UNIVERSITY OF IOWA
3 1858 044 435 760

AF566820

INTELLIGENT COMPUTER BASED ENGINEERING THERMODYNAMICS AND CYCLE ANALYSIS

Intelligent Computer Based Engineering Thermodynamics and Cycle Analysis

Chih Wu

Nova Science Publishers, Inc.
New York

Senior Editors: Susan Boriotti and Donna Dennis
Coordinating Editor: Tatiana Shohov
Office Manager: Annette Hellinger
Graphics: Wanda Serrano
Editorial Production: Matthew Kozlowski and Maya Columbus
Circulation: Ave Maria Gonzalez, Vera Popovic and Vladimir Klestov
Communications and Acquisitions: Serge P. Shohov
Marketing: Cathy DeGregory

Library of Congress Cataloging-in-Publication Data
Available upon request.
ISBN 1-59033-359-4

400 Oser Ave, Suite 1600
Hauppauge, New York 11788-3619
Tele. 631-231-7269 Fax 631-231-8175
e-mail: Novascience@earthlink.net
Web Site: http://www.novapubishers.com

Printed in the United States of America

To My Wife, Hoying Tsai Wu, and my children,
Anna, Joy, Sheree and Patricia

CONTENTS

PREFACE

Two semester courses in engineering thermodynamics typically harbor a dissatisfying feature. Development of classical thermodynamics is, logically and traditionally, aimed at analysis of cycles. First semester lay the ground work: First and Second Laws, thermodynamic state and process, approximation hierarchies used in modeling, properties of working fluids, heat, work, entropy and Carnot cycles. When cycles take the forefront in second semester, the course hits an anticlimax. Computational effort imposes harsh constraints on the kinds and amounts of cycle analyses which can reasonably be attempted. Cycle simulations cannot approach realistic complexity. Even relative sensitivity analyses based on grossly simplified cycle models are computationally taxing compared to their pedagogical benefits.

Desire for more design content in undergraduate courses is a familiar theme emanating from engineering education oversight bodies. By no means are thermodynamic cycles sterile ground for engineering design. Niches for innovative power generating systems, for example, are being created by deregulation, co-generation, choppy fuel costs and concern over global warming. Computerized look-up tables do reduce computational labor somewhat, but modeling cycles with many interactive loops still lies well outside of undergraduate time budgets.

This book utilizes an intelligent computer software called CyclePad for classroom purposes. CyclePad is developed by Professor K. Forbus of Northwestern University and evaluated by Professor Wu. It is an intelligent, powerful, mature, user friendly software package developed expressly to simulate thermodynamic devices and cycles. It cuts by about a hundred fold the computational effort involved in modeling realistically complex systems and cycles. It thus makes it feasible for students to run meaningful sensitivity analyses, to consider combinations of design modifications, to make engineering cost-benefit analyses and to include refinements such as accounting for pressure changes and heat transfers occurring between major cycle components.

This book and the intelligent computer software are intended to comprise the cycle portion of a second semester in thermodynamics or to support cost-benefit analyses or direct design projects in a third semester course or seminar. This package is strictly design and problem-solving oriented. Release from the drain of repetitive and iterative hand calculation, students can be held to a far wider and deeper study of cycles than has been possible previously. Homework assignments, saved to CD's, can be submitted very frequently for grading, an effective tool for keeping students abreast. Immersion in a veritable sea of problems helps students to “think like cycles” in the same sense that a successful first semester gets them “think like working fluids”

Classical thermodynamics is based on the concept of “equilibrium”. Time is not involved in the conventional engineering thermodynamic textbooks. Heat transfer deals with rate of energy transfer, but does not cover cycles. There is a gap between thermodynamics and heat transfer. There is a chapter called “Finite-time thermodynamics” in this book which tends to bridge the gap between thermodynamics and heat transfer.

Some attitudinal benefits have been noted by Professor Wu during his eight semesters of CyclePad assisted teaching of thermodynamics at the U.S. Naval Academy and at the Johns Hopkins University. Students tend from the outset to be more positively disposed towards computer assisted learning of thermodynamics, quite a few describing it as “fun”. (Not surprisingly, higher education is increasingly embracing computer assisted course work.) Material that is more positively regarded tends to be better retained. Further, an ability to execute realistically complicated cycle simulations builds students confidence and sense of professionalism.

Both CyclePad and the text contain pedagogical aids. The intelligent computer software switches to a warning-tutoring mode when students attempt to impose erroneous assumptions or perform inappropriate operation during cycle analyses. Objectives are listed at the start of each text chapter to highlight and preview upcoming material. Chapter summaries review the more salient points and provide cohesion. Homework problems and worked examples appear liberally throughout the text.

Both SI and English unit systems are used in the book.

ACKNOWLEDGEMENTS

I wish to acknowledge the following individuals who assisted in the text preparation: Dr. Susan Chipman of Naval Office of Research, Professor Ken Forbus of Northwestern University, Professor Myron Miller of the Johns Hopkins University, Professor Al Adams of the U S Naval Academy, Professor Joe Gillerlain, Jr. of the U S Naval Academy, Associate Professor Karen Flack of the U S Naval Academy, Ensign Donald Sherrill of the U S Naval Academy, and Mr. Doug Richardson of the U S Naval Academy.

NOMENCLATURE

c	Specific heat, kJ/[kg(K)]
c_p	Specific heat at constant pressure process, kJ/[kg(K)]
c_v	Specific heat at constant volume process, kJ/[kg(K)]
COP	Coefficient of performance
COP_R	Coefficient of performance of a refrigerator
COP_{HP}	Coefficient of performance of a heat pump
e	Specific energy, kJ/kg
E	Energy, kJ
e_k	Specific kinetic energy, kJ/kg
E_k	Kinetic energy, kJ
e_p	Specific potential energy, kJ/kg
E_p	Potential energy, kJ
g	Gravitational acceleration, m/s^2
h	Specific enthalpy, kJ/kg
H	Enthalpy, kJ
h_f	Specific enthalpy of saturated liquid, kJ/kg
h_g	Specific enthalpy of saturated vapor, kJ/kg
h_{fg}	Difference of specific enthalpy of saturated vapor and specific enthalpy of saturated liquid, kJ/kg
k	Specific heat ratio
L	Length, m
m	Mass, kg
mdot	Mass flow rate, kg/s
MEP	Mean effective pressure, kPa
n	Polytropic process exponent
p	Pressure, kPa
q	Specific heat transfer, kJ/kg
Q	Heat transfer, kJ
Q_H	Heat transfer with high temperature thermal reservoir, kJ
Q_L	Heat transfer with low temperature thermal reservoir, kJ
Qdot	Heat transfer rate, kW
r	Compression ratio
r_c	Cut-off ratio

r_p	Pressure ratio
R	Gas constant, kJ/[kg(K)]
R_u	Universal gas constant, kJ/[kmol(K)]
s	Specific entropy, kJ/[kg(K)]
S	Entropy, kJ/K
s_{gen}	Specific entropy generation, kJ/[kg(K)]
S_{gen}	Entropy generation, kJ/K
t	Time, s
T	Temperature, K
T_H	Temperature of high temperature thermal reservoir, K
T_L	Temperature of low temperature thermal reservoir, K
T_o	Temperature of surroundings, K
u	Specific internal energy, kJ/kg
U	Internal energy, kJ
v	Specific volume, m^3/kg
V	Volume, m^3
V	Velocity, m/s
Vdot	Rate of volumetric flow, m^3/s
w	Specific work, kJ/kg
W	Work, kJ
Wdot	Power, kW
W_{in}	Work input, kJ
W_{irr}	Irreversible work, kJ
W_{out}	Work output, kJ
W_{rev}	Reversible work, kJ
x	Quality
z	Elevation, m

GREEK LETTERS

β	Coefficient of performance
β_R	Coefficient of performance of a refrigerator
β_{HP}	Coefficient of performance of a heat pump
Δ	Finite change in a quantity
δ	Differential change of a path function
η	Heat engine efficiency
η_{Car}	Carnot heat engine efficiency
ρ	Density, kg/m^3

SUBSCRIPTS

abs	Absolute
act	Actual
atm	Atmospheric

e	Exit section
f	Saturated liquid
fg	Difference between saturated vapor and saturated liquid
g	Saturated vapor
gen	Generation
H	High temperature
i	Inlet section
int	Internally
irrev	Irreversible
L	Low temperature
rev	Reversible
s	Isentropic
surr	Surroundings
sys	System
1	Initial state or inlet state
2	Final state or exit state

Chapter 1

BASIC CONCEPTS

OBJECTIVES

After reading and studying the material in this chapter, you should be able to:

1. Define thermodynamics.
2. Know the statements of the four thermodynamic laws.
3. Know both the SI and English unit systems and know how to convert from one system to another system by using CyclePad.
4. Know the difference between a control mass or closed system and a control volume or open system.
5. Use properties to describe a system.
6. Understand point functions.
7. Understand extensive and intensive properties such as pressure, temperature, specific volume, internal energy, enthalpy, specific heats, etc., and know the nomenclatures of properties.
8. Know gage pressure and absolute pressure, and metric temperature and absolute temperature.
9. Understand the term flow energy and apply it to a control volume system.
10. Understand the concept of equilibrium.
11. Know the minimum number of independent intensive properties required to define a state.
12. Know the concepts of process and cycle.
13. Know the three modes of CyclePad.

1.1 THERMODYNAMICS

The field of science dealing with the relationships of heat, work, and properties of systems is called *thermodynamics*. A macroscopic approach to the study of thermodynamics is called *classical thermodynamics*. In engineering fields, a substance is considered to be in continuum, that is, it is continuously distributed throughout. Such a postulate can allow engineers to easily describe a system using only a few properties. Engineering thermodynamics is based on this macroscopic point of view. If the *continuum* assumption is not valid, a statistical method based on microscopic molecular activity may be used to describe a system. Such a microscopic approach to the study of thermodynamics is called *statistical thermodynamics*.

Most engineering activity involves interactions of energy, entropy, heat, work, and matter. Thermodynamics likewise covers broad and diverse fields. Practical uses of thermodynamics are unlimited. Traditionally, the study of applied thermodynamics is emphasized in the analysis or design of large scale systems such as heat engines, refrigerators, air conditioners, and heat pumps.

Homework 1.1 Thermodynamics

1. What is thermodynamics?
2. Distinguish clearly between statistical (microscopic) and classical engineering (macroscopic) thermodynamics.

1.2 Basic Laws

Thermodynamics studies the transformation of energy from one form to another and the interaction of energy with matter. It is a protracted and deductive science based on four strict laws. These laws bear the titles: " The Zeroth Law of Thermodynamics", "The First Law of Thermodynamics", "The Second Law of Thermodynamics", and "The Third Law of Thermodynamics". Laws are statements in agreement with all human experience and are always assumed to be true. Thermodynamics is a science built and based on these four basic laws. Among the four basic laws, The First Law of Thermodynamics and The Second Law of Thermodynamics are the two most useful to thermodynamicists.

Zeroth law: Two systems which are each in thermal equilibrium with a third system are in thermal equilibrium with each other.

First law: Energy can neither be created nor destroyed.

Second law: Heat cannot flow spontaneously from a cold body to a hot body.

Third law: The entropy of all pure substances in thermodynamic equilibrium approaches zero as the temperature of the substance approaches absolute zero.

Homework 1.2 Basic Laws

1. What is a law? Can laws be ever violated?
2. What are the basic laws of thermodynamics?
3. Why does a bicyclist pick up speed on a downhill road even when he is not pedaling? Does this violate the first law of thermodynamics?
4. A man claims that a cup of cold coffee on his table warmed up to 90°C by picking up energy from the surrounding air, which is at 20°C. Does this violate the second law of thermodynamics?

5. Consider two bodies A and B. Body A contains 10,000 kJ of thermal energy at 37°C whereas Body B contains 10 kJ of thermal energy at 97°C. Now the bodies are brought into contact with each other. Determine the direction of the heat transfer between the two bodies.

1.3 DIMENSIONS AND UNITS

A *dimension* is a character to any measurable quantity. For example, the distance between two points is the dimension called length. A *unit* is a quantitative measure of a dimension. For example, the unit used to measure the dimension of length is the meter. The two most widely used systems are the *English* unit system and the *SI* (Standard International) unit system. The SI unit system is based on a decimal relationship among the various units. The decimal feature of the SI system has made it well-suited for use by the engineering world, with the single major exception of the United States.

Those units for which reproducible standards are maintained are called *basic units*. The four basic SI and English system units used in engineering thermodynamics are meter (m) and foot (ft) in length, kilogram (kg) and pound (lbm) in mass, second (s) in time, and degree of Kelvin (K) and Rankine (°R) in temperature. Not all units are independent of each other. Those units that are related to the basic units through defining equations are called *secondary units*. For example, the English unit of area, the acre, is related to the basic unit of length, the foot.

The conversion from English units to SI units, or vice versa, is built into the CyclePad software. One can change the unit system from one to the other by reviewing the following example.

Example 1.3.1 Convert the following quantities from the SI unit system to the English unit system:

(a) Temperature (T) 460°C, (b)pressure (p) 1200 kPa, (c) specific volume (v) 2.4 m^3/kg, (d) specific internal energy (u) 1500 kJ/kg, (e) specific enthalpy (h) 1600 kJ/kg, (f) specific entropy (s) 6.2 kJ/[kg(K)], (g) mass flow rate (mdot) 2.3 kg/s, (h) volumetric flow rate (Vdot) 5.52 m^3/s, (i) Rate of internal energy (Udot) 3450 kW, (j) rate of enthalpy (Hdot) 3680 kW, and (k) rate of entropy (Sdot) 14.26 kW/K.

To solve this problem by CyclePad, we take the following steps:

1. Build
   (A) Take a source and a sink from the open-system inventory shop and connect them.
   (B) Switch to analysis mode.
2. Analysis
   (A) Input the given information: (a) 460°C, 1200 kPa., etc. (b) Edit, (c) Preference, (d) Unit, and (e) English unit.
3. Display results
   (A) Display the results. The answers are: 860°F, 174 psia, 38.44 ft^3/lbm, 644.9 Btu/lbm, 687.9 Btu/lbm, 1.48 Btu/[lbm(°R)], 5.07 lbm/s, 194.9 ft^3/s, 4627 hp, 4935 hp, and 24.33 Btu/[s(°R)].

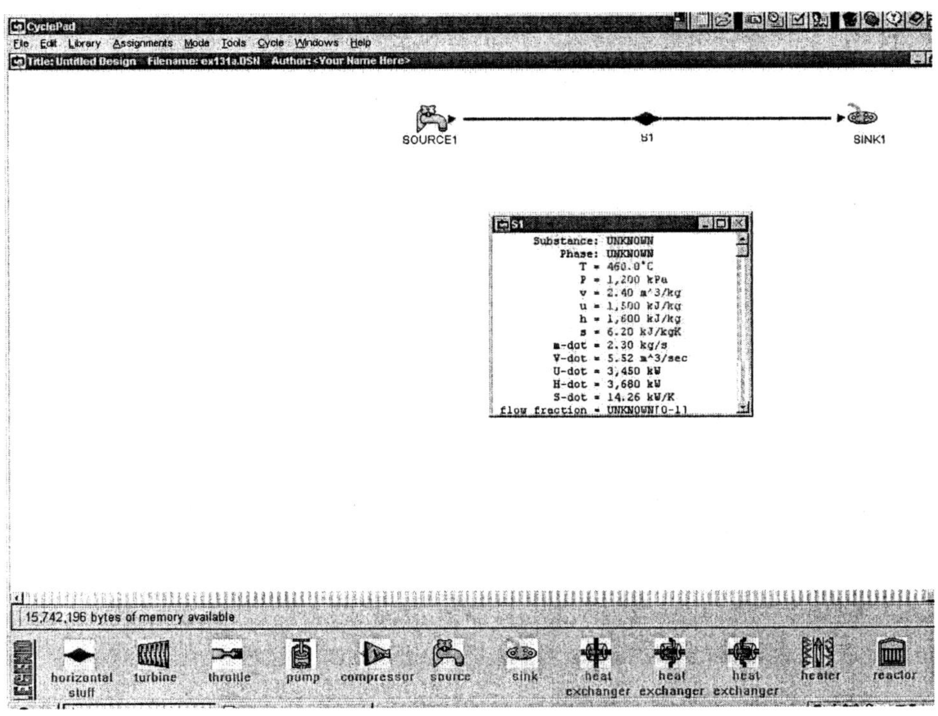

Figure Example 1.3.1 a Conversion (SI unit)

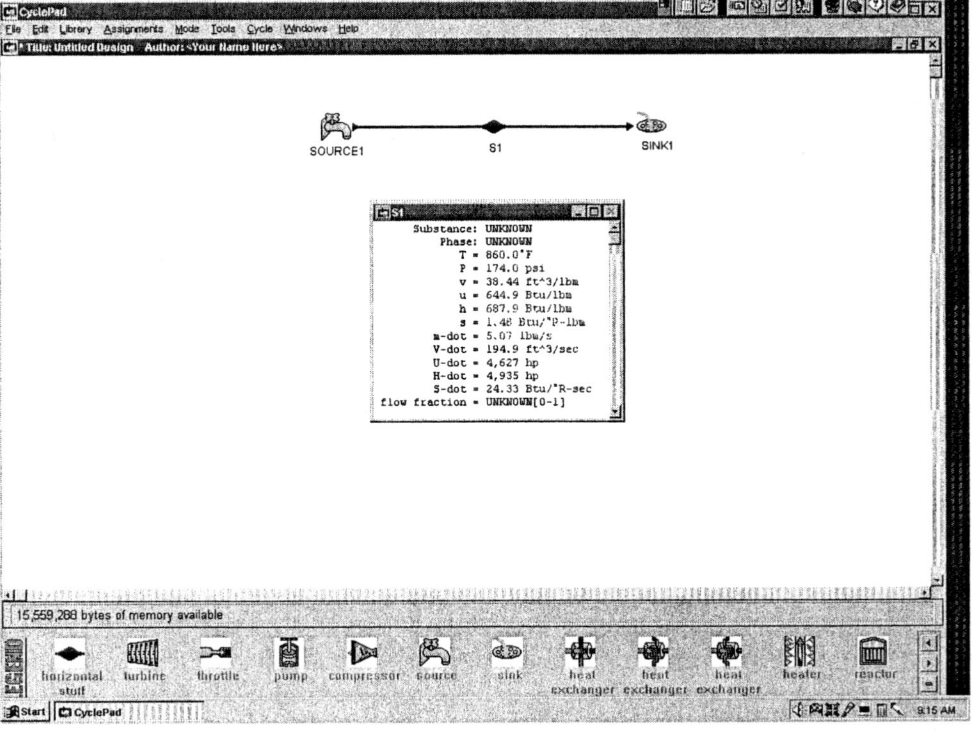

Figure Example 1.3.1 b Conversion (English unit)

Example 1.3.2 Convert the following quantities from the English unit system to the SI unit system:

(a) T 460°F, (b) p 120 psia,(c) v 2.4 ft^3/lbm, (d) u 1500 Btu/lbm, (e) h 1600 Btu/lbm, (f) s 6.2 Btu/[lbm(R)], (g) 2.3 lbm/s, (h) *Vdot* 5.52 ft^3/s (i) *Udot* 4881 hp, (j) *Hdot* 5207 hp, and (k) *Sdot* 46.2 Btu/R-s.

To solve this problem by CyclePad, we take the following steps:

1. Build
   (A) Take a source and a sink from the open-system inventory shop and connect them.
   (B) Switch to analysis mode.
2. Analysis
   (A) Input the given information: (a) 460°F, 120 psia, etc. (b) Edit, (c) Preference, (d) Unit, and (e) SI unit.
3. Display results
   (A) Display the results. The answers are: 237.8°C, 827.4 kPa, 0.1498 m^3/kg, 3489 kJ/kg, 3722 kJ/kg, 25.96 kj/[kg(K)], 1.04 kg/s, 0.1563 m^3/s, 3640 kW, 3883 kW, and 27.08 kW/K.

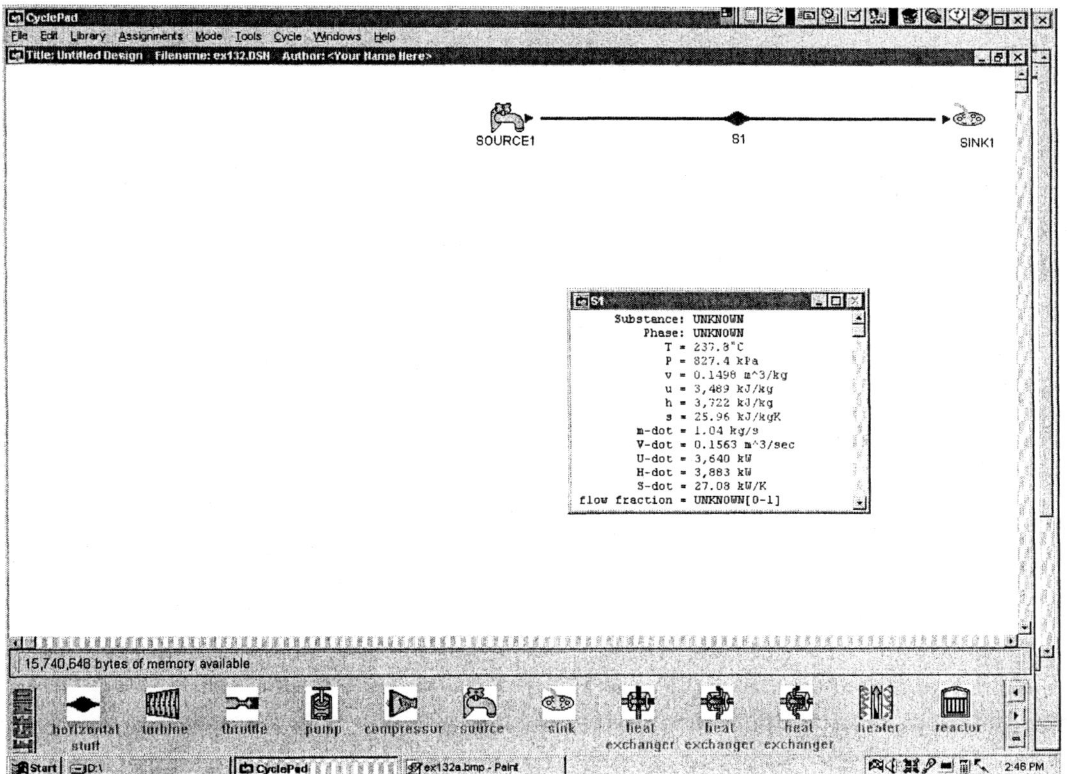

Figure Example 1.3.2 a Conversion from the English unit system to the SI unit system

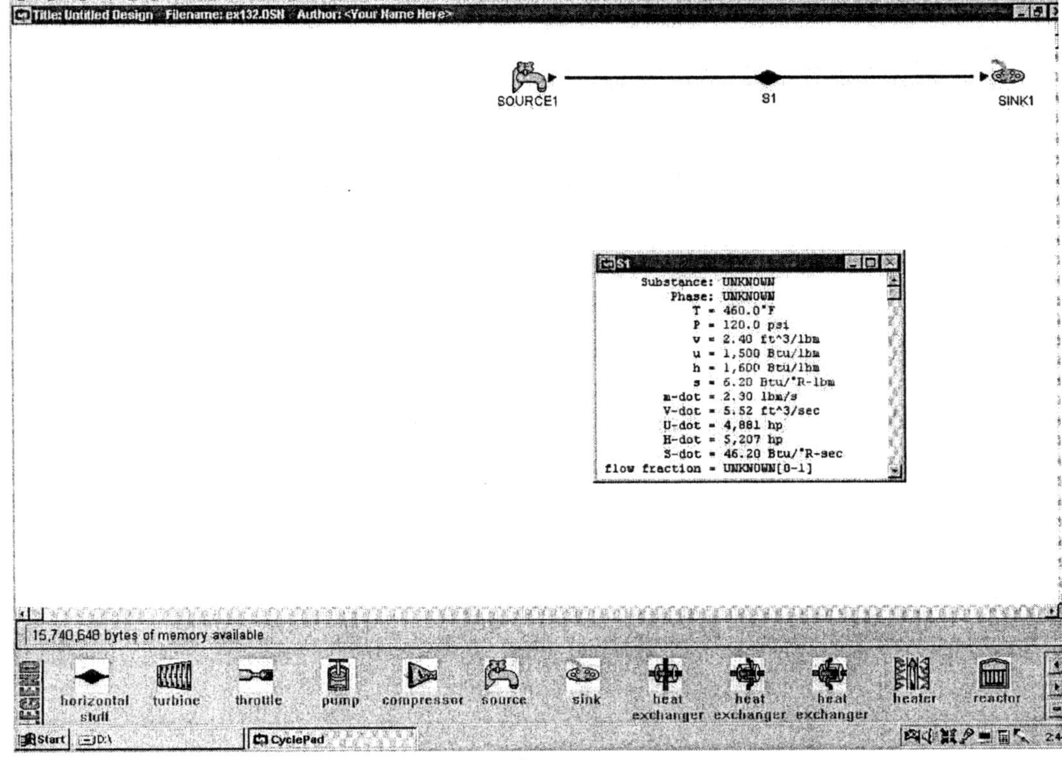

Figure Example 1.3.2b Conversion from the English unit system to the SI unit system

Homework 1.3 Dimensions and Units

1. Express the following secondary dimensions in terms of basic dimensions:
   (A) Volume
   (B) Velocity
   (C) Acceleration
   (D) Force
   (E) Pressure
   (F) Energy
   (G) Work
   (H) Power
   (I) Specific heat
2. Convert the following quantities from English units to SI units using CyclePad:
   (A) 500 psi
   (B) 400°F
   (C) 15 ft^3/lbm
   (D) 2000 Btu/lbm
   (E) 1.6 Btu/[lbm(°F)]
   (F) 10 lbm/s
   (G) 35 Btu
   (H) 10,000 Btu/hr

(I) 30,000 ft(lbf)/s
(J) 500 hp

3. What is the difference between lbm and lbf?
4. List the basic dimensions and state the units of each in the SI system.

1.4 Systems

A *system* may consist of a collection of matter or space. For example, a metal bar or a section of pipe can be considered a system. The surface, imaginary or real, enclosing the system is called the *boundary*. Everything outside the boundary which might affect the behavior of the system is called the *surroundings* of the system. Thermodynamics is concerned with the interactions of a system and its surroundings or one system interacting with another. If a system does not interact with its surroundings in mass, it is called a clos*ed* system. If a system does not interact with its surroundings in heat, it is called an *adiabatic* system. If a system does not interact with its surroundings, it is called an *isolated* system.

In many cases, a thermodynamic analysis is simplified if attention is focused on a mass without substance flow. Such a mass is called a *control mass* or a *closed system*. Water in a rigid tank and gas in a piston-cylinder apparatus are examples of control masses.

On the other hand, attention can be focused on a volume in space into which, and/or from which, a substance flows. Such a volume is called a *control volume* or *open system*. A turbine, a compressor, a boiler, a condenser, and a pump are examples of control volumes.

Homework 1.4 Systems

1. Explain the following concepts:
   (A) System, boundary, and surroundings
   (B) Closed system (control mass) and open system (control volume)
   (C) Adiabatic and isolated system
2. In which of the following processes would it be more appropriate to consider a closed system rather than a control volume?
   (A) Steady flow discharge of steam from a nozzle
   (B) Freezing a given mass of water
   (C) Stirring of air contained in a rigid tank using a mechanical agitator
   (D) Expansion of air contained in a piston and cylinder device
   (E) Heating of a metal bar in a furnace
   (F) Mixing of high pressure and low pressure air initially contained in two separate tanks connected by a pipe and valve
3. Identify the system, surroundings, and boundary you would use to describe the following processes:
   (A) Expansion of hot gas in the cylinder of an automobile engine
   (B) Evaporation of water from an open pot
   (C) Cooling of a steel rod
   (D) Cooking of an egg
4. Is a fixed mass usually treated as a closed or an open system?

1.5 PROPERTIES OF A SYSTEM

Once a system has been selected for analysis, it can be further described in terms of its properties. A *property* is a characteristic of a system and its value is independent of the history of the system. Some properties are directly or indirectly measurable, such as pressure, temperature, volume, specific heat at constant pressure, and specific heat at constant volume. Other properties, such as enthalpy, can be defined by mathematically combining other properties. The value of a property is unique at a fixed state.

Properties are classified as either extensive or intensive. A property is *extensive* if its value for the whole system is the sum of its value for the various parts of the system. Examples of an extensive property include volume (V) and energy (E). Generally, upper case letters denote extensive properties, with a few exceptions, such as mass (m). Extensive properties per unit mass are called intensive or specific properties, such as specific volume (v=V/m). An *intensive* property has the same value independent of the size of a system, such as specific volume (v) and specific energy (e=E/m). Generally, lower case letters denote intensive properties, with a few exceptions, such as temperature (T).

1.5.1 Volume (V)

The *volume* can be easily measured. It is a macroscopic property associated with thermodynamic boundary work. Volume is therefore called displacement of thermodynamic boundary work.

1.5.2 Density (ρ)

The *density* is the mass per unit volume. Density is defined by the equation

$$\rho = \lim(\Delta m/\Delta V). \quad (1.5.1)$$

where Δm is the finite mass contained in the finite volume ΔV.

In engineering thermodynamics, materials are considered to be in *continuum*. Therefore, ΔV cannot be allowed to shrink to zero. If ΔV became extremely small, Δm would vary discontinuously, depending on the number of molecules in ΔV. We must choose a ΔV sufficiently small but large enough to eliminate microscopic molecular effects.

The inverse of ρ, volume per unit mass (m), is called *specific volume* (v).

$$v = V/m. \quad (1.5.2)$$

For example, 2 kg of air contained in a 4 m^3 tank has a specific volume of 2 m^3/kg and a density of 0.5 kg/m^3.

The *specific weight* of a substance, denoted by γ, is its weight per unit volume. Density and specific weight are simply related by gravity:

$$\gamma = \rho g. \quad (1.5.3)$$

1.5.3 Specific Weight (γ)

Specific weight is the only thermodynamic specific property expressed in terms of a substance's unit volume rather than its unit mass. The units of γ are lbf/ft^3 or N/m^3. For example, 2 kg of air contained in a 4 m^3 tank has a specific weight of 0.5(9.8)=4.9 N/m^3.

1.5.4 Specific gravity (SG)

Specific gravity is the ratio of a fluid's density to the density of a standard reference fluid, which is water for liquids and air for gases. It is the only specific property expressed in dimensionless form. For example, the specific gravity of mercury (Hg) is SG_{Hg}=13580/998=13.6. This dimensionless ratio is easier to remember than the numerical value of density for a variety of fluids.

Example 1.5.1. 2 kg of a gas is contained in a 1 m^3 tank. Determine the specific volume of the gas.

To solve this problem by CyclePad, we take the following steps:

1. Build
   (A) Take a BEGIN and an END from the closed-system inventory shop and connect them.
   (B) Switch to analysis mode.
2. Analysis
   (A) Input the given information: (a) mass is 2 kg and volume is 1 m^3.
3. Display results
   (A) Display the results. The answer is 0.5 m^3/kg.

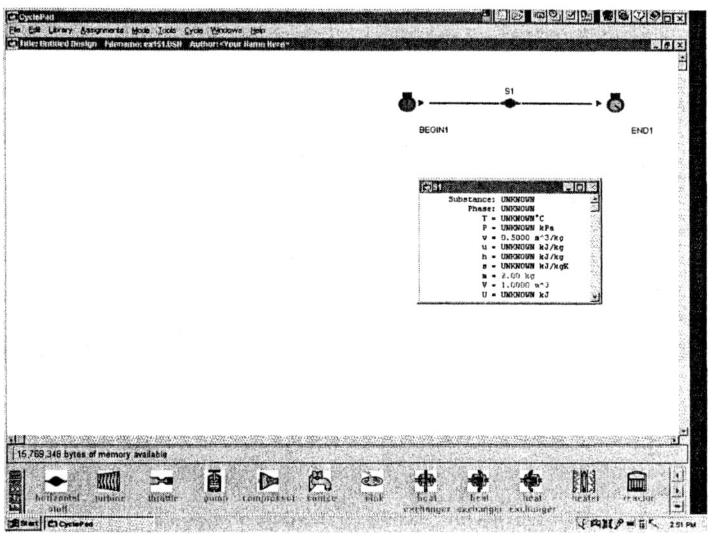

Figure Example 1.5.1 Determine the specific volume, specific weight and density

1.5.5 Pressure (p)

The normal force exerted by a system on a unit area of its surroundings is called the *pressure* (p) of the system. Pressure is a macroscopic property associated with thermodynamic boundary work. Pressure is therefore called the driving force of thermodynamic boundary work.

Two different pressures are common in engineering practice: gage pressure and absolute pressure. The difference between gage and absolute pressure should be understood. *Absolute pressure* (p_{abs}) is the amount of force per unit area exerted by a system on its boundaries. *Gage pressure* (p_{gage}) is the value measured by a pressure gauge, which indicates the pressure difference between a system

and its ambient, usually the atmosphere. The *atmospheric pressure* (p_{atm}) is due to the weight of the air per unit horizontal area in the earth's gravitational field. Hence,

$$p_{gage} = p_{abs} - p_{atm} \qquad (1.5.4)$$

The units of pressure commonly used are in or mm of mercury (Hg), kPa, Mpa, bar, psi, psf, etc. The standard atmospheric pressure at sea level is 29.92 in. Hg, 760 mmHg, 101.3 kPa, 0.1013 MPa, 1.013 bar, 14.69 psia, or 2117 psfa depending upon the units used. Very often atmospheric pressure is assumed to be 100 kPa and 14.7 psia for simplicity. Barometers are used to measure atmospheric pressure, and usually use mercury as a manometer fluid. Common devices for measuring pressures are a Bourdon gage, shown in Figure 1.5.1, and a manometer, shown in Figure 1.5.2.

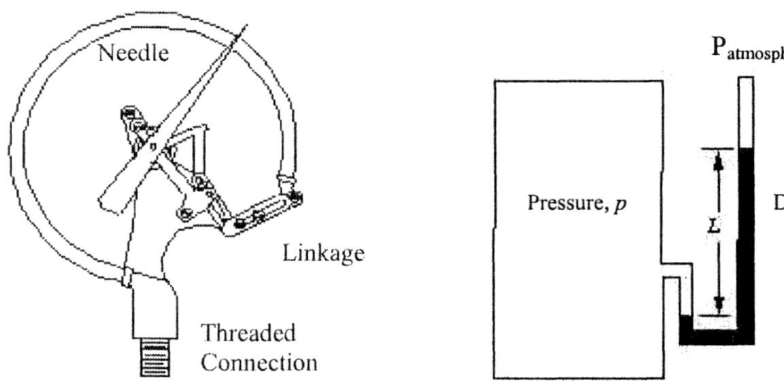

Figure 1.5.1 Bourdon gage **Figure 1.5.2 manometer**

A *manometer* is used to measure the system pressure in a container. If the system has a pressure p, the fluid in the manometer has a density ρ, and the surroundings are atmospheric with pressure p_{atm}, then the difference in pressure between the system and the surroundings is able to support the fluid in the manometer for a deflection L. This may be expressed by

$$p - p_{atm} = \rho L g \qquad (1.5.5)$$

where g is the gravitational acceleration.

Absolute pressures are always positive, while gauge pressures can be either positive or negative. Negative gauge pressures indicate pressures below atmospheric pressure. Pressures below atmospheric pressure are called vacuum pressures.

In the text if a pressure is not explicitly stated as being either gauge or absolute pressure, the implication is that the value is an absolute pressure.

Figure 1.5.3 depicts the various pressures in graphical form.

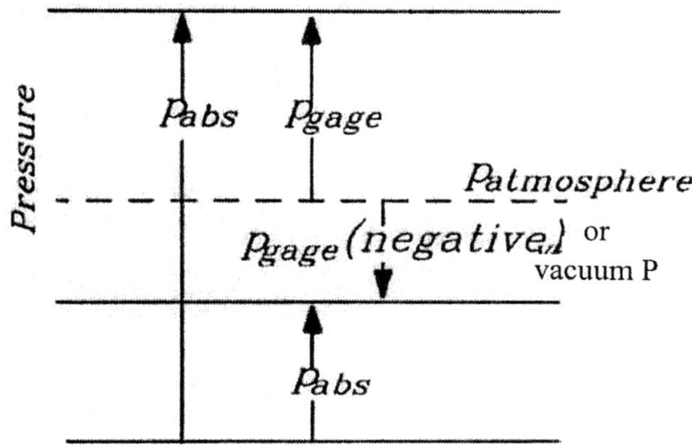

Figure 1.5.3 Graphical representation of pressure

It should be noted that when a system is subdivided, the pressure is not subdivided. This is a characteristic of an intensive property.

Example 1.5.2. Steam is exhausted from a turbine at an absolute pressure of 2 psia. The barometer reads 14.7 psia. Determine the gage pressure and vacuum pressure in psig at the turbine exhaust.

Solution: Eq. (1.5.4) gives p_{gage}=2-14.7 psia=-12.7 psig, and p_{vacuum}=12.7 psi.

Example 1.5.3. Convert 40 kPa gage pressure to absolute pressure. The barometer reads 101 kPa.

Solution: Eq. (1.5.4) gives p_{abs}=101+40 kPa Abs.=141 kPa Abs.

1.5.6 Temperature (T)

Temperature is often thought of as being a measure of the "hotness" of a substance. This statement is not exactly a good definition of temperature because the word hot is a relative rather than a quantitative term. In thermodynamics, *temperature* is defined to be the property having equal magnitude in systems that are in thermal equilibrium. Temperature is a microscopic property associated with heat. Temperature is therefore called the driving force of heat.

The *absolute temperature* scale is defined such that a temperature of zero corresponds to a theoretical state of no molecular movement of the substance. Negative absolute temperature is impossible. In the English unit system and SI unit system, the absolute temperature scales are the Rankine (°R) scale and the Kelvin (K) scale, respectively.

Metric temperature scales are made by arbitrarily selecting reference temperatures corresponding to reproducible state points (ice point and steam point). In the English unit system and SI unit system, the metric temperature scales are the Fahrenheit (°F) scale and the Celsius (°C) scale respectively. Negative temperatures exist for the metric temperature scale. The selection of reference temperatures allows us to write the relationships:

$$°F = (9/5)°C + 32$$

and

$$°C = (5/9)(°F - 32)$$

For example, 50°F is 10°C and 40°C is 104°F.

The absolute temperature scale is related to the metric temperature scale by the relationships:

$$K = °C + 273°$$

and

$$°R = °F + 460°$$

For example, 20°C is 293 K and 40°F is 500°R.

Figure 1.5.4 depicts the various temperatures in graphical form.

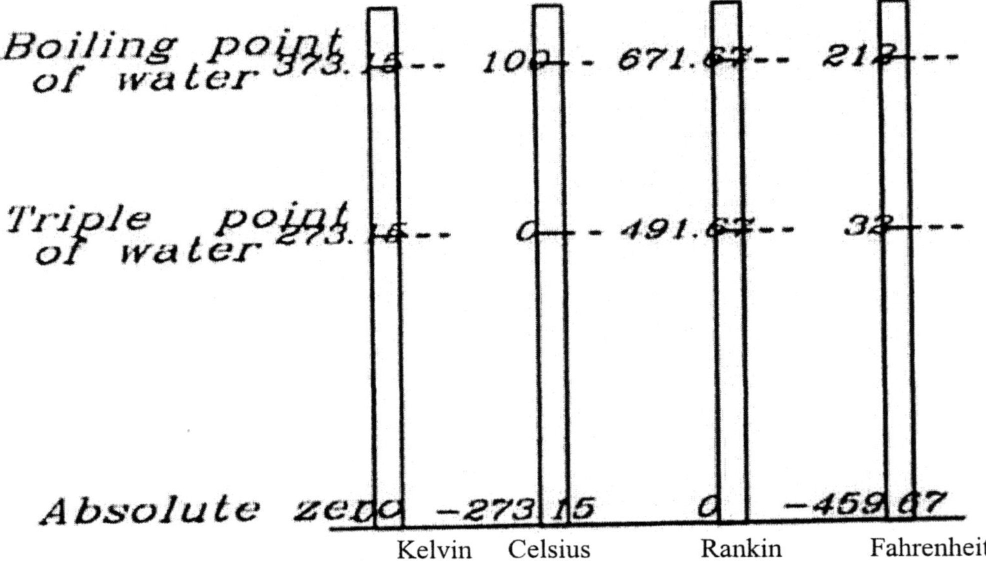

Figure 1.5.4 Graphical representation of temperature

Example 1.5.4. Convert 560°F to degree of Rankine, degree of Kelvin, and degree of Centigrade.

To solve this problem by CyclePad, we take the following steps:

1. Build
   (A) Take a source and a sink from the open-system inventory shop and connect them.
   (B) Switch to analysis mode.
2. Analysis
   (A) Input the given information: (a) 560°F (b) edit, (c) preference, (d) unit, (e) change unit, and (f) SI.
3. Display results
   (A) Display the results. The answers are 1020°R, 293.3°C, 566.5 K.

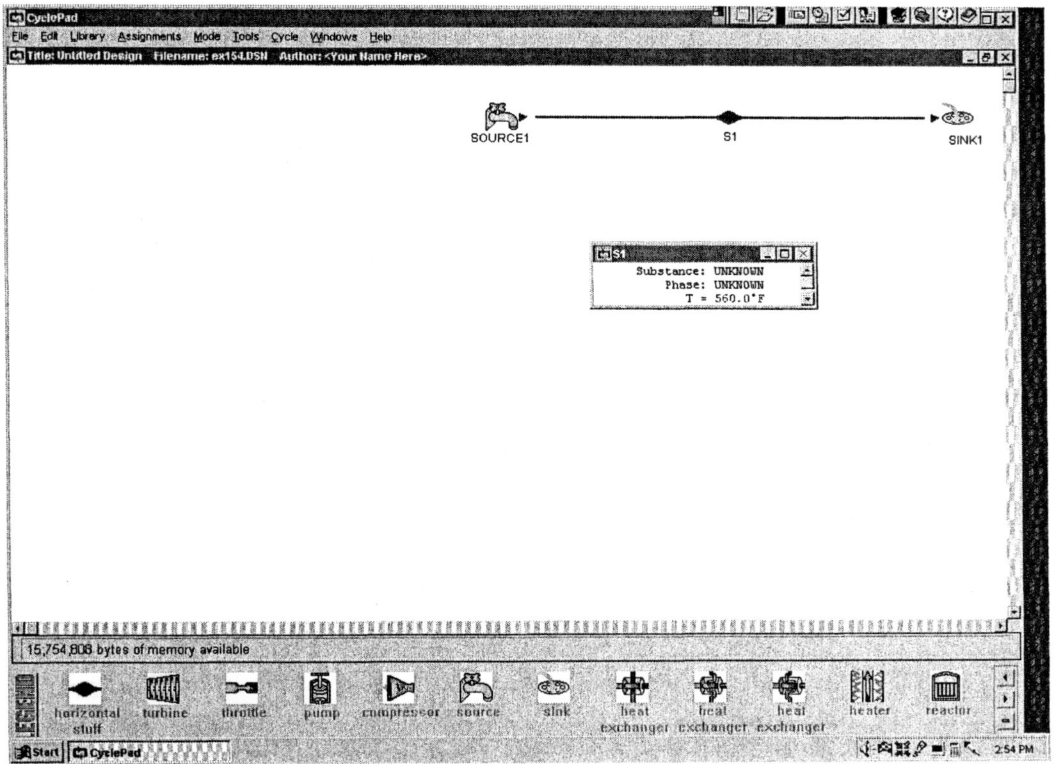

Figure Example 1.5.4a Temperature conversion

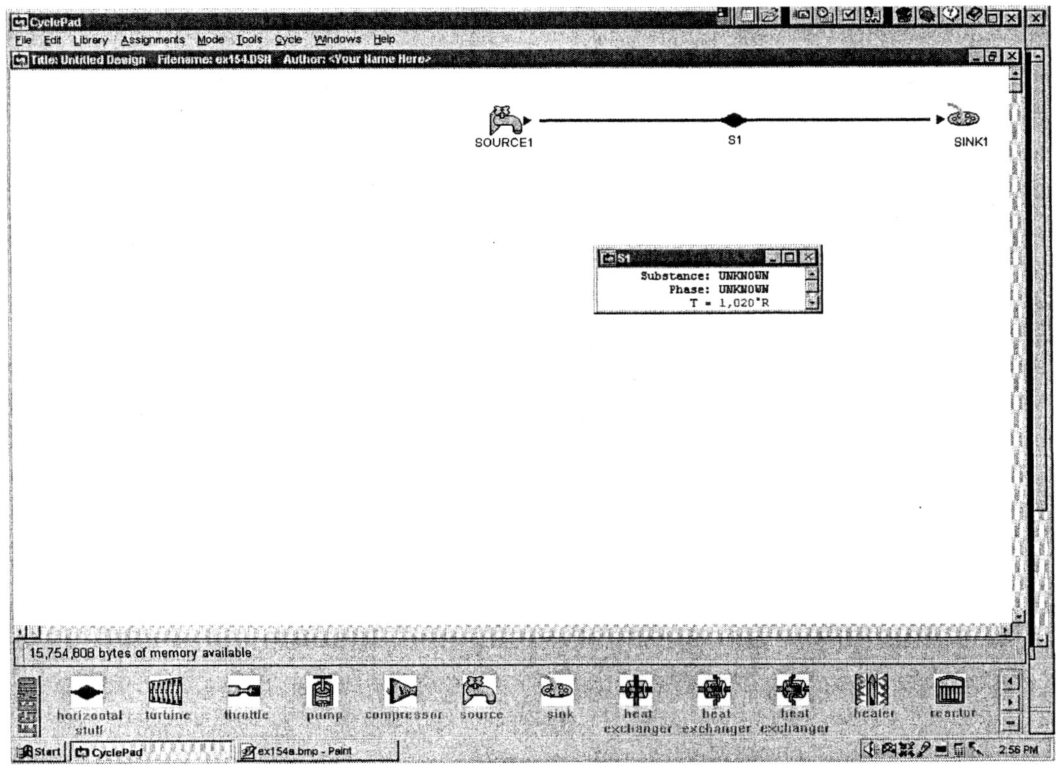

Figure Example 1.5.4b Temperature conversion

1.5.7 Energy (E)

Matter can store *energy*. Energy held within a system is associated with the matter of the system. The amount of energy of a system is reflected in properties such as temperature, velocity, or position in a gravitational field. As the amount of stored energy changes, the value of these properties change. An important characteristic of classical thermodynamics is that it deals with the changes in the amount of energy in a system and not with the system's absolute energy.

A system stores energy within and between its constituent molecules. This deeply stored microscopic energy is called *internal energy*. The symbol U is used to represent internal energy, and u is used to represent specific internal energy. Any change in molecular velocity, vibration rate in the bonds and forces between molecules, or in the number and kind of molecules, changes the internal energy. The change in internal energy is denoted by ΔU. The internal energy is not directly measurable, however, it is related to other measurable properties. *Kinetic energy* (E_k) is the stored macroscopic energy that a body of mass m has when it possesses a velocity V. The change in kinetic energy of a system when its velocity changes from V_1 to V_2 is $\Delta E_k = m(V_2^2 - V_1^2)/2$. The change in specific kinetic energy of a system when its velocity changes from V_1 to V_2 is $\Delta e_k = (V_2^2 - V_1^2)/2$.

Potential energy (E_p) is the stored macroscopic energy that a body of mass m has by virtue of its elevation (z) above ground level in a gravitation field whose acceleration is constant and equals to g. The potential energy is E_p=mgz. The change of potential energy from level z_1 to z_2 is $\Delta E_p = mg(z_2-z_1)$. The change of specific potential energy from level z_1 to z_2 is $\Delta e_p = g(z_2-z_1)$.

Flow energy (pV) is the energy required to push a volume V of a flowing substance through a boundary surface inlet section into the system from the surroundings by a pressure p, or to push a volume V of a flowing substance through a boundary surface exit section out from the system to the surroundings by a pressure p. Flow energy occurs only when there is a mass flow into the system or out from the system. If there is no mass flow into the system or out from the system, there is no flow energy.

1.5.8 Enthalpy (H)

Enthalpy is a not a directly measurable property. It is a synthetic combination of the internal energy and the flow energy exchanged with the surroundings. Enthalpy and specific enthalpy are symbolized by H and h, and are defined by

$$H = U + pV \tag{1.5.6}$$

$$h = u + pv \tag{1.5.7}$$

1.5.9 Specific Heat (c, c_p and c_v)

The quantity $C = \delta Q/dT$ is called the *specific heat* or *heat capacity*. It is a measure of the heat added to a mass of a system to produce a unit increase in temperature. For example, the specific heat of water at 25°C and 101.3 kPa is 4.18 kJ/[kg(K)], which means 4.18 kJ of heat added is required to a kg mass of water in order to raise its temperature by 1 K. The most commonly used specific heats are *specific heat at constant pressure* (c_p) and *specific heat at constant volume* (c_v); c_p and c_v are defined in the following equations:

$$c_p = (\partial h/\partial T)_P \tag{1.5.8}$$

$$c_v = (\partial u/\partial T)_v \tag{1.5.9}$$

Both c_p and c_v are measurable properties and are measured on a constant pressure process and a constant volume process for a closed system, respectively. Values of c_p and c_v can be obtained by measuring the heat transfer required to raise the temperature of a unit mass of substance by one degree, while holding the pressure and volume constant, respectively. The unit of c (c_p or c_v) is kJ/[kg(K)] in SI system and Btu/[lbm(°R)] in English system.

The heat capacities of gases other than c_p and c_v for an arbitrary process can also be defined (Reference 1, Chen J. and C. Wu, The heat capacities of gases in arbitrary process, *The International Journal of Mechanical Engineering Education,* vol. 29, no. 3, pp. 227-232, 2001.).

Since c_p and c_v are measurable properties, we therefore have a method of calculating the internal energy and enthalpy for any process if we know the end states.

1.5.10 Ratio of the Specific Heats (k)

A dimensionless property denoted by k is the *ratio of the specific heats*, $k=c_p/c_v$. It is extensively used in thermodynamics.

A liquid and vapor two-phase state is a mixture of liquid and vapor. *Quality* of vapor, usually represented by the symbol x, is defined as the vapor mass fraction of the total mixture. *Moisture content*, usually represented by (1-x), is defined as the liquid mass fraction of the total mixture. That is,

$$x = m_{vapor}/m_{total} \tag{1.5.10}$$

$$1\text{-}x = m_{liquid}/m_{total} \tag{1.5.11}$$

and

$$m_{total} = m_{vapor} + m_{liquid} \tag{1.5.12}$$

For example, 5 kg of saturated water liquid and vapor mixture consisting of 3 kg of saturated steam vapor and 2 kg of saturated liquid water has a quality of 0.6 and a moisture content of 0.4.

Example 1.5.5. 10 kg of water is contained in a tank. If 8 kg of the water is in vapor form and rest is in liquid form. Determine the quality and moisture content of the water.

Solution: Eq. (1.5.10) and Eq. (1.5.11) give

$$x=8/10=0.8$$

$$1\text{-}x=(10\text{-}8)/10=0.2$$

1.5.11 Entropy (S)

Entropy is a microscopic property associated with the microscopic energy transfer called heat (Q), between the system and its surroundings. Entropy is also called displacement of heat. It is not directly measurable, but can be related to other properties. Entropy is a measure of the level of irreversibility associated with any process. Unlike energy, it is non-conservative. It is a very important property in thermodynamics and will be discussed later in chapter 5.

1.5.12 Point Function

If the change of a function (quantity) of a system for a process between an initial state 1 and a final state 2 depends only on the two end states only, the function is said to be a *point function*. Otherwise, it is a path function. For example if T is a point function, then the change of T, ΔT, from state 1 to state 2 is

$$\Delta T = \int dT = T_2 - T_1 \qquad (1.5.13)$$

All properties are point functions.

Homewok 1.5 Properties

1. Which of the following are properties of a system: pressure, temperature, density, energy, work, heat, volume, specific heat, and power?
2. List at least three measurable properties of a system.
3. Distinguish clearly between intensive and extensive properties? Give three examples of each type.
4. What is a specific property?
5. Is the property denoted by H an intensive or extensive property ? How about the property denoted by h?
6. How are volume and specific volume related? What are the notations used for volume and specific volume?
7. What property is the sum of internal energy and flow energy?
8. Does a system possesses flow energy without mass flow in or out of the system?
9. What is meant by flow energy?
10. Define the property c_p.

Specific volume

1. Two pounds of air occupies a volume of 30 ft^3. Calculate the specific volume in ft^3/lbm and specific weight in lbf/ft^3.

Pressure

1. Is pressure an extensive property?
2. Define the units psi and kPa.
3. What is the difference between gage pressure and absolute pressure?
4. A pressure gage at a turbine inlet reads 500 psi and a vacuum gage at the turbine exhaust reads 20 psi. The corresponding barometer reading is 14.7 psi. What are the turbine inlet and exhaust pressures in psia?
5. The pressure of a system drops by 20 psi during an expansion process. Express this drop in psig and psia.
6. Standard atmospheric pressure at sea level is 14.7 psia. Convert it to kPa and bars using CyclePad.

Temperature

1. The body temperature of a healthy person is 37°C. What is it in Kelvin, in Fahrenheit and in Rankine scale?
2. Are the boiling pressure and boiling temperature of water dependent or independent?
3. Air temperature rises 400°C during a heating process. What is the air temperature rise in Kelvin during the heating process?
4. Copper block A contains 1000 kJ of thermal energy at 50°C and copper block B contains 100 kJ of thermal energy at 500°C. The two blocks are brought into contact with each other. Is heat transfer flow from block A to block B, or from block B to block A?

Energy

1. Define flow energy. Does a substance possesses flow energy when at rest?
2. What is the internal energy of a system?

1.6 Equilibrium State

A system is defined to be at *equilibrium* if it does not tend to undergo any further change of its own accord; any further change must be produced by external means.

Traditionally, classical thermodynamics has been able to deal with systems only when they were at rest, or at equilibrium. Therefore classical thermodynamics is also called equilibrium thermodynamics.

But thermodynamic systems are not always at equilibrium, and a method has been developed to apply it to non-equilibrium systems which conduct heat, matter, or work at a steady state. This relatively new field is called *FINITE-TIME THERMODYNAMICS* (Reference 2, Wu, C., L. Chen and J. Chen, Recent advances in finite time thermodynamics, Nova Science Publishers, Inc., ISBN 1-56077-664-4, New York, 1999.).

The *state* of a system is the condition of a system at any particular moment and can be identified by a statement of the properties of the system. A specification of a state describes a system completely. The number of independent intensive properties needed to define a state depends on the number of work modes of the substance. Substances used in engineering thermodynamics are usually restricted to simple compressible substances, which have only one work mode called compressible work. The state of a simple compressible substance can therefore be completely specified by two independent, intensive properties.

For example, a state of superheated vapor H_2O can be defined by a pressure of 100 kPa and a temperature of 400°C, because both pressure and temperature are intensive properties and are independent in one-phase region. However, a state of saturated mixture H_2O can not be defined by a pressure of 14.7 psia and a temperature of 212°F, because pressure and temperature are dependent in the two-phase region.

Similarly, a state of air cannot be defined by a specific volume of 5 m^3/kg and a density of 0.2 kg/m^3. because specific volume and density are dependent.

Homework 1.6 Equilibrium State

1. Explain the concept of equilibrium.
2. Which of the following systems are not in equilibrium?
   (A) Compressed air in a tank
   (B) A mixture of ice and water at 32°F
   (C) A burning log
   (D) A copper rod with one end immersed in a beaker of boiling water and the other in an ice bath
   (E) Steam, water and ice in a closed vessel at 0.01°C
   (F) Gasoline and oxygen in a closed vessel
   (G) Ice cubes floating in water at 50°F
   (H) A copper rod immersed in a beaker of boiling water
3. What is the minimum number of independent specific properties needed to define a thermodynamic state?
4. Is a state defined by given the boiling pressure and boiling temperature of water?
5. Is the state of air in your classroom completely specified by the temperature and pressure? Why?
6. How many different values does a property possess at a fixed state?
7. Does a state change if just one of its many properties changed?

1.7 PROCESSES AND CYCLES

A series of infinitesimal changes in a system between an initial state and a final state is called a *process*, or path change of state. The method by which the path is described is called a process. A process can be described only in terms of the properties of the system or a functional relation between the properties. A property of a system depends only on the state of the system and not on how a state is arrived at. Thus, a change in property is independent of path.

A differential change in property such as temperature (T) is written as dT. The change of temperature for a process between an initial state 1 and a final state 2 is

$$T_2 - T_1 = \int dT \qquad (1.7.1)$$

When a process proceeds in a manner in which the system remains too infinitesimally close to an equilibrium state at all times, it is called a *quasi-equilibrium* process or an ideal process. An *ideal process* is not a true representation of an actual process, which occurs at a faster rate. However, ideal processes are easy to analyze. Therefore, an actual process can be modeled as an ideal process with negligible error.

Important thermodynamic processes include isothermal, isobaric, isochoric, adiabatic, isentropic, throttling, and polytropic processes.

The prefix iso- is often used to designate a process for which a particular property remains constant. An *isothermal process* is a constant temperature process during which the temperature (T) remains constant; an *isobaric process* is a constant pressure process during which the pressure (p) remains constant; and an *isochoric* process is a constant volume process during which the volume (V) remains constant. An *adiabatic* process is a process during which the system does not exchange heat (Q=0) with its surroundings. A *constant internal energy* process is a process during which the internal energy (U) remains constant. An *isentropic* process is a constant entropy process during which the entropy (S) remains constant. A *throttling* process is a constant enthalpy process during which the enthalpy (H) remains constant. A *polytropic* process is a constant pV^n process, where n is a constant.

A cyclic process, or *cycle*, is a process for which the end states are identical (initial state = final state). The change in the value of any property of the system for a cycle is zero.

Homework 1.7 Processes and Cycles

1. What is a process?
2. Explain the concept of quasi-equilibrium
3. A system undergoes a process with an initial temperature of 700 K and a final temperature of 400 K, but the temperatures of the intermediate states are unknown. Can we determine the temperature difference between the final state and the initial state? If yes, what is the temperature difference in K and what is the temperature difference in °C?
4. Water is heated in an open container. After some time, the water starts to boil. Which of the following correctly describes the entire process?
   (A) Isothermal process
   (B) Adiabatic process
   (C) Isobaric process
   (D) Isochoric process
5. Does an adiabatic process mean that the net heat transfer is zero, or that there is no transfer heat at all?
6. What is a cycle?
7. What are the quantity changes of properties for a complete cycle?
8. Does the total quantity change of heat must equal to zero for a complete cycle?
9. Does the total quantity change of work must equal to zero for a complete cycle?
10. Steam is expanded through a turbine. Is this a process or a cycle?
11. Air is compressed through a compressor. Is this a process or a cycle?

1.8 CyclePad

1.8.1 Download

CyclePad can be downloaded free from Northwestern University's web page at Http://www.qrg.ils.northwestern.edu or from the U.S. Naval Academy's web page at http://web.usna.navy.mil/~cadigweb. CyclePad can be downloaded by the following steps:

Internet Explorer
Http://www.qrg.ils.northwestern.edu

Software
Dowload CyclePad v2.0
Fill in form - Submit form
License Agreement - Dowload Software
Save this program to disk - OK
Choose a location - Save
Close Internet Explorer
Open my computer
Go to the location you save the file to
Double click on webcpad - #######.exe
Double click on Setup.exe
Install
Yes
OK
OK - CyclePad is downloaded

1.8.2 Installation onto Your Own PC

CyclePad can be installed onto your own PC by the following steps:

Start up your computer normally
Insert the CyclePad first disk into your (3.5" or zip floppy disk) drive
Go to file manager and bring up the a:\ or e:\ drive window
You should see a file called **setup.exe**
Click on setup.exe and follow the on screen prompts
When you have successfully installed CyclePad, you should come to a screen that says
Congratulations! CyclePad has been installed successfully
You may now run CyclePad

Or,

Make new folder
Temp-CyclePad
Copy files from Zip to new folder Temp-CyclePad
Go to Temp-CyclePad
Double clip–webcpad- 20000504 20000504(date)
Double clip–cpadinst
Double clip–setup
Install
Yes
Yes
C:\CyclePad
OK
OK

1.8.3 Contents

An intelligent computer software called CyclePad has been co-developed and evaluated by Prof. C. Wu in the Mechanical Engineering Department at the U.S. Naval Academy (USNA) with Oxford University and Northwestern University since 1995. The software has been incorporated in three thermodynamic courses at USNA for six semesters.

CyclePad is designed to help with the learning and conceptual design of thermodynamic cycles. It works in two phases, build mode and analyze mode.

1.8.4 Modes

Build

In the build mode, the user uses a graphical editor to place components out from a thermodynamic inventory shop and connects them to form a state or several states, a process or several processes, or a cycle or several cycles. While the user can always quit CyclePad at any time, you can only proceed to the next phase (analysis) when CyclePad is satisified that your design is fully laid out; that is, when every component is connected via some other components via states, and every state has been used as both an input and an output for components in the design.

Analysis

In the analyze mode, the user chooses a working fluid, processes on assumption for each component, and inputs numerical property values. As soon as you give CyclePad some information, it draws as many conclusions as it can about your design, based on everything you have told it so far. All the calculations are then quickly done by the software and displayed. The user is free to inquire about how values were derived and how one might proceed at any time, using a hypertext query system. At any time you can save your design to a file so that you can continue working on it later, and generate reports describing the state of your analysis of the design.

There is a sensitivity tool which makes cycle performance parameter effects easy and quick, and generates the effects in graph form. Such a sensitivity analysis is quite tedious to do by hand.

Contradiction

It is possible to make assumptions that conflict with each other. In such a case, CyclePad cannot continue to analyze the design until one or more of the conflicting assumptions has been retracted. When Cycle Pad detects a conflict, it enters contradiction mode. The contradiction resolution window appears on the screen to inform you that there is a contradiction. There is a coach (instructor) in the software. If there is a mistake or a contradiction made by the user, the coach will show up, display the contradiction, and suggest ways to solve the contradiction.

1.10 SUMMARY

In this chapter, the concepts and basic laws of thermodynamics are introduced. Engineering thermodynamics is a macroscopic science that deals with heat, work, properties, and their relationships. The first law of thermodynamics is the principle of energy conservation. The second law of thermodynamics indicates the direction of a process. English and SI unit systems are introduced. A system of fixed mass is called a closed system or control mass. A system of fixed volume with mass flow is called an open system or control volume. The mass dependent properties of a system are called extensive properties and are usually denoted by upper class letters. The mass independent properties of a system are called intensive properties and are usually denoted by lower class letters. Specific volume, pressure and temperature are the three most important thermodynamic properties because they are directly measurable. Properties can be measurable or non-measurable. A state is a system at equilibrium. A process is a change of state. A cycle is a process with identical initial and final states. The state of a simple compressible substance is completely specified by two independent intensive properties. Relationships among the properties are called equations of state. An intelligent computer software called CyclePad is introduced. The procedures to download the software and install it into one's own PC are described.

Chapter 2

PROPERTIES OF THERMODYNAMIC SUBSTANCES

OBJECTIVES

After reading and studying the material in this chapter, you should be able to:

1. Know the three different types of thermodynamic substances.
2. Understand the p-v-T phase property diagram of pure substances.
3. Draw the (p or T) vs (v or h or u) liquid-vapor two phase diagram and identify the three regions.
4. Know how to find pure substance properties from steam tables and Mollier diagram.
5. Understand ideal gases.
6. Know how to find ideal gas properties from air tables.
7. Know how to find liquid and solid properties from equations.
8. Find properties of a state using CyclePad.

2.1 THERMODYNAMIC SUBSTANCES

In the analysis and design of thermodynamic processes, devices, and systems, we encounter many different types of thermodynamic *substances*. Among the many types, the three most important and frequently used types are pure substance, ideal gas, and incompressible substance.

A *pure substance* is a simple substance that has a homogeneous and invariable chemical composition and has only one relevant simple compressible work mode.

An *ideal gas* is mathematically defined as one whose thermodynamic equation of state is given by pv=RT, where p is the absolute pressure, v is the specific volume, R is the gas constant, and T is the absolute temperature of the gas, respectively.

An *incompressible substance* is a substance whose specific volume remains nearly constant during a thermodynamic process. Most liquids and solids can be assumed to be incompressible without much loss in accuracy.

Since system performance characteristics depend on the properties of the working substance used, it is essential that we have a good understanding of the thermodynamic behavior of a substance and know how to find properties of these substances.

Homework 2.1 Thermodynamic Substances

1. What are the three most important and frequently used types of thermodynamic substances?

2.2 Pure Substances

A substance can exist in solid, liquid, or gas phase. At normal room pressure and temperature, copper is a solid, water is a liquid, and nitrogen is a gas; but each of these substances can appear in a different phase if the pressure or temperature is changed sufficiently.

A *phase* is any homogeneous part of a system that is physically distinct and is separated from other parts of the system by a definite bounding surface. Solid ice, liquid water, and water vapor constitute three separate phases of the substance H_2O.

A pure substance is a simple substance that has a homogeneous and invariable chemical composition and has only one relevant work mode. A pure substance is known as a pure compressible substance when the only relevant work mode is the compression work (pdV work) only. Water is a pure substance even if it exists as a mixture of liquid and vapor, or as a mixture of liquid, vapor, and solid, since its chemical composition is the same in all phases. Atmospheric air, which is essentially a mixture of nitrogen and oxygen, may be treated as a pure substance as long as it remains in the gaseous state. A mixture of gaseous air in equilibrium with liquid air is not a pure substance since the chemical composition in the gaseous state is not the same as that in the liquid phase. A mixture of oil and water is also not a pure substance since the chemical composition is not homogeneous throughout the system.

Primary thermodynamic practical interest is in situations involving the liquid, liquid-vapor, and vapor regions. Consider a pure substance, water (H_2O), contained in a piston-cylinder arrangement in Figure 2.2.1 be the system. The change of a subcooled H_2O liquid state to a superheated H_2O vapor state by constant-pressure heat addition process can be demonstrated by the following simple experiment.

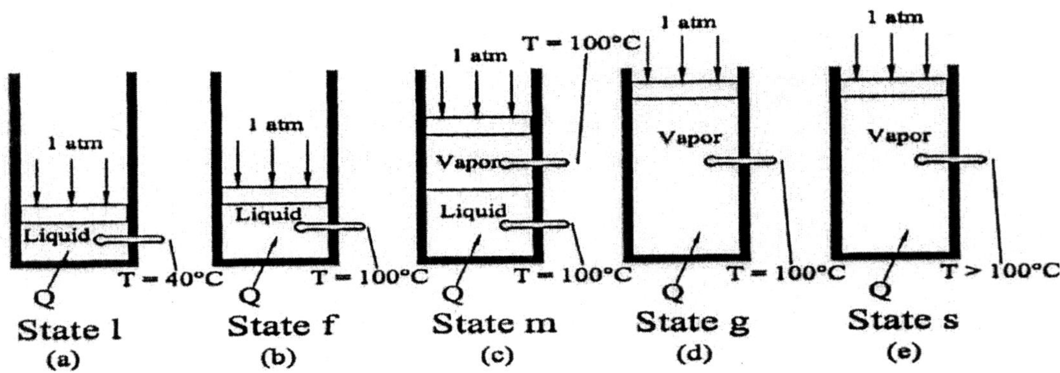

Figure 2.2.1 Liquid-vapor transition

1. Heat is added at constant pressure of 101.3 kPa (corresponding boiling temperature at this pressure is 100°C) to the liquid water initially at 40°C (point 1, Figure 2.2.2a). State 1 is in a region called the *sub-cooled* region because the state temperature is lower than the boiling temperature for this pressure at 101.3 kPa. The region is also called the *compressed-liquid* region because the state pressure is higher than the boiling pressure for this temperature at 40°C.
2. As heat is added, the water system temperature rises until it reaches 100°C (point f, Figure 2.2.2b), state f is called a *saturated liquid* and denoted by a subscript f, this means that it is at the highest temperature at which, for this pressure, it can remain liquid. The water starts to boil.
3. As more heat is added, the system temperature remains the same at 100°C during the boiling process. Some of the liquid changes to vapor, and a mixture of vapor and liquid occurs such as state m (point m, Figure 2.2.2c). This state is in a two phase region called *saturated mixture* region.
4. As more heat is added, the system temperature remains the same at 100°C and all the liquid in the mixture changes to vapor (point g, Figure 2.2.2d), state g is called a *saturated vapor* and denoted by a subscript g, this means that it is at the lowest temperature at which, for this pressure, it can remain vapor.
5. Any more heat addition to the system results in temperature rises over the boiling temperature (point s, Figure 2.2.2e), state s is called a superheated vapor. State s is in a region called *superheated vapor* region because the state temperature is higher than the boiling temperature for this pressure.

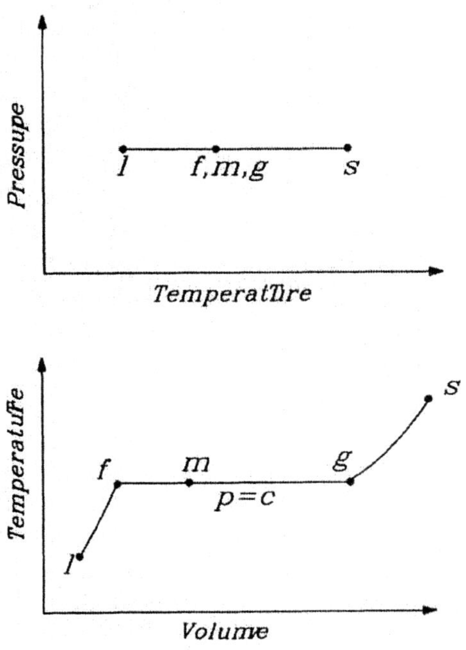

Figure 2.2.2. T-v diagram

Repeat the same isobaric process test but at different pressures. Plot the results on a T-v diagram. Connect the locus of point f and the locus of point g. These two lines are called the saturated liquid line and the saturated vapor line, respectively. The two lines merge at a point c called the *critical point*. Point c has a unique temperature called critical temperature and a unique pressure called critical pressure. The T-v diagram for water at a pressure process with the saturation lines is shown in Figure 2.2.3. The T-v diagram for water at various pressures is shown in Figure 2.2.3.

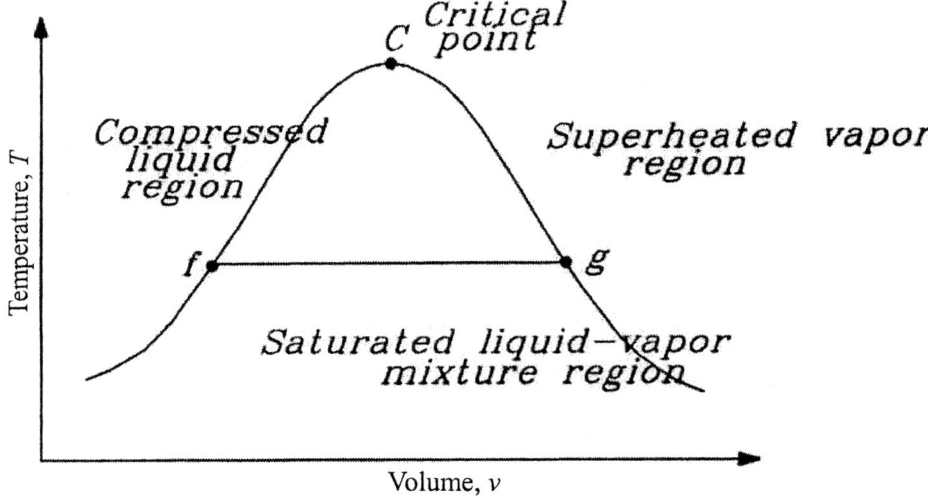

A *T-v* diagram illustrating three regions included in the steam tables

Figure 2.2.3. T-v diagram

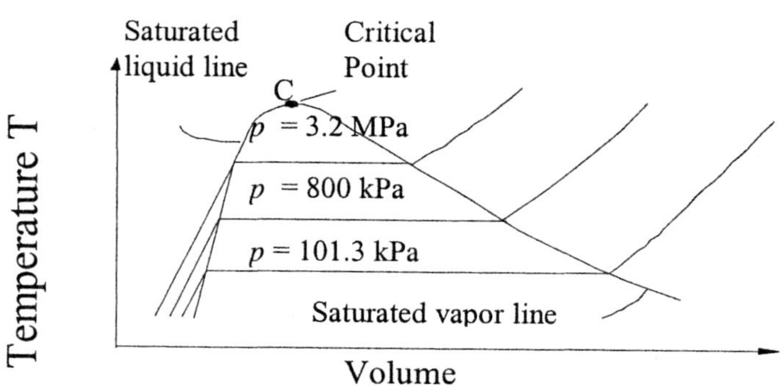

The T- v diagram for water at various pressures.

Figure 2.2.4. T-v diagram

As illustrated in Figure 2.2.3, there are three regions called sub-cooled (or compressed) liquid region, saturated mixture region and superheated region separated by the two saturated lines. At pressures higher than the critical pressure, the liquid could be heated from a low temperature to a high temperature without a phase transition occuring.

Referring to Figure 2.2.3, it is possible to locate a state (point) of water by knowing the temperature and pressure if the state is in the sub-cooled liquid or superheated vapor region. However, it is not possible to locate a state (point) of water by knowing the temperature and pressure if the state is in the saturated mixture region. In the saturated mixture region, temperature and pressure are not independent. In order to define a state in the saturated mixture region, another property such as quality is required so that the fraction of the vapor in the mixture would be known. It is important to realize that at least two independent intensive properties are needed to determine the state of a pure substance.

Similar experiments can be done for solid-liquid transition and solid-vapor transition for a pure substance. The results are plotted on a p-T diagram, Figure 2.2.5. There are three two-phase lines called freezing (solid and liquid) line, boiling (vapor and liquid) line, and sublimation (solid and vapor) line on the diagram. There are three phase (solid, liquid and vapor) regions separated by the three lines. The three lines intersect at a point where all three phases can coexist. This point is called triple point.

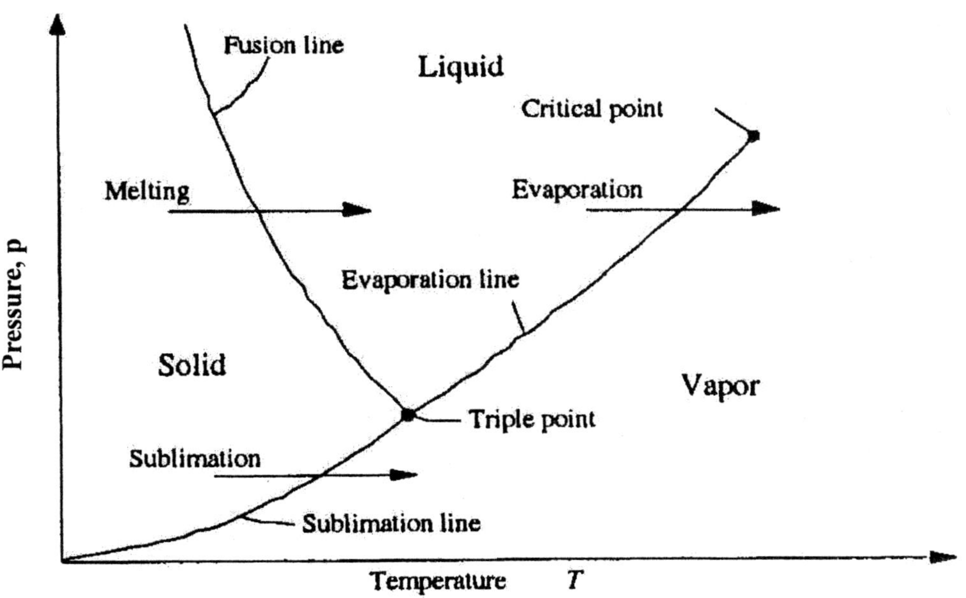

A *p-T* diagram showing phase equilibrium lines, the triple point, and the critical point

Figure 2.2.5 Three phase p-T diagram

The relationships among thermodynamic properties of the working substance at an equilibrium state are called equations of state. These equations in general are rather complicated and cumbersome to handle. It certainly would be convenient if tables or charts existed listing the values of the thermodynamic functions. Fortunately, tables and charts for many substances are available.

Thermodynamic data properties may be presented in the form of diagrams such as Mollier steam diagram (Figure 2.2.6). The Mollier diagram is a h-s diagram with constant pressure, constant temperature and constant quality lines. Notice that specific volume and internal energy values can not be read directly from the Mollier steam diagram.

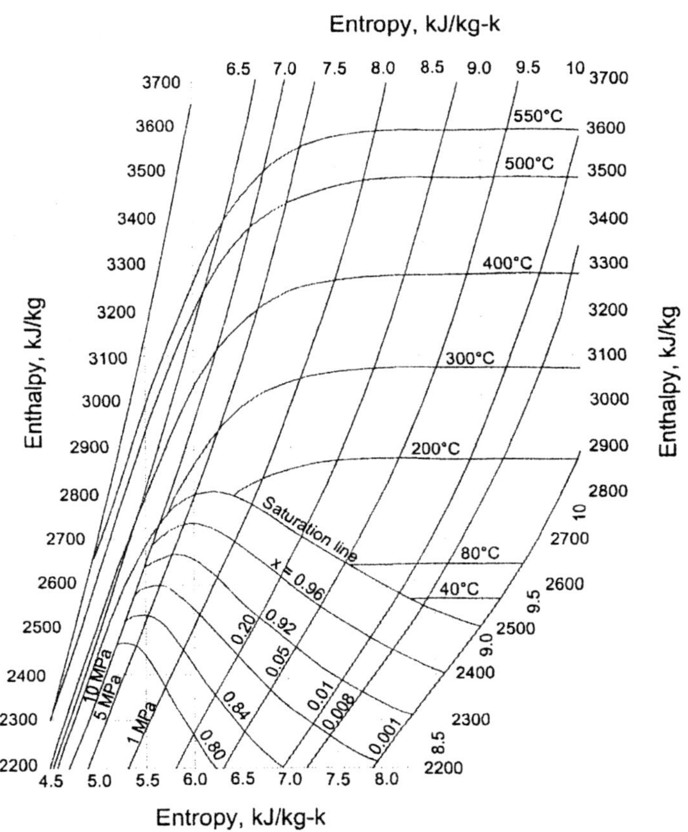

Figure 2.2.6 Mollier steam property diagram

Example 2.2.1. 3 kg of water is contained in a tank at (a) 0.1 Mpa and a quality of 0.9, and (b) 0.5 Mpa and 500°C. Determine the specific enthalpy and specific entropy of the water.

Solution: Mollier steam property diagram reading gives: (a) h=2450 kJ/kg and s=6.55 kJ/[kg(K)], and (b) h=3490 kJ/kg and s=8.3 kJ/[kg(K)].

However, the conventional way of presenting such data is in the form of saturation (saturated mixture) tables, superheated vapor tables, and compressed liquid tables. Typically, these tables give list values for p (pressure), T (temperature), v (specific volume), v_f (specific volume of saturated liquid), v_g (specific volume of saturated vapor), v_{fg} (difference between specific volume of saturated vapor and specific volume of saturated liquid, v_g-v_f), u (specific internal energy), u_f (specific internal energy of saturated liquid), v_g (specific internal energy of saturated vapor), u_{fg} (difference between specific internal energy of saturated vapor and specific internal energy of saturated liquid, u_g-u_f), h (specific enthalpy), h_f (specific enthalpy of saturated liquid), h_g (specific enthalpy of saturated vapor), h_{fg} (difference between specific enthalpy of saturated vapor and specific enthalpy of saturated liquid, h_g-h_f), s (specific entropy), s_f (specific entropy of saturated liquid), s_g (specific entropy of saturated vapor), and s_{fg} (difference between specific entropy of saturated vapor and specific entropy of saturated liquid, s_g-s_f).

Notice that the compressed liquid table for refrigerants is usually not available. In the absence of compressed liquid table, the following approximate equations are used to calculate v, h and u at a state with given pressure and temperature in the compressed liquid region.

$$v=v_f \text{ at the state temperature} \tag{2.2.1}$$

$$u=u_f \text{ at the state temperature} \tag{2.2.2}$$

$$h=h_f \text{ at the state temperature} \tag{2.2.3}$$

In the saturation mixture region, v, u, h and s are expressed in terms of the quality or called dryness (x) by the following equations:

$$v= (1-x)v_f + xv_g = v_f + xv_{fg} \tag{2.2.4}$$

$$u= (1-x)u_f + xu_g = u_f + xu_{fg} \tag{2.2.5}$$

$$h= (1-x)s_f + xs_g = s_f + xs_{fg} \tag{2.2.6}$$

and

$$s= (1-x)s_f + xs_g = s_f + xs_{fg} \tag{2.2.7}$$

The quality (dryness) of the mixture is only defined in the mixture region. This property does not apply to superheated vapor nor compressed liquid regions. It can be written as

$$x = (v-v_f)/v_{fg} \tag{2.2.8}$$

$$x = (u-u_f)/u_{fg} \tag{2.2.9}$$

$$x = (h-h_f)/h_{fg} \tag{2.2.10}$$

and

$$x = (s-s_f)/s_{fg} \tag{2.2.11}$$

A part of the typical steam saturation tables (Table 2.2.1 and Table 2.2.2), and superheated vapor table (Table2.2.3) are given in the following tables. Notice that Table 2.2.1 and Table 2.2.2 are the same, except that Table2.2.1 is based on boiling temperature and Table 2.2.1 is based on boiling pressure. Compressed or sub-cooled liquid water table is also available. However, the compressed or sub-cooled liquid water table is not very useful because it is large and can be replaced by a set of equations [Eq. (2.2.1), (2.2.2) and (2.2.3)].

Example 2.2.2. 8 kg of water is contained in a tank at 0.1 Mpa (1 Bar) and a quality of 0.9. Determine the temperature, specific volume, specific internal energy, specific enthalpy and specific entropy of the water.

Solution: Quality is only defined in the saturated mixture region. Steam saturation property table reading gives:

T=99.63°C, v_f=0.001043 m^3/kg, v_g=1.6940 m^3/kg, u_f=417.36 kJ/kg, u_{fg}=2088.7 kJ/kg, u_g=2506.1kJ/kg, h_f=417.46 kJ/kg, h_{fg}=2258.0 kJ/kg, s_f=1.3026 kJ/[kg(K)], s_{fg}=6.0568 kJ/[kg(K)], and s_g=7.3594 kJ/[kg(K)].

Eqs. (2.2.4), (2.2.5), (2.2.6), and (2.2.7) yields

$$v=(1-x)\,v_f + xv_g = (0.1)0.001043+ (0.9)1.6940 =1.5247 \text{ m}^3/\text{kg}$$

$$u= u_f + xu_{fg} =417.36+(0.9)2088.7 =2297.2 \text{ kJ/kg}$$

$$h= s_f + xs_{fg} =417.46+ (0.9)2258.0 =2449.7 \text{ kJ/kg}$$

and

$$s= s_f + xs_{fg} =1.3026+ (0.9)6.0568 =6.7537 \text{ kJ/[kg(K)]}$$

Example 2.2.3. 8 kg of water is contained in a tank at 0.5 Mpa (5 Bar) and 320°C. Determine the temperature, specific volume, specific internal energy, specific enthalpy and specific entropy of the water.

Solution: If one is not sure where the state is, always check first at the saturation mixture region. At 0.5 Mpa, the corresponding boiling temperature is 81.33°C. Since the given state temperature (320°C) is higher than 81.33°C, the state is in the superheated vapor region. Steam superheated vapor property table reading gives: v =0.5416 m^3/kg, u =2834.7 kJ/kg, h =3105.6 kJ/kg, and s =7.5308 kJ/[kg(K)].

Example 2.2.4. 8 kg of water is contained in a tank at 10 Mpa (100 Bar) and 120°C. Determine the temperature, specific volume, specific internal energy, specific enthalpy and specific entropy of the water.

Solution: If one is not sure where the state is, always check first at the saturation mixture region. At 120°C, the corresponding boiling pressure is 0.1985 Bar (1.985 Mpa). Since the given state pressure (10 Mpa) is higher than 1.985 Mpa, the state is in the compressed liquid region. Water compressed liquid property table is not needed. Using saturation table at 120°C, reading gives: v = 0.0010603 m^3/kg, u =503.5 kJ/kg, h =503.71 kJ/kg, and s =1.5276 kJ/[kg(K)].

Table 2.2.1 Steam Saturation Table

Properties of water—saturation-temperature table (SI units)*

v in cm³/g, 1 cm³/g = 10^{-3} m³/kg; h and u in kJ/kg; s in kJ/kg·K; p in bars, 1 bar = 10^5Pa

		Specific volume		Internal energy		Enthalpy			Entropy	
Temp. °C T	Press. bars P	Sat. liquid v_f	Sat. vapor v_g	Sat. liquid u_f	Sat. vapor u_g	Sat. liquid h_f	Evap. h_{fg}	Sat. vapor h_g	Sat. liquid s_f	Sat vapor s_g
0	0.00611	1.0002	206278	−0.03	2375.4	−0.02	2501.4	2501.3	−0.0001	9.1565
5	0.00872	1.0001	147120	20.97	2382.3	20.98	2489.6	2510.6	0.0761	9.0257
10	0.01228	1.0004	106379	42.00	2389.2	42.01	2477.7	2519.8	0.1510	8.9008
15	0.01705	1.0009	77926	62.99	2396.1	62.99	2465.9	2528.9	0.2245	8.7814
20	0.02339	1.0018	57791	83.95	2402.9	83.96	2454.1	2538.1	0.2966	8.6672
25	0.03169	1.0029	43360	104.88	2409.8	104.89	2442.3	2547.2	0.3674	8.5580
30	0.04246	1.0043	32894	125.78	2416.6	125.79	2430.5	2556.3	0.4369	8.4533
35	0.05628	1.0060	25216	146.67	2423.4	146.68	2418.6	2565.3	0.5053	8.3531
40	0.07384	1.0078	19523	167.56	2430.1	167.57	2406.7	2574.3	0.5725	8.2570
45	0.09593	1.0099	15258	188.44	2436.8	188.45	2394.8	2583.2	0.6387	8.1648
50	0.1235	1.0121	12032	209.32	2443.5	209.33	2382.7	2592.1	0.7038	8.0763
55	0.1576	1.0146	9568	230.21	2450.1	230.23	2370.7	2600.9	0.7679	7.9913
60	0.1994	1.0172	7671	251.11	2456.6	251.13	2358.5	2609.6	0.8312	7.9096
65	0.2503	1.0199	6197	272.02	2463.1	272.06	2346.2	2618.3	0.8935	7.8310
70	0.3119	1.0228	5042	292.95	2469.6	292.98	2333.8	2626.8	0.9549	7.7553
75	0.3858	1.0259	4131	313.90	2475.9	313.93	2321.4	2635.3	1.0155	7.6824
80	0.4739	1.0291	3407	334.86	2482.2	334.91	2308.8	2643.7	1.0753	7.6122
85	0.5783	1.0325	2828	355.84	2488.4	355.90	2296.0	2651.9	1.1343	7.5445
90	0.7014	1.0360	2361	376.85	2494.5	376.92	2283.2	2660.1	1.1925	7.4791
95	0.8455	1.0397	1982	397.88	2500.6	397.96	2270.2	2668.1	1.2500	7.4159
100	1.014	1.0435	1673.	418.94	2506.5	419.04	2257.0	2676.1	1.3069	7.3549
110	1.433	1.0516	1210.	461.14	2518.1	461.30	2230.2	2691.5	1.4185	7.2387
120	1.985	1.0603	891.9	503.50	2529.3	503.71	2202.6	2706.3	1.5276	7.1296
130	2.701	1.0697	668.5	546.02	2539.9	546.31	2174.2	2720.5	1.6344	7.0269
140	3.613	1.0797	508.9	588.74	2550.0	589.13	2144.7	2733.9	1.7391	6.9299
150	4.758	1.0905	392.8	631.68	2559.5	632.20	2114.3	2746.5	1.8418	6.8379
160	6.178	1.1020	307.1	674.86	2568.4	675.55	2082.6	2758.1	1.9427	6.7502
170	7.917	1.1143	242.8	718.33	2576.5	719.21	2049.5	2768.7	2.0419	6.6663
180	10.02	1.1274	194.1	762.09	2583.7	763.22	2015.0	2778.2	2.1396	6.5857
190	12.54	1.1414	156.5	806.19	2590.0	807.62	1978.8	2786.4	2.2359	6.5079
200	15.54	1.1565	127.4	850.65	2595.3	852.45	1940.7	2793.2	2.3309	6.4323
210	19.06	1.1726	104.4	895.53	2599.5	897.76	1900.7	2798.5	2.4248	6.3585
220	23.18	1.1900	86.19	940.87	2602.4	943.62	1858.5	2802.1	2.5178	6.2861
230	27.95	1.2088	71.58	986.74	2603.9	990.12	1813.8	2804.0	2.6099	6.2146
240	33.44	1.2291	59.76	1033.2	2604.0	1037.3	1766.5	2803.8	2.7015	6.1437

Note: In case there is no compressed liquid property table available, one can use the saturated mixture property table to locate temperature at 120°C (the nearest value on Table 2.2.1 is 120.23°C). Then, Eqs. (2.2.1), (2.2.1) and (2.2.1) are used to find the approximate values of v, h and u at a state with given pressure and temperature in the compressed liquid region.

Table 2.2.2 Steam Saturation Table

Properties of water—saturation-pressure table (SI units)*

v in cm³/g, 1 cm³/g = 10^{-3} m³/kg; h and u in kJ/kg; s in kJ/kg·K; p in bars, 1 bar = 10^5 Pa

		Specific volume		Internal energy		Enthalpy			Entropy	
Press. bars P	Temp. °C T	Sat. liquid v_f	Sat. vapor v_g	Sat. liquid u_f	Sat. vapor u_g	Sat. liquid h_f	Evap. h_{fg}	Sat. vapor h_g	Sat. liquid s_f	Sat. vapor s_g
0.040	28.96	1.0040	34800.	121.45	2415.2	121.46	2432.9	2554.4	0.4226	8.4746
0.060	36.16	1.0064	23729.	151.53	2425.0	151.53	2415.9	2567.4	0.5210	8.3304
0.080	41.51	1.0084	18103.	173.87	2432.2	173.88	2403.1	2577.0	0.5926	8.2287
0.10	45.81	1.0102	14674.	191.82	2437.9	191.83	2392.8	2584.7	0.6493	8.1502
0.20	60.06	1.0172	7649.	251.38	2456.7	251.40	2358.3	2609.7	0.8320	7.9085
0.30	69.10	1.0223	5229.	289.20	2468.4	289.23	2336.1	2625.3	0.9439	7.7686
0.40	75.87	1.0265	3993.	317.53	2477.0	317.58	2319.2	2636.8	1.0259	7.6700
0.50	81.33	1.0300	3240.	340.44	2483.9	340.49	2305.4	2645.9	1.0910	7.5939
0.60	85.94	1.0331	2732.	359.79	2489.6	359.86	2293.6	2653.5	1.1453	7.5320
0.70	89.95	1.0360	2365.	376.63	2494.5	376.70	2283.3	2660.0	1.1919	7.4797
0.80	93.50	1.0380	2087.	391.58	2498.8	391.66	2274.1	2665.8	1.2329	7.4346
0.90	96.71	1.0410	1869.	405.06	2502.6	405.15	2265.7	2670.9	1.2695	7.3949
1.00	99.63	1.0432	1694.	417.36	2506.1	417.46	2258.0	2675.5	1.3026	7.3594
1.50	111.4	1.0528	1159.	466.94	2519.7	467.11	2226.5	2693.6	1.4336	7.2233
2.00	120.2	1.0605	885.7	504.49	2529.5	504.70	2201.9	2706.7	1.5301	7.1271
2.50	127.4	1.0672	718.7	535.10	2537.2	535.37	2181.5	2716.9	1.6072	7.0527
3.00	133.6	1.0732	605.8	561.15	2543.6	561.47	2163.8	2725.3	1.6718	6.9919
3.50	138.9	1.0786	524.3	583.95	2548.9	584.33	2148.1	2732.4	1.7275	6.9405
4.00	143.6	1.0836	462.5	604.31	2553.6	604.74	2133.8	2738.6	1.7766	6.8959
4.50	147.9	1.0882	414.0	622.77	2557.6	623.25	2120.7	2743.9	1.8207	6.8565
5.00	151.9	1.0926	374.9	639.68	2561.2	640.23	2108.5	2748.7	1.8607	6.8213
6.00	158.9	1.1006	315.7	669.90	2567.4	670.56	2086.3	2756.8	1.9312	6.7600
7.00	165.0	1.1080	272.9	696.44	2572.5	697.22	2066.3	2763.5	1.9922	6.7080
8.00	170.4	1.1148	240.4	720.22	2576.8	721.11	2048.0	2769.1	2.0462	6.6628
9.00	175.4	1.1212	215.0	741.83	2580.5	742.83	2031.1	2773.9	2.0946	6.6226
10.0	179.9	1.1273	194.4	761.68	2583.6	762.81	2015.3	2778.1	2.1387	6.5863
15.0	198.3	1.1539	131.8	843.16	2594.5	844.89	1947.3	2792.2	2.3150	6.4448
20.0	212.4	1.1767	99.63	906.44	2600.3	908.79	1890.7	2799.5	2.4474	6.3409
25.0	224.0	1.1973	79.98	959.11	2603.1	962.11	1841.0	2803.1	2.5547	6.2575
30.0	233.9	1.2165	66.68	1004.8	2604.1	1008.4	1795.7	2804.2	2.6457	6.1869
35.0	242.6	1.2347	57.07	1045.4	2603.7	1049.8	1753.7	2803.4	2.7253	6.1253
40.0	250.4	1.2522	49.78	1082.3	2602.3	1087.3	1714.1	2801.4	2.7964	6.0701
45.0	257.5	1.2692	44.06	1116.2	2600.1	1121.9	1676.4	2798.3	2.8610	6.0199
50.0	264.0	1.2859	39.44	1147.8	2597.1	1154.2	1640.1	2794.3	2.9202	5.9734
60.0	275.6	1.3187	32.44	1205.4	2589.7	1213.4	1571.0	2784.3	3.0267	5.8892
70.0	285.9	1.3513	27.37	1257.6	2580.5	1267.0	1505.1	2772.1	3.1211	5.8133
80.0	295.1	1.3842	23.52	1305.6	2569.8	1316.6	1441.3	2758.0	3.2068	5.7432
90.0	303.4	1.4178	20.48	1350.5	2557.8	1363.3	1378.9	2742.1	3.2858	5.6772
100.	311.1	1.4524	18.03	1393.0	2544.4	1407.6	1317.1	2724.7	3.3596	5.6141
110.	318.2	1.4886	15.99	1433.7	2529.8	1450.1	1255.5	2705.6	3.4295	5.5527
120.	324.8	1.5267	14.26	1473.0	2513.7	1491.3	1193.6	2684.9	3.4962	5.4924
130.	330.9	1.5671	12.78	1511.1	2496.1	1531.5	1130.7	2662.2	3.5606	5.4323
140.	336.8	1.6107	11.49	1548.6	2476.8	1571.1	1066.5	2637.6	3.6232	5.3717
150.	342.2	1.6581	10.34	1585.6	2455.5	1610.5	1000.0	2610.5	3.6848	5.3098
160.	347.4	1.7107	9.306	1622.7	2431.7	1650.1	930.6	2580.6	3.7461	5.2455
170.	352.4	1.7702	8.364	1660.2	2405.0	1690.3	856.9	2547.2	3.8079	5.1777
180.	357.1	1.8397	7.489	1698.9	2374.3	1732.0	777.1	2509.1	3.8715	5.1044
190.	361.5	1.9243	6.657	1739.9	2338.1	1776.5	688.0	2464.5	3.9388	5.0228
200.	365.8	2.036	5.834	1785.6	2293.0	1826.3	583.4	2409.7	4.0139	4.9269
220.9	374.1	3.155	3.155	2029.6	2029.6	2099.3	0	2099.3	4.4298	4.4298

Table 2.2.3 Steam Superheated Vapor Table

Properties of water – Superheated table (SI units)
v in cm^3/g, 1 cm^3/g=0.001 m^3/kg;; h and u in kJ/kg; s in kJ/[(K)kg]; p in bars, 1 bar=10^2 kPa.

Temp. °C	*v*	*u*	*h*	*s*	*v*	*u*	*h*	*s*
	500 kPa (151.86°C)				700 kPa (164.97°C)			
Sat.	374.9	2561.2	2748.7	6.8213	272.9	2572.5	2763.5	6.7080
180	404.5	2609.7	2812.0	6.9656	284.7	2599.8	2799.1	6.7880
200	424.9	2642.9	2855.4	7.0592	299.9	2634.8	2844.8	6.8865
240	464.6	2707.6	2939.9	7.2307	329.2	2701.8	2932.2	7.0641
280	503.4	2771.2	3022.9	7.3865	357.4	2766.9	3017.1	7.2233
320	541.6	2834.7	3105.6	7.5308	385.2	2831.3	3100.9	7.3697
360	579.6	2898.7	3188.4	7.6660	412.6	2895.8	3184.7	7.5063
400	617.3	2963.2	3271.9	7.7938	439.7	2960.9	3268.7	7.6350
440	654.8	3028.6	3356.0	7.9152	466.7	3026.6	3353.3	7.7571
500	710.9	3128.4	3483.9	8.0873	507.0	3126.8	3481.7	7.9299
600	804.1	3299.6	3701.7	8.3522	573.8	3298.5	3700.2	8.1956
700	896.9	3477.5	3925.9	8.5952	640.3	3476.6	3924.8	8.4391
	1.0 MPa (179.91°C)				1.5 MPa (198.32°C)			
Sat.	194.4	2583.6	2778.1	6.5865	131.8	2594.5	2792.2	6.4448
200	206.0	2621.9	2827.9	6.6940	132.5	2598.1	2796.8	6.4546
240	227.5	2692.9	2920.4	6.8817	148.3	2676.9	2899.3	6.6628
280	248.0	2760.2	3008.2	7.0465	162.7	2748.6	2992.7	6.8381
320	267.8	2826.1	3093.9	7.1962	176.5	2817.1	3081.9	6.9938
360	287.3	2891.6	3178.9	7.3349	189.9	2884.4	3169.2	7.1363
400	306.6	2957.3	3263.9	7.4651	203.0	2951.3	3255.8	7.2690
440	325.7	3023.6	3349.3	7.5883	216.0	3018.5	3342.5	7.3940
500	354.1	3124.4	3478.5	7.7622	235.2	3120.3	3473.1	7.5698
540	372.9	3192.6	3565.6	7.8720	247.8	3189.1	3560.9	7.6805
600	401.1	3296.8	3697.9	8.0290	266.8	3293.9	3694.0	7.8385
640	419.8	3367.4	3787.2	8.1290	279.3	3364.8	3783.8	7.9391
	2.0 MPa (212.42°C)				3.0 MPa (233.90°C)			
Sat.	99.6	2600.3	2799.5	6.3409	66.7	2604.1	2804.2	6.1869
240	108.5	2659.6	2876.5	6.4952	68.2	2619.7	2824.3	6.2265
280	120.0	2736.4	2976.4	6.6828	77.1	2709.9	2941.3	6.4462
320	130.8	2807.9	3069.5	6.8452	85.0	2788.4	3043.4	6.6245
360	141.1	2877.0	3159.3	6.9917	92.3	2861.7	3138.7	6.7801
400	151.2	2945.2	3247.6	7.1271	99.4	2932.8	3230.9	6.9212
440	161.1	3013.4	3335.5	7.2540	106.2	3002.9	3321.5	7.0520
500	175.7	3116.2	3467.6	7.4317	116.2	3108.0	3456.5	7.2338
540	185.3	3185.6	3556.1	7.5434	122.7	3178.4	3546.6	7.3474
600	199.6	3290.9	3690.1	7.7024	132.4	3285.0	3682.3	7.5085
640	209.1	3362.2	3780.4	7.8035	138.8	3357.0	3773.5	7.6106
700	223.2	3470.9	3917.4	7.9487	148.4	3466.5	3911.7	7.7571

$v=v_f$ at the state temperature $=0.001061\ m^3/kg$

$u=u_f$ at the state temperature $=504.49$ kJ/kg

$h=h_f$ at the state temperature $=504.70$ kJ/kg

It should be pointed out that the values of u, h, and s in all tables are not absolute values. Each is the difference between the value at any state and the value of the respective property at a reference state. But it makes no difference, since we are only interested in changes of u, h and s.

Searching property values from the tables is tedious and long. The available software CyclePad saves users time and energy to spend on more creative activities. It is important for users to become familiar with the use of CyclePad.

There are nine thermodynamic working substances listed on the menu of CyclePad. Among the nine substances listed on the menu, ammonia, methane, refrigerant 12, refrigerant 22, refrigerant 134a, and water are pure substances. Among the most popular used pure working substances in thermodynamic application are refrigerants and water.

Example 2.2.5. Find the properties and determine whether water at each of the following states is a compressed liquid, a saturated mixture, or a superheated vapor.

(a) 10 Mpa and 0.0037 m^3/kg, (b) 10 kPa and 12°C, (c) 1 Mpa and 192°C, (d) 200 kPa and 132°C, (e) 120°C and 7.5 kJ/[(kg)K], and (f) 1 Mpa and x=0.7629.

To solve this problem by CyclePad, we take the following steps:

1. Build
   (A) Take a BEGIN and an END from the closed-system inventory shop and connect them.
   (B) Switch to analysis mode.
2. Analysis
   (A) Input the given information: (a) substance is water, and (b) pressure is 10 MPa and sp. volume is 0.0037 m^3 /kg.
3. Display results
   (A) Display the results. The answers are: (a) saturated mixture, x=0.1356, T=311°C, u=1549 kJ/kg, h=1586 kJ/kg, s=3.67 kJ/[(kg)K]; (b) compressed liquid, v=0.0010 m^3/kg, u=50.36 kJ/kg, h=50.37 kJ/kg, s=0.1803 kJ/[(kg)K]; (c) compressed liquid, v=0.2014 m^3/kg, u=2607 kJ/kg, h=2808 kJ/kg, s=6.65 kJ/[(kg)K]; (d) superheated vapor (gas), v=0.9153 m^3/kg, u=2548 kJ/kg, h=2731 kJ/kg, s=7.19 kJ/[(kg)K]; (e) superheated vapor (gas), p=94.92 kPa, v=0.9153 m^3/kg, u=2537 kJ/kg, h=2717 kJ/kg; (f) saturated mixture, T=179.9°C, v=0.1486 m^3/kg, u=2151 kJ/kg, h=2300 kJ/kg

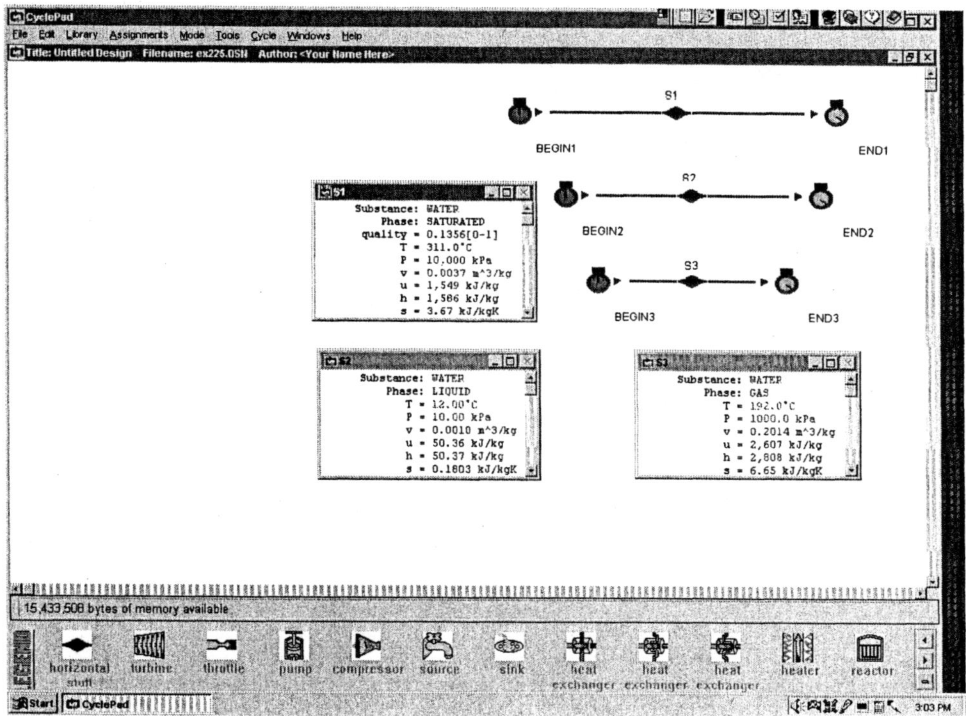

Figure Example 2.2.5a Water property relationships

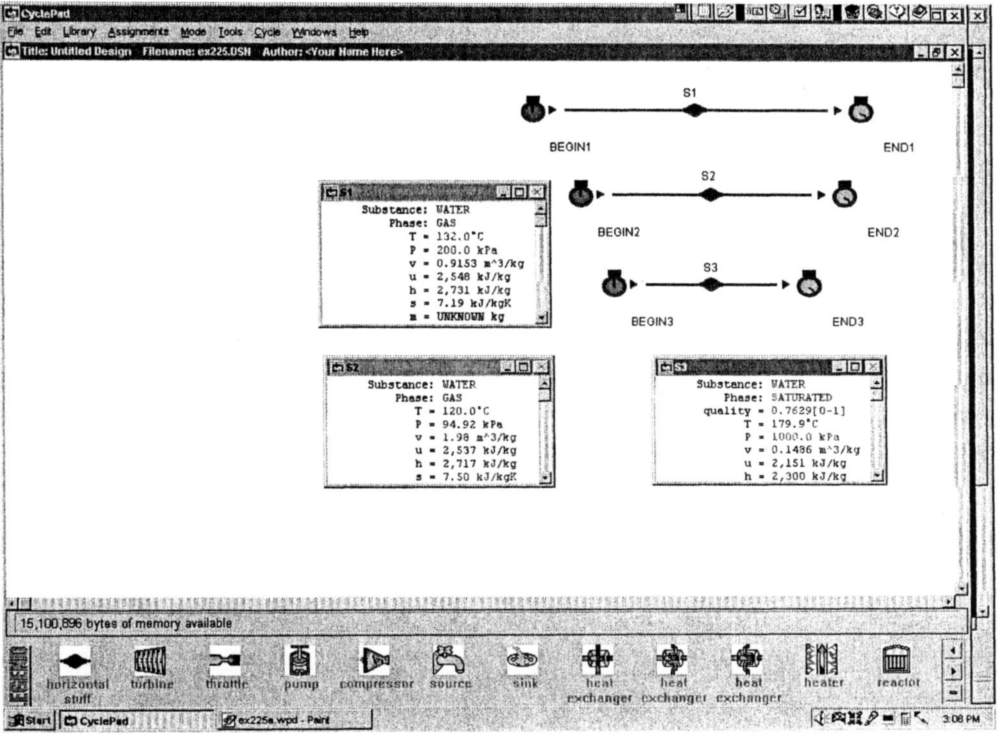

Figure Example 2.2.5b Water property relationships

Example 2.2.6. Find the properties and determine whether water at each of the following states is a compressed liquid, a saturated mixture, or a superheated vapor.

(a) 248°F and x=0.74, (b) 145 psia and h=1000 Btu/lbm, (c) 20 psia and 600°F, (d) 248°F and s=1.8 Btu/[lbm(R)], (e) 145 psia and 355.8°F, and (f) 2000 psia and 600°F.

To solve this problem by CyclePad, we take the following steps:

1. Build
   (A) Take a BEGIN and an END from the closed-system inventory shop and connect them.
   (B) Switch to analysis mode.
2. Analysis
   (A) Input the given information: (a) substance is water, and (b) temperature is 248°F and phase is saturated with x=0.74.
3. Display results
   (A) Display the results. The answers are: (a) saturated mixture, p=28.79 psia, v=10.58 ft^3/lbm, u=860.9 Btu/lbm, h=917.3 Btu/lbm, s=1.36 Btu/[lbm(R)]; (b) saturated mixture, x=0.7758, T=355.8 °F, v=2.42 ft^3/lbm, u=935.1 Btu/lbm, s=1.33 Btu/[lbm(R)]; (c) superheated vapor (gas), v=35.18 ft^3/lbm, u=1218 Btu/lbm, h=917.3 Btu/lbm, s=1.36 Btu/[lbm(R)]; (d) superheated vapor (gas), p=12.97 psia, v=34.87 ft^3/lbm, u=1091 Btu/lbm, h=1168 Btu/lbm, s=1.36 Btu/[lbm(R)]; (e) saturated mixture, x=0.7758, v=2.42 ft^3/lbm, u=935.1 Btu/lbm, h=1000 Btu/lbm, s=1.33 Btu/[lbm(R)]; (f) compressed liquid, v=0.0233 ft^3/lbm, u=605.4 Btu/lbm, h=614.0 Btu/lbm, s=0.8086 Btu/[lbm(R)].

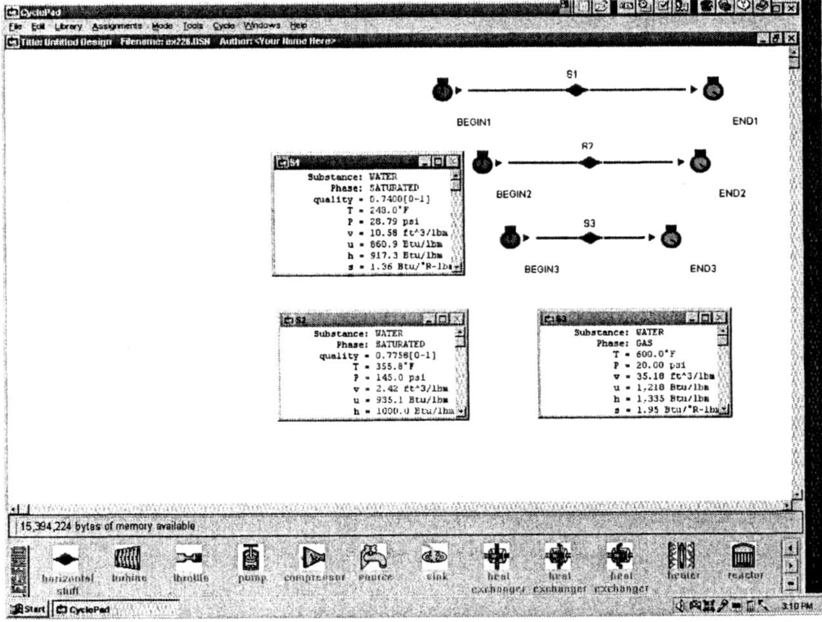

Figure Example 2.2.6a Water property relationships

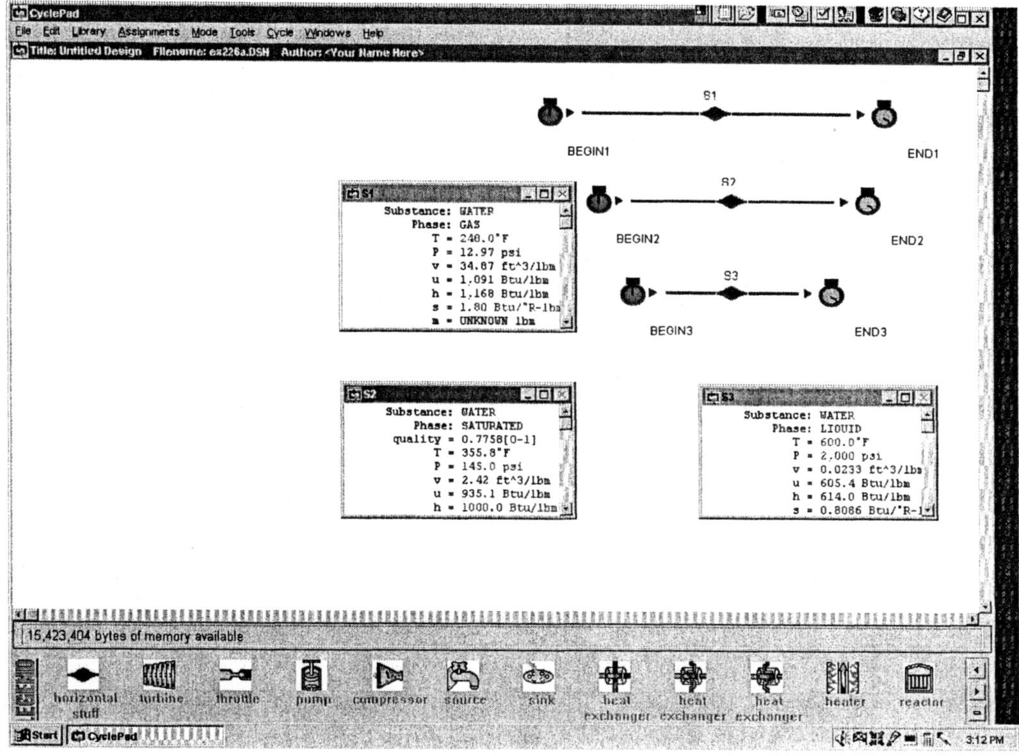

Figure Example 2.2.6b Water property relationships

Example 2.2.7. Find the properties and determine whether Refrigerant-134A at each of the following states is a compressed liquid, a saturated mixture, or a superheated vapor.

(a) 20°C and x=0.74, (b) 1000 kPa and h=450 kJ/kg, and (c) 0.15 Mpa and 40°C.

To solve this problem by CyclePad, we take the following steps:

1. Build
   (A) Take a BEGIN and an END from the closed-system inventory shop and connect them.
   (B) Switch to analysis mode.
2. Analysis
   (A) Input the given information: (a) substance is Refrigerant-134A, and (b) T=20°C, saturated mixture phase and x=0.74.
3. Display results
   (A) Display the results. The answers are: (a) saturated mixture, p=572.8 kPa, v=0.0269 m^3/kg, u=362.4 kJ/kg, h=362.4 kJ/kg, s=1.56 kJ/[(kg)K]; (b) superheated vapor (gas), T=67.79°C, v=0.0240 m^3/kg, u=426.0 kJ/kg, s=1.81 kJ/[(kg)K]; (c) superheated vapor (gas), v=0.1659 m^3/kg, u=411.6 kJ/kg, h=436.5 kJ/kg, s=1.91 kJ/[(kg)K]

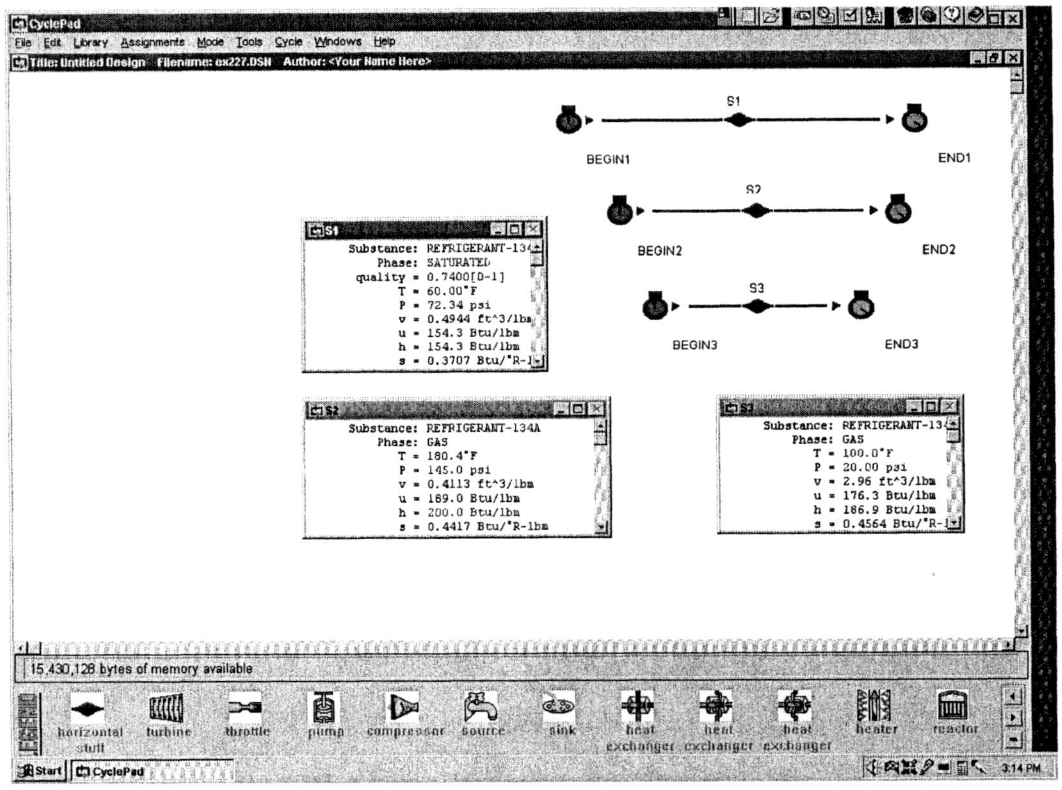

Figure Example 2.2.7 Refrigerant-134A property relationships

Example 2.2.8. Find the properties and determine whether ammonia at each of the following states is a compressed liquid, a saturated mixture, or a superheated vapor.

(a) 20°C and x=0.74, (b) 1000 kPa and h=450 kJ/kg, and (c) 0.15 Mpa and 40°C.

To solve this problem by CyclePad, we take the following steps:

1. Build
   (A) Take a BEGIN and an END from the closed-system inventory shop and connect them.
   (B) Switch to analysis mode.
2. Analysis
   (A) Input the given information: (a) substance is ammonia, and (b) T=20°C, saturated mixture phase and x=0.74.
3. Display results
   (A) Display the results. The answers are: (a) saturated mixture, p=857.2 kPa, v=0.1109 m^3/kg, u=1150 kJ/kg, h=1152 kJ/kg, s=4.03 kJ/[(kg)K]; (b) saturated mixture, x=0.1306, T=24.89°C, v=0.0182 m^3/kg, u=448.3 kJ/kg,; (c) superheated vapor (gas), v=1.01 m^3/kg, u=1406 kJ/kg, h=1557 kJ/kg, s=6.22 kJ/[(kg)K];

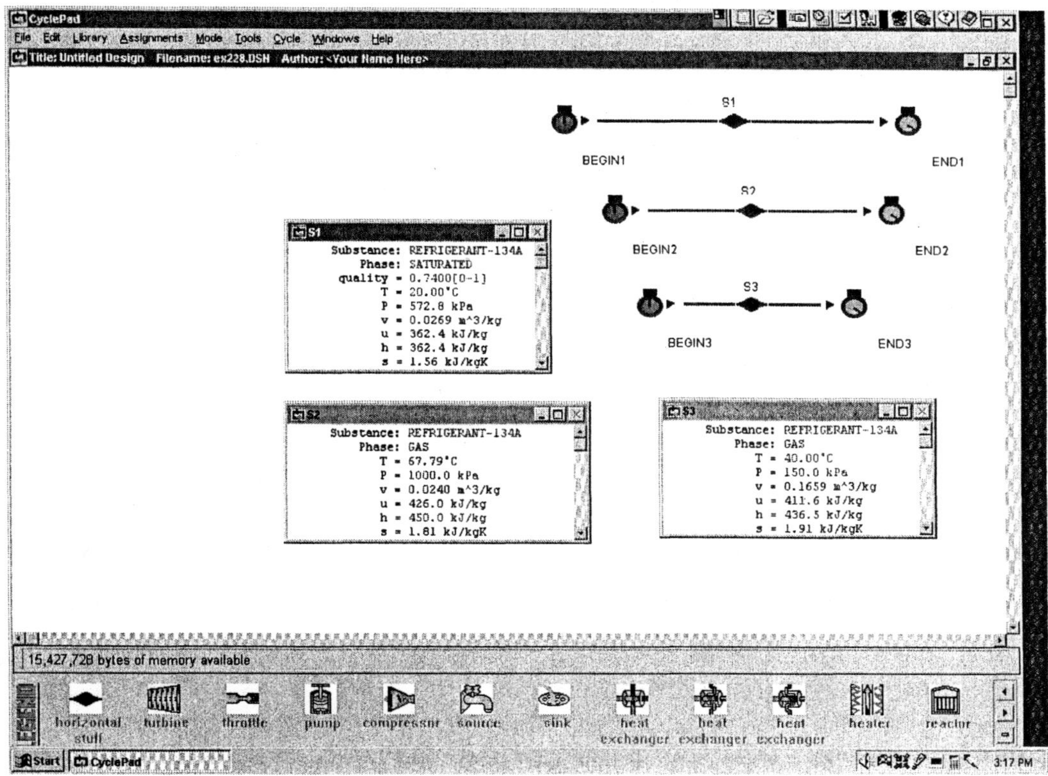

Figure Example 2.2.8 Ammonia property relationships

Example 2.2.9. Find the properties and determine whether Refrigerant-22 at each of the following states is a compressed liquid, a saturated mixture, or a superheated vapor.

(a) 20°C and x=0.74, (b) 1000 kPa and 20°C, and (c) 0.15 Mpa and 40°C.

To solve this problem by CyclePad, we take the following steps:

1. Build
   (A) Take a BEGIN and an END from the closed-system inventory shop and connect them.
   (B) Switch to analysis mode.
2. Analysis
   (A) Input the given information: (a) substance is Refrigerant-22, and (b) T=20°C, saturated mixture phase and x=0.74.
3. Display results
   (A) Display the results. The answers are: (a) saturated mixture, p=909.9 kPa, v=0.195 m^3/kg, u=206.9 kJ/kg, h=207.7 kJ/kg, s=0.7331 kJ/[(kg)K]; (b) compressed liquid, v=0.0 m^3/kg, u=68.67 kJ/kg, s=0.2588 kJ/[(kg)K]; (c) superheated vapor (gas), v=0.1970 m^3/kg, u=253.9 kJ/kg, h=283.5 kJ/kg, s=1.15 kJ/[(kg)K];

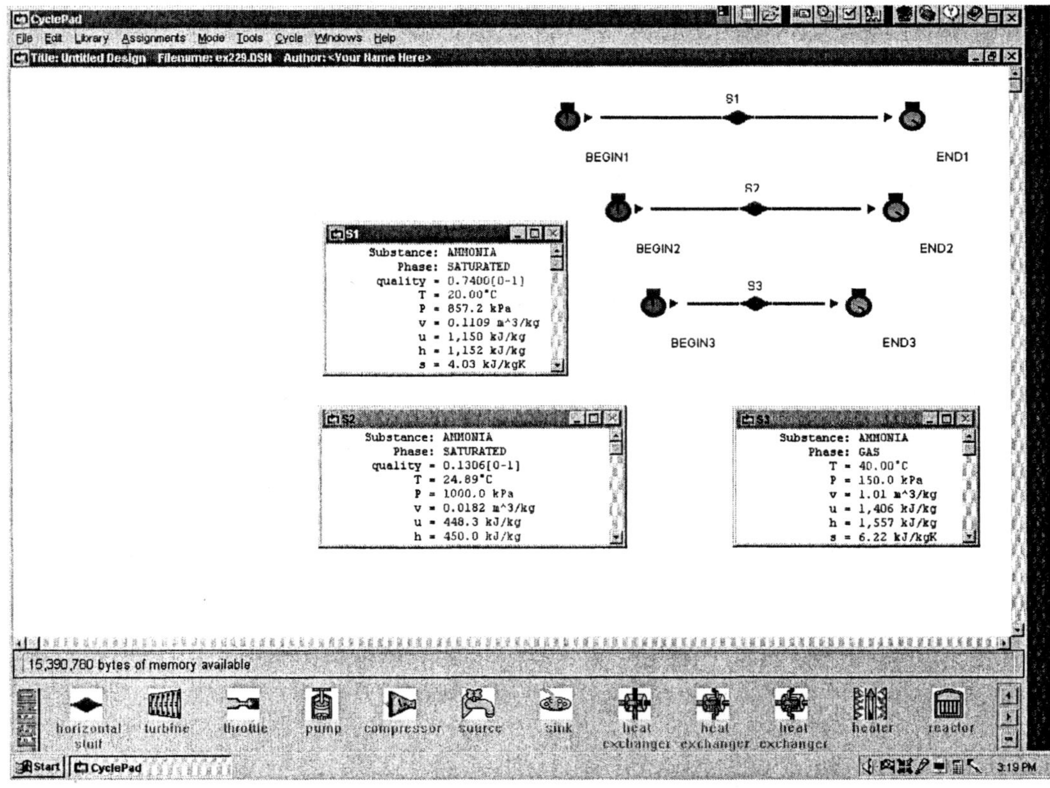

Figure Example 2.2.9 Refrigerant-22 property relationships

Example 2.2.10. Find the properties and determine whether Refrigerant-12 at each of the following states is a compressed liquid, a saturated mixture, or a superheated vapor.

(a) 60°F and x=0.74, (b) 60 psia and v=0.3 ft^3/lbm, and (c) 20 psia and 100°F.

To solve this problem by CyclePad, we take the following steps:

1. Build
   (A) Take a BEGIN and an END from the closed-system inventory shop and connect them.
   (B) Switch to analysis mode.
2. Analysis
   (A) Input the given information: (a) substance is water, and (b) temperature is 60°F and phase is saturated with x=0.74.
3. Display results
   (A) Display the results. The answers are: (a) saturated mixture, p=72.47 psia, v=0.4164 ft^3/lbm, u=61.8 Btu/lbm, h=67.38 Btu/lbm, s=0.134 Btu/[lbm(R)]; (b) saturated mixture, x=0.4366, T=48.57 °F, u=43.42 Btu/lbm, h=46.74 Btu/lbm; (c) superheated vapor (gas), v=2.50 ft^3/lbm, u=83.79 Btu/lbm, h=92.76 Btu/lbm, s=0.2010 Btu/[lbm(R)];

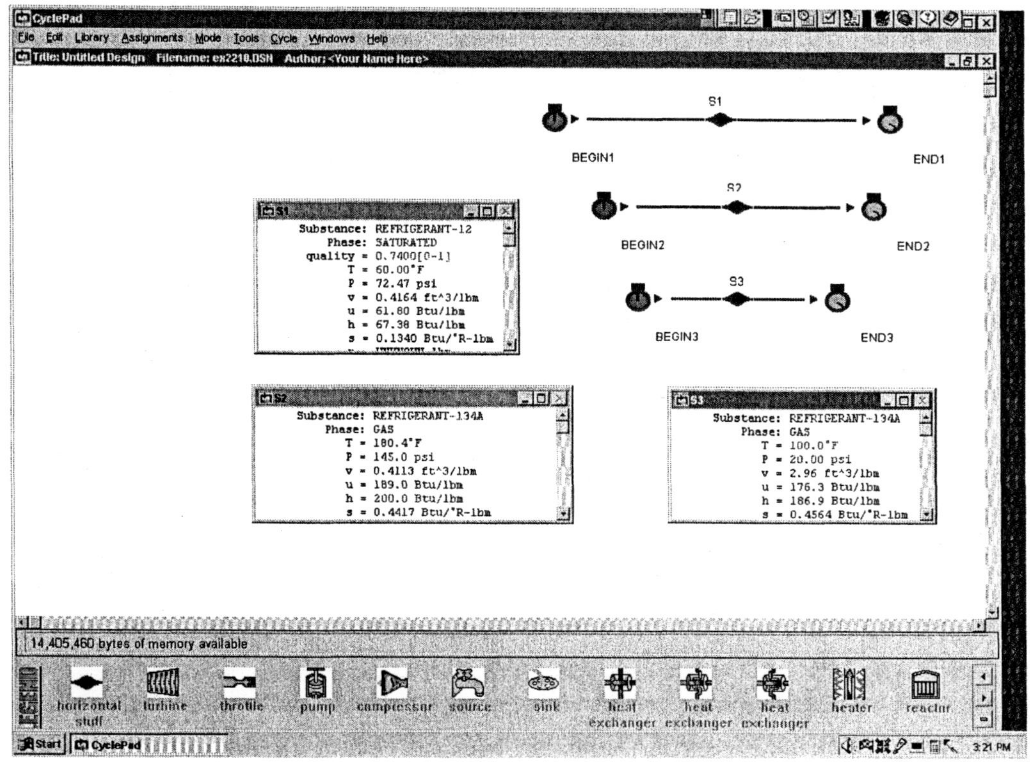

Figure Example 2.2.10 Refrigerant-12 property relationships

Homework 2.2 Pure Substances

1. Show the critical point, superheated vapor region, compressed liquid region, saturated mixture region, saturated liquid line, and saturated vapor line on a T-v phase diagram for a pure substance.
2. Draw a constant pressure heating process from compressed liquid region to superheated vapor region on a T-v phase diagram for a pure substance.
3. Is water a pure substance? Why?
4. What is the difference between saturated liquid and compressed liquid?
5. What is quality and what is moisture content of a saturated mixture of water? Is quality defined in the compressed liquid region?
6. Why water boils at lower temperature on the mountain than at sea level?
7. Are the temperature and pressure dependent properties in the saturated mixture region?
8. What are u_f, u_g and u_{fg}?
9. Can u ever be larger than h?
10. Complete the following table for water using CyclePad:

T(°C)	p(kPa)	v(m³/kg)	u(kJ/kg)	x(%)
50		5		
	200			100
200	400			

600	6000		
	500		50
	100	2000	
	20		97

11. Complete the following table for water using CyclePad:

T(°F)	p (psia)	v (ft^3/lbm)	h (Btu/lbm)	X(%)	
100				1000	
	100		2		
500	400				
270					43
	5				87
	20			1100	
100	2				
600	600				

12. Steam has an entropy of 1.325 Btu/lbm(R) at 100 psia. Determine the quality and moisture content of the steam.
13. Steam at a pressure of 600 psia has an enthalpy value of 1295 Btu/lbm. Determine the temperature of the steam.
14. Complete the following steam table using CyclePad.

State	p psia	T °F	v ft^3/lbm	u Btu/lbm	h Btu/lbm	s Btu/lbm(R)	x %	
1		212						0
2	90							60
3	120				910			
4	800	1000						
5	2000	600						
6	30		13.72					
7	50	370						
8	1000	500						
9		300						20
10	80							100

15. Complete the following steam table using CyclePa.

State	p psia	T °F	v ft^3/lbm	u Btu/lbm	h Btu/lbm	s Btu/lbm(R)	x %
1	4					1.8	
2	80						82
3	100	600					
4		400					90
5		600				1.9	
6	600				1100		
7	550			1300			
8		500	3				
9		600			1150		
10	80		2.41				

16. A saturated mixture of H_2O in a 5.01 ft^3 container is at 200°F. 5 ft^3 of the container volume is filled with vapor, and the remainder is filled with liquid. Determine the mass, quality, pressure, and internal energy of the mixture.
17. At 14.7 psia, steam has a specific volume of 30 ft^3/lbm. Determine the quality, internal energy and enthalpy of the steam at this state.
18. Seven pounds of steam with a pressure of 30 psia and a specific volume of 25 ft^3/lbm is flowing in a system. Determine the total enthalpy of the steam associated with this state.
19. At a vacuum of 2 psia, steam has a specific volume of 101.7 ft^3/lbm. Determine the quality, internal energy and enthalpy of the steam at this state.

2.3 IDEAL GASES

An *ideal gas* is a hypothetical substance. The mathematic definition of an ideal gas is that the ratio pv/T is exactly equal to R at all pressures and temperatures, otherwise it is called a real gas (note that when using this equation, the pressure and temperature must be expressed as absolute quantities). At low densities, the ratio pv/T has, by experiment, the same value of R (*gas constant*) for a specific gas. R is the gas constant of a gas. Different gases have different gas constants.

How close is a real gas to an ideal gas? This is an important question. It is convenient to use such a simple equation of state, but we must know how much accuracy we sacrifice for the convenience. In general a gas with a low pressure and high temperature is considered to be an ideal gas. But high and low are relative terms. For example, 100 kPa may be considered a high pressure, if very good accuracy is required. On the other hand, 100 kPa may be considered a low pressure, if good accuracy is not required.

The relation pv=RT is known as the ideal-gas equation of state. The gas constant R is given by $R=R_u/M$, where R_u is the *universal gas constant*, which has the same value for all gases, and M is the *molar mass*. The value of R_u, expressed in various units, is

$$R_u=8.314 \text{ kJ/kmol(K)}=1.986 \text{ Btu/lbmol(R)}=1545 \text{ ft(lbf)/lbmol(R)}$$

Experiment by Joule shows that internal energy (u) of an ideal gas is a function of temperature (T) only. Therefore, enthalpy (h=u+pv=u+RT), specific heat at constant pressure (c_p) and specific heat at constant volume (c_v) of an ideal gas must also each be a function of temperature (T) only.

The change of internal energy and change of enthalpy are given by the following equations:

$$\Delta u=\int c_v dT \tag{2.3.1}$$

and

$$\Delta h=\int c_p dT \tag{2.3.2}$$

The change of entropy is given by the following equations:

$$\Delta s=\int c_v dT/T + R \int dv/v \tag{2.3.3}$$

and

$$\Delta s=\int c_p dT/T - R \int dp/p \tag{2.3.4}$$

It is possible, but long and tedious, to calculate property changes of an ideal gas in a straight forward manner using above equations. It certainly would be convenient if tables or charts existed listing the values of the thermodynamic properties. Fortunately, tables and charts for many ideal gases are available.

The specific heats (both c_p and c_v) of a gas vary with temperature. The functional relationships denoting these variations are determined from experimental data. If accuracy is desired, these equations must be used. In many applications an average value of specific heat is used (based on the temperature range under consideration) for quick results. The internal energy change, enthalpy change and entropy change of the ideal gas from state 1 to state 2 can be simplified from Eqs. (2.3.1), (2.3.2), (2.3.3) and (2.3.4) and expressed as:

$$\Delta u=c_v(\Delta T) \tag{2.3.5}$$

$$\Delta h=c_p(\Delta T) \tag{2.3.6}$$

$$\Delta s=c_v [\ln(T_2/T_1)] + R [\ln(v_2/v_1)] \tag{2.3.7}$$

and

$$\Delta s=c_p [\ln(T_2/T_1)] - R [\ln(p_2/p_1)] \tag{2.3.8}$$

Air is the most important gas used in engineering thermodynamic application. The gas constant (R), specific heats (c_p and c_v) and specific heat ratio (k) of air at room temperature have the following numerical values:

$$R_{air}=0.06855 \text{ Btu/[lbm(R)]}=0.3704 \text{ [psia(ft}^3\text{)]/[lbm(R)]}=0.2870 \text{ kJ/[kg(K)]}$$

$$(c_p)_{air}=0.240 \text{ Btu/[lbm(R)]}=1.004 \text{ kJ/[kg(K)]}$$

$$(c_v)_{air}=0.171 \text{ Btu/[lbm(R)]}=0.718 \text{ kJ/[kg(K)]}$$

$$k_{air}=1.4$$

For air and many other gases, over the range of pressures and temperatures we commonly deal with, the assumption of ideal gas behavior yields a very excellent engineering approximation. However, as we get to high pressures, deviations from ideal gas behavior may be large in magnitude. In these cases, use of the ideal gas law will depend on the degree of accuracy required for a particular problem.

The tabulation of h, u and other properties of air, using the temperature as the argument has been made and is given by Keenan and Kaye. A part of the air table is given in Table 2.3.1.

Table 2.3.1 Air table

Thermodynamic properties of air at low pressure (SI units)

Reference level: 0 K and 1 atm

Absolute entropies may be calculated from

$$s^\circ = \phi - R \ln (p/p_0) + 4.1869 \text{ kJ/kg·K}$$

where p_0 is the reference pressure of 1 atm

T, K	h, kJ/kg	P_r	u, kJ/kg	v_r	ϕ kJ/kg K
100	99.76	0.029 90	71.06	2230	1.4143
110	109.77	0.041 71	78.20	1758.4	1.5098
120	119.79	0.056 52	85.34	1415.7	1.5971
130	129.81	0.074 74	92.51	1159.8	1.6773
140	139.84	0.096 81	99.67	964.2	1.7515
150	149.86	0.123 18	106.81	812.0	1.8206
160	159.87	0.154 31	113.95	691.4	1.8853
170	169.89	0.190 68	121.11	594.5	1.9461
180	179.92	0.232 79	128.28	515.6	2.0033
190	189.94	0.281 14	135.40	450.6	2.0575
200	199.96	0.3363	142.56	396.6	2.1088
210	209.97	0.3987	149.70	351.2	2.1577
220	219.99	0.4690	156.84	312.8	2.2043
230	230.01	0.5477	163.98	280.0	2.2489
240	240.03	0.6355	171.15	251.8	2.2915
250	250.05	0.7329	178.29	227.45	2.3325
260	260.09	0.8405	185.45	206.26	2.3717
270	270.12	0.9590	192.59	187.74	2.4096
280	280.14	1.0889	199.78	171.45	2.4461
290	290.17	1.2311	206.92	157.07	2.4813
300	300.19	1.3860	214.09	144.32	2.5153
310	310.24	1.5546	221.27	132.96	2.5483
320	320.29	1.7375	228.45	122.81	2.5802
330	330.34	1.9352	235.65	113.70	2.6111
340	340.43	2.149	242.86	105.51	2.6412
350	350.48	2.379	250.05	98.11	2.6704
360	360.58	2.626	257.23	91.40	2.6987
370	370.67	2.892	264.47	85.31	2.7264
380	380.77	3.176	271.72	79.77	2.7534
390	390.88	3.481	278.96	74.71	2.7796
400	400.98	3.806	286.19	70.07	2.8052
410	411.12	4.153	293.45	65.83	2.8302
420	421.26	4.522	300.73	61.93	2.8547
430	431.43	4.915	308.03	58.34	2.8786
440	441.61	5.332	315.34	55.02	2.9020

Example 2.3.1. Find the properties (h, u, and v) of air at 100 kPa and 400 K using the ideal gas equation and air table.

The gas constant of air is 0.287 kJ/[kg(K)].

Solution: Using the ideal gas equation, we have v = RT/p = {0.287 kJ/[kg(K)]}(400 K)/(100 kPa) =1.148 m^3/kg.

Air table reads (at 300 K), h=300.19 kJ/kg, and u=214.09 kJ/kg.

Example 2.3.2. Find the changes in enthalpy, internal energy and entropy of air if the air is heated from 100 kPa and 400 K to 100 kPa and 440 K using equations and the air table. Consider c_p and c_v are constant during this change of states. c_p of air is 1.004 kJ/[kg(K)] and c_v is 0.718 kJ/[kg(K)].

Solution: Using Eqs. (2.3.5), (2.3.6), and (2.3.8) we have

$$\Delta u=0.718(440-400)=28.72 \text{ kJ/kg}$$

$$\Delta h=1.004(440-400)=40.16 \text{ kJ/kg}$$

$$\Delta s=1.004[\ln(440/400)]-0.287[\ln(100/100)]=0.09569 \text{ kJ/[kg(K)]}$$

Again, searching property values from the tables or calculation from equations is tedious and long. It is time to abandon the time consuming use of property tables or calculation and move over entirely to the use of computer database software such as CyclePad. Among the nine substances listed on the CyclePad menu: air, helium, and carbon dioxide are ideal gases.

Notice that CyclePad databases account for the variation of specific heat with temperature for an ideal gas. Therefore equations using constant specific heats cannot be used in the thermodynamic analysis in this book, since the ideal gas substance using CyclePad is considered to be a gas with variable specific heats.

Example 2.3.3. Find the properties of air at the following states: (a) 14.7 psia and 1200 R, (b) v=20 ft^3/lbm and 1500 R, and (c) 10 psia and v=16 ft^3/lbm using CyclePad.

To solve this problem by CyclePad, we take the following steps:

1. Build
   (A) Take a BEGIN and an END from the closed-system inventory shop and connect them.
   (B) Switch to analysis mode.
2. Analysis
   (A) Input the given information: (a) substance is air, and (b) 14.7 psia and 1200 R.
3. Display results
   (A) Display the results. The answers are: (a) v=30.2 ft^3/lbm, u=205.4 Btu/lbm, h=287.6 Btu/lbm, s=0.7691 Btu/[lbm(R)]; (b) p=27.75 psia, u=256.8 Btu/lbm, h=359.5 Btu/lbm, s=0.7791 Btu/[lbm(R)]; and (c) T=432.4 R, u=74.03 Btu/lbm, h=103.6 Btu/lbm, s=0.5509 Btu/[lbm(R)].

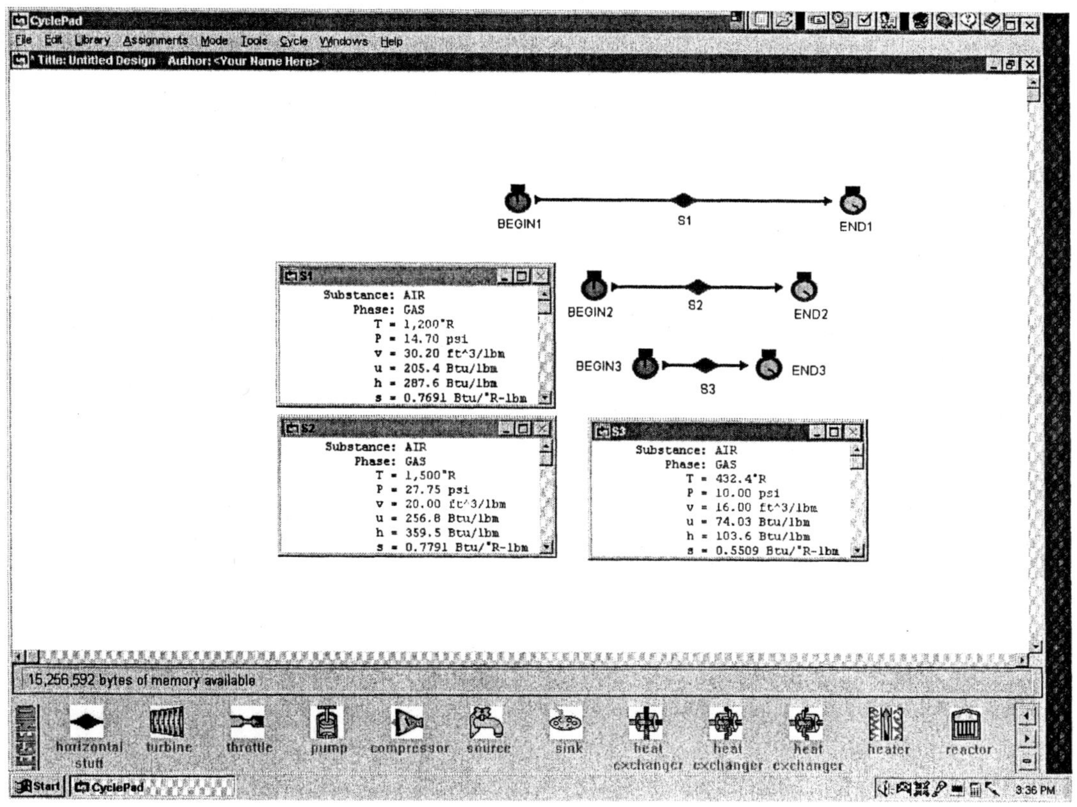

Figure Example 2.3.3 Air property relationships

Example 2.3.4. Find the properties of helium at the following states: (a) 100 kPa and 700 K, (b) v=1.23 m^3/kg and 800 K, and (c) 80 kPa and v=1 m^3/kg, using CyclePad.

To solve this problem by CyclePad, we take the following steps:

1. Build
   (A) Take a BEGIN and an END from the closed-system inventory shop and connect them.
   (B) Switch to analysis mode.
2. Analysis
   (A) Input the given information: (a) substance is helium, and (b) 100 kPa and 700 K.
3. Display results
   (A) Display the results. The answers are: (a) v=14.54 m^3/kg, u=2170 kJ/kg, h=3624 kJ/kg, s=10.00 kJ/[kg(K)]; (b) p=1351 kPa, u=2480 kJ/kg, h=4141 kJ/kg, s=5.29 kJ/[kg(K)]; and (c) T=38.52 K, u=119.4 kJ/kg, h=199.4 kJ/kg, s=-4.55 kJ/[kg(K)].

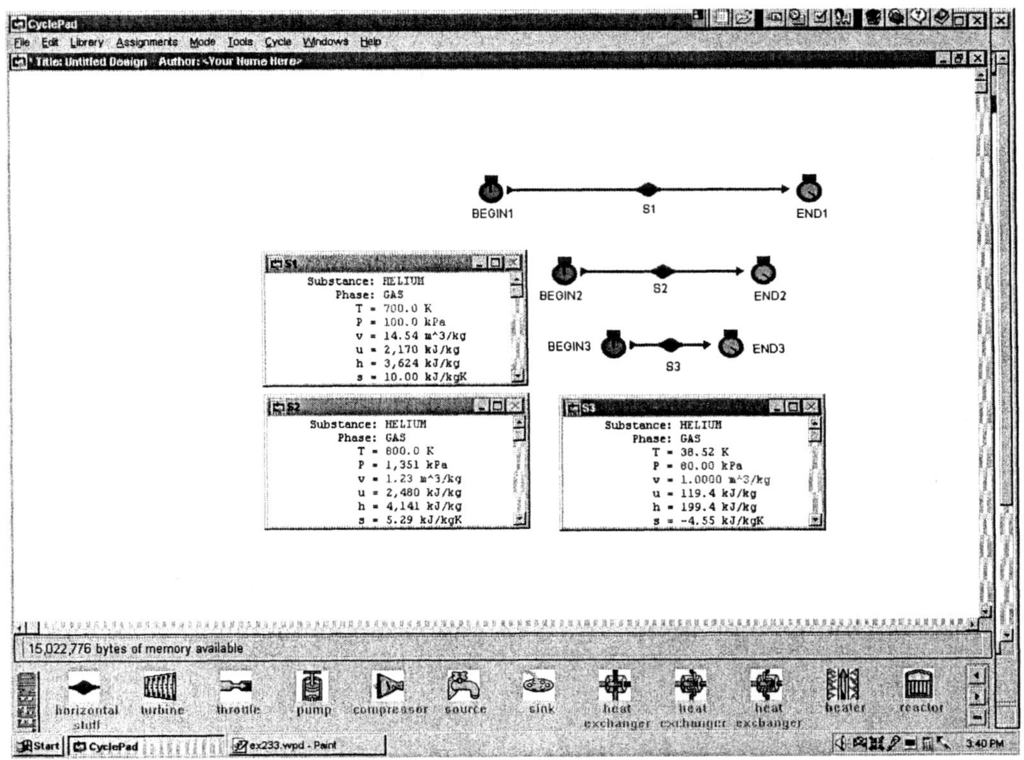

Figure Example 2.3.4 Helium properties

Example 2.3.5. Find the properties of carbon dioxide (CO_2) at the following states: (a) s=1.5 kJ/[kg(K)] and 700 K, (b) v=1.23 m^3/kg and u=500 kJ/kg, and (c) 80 kPa and s=3 kJ/[kg(K)], using CyclePad.

To solve this problem by CyclePad, we take the following steps:

1. Build
   (A) Take a BEGIN and an END from the closed-system inventory shop and connect them.
   (B) Switch to analysis mode.
2. Analysis
   (A) Input the given information: (a) substance is carbon dioxide (CO_2), and (b) s=1.5 kJ/[kg(K)] and 700 K.
3. Display results
   (A) Display the results. The answers are: (a) p=1612 Mpa, v=0 m^3/kg, u=456 kJ/kg, h=588.2 kJ/kg,; (b) p=117.9 kPa, T=767.5 K, h=645.0 kJ/kg, s=3.38 kJ/[kg(K)]; and (c) T=449.4 K, v=1.06 m^3/kg, u=292.8 kJ/kg, h=377.7 kJ/kg.

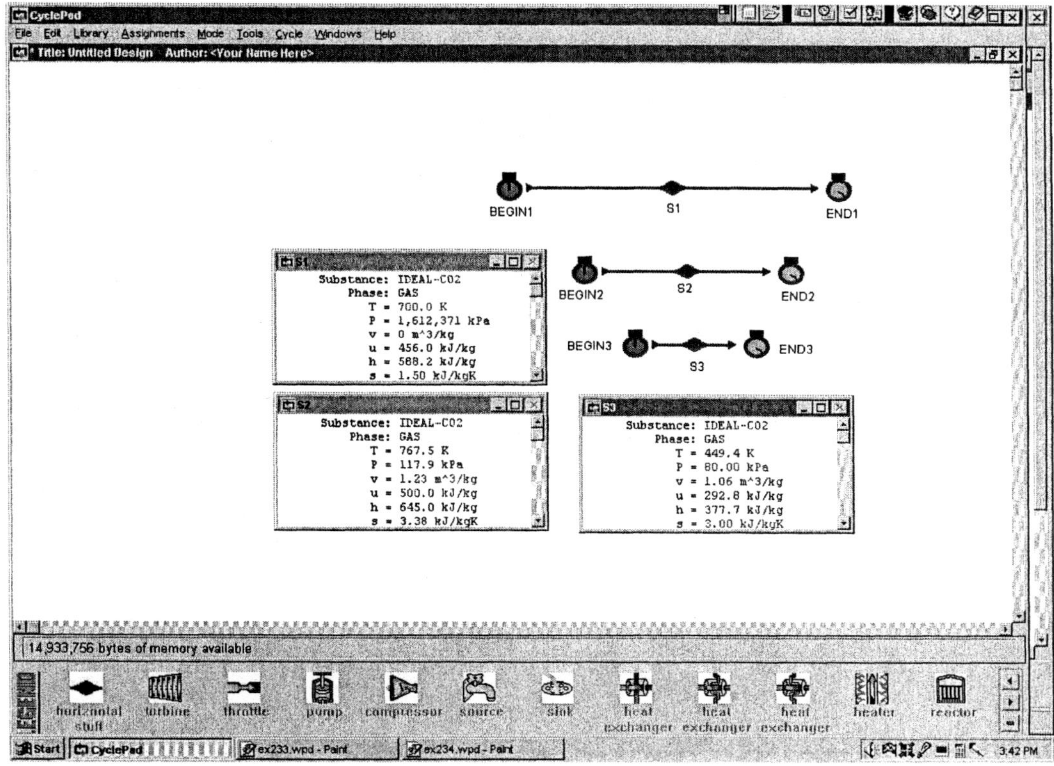

Figure Example 2.3.5 Carbon dioxide (CO_2) properties

Example 2.3.6. Find the changes in volume, enthalpy, internal energy and entropy of air if the air is heated from 14.7 psia and 1200 R to 14.7 psia and 1600 R using CyclePad.

To solve this problem by CyclePad, we take the following steps:

1. Build
   (A) Take a BEGIN, a combustor and an END from the closed-system inventory shop and connect them.
   (B) Switch to analysis mode.
2. Analysis
   (A) Input the given information: (a) substance is air, and (b) state 1 (input 14.7 psia and 1200 R), and state 1 (input 14.7 psia and 1600 R).
3. Display results
   (A) Display the results. The answers are: (a) Δv=40.27-30.2=10.07 ft^3/lbm, Δu=273.9-205.4 =68.5 Btu/lbm, Δh=383.5-287.6=95.9 Btu/lbm, Δs=0.8380-0.7691=0.0689 Btu/[lbm(R)].

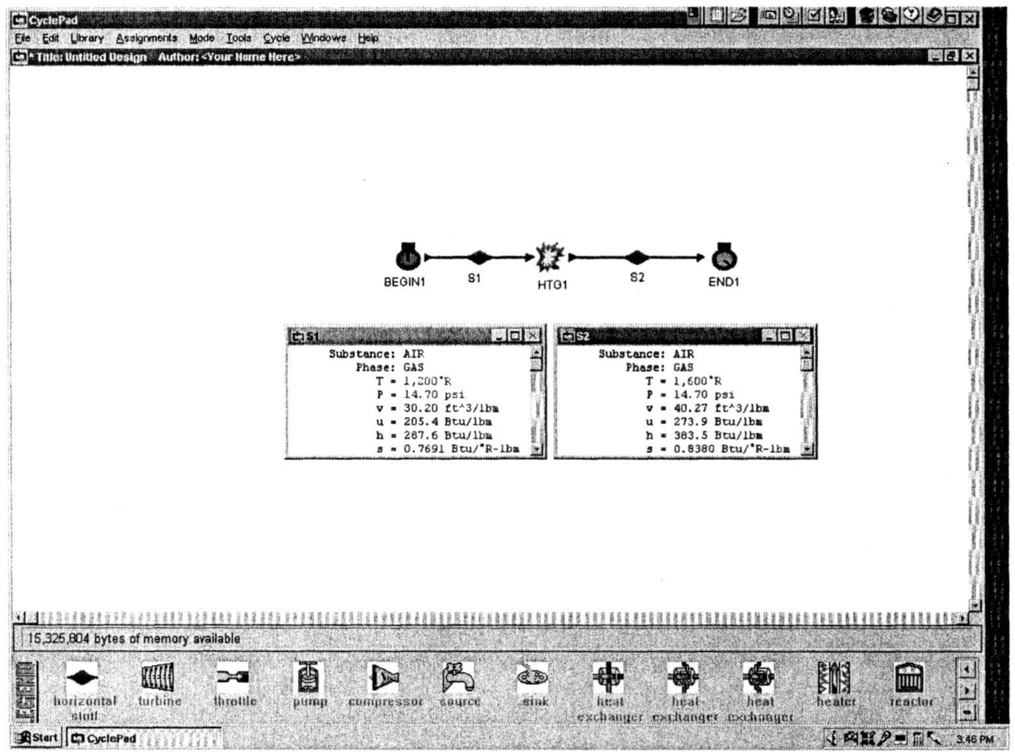

Figure Example 2.3.6 Air property changes

Example 2.3.7. Find the changes in volume, enthalpy, internal energy and entropy of air if the air state is changed from 14.7 psia and 1200 R to 1470 psia and 1200 R using CyclePad.

To solve this problem by CyclePad, we take the following steps:

1. Build
   (A) Take a BEGIN, a combustor and an END from the closed-system inventory shop and connect them.
   (B) Switch to analysis mode.
2. Analysis
   (A) Input the given information: (a) substance is air, and (b) state 1 (input 14.7 psia and 1200 R), and state 1 (input 1470 psia and 1200 R).
3. Display results
   (A) Display the results. The answers are: (a) Δv=0.3020-30.2=-29.90 ft^3/lbm, Δu=205.4-205.4 =0 Btu/lbm, Δh=287.6-287.6=0 Btu/lbm, Δs=0.4537-0.7691=-0.3154 Btu/[lbm(R)].

Comment: For ideal gases, h and u are functions of temperature only, but v and s are not functions of temperature only.

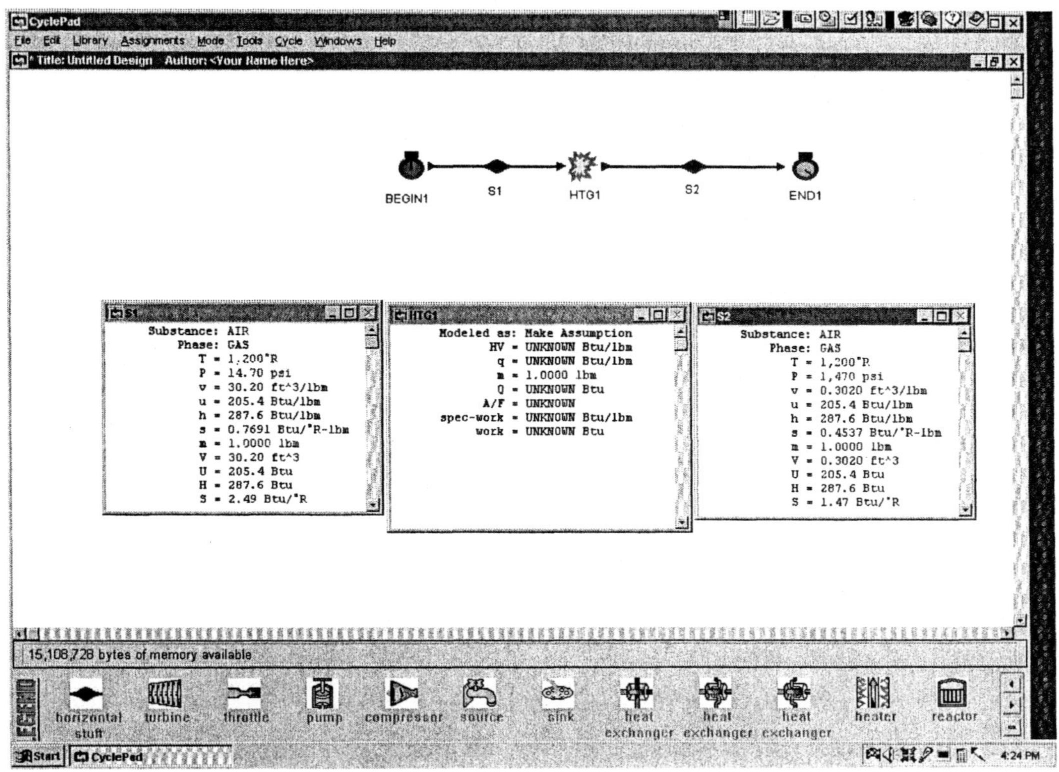

Figure Example 2.3.7 Air property changes

Homework 2.3 Ideal Gases

1. Under what conditions would it be reasonable to treat the water as an ideal gas?
2. Define the specific heat at constant pressure process.
3. What is the specific volume of carbon dioxide at 20 psia and 180°F?
4. 2 lbm of air at 100 psia and a specific volume of 3 ft^3/lbm is contained in a tank. Find the internal energy and enthalpy of the air in Btu/lbm.
5. The internal energy in 2 lbm of a gas at 100 psia and a specific volume of 3 ft^3/lbm is 138.8 Btu/lbm. Find the enthalpy of one pound mass of the gas.
6. A closed rigid tank contains 100 lbm of air at 200 psia and 180°F. The air is subsequently cooled to 80°F. Determine (a) the initial specific volume of the gas, (b) the volume of the tank, (c) the final specific volume of the gas, and (d) the final pressure of the gas.
7. A pressure vessel with a volume of 1.2 m^3 contains air at 400 kPa and 50°C. Determine the mass of air in the vessel.
8. A tank is filled with air at a pressure of 42 psia and a temperature of 180°F. What is the specific volume of the air in the tank?
9. A pressure vessel with a volume of 87 ft^3 contains air at 168 psia and 250°F. Determine the mass of air in the vessel.
10. Find the properties of air at the following states: (a) s=1.5 kJ/[kg(K)] and 700 K, (b) v=1.23 m^3/kg and u=500 kJ/kg, and (c) 80 kPa and s=3 kJ/[kg(K)], using CyclePad.

11. Find the properties of air at the following states: (a) 100 kPa and 700 K, (b) v=1.23 m^3/kg and 800 K, and (c) 80 kPa and v=1 m^3/kg, using CyclePad.
12. Find the properties of air at the following states: (a) 14.7 psia and 1700 R, (b) v=20 ft^3/lbm and 1200 R, and (c) 10 psia and v=11 ft^3/lbm using CyclePad.
13. Find the properties of carbon dioxide (CO_2) at the following states: (a) 14.7 psia and 1200 R, (b) v=20 ft^3/lbm and 1200 R, and (c) 10 psia and v=10 ft^3/lbm using CyclePad.
14. Find the properties of helium at the following states: (a) 14.7 psia and 1210 R, (b) v=21 ft^3/lbm and 1508 R, and (c) 10 psia and v=11 ft^3/lbm using CyclePad.

2.4 Incompressible Substances

Many liquids and solids are considered as incompressible substances whose specific volume remains constant regardless of changes in other properties.

Several useful properties relationships (equations of state) of the incompressible substances are given in the following.

The difference between the constant-pressure specific heat and the constant-volume specific heat is zero. The subscripts are often dropped and the specific heat of an incompressible substances is simply designated by c:

$$c_p = c_v = c \tag{2.4.1}$$

The internal energy change, enthalpy change and entropy change of an incompressible substances depend on temperature changes only:

$$\Delta u = \int c(dT) \tag{2.4.2}$$

$$\Delta h = \int c(dT) \tag{2.4.3}$$

$$\Delta s = \int c(dT)/T \tag{2.4.4}$$

For small temperature changes, the specific heat can be approximately as a constant. Equation (2.4.2) can be written as

$$u_2 - u_1 = c\ (T_2 - T_1) \tag{2.4.5}$$

$$\Delta h = \Delta u + v(\Delta p) \tag{2.4.6}$$

$$\Delta s = s_2 - s_1 = c[\ln(T_2 / T_1)] \tag{2.4.7}$$

Water is the most used incompressible working fluid in engineering thermodynamic application. The heat capacity and specific volume of water at room temperature are c= 4.18 kJ/[kg(K)]=1.0 Btu/[lbm(R)] and v=0.0010 m^3/kg=0.0160 ft^3/lbm, respectively.

Example 2.4.1. Find the change of properties (h, u, and v) of water from 14.7 psia and 42°F to 14.7 psia and 52°F using the Eqs. (2.4.5), (2.4.6) and (2.4.7).

Solution: Using the equations, we have $u_2 - u_1 = c(T_2 - T_1) = 1.0(52-42) = 10$ Btu/lbm

$\Delta h = \Delta u + v(\Delta p) = 10 + 0.0160(0) = 10$ Btu/lbm

$\Delta s = s_2 - s_1 = c[\ln(T_2/T_1)] = 1.0[\ln(460+52)/(460+42)] = 0.01972$ Btu/[lbm(R)].

Example 2.4.2. Find the properties of water (incompressible substance) at (a) 20°C and 100 kPa, (a) 20°C and 5000 kPa, and (a) 30°C and 100 kPa using CyclePad.

To solve this problem by CyclePad, we take the following steps:

1. Build
   (A) Take a BEGIN and an END from the closed-system inventory shop and connect them.
   (B) Switch to analysis mode.
2. Analysis
   (A) Input the given information: (a) substance is water, and (b) 20°C and 100 kPa.
3. Display results
   (A) Display the results. The answers are: (a) v=0.0010 m^3/kg, u=83.83 kJ/kg, h=83.93 kJ/kg, s=0.2962 kJ/[kg(K)]; (b) v=0.0010 m^3/kg, u=83.53 kJ/kg, h=88.52 kJ/kg, s=0.2951 kJ/[kg(K)]; and (c) v=0.0010 m^3/kg, u=125.7 kJ/kg, h=125.8 kJ/kg, s=0.4365 kJ/[kg(K)].

Comments: 1. v of incompressible substance is constant; 2. u of incompressible substance is a strong function of T and a weak function of p; 3. h of incompressible substance is a strong function of T and a weak function of p; and 4. h and u of incompressible substance are almost equal.

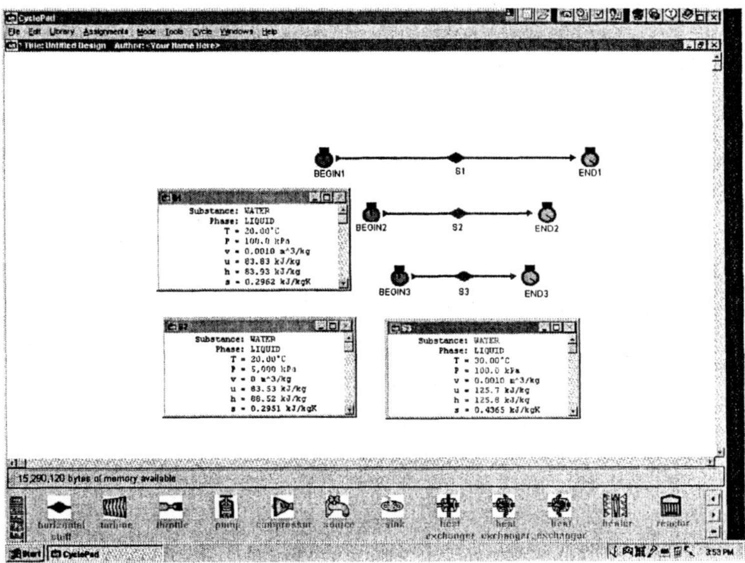

Figure Example 2.4.2 Compressed (Subcooled) Water properties

Example 2.4.3. Find the change of properties (h, u, and v) of water from 14.7 psia and 42°F to 14.7 psia and 52°F using CyclePad.

To solve this problem by CyclePad, we take the following steps:

1. Build
   (A) Take a BEGIN, a combustor and an END from the closed-system inventory shop and connect them.
   (B) Switch to analysis mode.
2. Analysis
   (A) Input the given information: (a) substance is water, and (b) state 1 (input 14.7 psia and 42°F), and state 2 (input 14.70 psia and 52°F).
3. Display results
   (A) Display the results. The answers are: The answers are: Δv=0.00160-0.0160=0 ft^3/lbm Δu=20.05-10.04=10.01 Btu/lbm, Δh=20.09-10.08=10.01 Btu/lbm, Δs=0.0399-0.0202=0.0197 Btu/[lbm(R)].

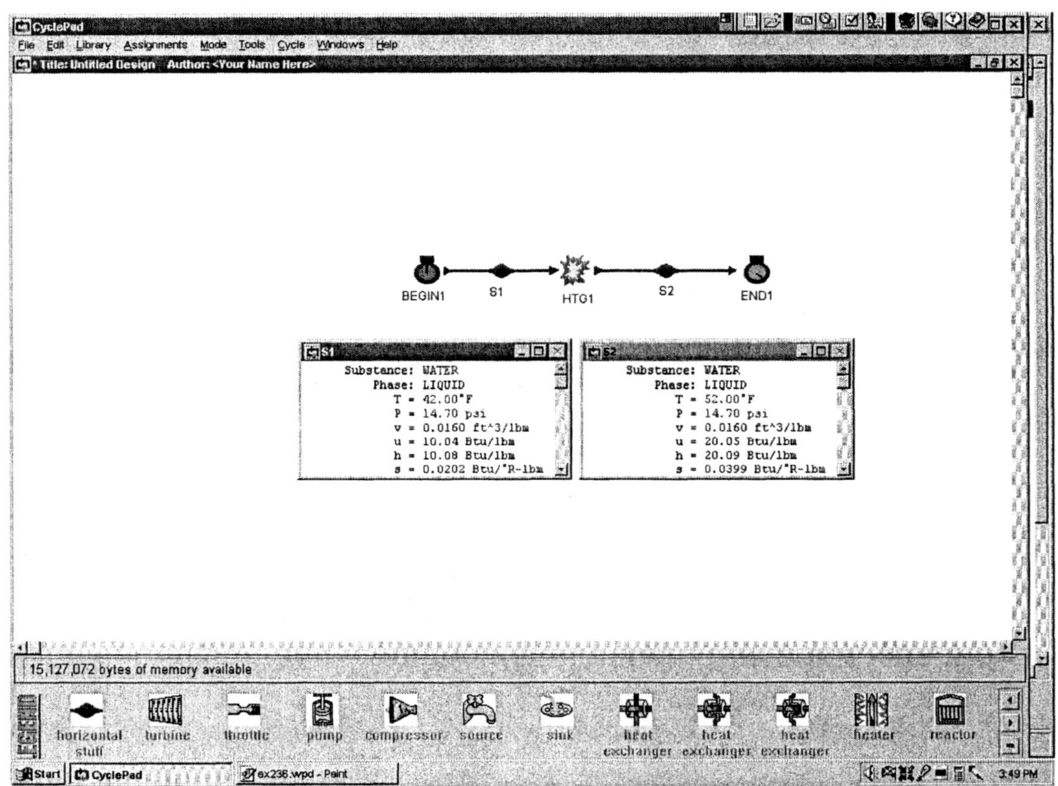

Figure Example 2.4.3 Compressed (Subcooled) water properties

Homework 2.4 Incompressible Substances (Liquids and Solids)

1. Water is compressed from 5 psia and 101.7 F to 600 psia and 101.7 F. Find the internal energy change and enthalpy change of the water.
2. In the absence of compressed liquid table, how do you find the specific volume and specific internal energy of refrigerant R-134a at 100 psia and 10°F without using CyclePad?
3. Find the properties of water (incompressible substance) at (a) 15°C and 100 kPa, (b) 21°C and 5000 kPa, and (c) 33°C and 100 kPa using CyclePad.
4. Find the properties of water (incompressible substance) at (a) 35°F and 100 psia, (b) 81°F and 500 psia, and (c) 53°F and 14.7 psia using CyclePad.
5. Find the change of properties (h, u, and v) of water from 14.7 psia and 42°F to 14.7 psia and 52°F using CyclePad.

2.5 SUMMARY

There are three types of thermodynamic substances: pure substances, ideal gases and incompressible substances.

A pure substance has a fixed chemical composition throughout all different phases. The liquid and gas phases of the pure substances are important in thermodynamic application. The two phases are separated by a saturated mixture dome on a T-v diagram. The liquid region is called the compressed or subcooled liquid region. The gas region is called the superheated vapor region. The two phase mixture region is called saturated mixture region. Temperature and pressure are dependent in the saturated mixture region. Quality is defined as the vapor mass fraction of the saturated mixture, and is only defined in the saturated mixture region. Properties among T, p, v, u, h, and s relationships are determined and listed on conventional steam and refrigerant tables. In the case of compressed or subcooled liquid table is not available, a general approximation is to treat the compressed or subcooled liquid as a saturated liquid (x=0) at the given state temperature.

An ideal gas is a low density gas and follows the equation pv=RT. The pressure and temperature must be expressed as absolute quantities in the pv=RT equation. It is a fictitious substance but the simplest to use. Specific heats, u and h of ideal gases are functions of temperature only. Specific heats may be treated as constants over a small temperature range. Air and other gas tables, assuming variable specific heats are commercially available.

Liquids and solids are considered as incompressible substances whose specific volumes are constant. Specific heat, u and h of incompressible substances are considered as functions of temperature only.

Properties of nine different thermodynamic substances including air and water are built in the software of CyclePad.

Chapter 3

FIRST LAW OF THERMODYNAMICS FOR CLOSED SYSTEMS

OBJECTIVES

After reading and studying the material in this chapter, you should be able to:

1. State the definitions of work, energy and heat.
2. Understand that both work and heat are energy of transition without mass transfer.
3. Know the sign conventions of work and heat.
4. Know that work is the area under the p-V process curve.
5. Understand that both work and heat are path functions.
6. State the first law of thermodynamics, or energy conservation for a closed system.
7. Apply the first law of thermodynamics to closed systems.
8. Solving closed system problems using CyclePad.

3.1 INTRODUCTION

A closed system is defined as a particular quantity of matter. The closed system always contains the same matter, and no matter crosses the system boundary. The mass of the closed system is constant. Although a closed system does not interact mass with its surroundings, it does interact energy with its surroundings.

Certain important energy forms cannot be stored within a system. They exist only in transit between the system and whatever interacts with the system. These forms are called transitory energy forms. Transitory energy is the energy entering or leaving a system without accompanying matter. It appears in two forms, microscopic as heat and macroscopic as work.

Homework 3.1 Introduction

1. What is a closed system?
2. Can a closed system interact mass with its surroundings?
3. Can a closed system interact energy with its surroundings?

4. What are the two transitory energy forms?
5. Both work and flow energy can flow in to a system from its surroundings. What is the difference between work and flow energy?

3.2 WORK

If a system undergoes a displacement due to the application of a force, work is said to be done. The amount of work is equal to the product of the force and the component of the displacement in the direction of the force. Work is a macroscopic transitory energy; it is never contained in a body. Transitory energy is pure energy not associated with matter in transit between the system and its surroundings. Since a system does not possess work, work is not a property of the system.

In thermodynamics, interest lies in devices that do work. It thus has become customary to refer to work done by a system as positive. As a convention of sign, work input to a system is taken as negative, and work done by the system is positive. The notation of work is W.
Since work is not a property, the derivative of work is written as δW. The amount of work transferred to a system from the surroundings between an initial state i and a final state f is

$$\int \delta W = W_{if}. \tag{3.2.1}$$

rather than $(W_f - W_i)$. There is no such thing as W_f or W_i.

Three commonly used units of work are ft-lbf and Btu in English unit system and kJ in SI unit system.
There are many modes of work. The two major work modes in thermodynamics are: simple compressible boundary work or simply boundary work, and shaft work. Neglecting other modes of work, the total work can be written as

$$W = W_{boundary} + W_{shaft} \tag{3.2.2}$$

Consider a system enclosed by a piston and cylinder. As the piston moves over a displacement distance dx in the direction of the driving force F acting on the piston, the boundary work done by this infinitesimal motion can be expressed as

$$\delta W_{boundary} = F\,dx = (F/A)\,(Adx) = p\,dV. \tag{3.2.3}$$

A system whose work is represented by (p dV) is called simple compressible boundary work or boundary work. The boundary work can be evaluated as

$$W_{boundary} = \int \delta W_{boundary} = \int p\,dV, \tag{3.2.4}$$

Where p is the driving force for the boundary work, and θ is the displacement for the boundary work.

In order to integrate Eq. (3.2.3), the process or functional relations between p and V must be known. Using p and V as thermodynamic coordinates, the area underneath the process represents the boundary work of the process as illustrated by Fig. 3.2.1. Boundary work is therefore a path function. This can be easily seen by referring to a p-V diagram as shown by Fig. 3.2.2. In the process from the initial state i to the final state f, regardless of the path between i and f, the change in volume is ($V_f - V_i$), and change in pressure is ($p_f - p_i$); V and p are point functions and properties. The boundary work of the process i-a-f is greater than that of i-b-f. Work is therefore clearly a path function and is not a property.

Similarly, the shaft work can be evaluated as

$$W_{shaft} = \int \delta W_{shaft} = \int \tau \, d\theta, \tag{3.2.5}$$

Where τ is the driving force for the shaft work, and θ is the displacement for the shaft work.

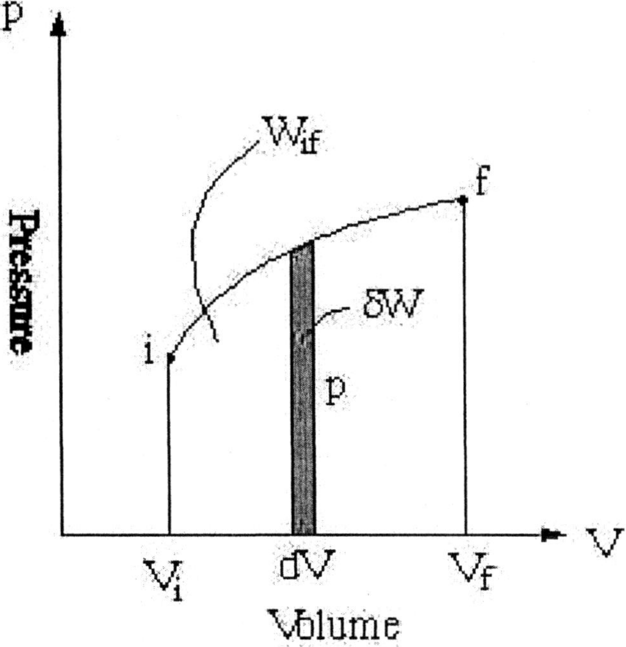

Fig. 3.2.1. p-V diagram

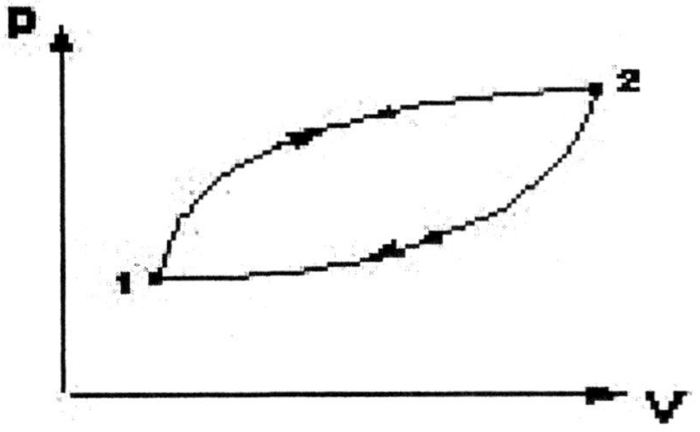

Fig. 3.2.2. Work of different processes on p-V diagram

Example 3.2.1. A piston-cylinder device contains 0.1 kg of air at 150 kPa, 0.09043 m^3 and 200°C. Consider the air is the system, find the boundary work of the following processes and draw the processes on a p-V diagram:

(A) Heat is added to the air in a constant volume (isochoric) process until the temperature reaches 300°C. During this process the piston does not move.
(B) Heat is added in a constant pressure (isobaric) process until the volume is doubled.
(C) Air is expanded in a linear process until the pressure reaches 300 kPa and volume is doubled.

Solution:

(A) W=0. There is no area under the constant volume process, 1-2A.
(B) $W=\int pdV=p\int dV=p\ (V_2-V_1)=150[2x\ 0.09043-0.09043]=13.56$ kJ, which is the rectangular area under the constant volume process, 1-2B.
(C) The final volume and pressure are 0.1808 m^3 and 300 kPa. Since the process is a linear line, the average pressure is $p_{average}=(150+300)/2=225$ kPa. $W=\int pdV=p_{average}\int dV=p_{average}(V_2-V_1)=225(0.1808-0.09043)=20.35$ kJ, which is the triangle area under the process 1-2C, as shown in Figure Example 3.2.1.

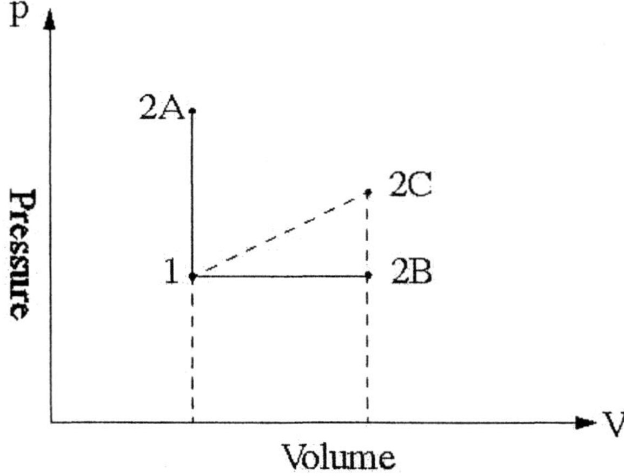

Figure Example 3.2.1 Work of different processes on p-V diagram

Example 3.2.2. A piston-cylinder device contains 0.1 kg of air at 150 kPa and 200°C. Consider the air is the system, find the boundary work of the following processes and draw the processes on a p-V diagram:

(A) Heat is added to the air in a constant volume (isochoric) process until the temperature reaches 300°C. During this process the piston does not move.
(B) Heat is added in a constant pressure (isobaric) process until the volume is doubled.
(C) Air is expanded in a constant temperature (isothermal) process until the volume is doubled.

To solve this problem by CyclePad, we take the following steps:

1. Build
   (A) Take a begin, a heating-combustion device in case (a) and case (b) or an expansion device in case (c), and an end from the closed-system inventory shop and connect them.
   (B) Switch to analysis mode.
2. Analysis
   (A) Assume the heating-combustion is isochoric in case (a) or isobaric in case (b).
   (B) Input the given information: (i) working fluid is air, (ii) the mass, initial air pressure and temperature of the process are 0.1 kg, 150 kPa and 200°C, (iii) the final air temperature of the process is 300°C in case (a), or the final specific volume is twice the initial specific volume in case (b) and case (c).
3. Display results
   (A) Display the expansion device results. The answers are: case (a) W=0 kJ, case (b) W=13.56 kJ, and case (c) W=9.40 kJ,

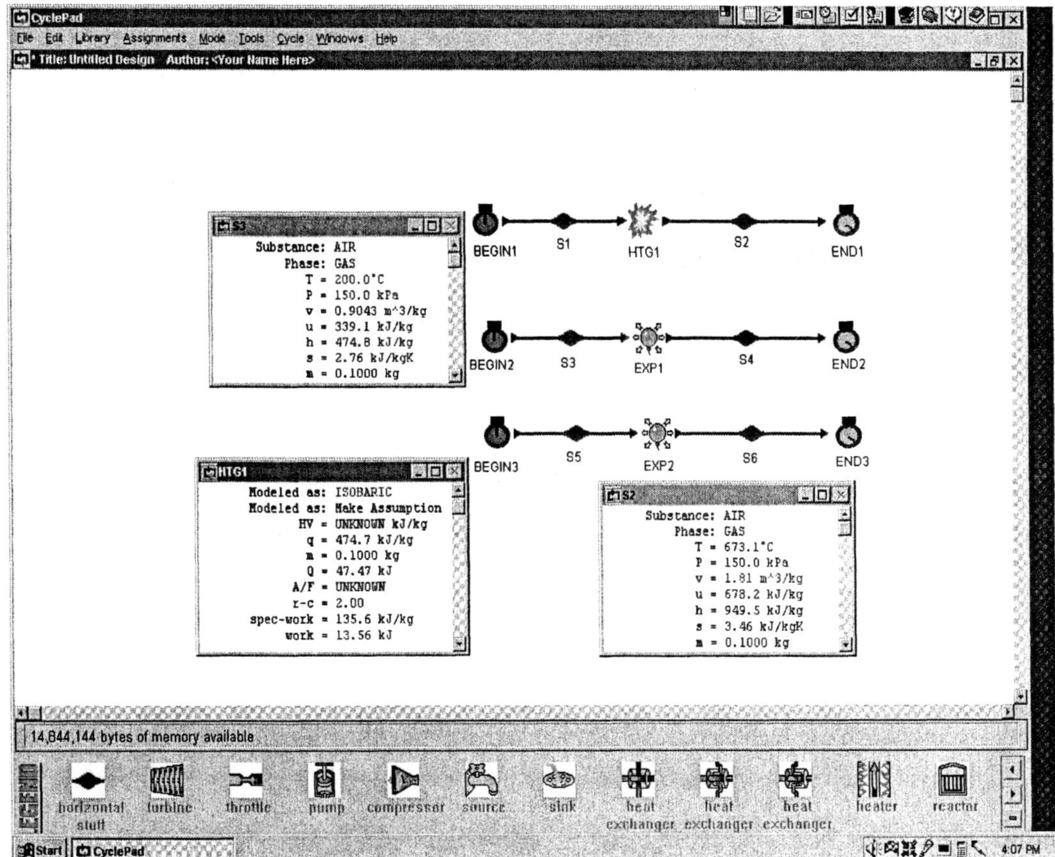

Figure Example 3.2.2a Work of different processes

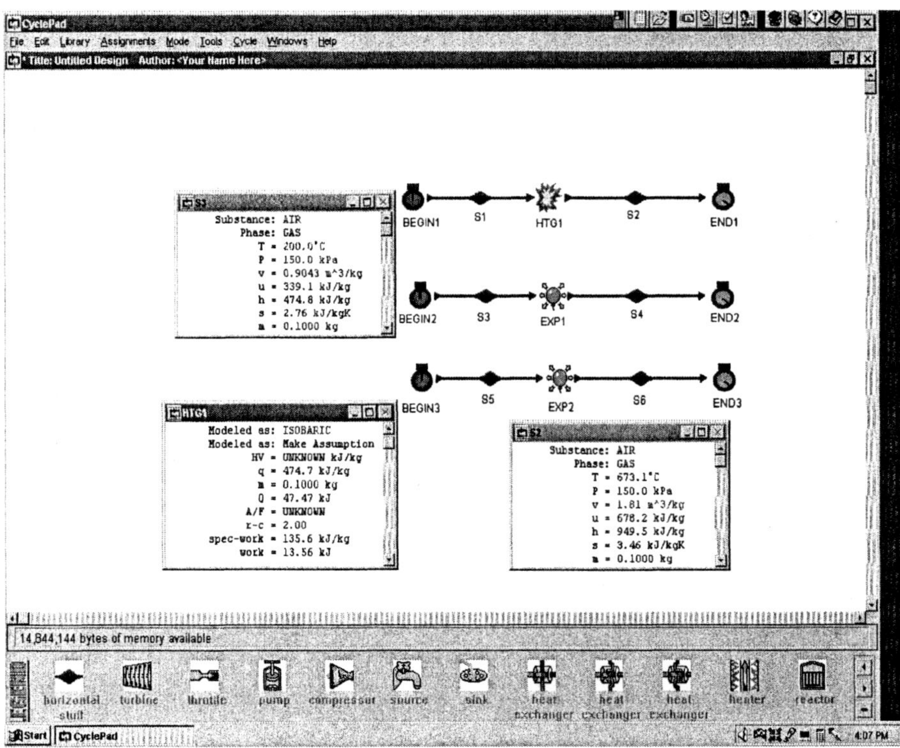

Figure Example 3.2.2b Work of different processes

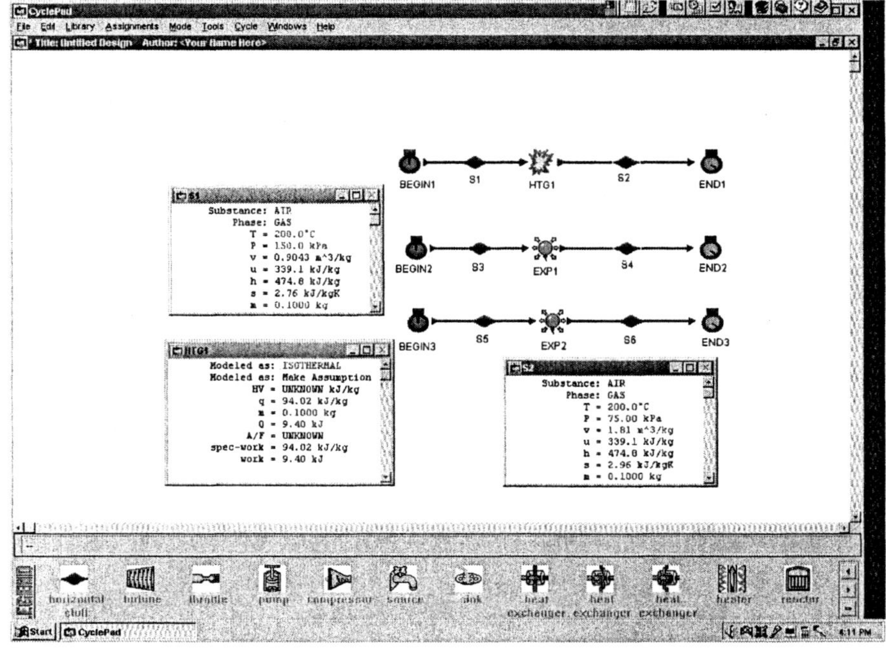

Figure Example 3.2.2c Work of different processes

Example 3.2.3. A frictionless piston-cylinder device contains 2.4 kg of helium at 100 kPa and 300 K. Air is now compressed adiabatically, slowly according to the relationship pV^n =constant until it reaches a final temperature and pressure of 400 K and 200 kPa. Find the work done during this process and n.

To solve this problem by CyclePad, we take the following steps:

1. Build
   (A) Take a begin, a compression device, and an end from the closed-system inventory shop and connect them.
   (B) Switch to analysis mode.
2. Analysis
   (A) Assume the compression as an adiabatic process.
   (B) Input the given information: (a) working fluid is helium, (b) the initial helium pressure and temperature of the process are 100 kPa and 300 K, (c) the final helium pressure and temperature of the process are 200 kPa and 400K.
3. Display results
   (A) Display the compression device results. The answers is W=-702.5 kJ and n=1.71.

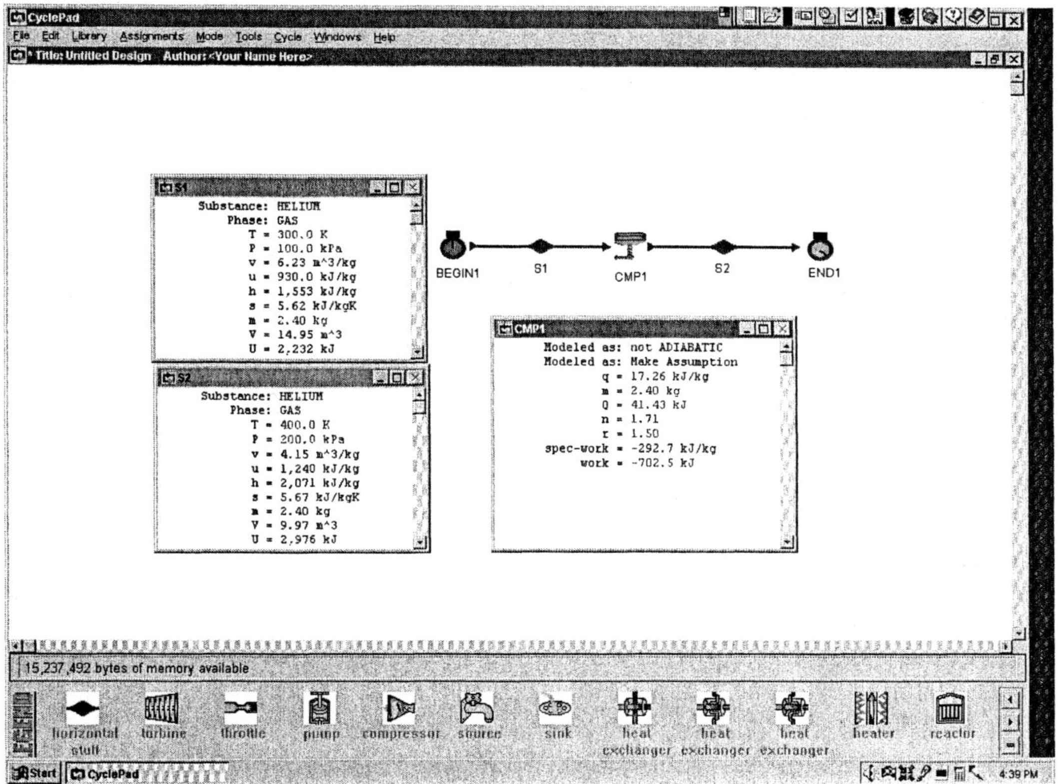

Figure Example 3.2.3 Polytropic process

Power is defined as the work per unit time crossing the boundary of the system. The symbol for power is Wdot, where the dot notation is used to signify a rate quantity. The units of power are W, kW, MW, Btu/s, Btu/h, ft-lbf/s, horsepower, etc. For example, 500 kW is produced by a heat engine which produced 500 kJ of work in each second.

$$Wdot=\delta W/dt=W/t \tag{3.2.6}$$

Homework 3.2 Work

1. Is work a property?
2. Does a system possess work?
3. What are the two major modes of work in thermodynamics?
4. What is the driving force and its corresponding displacement of boundary work?
5. What is the driving force and its corresponding displacement of shaft work?
6. A closed system consisting of an unknown gas undergoes a constant volume process from 1 Mpa to 6 Mpa. What is the boundary work done by the gas?
7. Is work added to or done by a closed system during an adiabatic expansion process?
8. Air in an internal combustion engine undergoes an expansion process ($pv^{1.45}=p_1(v_1)^{1.45}=p_2(v_2)^{1.45}=c$)from an initial state at 0.007 ft^3, 2400 F and 2000 psia to a final state at 0.045 ft^3. Find the final air pressure and the specific work done by the air.
9. 10 lbm of air undergoes a constant pressure process from 88.2 psia and 70 F to 100 F. Determine q, w and Δu for this process.
10. Air is compressed by a piston in a cylinder from 10 psia and 40 ft^3 to 20 psia. During the compression process, the product pV remains a constant. Determine the work done on the gas, the internal energy change of the gas and the heat transfer to the gas.
11. Carbon dioxide is compressed in a cylinder in a polytropical process $pV^{1.22}$=constant from p_1=100 kPa and V_1=0.004 m^3 to p_2=500 kPa. Compute the heat removed and work done or by the gas.
12. During a non-flow process 120 Btu of heat is removed from each lbm of the working substance while the internal energy of the working substance decreases by 85.5 Btu/lbm. Determine how much work is involved in the process and indicate whether the work is done on or done by the working substance.
13. During a non-flow process 1200 kJ of heat is removed from each kg of the working substance while the internal energy of the working substance decreases by 65 kJ/kg from an initial internal energy of 165 kJ/kg. Determine how much work is involved in the process and indicate whether the work is done on or done by the working substance.

3.3 Heat

When two bodies at different temperatures are brought into contact, heat flows from the higher temperature body to the lower temperature body. This flow of heat ceases when thermal equilibrium between the two bodies is reached. Heat is a microscopic transitory quantity; it is never contained in a body. Heat is pure energy in transit between a system and its surroundings, not associated with matter. A system does not possess heat. Heat is not a property of the system.

As a convention of sign, heat input to a system is taken as positive, and heat leaving the system is negative. The notation of heat is Q.

Since heat is not a property, the derivative of heat is written as δQ. For a reversible process, the amount of heat transferred to a system from the surroundings between an initial state i and a final state f is Q_{if} rather than $(Q_f - Q_i)$. There is no such thing as Q_f or Q_i.

$$\int \delta Q = Q_{if} = \int T\, dS \tag{3.3.1}$$

where temperature (T) is the driving force and entropy (S) is the displacement for the heat transfer.

Notice that Equation (3.3.1) for heat is similar to Equation (3.2.4) for boundary work. Two commonly used units of heat are Btu in English unit system and kJ in SI unit system.

Heat and work are the only energy forms that a system can give to or take from its surroundings without transferring matter. Heat and work change to some form of accumulated energy as soon as it crosses the boundaries of a system, much as pedestrians change into passengers as soon as they enter a bus.

Homework 3.3 Heat

1. Is it correct to say that a system possesses 5 kJ of heat?
2. Can heat be added to a system during a constant temperature process?
3. What is the driving force and its corresponding displacement of heat?
4. Block A and block B are at different temperatures ($T_A > T_B$). Five sides of each block are well insulated. The sides of each block that are not insulated are placed together. Keeping the definition of heat in mind, by prescribing the proper boundaries,
   (A) Select a system in such a manner that there is no transfer of heat
   (B) Select a system that receives energy by the mode of heat
   (C) Select a system that rejects energy by the mode of heat
5. A closed system receives 100 kJ of energy from its surroundings. Is this energy transferred by the mode of heat or by the mode of work?

3.4 First Law of Thermodynamics for a Closed System

The energy (E) content of a closed system may be changed by heat (Q) or work (W) or both from its surroundings without mass transfer. Let the initial energy content of a closed system at time t_1 be E_1; the final energy content of a closed system at time t_2 be E_2; Heat added to the system during the time interval from t_1 to t_2 is Q_{12}; and work added to the system during the time interval from t_1 to t_2 is W_{12}. Applying the energy conservation to the closed system under these conditions becomes

$$E_2 - E_1 = Q_{12} - W_{12} \tag{3.4.1}$$

Equation (3.4.1) is the First law of thermodynamics for a closed system. In using Equation (3.4.1), we must always remember the important sign convention that we have adopted on heat and work. Heat added to the system is positive and work added to the system is negative. We must also always remember the important subscripts 1 and 2 convention that we have adopted on Equation (3.4.1). 1 refers to time 1 (t_1) and 2 refers to time 2 (t_2).

The First law of thermodynamics for a closed system [Equation (3.4.1)] expressed in words is

[time change of the energy contained within the closed system] = [net heat added to the system] - [net work added to the system]

Various special forms of the First law of thermodynamics for a closed system can be written. For examples

$$e_2 - e_1 = q_{12} - w_{12} \tag{3.4.2}$$

Where e is the specific energy, q is heat per unit mass and w is work per unit mass.

Equation (3.4.2) is the specific form of Equation (3.4.1), where e=E/m, q=Q/m and w=W/m, respectively.

$$\int dE = \int \delta Q - \int \delta W \tag{3.4.3}$$

$$\int de = \int \delta q - \int \delta w \tag{3.4.4}$$

Equation (3.4.3) and Equation (3.4.4) are the integral forms of Equation (3.4.1) and Equation (3.4.2).

$$dE = \delta Q - \delta W \tag{3.4.5}$$

$$de = \delta q - \delta w \tag{3.4.6}$$

Equation (3.4.5) and Equation (3.4.6) are the differential forms of Equation (3.4.1) and Equation (3.4.2).

$$\Delta E/\Delta t = Q_{12}/\Delta t - W_{12}/\Delta t \tag{3.4.7}$$

$$\Delta e/\Delta t = q_{12}/\Delta t - w_{12}/\Delta t \tag{3.4.8}$$

Equation (3.4.7) and Equation (3.4.8) are the finite forms of Equation (3.4.1) and Equation (3.4.2), where $\Delta E = E_2 - E_1$, $\Delta e = e_2 - e_1$, and $\Delta t = t_2 - t_1$, respectively.

$$dE/dt = Qdot - Wdot \tag{3.4.9}$$

$$de/dt = qdot - wdot \tag{3.4.10}$$

Equation (3.4.9) and Equation (3.4.10) are the time rate forms of Equation (3.4.1) and Equation (3.4.2). The time rate of the first law of thermodynamics for a closed system [Equation (3.4.9)] expressed in words is

> [time rate of change of the energy contained within the closed system at time t] = [net rate of heat added to the system at time t] - [net rate of work added to the system at time t]

Example 3.4.1. 6 Btu of heat is added to 2 lbm of copper. Neglect the kinetic and potential energy change, find the change in internal Energy.

Solution: The copper is an incompressible substance with constant volume. Applying the definition of boundary work and the first law of thermodynamics for a closed system, we have

$$W = \int p\, dV = 0$$

and

$$\Delta E = \Delta U = Q - W = 6 - 0 = 6 \text{ Btu.}$$

Example 3.4.2. 1 kg of helium initially at 101.3 kPa and 20°C is contained within a cylinder and piston set up. The helium undergoes an isobaric heating process. The volume of the helium is doubled at the end of this process. Determine (a) the work done by the helium, (b) volume change, enthalpy change, entropy change, and internal energy change of the helium. Ignore the kinetic and potential energy changes.

To solve this problem by CyclePad, we take the following steps:

1. Build
   (A) Take a begin, a heating-combustion device, and an end from the closed-system inventory shop and connect them.
   (B) Switch to analysis mode.
2. Analysis
   (A) Assume the heating-combustion is isobaric.
   (B) Input the given information: (i) working fluid is helium, (ii) the helium mass, initial pressure and temperature of the process are 0.1 kg, 101.3 kPa and 20°C, (iii) the final specific volume is twice the initial specific volume.
3. Display results
   (A) Display the expansion device results. The answers are: (a) W=608.8 kJ ; (b) $\Delta V = 12.02 - 6.01 = 6.01 m^3$, $\Delta H = \Delta h = 3035 - 1518 = 1517$ kJ, $\Delta S = \Delta s = 9.06 - 5.47 = 3.59$ kJ/[kg(K)]. and $\Delta U = \Delta u = 1817 - 908.7 = 908.3$ kJ.

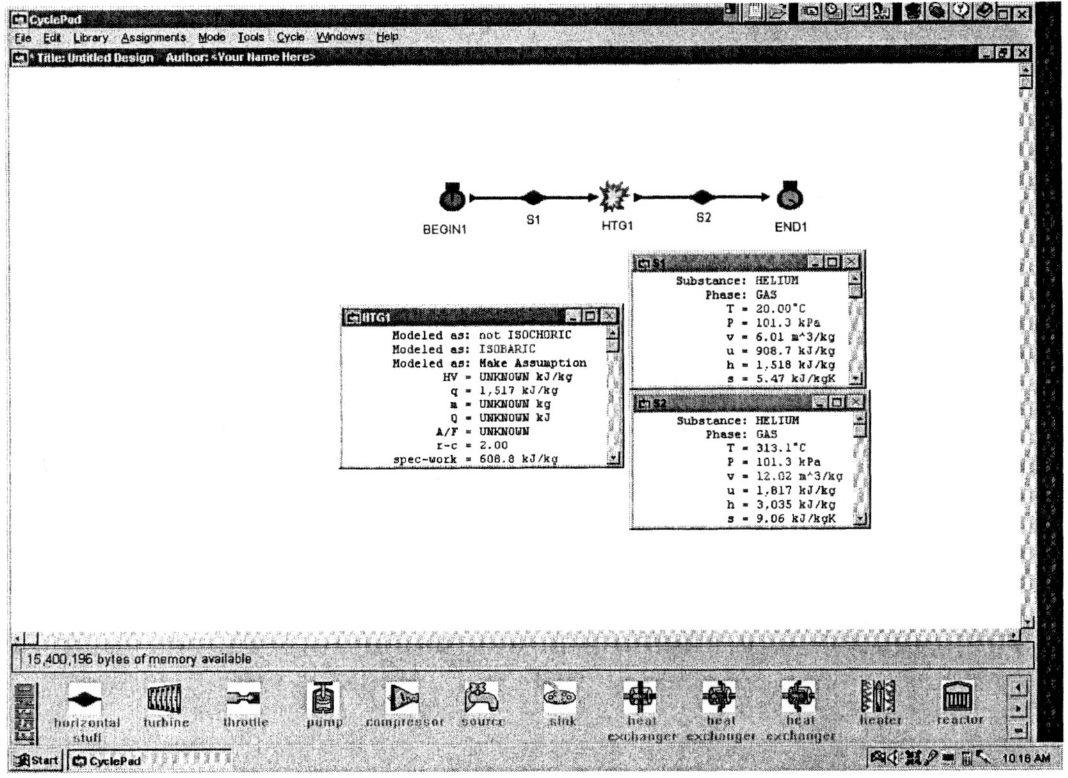

Figure Example 3.4.2 Isobaric process

Homework 3.4 First law of Thermodynamics for a Closed System

1. Fill in the missing data for each of the following processes of a closed system between the initial state and final state. ($\Delta E = E_{final} - E_{initial}$)
   a. Q=18 kJ, W=6 kJ, ΔU=?, U_{final}=34 kJ.
   b. Q=? kJ, W=25 kJ, ΔU=18 kJ.
   c. Q=44 kJ, W=?, ΔU=18 kJ.
   d. Q=?, W=12 kJ, ΔU=0 kJ.
2. 100 pounds of nitrogen undergoes a process. The internal energy at the start of the process is 90 Btu/lbm and at the completion of the process is 150 Btu/lbm. A total of 350 Btu of work is done on the gas. How much heat was added or subtracted from each pound of the gas?
3. Determine the change in specific internal energy for nitrogen contained in a closed tank while 12 Btu/lbm of heat and 7.64 Btu/lbm of work are added to the gas.
4. Hw3. The final internal energy of nitrogen contained in a closed tank is 600 Btu/lbm. 200 Btu/lbm of heat was removed from the gas and 100 Btu/lbm of work was done on the gas. What was the initial internal energy?
5. The internal energy of water at the start of a process is 1300 Btu/lbm. During the process 50 Btu/lbm of heat is added to the water and 10 Btu of work is performed by the water. Determine the final internal energy of the water.

3.5 First Law of Thermodynamics for a Closed System Applied to CyclePad Devices

Let a closed system operate in a cycle consisting of three processes 1-2, 2-3, and 3-1as shown in Figure 3.5.1.

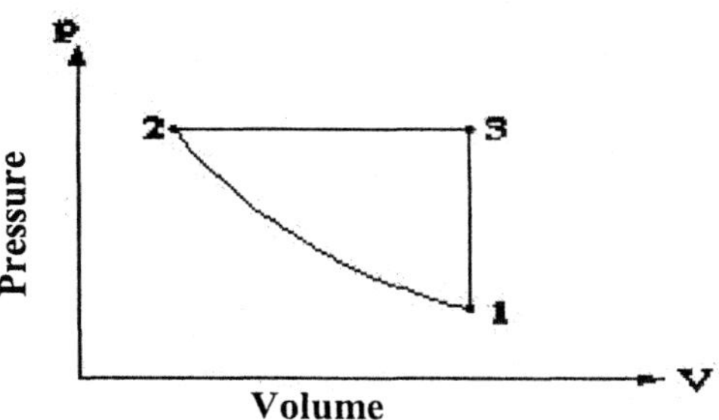

Figure 3.5.1. cycle

The initial state (1) of the cycle is the same as the finite state (also 1). Applying Equation (3.4.1) to the three processes 1-2, 2-3, and 3-1, we have

$$E_2 - E_1 = Q_{12} - W_{12} \tag{3.5.1}$$

$$E_3 - E_2 = Q_{23} - W_{23} \tag{3.5.2}$$

and

$$E_1 - E_3 = Q_{31} - W_{31} \tag{3.5.3}$$

For the cycle, which consists of three processes 1-2, 2-3, and 3-1, we have

$$\int_{Cycle} dE = (E_2 - E_1) + (E_2 - E_1) + (E_2 - E_1) = 0 \tag{3.5.4}$$

$$\int_{Cycle} \delta Q = Q_{12} + Q_{23} + Q_{31} = Q_{net} \tag{3.5.5}$$

$$\int_{Cycle} \delta W = W_{12} + W_{23} + W_{31} = W_{net} \tag{3.5.6}$$

Applying Equation (3.4.3) yields

$$\int_{Cycle} \delta Q = \int_{Cycle} \delta W \tag{3.5.7}$$

or

$$Q_{12} + Q_{23} + Q_{31} = W_{12} + W_{23} + W_{31}, \text{ or } Q_{net} = W_{net} \tag{3.5.8}$$

where $\int_{Cycle}$ is the summation around the cycle, $\int_{Cycle} \delta Q$ is the net heat added to the cycle and $\int_{Cycle} \delta W$ is the net work produced by the cycle, respectively.

In words, Equation (3.5.7) states that in a cycle the net heat addition to the cycle is equal to the net work removed from the cycle.

Example 3.5.1. An inventor claims to have developed a work-producing closed system cycle which receives 2000 kJ of heat from a heat source and rejects 800 kJ of heat to a heat sink. It produces a net work of 1200 kJ. How do we evaluate his claim?

Solution: Let us check if the first law of thermodynamics for a closed system cycle is satisfied or not.

$$\int_{Cycle} \delta Q = Q_{net} = +2000 + (-800) = 1200 \text{ kJ}$$

$$\int_{Cycle} \delta W = W_{net} = +1200 \text{ kJ}$$

Since $W_{net} = Q_{net}$, the claim is valid as far as the first law of thermodynamics is concerned.

Homework 3.5 First Law of Thermodynamics for a Closed System Applyto Cycles

1. What is the change of temperature (or any property) of a cycle?
2. Is it possible that the net heat added to a cycle is negative?
3. An inventor claims to have developed a heat-producing closed system cycle which receives 2000 kJ of heat from a low-temperature heat source and rejects 3200 kJ of heat to a high-temperature heat sink. It consumes a net work of 1200 kJ. How do we evaluate his claim?

3.6 Closed System for Various Processes

Processes encountered in engineering are of many varieties. It is necessary for one to acquire the ability to analyze simple processes before one can acquire the ability to analyze complex and multi-processes. In this chapter, we shall apply the fundamental first law of thermodynamics to the study of several common types of closed system processes.

For a closed system process, the mass is constant. Without shaft work, the boundary work and the First law of thermodynamics of the closed system for a process are:

$$W_{12} = \int pdV \tag{3.6.1}$$

and

$$Q_{12} - W_{12} = E_2 - E_1 \tag{3.6.2}$$

In many engineering applications, the kinetic energy change and potential energy change can be neglected. Eq. (3.6.2) is reduced to

$$Q_{12} - W_{12} = U_2 - U_1 \tag{3.6.3}$$

It will be seen that although the basic principles are identical for all processes, the analyses will depend on the kind of working substance involved and the state properties.

3.6.1 Constant Volume (Isochoric or Isometric) Process

If the volume of a process is constant, it is called *isochoric* or *isometric* process. The final volume (or specific volume) of the process is the same as its initial volume (or specific volume). Without shaft work, the boundary work and the first law of thermodynamics of the closed system for the constant volume process are:

$$W_{12} = \int p dV = 0, \qquad (3.6,1.1)$$

and

$$Q_{12} - W_{12} = U_2 - U_1 \qquad (3.6.1.2)$$

Example 3.6.1.1. The radiator (rigid-tank closed system) of a steam heating system has a mass of 0.02 kg and is filled with superheated vapor at 300 kPa and 280°C. At this moment both the inlet and exit valves to the radiator are closed. Determine (a) the final temperature and quality of the steam, (b) the change of internal energy and entropy, and (c) the amount of work and heat that will be transferred to the room when the steam pressure drops to 100 kPa.

To solve this problem by CyclePad, we take the following steps:

1. Build
   (A) Take a begin, a cooler, and an end from the closed-system inventory shop and connect them.
   (B) Switch to analysis mode.
2. Analysis
   (A) Assume the cooler as a isochoric process, W=0.
   (B) Input the given information: (a) working fluid is water, (b) the initial water pressure and temperature of the process are 300 kPa and 280°C, and (c) the final water pressure of the process is 100 kPa.
3. Display results
   (A) Display the cooler result. The answers are: (a) T=99.63°C and x=0.4977; (b) Δu=1457-2775=-1318 kJ/kg and Δs=4.32-7.63=-3.31 kJ/[kg(K)]; (c) W=0 and Q=-26.36 kJ.

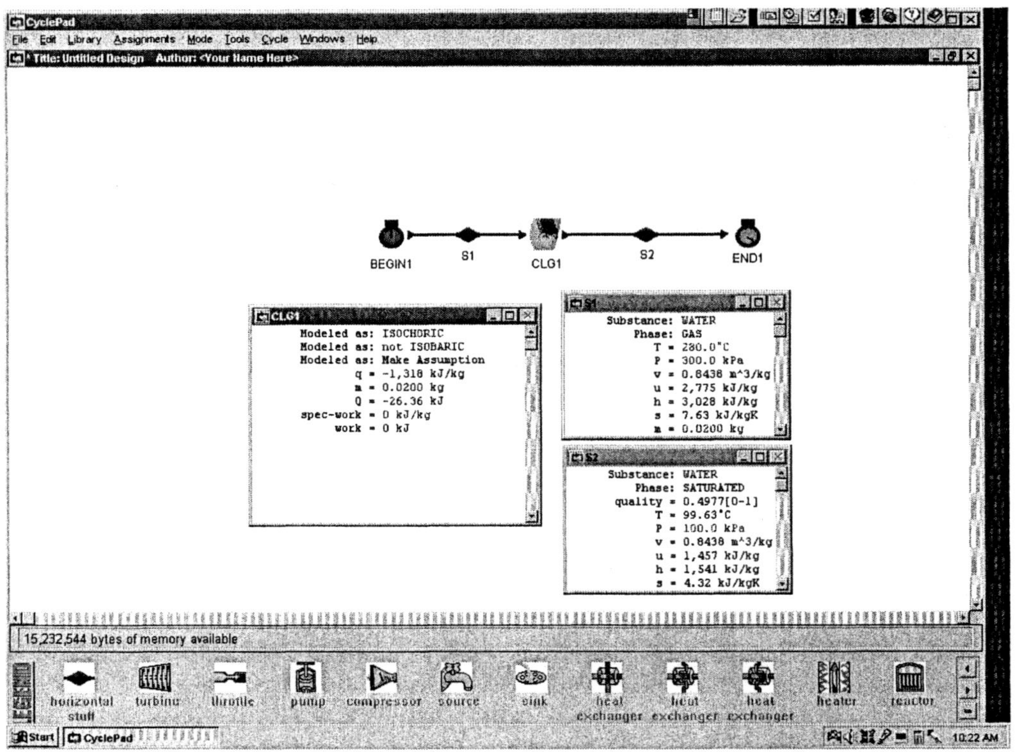

Figure Ex. 3.6.1.1 Isobaric cooling

Example 3.6.1.2. A rigid tank contains 3 lbm of air at 50 psia and 70°F. The air is now cooled until its pressure is reduced to 25 psia. Determine (a) the initial specific volume of the air, (b) the final temperature of the air, (c) the change of internal energy and entropy, and (c) the amount of work and heat.

To solve this problem by CyclePad, we take the following steps:

1. Build
   (A) Take a begin, a cooler, and an end from the closed-system inventory shop and connect them.
   (B) Switch to analysis mode.
2. Analysis
   (A) Assume the cooler as an isochoric process.
   (B) Input the given information: (a) working fluid is air, (b) the initial air mass, pressure and temperature of the process are 3 lbm, 50 psia and 70°F, and (c) the final air pressure of the process is 25 psia.
3. Display results
   (A) Display the cooler and final state result. The answers are: (a) v=3.92 ft^3/lbm; (b) T=-194.8 °F; (c) Δu=45.34-90.67=-45.33 Btu/lbm, and Δs=0.3706-0.4893=-0.1187 Btu/[lbm(R)]; (d) W=0 and Q=-136 kJ.

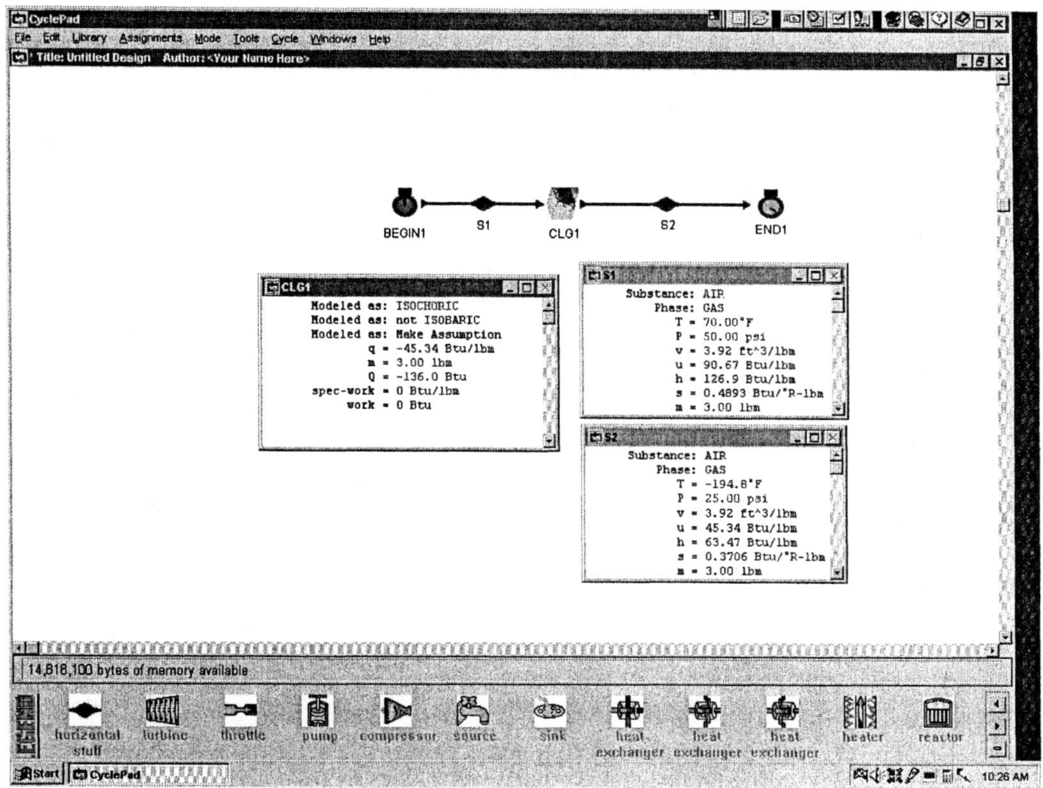

Figure Example 3.6.1.2 Constant volume process

Homework 3.6.1 Constant Volume

1. A rigid, closed vessel contains 2 lbm of steam initially at 100 psia, 350°F. It is cooled to 200°F. What is the size of the vessel? Determine the amount of heat transferred to the vessel.
2. Saturated steam vapor (x=1) at 14.7 psia is contained inside a rigid vessel having a volume of 50 ft^3. Heat is removed until the temperature drops to 180 F. Find the amount of heat removed, work added, and the quality at the final state.
3. A 5 m^3 rigid tank contains a quality 0.05745 steam (0.05 m^3 of saturated liquid water and 4.95 m^3 of saturated water vapor) at 0.1 Mpa. Heat is transferred until the pressure reaches 150 kPa. Determine the initial amount of water in the system, final quality of the steam, and heat transfer added to the system.
4. A closed rigid vessel contains 1.5 lbm of liquid water and 1.5 lbm of water vapor at 100 psia. Heat added to the mixture causes the H_2O to reach a pressure of 300 psia. Determine the final temperature of the H_2O, work and heat added during this process.
5. One kg of air at 350 kPa is confined to a 0.2 m^3 rigid tank. If 120 kJ of heat are supplied to the gas, the temperature increases to 411.5°C. Find (a) the work done, (b) the heat added, and (c) the final pressure of the air.

3.6.2 Constant Pressure (Isobaric) Process

If the pressure of a process is constant, it is called *isobaric* process. The final pressure of the process is the same as its initial pressure. Without shaft work, the boundary work and the first law of thermodynamics of the closed system for the constant pressure process 1-2 are:

$$W_{12} = \int pdV = p(V_2 - V_1), \qquad (3.6.2.1)$$

and

$$Q_{12} = W_{12} + U_2 - U_1 = H_2 - H_1 \qquad (3.6.2.2)$$

Example 3.6.2.1. 0.005 kg of air contained inside a piston-cylinder set up is initially at 128°C and 900 kPa. Heat is added at a constant pressure until its volume is doubled. Determine (a) the final temperature of the air, (b) the change of internal energy and entropy, and (c) the amount of work and heat.

To solve this problem by CyclePad, we take the following steps:

1. Build
   (A) Take a begin, a heater, and an end from the closed-system inventory shop and connect them.
   (B) Switch to analysis mode.
2. Analysis
   (A) Input the given information: (a) the process is isobaric, (b) working fluid is air and mass is 0.005 kg, and (c) the initial temperature and pressure of the process are 128°C and 900 kPa.
3. Display results
   (A) Display the heater and the final state results. The answers are: (a) T=529.3°C; (b) Δu=575.1-287.5=287.6 kJ/kg and Δs=2.78-2.08=0.70 kJ/[kg(K)]; (c) W=0.5752 kJ and Q=2.01 kJ.

Homework 3.6.2 Isobaric Process

1. 50 Btu's of heat are added to 1 lbm of air during an isobaric process with a pressure of 15 psia starting at a temperature of 100°F. Find the (a) final temperature and (b) work done.
2. Hw56. Half pound of CO_2 undergoes an isobaric process with a pressure of 50 psia while the volume increases from 4 ft^3 to 8 ft^3. Find the work done and heat added.
3. Air at 100 kPa, 27°C occupies a 0.01 m^3 piston-cylinder device that is arranged to maintain constant air pressure. This device is now heated until its volume is doubled. Determine the heat transfer to the air and work produced by the air.
4. A cylinder fitted with a piston has an initial volume of 0.1 m^3 and contains 0.1618 kg of air at 150 kPa, 50 C. The piston is moved, expanding the air isobarically (constant pressure) until the temperature is 150 C. Determine work added, and heat added during this process.

5. A closed system holds a mixture of 1 kg of liquid water and 1 kg of water vapor at 700 kPa. (A) Determine the initial temperature. (B) Heat transfer to the content occurs until the temperature reaches 360 C. The pressure is maintained constant during the process. Determine the work and heat added to the system.
6. 0.1 kg of air is expanded from 3 m^3 to 8 m^3 in an isobaric process with a constant pressure of 200 kPa. How much heat is added and work is done by the gas?
7. Five pounds of CO_2 is compressed in a piston-cylinder set up from a volume of 23 ft^3 to a volume of 13 ft^3. During the process the pressure remains constant at 20 psia. What is the work done and heat transferred?

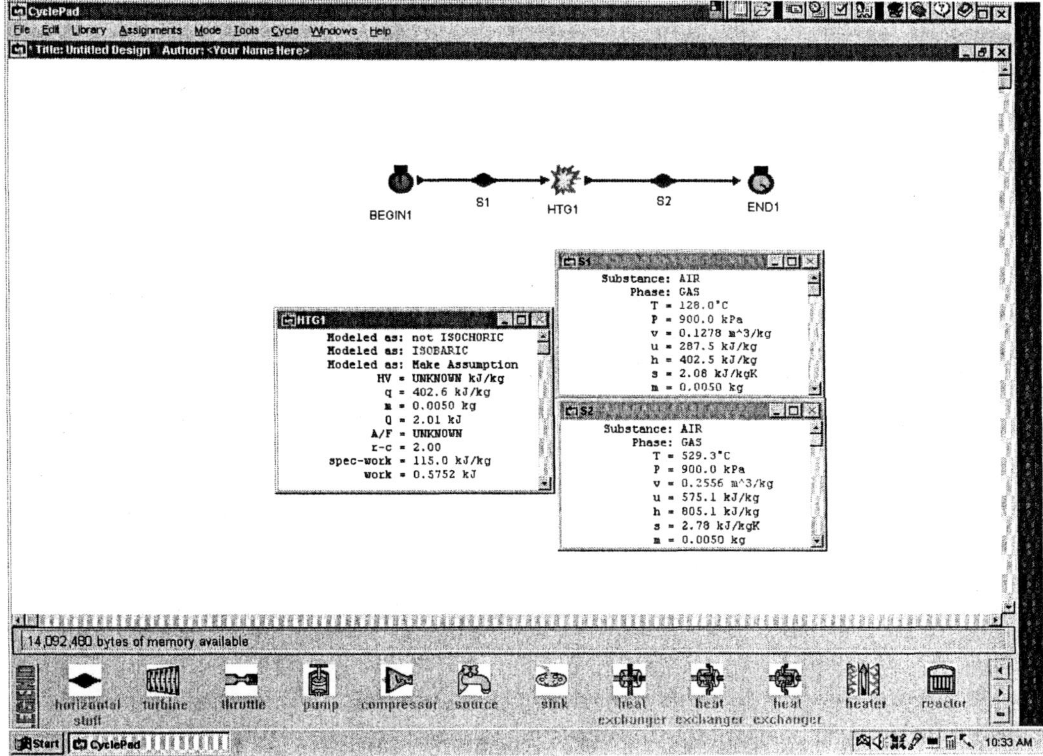

Example 3.6.2.1 Constant pressure process

3.6.3 Constant Temperature (Isothermal) Process

If the temperature of a process is constant, it is called *isothermal*. The final temperature of the process is the same as its initial temperature. The First law of thermodynamics of the closed system for the constant temperature process 1-2 are:

$$Q_{12} - W_{12} = U_2 - U_1 \tag{3.6.3.1}$$

For an ideal gas, $U_2 - U_1 = 0$. Therefore $Q_{12} = W_{12}$ for the isothermal process.

Example 3.6.3.1. A piston-cylinder device contains 0.1 m^3 of carbon dioxide at 100 kPa and 85°C. The carbon dioxide is compressed isothermally until the volume becomes 0.01 m^3. Determine (a) the final pressure of the carbon dioxide, (b) the change of internal energy and entropy, and (c) the amount of work and heat.

To solve this problem by CyclePad, we take the following steps:

1. Build
   (A) Take a begin, a heater, and an end from the closed-system inventory shop and connect them.
   (B) Switch to analysis mode.
2. Analysis
   (A) Input the given information: (a) the process is isothermal, (b) working fluid is carbon dioxide, (c) the initial volume, temperature and pressure of the process are 0.1 m^3, 85°C and 100 kPa, and (d) the final volume is 0.01 m^3.
3. Display results
   (A) Display the heater and the final state results. The answers are: (a) p=1000 kPa; (b) Δu=233.3-233.3=0 kJ/kg and Δs=2.33-2.77=-0.44 kJ/[kg(K)]; (c) W=-23.03 kJ and Q=-23.03 kJ.

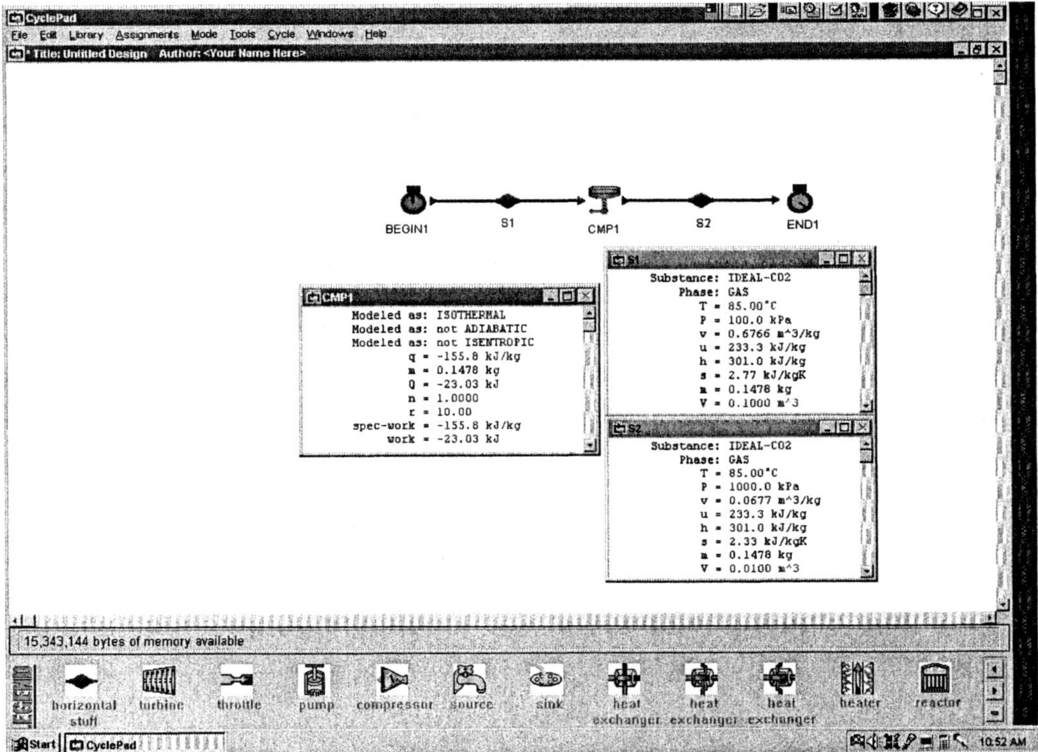

Figure Example 3.6.3.1. Isothermal process

Homework 3.6.3 Constant Temperature Process

1. 0.25 lbm of air is compressed by a piston in a cylinder from 10 psia and 40 ft^3 to 20 psia. During the compression process, the product pV remains a constant. Determine the work done on the gas, and the heat transfer to the gas.
2. 1.2 kg of air occupying a volume of 0.2 m^3 at a pressure of 1.5 Mpa is heated and expands isothermally (T=c) to a volume of 0.5 m^3. Find the work added and heat added during this process.
3. 3 ft^3 of air are expanded in a piston-cylinder set up from 700 psia and 2900°F to a final volume of 6 ft^3. Assuming the process is isothermal, determine the (a) heat transferred and work done.

3.6.4 Adiabatic Process

If a process is adiabatic, the heat transfer to the system is equal to zero. The adiabatic process may be reversible or irreversible.

$$Q_{12} = 0, \tag{3.6.14.1}$$

The first law of thermodynamics of the closed system for the adiabatic process 1-2 is:

$$-W_{12} = U_2 - U_1 \tag{3.6.4.2}$$

Example 3.6.4.1. A frictionless piston-cylinder device contains 2 kg of air at 100 kPa and 290K. Air is now compressed adiabatically to 1800 kPa and 662.3 K. Find the work added to the device.

To solve this problem by CyclePad, we take the following steps:

1. Build
   (A) Take a begin, a compression device, and an end from the closed-system inventory shop and connect them.
   (B) Switch to analysis mode.
2. Analysis - Expansion device
   (A) Assume the compression as an adiabatic process.
   (B) Input the given information: (a) working fluid is air, (b) the initial air mass, pressure and temperature of the process are 2 kg, 100 kPa and 290K, and (c) the final air pressure and temperature of the process are 1800 kPa and 662.3 K.
3. Display results
   (A) Display the compression device results. The answers is Q=0 kJ, and W=-533.7 kJ.

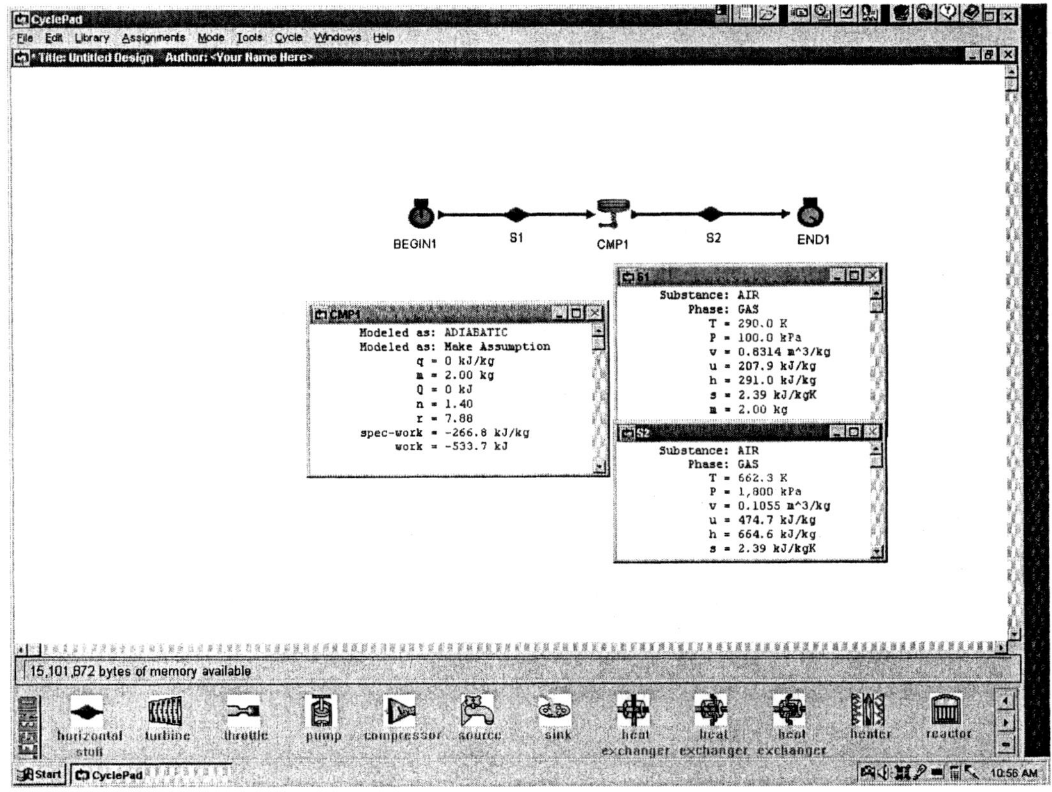

Figure Example 3.6.4.1. Adiabatic process

Example 3.6.4.2. A frictionless piston-cylinder device contains 5 lbm of air at 14.7 psia and 80°F. Air is now expanded adiabatically to 200 psia. 511.9 Btu of work is added to the process. Find the Final temperature and volume of the air.

To solve this problem by CyclePad, we take the following steps:

1. Build
   (A) Take a begin, an expansion device, and an end from the closed-system inventory shop and connect them.
   (B) Switch to analysis mode.
2. Analysis
   (A) Assume the expansion as an adiabatic process, W added is -511.9 Btu.
   (B) Input the given information: (a) working fluid is air, (b) the initial air mass, pressure and temperature of the process are 5 lbm of air at 14.7 psia and 80°F, and (c) the final air pressure is 200 psia.
3. Display results
   (A) Display the expansion device results. The answers are T=678.1°F and V=10.52 ft^3.

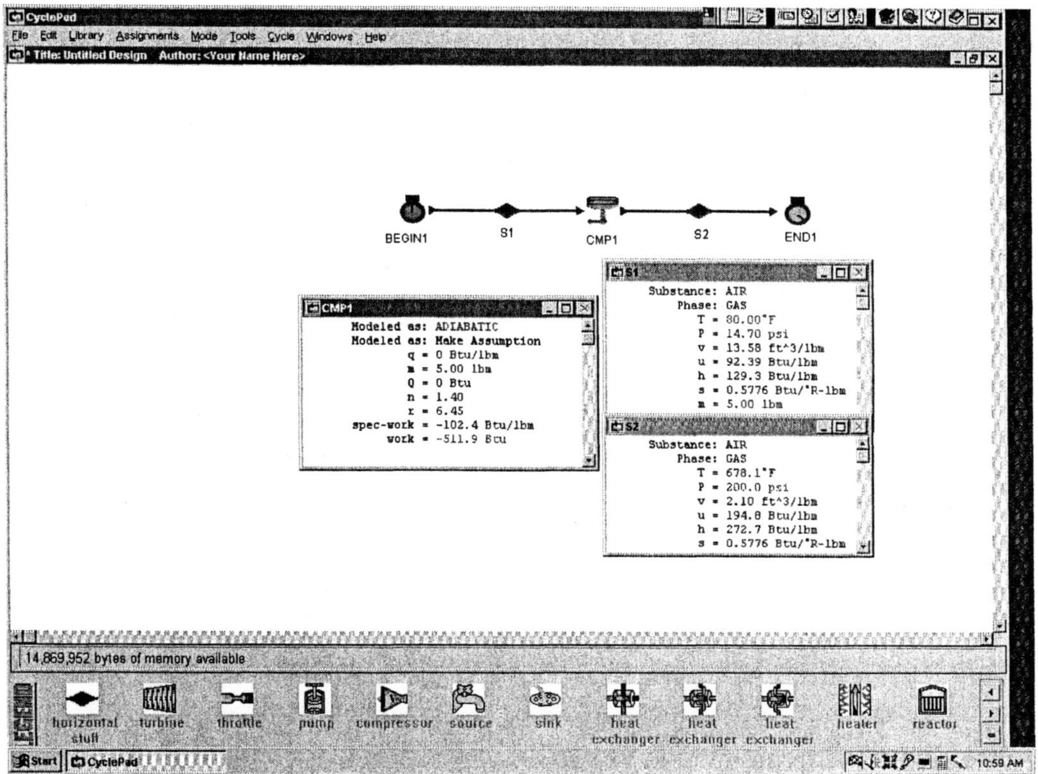

Figure Example 3.6.4.2. Adiabatic process

Homework 3.6.4 Adibatic Process

1. 0.1 kg of air is expanded adiabatically from 5000 kPa and 2000 C to 500 kPa and 906 C. Determine the work done by the air.
2. 3 kg of air is compressed adiabatically from 100 kPa and 40 C to 4000 kPa and 627 C. Determine the work added to the air.

3.6.5 Constant Entropy (Isentropic) Process

If the entropy of a process is constant, it is called *isentropic*. The heat transfer to the system is equal to zero and the final entropy of the process is the same as its initial entropy.

$$Q_{12} = 0, \text{ and } S_2 = S_1 \tag{3.6.5.1}$$

The first law of thermodynamics of the closed system for the isentropic process 1-2 is:

$$-W_{12} = U_2 - U_1 \tag{3.6.5.2}$$

Example 3.6.5.1. Helium undergoes an isentropic expansion process from an initial volume of 0.007 ft^3 and 2000 psia to 0.045 ft^3. Find the work performed by the expanding air. Determine the work added or removed from the air during this process. Also determine the final temperature, the internal energy change, and entropy change of the gas.

To solve this problem by CyclePad, we take the following steps:

1. Build
   (A) Take a begin, an expansion device, and an end from the closed-system inventory shop and connect them.
   (B) Switch to analysis mode.
2. Analysis
   (A) Assume the expansion as an isentropic process.
   (B) Input the given information: (a) working fluid is helium, (b) the initial helium volume, pressure and temperature of the process are 0.007 ft^3, 2000 psia and 300°F, and (c) the final helium volume is 0.045 ft^3.
3. Display results
   (A) Display the expansion device results. The answers are W=2.76 Btu, T=-241.3°F, Δu=161.7-562.5=-400.8 Btu/lbm, and Δs=-0.6805-(-0.6805)=0 Btu/[lbm(R)].

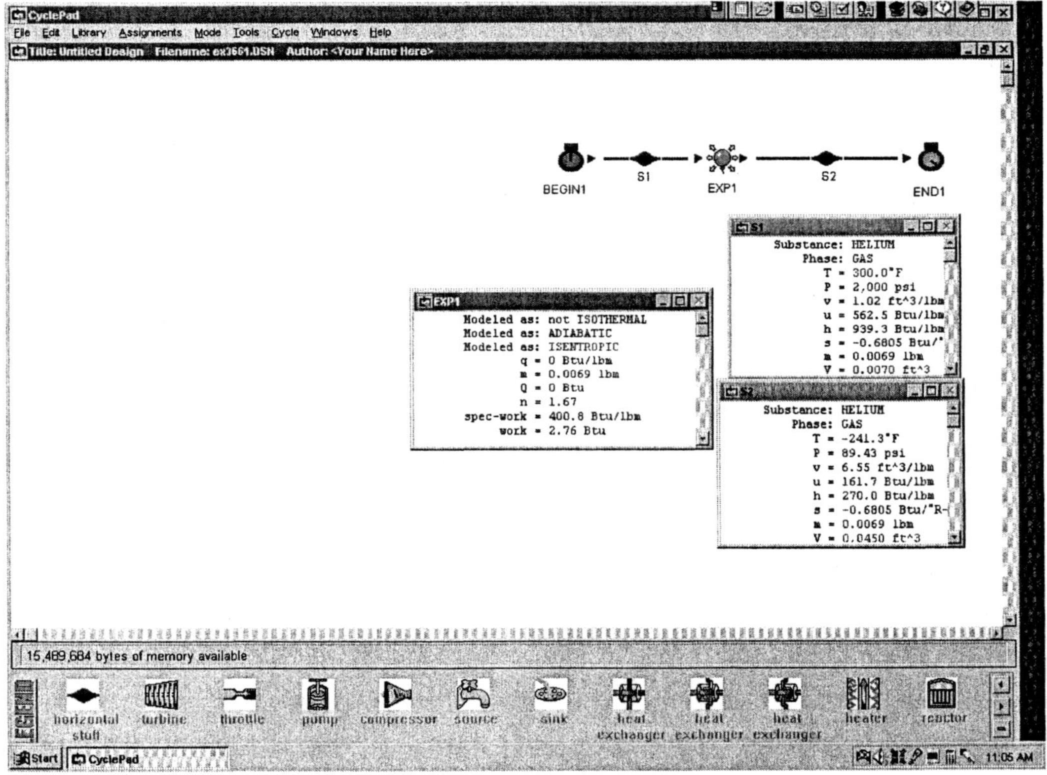

Figure Example 3.6.5.1. Isentropic process

Example 3.6.5.2. 2.2 lbm of water undergoes an isentropic expansion process from an initial 600 psia and 600°F to 100 psia. Find the work performed by the expanding water. Determine the work added or removed from the water during this process. Also determine the final temperature and quality, the internal energy change, and entropy change of the gas.

To solve this problem by CyclePad, we take the following steps:

1. Build
   (A) Take a begin, an expansion device, and an end from the closed-system inventory shop and connect them.
   (B) Switch to analysis mode.
2. Analysis
   (A) Assume the expansion as an isentropic process.
   (B) Input the given information: (a) working fluid is water, (b) the initial pressure and temperature of the process are 600 psia and 600°F, (c) the mass is 2.2 lbm, and (d) the final pressure is 100 psia.
3. Display results
   (A) Display the expansion device results. The answers are W=293.4 Btu (removed), T=327.9°F, x=0.9637; Δu=1055-1184=-129 Btu/lbm, and Δs=1.53-1.53=0 Btu/[lbm(R)].

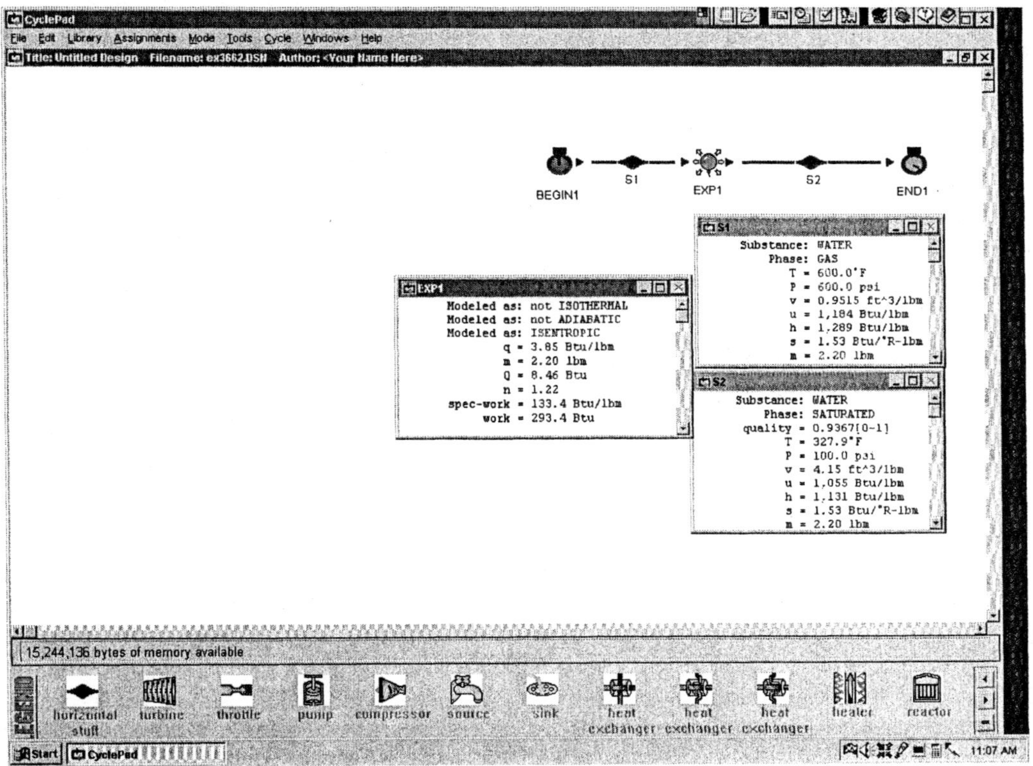

Figure Example 3.6.5.2. Isentropic process

Homework 3.6.5 Isentropic Process

1. Air at 2900°F and 550 psia contained in a tank expands isentropically. The final volume is 5 times the initial volume. Find the (a) final pressure, (b) final temperature, (c) heat transferred, (d) work done, and (e) change in entropy.
2. 1.3 kg of ir at 2400°C and 1 Mpa expands isentropically. The final volume is 5 times the initial volume. Find the (a) final pressure, (b) final temperature, (c) heat transferred, (d) work done, and (e) change in entropy.
3. Water is expanded isentropically from 1 Mpa and 400 C to 200 kPa. Find the final temperature of the water and the work done by the water.

3.6.6 Polytropic Process

Many practical thermodynamic applications undergo processes described by a specific relationship between pressure and volume. A process described by pV^n=constant is called a polytropic process. The exponential index n is either known or is obtained from experimental data. The polytropic process is a generalization of the isentropic and other processes that are used when the working fluid is an ideal gas. It is quite useful for analyzing ideal gas processes.

The boundary work and the first law of thermodynamics of the closed system for the constant pressure process 1-2 are:

$$W_{12} = \int pdV, \tag{3.6.6.1}$$

and

$$Q_{12} = W_{12} + U_2 - U_1 \tag{3.6.6.2}$$

Example 3.6.6.1. 2 lbm of water undergoes an polytropic compression process from an initial 100 psia and 600°F to 300 psia and 620°F. Find the work added to the water. Determine the heat added or removed from the water during this process. Also determine the exponential index n of the polytropic process, the internal energy change, and entropy change of the water.

To solve this problem by CyclePad, we take the following steps:

1. Build
   (A) Take a begin, a compression device, and an end from the closed-system inventory shop and connect them.
   (B) Switch to analysis mode.
2. Analysis
   (A) Input the given information: (a) working fluid is water, (b) the initial pressure and temperature of the process are 100 psia and 600°F, (c) the mass is 2 lbm, and (d) the final pressure and temperature of the process are 300 psia and 620°F.
3. Display results

(A) Display the compression device results. The answers are W=-252.9 Btu (added), Q=-258.3 Btu (removed), n=1.00,; Δu=1211-1214=-3 Btu/lbm, and Δs=1.64-1.76=-0.08 Btu/[lbm(R)].

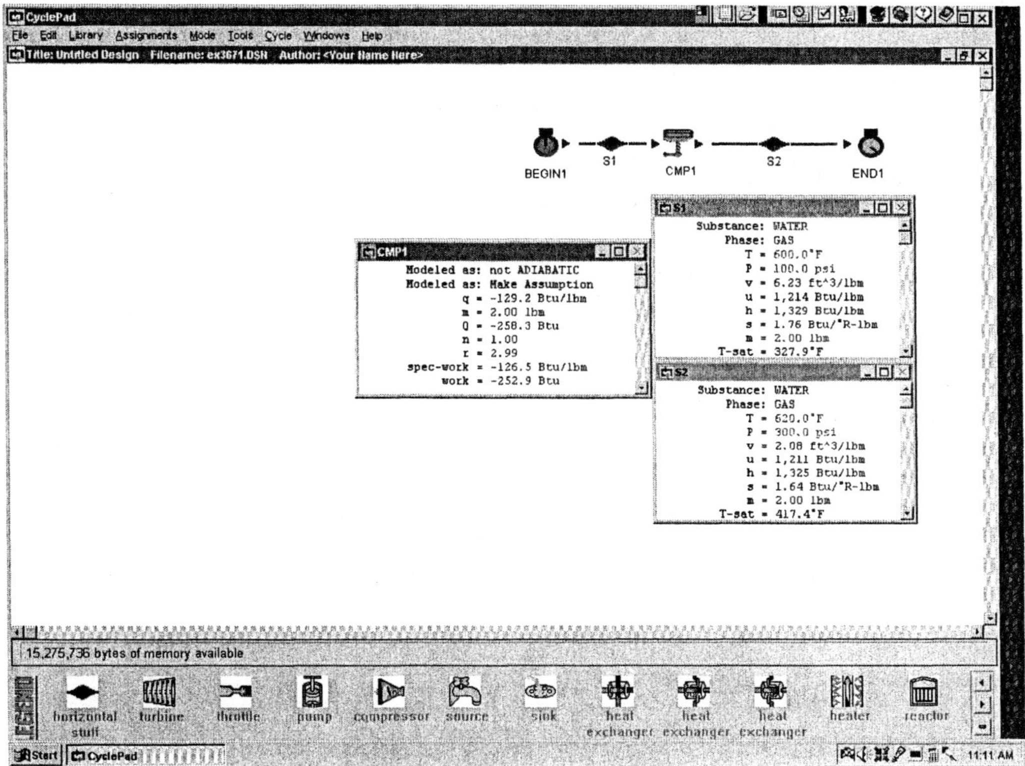

Figure Example 3.6.6.1. Polytropic process

Example 3.6.6.2. 0.34 lbm of carbon dioxide at 900°F and 300 psia contained in a tank expands polytropically to 100 psia and 300°F. Determine the heat added or removed from the gas during this process. Also determine the exponential index n of the polytropic process, the internal energy change, and entropy change of the gas.

To solve this problem by CyclePad, we take the following steps:

1. Build
   (A) Take a begin, an expansion device, and an end from the closed-system inventory shop and connect them.
   (B) Switch to analysis mode.
2. Analysis
   (A) Input the given information: (a) working fluid is carbon dioxide, (b) the initial pressure and temperature of the process are 550 psia and 900°F, (c) the mass is 0.34 lbm, and (d) the final pressure and temperature of the process are 100 psia and 300°F.
3. Display results

(A) Display the expansion device results. The answers are W=14.45 Btu (removed), Q=-4.06 Btu, n=1.37, Δu=157.1- 211.6=-54.5 Btu/lbm, and Δs=0.6638-0.6740=-0.0102 Btu/[lbm(R)].

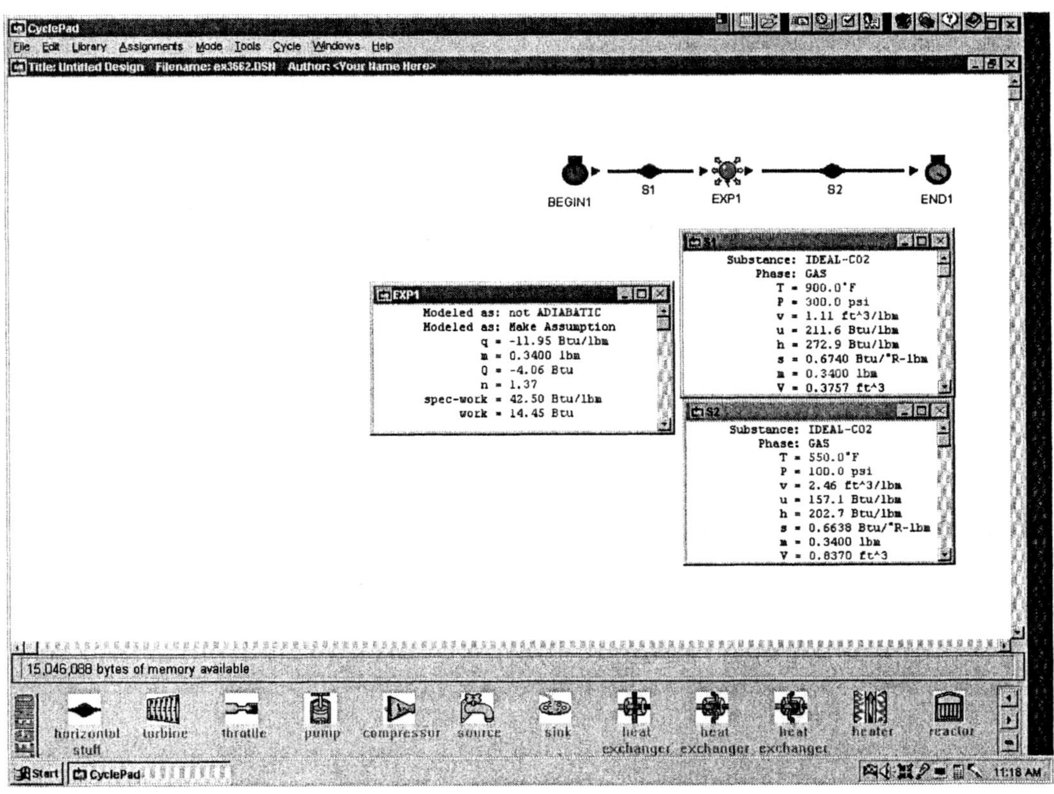

Figure Example 3.6.6.2. Polytropic process

Homework 3.6.6 Polytropic

1. Helium undergoes an expansion process $[pv^2=p_1(v_1)^2=p_2(v_2)^2=\text{constant}]$ from an initial state at 0.01 ft^3, 80 F and 1000 psia to a final state at 0.1 ft^3. Find the helium mass, temperature, pressure, specific heat added and the specific work done by the helium.
2. Carbon dioxide undergoes an expansion process $[pv^3=p_1(v_1)^3=p_2(v_2)^3=\text{constant}]$ from an initial state at 0.04 ft^3, 80 F and 1000 psia to a final state at 0.1 ft^3. Find (a) the air mass, (b) final temperature and pressure, (c) specific heat added and (d) the specific work done by the carbon dioxide.
3. 0.08 lb of helium in a cylinder fitted with a piston is compressed from 14.3 psia and 75 F to 2000 psia in a polytropic process, $pv^{1.3}$=constant. Determine the specific work and heat of the compression process.
4. 1 kg of helium is compressed in a polytropic process ($pv^{1.3}$=constant). The initial pressure, temperature and volume are 620 kPa, 715.4 K and 0.15 m^3. The final volume is 0.1 m^3. Find (a) the final temperature and pressure, (b) the work done, and (c) the heat interaction.

5. Air initially at 65°F and 75 psia is compressed to a final pressure of 300 psia and temperature of 320°F. Find the value of the polytropic exponent for this process.
6. Air in a piston-cylinder set up expands from 30 psia and 12 ft^3/lbm to 22 psia and 18 ft^3/lbm. Find the work done for the processes.
7. Air initially at 15 psia and 250°F is expanded from 13 ft^3 to 20 ft^3. If the final temperature of the air is 50°F, what is the final pressure? What is the heat added? What is the work added?

3.6.7 Heating and Cooling Processes

Many practical thermodynamic applications undergo heating and cooling processes. The heating and cooling processes can be isobaric, or isochoric, or other processes. In the software CyclePad, there is a heating device and a cooling device in the closed system inventory shop. Without the shaft work, the boundary work and the first law of thermodynamics of the closed system for the heating and cooling process 1-2 are:

$$W_{12} = \int p dV, \tag{3.6.7.1}$$

and

$$Q_{12} = W_{12} + U_2 - U_1 \tag{3.6.7.2}$$

Example 3.6.7.1. 0.4 lbm of helium is cooled in a rigid tank from 525 psia and 870°F to 220°F. Determine the heat removed, final pressure of helium, change of internal energy, and change of entropy.

To solve this problem by CyclePad, we take the following steps:

1. Build
   (A) Take a begin, a cooler, and an end from the closed-system inventory shop and connect them.
   (B) Switch to analysis mode.
2. Analysis
   (A) Assume the cooler as an isochoric process, W added is 0.
   (B) Input the given information: (a) working fluid is helium and mass is 0.4 lbm, (b) the initial helium pressure and temperature of the process are 525 psia and 870°F, and (d) the final temperature of the process is 220°F.
3. Display results
   (A) Display the cooling device results. The answers are Q=-192.5 Btu, p=268.4 psia, Δu=503.2-984.5=-481.3 Btu/lbm, and Δs=0.1783-0.6752=-0.4969 Btu/[lbm(R)].

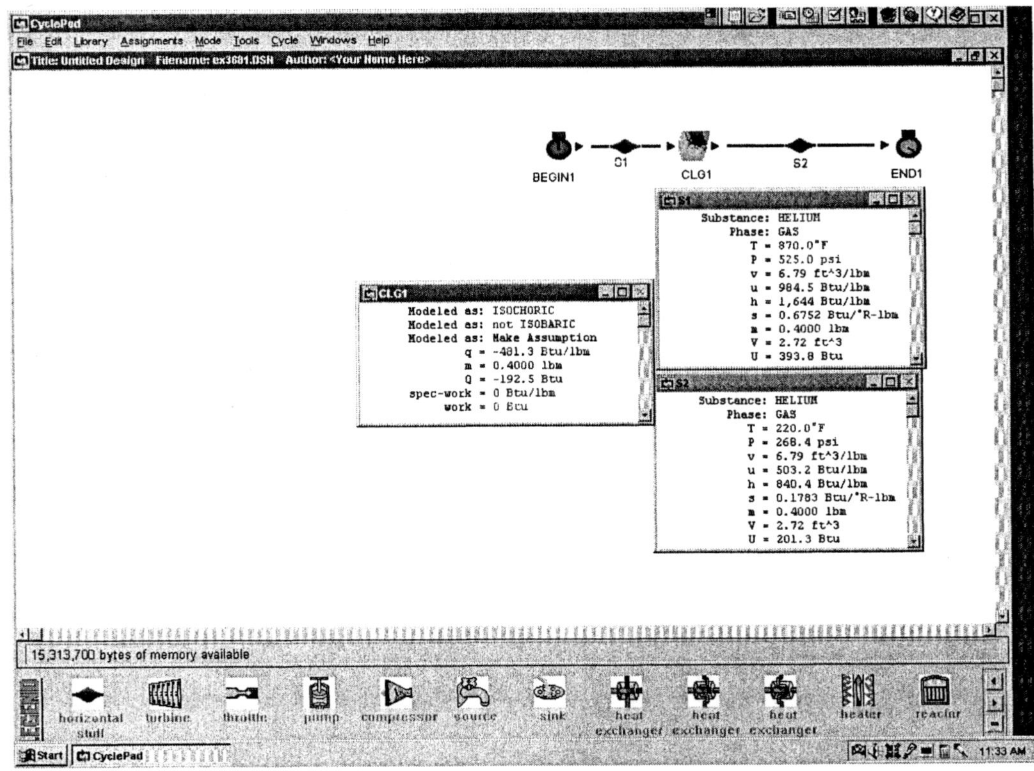

Figure Example 3.6.7.1. Cooling process

Example 3.6.7.2. 0.2 kg of R-134a is heated from a quality of 0.58 and 8°C to saturated vapor at 410 kPa. The heat added to the R-134a is 90 kJ. Determine the work added or removed, final temperature of, change of internal energy, and change of entropy.

To solve this problem by CyclePad, we take the following steps:

1. Build
   (A) Take a begin, a heater, and an end from the closed-system inventory shop and connect them.
   (B) Switch to analysis mode.
2. Analysis
   (A) Q added to the heater is 90 kJ.
   (B) Input the given information: (a) working fluid is R-134a and mass is 0.2 kg, (b) the initial R-134a quality and temperature of the process are 0.58 and 8°C, and (d) the final quality and pressure of the process are 1 and 410 kPa.
3. Display results
   (A) Display the heating device results. The answers are W=73.67 kJ (removed), T=9.55°C, Δu=403.9-322.3=81.6 kJ, and Δs=1.72-1.44=0.28 kJ/[kg(K)].

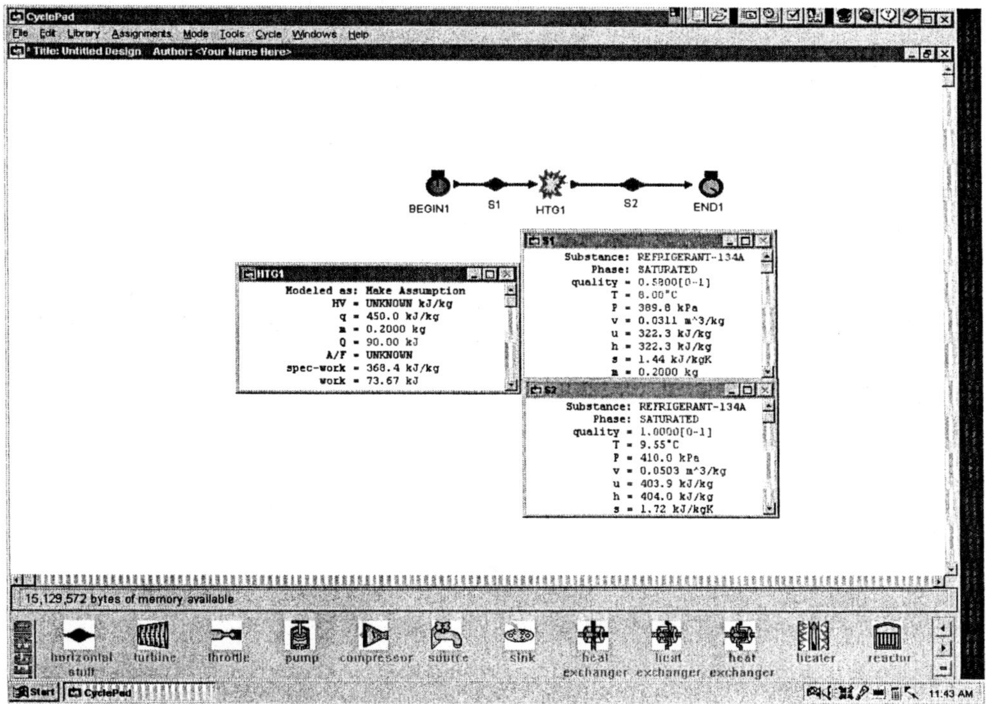

Figure Example 3.6.7.2. Heating process

Homework 3.6.7 Heating and Cooling Process

1. 4 lbm of helium is cooled in a rigid tank from 25 psia and 810°F to 250°F. Determine the work added, heat removed, and final pressure of helium.
2. 2 kg of R-22 is heated from a quality of 0.8 and 9°C to saturated vapor at 400 kPa. The heat added to the R-12 is 94 kJ. Determine the work added or removed, and final temperature of R-22.
3. 2.4 kg of water is heated from a quality of 0.98 and 59°C to saturated vapor at 710 kPa. The heat added to the water is 940 kJ. Determine the work added or removed, and final temperature of water.

3.6.8 Compression and Expansion Processes

Many practical thermodynamic applications undergo compression and expansion processes. The compression and expansion processes can be adiabatic, isentropic, polytropic, or other processes. In the software CyclePad, there is a compression device and an expansion device in the closed system inventory shop.

The boundary work and the first law of thermodynamics of the closed system for the compression and expansion process 1-2 are:

$$W_{12} = \int p dV, \tag{3.6.8.1}$$

and

$$Q_{12} = W_{12} + U_2 - U_1 \tag{3.6.8.2}$$

Example 3.6.8.1. 2.3 lbm of air at 290°F and 100 psia expands isentropically. The final pressure is 25 psia. Find the (a) final temperature, (b) heat transferred, (c) work done, and (d) change in entropy.

To solve this problem by CyclePad, we take the following steps:

1. Build
   (A) Take a begin, an expansion device, and an end from the closed-system inventory shop and connect them.
   (B) Switch to analysis mode.
2. Analysis
   (A) Assume the expansion as an isentropic process.
   (B) Input the given information: (a) working fluid is air, (b) the initial temperature and pressure of the process are 290°F and 100 psia, (c) the mass is 2.3 lbm, and (d) the final pressure is 25 psia.
3. Display results
   (A) Display the expansion device results. The answers are T=44.82°F, Q=0 Btu, W=96.54 Btu (removed), and Δs=0.525-0.525=0 Btu/[lbm(R)].

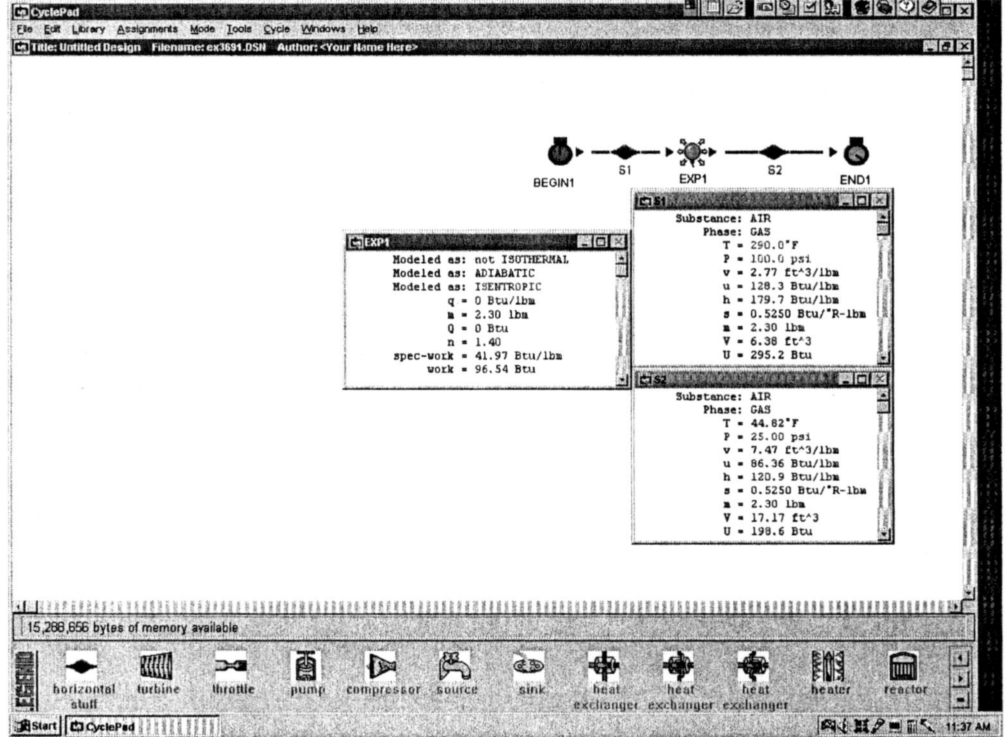

Figure Example 3.6.8.1. Expansion process

Example 3.6.8.2. 0.0804 kg of helium in a cylinder fitted with a piston is compressed from 103 kPa and 0.6 m^3 to 443 kPa in a polytropic process with n=1.5. Determine the final temperature, specific internal energy change, specific internal entropy change of the helium, work and heat of the compression process.

To solve this problem by CyclePad, we take the following steps:

1. Build
   (A) Take a begin, an expansion device, and an end from the closed-system inventory shop and connect them.
   (B) Switch to analysis mode.
2. Analysis
   (A) Assume the expansion as a polytropic process, n=1.5.
   (B) Input the given information: (a) working fluid is helium, (b) the initial helium mass, pressure and volume of the process are 0.0804 kg, 103 kPa and 0.6 m^3, and (c) the final helium pressure is 443 kPa.
3. Display results
   (A) Display the expansion device results. The answers are T=210.7°C, Δu=1500-922.4=577.6 kJ/kg, Δs=5.00-5.51=-0.51 kJ/[kg(K)], W=-62.23 kJ (added), and Q=-15.79 kJ.

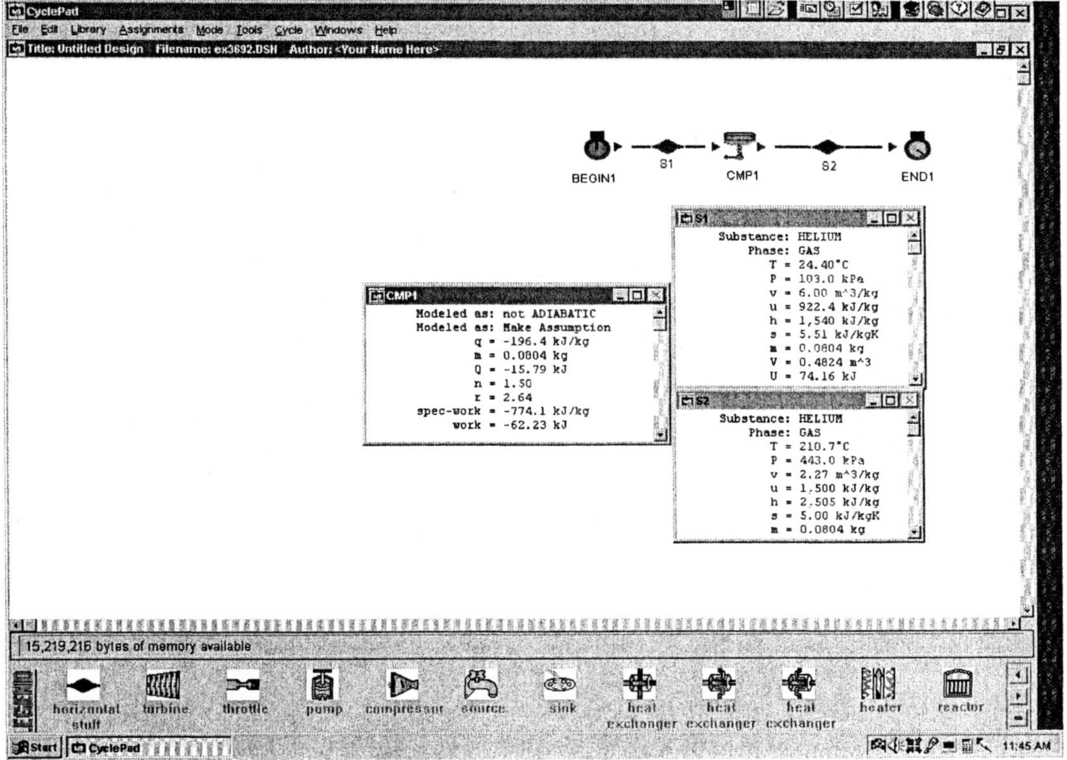

Figure Example 3.6.8.2 Compression process

Homework 3.6.8 Compression and Expansion Processes

1. Carbon dioxide expands in a polytropic process with n=1.3. The initial pressure, temperature and volume are 620 kPa, 715.4 K and 0.15 m^3. The final volume is 1 m^3. Find (a) the final temperature, (b) the work done, (c) the mass of water.
2. Air initially at 65°F and 75 psia is compressed to a final pressure of 300 psia in a polytropic process with n=1.57. Find (a) the final temperature, (b) the work done, (c) the change of internal energy, and (d) the heat interaction.
3. 0.023 lbm of air initially at 15 psia and 250°F is expanded adiabatically to 0.92 ft^3. If the final temperature of the air is 50°F, what is the final pressure? What is the heat added? What is the work added or removed?

3.7 MULTI- PROCESS

We have studied several processes separately. These processes can be combined to perform an engineering task. The multi-process analysis is illustrated by the following examples.

Example 3.7.1. 0.023 lbm of air initially at state 1 (80°F and 100 psia) in a piston-cylinder set up undergoes the following two processes:

1-2 isometric (constant volume) process where q_{12}=300 Btu/lbm(heat added)
2-3 isentropic expansion to a pressure of 25 psia

Find (a) the heat transferred and work done for process 1-2, and (b) the heat transferred and work done for process 2-3, and (c) the heat transferred and work done for process 1-2-3, (d) pressure and temperature at state 2, (e) temperature at state 3, (f) specific internal energy change 1-3, and (g) specific entropy change 1-3.

To solve this problem by CyclePad, we take the following steps:

1. Build
   (A) Take a begin, a heating device, an expansion device, and an end from the closed-system inventory shop and connect them.
   (B) Switch to analysis mode.
2. Analysis
   (A) Assume the heater is isometric and expansion device as an isentropic process.
   (B) Input the given information: (a) working fluid is air, (b) the initial mass, temperature and pressure of the process are 0.023 lbm, 290°F and 100 psia, and (d) the final pressure at state 3 is 25 psia.
3. Display results
   (A) Display state 1, state 2, state 3, the heating device and expansion device results. The answers are: (a) Q_{12}=6.9 Btu, W_{12}=0 Btu; (b) Q_{23}=0 Btu, W_{23}=5.01 Btu; (c) $Q_{13}=Q_{12}+Q_{23}$=6.9+0=6.9 Btu, $W_{13}=W_{12}+W_{23}$=0+5.01=5.01 Btu, (d) p_2=424.7 psia, T_2=1832°F; (e) T_3=560.7°F; (f) $\Delta u=u_3-u_1$=174.7-92.39=82.31 Btu/lbm; (g) $\Delta s=s_3-s_1$=0.6939-0.4463=0.2476 Btu/[lbm(R)].

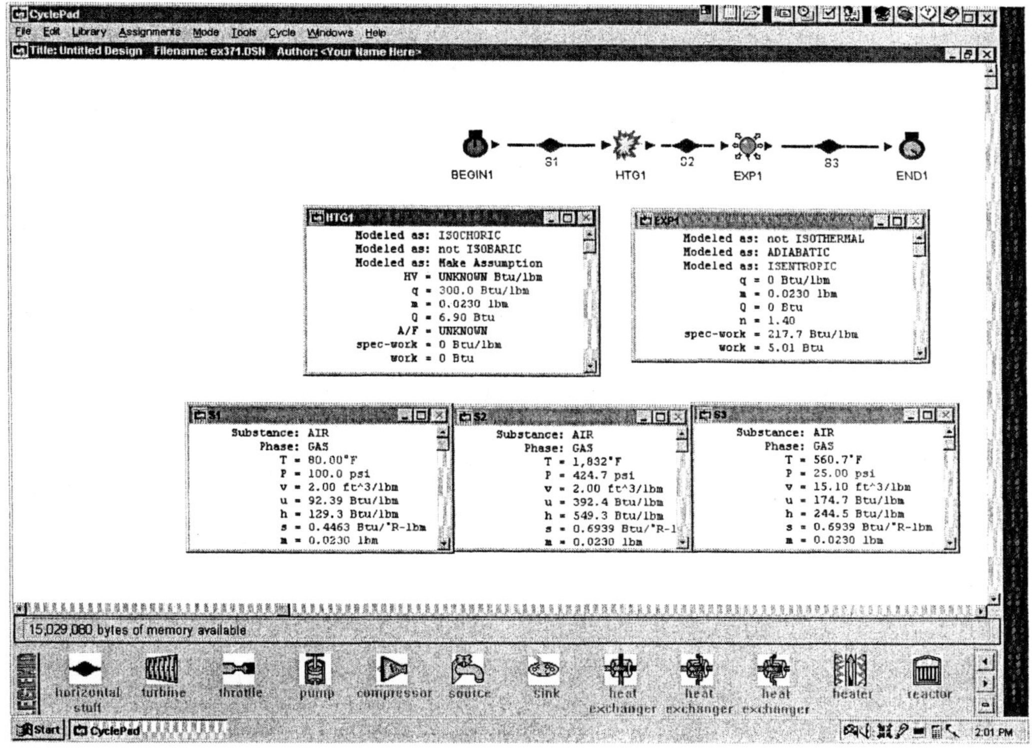

Figure Example 3.7.1. Multi process

Example 3.7.2. Air initially at state 1 (20°C, 0.0024 m^3 and 100 kPa) in a piston-cylinder set up undergoes the following two processes:

1-2 isentropic compression to a volume of 0.0003 m^3.
2-3 constant volume heating with heat added 1.5 kJ.

Find (a) the heat transferred and work done for process 1-2, and (b) the heat transferred and work done for process 2-3, and (c) the heat transferred and work done for process 1-2-3, (d) pressure and temperature at state 2, (e) pressure and temperature at state 3, (f) specific internal energy change 1-3, and (g) specific entropy change 1-3.

To solve this problem by CyclePad, we take the following steps:

1. Build
   (A) Take a begin, a compression device, a heating device, and an end from the closed-system inventory shop and connect them.
   (B) Switch to analysis mode.
2. Analysis
   (A) Assume the compression device as an isentropic and heater as isometric process. Q_{23} added in the heater is 1.5 kJ.

(B) Input the given information: (a) working fluid is air, (b) the initial temperature, volume and pressure of the process are 20°C, 0.0024 m^3 and 100 kPa, and (d) the volume at state 2 is 0.0003 m^3.

3. Display results

(A) Display state 1, state 2, state 3, the compression device and heating device results. The answers are: (a) Q_{12}=0 kJ, W_{12}=-0.7784 kJ; (b) Q_{23}=1.5 kJ, W_{23}=0 kJ; (c) $Q_{13}=Q_{12}+Q_{23}$=0+1.5=1.5 kJ, $W_{13}=W_{12}+W_{23}$=-0.7784+0=-0.7784 kJ; (d) p_2=1838 kPa, T_2=400.3°C; (e) p_3=3838 kPa and T_3=1133°C; (f) $\Delta u=u_3-u_1$=1008-210.1=797.9 kJ; (g) $\Delta s=s_3-s_1$=2.93-2.40=0.53 kJ/[kg(K)].

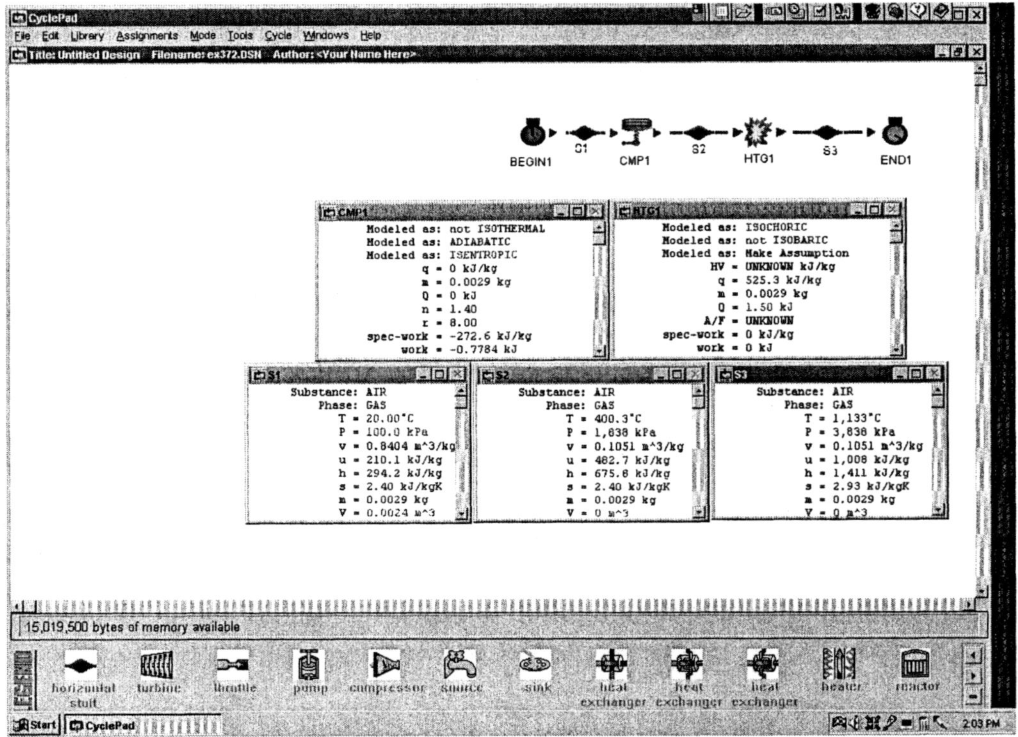

Figure Example 3.7.2. Multi process

Homework 3.7 Multi-Process

1. 1.2 lbm of air initially at 80°F and 100 psia in a piston-cylinder set up undergoes the following three reversible processes:

   1-2 isometric process where q_{12}=300 Btu/lbm

   2-3 isentropic expansion

   3-4 isothermal expansion

   The pressure and temperature of air at state 4 are 1 psia and 80 F.

   Find the (a) heat transferred and work done for process 1-2, (b) heat transferred and work done for process 2-3, and (c) heat transferred and work done for process 3-1.

2. Air initially at state 1 (20°C, 0.0024 m^3 and 100 kPa) in a piston-cylinder set up undergoes the following two processes:

   1-2 isentropic compression to a volume of 0.0003 m^3.
   2-3 constant pressure heating with heat added 1.5 kJ.

   Find (a) the heat transferred and work done for process 1-2, and (b) the heat transferred and work done for process 2-3, and (c) the heat transferred and work done for process 1-2-3, (d) pressure and temperature at state 2, (e) pressure and temperature at state 3, (f) specific internal energy change 1-3, and (g) specific entropy change 1-3.
3. 0.23 lbm of air in a piston-cylinder set up is heated at 100 psia constant pressure from a temperature of 100°F to a temperature of 200°F. The air is then expanded isentropically until the volume doubles. Find (a) the pressure, temperature and specific volume of the air at the final state, (b) heat transferred and work done for the heating process, (c) heat transferred and work done for the isentropic process, and (d) total heat transferred and work done for the heating and the isentropic processes.
4. Air at 100 kPa, 20 C and 0.004 m^3 in a piston-cylinder set up is compressed isentropically to one-fourth its original volume. It is then cooled at constant volume to 175°C. Find (a) the pressure and temperature of the air at the final state of the isentropic process, (b) the pressure of the air at the final state of the cooling process, (c) heat transferred and work done for the isentropic process, (d) heat transferred and work done for the cooling process, and (e) total heat transferred and work done for the cooling and the isentropic processes.

3.8 SUMMARY

Heat and work are the microscopic and macroscopic energy transfers across the boundary surface from the surroundings to the system or vice versa without mass transfer. Both work and heat are path functions. There are two types of important thermodynamic work: boundary work and shaft work. Boundary work is given by the expression $\int pdV$, where p is a function of V. Work produced by the system is positive and heat added to the system is positive. The energy balance for a closed system, called the First law of thermodynamics for a closed system, can be expressed as $Q-W=\Delta E$, where $E=E_k+E_p+U$. Notice that flow energy is equal to zero because there is no mass transfer across the boundary surface of the system. Application of the First law of thermodynamics to various processes and devices using CyclePad are illustrated.

Chapter 4

FIRST LAW OF THERMODYNAMICS FOR OPEN SYSTEMS

OBJECTIVES

After reading and studying the material in this chapter, you should be able to:

1. State the general continuity equation for an open system.
2. Understand steady flow and state the steady flow continuity equation.
3. Solving simple open system mass conservation problems by hand.
4. State the general and steady flow first law of thermodynamics, or energy conservation for an open system.
5. Know the commonly used open system devices in engineering application.
6. Set up the first law of thermodynamics and continuity equation to open systems.
7. Solving open system problems using CyclePad.

4.1 INTRODUCTION

In Chapter 3, several processes, by focusing attention on a fixed mass (control mass or closed system), are studied. In many engineering applications, it is more convenient to draw the system boundary around a fixed space (control volume or open system) through which working fluid may flow. In this chapter, many applications using the open system approach will be studied. The principles used in the study of open systems are the conservation of mass (mass balance) and the first law of thermodynamics (energy balance).

4.2 CONSERVATION OF MASS

4.2.1 General Case

The law of the conservation of mass states that mass can not be created nor destroyed. In thermodynamics there are two kinds of systems: open and closed. For a closed system, that is, a fixed mass, the conservation of mass is true. No equation is necessary. However, for an open system, that is, a fixed space, an expression for the conservation of mass needs to be developed.

The mass (m) content of an open system may be changed by mass flow in (m_i) across the boundary of the open system from the surroundings or mass exit (m_e) across the boundary of the open system to the surroundings, or both. Let the initial mass content of system at time t_1 is m_1; the final mass content of system at time t_2 is m_2; Mass flow in across the boundary of the open system from the surroundings from t_1 to t_2 is m_i; mass exit across the boundary of the open system to the surroundings from t_1 to t_2 is m_e. Applying the *conservation of mass* to the open system under these conditions becomes

$$m_2 - m_1 = m_i - m_e \tag{4.2.1.1}$$

Equation (4.2.1.1) is the law of conservation of mass for an open system. In using Equation (4.2.1.1), we must always remember the important subscripts that 1 refers to time 1 (t_1) and 2refers to time 2 (t_2); and that i and e refer to locations; i refers to inlet section i and e refers to exit section e, respectively.

The law of conservation of mass for an open system [Equation (4.2.1.1)] expressed in words is

> [time change of the mass contained within the open system] = [net mass flow in across the boundary of the open system from the surroundings] - [net mass exit across the boundary of the open system to the surroundings]

Various special forms of the law of conservation of mass for an open system can be written. For examples

$$m_2 - m_1 = (\rho_i V_i A_i - \rho_e V_e A_e)\Delta t \tag{4.2.1.2}$$

$$\int_{\text{Time from t1 to t2}} dm = \int_{\text{inlet sections}} dm_i - \int_{\text{exit sections}} dm_e \tag{4.2.1.3}$$

$$\int_{\text{Time from t1 to t2}} dm = \int_{\text{inlet sections}} \rho_i V_i dA_i - \int_{\text{exit sections}} \rho_e V_e dA_e \tag{4.2.1.4}$$

Equation (4.2.1.2) is the algebraic form of Equation (4.2.1.1); Equation (4.2.1.3) and Equation (4.2.1.4) are the integral forms of Equation (4.2.1.1), respectively.

$$dm = dm_i - dm_e \tag{4.2.1.5}$$

Equation (4.2.1.5) is the differential forms of Equation (4.2.1.1).

$$\Delta m/\Delta t = m_i/\Delta t - m_e/\Delta t \tag{4.2.1.6}$$

Equation (4.2.1.6) is the finite time rate form of Equation (4.2.1.1), where $\Delta m = m_2 - m_1$, and $\Delta t = t_2 - t_1$, respectively.

$$dm/dt = mdot_i - mdot_e \tag{4.2.1.7}$$

Equation (4.2.1.7) is the differential time rate form of Equation (4.2.1.1). The time rate of the law of conservation of mass for an open system [Equation (4.2.1.7)] expressed in words is

[time rate of change of the mass contained within the open system at time t] = [net rate of mass flow in across the boundary of the open system from the surroundings at time t] - [net rate of mass exit across the boundary of the open system to the surroundings at time t]

4.2.2 Steady Flow Case

Many engineering processes that operate under a steady condition for long period of time are called steady-flow processes. A *steady system* is one whose quantities such as mass and properties within the system do not change with respect to time. A *steady-flow* process is a process whose flow quantities at inlet and exit boundary sections do not change with respect to time. Equation (4.2.1.2), the law of conservation of mass for an open system becomes

$$m_2 - m_1 = 0 \quad (4.2.2.1)$$

$$mdot_i = mdot_e \quad (4.2.2.2)$$

and

$$\rho_i V_i A_i = \rho_e V_e A_e \quad (4.2.2.3)$$

The law of conservation of mass for a steady flow open system [Equation (4.2.2.2)] expressed in words is

[total rate of mass flow in across the boundary of the open system from the surroundings] = [total rate of mass exit across the boundary of the open system to the surroundings at time t]

Equation (4.2.2.3) is called *continuity equation.*

Example 4.2.1. A hot-air stream at 200°C and 200 kPa with mass flow rate of 0.1 kg/s enters a mixing chamber where it is mixed with a stream of cold-air at 200 kPa and 10°C with a volumetric rate flow of 1 m^3/s. The mixture leaves the mixing chamber at 200 kPa. Determine the mass flow rate and the volumetric rate flow of air leaving the chamber.

To solve this problem by CyclePad, we take the following steps:

1. Build
   (A) Take two sources, a mixing chamber, and a sink from the open-system inventory shop and connect them.
   (B) Switch to analysis mode.
2. Analysis

(A) Assume the mixing chamber is isobaric.

(B) Input the given information: (a) working fluids are air, (b) mass flow rate, pressure and temperature of the hot air at the inlet are 0.1 kg/s, 200 kPa and 200°C, (c) volumetric flow rate, pressure and temperature of the cold air at the other inlet are 1 m^3/s, 200 kPa and 10°C.

3. Display results

(A) Display the outlet states results. The answers are mdot,mix=2.56 kg/s and Vdot,mix=1.07 m^3/s.

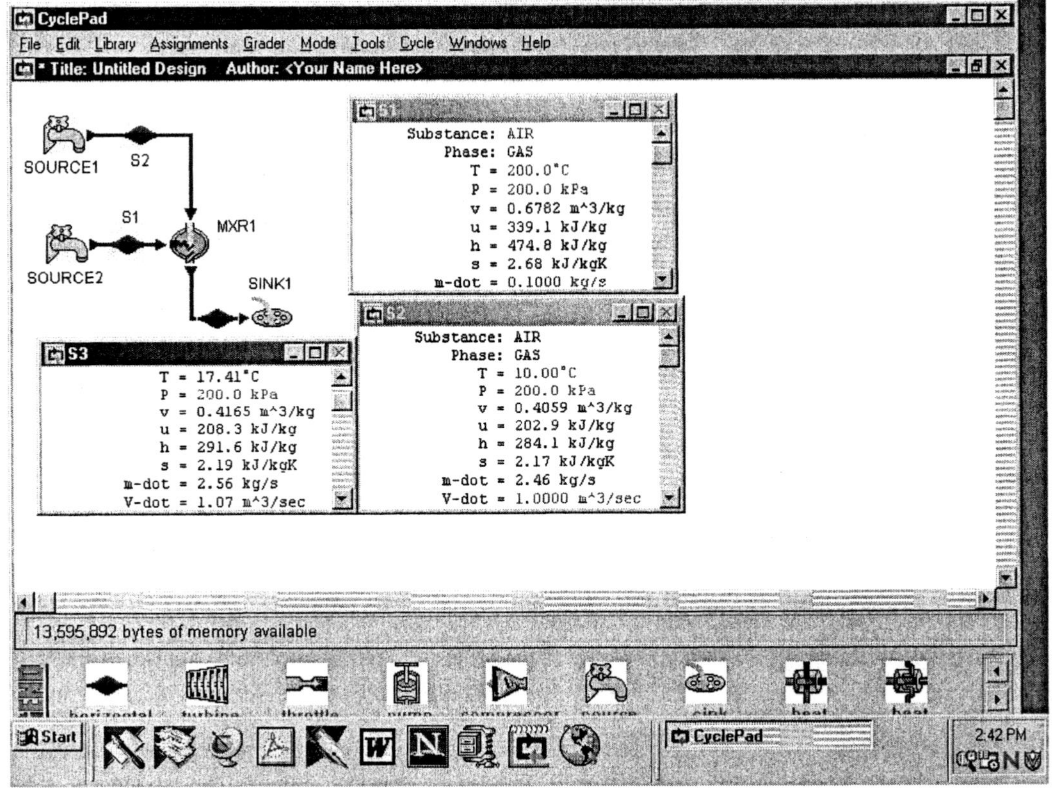

Figure Example 4.2.1 Mixing chamber

Homework 4.2 Mass Conservation

1. What is an open system?
2. Water flows through a steam power plant, which consists of a pump, a boiler, a turbine and a condenser. Is the steam power plant an open system? Is the turbine of the steam power plant an open system?
3. Define mass flow rate. How is it differ from mass?
4. Describe mass conservation in words for an open system.
5. In the mass balance equation, $m_2 - m_1 = m_i - m_e$, what are the subscripts 1, 2, i and e refer to?

6. In the mass balance equation, $m_2 - m_1 = m_i - m_e$, state the physical meaning of each term.
7. What is a steady flow?
8. Does mass flow rate proportional to fluid flowing velocity? Does mass flow rate proportional to fluid specific volume?
9. Describe mass conservation in words for pumping air into your car tire.
10. A 1 kg/s stream of refrigerant R-134a at 1 Mpa and 10°C is mixed with another stream at 1 Mpa and 50°C in a mixing chamber. If the mass flow rate of the hot stream is twice that of the cold one, determine the mass and volumetric rate flow of the refrigerant at the exit of the mixing chamber?

4.3 First Law of Thermodynamics

4.3.1 General Case

The energy (E) content of an open system may be changed by energy flow in (E_i) with mass flow in across the boundary of the open system from the surroundings, or energy exit (E_e) with mass exit across the boundary of the open system to the surroundings, or heat (Q), or work (W) or any combination of the four quantities. Let the initial energy content of an open system at time t_1 is E_1; the final energy content of an open system at time t_2 is E_2; Heat added to the system during the time from t_1 to t_2 is Q_{12}; and work added to the system during the time from t_1 to t_2 is W_{12}; Energy flow in across the boundary of the open system from the surroundings from t_1 to t_2 is E_i; Energy exit across the boundary of the open system to the surroundings from t_1 to t_2 is E_e. Apply the *energy conservation* to the open system under these conditions becomes

$$E_2 - E_1 = Q_{12} - W_{12} + E_i - E_e \qquad (4.3.1.1)$$

where E is the total energy.

The total energy (E) is made of potential energy (E_p), kinetic energy (E_k), internal energy (U) and flow energy (pV), i.e. $E = E_p + E_k + U + pV$. Notice that flow energy (or called flow work in some textbooks) is the energy required to push a volume in (or out) by pressure across the boundary surface of the open system. Thus energy contained within the open system does not have flow energy. Therefore, $E_1 = (E_p)_1 + (E_k)_1 + U_1$, $E_2 = (E_p)_2 + (E_k)_2 + U_2$, $E_i = (E_p)_i + (E_k)_i + U_i + pV_i = (E_p)_i + (E_k)_i + H_i$ and $E_e = (E_p)_e + (E_k)_e + U_e + pV_e = (E_p)_e + (E_k)_e + H_e$.

Also notice that the boundary work is zero in the case of the open system, which is considered to be a volume of fixed identity, i.e., dV=0. Therefore $W_{boundary} = \int pdV = 0$. Hence, the total work (W) in Equation (4.3.1.1) is made of shaft work (W_{shaft}) only if other modes of work are not presented.

$$W = W_{boundayr} + W_{shaft} = W_{shaft}$$

Equation (4.3.1) is the *first law of thermodynamics* for an open system.

The first law of thermodynamics for an open system [Equation (4.3.1.1)] expressed in words is

> [time change of the energy contained within the open system] = [net heat added to the system] - [net work added to the system] +[energy flow in with mass flow in across the boundary of the open system from the surroundings] - [energy exit (E_e) with mass exit across the boundary of the open system to the surroundings]

Various special forms of the first law of thermodynamics for an open system can be written. For example, the amount of energy change (∫dE) from time t_1 to time t_2 can be expressed as

$$\int dE=\int \delta Q-\int \delta W+\int_{\text{inlet sections}} dE_i - \int_{\text{exit sections}} -dE_e \qquad (4.3.1.2)$$

Equation (4.3.1.2) is the integral form of Equation (4.3.1.1).

$$dE=\delta Q-\delta W + dE_i -dE_e \qquad (4.3.1.3)$$

Equation (4.3.1.3) is the differential form of Equation (4.3.1.1).

$$\Delta E/\Delta t=Q_{12}/\Delta t-W_{12}/\Delta t + \Delta E_i/\Delta t - \Delta E_e/\Delta t \qquad (4.3.1.4)$$

Equation (4.3.1.4) is the finite time rate form of Equation (4.3.1.1), where $\Delta E=E_2 -E_1$, and $\Delta t=t_2 -t_1$, respectively.

$$dE/dt=Qdot-Wdot+ Edot_i -Edot_e \qquad (4.3.1.5)$$

Equation (4.3.1.5) is the differential time rate form of Equation (4.3.1.1). The time rate of the first law of thermodynamics for an open system [Equation (4.3.1.5)] expressed in words is

> [time rate of change of the energy contained within the open system at time t] = [net rate of heat added to the system at time t] - [net rate of work added to the system at time t] +[rate of energy flow in with mass flow in across the boundary of the open system from the surroundings at time t] - [rate of energy exit (E_e) with mass exit across the boundary of the open system to the surroundings at time t]

4.3.2 Steady Flow

For steady flow, the change in mass and energy within the open system are both zero.

$$E_2 -E_1= 0 \text{ and } m_2 -m_1= 0 \qquad (4.3.2.1)$$

Thus the various forms of the first law of thermodynamics of an open system [Equations (4.3.1.1), (4.3.1.2), (4.3.1.3), (4.3.1.4), and (4.3.1.5)] are reduced to:

$$0 = Q_{12} - W_{12} + E_i - E_e \qquad (4.3.2.2)$$

$$0 = \int \delta Q - \int \delta W + \int_{\text{inlet sections}} dE_i - \int_{\text{exit sections}} -dE_e \qquad (4.3.2.3)$$

$$0 = \delta Q - \delta W + dE_i - dE_e \qquad (4.3.2.4)$$

$$0 = Q_{12}/\Delta t - W_{12}/\Delta t + \Delta E_i/\Delta t - \Delta E_e/\Delta t \qquad (4.3.2.5)$$

and

$$0 = \text{Qdot} - \text{Wdot} + \text{Edot}_i - \text{Edot}_e \qquad (4.3.2.6)$$

The specific form of the first law of thermodynamics of an open system [Equations (4.3.2.2)] becomes

$$0 = q_{12} - w_{12} + e_i - e_e \qquad (4.3.2.7)$$

The first law of thermodynamics for an open system [Equation (4.3.2.2)] expressed in words is

0 = [net heat added to the system] - [net work added to the system] +[energy flow in with mass flow in across the boundary of the open system from the surroundings] + [energy exit (E_e) with mass exit across the boundary of the open system to the surroundings]

In many occasions, the changes of kinetic energy and potential energy are relatively small compared with the change of internal energy and flow energy, Equation (4.3.2.2) is reduced to:

$$0 = q_{12} - w_{12} + h_i - h_e \qquad (4.3.2.8)$$

Homework 4.3 First Law of Thermodynamics

1. What is the boundary work of an open system?
2. What is flow energy? Do substances at rest possess any flow energy?
3. What is enthalpy? How does it relate to internal energy and flow energy?
4. Describe energy conservation in words for an open system.
5. In the energy balance equation, $E_2 - E_1 = Q_{12} - W_{12} + E_i - E_e$, what are the subscripts 1, 2, i and e refer to?
6. In the energy balance equation, $E_2 - E_1 = Q_{12} - W_{12} + E_i - E_e$, state the physical meaning of each term.

4.4 CYCLEPAD OPEN SYSTEM DEVICES

Since the majority of engineering devices may be modeled as operating under steady state, steady flow conditions; the major effort of CyclePad is devoted to the analysis of systems with steady state, steady flow processes. The following devices in the open system inventory shop of CyclePad are used to illustrate the usefulness of the general principles and to strengthen our understanding of the basic thermodynamic principles.

4.4.1. Heater (Including Boiler, Combustion Chamber, Evaporator, Reheater, Preheater and Open Feed Water Heater)

A *heater* is a simple single stream fluid flow through a device where heat is transferred to the fluid from the surroundings. The fluid is heated and may or may not change phases. The heating process tends to occur at constant pressure, since a fluid flowing through the device usually undergoes only a small pressure drop due to fluid friction at the walls. There is no means for doing any shaft or electric work, and changes in kinetic and potential energies are commonly negligible small.

Application of the first law of thermodynamics to a heater gives

$$W=0 \text{ and } Q = m(h_e-h_i) \qquad (4.4.1.1)$$

Heaters at various applications are called boiler, combustion chamber, evaporator, reheater, preheater, open feed water heater, etc.

Example 4.4.1.1. Water at a mass flow rate of 2 lbm/s is heated in a steam boiler from 400 psia and 100°F to 400 psia and 400°F. Find the rate of heat added to the water in the boiler, specific enthalpy change and specific entropy change of the water,

To solve this problem by CyclePad, we take the following steps:

1. Build
   (A) Take a source, a boiler (heater), and a sink from the open-system inventory shop and connect them.
   (B) Switch to analysis mode.
2. Analysis
   (A) Assume the heater as an isobaric process.
   (B) Input the given information: (a) working fluid is water, (b) the inlet pressure and temperature of the boiler are 400 psia and 100°F, (c) the outlet temperature of the boiler is 400°F, and (d) the mass flow rate is 2 lbm/s.
3. Display results
   (A) Display the boiler results. The answers are Qdot=612.4 Btu/s, Δh=375.3-69.07=306.2 Btu/lbm, and Δs=0.5663-0.1293=0.4370 Btu/[lbm(R)].

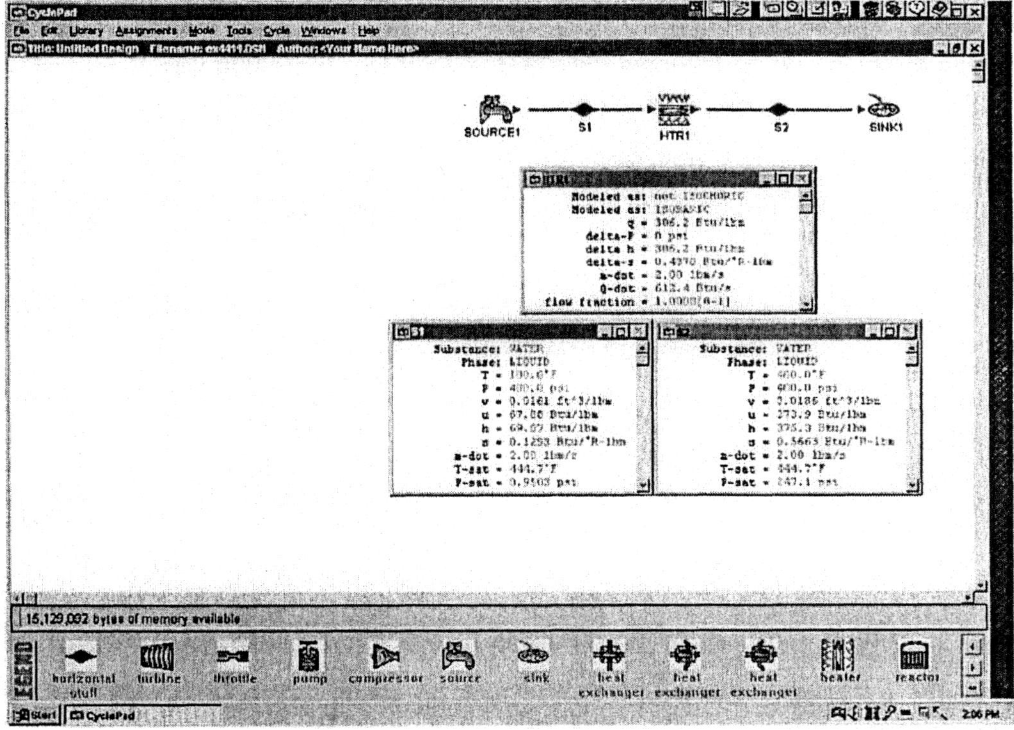

Figure Example 4.4.1.1 Heater

Example 4.4.1.2. Air at a mass flow rate of 0.01 lbm/s is heated in a combustion chamber from 150 psia and 100°F to 150 psia and 1200°F. Find the rate of air flow at the exit section, the rate of heat added to the air in the combustion chamber, specific enthalpy change and specific entropy change of the air.

To solve this problem by CyclePad, we take the following steps:

1. Build
   (A) Take a source, a combustion chamber (heater), and a sink from the open-system inventory shop and connect them.
   (B) Switch to analysis mode.
2. Analysis
   (A) Assume the heater as an isobaric process.
   (B) Input the given information: (a) working fluid is air, (b) the inlet pressure and temperature of the combustion chamber are 150 psia and 100°F, (c) the outlet temperature of the combustion chamber is 1200°F, and (d) the mass flow rate is 0.01 lbm/s.
3. Display results
   (A) Display the combustion chamber results. The answers are Vdot=0.0409 ft^3/s, Qdot=2.64 Btu/s, Δh=397.8-134.1=263.7 Btu/lbm, and Δs=0.6878-0.4272=0.2606 Btu/[lbm(R)].

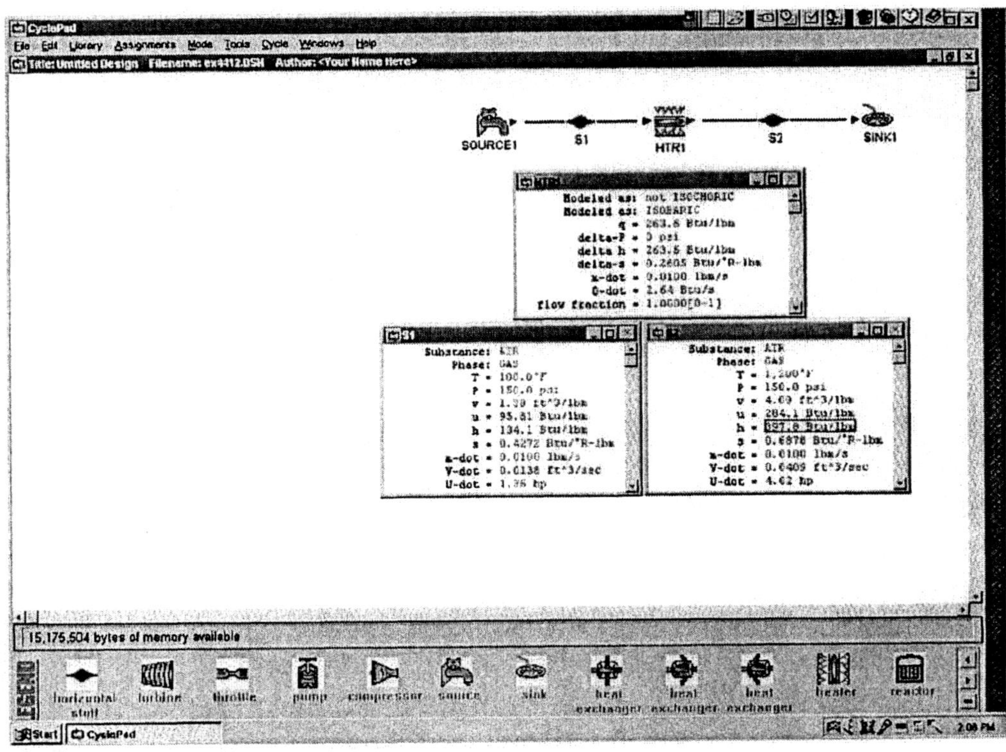

Figure Example 4.4.1.2. Heater

Example 4.4.1.3. Freon R-12 at a mass flow rate of 0.001 lbm/s is heated in an evaporator from 36 psia and a quality of 0.2 to saturated vapor. Find the rate of the freon flow at the exit section, the rate of heat added to the freon in the combustion chamber, specific enthalpy change and specific entropy change of the freon.

To solve this problem by CyclePad, we take the following steps:

1. Build
   (A) Take a source, an evaporator (heater), and a sink from the open-system inventory shop and connect them.
   (B) Switch to analysis mode.
2. Analysis
   (A) Assume the heater as an isobaric process.
   (B) Input the given information: (a) working fluid is R-12, (b) the inlet pressure and quality of the evaporator are 36 psia psia and 0.2, (c) the outlet quality of the boiler is 1, and (d) the mass flow rate is 0.001 lbm/s.
3. Display results
   (A) Display the evaporator results. The answers are Vdot=0.0011 ft^3/s, Qdot=0.0532 Btu/s, Δh=79.41-26.22=53.19 Btu/lbm, and Δs=0.1672-0.0563=0.1109 Btu/[lbm(R)].

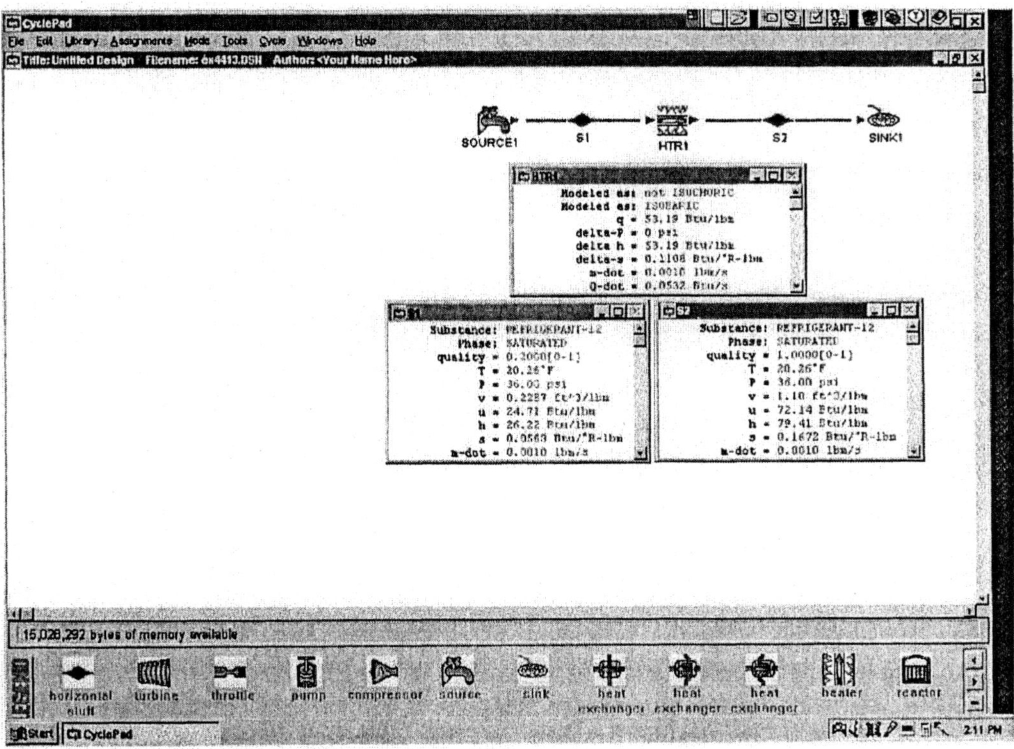

Figure Example 4.4.1.3. Heater

Homework 4.4.1 Heater

1. 0.1 kg/s of water enters a boiler (heater) at 323 K and 5 Mpa, and leaves as steam at 673 K and 5 Mpa. For steady flow, what is the heat transferred and work added to the boiler?
2. 0.01 kg/s of air enters a combustion chamber. 1000 kJ/kg of heat transfer is added to the isobaric combustion chamber. The inlet pressure and temperature of the air are 1000 kPa and 50 C and the exit air pressure is 1000 kPa.
3. For steady flow, what is the exit temperature and work added to the combustion chamber? What is the specific entropy change of the air between the inlet and exit section?

4.4.2. Cooler (Including Condenser, Intercooler)

A *cooler* is a simple single stream fluid flow through a device where heat is removed from the fluid to the surroundings. The fluid is cooled and may or may not change phases. The cooling process tends to occur at constant pressure, since a fluid flowing through the device usually undergoes only a small pressure drop due to fluid friction at the walls. There are no means for doing any shaft or electric work, and changes in kinetic and potential energies are commonly, negligibly small.

Application of the first law of thermodynamics to a steady flow steady system cooler gives

$$W=0 \text{ and } Q = m(h_e-h_i) \tag{4.4.2.1}$$

Coolers at various applications are called condenser, intercooler, etc.

Example 4.4.2.1. Water at a mass flow rate of 2 kg/s is condensed in a steam condenser from 20 kPa and a quality of 0.9 to saturated liquid. Find the rate of the water flow at the exit section, the rate of heat removed from the water in the condenser, specific enthalpy change and specific entropy change of the water.

To solve this problem by CyclePad, we take the following steps:

1. Build
   (A) Take a source, a condenser (cooler), and a sink from the open-system inventory shop and connect them.
   (B) Switch to analysis mode.
2. Analysis
   (A) Assume the condenser as an isobaric process.
   (B) Input the given information: (a) working fluid is water, (b) the inlet pressure and quality of the condenser are 20 kPa and 0.9, (c) the outlet quality of the condenser is 0, and (d) the mass flow rate is 2 kg/s.
3. Display results
   (A) Display the condenser results. The answers are Vdot=0.002 m^3/s, Qdot=-4243 kW, Δh=-2122 kJ/kg, and Δs=-6.37 kJ/[kg(K)].

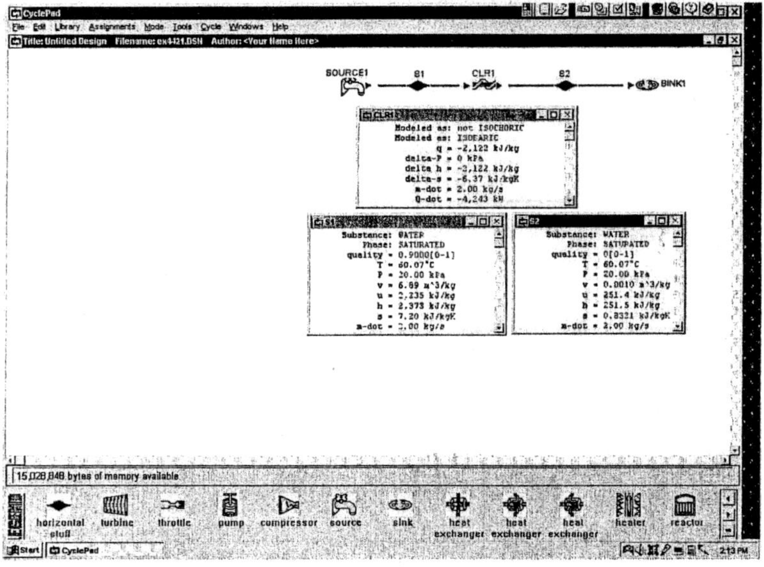

Figure Example 4.4.2.1. Cooler

Example 4.4.2.2. Freon R-12 at a mass flow rate of 0.02 kg/s is condensed in a condenser from 900 kPa and 40°C to saturated liquid. Find the rate of heat removed from the water in the condenser, specific enthalpy change and specific entropy change of the water.

To solve this problem by CyclePad, we take the following steps:

1. Build
   (A) Take a source, a condenser (cooler), and a sink from the open-system inventory shop and connect them.
   (B) Switch to analysis mode.
2. Analysis
   (A) Assume the condenser as an isobaric process.
   (B) Input the given information: (a) working fluid is R-12, (b) the inlet pressure and temperature of the condenser are 900 psia and 40°C, (c) the outlet quality of the condenser is 0, and (d) the mass flow rate is 0.02 kg/s.
3. Display results
   (A) Display the condenser results. The answers are Qdot=-2.64 kW, Δh=-132.2 kJ/kg, and Δs=-0.4347 kJ/[kg(K)].

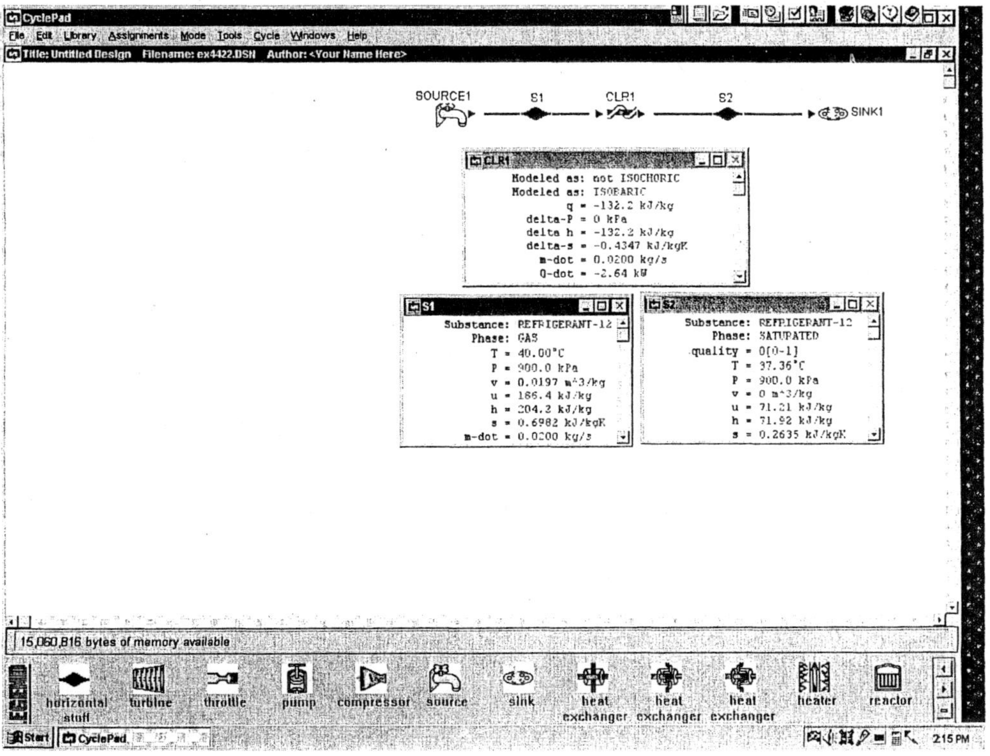

Figure Example 4.4.2.2. Cooler

Homework 4.4.2 Cooler

1. Ammonia enters a condenser operating at a steady state at 225 psia and 600 R and is condensed to saturated liquid at 225 psia; on the outside of tubes through which cooling water flows. The volumetric flow rate of the ammonia is 2 ft^3/s. Neglect heat transfer and kinetic energy effects. Determine (a) the mass flow rate of ammonia in lbm/s, and (b) the rate of energy transfer from the condensing ammonia to the cooling water, and (c) the specific entropy change of the ammonia between the inlet and exit section.
2. Steam enters a condenser (cooler) at a pressure of 1 psia and 90% quality. It leaves the condenser as a saturated liquid at 1 psia. For a flow rate of 75 lbm/s of steam, determine the heat removed from the steam, the temperature of the ammonia at the exit section, the specific enthalpy change of the ammonia between the inlet and exit section., and the specific entropy change of the ammonia between the inlet and exit section.

4.4.3. Compressor

The purpose of a *compressor* is to compress a gas or a vapor from a low pressure inlet state to a high pressure exit state. In general, this requires much more work than pumping liquid, and the working fluid experiences a significant increase in temperature. Compressors are not supposed to handle wet saturated vapors (saturated two-phaes vapor and liquid mixtures), as such fluids cause excessive wear. A compressor is usually modeled as adiabatic.

Application of the first law of thermodynamics to a steady flow steady system compressor gives

$$Q=0 \text{ and } W = m(h_i-h_e) \qquad (4.4.3.1)$$

The adiabatic assumption is reasonable because the heat transfer surface area of the compressor is relatively small, and the length of time required for the working fluid to pass through the compressor is short. Therefore the ideal compression process is considered to be a reversible and adiabatic or isentropic one. Comparison between the actual and the ideal compressor performance is given by the compressor efficiency, η. Since it requires more actual work to drive an actual compressor than the ideal work to drive an isentropic compressor, the compressor efficiency, η, is defined as

$$\eta=w_{isentropic}/w_{actual} \qquad (4.4.3.2)$$

Example 4.4.3.1. Freon R-134a at a mass flow rate of 0.015 kg/s enters an adiabatic compressor at 200 kPa and -10°C and leaves at 1 MPa and 70°C. The power input to the compressor is 1 kW. Determine the power required to drive the compressor, specific enthalpy change and specific entropy change of the R-134a.

To solve this problem by CyclePad, we take the following steps:

1. Build
   (A) Take a source, a compressor, and a sink from the open-system inventory shop and connect them.
   (B) Switch to analysis mode.
2. Analysis
   (A) Input the given information: (a) working fluid is R-134a, (b) the inlet pressure and temperature of the compressor are 200 kPa and -10°C, (c) the outlet pressure and temperature of the compressor are 1 MPa and 70°C, (d) the mass flow rate is 0.015 kg/s, and (e) the shaft power of the compressor is 1 kW.
3. Display results
   (A) Display the compressor results. The answers are Wdot=-0.9000 kW, Δh=60.00 kJ/kg, and Δs=0.0799 kJ/[kg(K)].

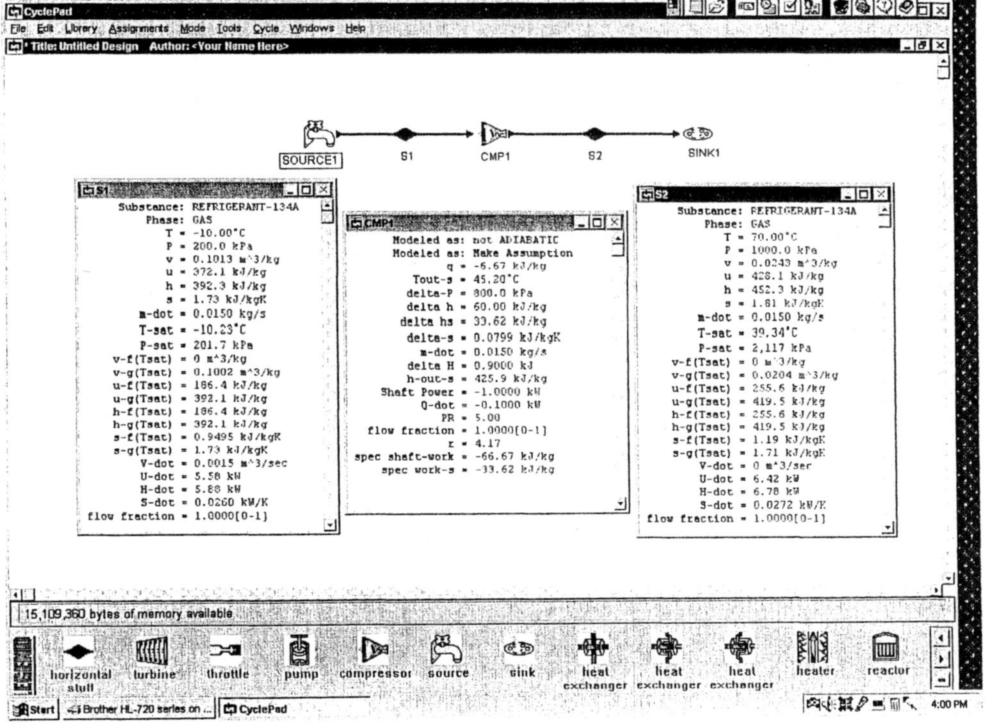

Figure Example 4.4.3.1. Freon Compressor

Example 4.4.3.2. Air at a mass flow rate of 0.15 kg/s enters an adiabatic compressor at 100 kPa and 10°C and leaves at 1 MPa and 280°C. Find the rate of the air flow at the exit section, the power added to the air, specific enthalpy change, specific entropy change of the air, and efficiency of the compressor.

Determine the power input to the compressor.

To solve this problem by CyclePad, we take the following steps:

1. Build
   (A) Take a source, a compressor, and a sink from the open-system inventory shop and connect them.
   (B) Switch to analysis mode.
2. Analysis
   (A) Assume the compressor is adiabatic.
   (B) Input the given information: (a) working fluid is air, (b) the inlet pressure and temperature of the compressor are 100 kPa and 10°C, (c) the outlet pressure and temperature of the compressor are 1 MPa and 280°C, and(d) the mass flow rate is 0.15 kg/s.
3. Display results
   (A) Display the compressor results. The answers are Vdot=0.0238 m^3/s, Wdot=-40.64 kW, Δh=270.9 kJ/kg, Δs=0.0118 kJ/[kg(K)], and η=97.60%,

Example 4.4.3.3. Air at a mass flow rate of 0.15 kg/s enters an isentropic compressor at 100 kPa and 10°C and leaves at 1 MPa. Find the temperature of air at exit section, the rate of the air flow at the exit section, the power added to the air, specific enthalpy change, specific entropy change of the air, and efficiency of the compressor.
Determine the power input to the compressor.

To solve this problem by CyclePad, we take the following steps:

1. Build
   (A) Take a source, a compressor, and a sink from the open-system inventory shop and connect them.
   (B) Switch to analysis mode.
2. Analysis
   (A) Assume the compressor is adiabatic and isentropic.
   (B) Input the given information: (a) working fluid is air, (b) the inlet pressure and temperature of the compressor are 100 kPa and 10°C, (c) the outlet pressure and temperature of the compressor are 1 MPa and 280°C, and(d) the mass flow rate is 0.15 kg/s.
3. Display results
   (A) Display the compressor results. The answers are T=273.5°C, Vdot=0.0235 m^3/s, Wdot=-39.66 kW, Δh=264.4 kJ/kg, Δs=0 kJ/[kg(K)],and η=100%.

Comments: Comparing Ex.4.4.3.3 and Ex.4.4.3.2, we see that:

1. $Wdot_{isentropic}$ (-39.66 kW) is less than $Wdot_{adiabatic}$(-40.64 kW);
2. $(Texit)_{sentropic}$(273.5°C) is lower than $(Texit)_{adiabatic}$(280°C)
3. $(\Delta s)_{isentropic}$=0 and $(\Delta s)_{adiabatic}$=0.0118 kJ/[kg(K)]
4. $\eta_{isentropic}$(100%) is larger than $\eta_{isentropic}$(97.60%).

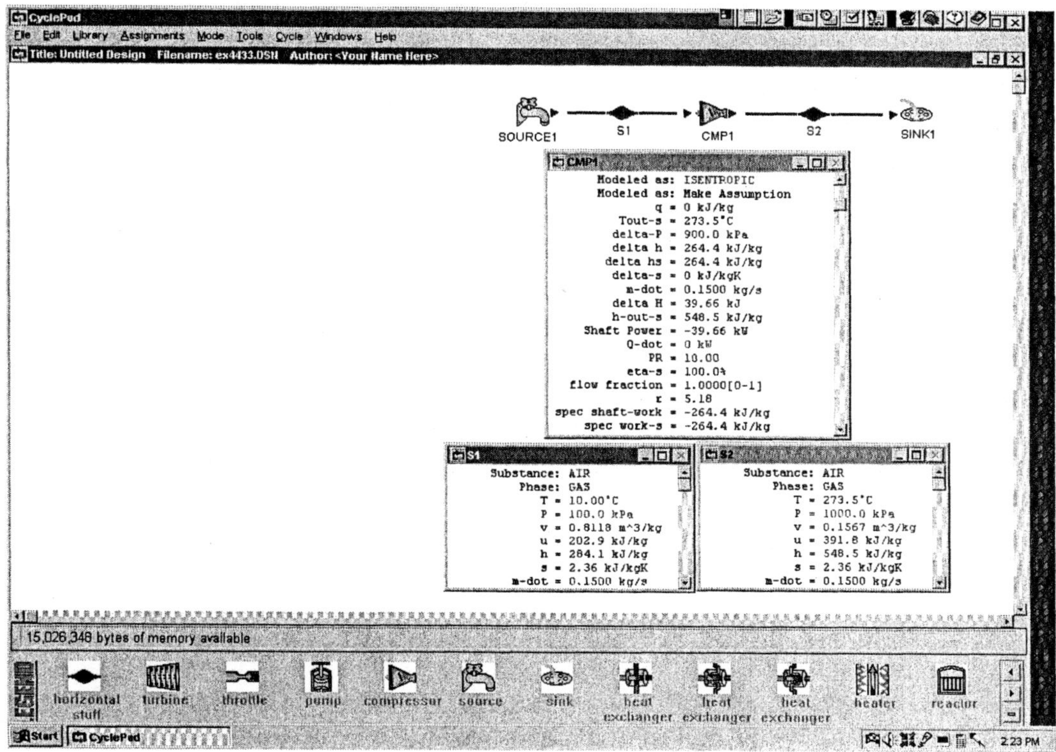

Figure Example 4.4.3.3. Compressor

Homework 4.4.3 Compressor

1. What is the function of a compressor?
2. What is the difference between a compressor and a pump?
3. Do you expect the pressures of air at the compressor inlet and exit to be the same?
4. Do you expect the temperatures of air at the compressor inlet and exit to be the same? Why?
5. Do you expect the specific volumes of air at the compressor inlet and exit to be the same? Why?
6. Does the mass rate flow at the inlet of a steady flow compressor the same as that at the exit of the compressor?
7. Refrigerant R-134a enters an adiabatic compressor as saturated vapor at -20°C and leaves at 0.7 Mpa and 70°C. The mass flow rate of the refrigerant is 2.5 kg/s. Determine (A) the power input to the compressor and (B) the volume flow rate at the compressor inlet.
8. A compressor receives air at 100 kPa and 300 K and discharges it to 400 kPa and 480 K. The mass flow rate of the air is 15 kg/s. The heat transfer from the compressor to its surroundings is 8.5 kW. Determine the specific work and power required to run this compressor.

9. A well insulted compressor takes in air at 520 R and 14.7 psia with a volumetric flow rate of 1200 ft^3/min, and compresses it to 960 R and 120 psia. Determine the compressor power and the volumetric flow rate at the exit.
10. A compressor increases the pressure of 3 lbm/s of air from 15 psia to 150 psia. If the specific enthalpy of the entering air is 150 Btu/lbm and that of the exiting air is 300 Btu/lbm. Determine how much power is required to drive the compressor if 1.53 Btu/lbm of heat is lost in the process.
11. Air enters a compressor at 100 kPa and an specific enthalpy of 240 kJ/kg. The air leaves the compressor at 400 kPa and an enthalpy of 432 kJ/kg. If the heat loss to the atmosphere is 2 kJ/kg of air, how much work is required per kg of air.

4.4.4. Turbine

Turbines are high rotating speed devices used to produce work. Turbines consist of a set of rotor blades interleaved with a set of stationary blades or stators. Working fluid in a turbine is expanded from a high pressure and high temperature inlet state to a low pressure exit state. The temperature of the working fluid also drops during the expansion process. The fluid entering the turbine must be either dry saturated steam or gas. Wet saturated vapors (saturated two-phase vapor and liquid mixtures) will seriously erode the turbine blades. A turbine is usually modeled as adiabatic.

Application of the first law of thermodynamics to a steady flow steady system compressor gives

$$Q=0 \text{ and } W = m(h_i - h_e) \quad (4.4.4.1)$$

The adiabatic assumption is reasonable because the heat transfer surface area of the turbine is relatively small, and the length of time required for the working fluid to pass through the turbine is short. Therefore the ideal expansion process is considered to be a reversible and adiabatic or isentropic one. Comparison between the actual and the ideal turbine performance is given by the turbine efficiency, η. Since ideal turbine produces more isentropic work than the actual work by an actual adiabatic turbine, the turbine efficiency, η, is defined as

$$\eta = w_{actual} / w_{isentropic} \quad (4.4.4.2)$$

Example 4.4.4.1. Air at a mass flow rate of 0.15 kg/s enters an adiabatic turbine at 1000 kPa and 1100°C and leaves at 100 kPa and 500°C. Determine the enthalpy change and entropy change of the air. Find the efficiency and power ouput to the turbine.

To solve this problem by CyclePad, we take the following steps:

1. Build
   (A) Take a source, a turbine, and a sink from the open-system inventory shop and connect them.
   (B) Switch to analysis mode.

2. Analysis
   (A) Assume the turbine is isentropic and adiabatic.
   (B) Input the given information: (a) working fluid is air, (b) the inlet pressure and temperature of the turbine are 1000 kPa and 1100°C, (c) the outlet pressure and temperature of the turbine are 100 kPa and 500°C, and(d) the mass flow rate is 0.15 kg/s.
3. Display results
   (A) Display the turbine results. The answers are Δh=-602.0 kJ/kg, Δs=3.37-3.29=0.08 kJ/[kg(K)], η=90.64% and Wdot=90.31 kW.

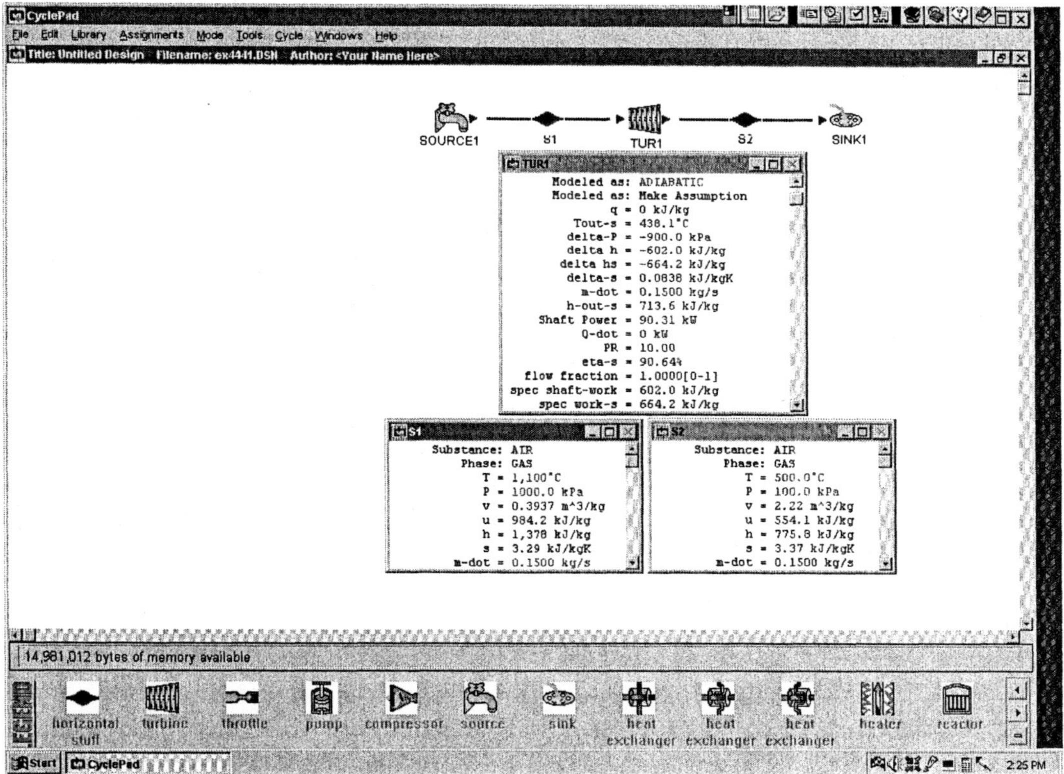

Figure Example 4.4.4.1. Turbine

Example 4.4.4.2. Air at a mass flow rate of 0.15 kg/s enters an isentropic turbine at 1000 kPa and 1100°C and leaves at 100 kPa. Determine the enthalpy change and entropy change of the air. Find the efficiency and power ouput to the turbine.

To solve this problem by CyclePad, we take the following steps:

1. Build
   (A) Take a source, a turbine, and a sink from the open-system inventory shop and connect them.
   (B) Switch to analysis mode.
2. Analysis

(A) Assume the turbine is isentropic.

(B) Input the given information: (a) working fluid is air, (b) the inlet pressure and temperature of the turbine are 1000 kPa and 1100°C, (c) the outlet pressure and temperature of the turbine are 100 kPa and 500°C, and(d) the mass flow rate is 0.15 kg/s.

3. Display results

(A) Display the turbine results. The answers are Δh=-664.2 kJ/kg, Δs=3.29-3.29=0 kJ/[kg(K)], η=100% and Wdot=99.63 kW.

Comments: Comparing Ex.4.4.4.1 and Ex.4.4.4.2, we see that:

1. $Wdot_{isentropic}$ (99.63 kW) is more than $Wdot_{adiabatic}$(90.31 kW);
2. $(Texit)_{sentropic}$(438.1°C) is lower than $(Texit)_{adiabatic}$(500°C)
3. $(\Delta s)_{isentropic}$=0 and $(\Delta s)_{adiabatic}$=0.08 kJ/[kg(K)]
4. $\eta_{isentropic}$(100%) is larger than $\eta_{isentropic}$(97.60%).

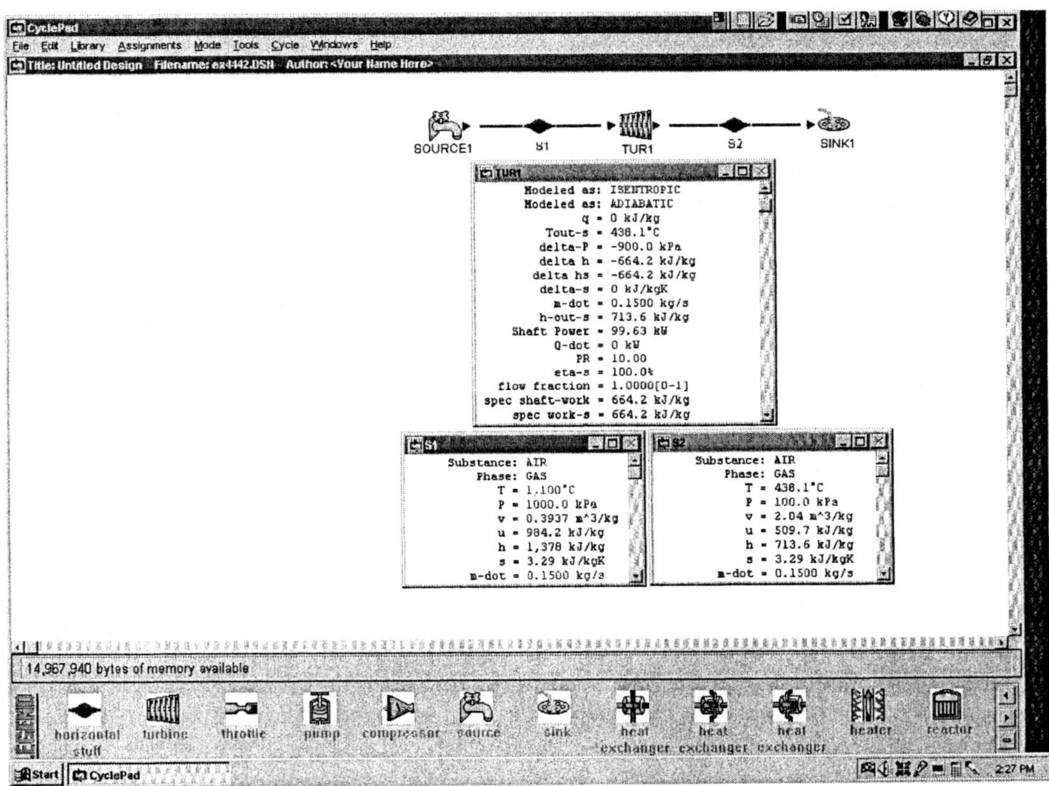

Figure Example 4.4.4.2. Turbine

Example 4.4.4.3. The shaft power produced by an adiabatic steam turbine is 100 Mw. Steam enters the adiabatic turbine at 4000 kPa and 500°C and leaves at 10 kPa and a quality of 0.9. Determine the rate of steam flow through the turbine, the entropy change of the steam, and efficiency of the turbine.

To solve this problem by CyclePad, we take the following steps:

1. Build
   (A) Take a source, a turbine, and a sink from the open-system inventory shop and connect them.
   (B) Switch to analysis mode.
2. Analysis
   (A) Assume the turbine is adiabatic, and the shaft power is 100,000 kW,
   (B) Input the given information: (a) working fluid is water, (b) the inlet pressure and temperature of the turbine are 4000 kPa and 500°C, and (c) the outlet pressure and quality of the turbine are 10 kPa and 0.9.
3. Display results
   (A) Display the turbine results. The answers are mdot=90.84 kg/s, Δs=7.40-7.09=0.31 kJ/[kg(K)], η=91.80%.

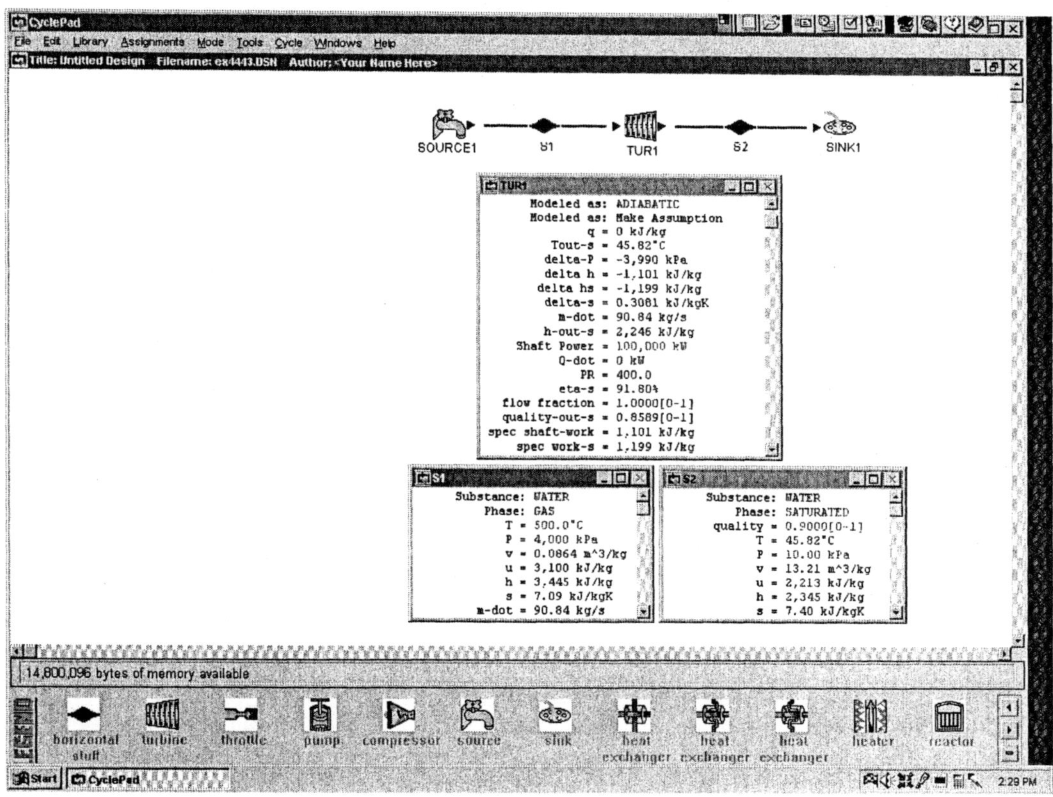

Figure Example 4.4.4.3. Turbine

Example 4.4.4.4. The shaft power produced by an isentropic steam turbine is 100 Mw. Steam enters the turbine at 4000 kPa and 500°C and leaves at 10 kPa. Determine the quality of steam at the exit section, rate of steam flow through the turbine, the entropy change of the steam, and efficiency of the turbine.

To solve this problem by CyclePad, we take the following steps:

1. Build
   (A) Take a source, a turbine, and a sink from the open-system inventory shop and connect them.
   (B) Switch to analysis mode.
2. Analysis
   (A) Assume the turbine is isentropic and adiabatic, and the shaft power is 100,000 kW.
   (B) Input the given information: (a) working fluid is water, (b) the inlet pressure and temperature of the turbine are 4000 kPa and 500°C, and (c) the outlet pressure of the turbine are 10 kPa.
3. Display results
   (A) Display the turbine results. The answers are x=0.8589, mdot=83.40 kg/s, Δs=7.09-7.09=0 kJ/[kg(K)], and η=100%.

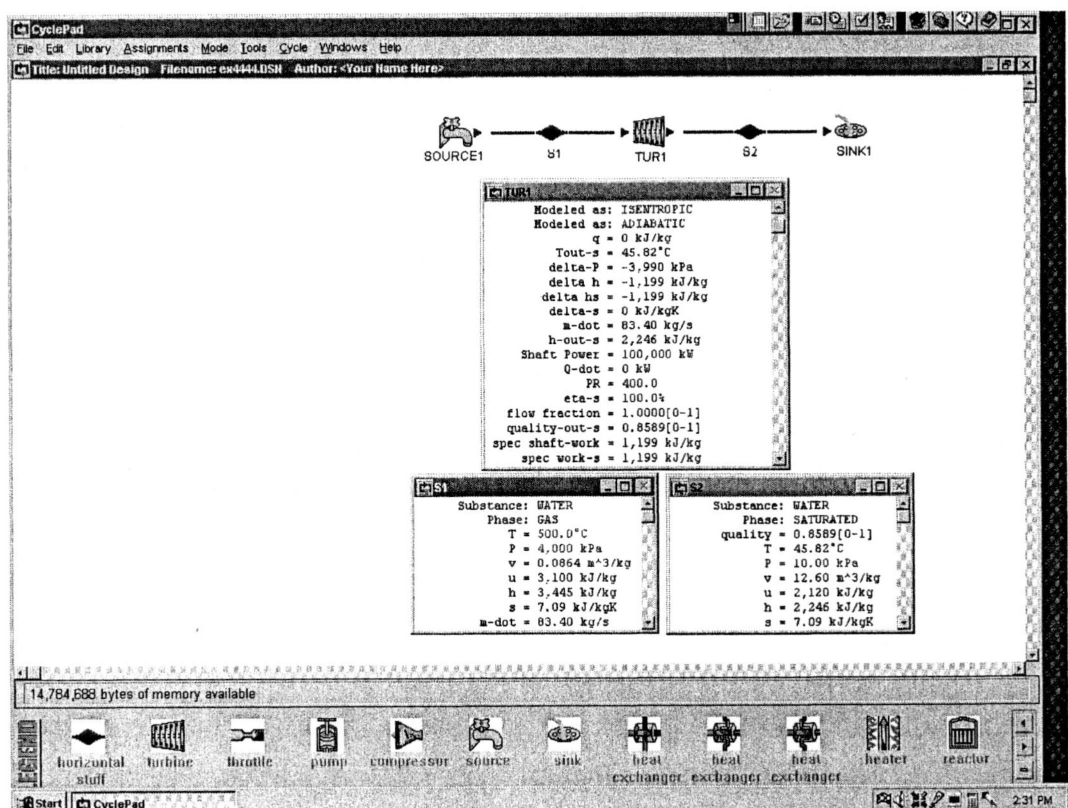

Figure Example 4.4.4.4 Turbine

Homework 4.4.4 Turbine

1. 3 lbm/s of steam enters a steady-state steady-flow adiabatic turbine at 3000 psia and 1000°F. It leaves the turbine at 1 psia with a quality of 0.74. Determine the power produced by the turbine.
2. A steam turbine is designed to have a power output of 9 MW for a mass flow rate of 17 kg/s. The inlet state is 3 Mpa and 450 C, and the outlet state is 0.5 Mpa and saturated vapor. What is the heat transfer for this turbine?
3. Steam enters a turbine at steady state with a mass flow rate of 2 kg/s. The turbine developes a power output of 1000 kW. At the inlet, the temperature is 400°C and the pressure is 6000 kPa. At the exit, the pressure is 10 kPa and the quality is 90%. Calculate the rate of heat transfer between the turbine and its surroundings.
4. 50 lbm/s of steam enters a turbine at 700°F and 600 psia, and leaves at 0.6 psia with a quality of 90%. The heat transfer from the turbine to the surroundings is 2.5×10^6 Btu/hr. Determine the power developed by the turbine.
5. The mass rate of flow into a steam turbine is 1.5 kg/s, and the heat transfer from the turbine to its surroundings is 8.5 kW. The steam is 2 Mpa and 350°C at the inlet, and 0.1 Mpa and saturated (100% quality) at the exit. Determine the specific work and power produced by the turbine.
6. Air expands through a turbine from 1000 kPa and 900 K to 100 kPa and 500K. The turbine operates at steady state and develops a power output of 3200 kW. Neglect heat transfer and kinetic energy effects. Determine the mass flow rate in kg/s.
7. A ship's steam turbine receives steam at a pressure of 900 psia and 1000°F. The steam leaves the turbine at 2 psia and 114°F. Find the power produced if the mass flow rate is 3 lbm/s and the heat loss from the turbine is 12 Btu/s.
8. A steam turbine produces 6245 hp. At the entrance to the turbine, the pressure is 1000 psia and specific volume is 0.667 ft^3/lbm. At the exit of the turbine, the pressure is 5 psia and specific volume is 40 ft^3/lbm. If 0.578 Btu/lbm of heat is lost to the environment through the turbine casing, determine the required steam flow rate.
9. Superheated steam enters a turbine at 700 psia and 700°F, and leaves as a dry saturated vapor at 10 psia. The mass flow rate is 20 lbm/s and the heat loss from the turbine is 14 Btu/lbm. Find the power output of the turbine.
10. Steam initially at a pressure of 500 psia and a quality of 92% expands isentropically to a pressure of 1 psia. Determine the final quality and moisture content of the steam, and the specific work of the turbine.
11. Steam at a pressure of 800 psia and 650°F is expanded isentropically to a pressure of 14.7 psia. Determine the exit enthalpy of the steam, and the specific work of the turbine.
12. Steam at a pressure of 1200 psia and 900°F is expanded isentropically to a pressure of 2 psia. Determine the final enthalpy and temperature of the steam.
13. Superheated steam at 400 psia and 700°F is expanded in an isentropic turbine to a pressure of 100 psia. Determine the final enthalpy and temperature of the steam.

4.4.5. Pump

The purpose of a *pump* is to compress a liquid from a low pressure inlet state to a high pressure exit state. Pumping a liquid is more efficient than compressing a gas, because one can pump a great deal more liquid per unit volume, and liquids, being basically incompressible, do not gain an appreciable amount of heat during the pumping process. A pump can not handle gases or saturated vapors, because such fluids tend to cavitate, or boil. Cavitation causes excessive shocks within the pump and can rapidly lead to pump failure. A pump is usually modeled as adiabatic.

Application of the first law of thermodynamics to a steady flow steady system pump gives

$$Q=0 \text{ and } W = m(h_i-h_e) \tag{4.4.5.1}$$

Since liquid is an incompressible substance and liquid temperature change across the pump is negligible, Eq. (4.4.5.1) can be reduced to

$$W = mv(p_i-p_e) \tag{4.4.5.2}$$

The adiabatic assumption is reasonable because the heat transfer surface area of the pump is relatively small, and the length of time required for the working fluid to pass through the pump is short. Therefore the ideal pumping process is considered to be a reversible and adiabatic or isentropic one. Comparison between the actual and the ideal pump performance is given by the isentropic efficiency, η. Since it requires more actual work to drive an actual pump than the ideal work to drive an isentropic pump, the pump efficiency, η, is defined as

$$\eta=w_{isentropic}/w_{actual} \tag{4.4.5.3}$$

Example 4.4.5.1. Saturated water at a mass flow rate of 1 kg/s enters an adiabatic pump at 10 kPa and leaves at 4000 kPa and 46°C. Determine the entropy change of the water, efficiency of the pump and the power required for the pump.

To solve this problem by CyclePad, we take the following steps:

1. Build
   (A) Take a source, a pump, and a sink from the open-system inventory shop and connect them.
   (B) Switch to analysis mode.
2. Analysis
   (A) Assume the pump is adiabatic and isentropic.
   (B) Input the given information: (a) working fluid is water, (b) pressure and quality of the water at the pump inlet are 10 kPa and 0, (c) the pressure and of the water at the pump outlet are 4000 kPa and, and(d) the mass flow rate is 1 kg/s.
3. Display results
   (A) Display the pump results. The answers are Δs=0.6498-0.6493=0.0005 kJ/[kg(K)], η=95.89% and Wdot=-4.24 kW.

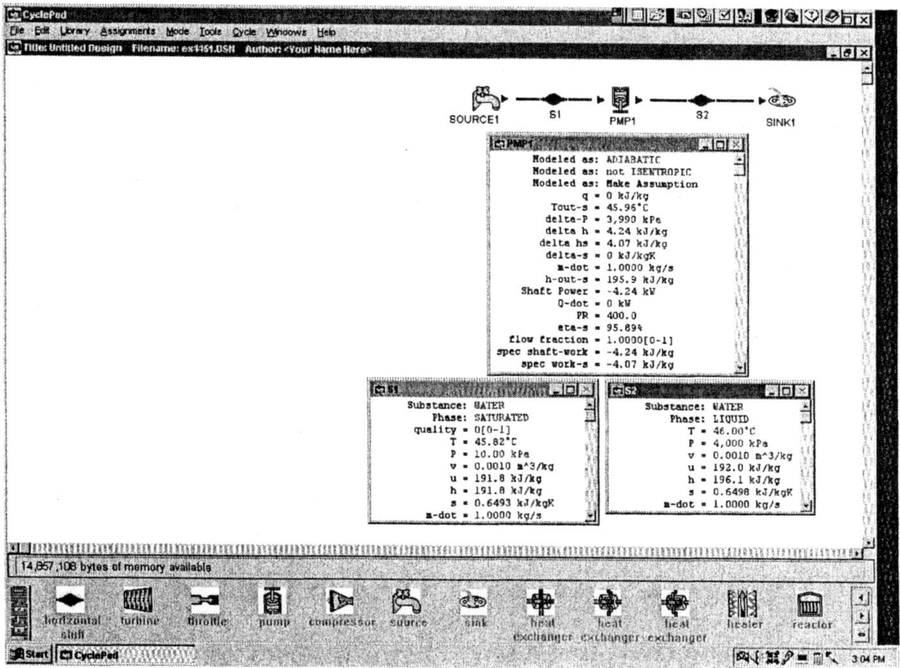

Figure Example 4.4.5.1. Pump

Example 4.4.5.2. Saturated water at a mass flow rate of 1 kg/s enters an isentropic pump at 10 kPa and leaves at 4000 kPa. Determine the entropy change of the water, the efficiency of the pump and the power required for the pump.

To solve this problem by CyclePad, we take the following steps:

1. Build
   (A) Take a source, a pump, and a sink from the open-system inventory shop and connect them.
   (B) Switch to analysis mode.
2. Analysis
   (A) Assume the pump is adiabatic.
   (B) Input the given information: (a) working fluid is water, (b) pressure and quality of the water at the pump inlet are 10 kPa and 0, (c) the pressure and of the water at the pump outlet are 4000 kPa and, and(d) the mass flow rate is 1 kg/s.
3. Display results
   (A) Display the pump results. The answers are Δs=0.6493-0.6493=0 kJ/[kg(K)], η=100% and Wdot=-4.07 kW.

Comments: Comparing Ex.4.4.5.1 and Ex.4.4.5.2, we see that:

1. $Wdot_{isentropic}$ (-4.07 kW) is less than $Wdot_{adiabatic}$(-4.24 kW);
3. $(\Delta s)_{isentropic}$=0 and $(\Delta s)_{adiabatic}$=0.005 kJ/[kg(K)]
4. $\eta_{isentropic}$(100%) is larger than $\eta_{isentropic}$(95.89%).

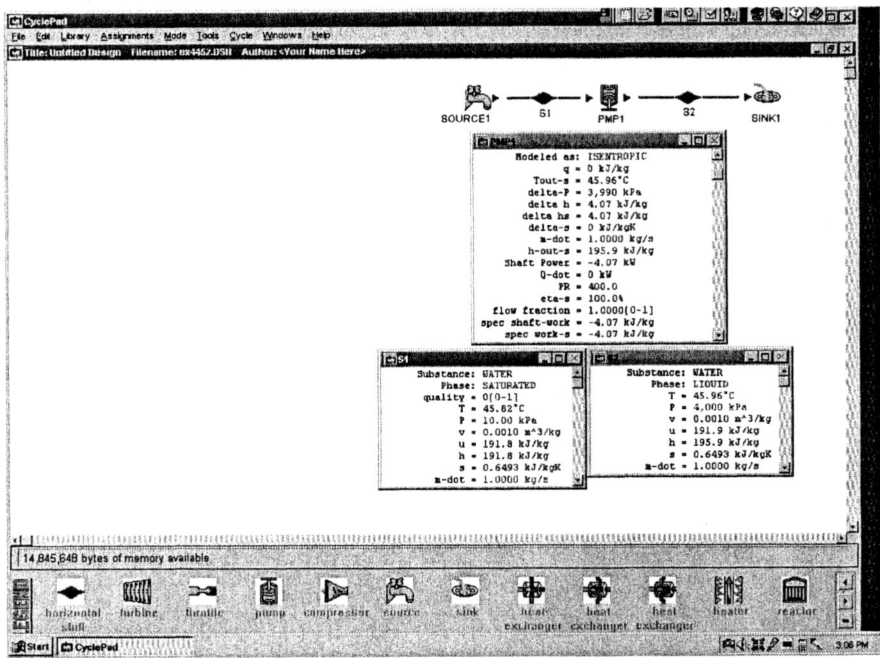

Figure Example 4.4.5.2. Pump

Homework 4.4.5 Pump

1. Water enters a steady-state steady-flow isentropic pump as saturated liquid at 40°C. It leaves the pump at 20 Mpa. Determine the pump specific work required.
2. An adiabatic pump pumps 10 kg/s of water from 100 kPa and 20°C to 1100 kPa and 20.03°C. Find the power input to the pump.
3. An adiabatic pump pumps 10 lbm/s of water from 14.7 psia and 60°F to 214.7 psia and 60.1°F. Find the power input to the pump.

4.4.6 Mixing Chamber

A *mixing chamber* is a device to mix two or more streams of fluids into one stream. The mixing chamber does not have to be a distinct chamber. An ordinary T-elbow in a shower, for example, serves as a mixing chamber for the cold- and hot-water stream. Mixing chambers are usually well insulated (Q=0) and do not involve electric or shaft work. Also, the kinetic and potential energies of the working fluid streams are usually negligible.

Application of the mass balance equation and the first law of thermodynamics to a steady flow steady state mixing chamber gives that the total mass rate flow in at the inlets is equal to the total mass rate flow out at the exit of the mixing chamber, and the total enthalpy rate flow in at the inlets is equal to the total enthalpy rate flow out at the exit of the mixing chamber.

$$\sum(\text{mdot})_{in} = \sum(\text{mdot})_{out} \qquad (4.4.6.1)$$

and

$$\sum[(mdot)_{in}\ (h_{in})]=\sum[(mdot)_{out}\ (h_{out})] \qquad (4.4.6.2)$$

Example 4.4.6.1. Consider an ordinary shower where hot water at 15 psia and 140°F is mixed with cold water at 15 psia and 60°F. It is desirable that a steady stream of warm water at 15 psia and 100°F, and 0.1 lbm/s be supplied, Determine the mass flow rates of the hot water and the cold water.

To solve this problem by CyclePad, we take the following steps:

1. Build
   (A) Take two sources, a mixing chamber, and a sink from the open-system inventory shop and connect them.
   (B) Switch to analysis mode.
2. Analysis
   (A) Assume the mixing chamber is isobaric.
   (B) Input the given information: (a) working fluids are water, (b) pressure and temperature of the hot water at the inlet are 15 psia and 140°F, (c) pressure and temperature of the cold water at the inlet are 15 psia and 60°F, (c) the mass flow rate, pressure and temperature of the mixed water at the mixing chamber outlet are 0.1 lbm/s, 15 psia and 100°F.
3. Display results
   (A) Display the inlet states results. The answers are mdot,hot=0.05 lbm/s and mdot,cold=0.05 lbm/s.

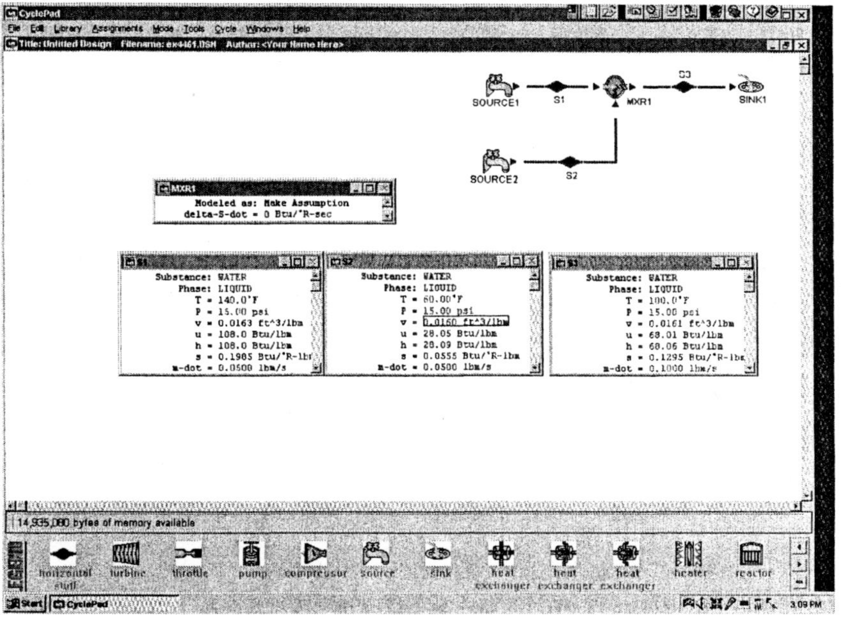

Figure Example 4.4.6.1. Mixing chamber

Example 4.4.6.2. A flow rate of 2 kg/s of hot water at 100 kPa and 60°C is mixed with 3 kg/s of cold water at 100 kPa and 15°C. Determine the mass flow rate and temperature of the mixed water.

To solve this problem by CyclePad, we take the following steps:

1. Build
   (A) Take two sources, a mixing chamber, and a sink from the open-system inventory shop and connect them.
   (B) Switch to analysis mode.
2. Analysis
   (A) Assume the mixing chamber is isobaric.
   (B) Input the given information: (a) working fluids are water, (b) mass flow rate, pressure and temperature of the hot water at the inlet are 2 kg/s, 100 kPa and 60°C, (c) mass flow rate, pressure and temperature of the cold water at the inlet are 3 kg/s, 100 kPa and 15°C.
3. Display results
   (A) Display the outlet states results. The answers are mdot,mix=5 kg/s and T=33°C.

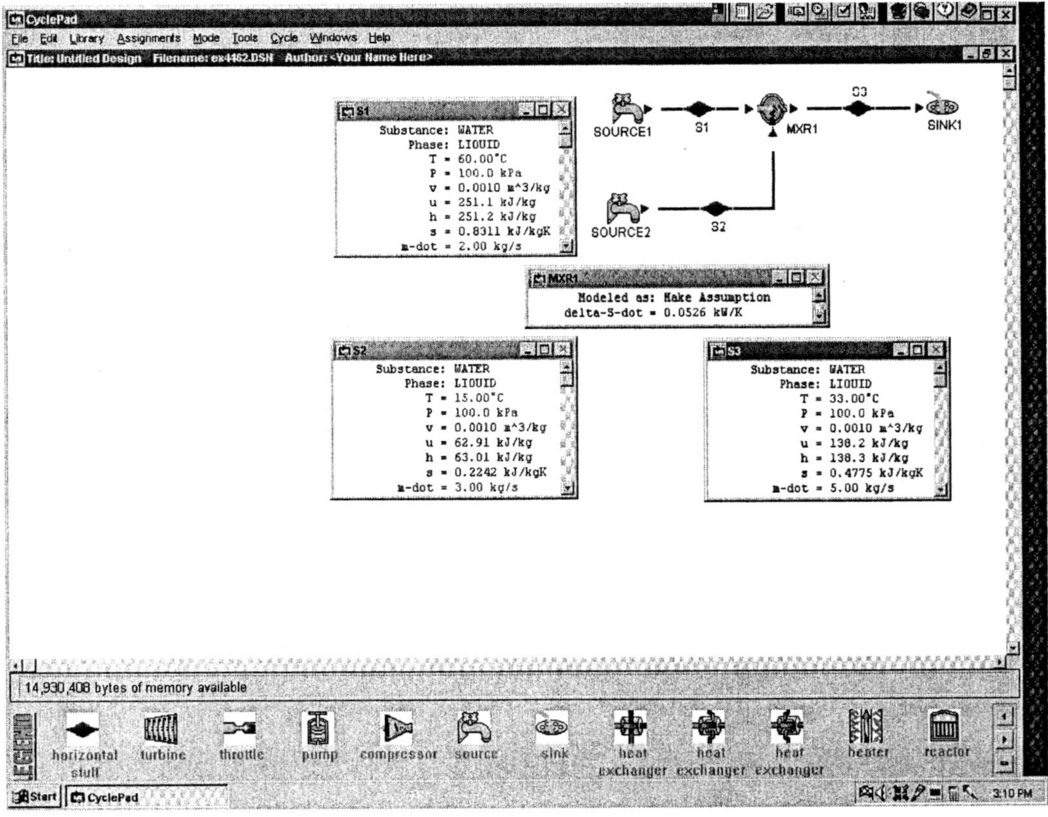

Figure Example 4.4.6.2. Mixing chamber

Example 4.4.6.3. A flow rate of 4.41 lbm/s of hot air at 15 psia and 540°F is mixed with 100 ft^3/s of cold air at 15 psia and 20°F. Determine the mass flow rate, volumetric flow rate and temperature of the mixed air.

To solve this problem by CyclePad, we take the following steps:

1. Build
   (A) Take two sources, a mixing chamber, and a sink from the open-system inventory shop and connect them.
   (B) Switch to analysis mode.
2. Analysis
   (A) Assume the mixing chamber is isobaric.
   (B) Input the given information: (a) working fluids are air, (b) mass flow rate, pressure and temperature of the hot air at the inlet are 2 kg/s, 15 psia and 540°F, (c) mass flow rate, pressure and temperature of the cold air at the inlet are 3 kg/s, 15 psia and 20°F.
3. Display results
   (A) Display the outlet states results. The answers are Vdot=208.7 ft^3/s, mdot,mix=12.86 lbm/s and T=198.3 °F.

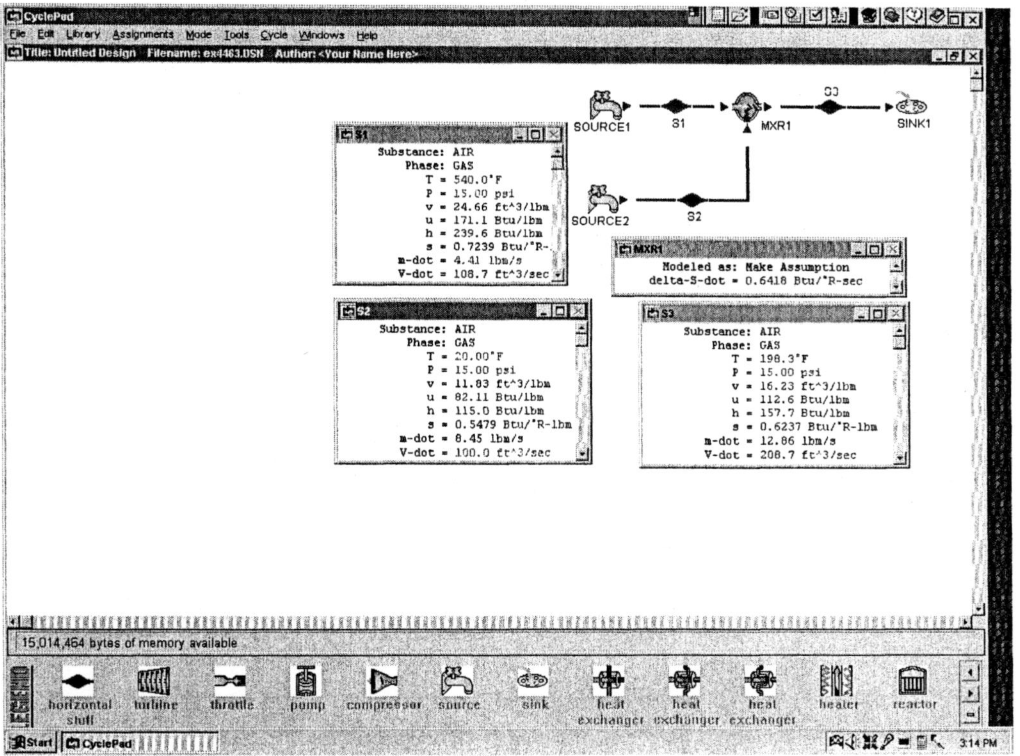

Figure Example 4.4.6.3. Mixing chamber

Homework 4.4.6 Mixing Chamber

1. Is a mixing process reversible or irreversible?
2. When two fluid streams are mixed in a mixing chamber without chemical reaction. Can the mixture temperature be higher than the temperature of both streams?
3. Water at 80°F and 50 psia is heated in a chamber by mixing it with saturated water vapor at 50 psia. If both streams enter the isobaric mixing chamber at the same flow rate, determine the temperature and quality of the exit steam,
4. 0.1 kg/s of saturated water (x=0) at 600 kPa is injected into 1 kg of saturated steam (x=1) at 1400 kPa in an adiabatic mixing process. Find the enthalpy of the mixture. If the mixed stream pressure is 1000 kPa, what is the temperature and quality of the mixture?
5. A mixing chamber operating at steady state has two inlets and one exit. At inlet 1, 40 kg/s of water vapor enters at 700 kPa and 200°C. At inlet 2, water enters at 700 kPa and 40°C. Saturated vapor at 700 kPa exits at exit 3. Determine the mass flow rate at inlet 2 and at the exit 3. Find the rate of entropy change of the mixing chamber.
6. A 5 lb/s stream of cold water at 40 F and 14.7 psia is to be heated by a hot water flow of 3 lb/s at 180 F and 14.7 psia in a steady flow mixing process. What is the resultant temperature and quality of the mixed stream?
7. A stream of water at 50 psia and 70°F is mixed in an adiabatic mixing chamber with steam at 50 psia, 200 lbm/s and 500°F. The mixture leaves the chamber at 50 psia and saturated vapor. Determine the mass flow rate of saturated vapor leaving the chamber.

4.4.7. Splitter

A *splitter* is a device to separate one stream fluid into two or more streams of fluids. The splitter does not have to be a distinct chamber. An ordinary Y-elbow in a shower, for example, serves as a splitter. Splitters are usually well insulated (Q=0) and do not involve shaft work. Also, the kinetic and potential energies of the working fluid streams are usually negligible.

Application of the mass balance equation and the first law of thermodynamics to a steady flow steady state splitter give the total mass rate flow in at the inlets equal to the total mass rate flow out at the exit of the splitter, and the total enthalpy rate flow in at the inlet equal to the total enthalpy rate flow out at the exits of the splitter mixing chamber.

$$\sum(\text{mdot})_{\text{in}} = \sum(\text{mdot})_{\text{out}} \qquad (4.4.7.1)$$

and

$$\sum[(\text{mdot})_{\text{in}}\,(h_{\text{in}})] = \sum[(\text{mdot})_{\text{out}}\,(h_{\text{out}})] \qquad (4.4.7.2)$$

Example 4.4.7.1. A geothermal saturated steam with a mass rate flow of 3 kg/s at a pressure of 1200 kPa and 90 percent quality is separated into saturated liquid and saturated vapor. Determine the mass flow rates of the saturated liquid and saturated vapor.

To solve this problem by CyclePad, we take the following steps:

1. Build
   (A) Take a source, a, splitter and two sinks from the open-system inventory shop and connect them.
   (B) Switch to analysis mode.
2. Analysis
   (A) Assume the splitter is isobaric.
   (B) Input the given information: (a) working fluids are water, (b) mass flow rate, pressure and quality of the water at the inlet are 3 kg/s, 1200 kPa and 0.9, (c) quality of the water at one outlet is 1, and (d) quality of the water at the other outlet is 0.
3. Display results
   (A) Display the outlet states results. The answers are mdot,liquid=0.3 kg/s and mdot,vapor=2.7 kg/s.

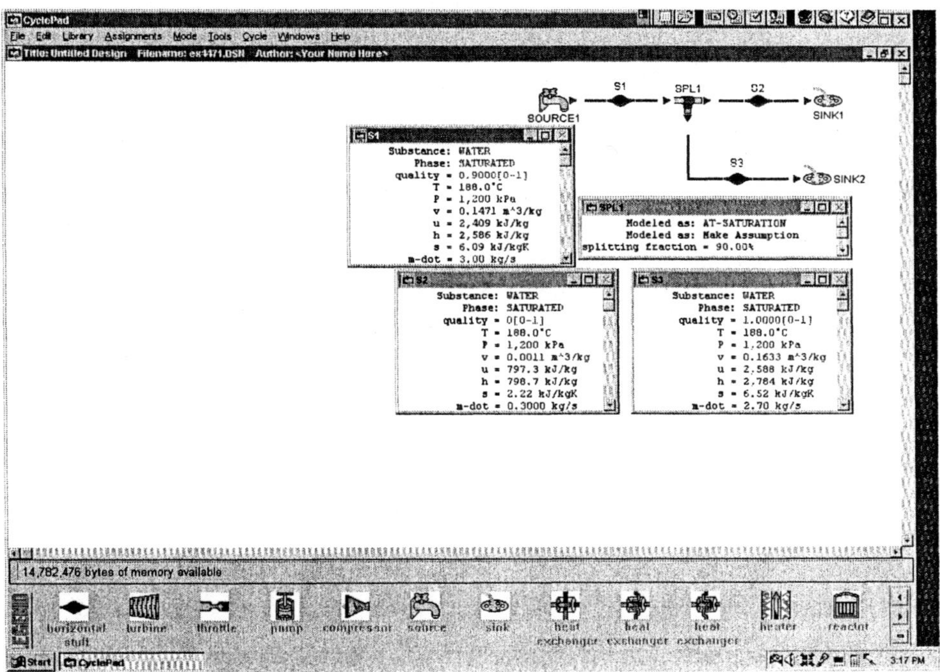

Figure Example 4.4.7.1. Splitter

Homework 4.4.7 Splitter

1. A 1.3 kg/s stream of geothermal hot water at 350 kPa flows into a separator to make saturated water and saturated steam at 350 kPa. If we want to produce a flow rate of 1 kg/s of 350 kPa saturated steam, determine the quality of the geothermal hot water.
2. A 1.7 kg/s stream of saturated liquid and vapor mixture R134a at 6°C flows into an isobaric separator to make saturated liquid and saturated vapor. If we want to produce a flow rate of 1 kg/s of saturated vapor, determine the quality of the two-phase mixture.

3. A 4 kg/s stream of saturated liquid and vapor mixture methane at 200 kPa flows into an isobaric separator to make two separate streams of mixture methane. If we want to produce a flow rate of 1 kg/s of 80% quality of saturated mixture in one stream and another stream of 50% quality of saturated mixture, determine the inlet quality of the two-phase mixture.
4. A 4 kg/s stream of saturated liquid and vapor mixture ammonia at 200 kPa flows into an isobaric separator to make two separate streams of mixture ammonia. If we want to produce a flow rate of 1 kg/s of 80% quality of saturated mixture in one stream and another stream of 50% quality of saturated mixture, determine the inlet quality of the two-phase mixture.
5. A 3 kg/s stream of steam at 200 kPa flows into an isobaric separator to make two separate streams of mixture steam. If we want to produce a flow rate of 1 kg/s of 80% quality of saturated mixture in one stream and another stream of 50% quality of saturated mixture, determine the inlet quality of the two-phase mixture.
6. A 2.5 kg/s stream of steam at 200 kPa flows into an isobaric separator to make two separate streams of mixture steam. If we want to produce a flow rate of 1 kg/s of 90% quality of saturated mixture in one stream and another stream of 20% quality of saturated mixture, determine the inlet quality of the two-phase mixture.

4.4.8. Heat Exchanger

A *heat exchanger* is a simple multi stream fluids (usually two streams) flow through a device where heat is transferred from one stream fluid at a higher temperature to the other stream fluid at a lower temperature. The fluids are heated or cooled and may or may not change phases. The heat exchanging process tends to occur at constant pressure, since the fluids flowing through the device usually undergo only small pressure drops due to fluid friction at the walls. There is no means for doing any shaft work, and changes in kinetic and potential energies are commonly negligible small. Energies are exchanged from one stream fluid to another stream fluid inside the heat exchanger. There is no heat interaction with the surroundings.

Application of the mass balance equation and the first law of thermodynamics to a steady flow steady state heat exchanger, the total mass rate flow in at the inlets is equal to the total mass rate flow out at the exit of the mixing chamber, and the total enthalpy rate flow in at the inlets is equal to the total enthalpy rate flow out at the exit of the mixing chamber.

$$\sum(\text{mdot})_{in} = \sum(\text{mdot})_{out} \tag{4.4.8.1}$$

and

$$\sum[(\text{mdot})_{in}\,(h_{in})] = \sum[(\text{mdot})_{out}\,(h_{out})] \tag{4.4.8.2}$$

Heat exchangers at various applications are called boiler, condenser, open feed water heater, etc.

Example 4.4.8.1. Steam enters a condenser (Heat exchanger) at a pressure of 1 psia and 90 percent quality. It leaves the condenser as saturated liquid at 1 psia. Cooling lake water available at 14.7 psia and 55°F is used to remove heat from the steam. For a flow rate of 5 lbm/s of steam, determine the flow rate of cooling water, if (a) the cooling water leaves the condenser at 60°F, and (b) the cooling water leaves the condenser at 70°F.

To solve this problem by CyclePad, we take the following steps:

1. Build
   (A) Take two sources, a heat exchanger, and two sinks from the open-system inventory shop and connect them. Please note that red indicates hot and blue indicates cold, respectively.
   (B) Switch to analysis mode.
2. Analysis
   (A) Assume the heat exchanger are isobaric on both hot and cold sides.
   (B) Input the given information: (a) working fluid are water for both streams, (b) mass flow rate, pressure and quality of the steam at the heat exchanger inlet are 5 lbm/s, 1 psia and 0.9, (c) pressure and temperature of the lake water at the heat exchanger inlet are 14.7 psia and 55°F, and (d) quality of the steam at the heat exchanger outlet is 0, and (e) temperature of the lake water at the heat exchanger outlet is 60°F (or 70°F).
3. Display results
   (A) Display the state of the lake water at the outlet heat exchanger result. The answer is (a) mdot=932.4 lbm/s, and (b) mdot=310.9 lbm/s.

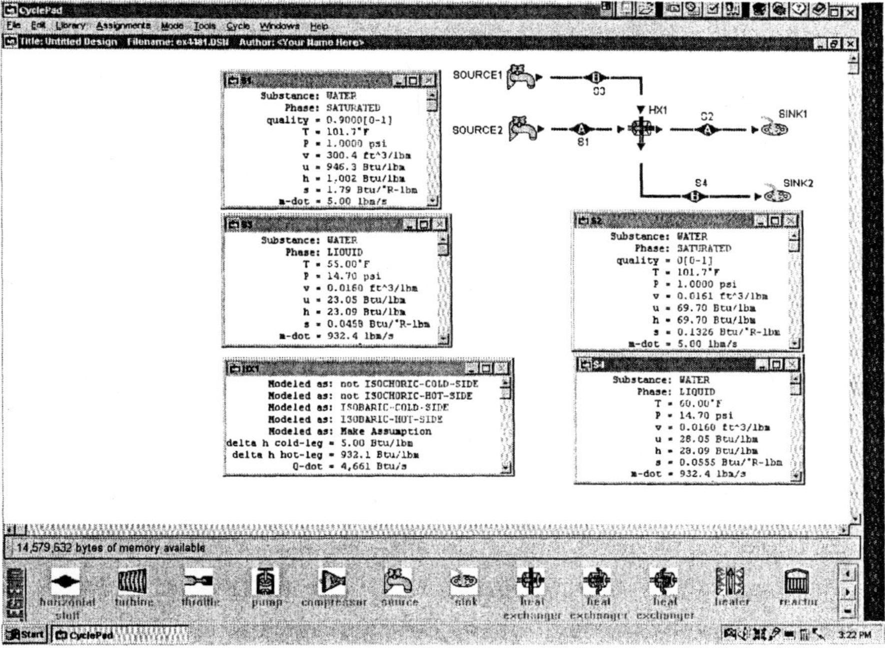

Figure Example 4.4.8.1. Heat exchanger

Example 4.4.8.2. Air with a mass rate flow of 0.1 kg/s enters a water-cooled condenser at 1000 kpa, 260°C and leaves at 1000 kpa and 85°C. Cooling lake water available at 100 kPa and 15°C is used to remove heat from the air. The lake water leaves at 100 kPa and 30°C. Determine the mass rate flow of the lake water required.

To solve this problem by CyclePad, we take the following steps:

1. Build
   (A) Take two sources, a heat exchanger, and two sinks from the open-system inventory shop and connect them.
   (B) Switch to analysis mode.
2. Analysis
   (A) Assume the heat exchanger are isobaric on both hot and cold sides.
   (B) Input the given information: (a) working fluid are air for one stream and water for the other stream, (b) mass flow rate, pressure and quality of the air at the heat exchanger inlet are 0.1 kg/s, 1 MPa and 260°C, (c) pressure and temperature of the lake water at the heat exchanger inlet are 100 kPa and 15°C, and (d) pressure and temperature of the lake water at the heat exchanger inlet are 100 kPa and 30°C.
3. Display results
   (A) Display the state of the lake water at the outlet heat exchanger result. The answer is mdot=0.2789 kg/s.

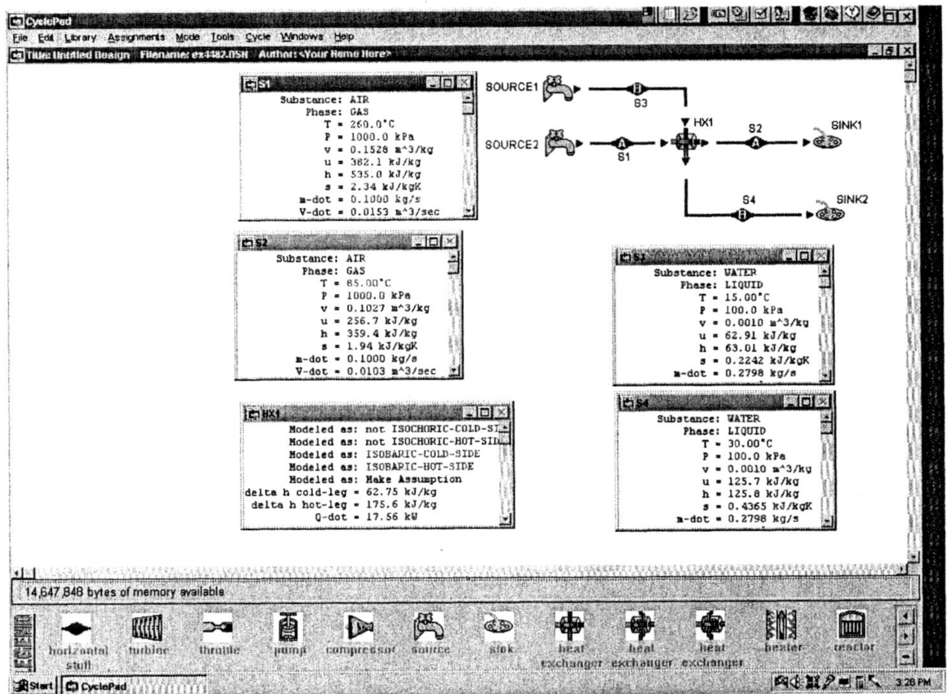

Figure Example 4.4.8.2. Heat exchanger

Example 4.4.8.3. Freon R-134a with a mass rate flow of 0.012 kg/s enters a water-cooled condenser at 1 Mpa, 60°C and leaves as a liquid at 1 Mpa and 35°C. Cooling lake water available at 100 kPa and 15°C is used to remove heat from the freon. The lake water leaves at 100 kPa and 20°C. Determine the mass rate flow of the lake water required.

To solve this problem by CyclePad, we take the following steps:

1. Build
   (A) Take two sources, a heat exchanger, and two sinks from the open-system inventory shop and connect them.
   (B) Switch to analysis mode.
2. Analysis
   (A) Assume the heat exchanger is isobaric on both hot and cold sides.
   (B) Input the given information: (a) working fluid are R-134a for one stream and water for the other stream, (b) mass flow rate, pressure and quality of the R-134a at the heat exchanger inlet are 0.014 kg/s, 1 MPa and 60°C, (c) pressure and temperature of the lake water at the heat exchanger inlet are 100 kPa and 15°C, and (d) pressure and temperature of the lake water at the heat exchanger inlet are 100 kPa and 20°C.
3. Display results
   (A) Display the state of the lake water at the outlet heat exchanger result. The answer is mdot=0.1111 kg/s.

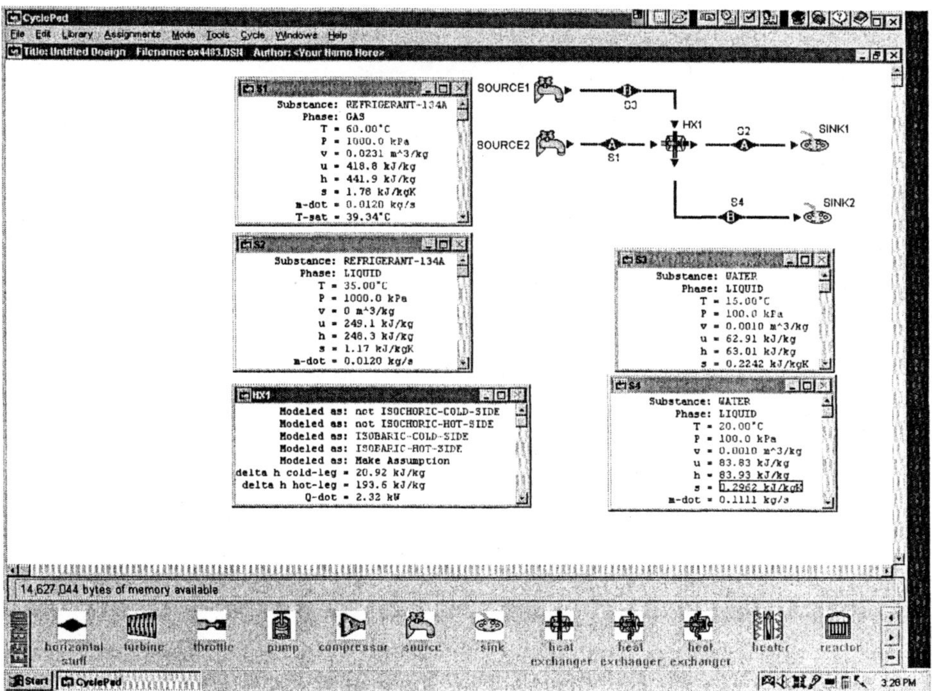

Figure Example 4.4.8.3. Heat exchanger

Example 4.4.8.4. Freon R-134a with a mass rate flow of 0.014 kg/s enters a water-cooled condenser at 1 Mpa, 60°C and leaves as a liquid at 1 Mpa and 35°C. Air available at 100 kPa and 15°C is used to remove heat from the freon. The air leaves at 100 kPa and 30°C. Determine the mass rate flow and volumetric rate flow of the air required.

To solve this problem by CyclePad, we take the following steps:

1. Build
   (A) Take two sources, a heat exchanger, and two sinks from the open-system inventory shop and connect them.
   (B) Switch to analysis mode.
2. Analysis
   (A) Assume the heat exchanger is isobaric on both hot and cold sides.
   (B) Input the given information: (a) working fluid are R-134a for one stream and water for the other stream, (b) mass flow rate, pressure and quality of the R-134a at the heat exchanger inlet are 0.014 kg/s, 1 MPa and 60°C, (c) pressure and temperature of the air at the heat exchanger inlet are 100 kPa and 15°C, and (d) pressure and temperature of the lake water at the heat exchanger inlet are 100 kPa and 30°C.
3. Display results
   (A) Display the state of the air at the outlet heat exchanger result. The answer is mdot=0.1801 kg/s and Vdot=0.1565 m^3/s.

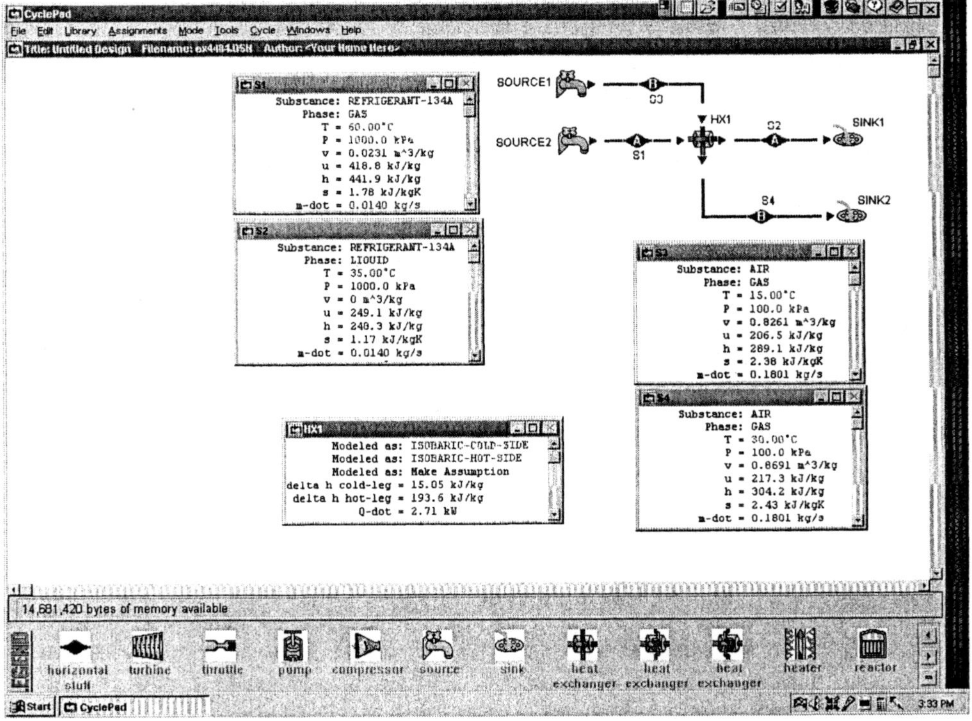

Figure Example 4.4.8.4. Heat exchanger

Example 4.4.8.5. We want to cool 50 lbm/s of air from 14.7 psia and 540°R to 14.7 psia and 440°R in a steady-state steady-flow heat exchanger. If 40 lbm/s nitrogen gas at 15 psia and 300°R is available, determine the temperature of the nitrogen at the outlet.

To solve this problem by CyclePad, we take the following steps:

1. Build
   (A) Take two sources, a heat exchanger, and two sinks from the open-system inventory shop and connect them.
   (B) Switch to analysis mode.
2. Analysis
   (A) Assume the heat exchanger is isobaric on both hot and cold sides.
   (B) Input the given information: (a) working fluids are air in one stream and nitrogen in another stream, (b) mass flow rate, pressure and temperature of the air at the heat exchanger inlet are 50 lbm/s, 14.7 psia and 540°R 10 kPa, (c) mass flow rate, pressure and temperature of the nitrogen at the heat exchanger inlet are 40 lbm/s, 15 psia and 300°R, and (d) temperature of the air at the heat exchanger outlet is 340°R. 4000 kPa,
3. Display results
   (A) Display the state of nitrogen at the outlet heat exchanger results. The answers is T=420.1 °R.

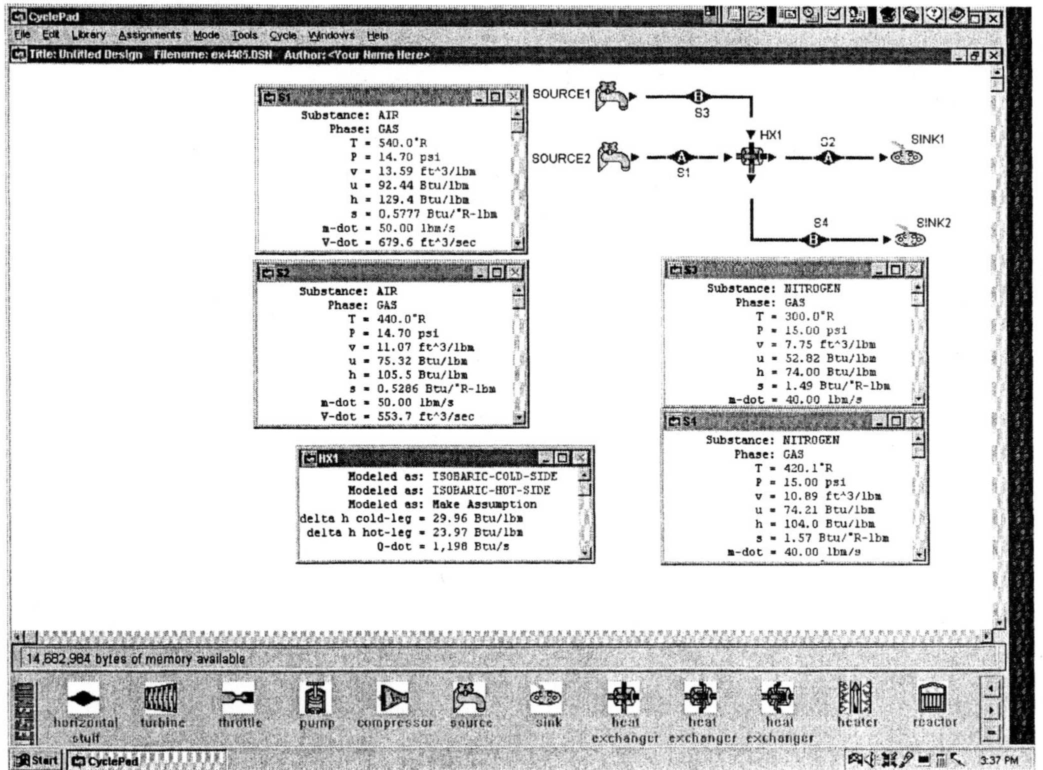

Figure Example 4.4.8.5. Heat exchanger

Homework 4.4.8 Heat Exchanger

1. Steam enters a constant pressure heat exchanger at 200 kPa and 200°C at a rate of 8 kg/s, and it leaves at 180 kPa and 100°C. Air enters at 100 kPa and 25°C and leaves at 100 kPa and 47°C. Determine the mass rate flow of air.
2. A heat exchanger is designed to use exhaust steam from a turbine to heat air in a manufacturing plant. Steam enters the well-insulated heat exchanger with a mass flow rate of 1.2 kg/s, 200 kPa and 200°C. The steam leaves the heat exchanger at 200 kPa and as saturated vapor. The air enters the heat exchanger at 20°C, 3 kg/s and 100 kPa and leaves at 100 kPa. Find the temperature of the air as it leaves the heat exchanger.
3. Water is used to cool a refrigerant in a condenser of a large refrigeration system. Cooling water flows through the condenser at a rate of 2 kg/s. The cooling water enters the condenser at 10 °C and exits at 20 °C. Refrigerant R134a flows through the condenser at a rate of 1 kg/s. The R134a enters the condenser at 40 °C as a two-phase saturated mixture and exits at 40 °C as a saturated liquid. Determine the rate at which heat is removed by the cooling water in kJ/s.
4. Hot gas enters the heat recovery steam generator of a cogeneration system at 500°C and 100 kPa and leaves at 150°C and 100 kPa. Water enters steadily at 100°C and 1,000 kPa and leaves as dry saturated steam at 1,000 kPa. For a mass flow rate of hot gas of 25 kg/s, determine the flow rate of water in kg/s. Find the rate of the gas flow leaving the heat exchanger.

4.4.9. Throttling Valve

A *throttling valve* is an inexpensive control device used to reduce the pressure of the working fluid, or to measure mass flow rates and quality of a mixture. Throttling processes occur in obstructed flow passages such as valves, flow meters, capillary tubes, and other devices that reduce the pressure of the working fluid without any shaft work or heat interactions. There is no work nor heat transfer interaction between the valve and its surroundings. Application of the first law of thermodynamics (neglecting kinetic and potential energy changes) to a steady flow steady state throttling valve gives

$$W=0,\ Q=0,\ \text{and}\ h_e=h_i \qquad (4.4.9.1)$$

Therefore, a *throttling process* is a constant enthalpy process. The inlet enthalpy of the working fluid of a throttling valve is equal to the exit enthalpy of the working fluid. Reducing the pressure of the working fluid in throttling manner involves considerable irreversibility. As a result, the throttling process is highly irreversible. The throttling valve is primarily used as a means to control a turbine, to reduce the pressure between the condenser and evaporator of a refrigerator or heat pump, or as a flow measurement device.

Example 4.4.9.1. Refrigerant R-12 enters a throttling valve as a saturated liquid at 50°C. It leaves at -5°C as a saturated mixture. The process is steady-state flow. Determine the quality of the refrigerant at the outlet of the throttling valve, the pressure drop and entropy change of the refrigerant R-12.

To solve this problem by CyclePad, we take the following steps:

1. Build
   (A) Take a source, a throttling valve, and a sink from the open-system inventory shop and connect them.
   (B) Switch to analysis mode.
2. Analysis
   (A) Input the given information: (a) working fluid is R-12, (b) the inlet temperature and quality of the throttling valve are 100°F and 0, (c) the outlet temperature of the throttling valve is -5°C, and the phase is saturated.
3. Display results
   (A) Display the outlet state results. The answers are x=0.3476, Δp=-959.3 kPa and Δs=0.0209 kj?[kg(K)].

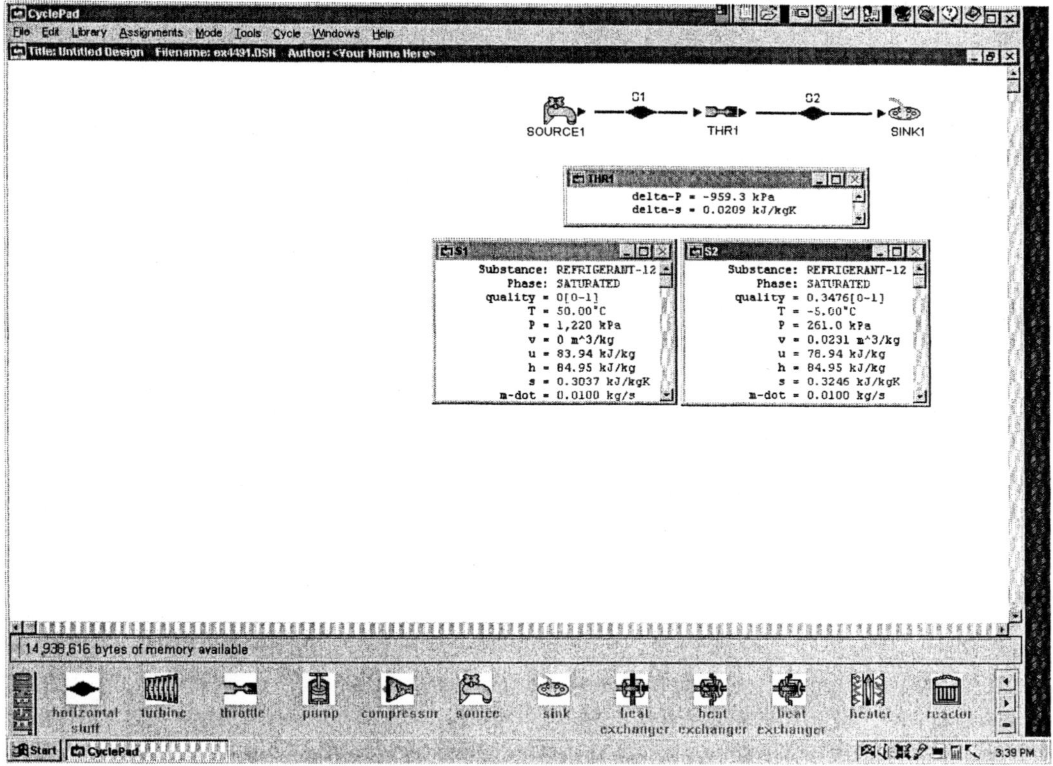

Figure Example 4.4.9.1. Throttling valve

Example 4.4.9.2. 0.1 lbm/s of air flow rate enters a throttling valve at 100 psia and 100°F. It leaves at 90 psia. The process is steady-state flow. Determine the pressure drop and entropy change of the air, and the temperature of the air at the outlet of the throttling valve.

To solve this problem by CyclePad, we take the following steps:

1. Build
   (A) Take a source, a throttling valve, and a sink from the open-system inventory shop and connect them.
   (B) Switch to analysis mode.
2. Analysis
   (A) Input the given information: (a) working fluid is air, (b) the inlet temperature and pressure of the throttling valve are 100°F and 100 psia, (c) the outlet pressure of the throttling valve is 90 psia.
3. Display results
   (a) Display the outlet state results. The answer answers are Δp=-10 psia, and Δs=0.0072 Btu/[lbm(R)], and T=100°F.

Comment: Since h of ideal gases is a function of temperature only, $h_e=h_i$ gives $T_e=T_i=100°F$.

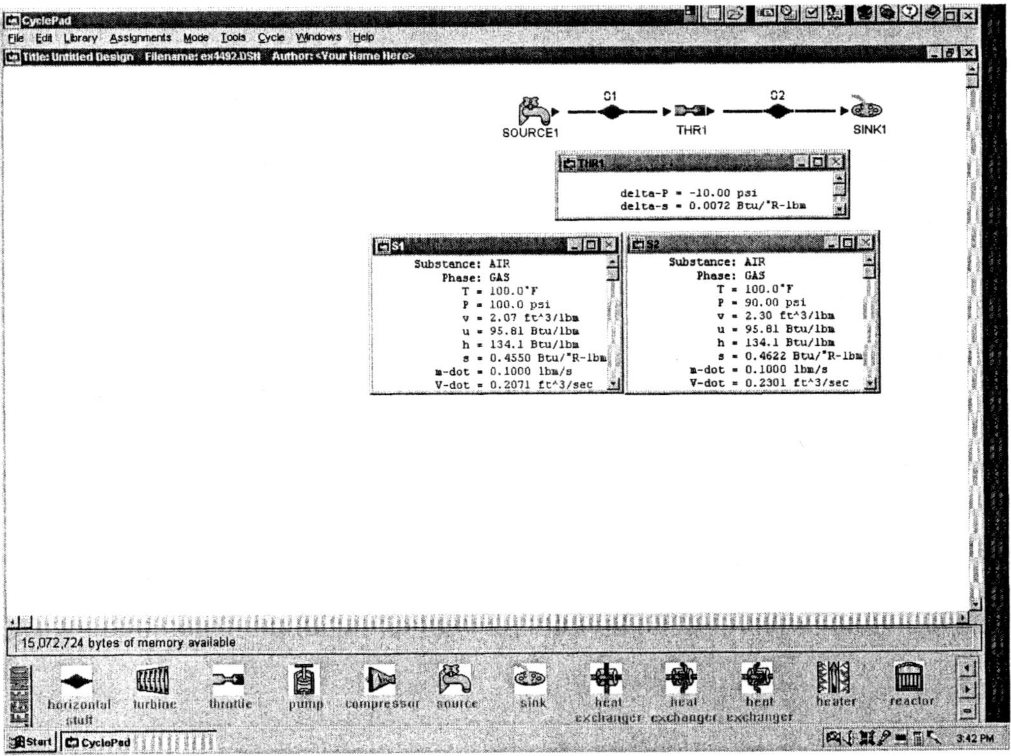

Figure Example 4.4.9.2. Throttling valve

Example 4.4.9.3. A throttling steam calorimeter (valve) is an instrument used for the determination of the quality of wet steam flowing in a steam main. It utilizes the fact that when wet steam is throttled sufficiently, superheated steam will form. If wet steam at 200 psia is throttled in a throttling steam calorimeter to 15 psia and 300 °F, Determine the pressure drop and entropy change of the water, and the quality of the wet steam.

To solve this problem by CyclePad, we take the following steps:

1. Build
   (A) Take a source, a throttling valve, and a sink from the open-system inventory shop and connect them.
   (B) Switch to analysis mode.
2. Analysis
   (A) Input the given information: (a) working fluid is water, (b) the inlet phase is saturated, and pressure of the throttling valve is 200 psia, (c) the outlet pressure and temperature of the throttling valve are 15 psia and 300 °F.
3. Display results
   (A) Display the outlet state results. The answers are Δp=-185 psia, and Δs=0.2759 Btu/[lbm(R)], and x=0.9922.

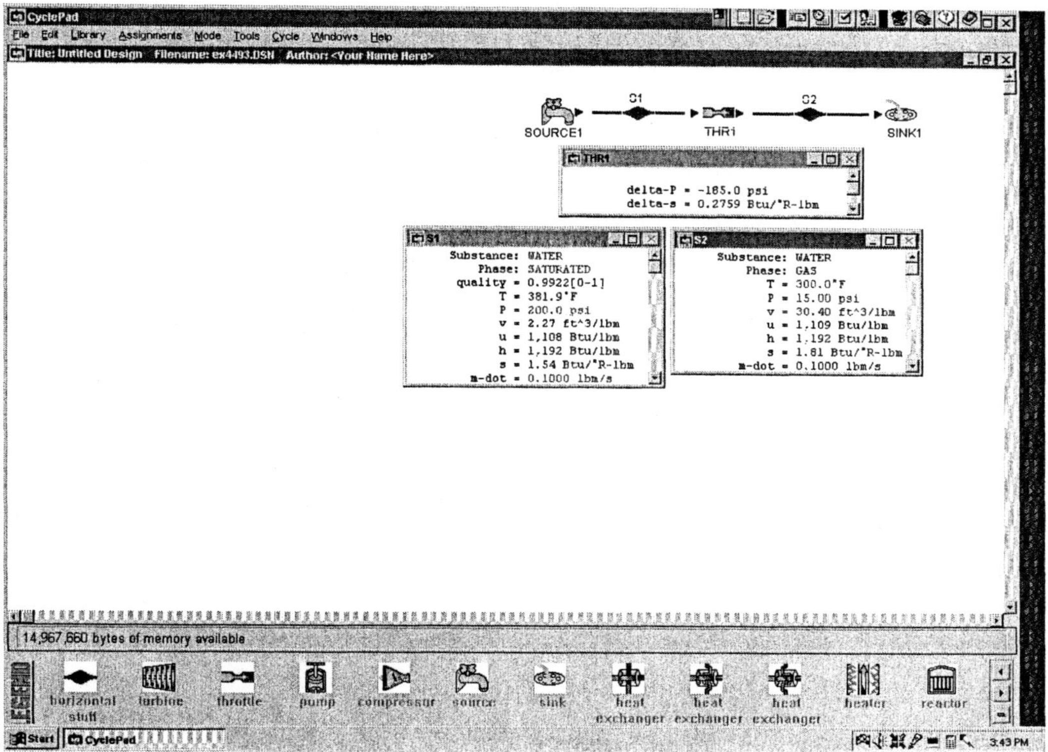

Figure Example 4.4.9.3. Throttling valve

Homework 4.4.9 Throttling Valve

1. Is the throttling process reversible?
2. Would you expect the pressure of steam to drop as it undergoes a steady flow throttling process?
3. Would you expect the temperature of steam to drop as it undergoes a steady flow throttling process?

4. Would you expect the pressure of air to drop as it undergoes a steady flow throttling process?
5. Would you expect the temperature of air to drop as it undergoes a steady flow throttling process?
6. Both turbines and throttling valves are expansion devices. Why are throttling valves used in refrigeration and heat pumps rather than turbines?
7. A throttling steam calorimeter is an instrument used for the determination of the quality of wet steam flowing in a steam main. It utilizes the fact that when wet steam is throttled sufficiently, superheated steam will form. If the wet steam at 200 psia is throttled in a calorimeter to 15 psia and 300°F, determine the quality of the steam,
8. Ammonia enters an expansion valve at 1.5 MPa and 32 C and exits at 268 kPa. Find the quality and temperature of the ammonia leaving the valve.
9. A steady flow of refrigerant-134a enters a throttling valve at 100°F and as saturated liquid, and leaves at 50 °F as a two-phase saturated vapor and liquid mixture. Determine the inlet and exit pressure and the exit quality.
10. A throttling calorimeter is connected to a saturated steam line. The line pressure is 400 psia, the calorimeter pressure is 14.7 psia and the temperature is 260°F. Determine the enthalpy and quality of the steam.
11. A throttling calorimeter is used to determine the quality of steam originating in a system maintained at 600 psia. What is the quality of the steam if the calorimeter temperature and pressure are 225 °F and 14.7 psia?

4.4.10 Reactor

A *reactor* is a heater in which a simple single stream fluid flows through a device where nuclear heat is transferred to the fluid. The fluid is heated and may or may not change phases. The heating process tends to occur at constant pressure, since a fluid flowing through the device usually undergoes only a small pressure drop due to fluid friction at the walls. There are no means for doing any shaft or electric work, and changes in kinetic and potential energies are commonly negligibly small.

Application of the first law of thermodynamics to a steady flow steady state reactor gives

$$W=0 \text{ and } Q = m(h_e-h_i) \tag{4.4.10.1}$$

Example 4.4.10.1. 1000 kW of nuclear heat is added to helium at a constant volume process in a nuclear reactor. The inlet temperature and pressure of the helium are 50°C and 300 kPa and the outlet temperature of the helium is 2000°C. Determine the mass flow rate, volumetric flow rate, and pressure of the helium at the exit section.

To solve this problem by CyclePad, we take the following steps:

1. Build
   (A) Take a source, a reactor, and a sink from the open-system inventory shop and connect them.
   (B) Switch to analysis mode.

2. Analysis
   (A) Input the given information: (a) working fluid is helium, (b) the inlet temperature and pressure of the helium are 50°C and 300 kPa, (c) the outlet temperature of the helium is 2000°C.
   (B) the model process of the reactor is constant volume and heat rate added is 1000 kW.
3. Display results
   (A) Display the outlet state results. The answers are mdot=0.0991 kg/s, Vdot=0.2216 m^3/s, p=2110 kPa and Δs=6.05 kJ/[kg(K)].

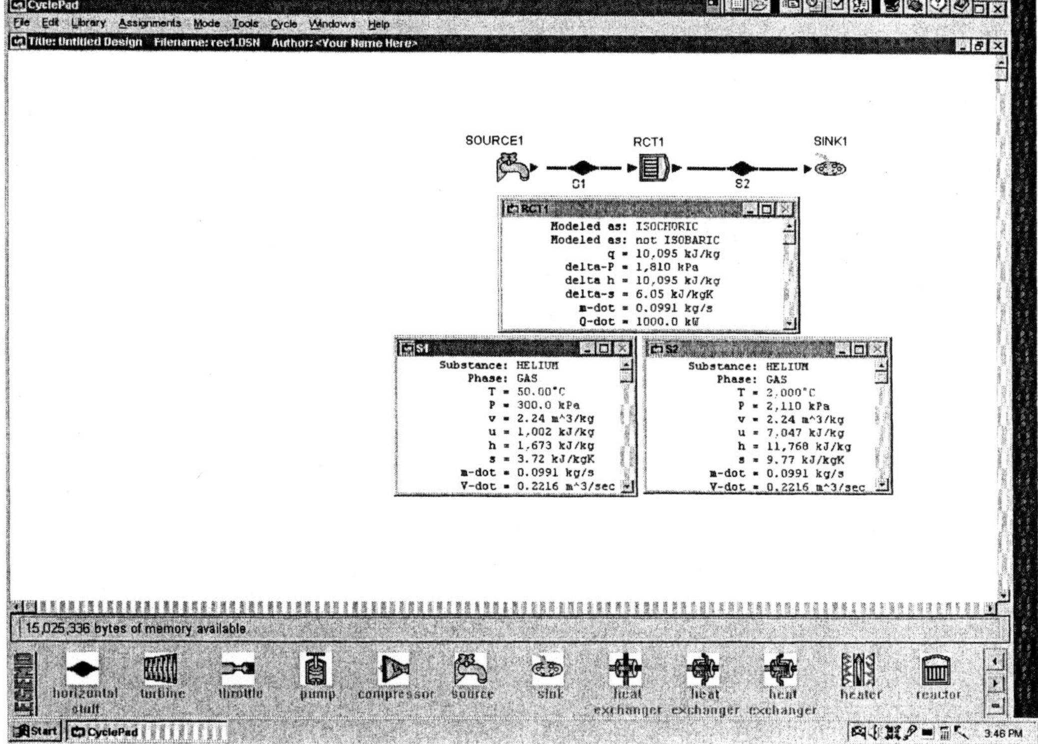

Figure Example 4.4.10.1. Reactor

Example 4.4.10.2. 1000 kW of nuclear heat is added to helium at a constant pressure process in a nuclear reactor. The inlet temperature and pressure of the helium are 50°C and 300 kPa and the outlet temperature of the helium is 2000°C. Determine the mass flow rate, volumetric flow rate, and pressure of the helium at the exit section.

To solve this problem by CyclePad, we take the following steps:

1. Build
   (A) Take a source, a reactor, and a sink from the open-system inventory shop and connect them.
   (B) Switch to analysis mode.
2. Analysis

(A) Input the given information: (a) working fluid is helium, (b) the inlet temperature and pressure of the helium are 50°C and 300 kPa, (c) the outlet temperature of the helium is 2000°C.

(B) the model process of the reactor is constant pressure and heat rate added is 1000 kW.

3. Display results

(A) Display the outlet state results. The answers are mdot=0.0991 kg/s, Vdot=1.56 m^3/s, p=300 kPa and Δs=10.10 kJ/[kg(K)].

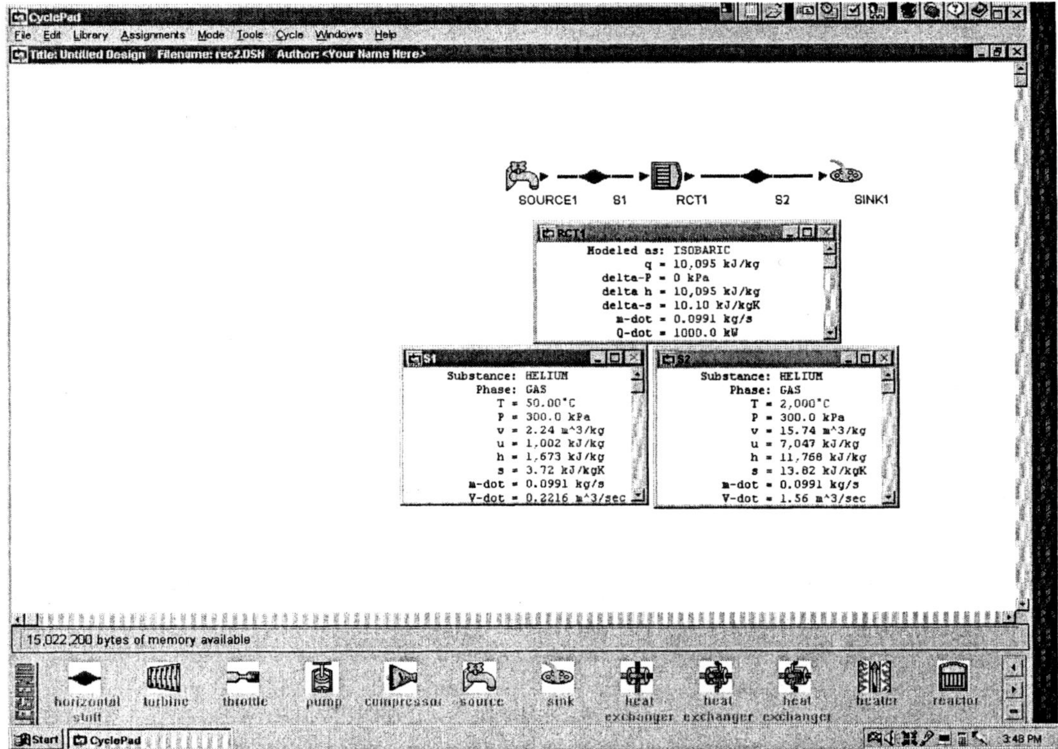

Figure Example 4.4.10.2. Reactor

4.5 Other Devices (Unable to Use Cyclepad)

4.5.1 Nozzle

A *nozzle* is a small device used to create a high velocity fluid stream at the expense of its pressure. There is no means to do work. A nozzle is usually modeled as adiabatic. This assumption is reasonable because the heat transfer surface area of the nozzle is very small, and the length of time required for the working fluid to pass through the nozzle is very short. Therefore the ideal process is considered to be a reversible and adiabatic or isentropic one.

Application of the first law of thermodynamics to a steady flow steady state nozzle gives

$$W=0,\ Q=0,\ \text{and}\ h_i + (V_i)^2/2 = h_e + (V_e)^2/2 \qquad (4.5.1.1)$$

Usually the kinetic energy of the fluid at the inlet of the nozzle is much smaller than the kinetic energy of the fluid at the exit, and would be neglected if its value is not known. Equation (4.5.1) is then reduced to

$$h_i = h_e + (V_e)^2/2 \qquad (4.5.1.2)$$

Comparison between the actual and the ideal nozzle performance is given by the isentropic efficiency, η. Since an ideal nozzle raises higher theoretical kinetic energy than the actual kinetic energy by an actual adiabatic nozzle at exit, the nozzle efficiency, η, is defined as

$$\eta=[(V_{actual})^2/2]/[(V_{isentropic})^2/2] \qquad (4.5.1.3)$$

Example 4.5.1. Air enters a nozzle at 100 psia and 200°F, and leaves at 15 psia and -40°F. The inlet velocity is 100 ft/s. Determine the air exit velocity.

Solution: Applying Eq. (4.5.1) gives $V_e = [2(h_i-h_e) + (V_e)^2]^{1/2} = [2c_p(T_i-T_e) + (V_e)^2]^{1/2} = [2(0.24)(200+40)25000+100^2]^{1/2} = 1702$ ft/s.
Notice that the conversion factor of 1 Btu/lbm=25000 (ft/s)^2.

4.5.2 Diffuser

A *diffuser* is a small device used to decelerate a high velocity fluid stream and raise the pressure of the fluid. There are no means to do work. A diffuser is usually modeled as adiabatic. This assumption is reasonable because the heat transfer surface area of the diffuser is very small, and the length of time required for the working fluid to pass through the diffuser is very short. Therefore the ideal process is considered to be a reversible and adiabatic or isentropic one.

Application of the first law of thermodynamics to a steady flow steady state diffuser gives

$$W=0,\ Q=0,\ \text{and}\ h_i + (V_i)^2/2 = h_e + (V_e)^2/2 \qquad (4.5.2.1)$$

Usually the kinetic energy of the fluid at the exit of the diffuser is much smaller than the kinetic energy of the fluid at the inlet, and would be neglected if its value is not known. Equation ( ) is then reduced to

$$h_i + (V_i)^2/2 = h_e \qquad (4.5.2.2)$$

Example 4.5.2.1. Air enters an isentropic diffuser at 250 m/s, 120 kPa and 40°C, and leaves at 90 m/s. Determine the air exit temperature.

Solution: Applying Eq. (4.5.4) gives $h_e - h_i = c_p(T_e-T_i) = (V_i)^2/2 - (V_e)^2/2$

Thus $T_e=T_i+[(V_i)^2/2 - (V_e)^2/2]/c_p=313+\{[250^2 -90^2]/2000\}/0.24=340.2$ K

Notice that the conversion factor of 1 kJ/kg=1000 $(m/s)^2$.

Homework 4.5 Nozzle and Diffuser

1. What is the function of a nozzle?
2. What is the function of a diffuser?

4.6 Systems Consisting of More Than one Open-System Device

We have studied several problems involving only one steady flow device in each case. In this section, we will study a few problems in which two or more devices are combined to perform an application task.

Example 4.6.1. Water is heated in an isobaric boiler from 6 Mpa and 90°C to 500°C (process 1-2). It is then expanded in an isentropic turbine to 50 kPa (process 2-3). The required turbine shaft power is 100,000 kW. Determine the quality and temperature of steam at the exit of the turbine, entropy change from state 1 to state 3, enthalpy change from state 1 to state 3, heat added in the boiler, work produced by the turbine, and mass rate of water.

To solve this problem by CyclePad, we take the following steps:

1. Build
   (A) Take a source, a heater, a turbine, and a sink from the open-system inventory shop and connect them.
   (B) Switch to analysis mode.
2. Analysis
   (A) Assume the heater is isobaric, turbine is isentropic.
   (B) Input the given information: (a) working fluid is water, (b) inlet pressure and temperature to the boiler are 6 Mpa and 90°C, (b) outlet temperature of the boiler is 500°C, (c) outlet pressure of the turbine is 50 kPa, and (d) the turbine turbine shaft power is 100,000 kW.
3. Display results
   (A) Display the states and the devices results. The answers are x=0.8904, T=81.34°C, Δs=6.88-1.19=5.69 kJ/[kg(K)], Qdot=295,363 kW.

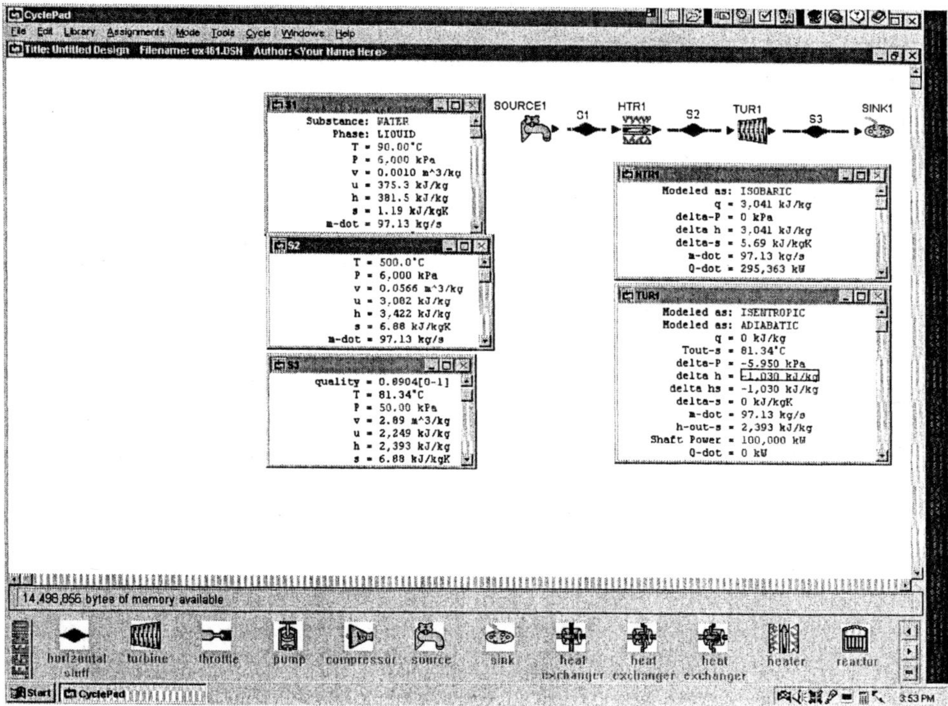

Figure Example 4.6.1. Multi-process

Example 4.6.2. Water is expanded in an adiabatic turbine from 6 Mpa and 500°C to 50 kPa and a quality of 0.93 (process 1-2). Water is then condensed in an isobaric condenser to saturated liquid (process 2-3). The water mass flow rate is 32 kg/s. Determine the temperature of steam at the exit of the turbine, entropy change from state 1 to state 3, enthalpy change from state 1 to state 3, rate of heat removed from the condenser, power produced by the turbine, and efficiency of the turbine.

To solve this problem by CyclePad, we take the following steps:

1. Build
   (A) Take a source, a heater, a turbine, and a sink from the open-system inventory shop and connect them.
   (B) Switch to analysis mode.
2. Analysis
   (A) Assume the heater is isobaric, turbine is isentropic.
   (B) Input the given information: (a) working fluid is water, (b) inlet pressure and temperature to the condenser are 6 Mpa and 90°C, (b) outlet temperature of the condenser is 500°C, (c) outlet pressure of the turbine is 50 kPa, and (d) the turbine shaft power is 100,000 kW.
3. Display results
   (A) Display the states and the devices results. The answers are T=81.34°C, Δs=1.09-6.88=-5.79 kJ/[kg(K)], Δh=340.5-3422=-3081.5 kJ/kg, Qdot=-68,590 kW, Wdot=30,027 kW, and η=91.14%.

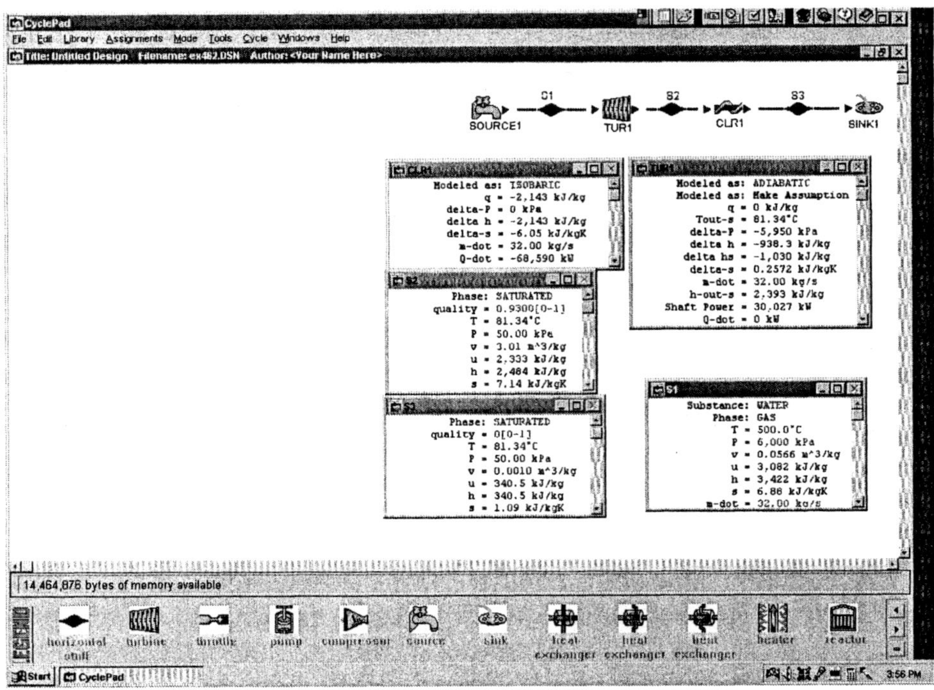

Figure Example 4.6.2. Device combination

Example 4.6.3. Refrigerant R-134a is compressed in an adiabatic compressor from 0.14 Mpa and -10°C to 0.8 MPa and 50°C (process 1-2). R-134a is then condensed in an isobaric condenser to saturated liquid (process 2-3). The R-134a mass flow rate is 0.02 kg/s. Determine the temperature of R-134a at the exit of the condenser, entropy change from state 1 to state 3, enthalpy change from state 1 to state 3, rate of heat removed from the condenser, power required by the compressor, and efficiency of the compressor.

To solve this problem by CyclePad, we take the following steps:

1. Build
   (A) Take a source, a heater, a compressor, and a sink from the open-system inventory shop and connect them.
   (B) Switch to analysis mode.
2. Analysis
   (A) Assume the heater is isobaric, compressor is adiabatic.
   (B) Input the given information: (a) working fluid is R-134a, (b) inlet pressure and temperature to the condenser are 6 Mpa and 90°C, (b) outlet temperature of the condenser is 500°C, (c) outlet pressure of the compressor is 50 kPa, and (d) the compressor shaft power is 100,000 kW.
3. Display results
   (A) Display the states and the devices results. The answers are T=31.24°C, Δs=1.15-1.77=-0.62 kJ/[kg(K)], Δh=243.6-394.1=-150.5 kJ/kg, Qdot=-3.83 kW, Wdot=-0.82 kW, and η=93.23%.

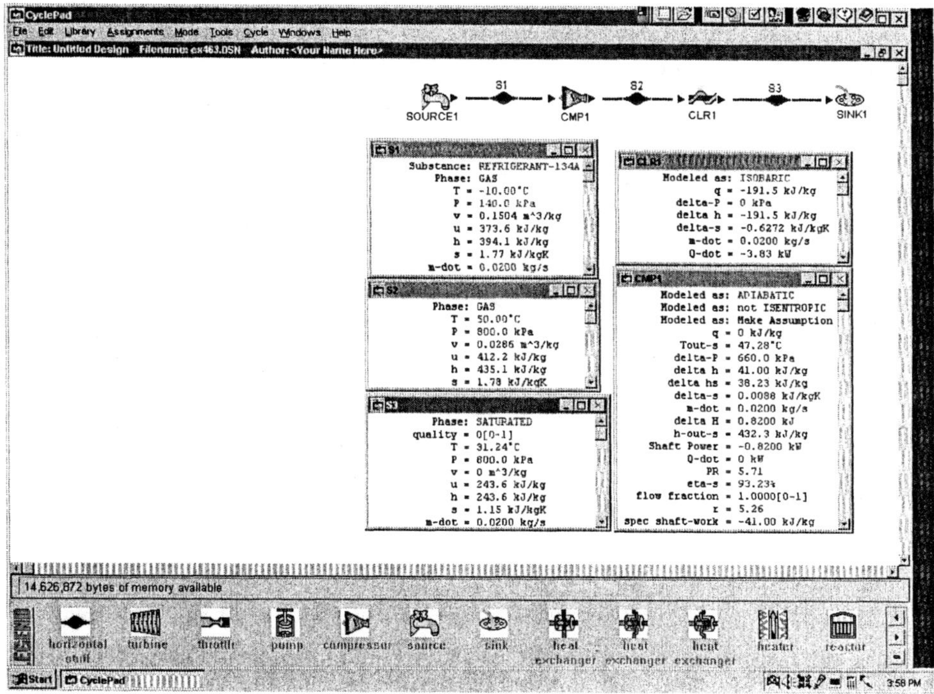

Figure Example 4.6.3. Device combination

Homework 4.6 Combination

1. 2.3 kg/s of air flows through a two-stage turbine with a reheater. Air flows into the high-pressure adiabatic turbine at 1400 K and 1200 kPa and leaves at 1100 K and 400 kPa. It enters the isobaric reheater and leaves at 1400 K. Air then enters the low-pressure adiabatic turbine and leaves at 800 K and 120 kPa. Determine (a) the power produced by the high-pressure turbine, (b) the heat transfer added by the reheater, and (c) the power produced by the low-pressure turbine.
2. 3.4 kg/s of steam flows through a two-stage turbine with a reheater. Steam flows into the high-pressure isentropic turbine at 800 K and 1200 kPa and leaves at 600 kPa. It enters the isobaric reheater and leaves at 800 K. Steam then enters the low-pressure isentropic turbine and leaves at 30 kPa. Determine (a) the power produced by the high-pressure turbine, (b) the heat transfer added by the reheater, and (c) the power produced by the low-pressure turbine.
3. 0.12 kg/s of refrigerant R12 enters an isentropic compressor at -25°C as saturated vapor and leaves at 1000 kPa. The refrigerant then enters an isobaric cooler and leaves as a saturated liquid. Determine the power required by the compressor and heat transfer removed from the cooler.
4. 0.012 kg/s of refrigerant R22 enters an adiabatic compressor at -25°C as saturated vapor and leaves at 1000 kPa and 52°C. The refrigerant then enters an isobaric cooler and leaves as a saturated liquid. Determine the power required by the compressor and heat transfer removed from the cooler.

4.7 SUMMARY

An open system (control volume) is a fixed volume system, which is different from a fixed mass system (control mass). The mass balance of the open system can be expressed as $\Delta m= m_i - m_e$. The energy balance of the open system called the first law of thermodynamics for open system can be expressed as $\Delta E=Q-W +E_i -E_e$, where energy flow in or out with mass across the boundary surface is $E= E_k +E_k+U +pV= E_k +E_k+H$. Boundary work for open systems is zero, because the volume of the system is fixed.

Steady flow and steady state is the most important case of engineering applications. The mass, energy and other quantities of the system are constant in the steady state, and the fluid flows through the system steadily. The mass and energy balances in this case can be expressed as: $m_i=m_e$, and $0=Q-W +E_i -E_e$.

Various engineering steady flow and steady state processes and devices analyzed by the mass and energy balances are illustrated using CyclePad.

Chapter 5

SECOND LAW OF THERMODYNAMICS

OBJECTIVES

After reading and studying the material in this chapter, you should be able to:

3. Understand the need of second law of thermodynamics.
4. Know the definitions of heat engine, refrigerator, and heat pump.
5. Know the difference between a refrigerator and heat pump.
6. Understand that heat engines, refrigerators, and heat pumps are cycles.
7. Find the heat engine efficiency, refrigerator COP, and heat pump COP by hand.
8. Know the two basic statements of second law of thermodynamics.
9. Understand reversible and irreversible processes.
10. Understand that all real processes are irreversible and cite some factors that cause irreversibility.
11. Know the four processes that constitute the Carnot cycle.
12. Know the three important equations concerning the limits of the efficiency or COP of heat engines, refrigerators, and heat pumps.
13. Know that the Carnot cycle is independent of devices and working fluids used in the cycle.

5.1 INTRODUCTION

Thus far we have analyzed thermodynamic systems according to the first law of thermodynamics and state properties relationships. Water does not flow up a hill, heat does not flow from a low temperature body to a high temperature body. Our experiences suggest that processes have a definite direction. A proposed thermodynamic system that does not violate the first law of thermodynamics does not ensure that the thermodynamic system will actually occur. The first law of thermodynamics does not give any information to the direction of the process. There is a need to place restrictions on the direction of flow of a process. The limited amount of energy that can be transformed from one form to another form and the direction of flow of heat and work have not been discussed. The second law of thermodynamics addresses these areas.

5.2 DEFINITIONS

5.2.1 Thermal Reservoirs

A *thermal reservoir* is any object or system which can serve as a heat source or sink for another system. Thermal reservoirs usually have accumulated energy capacities which are very, very large compared with the amounts of heat energy they exchange. Therefore the thermal reservoirs are considered to operate at constant temperatures. Examples of large capacity, constant temperature thermal reservoirs which make convenient heat sources and sinks are: ocean, atmosphere, etc.

5.2.2 Heat Engines

A *heat engine* is a continuous cyclic device which produces positive net work output by adding heat. The energy flow diagram of a heat engine and its thermal reservoirs are shown in Figure 5.2.2.1. Heat (Q_H) is added to the heat engine from a high-temperature thermal reservoir at T_H, output work (W) is done by the heat engine, and heat (Q_L) is removed from the heat engine to a low-temperature thermal reservoir at T_L.

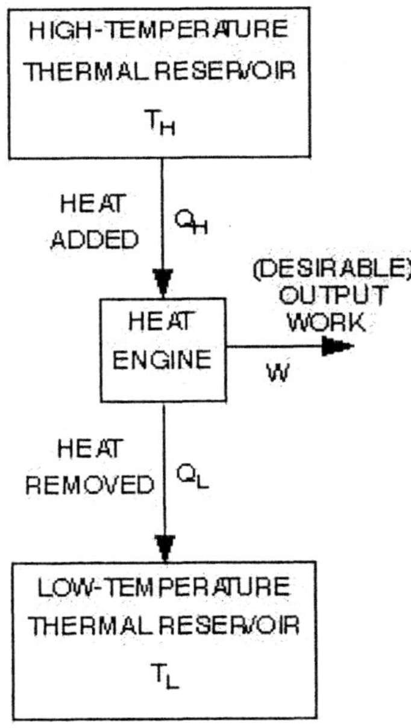

Figure 5.2.2.1 Heat engine

For example, a commercial central power station using a heat engine called a Rankine steam power plant is shown in Figure 5.2.2.2. The Rankine heat engine consists of a pump, a boiler, a turbine and a condenser. Heat (Q_H) is added to the working fluid (water) in the boiler from a high-temperature (T_H) flue gas by burning coal or oil. Output work (W_o) is done by the turbine. Heat (Q_L) is removed from the the working fluid in the condenser to low-temperature

(T_L) lake cooling water. Input work (W_i) is added to drive the pump. The net work (W) produced by the Rankine heat engine is $W=W_o-W_i$.

Another example of a heat engine is a nuclear helium gas power plant which is made of a nuclear reactor, a gas turbine, a cooler and a compressor as shown in Figure 5.2.2.3. Heat is added to the nuclear gas power plant in the nuclear reactor, work is produced by the turbine, heat is removed from the cooler, and work input is required to operate the compressor. A part of the work produced by the turbine is used to drive the pump. The net work output of the nuclear helium gas power plant is the difference between the work produced by the turbine and work required to operate the compressor.

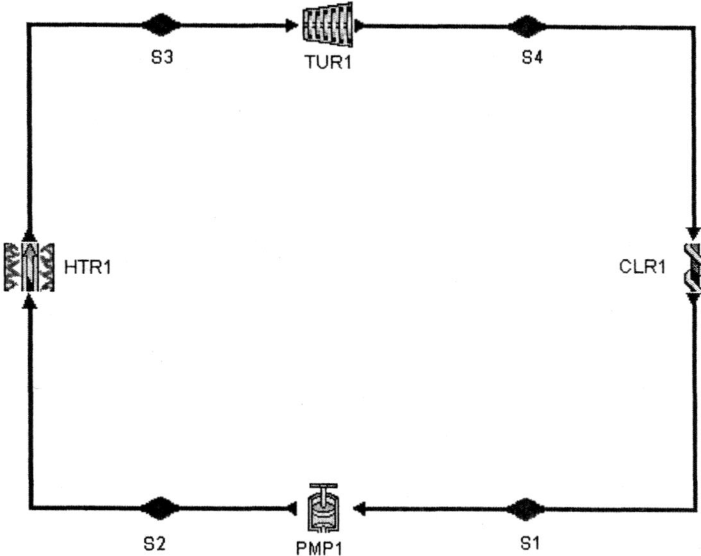

Figure 5.2.2.2. Rankine heat engine

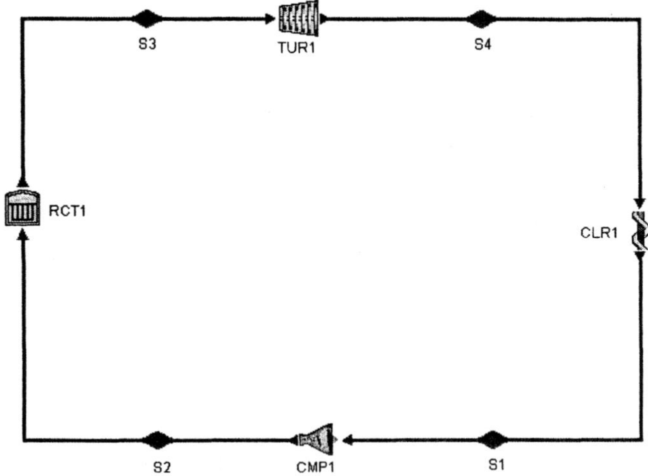

Figure 5.2.2.3 Nuclear helium gas heat engine

The measurement of performance for a heat engine is called the thermal *efficiency,* η. The thermal efficiency of a heat engine is defined as the ratio of the desirable net output work sought to the heat input of the engine:

$$\eta = W_{net}/Q_{input} \quad (5.2.2.1)$$

Example 5.2.2.1. Heat is transferred to a Rankine power plant at a rate of 80 MW. If the net power output of the plant is 30 MW. Determine the thermal efficiency of the power plant.

To solve this problem by CyclePad, we take the following steps:

1. Build
   (A) Go to LIBRARY, select Rankine cycle unsolved.
   (B) Switch to analysis mode.
2. Analysis
   (A) Go to cycle, then cycle properties.
   (B) Input the given information: (a) heat added is +80 MW, (b) net power output is +30 MW.
3. Display results
   (A) The answer is η=0.375.

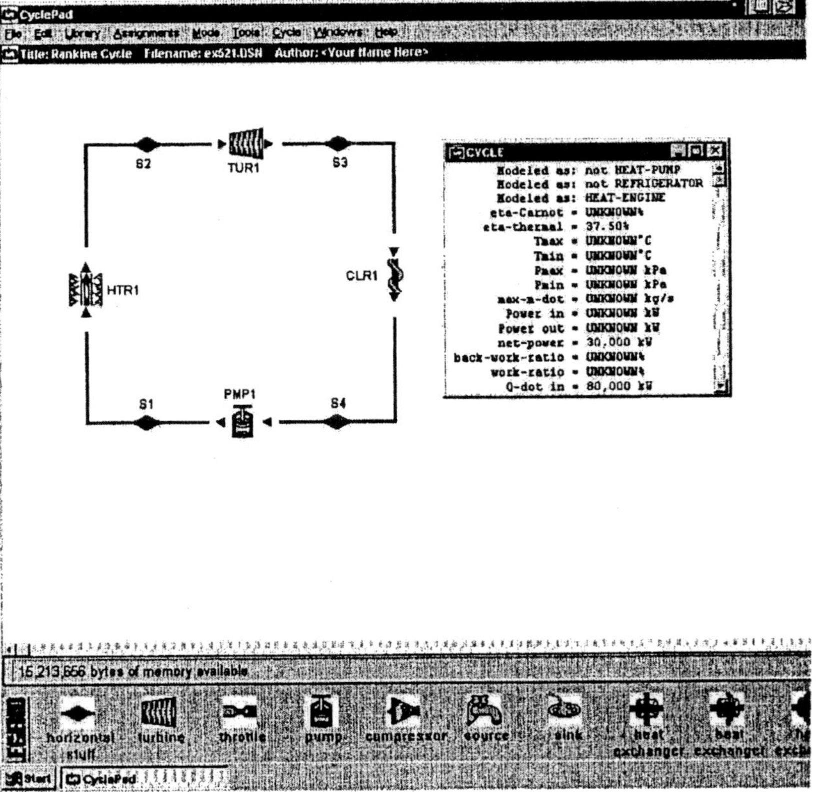

Figure Example 5.2.2.1. Heat engine efficiency

Example 5.2.2.2 A Rankine heat engine with a net power output of 85000 hp has a thermal efficiency of 35%. Determine the rate of heat added to the car engine.

To solve this problem by CyclePad, we take the following steps:

1. Build
   (A) Go to LIBRARY, select Rankine cycle unsolved.
   (B) Switch to analysis mode.
2. Analysis
   (A) Go to cycle, then cycle properties.
   (B) Input the given information: (a) net power output is +85000 hp, (b) cycle efficiency is 35%.
3. Display results
   (A) The answer is the rate of heat added=171657 Btu/s.

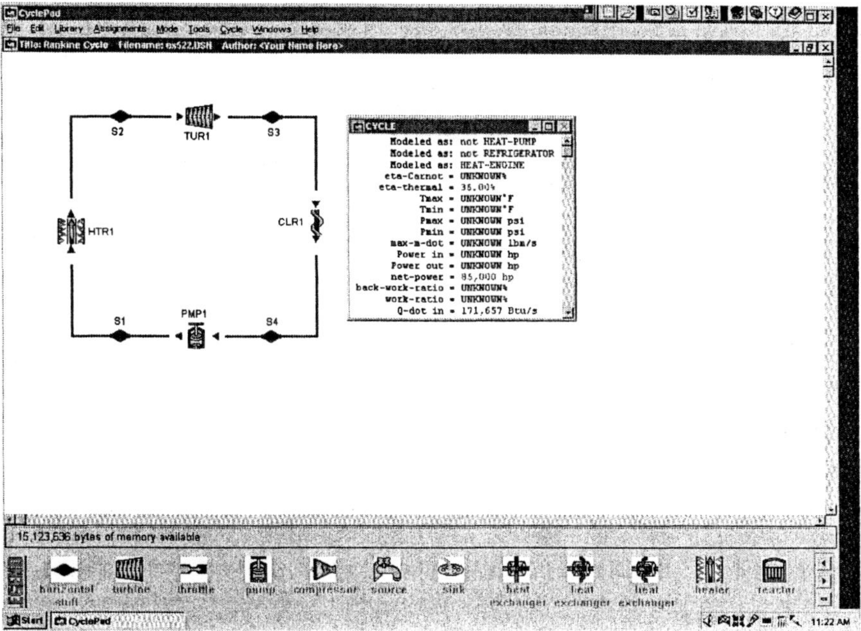

Figure Example 5.2.2.2 Heat engine

Homework 5.2.2 Heat Engines

1. What is a thermal reservoir?
2. Is a gas turbine a heat engine?
3. Is a steam power plant a heat engine?
4. Is it possible to have a heat engine with efficiency of 100%?
5. Can a heat engine operating with only one thermal reservoir?
6. Is a reversible heat engine more efficient than an irreversible heat engine when operating between the same two thermal reservoirs?

7. Give two expressions to calculate the efficiency of a heat engine.
8. A closed system cycle has a thermal efficiency of 28%. The heat supplied from the energy source is 1000 Btu/lbm of working substance. Determine (a) heat rejected, and (b) net work of the cycle.
9. A closed system undergoes a cycle in which 600 Btu of heat is transferred to the system from a source at 600 °F and 350 Btu of heat is rejected to a sink at 250 °F. From the start to end of the cycle,
   (a) What is the change in internal energy of the system (Btu)?
   (b) What is the net work of the system (Btu)?
   (c) Is this cycle possible? Justify your answer.
10. A simple steam power cycle plant receives 100,000 kJ/min as heat transfer from hot combustion gases at 3000 K and rejects 66,000 kJ/min as heat transfer to the environment at 300 K. Determine (a) the thermal efficiency of the cycle, (b) the net power produced by the cycle, and (c) the maximum possible thermal efficiency of any cycle operating between 3000 K and 300 K.

5.2.3 Refrigerators

A *refrigerator* is a continuous cyclic device which removes heat from a low temperature reservoir to a high temperature reservoir at the expense of work input. The energy flow diagram of a refrigerator and its thermal reservoirs are shown in Figure 5.2.3.1. Input work (W) is added to the refrigerator, desirable heat (Q_L) is removed from the low-temperature thermal reservoir at T_L, and heat (Q_H) is added to the high-temperature thermal reservoir at T_H.

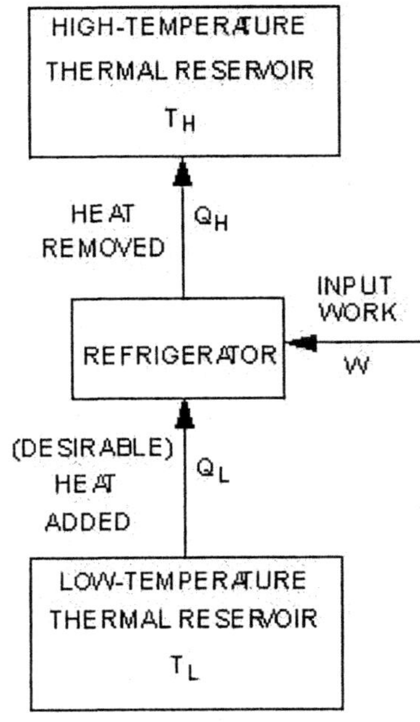

Figure 5.2.3.1. Refrigerator

An example of a refrigerator is a domestic refrigerator which is made up of a compressor, a condenser, an expansion valve and an evaporator. The domestic refrigerator is illustrated in Figure 5.2.3.2. Work (W) is added to drive the compressor by an electric motor, heat (Q_L) is added in the evaporator from the low temperature (T_L)refrigerator inner space by the working fluid (refrigerant), and heat (Q_H) is removed from the condenser from the working fluid to the high temperature (T_H) reservoir (kitchen).

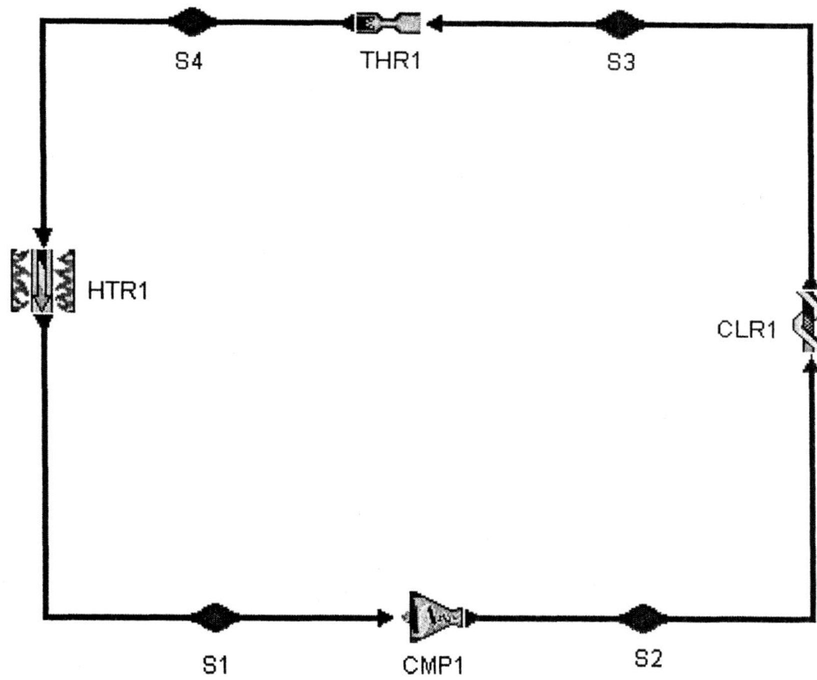

Figure 5.2.3.2. Domestic Refrigerator

The measurement of performance for a refrigerator is called the *coefficient of performance (COP)*, and is denoted by β_R. The coefficient of performance of a refrigerator is defined as the ratio of the desirable heat removed Q_L to the work input W of the refrigerator:

$$\beta_R = Q_L/W \tag{5.2.3.1}$$

Q_L is called *refrigerator capacity* and is usually expressed in tons of refrigeration. One *ton of refrigeration* is 3.516 kW or 12,000 Btu/h. The term "ton" is derived from the fact that the heat required to melt one ton of ice is about 12,000 Btu/h.

Notice that the coefficient of performance of a refrigerator may be larger or smaller than one.

Example 5.2.3.1. The inside space of a refrigerator is maintained at low temperature by removing heat (Q_L) from it at a rate of 6 kW. If the COP of the refrigerator is 1.5, determine the refrigerator capacity in tons of refrigeration and the required power input to the refrigerator.

To solve this problem by CyclePad, we take the following steps:

1. Build
   (A) Go to LIBRARY, select basic refrigerator cycle unsolved.
   (B) Switch to analysis mode.
2. Analysis
   (A) Go to cycle, then cycle properties.
   (B) Input the given information: (a) COP is 1.5, (b) heat removed from the refrigerator but added to the cycle is +6 kW.
3. Display results
   (A) The answers are: refrig. Capacity is 1.71 ton, and net power input is -4kW.

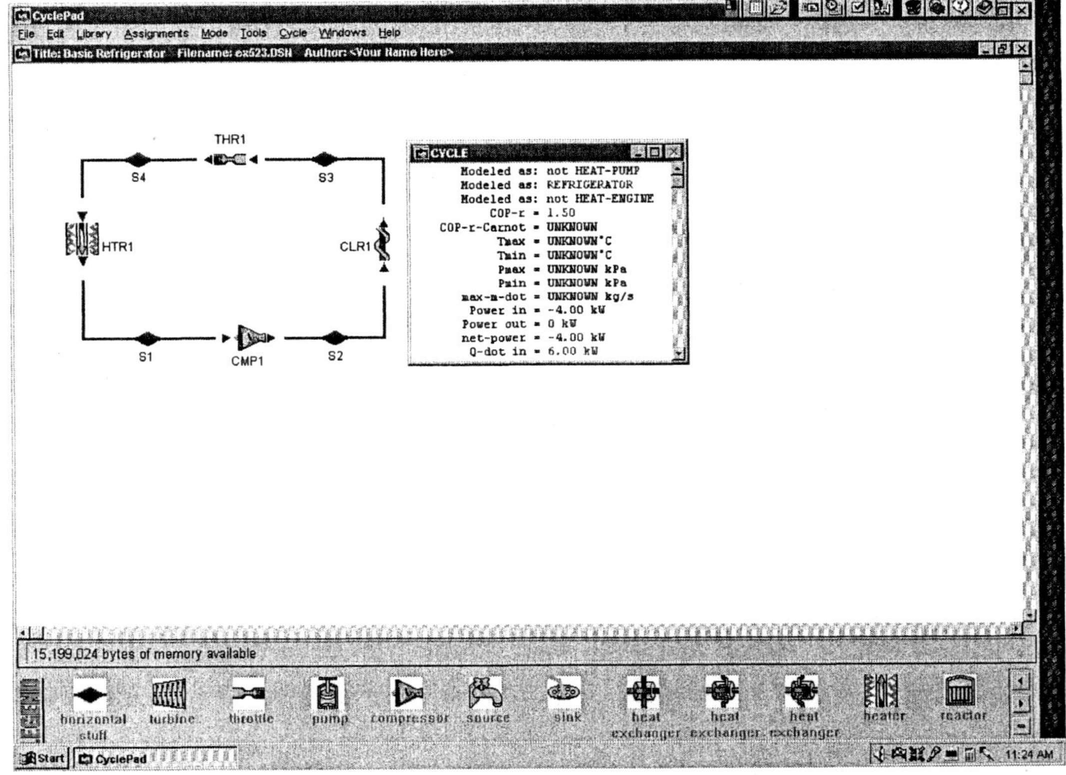

Figure Example 5.2.3.1. Refrigerator

Example 5.2.3.2. The inside space of a refrigerator is maintained at 3°C by removing heat from it at a rate of 5 kW. If the required power input to the refrigerator is 2 kW. Determine the COP of the refrigerator.

To solve this problem by CyclePad, we take the following steps:

1. Build
   (A) Go to LIBRARY, select basic refrigerator cycle unsolved.
   (B) Switch to analysis mode.

2. Analysis
   (A) Go to cycle, then cycle properties.
   (B) Input the given information: (a) power input is -2 kW, (b) heat removed from the refrigerator but added to the cycle is +5 kW.
3. Display results
   (A) The answer is COP=2.50.

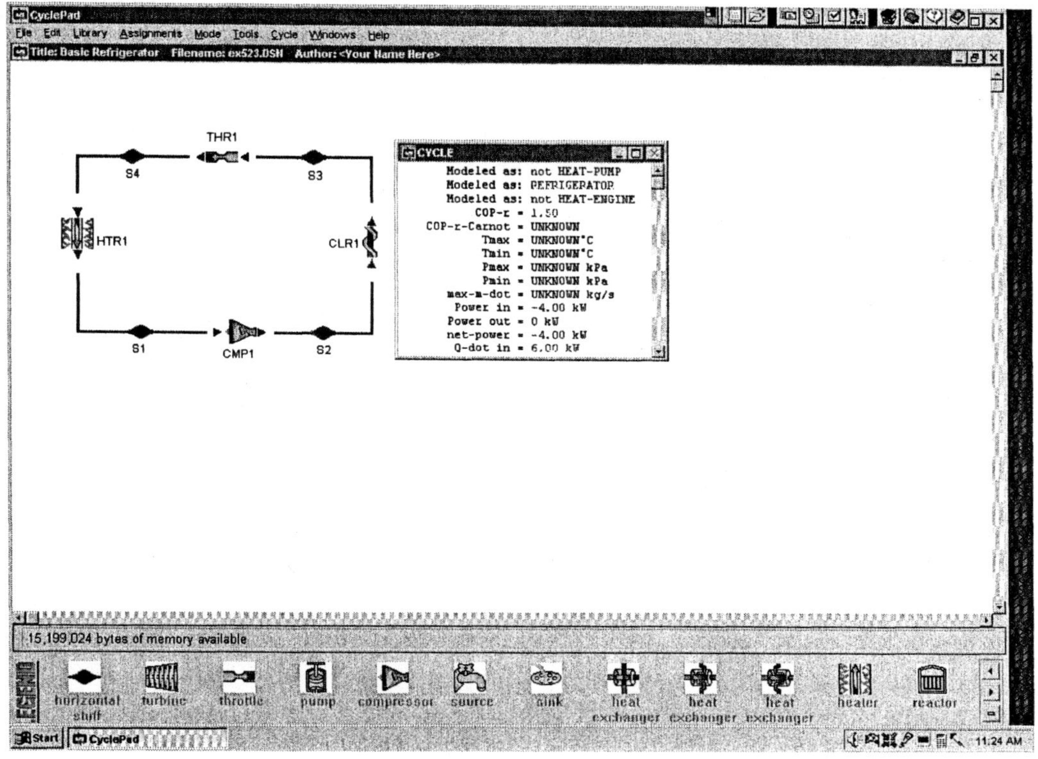

Figure Example 5.2.3.2. Refrigerator

Homework 5.2.3 Refrigerator

1. Is air conditioner a refrigerator?
2. Is it possible to have a refrigerator operating with one thermal reservoir?
3. Define the COP of a refrigerator.
4. Is the COP of a refrigerator always larger than 1?
5. What is a ton of refrigeration?
6. The refrigeration plant of an air conditioning system has a capacity of 12000 Btu/h. The COP of the refrigerator is 4. Determine the power required by the compressor.
7. The refrigeration plant of an air conditioning system has a cooling capacity of 2400 Btu/h. The compressor power added is 1200 Btu/h. What is the COP of the refrigerator? What is the amount of heat transfer to the atmosphere?

5.2.4 Heat Pumps

A *heat pump* is a continuous cyclic device which pumps heat to a high temperature reservoir from a low temperature reservoir at the expense of work input. The energy flow diagram of a heat pump and its thermal reservoirs are shown in Figure 5.2.4.1. Input work (W) is added to the heat pump, desirable heat (Q_H) is pumped to the high-temperature thermal reservoir at T_H, and heat (Q_L) is removed from the low-temperature thermal reservoir at T_L.

An example of a heat pump is a house heat pump which is made up of a compressor, a condenser, a throttling valve and an evaporator as illustrated in Figure 5.2.4.2. Heat (Q_H) is removed to the high-temperature house from the working fluid in the condenser, work is added to the compressor by an electric motor, and heat (Q_L) is added to the evaporator from the low-temperature outside atmospheric air in the winter season.

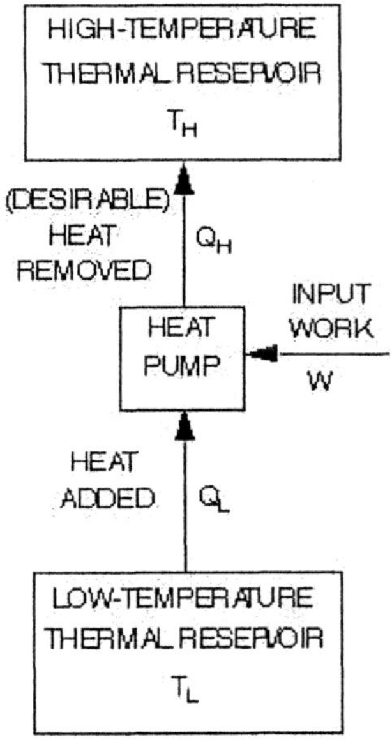

Figure 5.2.4.1. Heat pump

Notice that the energy flow diagram and hardware components of a heat pump are exactly the same as those of a refrigerator. The difference between the heat pump and the refrigerator is the function of the cyclic devices. A refrigerator is used to remove heat (Q_L) from a low temperature (T_L) thermal reservoir by adding work. A heat pump is used to add heat (Q_H) to a high temperature (T_H) thermal reservoir by adding work.

COP

The measurement of performance for a heat pump is called the coefficient of performance (COP), and is denoted by β_{HP}. The coefficient of performance of a heat pump is defined as the ratio of the desirable heat output Q_H to the work input W of the heat pump:

$\beta_{HP}=Q_H/W$ (5.2.4.1)

Notice that the coefficient of performance of a heat pump is always larger than one.

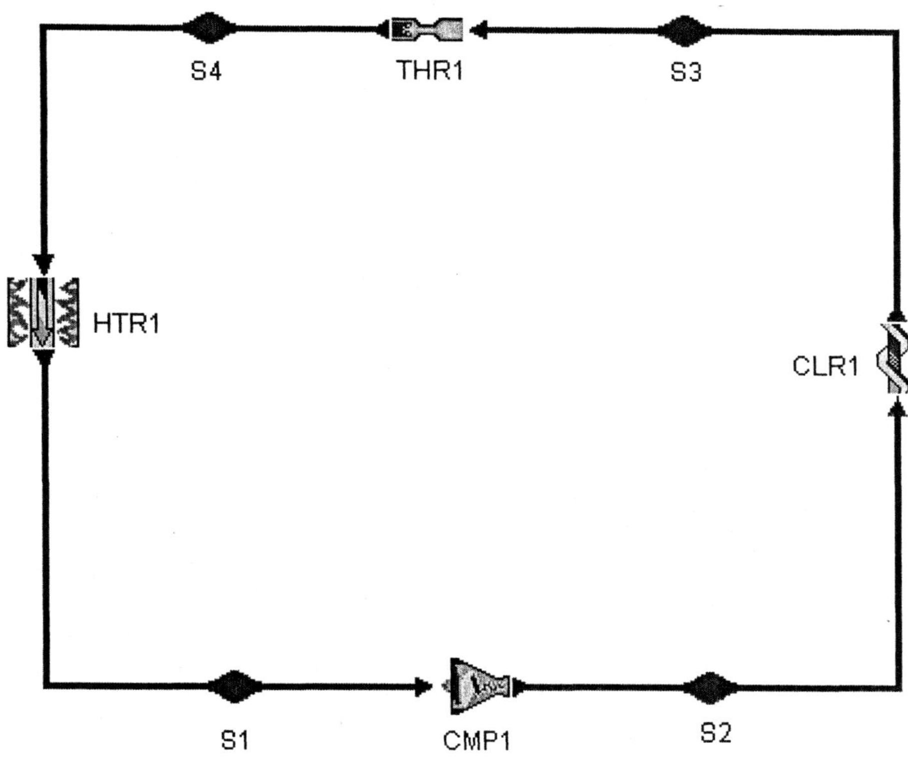

Figure 5.2.4.2. House Heat pump

Example 5.2.4.1. A heat pump with a COP of 2.8 is selected to meet the heating requirements of a house and maintain it at a comfortable temperature (T_H). Heat (Q_L) is pumped from the outdoor ambient air at low temperature (T_L). The house is estimated to lose heat (Q_H) at a rate of 1000 kW. Determine the electric motor power consumed by the heat pump.

To solve this problem by CyclePad, we take the following steps:

1. Build
   (A) Go to LIBRARY, select basic refrigeration cycle unsolved.
   (B) Switch to analysis mode.
2. Analysis
   (A) Go to cycle, then cycle properties.
   (B) Input the given information: (a) COP is 2.8, (b) heat removed is -1000 kW.
3. Display results
   (A) The answer is net power=357.1 kW.

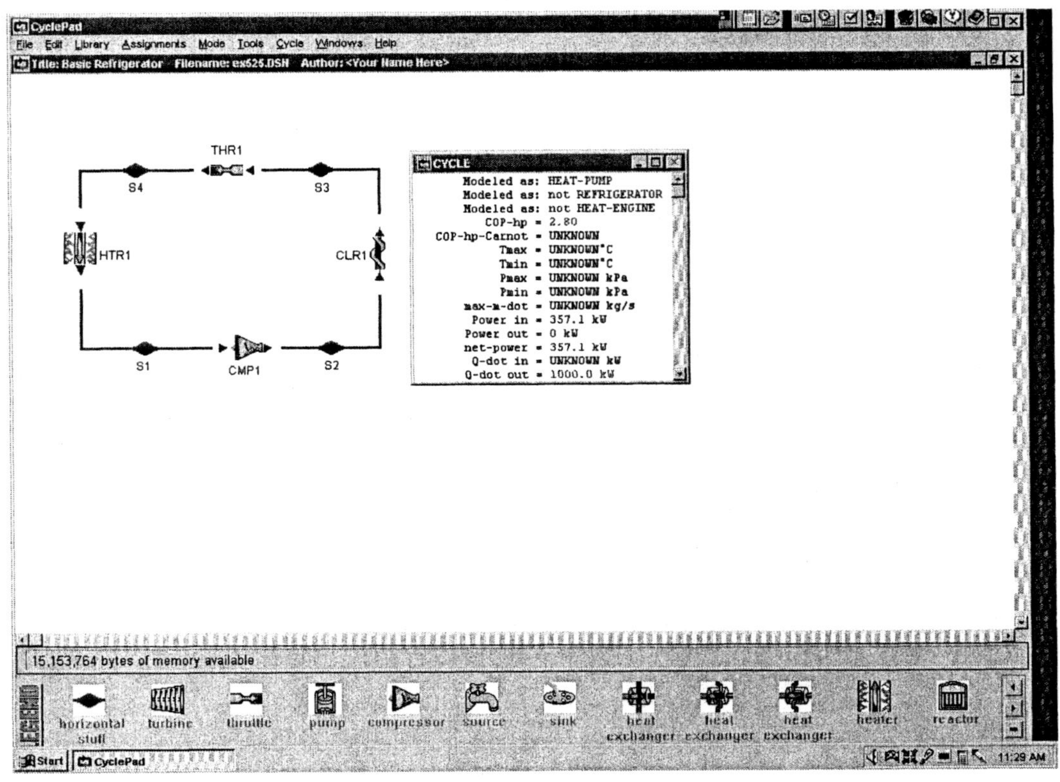

Figure Example 5.2.4.1. Heat pump

Example 5.2.4.2. A heat pump is selected to meet the heating requirements of a house and maintain it at 18°C. If the outdoor temperature is 2°C, the house is estimated to lose heat at a rate of 1000 kW and the net power input to the house is 680 kW. Determine the COP of the heat pump.

To solve this problem by CyclePad, we take the following steps:

1. Build
   (A) Go to LIBRARY, select basic refrigeration cycle unsolved.
   (B) Switch to analysis mode.
2. Analysis
   (A) Go to cycle, then cycle properties.
   (B) Input the given information: (a) net power input is -680 kW, (b) heat removed is -1000 kW.
3. Display results
   (A) The answer is COP=1.47.

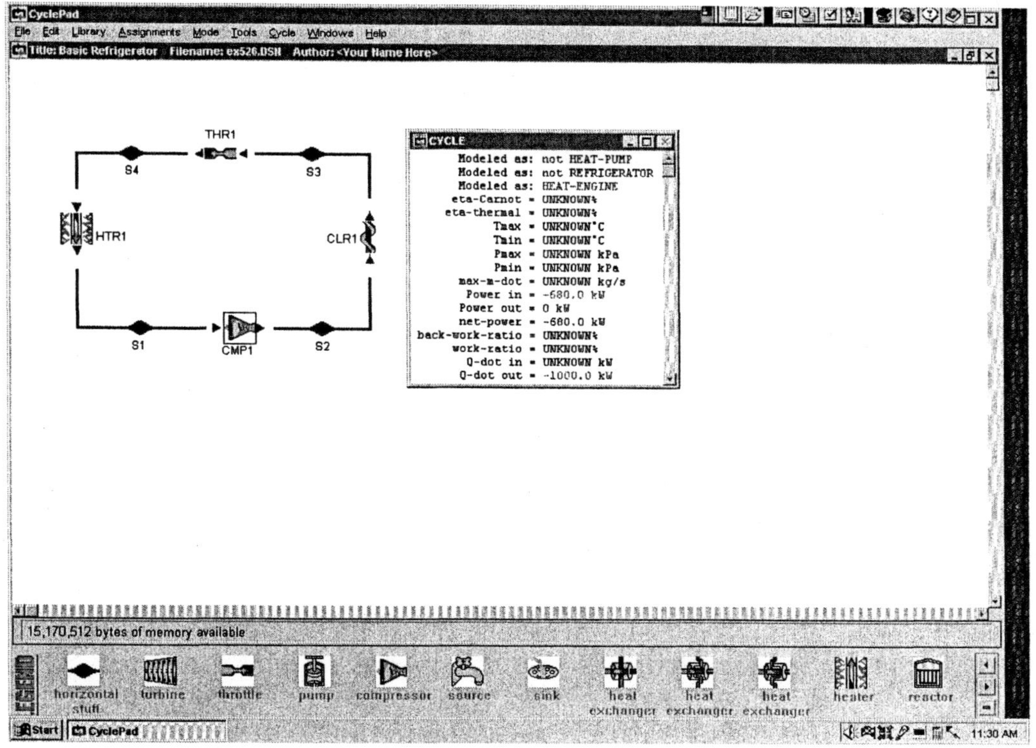

Figure Example 5.2.4.2. Heat pump

Homework 5.2.4 Heat Pump

1. What is the difference between a refrigerator and a heat pump?
2. A heat pump and a refrigerator are operating between the same two thermal reservoirs. Which one has a higher COP?
3. Define the COP of a heat pump.
4. Is the COP of a heat pump always larger than 1?
5. A heat pump picks up 1000 kJ of heat from well water at 10 C and discharges 3000 kJ of heat to a building to maintain it at 20 C. What is the COP of the heat pump? What is the required cycle net work?

5.3 SECOND LAW STATEMENTS

The principle underlying the directionality of spontaneous change and inefficiency of a heat engine is called the *second law of thermodynamics*. This law may be stated in various equivalent ways. Among the best known statements are the following two statements.

Kelvin-Planck statement: It is impossible to construct a heat engine that produces work with heat interaction only from a single thermal reservoir. This implies that a heat engine requires at least two thermal reservoirs with different temperatures, and that it is impossible to build a heat engine that has a thermal efficiency of 100%.

Clausius statement: It is impossible to construct a heat pump or a refrigerator which moves heat from a low temperature thermal reservoir to a high temperature thermal reservoir without adding work. This implies that the coefficient of performance of a heat pump or a refrigerator is always less than infinity.

Every relevant experiment that has been conducted verifies the second law of thermodynamics.

5.4 Reversible and Irreversible Processes

A *reversible process* is an idealized one that is performed in such a way that, at the conclusion of the process, both the system and its surroundings may be restored to their initial states without producing any changes in the rest of the universe. Any process that does not fulfill these stringent requirements is called an *irreversible process*. The factors that cause a process to be irreversible are called irreversibility factors. The irreversibility factors include friction, heat transfer across a finite temperature difference, free expansion, mixing of two gases, chemical reactions, etc.

A process is called *internally reversible* if no irreversibilities occur within the boundaries of the system during the process. A process is called *externally reversible* if no irreversibilities occur outside the boundaries of the system during the process. A process is called *totally reversible* if no irreversibilities occur neither within nor outside the boundaries of the system during the process. A process is called *endo-reversible* if no irreversibilities occur within but outside the boundaries of the system during the process.

Reversible processes actually do not occur in nature. They serve as idealized models in which actual processes can be compared. It is generally much easier to describe a system's behavior under reversible conditions rather than irreversible ones. Thus the assumption of reversibility, although not perfect, is often useful engineering approach.

5.5 Carnot Cycle

Considering the concepts of reversible processes, a reversible cycle can be carried out for given thermal reservoirs at temperatures T_H and T_L. The reversible cycle was introduced by a French engineer N.S. Carnot. The Carnot heat engine cycle on a p-v diagram as shown in Figure 5.5.1 is composed of the following four reversible processes:

1-2 reversible adiabatic (isentropic) compression
2-3 reversible isothermal heating at T_H
3-4 reversible adiabatic (isentropic) expansion
4-1 reversible isothermal cooling at T_L

Referring to Figure 5.5.1, the system undergoes a Carnot heat engine cycle in the following manner:

(A) During process 1-2, the system is thermally insulated and the temperature of the working substance is raised from the low temperature T_L to the high temperature T_H.

The process is an isentropic process. The amount of heat transfer during the process is $Q_{12}=\int Tds=0$, because there is no area underneath a constant entropy (vertical) line.

(B) During process 2-3, heat is transferred isothermally to the working substance from the high temperature reservoir at T_H. This process is accomplished reversibly by bringing the system in contact with the high temperature reservoir whose temperature is equal to or infinitesimally higher than the working substance. The amount of heat transfer during the process is $Q_{23}=\int Tds= T_H(S_3-S_2)$, which can be represented by the area 2-3-5-6-2. Q_{23} is the amount of heat added to the Carnot cycle from a high temperature thermal reservoir.

(C) During process 3-4, the system is thermally insulated and the temperature of the working substance is decreased from the high temperature T_H to the low temperature T_L. The process is an isentropic process. The amount of heat transfer during the process is $Q_{34}=\int Tds=0$, because there is no area underneath a constant entropy (vertical) line.

(D) During process 4-1, heat is transferred isothermally from the working substance to the low temperature reservoir at T_L. This process is accomplished reversibly by bringing the system in contact with the low temperature reservoir whose temperature is equal to or infinitesimally lower than the working substance. The amount of heat transfer during the process is $Q_{41}=\int Tds= T_L(S_1-S_4)$, which can be represented by the area 1-4-5-6-1. Q_{41} is the amount of heat removed from the Carnot cycle to a low temperature thermal reservoir.

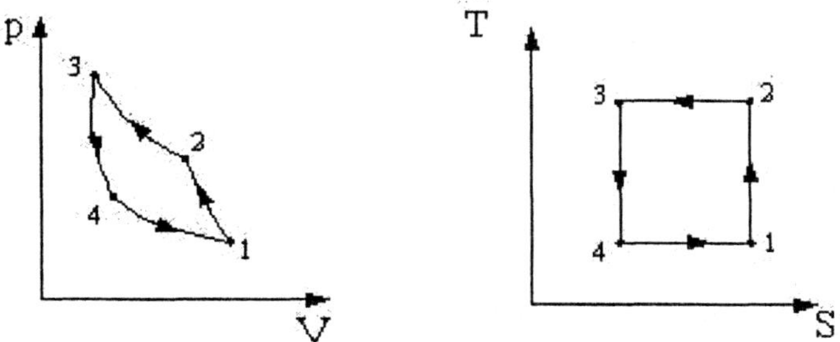

Figure 5.5.1 Carnot heat engine cycle on p-v and T-s diagram

The net heat added to the cycle is $Q_{net}=Q_{12}+Q_{23}+Q_{34}+Q_{41}=0+T_H(S_3-S_2)+0+T_L(S_1-S_4)=(T_H-T_L)(S_4-S_1)$=[Area 2-3-5-6-2]-[area 1-4-5-6-1]=Area 1-2-3-4-1. Notice that the area 1-2-3-4-1 is the area enclosed by the cycle.

The net work produced to the cycle is $W_{net}=Q_{net}$= Area 1-2-3-4-1.

According to the definition of a heat engine efficiency, the *efficiency of the Carnot heat engine* is $\eta_{Carnot}=W_{net\ output}/Q_{input}$= [Area 1-2-3-4-1]/[Area 2-3-5-6-2]= $(T_H-T_L)(S_4-S_1)/[T_H(S_4-S_1)]=(T_H-T_L)/T_H$.

Or $\eta_{Carnot}==1-T_L/T_H$ (5.5.1)

Example 5.5.1. Heat is transferred to a Carnot heat engine at a rate of 500 MW from a high-temperature source at 700 °C and rejects heat to a low-temperature sink at 40 °C. Determine the power produced and the efficiency of the Carnot heat engine.

To solve this problem by CyclePad, we take the following steps:

1. Build
   (A) Go to LIBRARY, select Rankine cycle unsolved.
   (B) Switch to analysis mode.
2. Analysis
   (A) Go to cycle, then cycle properties.
   (B) Input the given information: (a) Tmax is 700°C, (b) Tmin is 40°C, (c) rate of heat added is +500 MW.
   (C) After the Carnot cycle efficiency is posted, input eta-thermal=eta Carnot=67.82%.
3. Display results
   (A) The answer is η=67.82% and net-power=339,100 kW.

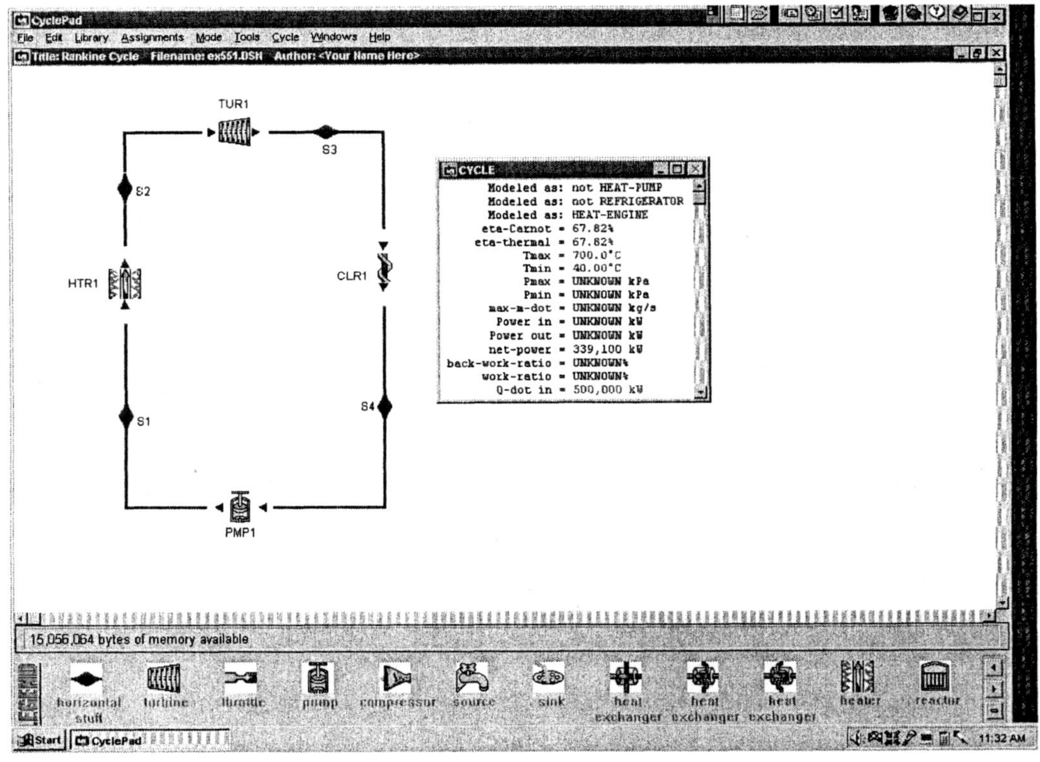

Figure Example 5.5.1. Carnot heat engine

Example 5.5.2. A Carnot heat engine operates between 1000°F and 100°F develops 5 hp. Determine the thermal efficiency and the rate of heat supplied.

If an inventor claims that he has developed a heat engine operating between the same temperature interval producing 5 hp and the rate of heat supplied to his engine is 10,000 Btu/h. Is it possible?

To solve this problem by CyclePad, we take the following steps:

1. Build
   (A) Go to LIBRARY, select Rankine cycle unsolved.
   (B) Switch to analysis mode.
2. Analysis
   (A) Go to cycle, then cycle result.
   (B) Input the given information: (a) Tmax is 1000°F, (b) Tmin is 100°F, (c) net power output is 5 hp.
   (C) After the Carnot cycle efficiency is posted, input eta-thermal=eta Carnot=61.66%.
3. Display results
   (A) The answer is rate of heat added for the Carnot cycle is 20,638 Btu/h. Since the Carnot cycle is the most efficient one among heat engines operating between the same two temperature reservoirs, the claim of having the rate of heat supplied to his engine is 10,000 Btu/h is impossible.

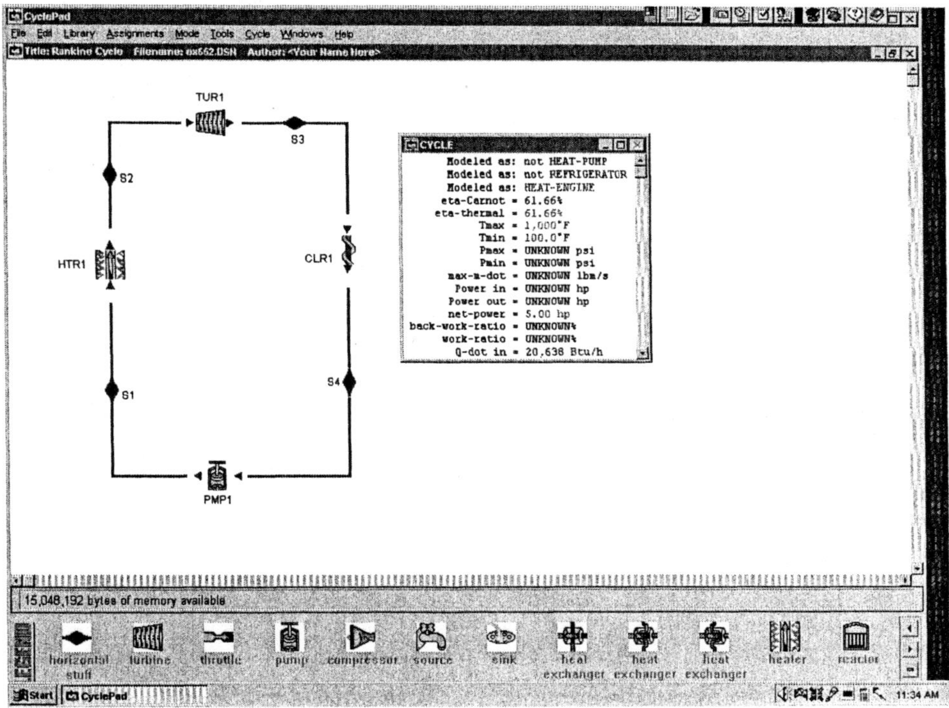

Figure Example 5.5.2. Carnot heat engine

Homework 5.5 Carnot cycle

Carnot heat engine

1. Does the Carnot heat engine efficiency depends on the working fluid used in the engine? Which working fluid used in the engine would be more efficient, air or water?
2. What kind of shape is the Carnot cycle illustrating on a T-s diagram?
3. What is the area enclosed by the cycle area of the Carnot cycle illustrating on a T-s diagram?

4. What is the area enclosed by the cycle area of the Carnot cycle illustrating on a p-V diagram?
5. Carnot heat engine A operates between 20°C and 520°C. Carnot heat engine B operates between 20°C and 820°C. Which Carnot heat engine is more efficient than the other.
6. What are the four processes of a Carnot heat engine?
7. A Carnot heat engine operates between a high temperature thermal reservoir at T_H and a low temperature thermal reservoir at T_L. What is the efficiency of the Carnot heat engine?
8. An inventor claims to have developed a heat engine with better efficiency than the Carnot heat engine efficiency when operating between the same two thermal reservoirs. Is it possible?
9. Consider the design of a power plant operating between a high-temperature reservoir at 1000°R and a low-temperature reservoir at 500°R. (a) What is the maximum possible efficiency of such a power plant? (b) for a production of 1,000,000 kW of power, what is the minimum rate of heat addition? (c)What is the maximum rate of heat rejection? (d) What is the rate of heat addition if the actual efficiency is 10 %?
10. A midshipman has invented an auto engine cycle that receives heat from combustion gases at 2500 R and rejects heat to ambient air at 500°R. He claims that for a steady fuel flow of 10 lbm/h, his engine can produce 60000 Btu/h. The heating value of the fuel is 20000 Btu/lbm. How do you evaluate his claim?
11. An inventor claims to have developed a heat engine when receives 800 kJ of heat from a source at 400K and produces 400 kJ of net work while rejecting the waste heat to a sink at 300 K. Is this a reasonable claim? Why? Show your justification in detail.

Carnot Heat Pump

If the Carnot cycle for a heat engine is carried out in the reverse direction, the result will be either a *Carnot heat pump* or a Carnot refrigerator. Such a cycle is shown in Figure 5.5.2. Using the same graphical explanation that was used in the Carnot heat engine, the heat added from the low temperature reservoir at T_L is area 1-4-5-6-1. Q_{41} is the amount of heat added to the Carnot cycle from a low temperature thermal reservoir.

The net heat added to the cycle is $Q_{net}=Q_{12}+Q_{23}+Q_{34}+Q_{41}=0+T_H(S_3-S_2)+0+T_L(S_1-S_4)=(T_H-T_L)(S_4-S_1)$=[Area 2-3-5-6-2]-[area 1-4-5-6-1]=Area 1-2-3-4-1. Notice that the area 1-2-3-4-1 is the area enclosed by the cycle.

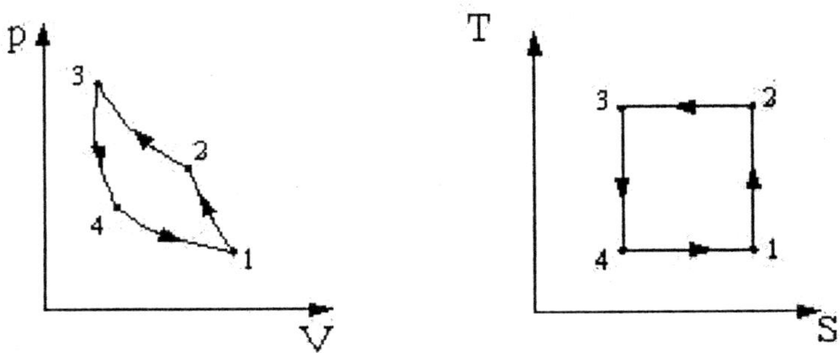

Figure 5.5.2 Carnot heat pump or Carnot refrigerator cycle on p-v and T-s diagram

The net work added to the cycle is $W_{net}=Q_{net}=$ Area 1-2-3-4-1. According to the COP (β) definition of a heat pump, the *COP of the Carnot refrigerator* is

$\beta_{Carnot,R}=Q_{desirable\ output}/W_{input}=$ [Area 4-1-5-6-4]/[Area 1-2-3-4-1]= $(T_L)(S_2-S_3)/[(T_H-T_L)(S_2-S_3)]=T_L/(T_H-T_L)$.

or

$$\beta_{Carnot,R}=1/(T_H/T_L-1) \tag{5.5.2}$$

The net work added to the cycle is $W_{net}=Q_{net}=$ Area 1-2-3-4-1. According to the COP (β) definition of a heat pump, the *COP of the Carnot heat pump* is

$\beta_{Carnot,HP}=Q_{desirable\ output}/W_{input}=$ [Area 2-3-5-6-2]/[Area 1-2-3-4-1]= $(T_H)(S_2-S_3)/[(T_H-T_L)(S_2-S_3)]=T_H/(T_H-T_L)$.

or

$$\beta_{Carnot,HP}=1/(1-T_L/T_H) \tag{5.5.3}$$

Referring to Figure 5.5.2, the system undergoes a Carnot heat pump or Carnot refrigerator cycle in the following manner:

(A) During process 1-2, the system is thermally insulated and the temperature of the working substance is raised from the low temperature T_L to the high temperature T_H.
(B) During process 2-3, heat is transferred isothermally from the working substance to the high temperature reservoir at T_H. This process is accomplished reversibly by bringing the system in contact with the high temperature reservoir whose temperature is equal to or infinitesimally lower than the working substance.
(C) During process 3-4, the system is thermally insulated and the temperature of the working substance is decreased from the high temperature T_H to the low temperature T_L.
(D) During process 4-1, heat is transferred isothermally to the working substance from the low temperature reservoir at T_L. This process is accomplished reversibly by bringing the system in contact with the low temperature reservoir whose temperature is equal to or infinitesimally higher than the working substance.

Example 5.5.2. An inventor claims to have developed a refrigerator that removes heat from a region at 270 K and transfer it to a thermal reservoir at 540 K while maintaining a COP of 8.5. Is this claim reasonable?

Solution:

To solve this problem, we take the following steps:

1. From Eq. (5.5.2), we have $\beta_{Carnot,R}=1/(T_H/T_L-1)=1/(540/270-1)=1$.
2. Since the Carnot refrigerator has the maximum COP, the claim of 8.5 is not valid.

Homework Carnot Heat Pump and Carnot Refrigerator

1. What is the relationship between the COP of a Carnot refrigerator and the COP of a Carnot heat pump when the two cycles are operating between the same two thermal reservoirs?
2. A reversible heat pump and a Carnot heat pump operate between the same two thermal reservoirs. Which heat pump has higher COP?
3. An irreversible heat pump and a Carnot heat pump operate between the same two thermal reservoirs. Which heat pump has higher COP?
4. A heat pump is used to heat a system and maintain it at 300 K. On a winter day when the outdoor air temperature is 270 K, the system is estimated to lose heat at a rate of 5,000 kW. Determine the minimum power required to operate this heat pump. What would be the COP at this condition.
5. Determine the maximum possible COP for a heat pump operating between an ambient temperature of -23°C and the interior of a house at 27°C.
6. A Carnot refrigerator operates in a room in which the temperature is 27°C. The refrigerator consumes 500 W of power when operating, and has a COP of 4. Determine (a) the rate of heat removal from the refrigerated space, and (b) the temperature of the refrigerated space.
7. An air conditioning system is used to maintain a house at 60°F when the outside air temperature is 100°F. If heat enters the house from the outside at a rate of 5,000 Btu/h, Find the COP of the Carnot refrigerator, and the minimum possible power (in Btu/h) required for the air-conditioner.
8. A heat pump is used to maintain a house at 60°F when the outside air temperature is 40°F. If heat leaves the house at a rate of 10,000 Btu/h, what is the minimum amount of power (in Btu/h) required to run the heat pump?
9. A Carnot power cycle using carbon mono-oxide as a working fluid has a thermal efficiency of 40 percent. At the beginning of the isothermal heating process, the temperature is 500 K. Determine (a) the temperature of the low temperature thermal reservoir, and (b) the COP (coefficient of performance) for the refrigerator obtained by reversing the power cycle.

5.6 CARNOT COROLLARIES

Six corollaries deduced from the Carnot cycle are of great use in comparing the performance of cycles. The corollaries are:

1. The efficiency of the Carnot heat engine operating between a fixed high-temperature heat source thermal reservoir at T_H and a fixed low-temperature heat sink thermal reservoir at T_L is irrespective of the working substance.

2. No heat engine operating between a fixed high-temperature heat source thermal reservoir and a fixed low-temperature heat sink thermal reservoir can be more efficient than a Carnot heat engine operating between the same two thermal reservoirs.
3. All reversible heat engines operating between a fixed high-temperature heat source thermal reservoir and a fixed low-temperature heat sink thermal reservoir have the same efficiency.
4. The COP (coefficient of performance) of the Carnot heat pump (or refrigerator) operating between a fixed high-temperature thermal reservoir at T_H and a fixed low-temperature thermal reservoir at T_L is irrespective of the working substance.
5. No heat pump (or refrigerator) operating between a fixed high-temperature thermal reservoir and a fixed low-temperature thermal reservoir can have higher COP (coefficient of performance) than a Carnot heat pump (or refrigerator) operating between the same two thermal reservoirs.
6. All reversible heat pump (or refrigerator) operating between a fixed high-temperature thermal reservoir and a fixed low-temperature thermal reservoir have the same COP (coefficient of performance).

These corollaries can be proved by demonstrating that the violation of any of the corollary results in the violation of the second law of thermodynamics.

5.7 The Thermodynamic Temperature Scale

A temperature scale that is independent of the properties of the substance that are used to measure temperature is called a thermodynamic temperature scale. With the aid of the first Carnot corollary, W.T. Kelvin devised such a temperature scale.

It was shown in previous sections that the efficiency of a Carnot heat engine operating between a fixed high-temperature heat source thermal reservoir at T_H and a fixed low-temperature heat sink thermal reservoir at T_L is a function of the temperatures of the thermal reservoirs only. The function was found by Kelvin as

$$\eta=1-T_L/T_H \tag{5.7.1}$$

This thermodynamic temperature scale is called the *Kelvin temperature* scale. On this scale, the temperature varies from zero to infinity.

Similarily, for a Carnot heat pump and a Carnot refrigerator operating between a fixed high-temperature thermal reservoir at T_H and a fixed low-temperature thermal reservoir at T_L, the COP (coefficient of performance) of a Carnot heat pump and a Carnot refrigerator are

$$\beta_R=T_L/(T_H-T_L) \tag{5.7.2}$$

$$\beta_{HP}=T_H/(T_H-T_L) \tag{5.7.2}$$

The three expressions give the maximum values for any cycle operating between two thermal reservoirs and can be used as standards for comparsion for actual cycles.

5.8 SUMMARY

The second law of thermodynamics is needed to give the direction of a process. A process will not occur unless both the first law of thermodynamics and the second law of thermodynamics are satisfied.

A thermal reservoir is a huge system which can absorb or reject energy without changing its temperature.

A heat engine is a continuous cyclic device which produces output work by adding input heat. The efficiency of a heat engine is $\eta=W_{output}/Q_{input}$.

A heat pump is a continuous cyclic device which pumps output heat to a high temperature reservoir from a low temperature reservoir by adding input work. The coefficient of performance (COP) of a heat pump is $\beta_{HP}=Q_{output}/W_{input}$.

A refrigerator is a continuous cyclic device which removes output heat from a low temperature reservoir to a high temperature reservoir by adding input work. The coefficient of performance (COP) of a refrigerator is $\beta_R=Q_{output}/W_{input}$.

A heat engine, heat pump and refrigerator all are required to have at least two thermal reservoirs with different temperature.

The statements of the second law of thermodynamics state that no heat engine can be 100% efficient, and that no heat pump nor refrigerator can have an infinity COP.

A reversible process is one that at the conclusion of the process, both the system and its surroundings may be restored to their initial states without producing any changes in the rest of the universe. Otherwise, it is an irreversible process.

The Carnot cycle, that is composed of four reversible processes, is the most efficient cycle operating between two fixed thermal reservoirs with different temperature. All reversible heat engines operating between two fixed thermal reservoirs with different temperature have the same efficient as the Carnot cycle efficiency.

The Carnot heat engine efficiency is $\eta_{Carenot}=1-T_L/T_H$.
The Carnot heat pump COP is $\beta_{Carenot}=1/(1-T_L/T_H)$.
The Carnot refrigerator COP is $\beta_{Carenot}=1/(T_H/T_L-1)$.

Chapter 6

ENTROPY

OBJECTIVES

After reading and studying the material in this chapter, you should be able to:

1. Know the relationship between entropy and heat.
2. Know the area under a process on a T-s diagram, and the area enclosed by a cycle.
3. Understand why T-s diagram is important in thermodynamics.
4. Understand that entropy is a microscopic property.
5. State the Second law of thermodynamics equation for a closed system.
6. State the Second law of thermodynamics equation for a steady flow open system.
7. Understand isentropic process.
8. Define isentropic efficiency for turbine, pump, and compressor.
9. Find the entropy change of both closed and open system using CyclePad.

6.1 CLAUSIUS INEQUALITY

When a system is carried through a complete cycle, the integral of $(\delta Q/T)$ around the cycle is less than or equal to zero. The statement is the *Clausius inequality*.

Applying Carnot engine efficiency to a reversible heat engine gives

$$\eta_R = (1\text{-}Q_L/Q_H)_R = 1\text{-}T_L/T_H \qquad (6.1.1)$$

Hence, $Q_L/T_L\text{-}Q_H/T_H = 0$ for a reversible cycle. Extending to integrals of infinitesimals, we have

$$\int_{Cycle} (\delta Q/T)_R = 0 \qquad (6.1.2)$$

For an irreversible cycle between the same T_L and T_H, chapter 5 states that

$$\eta_I < \eta_R, \qquad (6.1.3)$$

or

$$(1\text{-}Q_L/Q_H)_I < (1\text{-}T_L/T_H) \tag{6.1.4}$$

Hence, $Q_L/T_L\text{-}Q_H/T_H < 0$ for an irreversible cycle. Similarly, extending to integrals of infinitesimals, we have

$$\int_{Cycle} (\delta Q/T)_I < 0 \tag{6.1.5}$$

Homework 6.1 Clausius Inequality

1. A proposed steam power plant has the following data:
   heat added to the boiler (at 200°C)=2600 kJ/kg
   heat rejected from the condenser (at 50°C)=2263 kJ/kg
   adiabatic pump with 80% adiabatic efficiency
   adiabatic turbine with 90% adiabatic efficiency
   Does this plant violate the Clausius theorem? Show your detail justification.

6.2 Entropy and Heat

A property of a system is some calculable or measurable quantity of that system whose value depends only on the state of the system. If we reexamine the operation of a reversible heat engine, we shall find that the mathematical term $(\delta Q/T)_{reversible}$ behaves just this way:

$$\int_{cycle} (\delta Q/T)_{reversible}=0. \tag{6.2.1}$$

This property $(\delta Q/T)_{reversible}$, called *entropy* and denoted by S was discovered in 1862 by Rudolf Clausius who named it entropy. Entropy is a microscopic property associated with the microscopic energy transfer, Q. Entropy is similar to volume which is a macroscopic property associated with the macroscopic energy transfer, W. Phenomenologically, entropy is related to the molecular disorder of a system associated with the microscopic energy transfer, Q.

On a microscopic point of view, entropy can be considered as a measure of randomness or disorder. Consider two different inert gases that are initially separated from each other. If the partition separating the two gases is removed, the gases will quickly mix without outside influence (even if the gases are at the same temperature and pressure). Although the net energy content of this mixture has not changed, the entropy of the mixture is greater than the total entropy of the individual gases before mixing. The Second law indicates that the reverse process (i.e. the gases separating without outside influence) will never occur as this be a state of "less disorder" and lower entropy.

Let us consider another simple example. Water in solid phase is called ice and has a very regular, repeating arrangement of molecules. This arrangement is called a crystalline structure. We can examine an ice crystal and determine the location of each of the molecules. If heat is added, the ice is melted into liquid water. The molecular pattern is no longer regular, and our ability to state exactly the location of individual molecules is gone. The liquid has a more random structure than the ice, thus it has a higher entropy. If more heat is added, the liquid water is boiled and becomes the vapor phase called steam. This causes the molecules to move even faster and over greater distances, so we have even less of a chance to predict an individual molecule's location. Thus the steam has an even greater level of entropy than either the liquid water or the ice.

With these two examples, it is seen that both irreversibility within a system and heat transfer added to the system increase the entropy of the system. For a reversible process, the entropy change is caused by heat transfer added to the system only.

While energy is conserved and hence can not be used up or exhausted, the usefulness of a quantity of energy does decrease through usage. Energy is degraded through usage, eventually to the point of zero usefulness. Entropy is a negative measure associated with energy usefulness. An increase in entropy corresponds to a decrease in energy usefulness. Hence, entropy is a measure of dis-usefulness.

It is rather difficult to explain a microscopic property with a macroscopic approach. To fully understand the idea of entropy, one should learn it from microscopic thermodynamics or called statistical thermodynamics.

It can be proved that the quantity $\int(\delta Q/T)_{reversible}$ are identical for two different processes (1-A-2 and 1-B-2) with identical end states 1 and 2 as shown in Figure 6.2.1, i.e.

$$\int_{process\ 1\text{–}A\text{-}2} (\delta Q/T)_{reversible} = \int_{process\ 1\text{–}B\text{-}2} (\delta Q/T)_{reversible} \qquad (6.2.2)$$

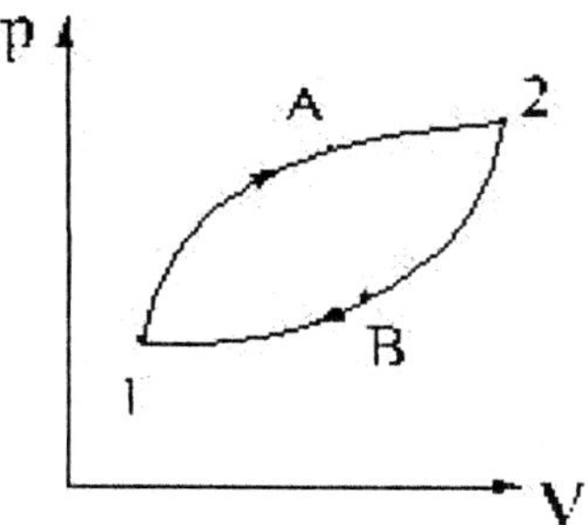

Figure 6.2.1 p-v diagram

Referring to the p-v diagram as shown in Figure 6.2.1, the two processes 1-A-2 and 1-B-2 are arbitrarily chosen. Therefore $\int_{process} (\delta Q/T)_{reversible}$ is independent of path and related to states 1 and 2 only. In order words, it is a property. The name entropy and the symbol S are given to this property. Restate Equations (6.2.1) and (6.2.2), we have

$$\int_{cycle} (\delta Q/T)_{reversible} = \int_{cycle}(dS) = 0. \qquad (6.2.3)$$

and

$$\int_{process\ 1\text{–}A\text{-}2} (\delta Q/T)_{reversible} = \int_{process\ 1\text{–}B\text{-}2} (\delta Q/T)_{reversible} = S_2 - S_1 \qquad (6.2.4)$$

The quantity $(S_2 - S_1)$ represents the entropy change of a system from state 1 to state 2. Note that the equality in the above equations holds for reversible cycles and processes only.

Equation (6.2.3) is the macroscopic definition of entropy. Note that entropy is defined only for reversible processes. and a change in entropy may be calculated with

$$\Delta S=S_2 - S_1=\int_{\text{process 1-2}} (\delta Q/T)_{\text{reversible}} \tag{6.2.5}$$

or, in differential form

$$dS= (\delta Q/T)_{\text{reversible}} \tag{6.2.6}$$

or,

$$(\delta Q)_{\text{reversible}} =Tds \tag{6.2.7}$$

Homework 6.2 Entropy and Heat

1. An insulated piston-cylinder device initially contains 10 ft^3 of air at 20 psia and 60°F. Air is now heated for 10 minutes by a 200 Btu/min electric heater placed inside the cylinder. The pressure of air is maintained constant during this process. Determine the entropy change of air.
2. Find the change in specific entropy as air is heated from 560 to 1260 R while pressure drops from 50 to 40 psia.

6.3 Heat and Work As Areas

Since $(\delta Q)_{\text{reversible}}$ =Tds, temperature and entropy are the natural thermodynamic coordinates for reversible heat. One of the great advantages of a T-s diagram is that the reversible heat transfer of a process may be presented by the area underneath the process as shown in the following T-s diagram.

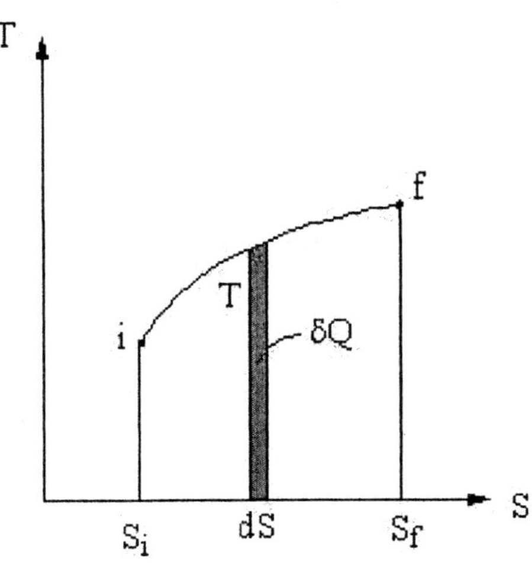

Figure 6.3.1 T-s diagram

6.4 CARNOT CYCLES

The temperature and entropy (T-s) diagram of the Carnot reversible heat engine cycle is shown in Figure 6.4.1. The Carnot cycle which operates between a high temperature (T_H) thermal reservoir and a low temperature (T_L) thermal reservoir is composed of the following four reversible processes:

1-2 isentropic compression
2-3 isothermal heating
3-4 isentropic expansion
4-1 isothermal cooling

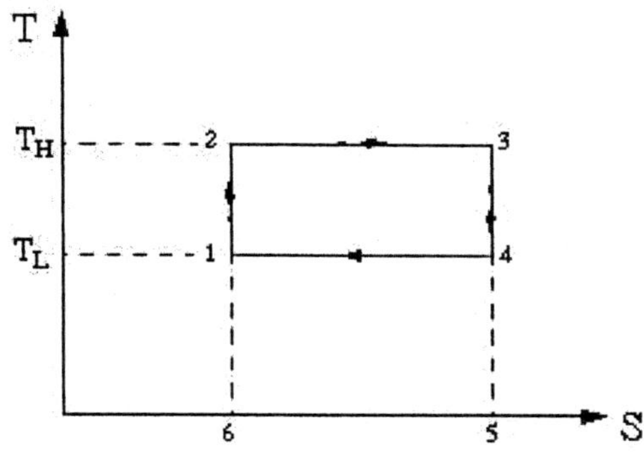

Figure 6.4.1 Carnot T-s diagram

Notice that since process 1-2 and process 3-4 are both isentropic, they appear as vertical lines on the diagram. The cycle has a rectangular shape regardless of the working fluid used in the heat engine. Since the heat transfer for each reversible process ie represented by the area under the process curve on the T-S diagram. Therefore, we have

$$Q_{12} = 0 \tag{6.4.1}$$

$$Q_{23} = Q_H = T_H (S_3 - S_2) \tag{6.4.2}$$

$$Q_{34} = 0 \tag{6.4.3}$$

and

$$Q_{41} = Q_L = T_L (S_1 - S_4) \tag{6.4.4}$$

The ratio of the heat transfer quantities can be written as

$$Q_H/Q_L = -T_H/T_L \tag{6.4.5}$$

The cycle net work (W_{net}), which is equal to the cycle net heat transfer (Q_{net}) can be written as

$$W_{net} = Q_{net} = Q_H + Q_L \tag{6.4.6}$$

The cycle net work (W_{net}) is the area enclosed by the cycle.

The Carnot cycle efficiency (η_{Carnot}) is therefore

$$\eta_{Carnot} = W_{net}/Q_H = 1 - T_L/T_H \tag{6.4.7}$$

The Carnot cycle efficiency (η_{Carnot}) is obtained directly using the proper thermodynamic T-S diagram. Again, it demonstrates that Equation (6.4.7) is valid regardless of the working fluid employed in the Carnot reversible heat engine cycle.

6.5 SECOND LAW OF THERMODYNAMICS FOR CLOSED SYSTEMS

It is difficult to explain entropy which is a microscopic property with a macroscopic point of view. A macroscopic analogy of entropy is herewith given in the following and hopefully will aid the understanding of entropy to the readers.

A person walks down from an initial position on a constant slope mountain path. Each step downward is equivalent to one foot downward in elevation. A position is a state. Elevation is a property, because it has a unique value at a certain position (state). A path is a process which is a change of positions. Step is a measurement quantity. Consider the following two cases:

Case A: The path is a firm path. If the person walks down 10 steps, his elevation changes is 10 feet. It will take the person walk back 10 steps to get back to his initial position without any surrounding help. This process is therefore a reversible process. The elevation change (10 feet) is equal to the measurement of steps (10 steps).

Case B: It snows and the path is covered by a layer of snow and becomes slippery. If the person walks down on the slippery path 10 steps, his elevation changes is going to be more than 10 feet. It will take the person walk back more than 10 steps to get back to his initial position without any surrounding help. This process is therefore an irreversible process. The elevation change (more than 10 feet) is larger than the measurement of steps (10 steps).

The conclusion of both cases is therefore

- Change of elevation is either equal or larger than the measurement steps which is corresponding to
- Change of property is either equal or larger than the measurement quantity or, thermodynamic specifically, equivalent to
- Change of entropy is either equal or larger than the measurement ($\delta Q/T$), i.e.

$$dS = (\delta Q/T)_{rev} \text{ or } dS > (\delta Q/T)_{rev} \quad (6.5.1)$$

Where T is the absolute temperature at the boundary of the system where the differential amount of heat (δQ) is transferred between the system and its surroundings, the inequality holds for an irreversible process and the equality holds for a reversible process.

The inequality in equation (6.5.1) reminds us that the entropy change of a closed system during an irreversible process is always greater than the entropy transfer. Therefore, some entropy must be generated during an irreversible process. The *entropy generation* ($S_{generation}$) during the process is created by irreversibilities. Equation (6.) can be rewritten as

$$\Delta S = S_2 - S_1 = \int_{process\ 1\text{-}2} (\delta Q/T) + S_{generation} \quad (6.5.2)$$

Where $S_{generation}$ is a non-negative term which depends on the irreversibilities during the process 1-2 and therefore is not a property of the system.

$S_{generation} > 0$ for irreversible processes, and $S_{generation} = 0$ for reversible processes.

Equation (6.) is the mathematical form of the *second law of thermodynamics for a closed system*. The second law of thermodynamics states that the change in entropy of a closed system is greater than or equal to the sum of the heat transfers divided by the corresponding absolute temperatures of the boundary.

We may reach the following conclusions regarding the entropy change of a closed system.

1. For an adiabatic (Q=0) closed system, the entropy will increase due to internal irreversibilities. The entropy is constant (S=constant, or $\Delta S=0$) only during a reversible adiabatic process.
2. A reversible adiabatic process is called an isentropic or a constant entropy process. Notice that an isentropic implies the process is adiabatic, but an adiabatic process is not always an isentropic process.
3. For an isolated system, which has no interactions with its surroundings including mass and heat transfers, the entropy will keep increasing ($\Delta S>0$) due to internal irreversibility activities within the system.
4. The entropy value of an isolated system reaches its maximum value ($S_{maximum}$) when there is no further internal irreversibility activity within the system.
5. A system may have several parts. The entropy value of an isolated system will no longer change once the system reaches its ultimate equilibrium state (equilibrium among all parts).
6. The universe is an isolated system. The entropy of the universe will keep increasing due to internal irreversibilities until the universe is dead.

Homework 6.5 Entropy and Second Law

1. An inventor claims to have developed an adiabatic device that executes a steady state expansion process in which the entropy of the surroundings decreases at 5 kJ/(Ksec). Is this possible? Why or why not?
2. Air is compressed in a piston-cylinder set up from 516.3 °R and 10 ft^3/lbm (state 1) to 1350 R, 500 psia and 1 ft^3/lbm (state 2). Heat in then added to the air in a constant volume process 2-3 until the pressure reaches 900 psia. More heat is added to the air in a constant pressure process 3-4 until the temperature reaches 5000°R.
Determine: (a) the entropy change of the air in the adiabatic process, (s_2-s_1) Btu/lbm(°R), (b) Is the process adiabatic? (c) the amount of work added to the air in process 1-2, w_{12}, (d) the entropy change of the air in the constant volume process, (s_3-s_2) in Btu/lbm(°R), (e) the entropy change of the air in the constant pressure process, (s_4-s_3) in Btu/lbm(°R), (f) the entropy change of the air from state 1 to state 4, (s_4-s_1) in Btu/lbm(°R), (g) the amount of work added to the air in process 2-3, w_{23} in Btu/lbm, (h) the amount of heat added to the air in process 2-3, q_{23} in Btu/lbm, and (i) the amount of heat added to the air in process 3-4, q_{34} in Btu/lbm.

6.6 Second Law of Thermodynamics for Open Systems

An open system permits mass and energy interactions between the system and its surroundings. The mass transfer carries the property entropy into and out of the system. Therefore the change in entropy within the open system is modified by the mass interaction. The statement of the *second law of thermodynamics for open systems* is:

The change in entropy within the open system minus the net entropy transported into the open system with the mass flow is greater than or equal to the sum of the heat transfer divided by the corresponding absolute temperatures.

$$(S_2 - S_1) - [\,(mdot)_i\,(s_i) - (mdot)_e\,(s_e)](\Delta t) = [\textstyle\int_{process\ 1-2} (\delta Qdot/T) + (Sdot)_{generation}](\Delta t) \qquad (6.6.1)$$

Where $(S_2 - S_1)$ is the change in entropy within the open system from time t_1 to t_2, $(mdot)_i\,(s_i)$ is the rate of entropy flow in with the mass at inlet section of the open system during time t_1 to t_2, $(mdot)_e\,(s_e)$ is the rate of entropy flow out with the mass at exit section of the open system during time t_1 to t_2, $[\,(mdot)_i\,(s_i) - (mdot)_e\,(s_e)](\Delta t)$ is the net entropy transported into the open system with the mass flow during time t_1 to t_2, $\int_{process\ 1-2} (\delta Qdot/T)(\Delta t)$ is the contribution to the entropy change due to the sum of the heat transfer divided by the corresponding absolute temperatures during time t_1 to t_2, $[(Sdot)_{generation}](\Delta t)$ is a non-negative contribution to the entropy change term which depends on the irreversibilities during time t_1 to t_2.

For steady state and steady flow, Equation (6.6.1) is reduced to

$$(mdot)_e\,(s_e) - (mdot)_i\,(s_i) = \textstyle\int_{process\ 1-2} (\delta Qdot/T) + (Sdot)_{generation} \qquad (6.6.2)$$

For no heat transfer across the boundary surface, Equation (6.6.2) is reduced to

$$s_e - s_i = (s)_{generation} \qquad (6.6.3)$$

Homework 6. 6 Second Law of Thermodynamics for Open Systems

1. Can $(s)_{generation}$ ever be negative?
2. The second law of thermodynamics for open systems is
   $(S_2 - S_1)$ -[$(mdot)_i$ (s_i) - $(mdot)_e$ (s_e)](Δt) =[$\int_{process\ 1-2}$ $(\delta Qdot/T)$ + $(Sdot)_{generation}$](Δt)
   What is the physical meaning of each term?
3. Is entropy change of an open system always non-negative?
4. Is entropy change of an adiabatic open system always non-negative?

6.7 Property Relationships

6.7.1 Pure Substance

Entropy has been defined as a property of a system, and the specific entropy (s) is tabulated in the conventional tables of thermodynamics. It is listed the same way as internal energy (u) and enthalpy (h). Again, the values listed on the tables are not absolute entropy values because we are interested only in the entropy changes.

Example 6.7.1. 1. A piston-cylinder device contains 1 kg of saturated R-134a vapor at -5°C. An amount work is added to compress the vapor until the pressure and temperature are 1 Mpa and 50°C. Determine the amount of work added, amount of heat added, and entropy change of the refrigerant during this processs.

To solve this problem by CyclePad, we take the following steps:

1. Build
   (A) Take a begin, a compression device, and an end from the closed-system inventory shop and connect them.
   (B) Switch to analysis mode.
2. Analysis
   (A) Input the given information: (a) working fluid is R-134a, (b) the initial R-134a mass, quality and temperature of the process are 1 kg, 1 and -5°C, and (c) the final pressure and temperature of the process are 1 Mpa and 50°C.
3. Display results
   (A) Display the compression device results. The answers are W=-27.94 kJ, Q=-13.86 kJ and ΔS=1.75-1.73=0.02 kJ/K.

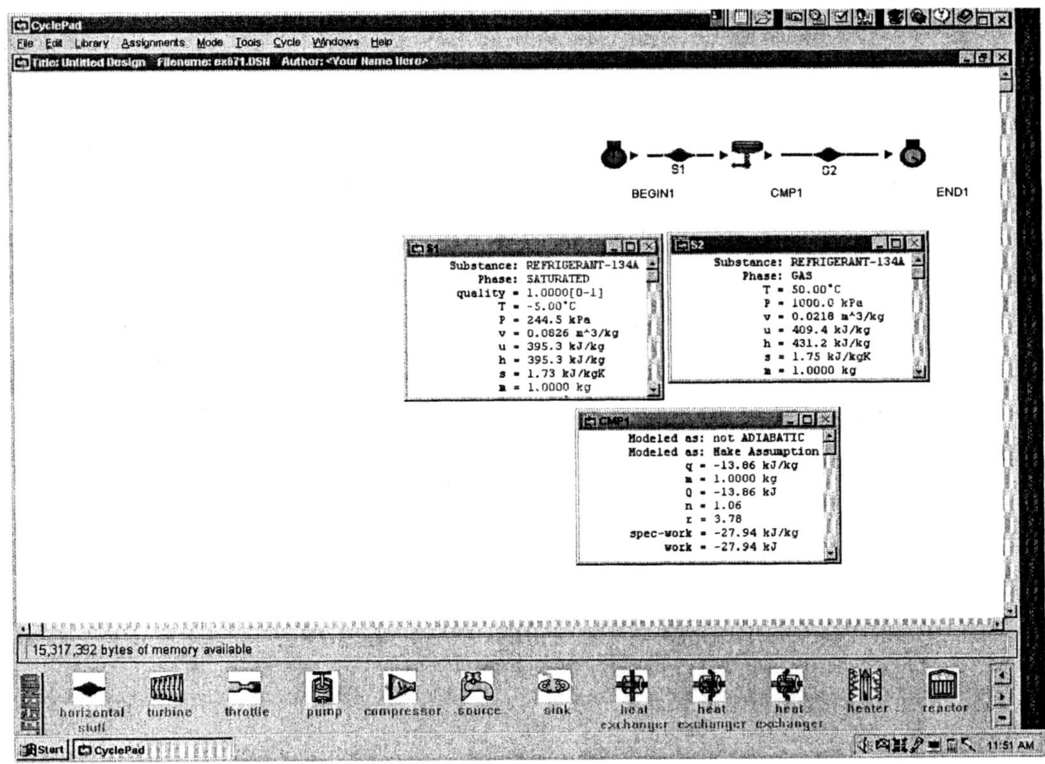

Figure Example 6.7.1.1. Entropy change

Example 6.7.1.2. A rigid tank contains 0.2 kg of water vapor at 100 kPa and 170°C. An amount of heat is added to heat the vapor until the pressure is 400 kPa. Determine the final temperature and entropy change of the water vapor, and the amount of work and heat added during this process.

To solve this problem by CyclePad, we take the following steps:

1. Build
   (A) Take a begin, a heating device, and an end from the closed-system inventory shop and connect them.
   (B) Switch to analysis mode.
2. Analysis
   (A) Input the given information: (a) working fluid is water, (b) the initial mass, pressure and temperature of the process are 0.2 kg, 100 kPa and 170°C, (c) the final pressure of the process is 400 kpa, (d) the process is constant volume (isochoric), and (e) W=0 kJ.
3. Display results
   (A) Display the heating device results. The answers are T=1487 °C ΔS=0.2(10.18-7.70)=0.496 kJ/K, W=0 kJ, and Q=498.5 kJ.

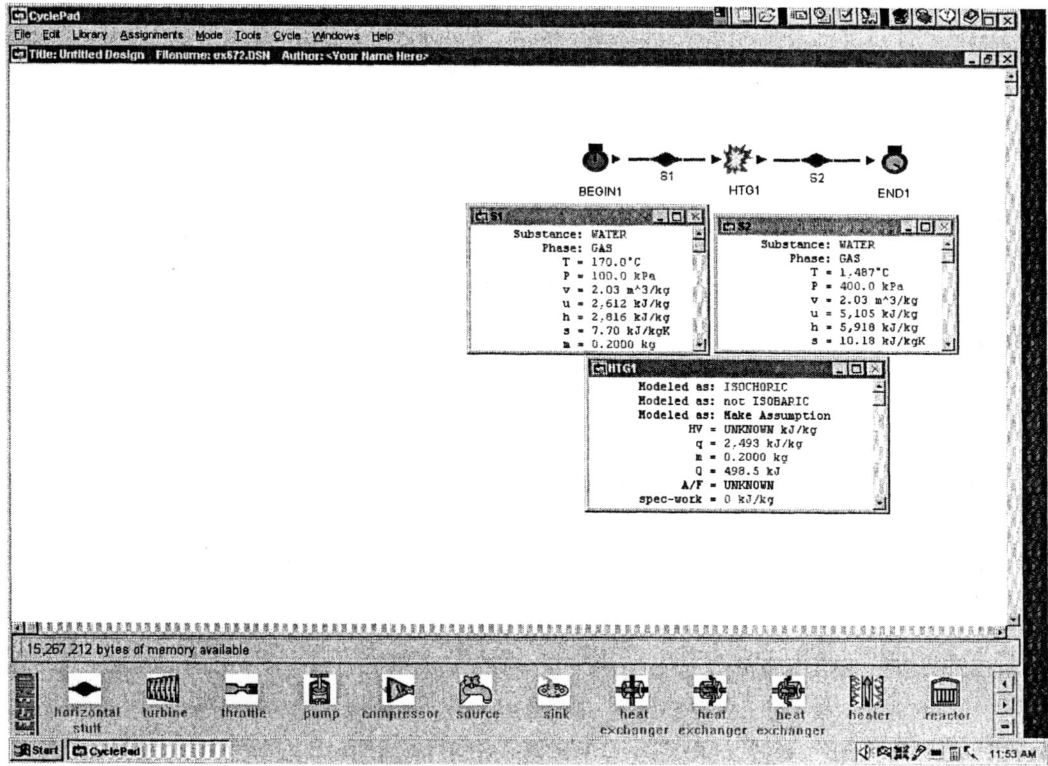

Figure Example 6.7.1.2.Entropy change

Example 6.7.1.3. Steam at a mass flow rate of 1 kg/s enters an adiabatic turbine at 4000 kPa and 500°C and leaves at 10 kPa and a quality of 0.9. Determine the power produced by the turbine and the rate of entropy change of the steam.

To solve this problem by CyclePad, we take the following steps:

1. Build
   (A) Take a source, a turbine, and a sink from the open-system inventory shop and connect them.
   (B) Switch to analysis mode.
2. Analysis
   (A) Assume the turbine is adiabatic.
   (B) Input the given information: (a) working fluid is water, (b) the inlet pressure and temperature of the turbine are 4000 kPa and 500°C, (c) the exit pressure and quality of the turbine are 10 kPa and 0.9, and (d) the mass flow rate is 1 kg/s.
3. Display results
   (A) Display the turbine results. The answers are Wdot=1101 kW and ΔSdot=1(0.3081)=0.3081 kJ/K.

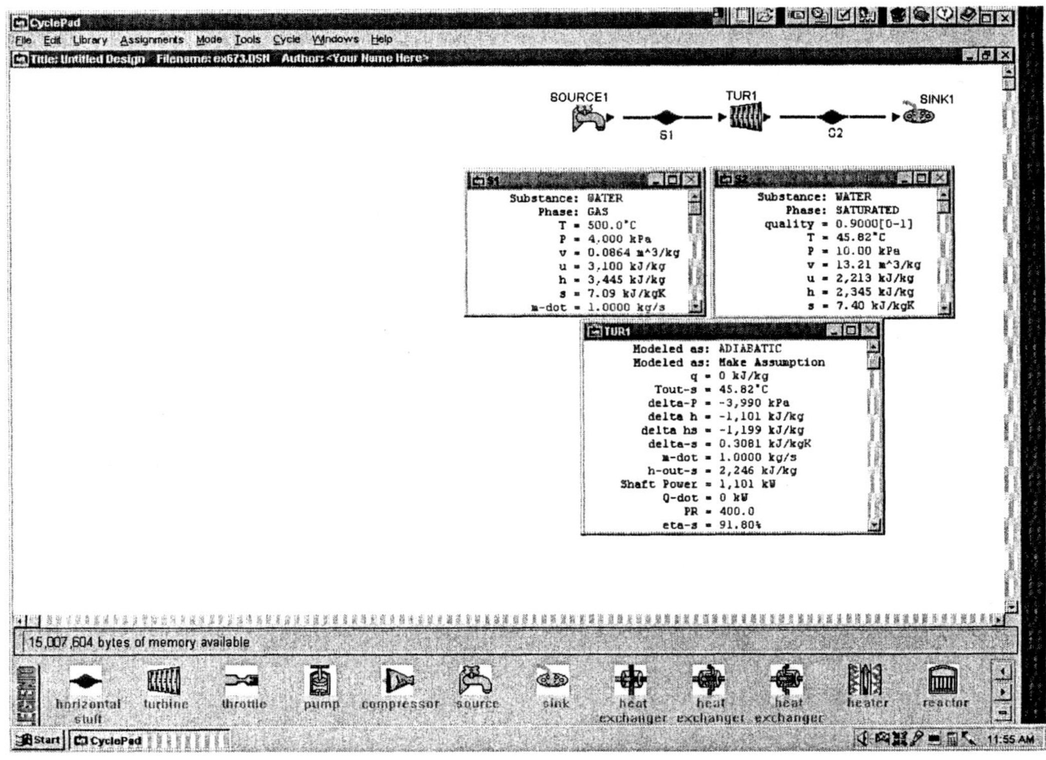

Figure Example 6.7.1.3 Entropy relationship

Example 6.7.1.4. Determine the rate of entropy change and the pump power input required to an adiabatic pump for a mass flow rate of 0.7 kg/s saturated water from 100 kPa to 2 Mpa and 101 °C.

To solve this problem by CyclePad, we take the following steps:

1. Build
   (A) Take a source, a pump, and a sink from the open-system inventory shop and connect them.
   (B) Switch to analysis mode.
2. Analysis
   (A) Assume the pump is adiabatic.
   (B) Input the given information: (a) working fluid is water, (b) the mass rate flow, quality and pressure at the inlet of the pump are 0.7 kg/s, 0 and 100 kPa, and (c) the exit pressure and temperature of the pump are 2 Mpa and 101 °C.
3. Display results
   (A) Display the pump result. The answers are ΔSdot=0.7(0.0139)=0.00973 kJ/[K(s)], Wdot=-5.04 kW.

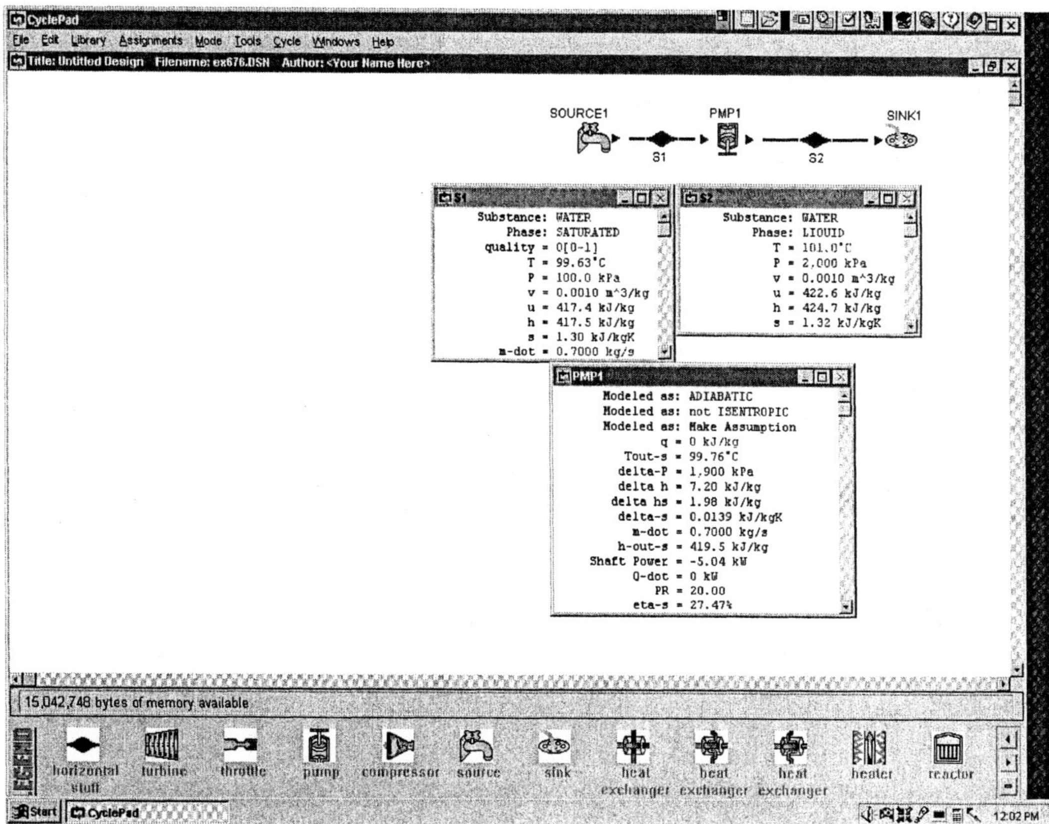

Figure Example 6.7.1.4. Entropy

6.7.2 Ideal Gas

The entropy change of an ideal gas during a process 1-2 are given by the following equations:

$$S_2 - S_1 = \int_{\text{process } 1\text{-}2} [C_v\,(dT)]/T + \int_{\text{process } 1\text{-}2} [p(dV)]/T \qquad (6.\ 7.2.1)$$

and

$$S_2 - S_1 = \int_{\text{process } 1\text{-}2} [C_p\,(dT)]/T - \int_{\text{process } 1\text{-}2} [V(dp)]/T \qquad (6.7.2.2)$$

Example 6.7.2.1. 0.25 kg of air is compressed from an initial state of 100 kPa and 19°C to a final state of 600 kPa and 60°C. Determine the heat added, work added, and the entropy change of the air during this compression process.

To solve this problem by CyclePad, we take the following steps:

1. Build
   (A) Take a begin, a compression device, and an end from the closed-system inventory shop and connect them.
   (B) Switch to analysis mode.
2. Analysis
   (A) Assume the compression is an adiabatic process.
   (B) Input the given information: (a) working fluid is air, (b) the initial air mass, pressure and temperature of the process are 0.25 kg, 100 kPa and 19°C, (c) the final pressure and temperature of the process are 600 kPa and 60°C.
3. Display results
   (A) Display the compression device result. The answers are Q=-29.81 kJ, W=-37.15 kJ and ΔS=0.25(2.01-2.40)=-0.0795 kJ/K.

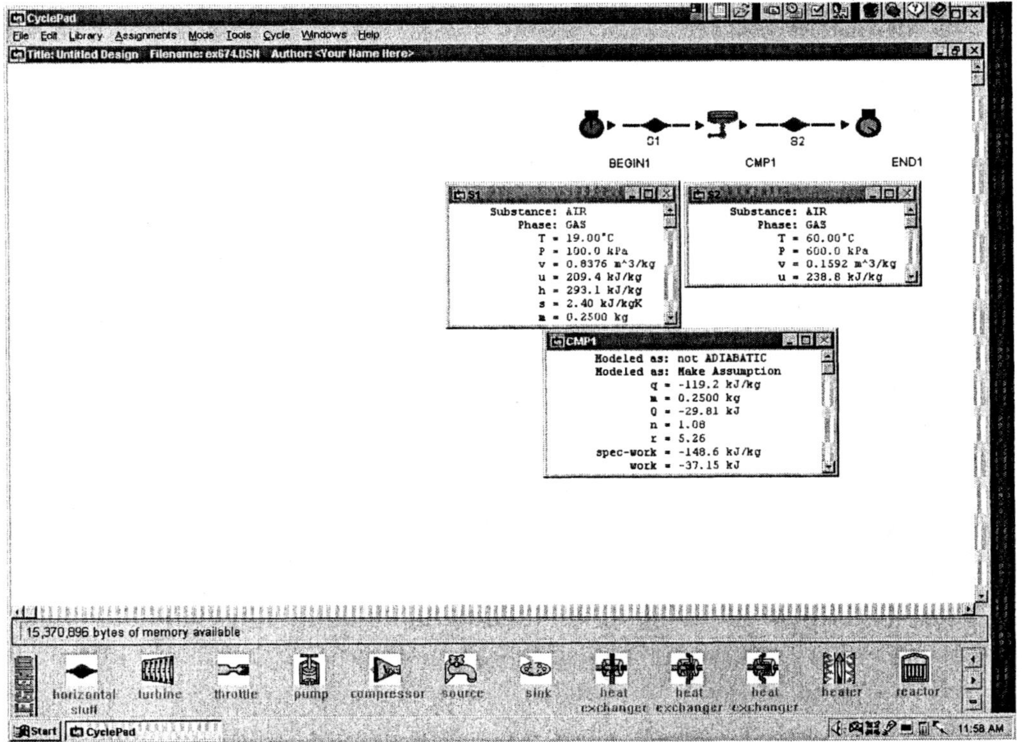

Figure Example 6.7.2.1 Entropy

Example 6.7.2.2. 5 ft^3/s of air is compressed in an adiabatic compressor from an initial state of 14.7 psia and 60°F to 40.5 psia and 350°F. Determine the power required, mass rate flow, and the rate of the entropy change of the air during this compression process.

To solve this problem by CyclePad, we take the following steps:

1. Build
   (A) Take a source, a compressor, and a sink from the open-system inventory shop and connect them.

(B) Switch to analysis mode.

2. Analysis
   (A) Assume the compressor is adiabatic.
   (B) Input the given information: (a) working fluid is air, (b) the flow rate, pressure and temperature at the inlet of the compressor turbine are 5 ft^3/s, 14.7 psia and 60°F, and (c) the exit pressure and temperature of the compressor is 40.5 psia and 350°F.
3. Display results
   (A) Display the exit state and the compressor results. The answers are Wdot=-37.59 hp, mdot=0.3823 lbm/s, and ΔSdot=0.3823(0.0369)=0.01411 Btu/[s(°R)].

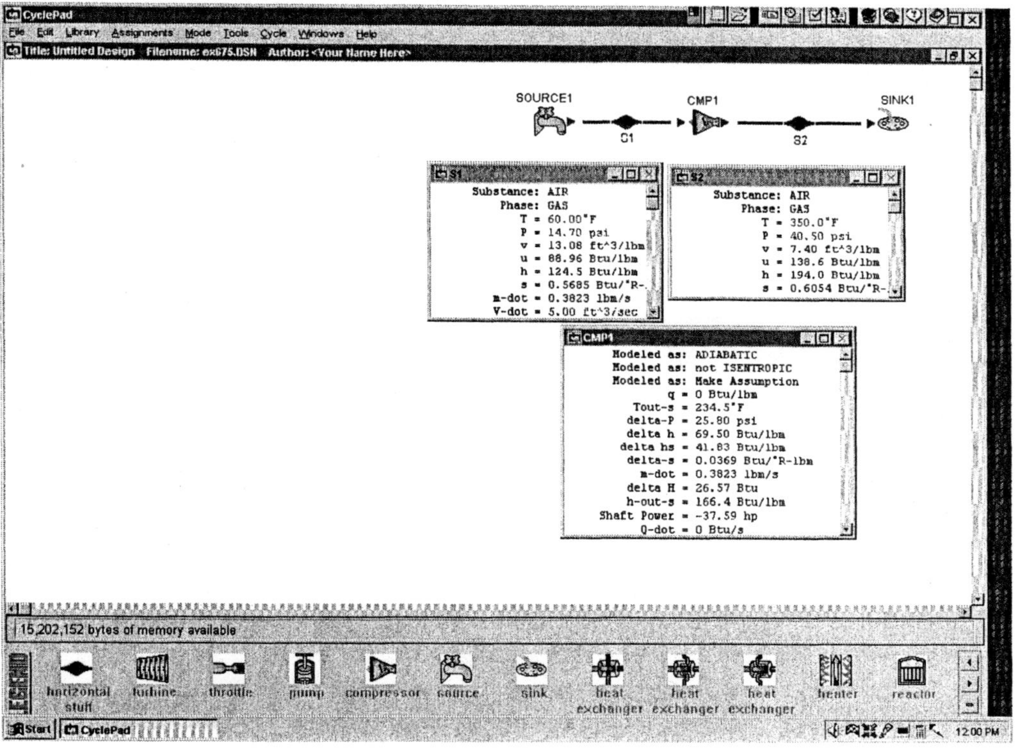

Figure Example 6.7.2.2. Entropy

6.7.3 Incompressible Liquid and Solid

The entropy change of an incompressible liquid or a solid during a process 1-2 are given by the following equations:

$$S_2 - S_1 = \int_{\text{process } 1\text{-}2} [C(dT)]/T \qquad (6.\ 7.3.1)$$

The temperature change of an incompressible liquid or a solid during an isentropic process 1-2 is zero.

Example 6.7.3.1. Determine the rate of entropy change and the pump power input required to an adiabatic pump for a mass flow rate of 0.7 kg/s saturated water from 100 kPa to 2 Mpa and 101 °C.

To solve this problem by CyclePad, we take the following steps:

1. Build
   (A) Take a source, a pump, and a sink from the open-system inventory shop and connect them.
   (B) Switch to analysis mode.
2. Analysis
   (A) Assume the pump is adiabatic.
   (B) Input the given information: (a) working fluid is water, (b) the mass rate flow, quality and pressure at the inlet of the pump are 0.7 kg/s, 0 and 100 kPa, and (c) the exit pressure and temperature of the pump are 2 Mpa and 101 °C.
3. Display results
   (A) Display the pump result. The answers are ΔSdot=0.7(0.0139)=0.00973 kJ/[K(s)], Wdot=-5.04 kW.

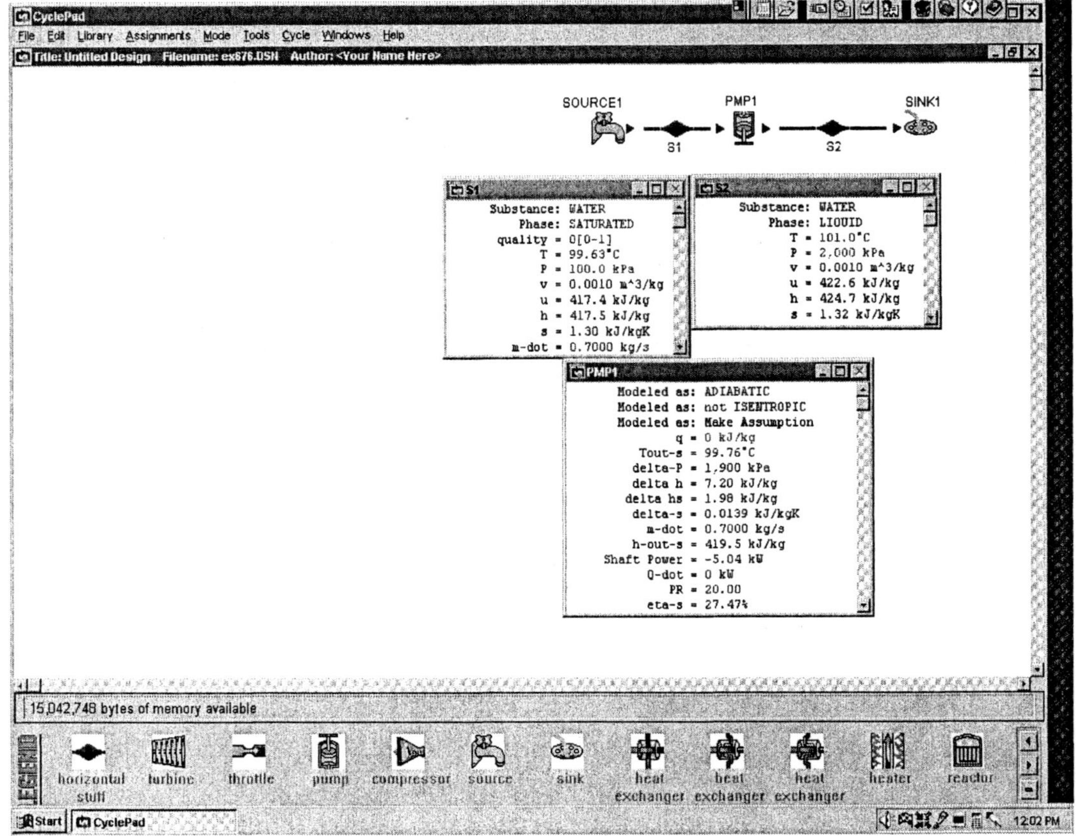

Figure Example 6.7.3.1. Entropy

Homework 6.7 Property Relationships

1. 2 kg/s of water at 100 kPa and 90°C are mixed with 3 kg/s of water at 100 kPa and 10°C to form 5 kg/s of water at 100 kPa in an adiabatic mixing chamber. Find the water temperature of the mixed 5 kg/s stream, and rate change of entropy due to the mixing process.
2. A mixing chamber has two inlets and one outlet. The following information is known:
   inlet one: liquid water at 10 Mpa, 260°C and 1.8 kg/s
   inlet two: steam in at 10 Mpa and 500°C
   outlet: saturated mixture at 10 Mpa with quality of 0.5.
   Find: (a) the mass flow rates for inlet two and for the outlet, and (b) the rate of entropy generation during the process.
3. A rigid container holds 5 lb_m of R134a. Initially, the refrigerant is 72% vapor by mass at 70°F. Heat is added until the R134a. temperature inside the container reaches 220°F. Determine: (a) the final pressure in the container, and (b) total entropy change of the R134a.
4. A rigid tank contains 1.2 kg of air at 350 K and 100 kPa. Heat is added to the air until the pressure reaches 120 kPa. Find (a) the heat added to the gas, (b) the final temperature of the gas, (c) the entropy generation for the air.
5. Steam initially at 500 kPa and 553 K is contained in a cylinder fitted with a frictionless piston. The initial volume of the steam is 0.057 m^3. The steam undergoes an isothermal compression process until it reaches a specific volume of 0.3 m^3/kg. Determine the total entropy change of the steam, and the magnitude and direction of the heat transfer and work.
6. A piston-cylinder device contains 1.2 kg of carbon dioxide at 120 kPa and 27°C. The gas is compressed slowly in a polytropic process during which $p_1(v_1)^{1.3}=p_2(v_2)^{1.3}$=constant. The process ends when the specific volume reaches 0.3 m^3/kg. Determine (a) the final volume of the gas, (b) final temperature of the gas, (c) entropy change of the gas. Also determine (d) the work added to the gas, and (e) the heat transfer during the process.
7. A piston-cylinder device contains 3 lbm of refrigerant-22 at 120 psia and 120°F. The refrigerant is cooled at constant pressure until it exists as a compressed liquid at 90 F. If the temperature of the surroundings is 70°F, determine (a) the work added to the refrigerant and (b) heat transfer removed from the system. Also determine (c) the total entropy change of the system.
8. A piston-cylinder device contains 3 lbm of refrigerant-12 at 120 psia and 120°F. The refrigerant is cooled at constant pressure until it exists as a compressed liquid at 90 F. If the temperature of the surroundings is 70°F, determine (a) the work added to the refrigerant and (b) heat transfer removed from the system. Also determine (c) the total entropy change of the system.
9. 2 kg of methane at 500 kPa and 100°C (state 1) undergoes a reversible polytropic expansion to a final (state 2) temperature and pressure of 20°C and 100 kPa. Find the work done, heat transferred, entropy change of the methane.
10. Air at 50 kPa and 400 K has 50 kJ of heat added to it in an internally-reversible isothermal process. Calculate the total entropy change of the air.

11. Helium at 50 kPa and 400 K has 50 kJ of heat added to it in an internally-reversible isothermal process. Calculate the total entropy change of the helium.
12. Helium is contained in a closed system initially at 135 kPa, 1 m^3 and 30°C and undergoes an isobaric process until the volume is doubled. Determine the mass, the final temperature and entropy change of the helium. Find the work done and heat transferred during this process.
13. An engineer in an industrial plant instruments a steady-flow mixing chamber and reports the following measurements: inlet steam at 200 kPa and 150°C, 1.5 kg/s; inlet water at saturated liquid, 15°C, 0.8 kg/s; exit mixture at 400 kPa and 2.3 kg/s; work added is 0; and heat transfer to environment is 0. Find the pressure of the inlet 15°C saturated liquid water, the rate of total entropy change, and the rate of entropy generation. It is known that an adiabatic mixing process is an irreversible process. Determine whether it is possible for these data to be correct.
14. Another engineer in the industrial plant instruments a steady-flow mixing chamber and reports the following measurements: inlet steam at 200 kPa and 150°C, 1.5 kg/s; inlet water at saturated liquid, 15°C, 0.8 kg/s; exit mixture at 100 kPa and 2.3 kg/s; work added is 0; and heat transfer to environment is 0. Find the pressure of the inlet 15°C saturated liquid water, the rate of total entropy change, and the rate of entropy generation. It is known that an adiabatic mixing process is an irreversible process. Determine whether it is possible for these data to be correct.

6.8 Isentropic Processes

An isentropic process is a constant entropy process. A constant entropy process implies a reversible and adiabatic process. It serves as an idealization of an adiabatic process for particular applications.

Example 6.8.1. 1 m^3 of air is compressed in an isentropic process from an initial state of 300 K, and 101 kPa to a final temperature of 870 K. Determine the pressure and specific volume of the air at the final state, and the work required.

To solve this problem by CyclePad, we take the following steps:

1. Build
   (A) Take a begin, a compression device, and an end from the closed-system inventory shop and connect them.
   (B) Switch to analysis mode.
2. Analysis
   (A) Assume the compression is an isentropic process.
   (B) Input the given information: (a) working fluid is air, (b) the initial air volume, pressure and temperature of the process are 1 m^3, 101 kPa and 300 K, (c) the final temperature of the process is 870 K.
3. Display results
   (A) Display the final state result. The answers is V=0.0595 m^3/kg, T=4195 kPa, and W=-479.8 kJ.

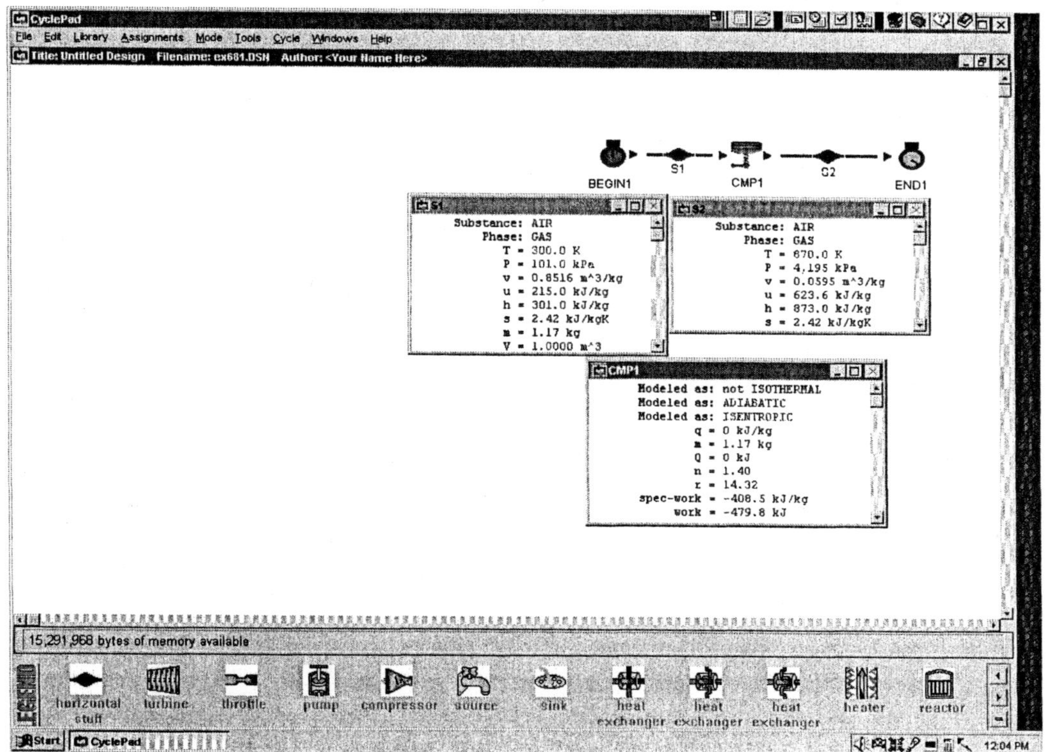

Figure Example 6.8.1. Entropy

Example 6.8.2. Steam at 1.5 Mpa and 400°C enters an adiabatic turbine and exhausts at 20 kPa. The mass rate flow of the steam is 0.24 kg/s. Determine the maximum power produced, exhaust temperature and quality of the steam at the maximum power condition.

To solve this problem by CyclePad, we take the following steps:

1. Build
   (A) Take a source, a turbine, and a sink from the open-system inventory shop and connect them.
   (B) Switch to analysis mode.
2. Analysis
   (A) Assume the turbine is adiabatic and isentropic.
   (B) Input the given information: (a) working fluid is steam, (b) the pressure and temperature at the inlet of the turbine are 1.5 Mpa and 400°C,(c) the exit pressure of the turbine is 20 kPa, and (d) the mass rate flow of the steam is 0.24 kg/s.
3. Display results
   (A) Display the turbine exit state results. The answers are Wdot=204.1 kW, T=60.07°C and x=0.9136.

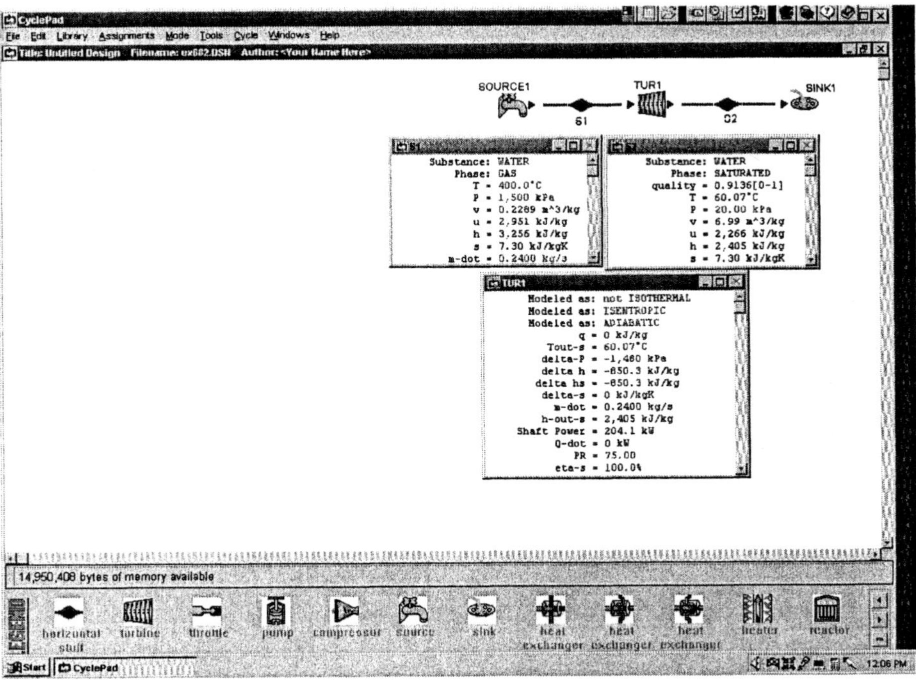

Figure Example 6.8.2. Entropy

Homework 6.8 Isentropic Processes

1. Air undergoes a change of state isentropically from (1500°R and 75 psia) to 16.5 psia. (a) Sketch the process on a T-s diagram. (b) Determine the temperature of the air at the final state. (c) Determine the entropy change due to change of state. (d) Determine the heat added and work added to the air. And (e) Determine the specific volume of air at the final state.
2. Air is compressed in a piston-cylinder device from 100 kPa and 17°C to 800 kPa in an isentropic process. Determine the final temperature and the work added per unit mass during this process.
3. 15 kg/s of steam enters an isentropic steam turbine steadily at 15 Mpa and 600°C and leaves at 10 kPa. Find the temperature and quality of the steam at exit, specific entropy change of the steam, and the shaft power produced by the turbine.
4. 15 kg/s of steam enters an adiabatic steam turbine with 87% efficiency steadily at 15 Mpa and 600°C and leaves at 10 kPa. Find the temperature and quality of the steam at exit, specific entropy change of the steam, and the shaft power produced by the turbine.
5. Water at 20 kPa and 38°C enters an isentropic pump and leaves at 16 Mpa. The water flow rate through the pump is 15 kg/s. Determine the temperature of the water at the exit, specific entropy change of the water, and the power required to drive the pump.
6. Water at 20 kPa and 38°C enters an adiabatic pump and leaves at 16 Mpa. The water flow rate through the pump is 15 kg/s. The adiabatic efficiency of the pump is 85%. Determine the temperature of the water at the exit, specific entropy change of the water, and the power required to drive the pump.

7. 0.2 kg/s of air (combustion gas) enters an isentropic gas turbine at 1200 K, 800kPa, and leaves at 400 kPa. Determine the temperature of the air at the exit, specific entropy change of the the air, and the power output of the turbine.
8. 0.2 kg/s of air (combustion gas) enters an adiabatic gas turbine at 1200 K, 800kPa, and leaves at 400 kPa. The gas turbine has an adiabatic efficiency of 86%. Determine the temperature of the air at the exit, specific entropy change of the the air, and the power output of the turbine.

6.9 Isentropic Efficiency

The *isentropic efficiency* is used to compare the actual adiabatic process to the isentropic process for many devices. The reduction of the irreversibilities of a device process is desired to increase the efficiency of the process. Therefore the device isentropic efficiency is the ratio of the actual adiabatic performance to the isentropic performance. The isentropic device is taken as the standard comparison of actual adiabatic operation.

6.9.1 Turbine Isentropic Efficiency

The isentropic efficiency ($\eta_{turbine}$) of a turbine is a comparison of the work produced by an actual adiabatic turbine (w_a) with the work produced by an isentropic turbine (w_s).

$$\eta_{turbine} = w_a/w_s \tag{6.9.1.1}$$

The inlet to the turbine is at a specified state, and the exit is at a specified pressure.

Example 6.9.1.1. An adiabatic turbine is designed to produce 10 Mw. It receives steam at 1 Mpa and 300°C and leaves the turbine at 15 kPa. Determine the minimum steam mass rate of flow required, the exit temperature and quality of steam.

To solve this problem by CyclePad, we take the following steps:

1. Build
   (A) Take a source, a turbine, and a sink from the open-system inventory shop and connect them.
   (B) Switch to analysis mode.
2. Analysis
   (A) Assume the turbine is adiabatic and isentropic.
   (B) Input the given information: (a) working fluid is steam, (b) the mass rate flow, pressure and temperature at the inlet of the turbine are 2 kg/s, 1 Mpa and 300°C, (c) the outlet pressure of the turbine is 15 kPa, and (d) the power output of the turbine is 1,000 kW.
3. Display results
   (A) Display the turbine results. The answers are mdot=13.48 kg/s, T=53.98°C and x=0.8780.

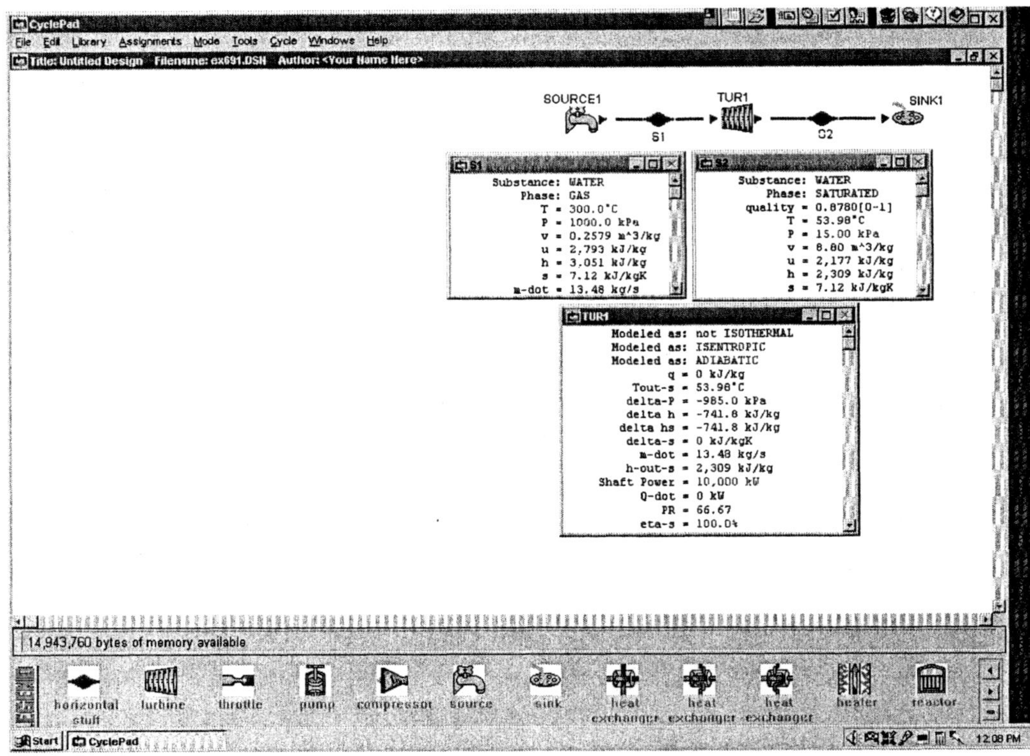

Figure Example 6.9.1. Turbine efficiency

Example 6.9.1.2. Air enters an adiabatic turbine with a mass flow rate of 0.5 kg/s at 1000 kPa and 1300°C and leaves the turbine at 100 kPa. The turbine efficiency is 88%. Determine the temperature at the exit and the power output of the turbine.

To solve this problem by CyclePad, we take the following steps:

1. Build
   (A) Take a source, a turbine, and a sink from the open-system inventory shop and connect them.
   (B) Switch to analysis mode.
2. Analysis
   (A) Assume the turbine is adiabatic.
   (B) Input the given information: (a) working fluid is air, (b) the mass rate flow, pressure and temperature at the inlet of the turbine are 0.5 kg/s, 1000 kPa and 1300°C, (c) the outlet pressure of the turbine is 100 kPa, and (d) the efficiency of the turbine is 0.88.
3. Display results
   (A) Display the inlet state and the turbine results. The answers are T=632.7°C, Wdot=334.8 kW.

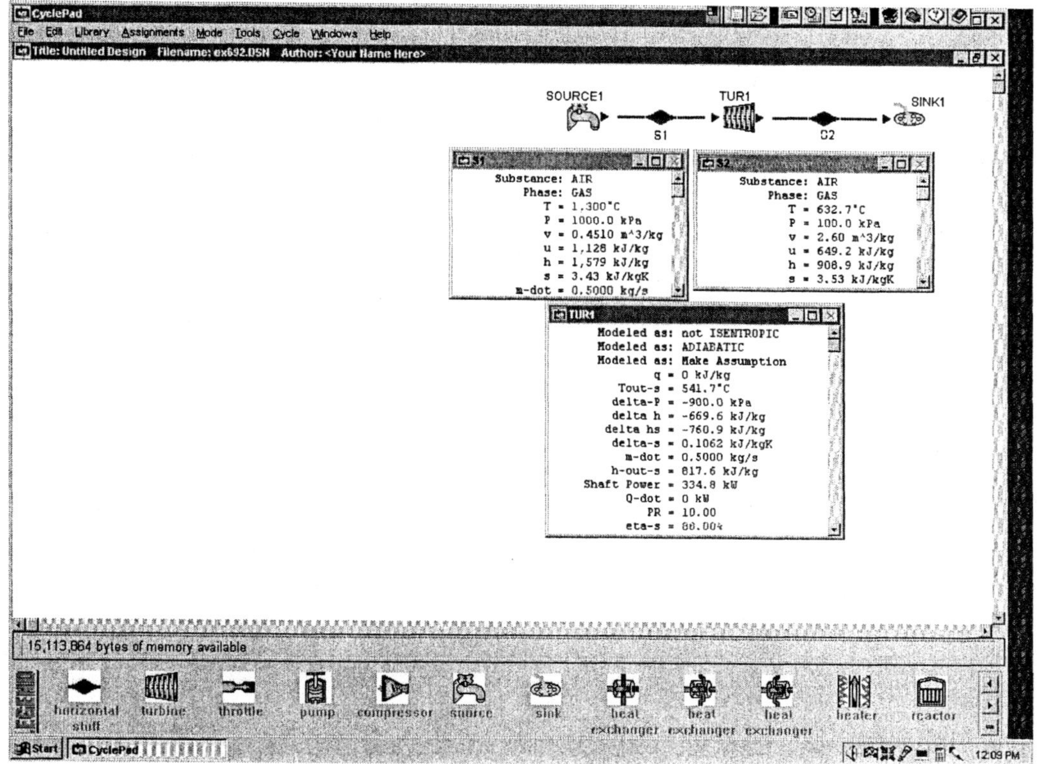

Figure Example 6.9.1.2. Turbine efficiency

6.9.2 Compressor Isentropic Efficiency

The isentropic efficiency ($\eta_{compressor}$) of a compressor is a comparison of the work required by an actual adiabatic compressor (w_a) with the work required by an isentropic turbine (w_s).

$$\eta_{compressor} = w_s/w_a \tag{6.9.2.1}$$

The inlet to the compressor is at a specified state, and the exit is at a specified pressure.

Example 6.9.2.1. 0.1 kg/s of air enters an adiabatic compressor at 100 kPa and 20°C and leaves the compressor at 800 kPa. The compressor efficiency is 87%. Determine the air exit temperature and the power required for the compressor.

To solve this problem by CyclePad, we take the following steps:

1. Build
   (A) Take a source, a compressor, and a sink from the open-system inventory shop and connect them.
   (B) Switch to analysis mode.

2. Analysis
   (A) Assume the compressor is adiabatic.
   (B) Input the given information: (a) working fluid is air, (b) the mass rate flow, pressure and temperature at the inlet of the compressor turbine are 0.1 kg/s, 100 kPa and 20°C, (c) the outlet pressure of the compressor is 800 kPa, and (d) the efficiency of the compressor turbine is 0.87.
3. Display results
   (A) Display the exit state and the compressor results. The answers are Wdot=-27.44 kW and T=293.4°C.

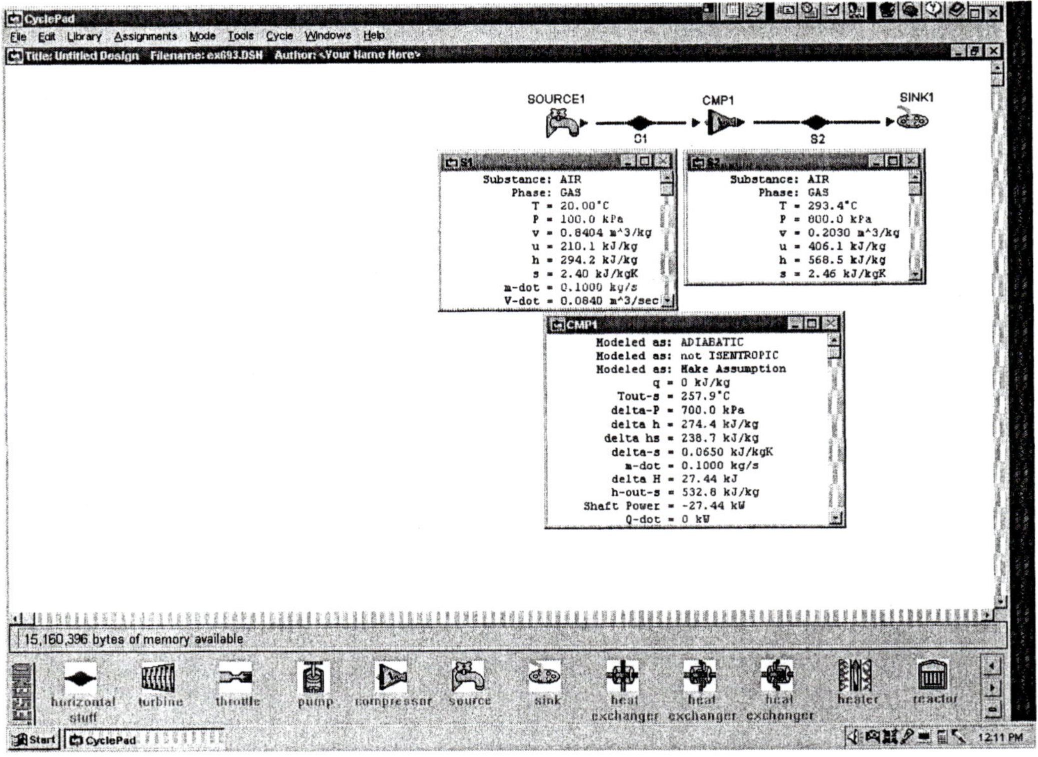

Figure Example 6.9.2.1. Compressor efficiency

Example 6.9.2.2. 0.01 kg/s of refrigerant R-12 enters an adiabatic compressor at 125 kPa and -10°C and leaves the compressor at 1.2 Mpa and 100°C. Determine the compressor efficiency and the power required for the compressor.

To solve this problem by CyclePad, we take the following steps:

1. Build
   (A) Take a source, a compressor, and a sink from the open-system inventory shop and connect them.
   (B) Switch to analysis mode.
2. Analysis
   (A) Assume the compressor is adiabatic.

(B) Input the given information: (a) working fluid is R-12, (b) the mass rate flow, pressure and temperature at the inlet of the compressor turbine are 0.01 kg/s, 125 kPa and -10°C, and (c) the exit pressure and temperature of the compressor are 1.2 Mpa and 100°C.

3. Display results

(A) Display the compressor results. The answers are Wdot=-0.6036 kW and η=72.58 %.

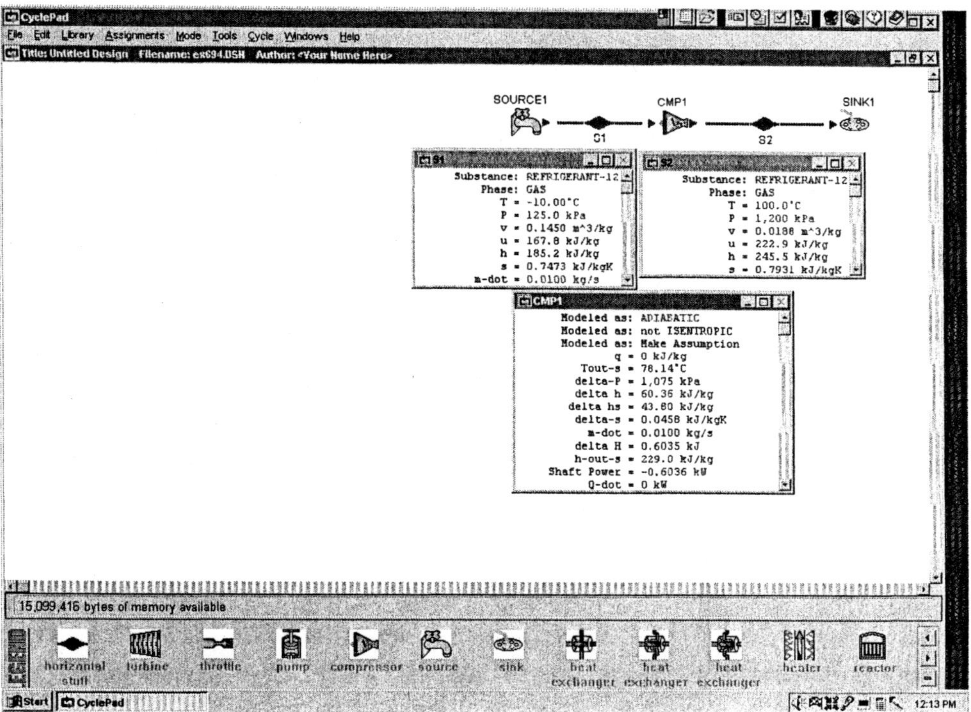

Figure Example 6.9.2.2. Compressor efficiency

6.9.3 Pump Isentropic Efficiency

The isentropic efficiency (η_{pump}) of a pump is a comparison of the work required by an actual adiabatic pump (w_a) with the work required by an isentropic pump (w_s).

$$\eta_{pump} = w_s/w_a \tag{6.9.3.1}$$

Example 6.9.3.1. 0.4 kg/s of saturated liquid water enters an adiabatic pump at 10 kPa and leaves the pump at 4 MPa. The pump efficiency is 83%. Determine the pump power required.

To solve this problem by CyclePad, we take the following steps:

1. Build

(A) Take a source, a pump, and a sink from the open-system inventory shop and connect them.

(B) Switch to analysis mode.

2. Analysis
   (A) Assume the pump is adiabatic.
   (B) Input the given information: (a) working fluid is water, (b) the mass rate flow and pressure at the inlet of the pump are 0.4 kg/s and 10 kPa, (c) the outlet pressure of the pump is 4 MPa, and (d) the efficiency of the pump compressor turbine is 0.83.
3. Display results
   (A) Display the pump result. The answers is Wdot=-1.96 kW.

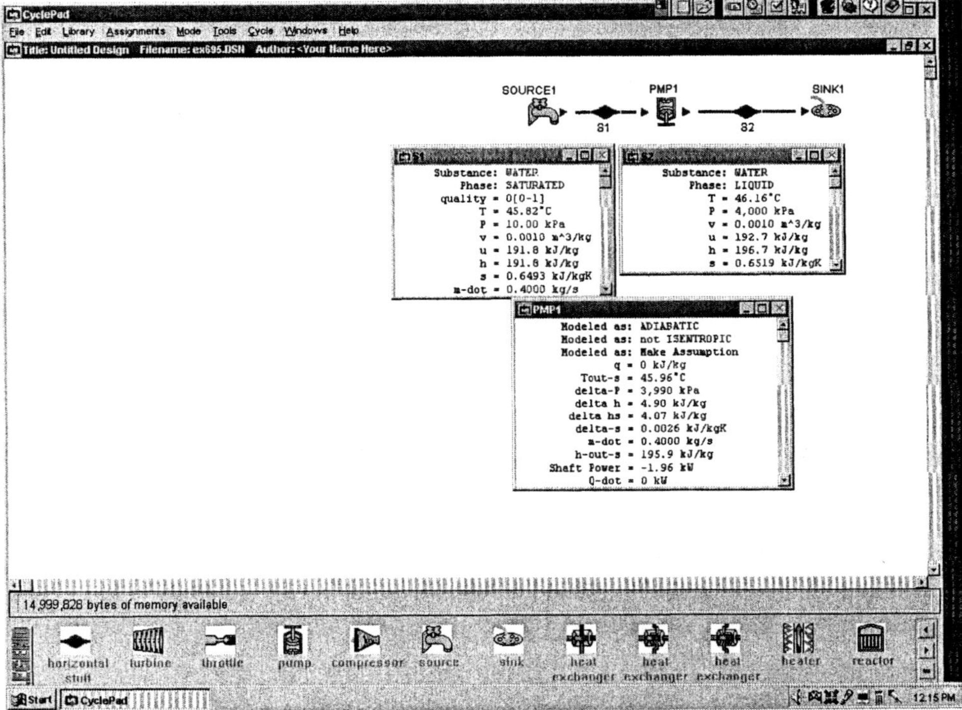

Figure Example 6.9.3.1. Pump efficiency

Homework 6.9 Isentropic Efficiency

1. Helium enters a steady-flow, adiabatic turbine at 1300 K and 2 Mpa and exhausts at 350 kPa. The turbine produces 1200 kW of power when the mass flow rate of helium is 0.4 kg/s. Determine the exit temperature of the helium, the specific entropy change of the helium and the efficiency of the turbine.
2. Steam is expanded from (1000 psia, 1000°F) in an adiabatic turbine to 1 psia. The efficiency of the turbine is 90 %. Determine (a) the actual specific work done by the turbine, (b) the mass flow rate required to produce an output of 100 MW, (c) the specific entropy generated by this process, and (d) the exhaust steam quality.
3. Steam enters an adiabatic turbine at 1000 psia and 900°F with a mass flow rate of 60 lbm/s and leaves at 5 psia. The adiabatic efficiency of the turbine is 90%. Determine the temperature of the steam at the turbine exit and the power output of the turbine in MW.

4. 0.23 kg/s of combustion air enters an adiabatic gas turbine with 86 % efficiency at 1200 K, 800 kPa, and leaves at 400 kPa. Determine (a) the power output of the turbine, and (b) the entropy change of the air.
5. Steam enters an adiabatic turbine at 1250 psia and 800°F and leaves at 5 psia with a quality of 90 %. Determine the mass flow rate required for a power output of 1 MW.
6. Combustion gases enter an adiabatic gas turbine at 1440°F and 120 psia and leaves at 68.5 psia. The power output from the turbine is 85 hp.
   (a) Find the isentropic efficiency of the turbine if the in-line compressor has a mass flow rate of 60 lbm/min, and (b) Find the entropy change across the turbine (Btu/lbm-°R).
7. Combustion gases enter an adiabatic gas turbine at 1440°F and 120 psia and leaves at 68.5 psia. The turbine has an isentropic efficiency of 85%.
   (a) Find the 'actual' power output of the turbine (horsepower) if the in-line compressor has a mass flow rate of 60 lbm/min, and (b) Find the entropy change across the turbine (Btu/lbm-°R).
8. Air at 800 kPa and 1200 K enters an actual adiabatic turbine at steady state. Pressure at 300.1 kPa is measured at the exit of the turbine. The turbine is known to have an isentropic efficiency of 85%. determine: (a) the actual temperature at the exit of the turbine, (b) the work output of the turbine in kJ/kg, and (c) the specific entropy generated by the turbine in kJ/[kg(K)].
9. Determine the minimum power input required to drive a steady-flow, adiabatic air compressor that compresses 2 kg/s of air from 105 kPa and 23°C to a pressure of 300 kPa.
10. A steady-flow, isentropic air compressor that compresses air at a rate of 6 kg/s from an inlet state of 100 kPa and 310 K to a discharge pressure of 350 kPa. Find the power required to drive the compressor.
11. An inventor claims that a steady-flow, adiabatic air compressor that compresses air at a rate of 6 kg/s from an inlet state of 100 kPa and 310 K to a discharge pressure of 350 kPa with a power input of 785 kW has been developed. Evaluate his claim.
12. Refrigerant-12 enters an adiabatic compressor as saturated vapor at 20 psia at a rate of 4 ft^3/s and exits at 100 psia pressure. If the adiabatic efficiency of the compressor is 80%, determine (a) the temperature of the refrigerant at the exit of the compressor and (b) the power input, in hp.
13. Refrigerant R134a enters an actual adiabatic compressor at 140 kPa and -10°C and exits at 1.4 Mpa and 80°C. Find (a) the efficiency of the compressor, (b) the actual work reqired by the compressor, and (c) the entropy generation.
14. Refrigerant-12 enters an adiabatic compressor with 80 % efficiency as saturated vapor at 120 kPa at a rate of 0.3 m^3/s and exits at 1 Mpa pressure. Determine (a) the temperature of the refrigerant at the exit of the compressor, (b) the power input in kW, and (c) the entropy change of the refrigerant.
15. 10 lbm/s of air is compressed through a compresser adiabatically from (15 psia and 70°F) to (60 psia and 350°F). Compute (a) the work input required to the compressor, (b) the efficiency of the compressor, and (c) the entropy change of the process.
16. Saturated liquid water enters an adiabatic pump at 100 kPa pressure at a rate of 5 kg/s and exits at 10 Mpa. Determine the minimum power required to drive the pump.

17. Saturated liquid water enters an adiabatic pump at 100 kPa pressure at a rate of 5 kg/s and exits at 10 Mpa. Determine the power required to drive the pump if the pump efficiency is (a) 80% and (b) 85%.
18. In a water pump, 13.9 lbm/s are raised from atmospheric pressure (14.7 psia) and 520°R to 100 psia. The isentropic pump efficiency is 80 percent. Determine the horsepower input to the pump.
19. 1.23 lbm/s of steam is expanded from (1000 psia, 1000°F) in a two-stage adiabatic turbine to 1 psia. The efficiency of the high-pressure stage turbine is 90 %. The efficiency of the low-pressure stage turbine is 85 %. The exit pressure of the high-pressure stage turbine is 300 psia. Determine (a) the actual power produced by the high-pressure stage turbine and the low-pressure stage turbine, (b) the specific entropy generated by the high-pressure stage and by the low-pressure stage turbine, (c) the steam temperature at the exit of the high-pressure stage turbine and the low-pressure stage turbine, and (d) the exhaust steam quality.
20. A steady-flow, two-stage air adiabatic compressor that compresses air at a rate of 6 kg/s from an inlet state of 100 kPa and 310 K to a low-pressure stage discharge pressure of 250 kPa, and to a high-pressure stage discharge pressure of 750 kPa. The efficiency of the high-pressure stage compressor is 90 %. The efficiency of the low-pressure stage compressor is 85 %. Determine (a) the actual power required by the high-pressure stage compressor and by the low-pressure stage compressor, (b) the specific entropy generated by the high-pressure stage compressor and by the low-pressure stage compressor, and (c) the air temperature at the exit of the low-pressure stage compressor, and at the exit of the high-pressure stage compressor.

6.10 Entropy Change of Irreversible Processes

Entropy can be generated by irreversibility factors such as friction, heat transfer, mixing, free expansion, combustion, etc. The following example illustrates the entropy generation due to these effects using CyclePad.

Example 6.10.1 A 0.5 m^3 rigid tank contains 0.87 kg of air at 200 kPa. Heat is added to the tank until the pressure in the tank rises to 300 kPa. Determine the entropy change of air and heat added to the tank during this process.

To solve this problem by CyclePad, we take the following steps:

1. Build
   (A) Take a begin, a heater, and an end from the closed-system inventory shop and connect them.
   (B) Switch to analysis mode.
2. Analysis
   (A) Assume the heater as a isochoric process, and work=0.
   (B) Input the given information: (a) working fluid is air, (b) the initial air mass, volume and pressure of the process are 0.87 kg, 0.5 m^3 and 200 kPa, and (c) the final air pressure of the process is 300 kPa.

3. Display results

   (A) Display the heater result. The answers is Q=125.0 kJ and ΔS=0.87(2.81-2.51)=0.261 kJ/K.

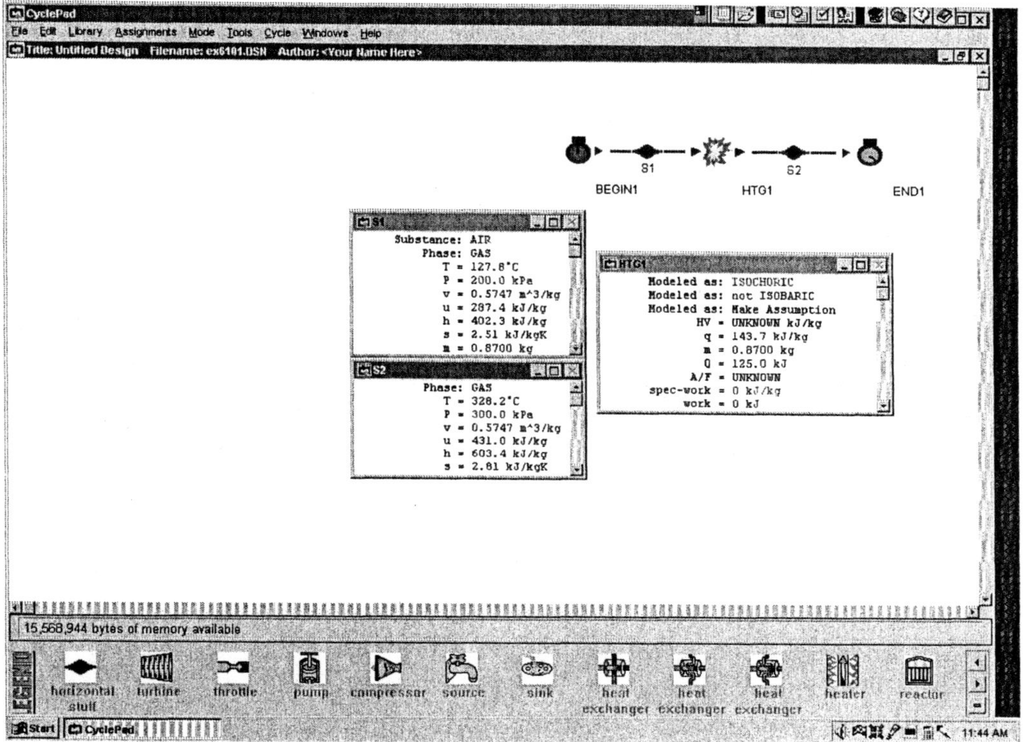

Figure Example 6.10.1. Entropy change of an isochoric process

Example 6.10.2 Steam at 4 MPa and 450°C is throttled in a valve to a pressure of 2 MPa during a steady flow process. Determine the specific entropy generated during this process.

To solve this problem by CyclePad, we take the following steps:

1. Build

   (A) Take a source, a throttling valve, and a sink from the open-system inventory shop and connect them.

   (B) Switch to analysis mode.

2. Analysis

   (A) Input the given information: (a) working fluid is water, (b) the inlet temperature and pressure of the throttling valve are 450°C and 4 MPa, (c) the outlet pressure of the throttling valve is 2 MPa.

3. Display results

   (A) Display the throttling valve results. The answer is Δs= kJ/K(kg).

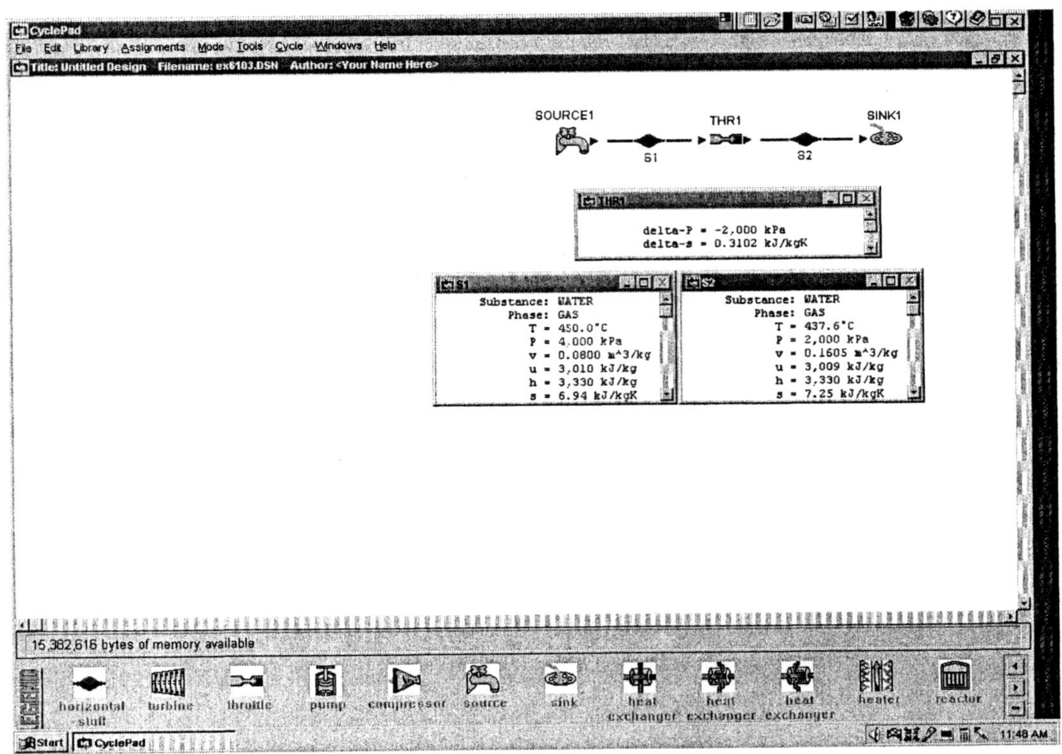

Figure Example 6.10.2 Entropy change of throttling process

6.11 The Increase of Entropy Principle

Entropy provides a means of determining whether a process can occur or not. The application of entropy is based on the *increase of entropy principle.* The principle states that the entropy change of an isolated system with respect of time is always non-negative. An isolated system is defined as a system with neither mass nor energy interaction with its surroundings. Thus, the principle means that changes of state in an isolated system can occur only in the direction of increasing entropy. This also implies that an isolated system will have attained equilibrium only when its entropy attained the maximum value.

Since a system and its surroundings include everything which is affected by a process, the combination of the system and its surroundings is called universe. The universe is an isolated system, because the surroundings of the universe is a null system, which does not have anything in it. Therefore, we conclude that the entropy change of the universe with respect of time is always non-negative until it reaches its equilibrium state. At its equilibrium state, the entropy value of the universe would attained its maximum value and there would not be any non-equilibrium activity occurs within the universe.

Homework 6.11 the Increase of Entropy Principle

1. What is the increase of entropy principle?

2. Is universe an isolated system? What is the surroundings of the universe?
3. When will the entropy value of the universe attained its maximum value?

6.12 Available and Unavailable Energy

The energy which is ideally available to do work is called *available energy*. In the Carnot cycle T-s diagram as shown in Figure 6.12.1, we see that a portion of the heat supplied (Area 623562) from a constant-temperature heat source at high-temperature T_H must be rejected to a constant-temperature heat sink at low-temperature T_L. The amount of heat rejected (Area 41654) for a given heat input is reduced by lowering the sink temperature T_L. However, in practice there is a minimum value of T_L which can be used. This normally corresponds to the ambient temperature (T_o) of the heat engine. The minimum energy which could be rejected by the heat engine is called *unavailable energy*. Unavailable energy is the portion of energy supplied from the heat source which is not converted into work. If the ambient temperature is T_L (T_L=T_o), it is apparently that the available energy is Area 12341 and the unavailable energy is Area 41654. The unavailable energy is always determined by $(T_o)\Delta S$, and is represented by a rectangular area. In general, many applications of interest involves a heat source of variable temperature T_H. The differential amount of available energy is $(T_H - T_o)dS$. The total amount of available energy is $\int(T_H - T_o)dS$.

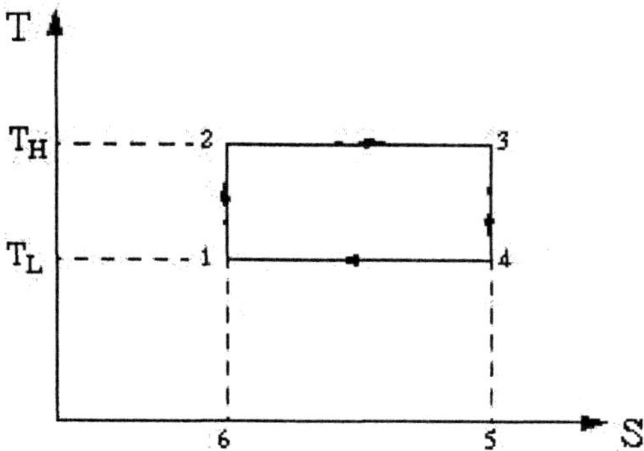

Figure 6.12.1 Carnot cycle T-s diagram

Homework 6.12 Available and Unavailable Energy

1. What are available and unavailable energy?
2. What is minimum temperature value of heat rejection T_L which can be used in real world?

6.13 Summary

A microscopic property called entropy is defined based on the Clausius inequality. Heat and entropy are related by the second law of thermodynamics for a closed system and is expressed as $\Delta S = \delta Q/T + S_{generation}$, where $S_{generation}$ is the non-negative entropy generated by irreversibility during a process. Heat can be shown by the area underneath a process on a T-s diagram.

The second law of thermodynamics for an open system is expressed as $\Delta S=\delta Q/T + S_i+S_e+S_{generation}$. For the case of steady flow and steady state process, The second law of thermodynamics for an open system is expressed as $0=\delta Q/T + S_i+S_e+S_{generation}$.

Entropy change of pure substances, ideal gases and incompressible substances are related to other properties. These relationships are listed on tables and are built in the software CyclePad.

An isentropic process is a constant entropy process. It is also a reversible adiabatic process. It serves as an idealization of an adiabatic process for many engineering applications. The isentropic or adiabatic efficiency for turbines, compressors and pumps are expressed as: $\eta_{turbine} = w_{actual\ turbine} / w_{isentropic\ turbine}$, $\eta_{compressor} = w_{isentropic\ compressor} / w_{actual\ compressor}$, and $\eta_{pump} = w_{isentropic\ pump} / w_{actual\ pump}$.

The increase of entropy principle states that the entropy change of an isolated system with respect of time is always non-negative. The universe is an isolated system, the entropy change of the universe with respect of time is always non-negative until it reaches its equilibrium state.

The energy which is ideally available to do work is called available energy. The minimum energy which could be rejected by the heat engine is called unavailable energy.

Chapter 7

VAPOR CYCLES

OBJECTIVES

After reading and studying the material in this chapter, you should be able to:

1. Understand the difference between vapor and gas cycles.
2. Know why the Carnot cycle is not used and how it is modified to the Rankine cycle.
3. Know the four basic components and their corresponding processes of the Rankine cycle.
4. Draw the Rankine cycle on a T-s Diagram.
5. Conclude that for the same pressures, the Rankine cycle efficiency is less than the Carnot cycle efficiency.
6. Sketch the reheat Rankine cycle on a T-s Diagram.
7. Understand that reheat does not greatly increase basic Rankine cycle efficiency, but does increase the net work and decrease the moisture content of the steam in the later stage of the turbine.
8. Sketch the regenerating Rankine cycle on a T-s Diagram.
9. Understand that regeneration increases basic Rankine cycle efficiency.
10. Quantitatively understand the construction and operation of the major pieces of equipment used in Rankine power plant.
11. Modify and analyze the Rankine cycle using CyclePad.
12. Know low-temperature Rankine cycles.
13. Know cascaded and combined cycles.
14. Know the working fluids for vapor cycles.
15. Obtain some design experience on vapor cycles.

7.1 CARNOT CYCLE

Theoretically, the Carnot vapor cycle is the most efficient vapor power cycle operating between two temperature reservoirs.

The *Carnot vapor cycle* as shown in Figure 7.1.1 is composed of the following four processes:

1-2 isentropic compression
2-3 isothermal heat addition
3-4 isentropic expansion
4-1 isothermal heat rejection

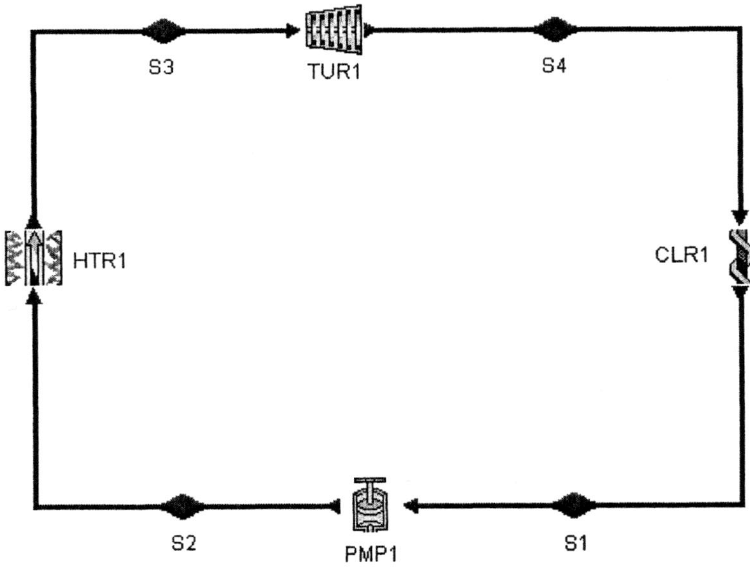

Figure 7.1.1 Carnot vapor cycle

In order to achieve the isothermal heat addition and isothermal heat rejection processes, the Carnot cycle must operate inside the vapor dome. The T-s diagram of a Carnot cycle operating inside the vapor dome is shown in Figure 7.1.2. Saturated water at state 1 is evaporated isothermally to state 2, where it is saturated vapor. The steam enters a turbine at state 2 and expands isentropically, producing work, until state 3 is reached. The vapor-liquid mixture would then be partially condensed isothermally until state 4 is reached. At state 4, a compressor would isentropically compress the vapor-liquid mixture to state 1.

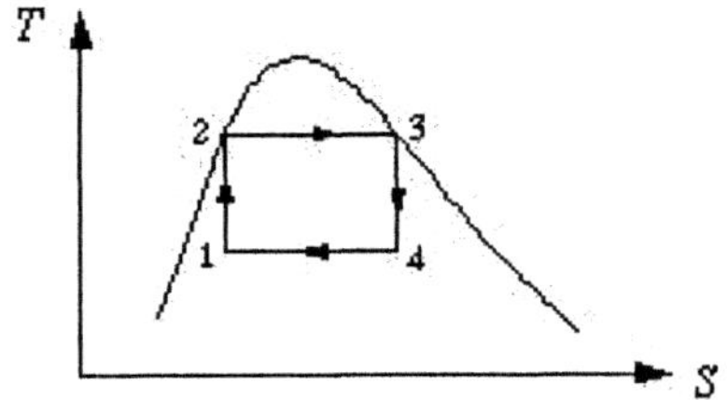

Figure 7.1.2 Basic vapor Carnot cycle T-S diagram

Applying the first law and second law of thermodynamics of the open system to each of the four processes of the Carnot vapor cycle yields:

$$Q_{12} = 0, \tag{7.1.1}$$

$$W_{12} = m(h_1 - h_2), \tag{7.1.2}$$

$$W_{23} = 0, \tag{7.1.3}$$

$$Q_{23} - 0 = m(h_3 - h_2), \tag{7.1.4}$$

$$Q_{34} = 0, \tag{7.1.5}$$

$$W_{34} = m(h_3 - h_4), \tag{7.1.6}$$

$$W_{41} = 0, \tag{7.1.7}$$

and

$$Q_{41} - 0 = m(h_1 - h_4), \tag{7.1.8}$$

The net work (W_{net}), which is also equal to net heat (Q_{net}), is

$$W_{net} = Q_{net} = Q_{23} + Q_{41} \tag{7.1.9}$$

The thermal efficiency of the cycle is

$$\eta = W_{net} / Q_{23} = Q_{net} / Q_{23} = 1 - Q_{41} / Q_{23} = 1 - (h_4 - h_1)/(h_3 - h_2) \tag{7.1.10}$$

Example 7.1.1. A steam Carnot cycle operates between 250°C and 100°C. Determine the pump work, turbine work, heat added, quality of steam at the exit of the turbine, quality of steam at the inlet of the pump, and cycle efficiency.

To solve this problem by CyclePad, we take the following steps:

1. Build
   (A) Take a pump, a boiler, a turbine and a condenser from the inventory shop and connect the four devices to form the basic Rankine cycle.
   (B) Switch to analysis mode.
2. Analysis
   (A) Assume a process each for the four devices: (a) pump as isentropic, (b) boiler as isothermal (isobaric inside the vapor dome), (c) turbine as isentropic, and (d) condenser as isothermal (isobaric inside the vapor dome).
   (B) Input the given information: (a) working fluid is water, (b) the inlet temperature and quality of the boiler are 400°C and 0, (c) the exit quality of the boiler is 1, and (d) the inlet temperature of the condenser is 100°C.
3. Display results
   (A) Display the T-s diagram and cycle properties results. The cycle is a heat engine. The answers are pump work=-111.9 kJ/kg, turbine work=603.7 kJ/kg, and η=28.67%, and
   (B) Display the sensitivity diagram of cycle efficiency vs boiler temperature.

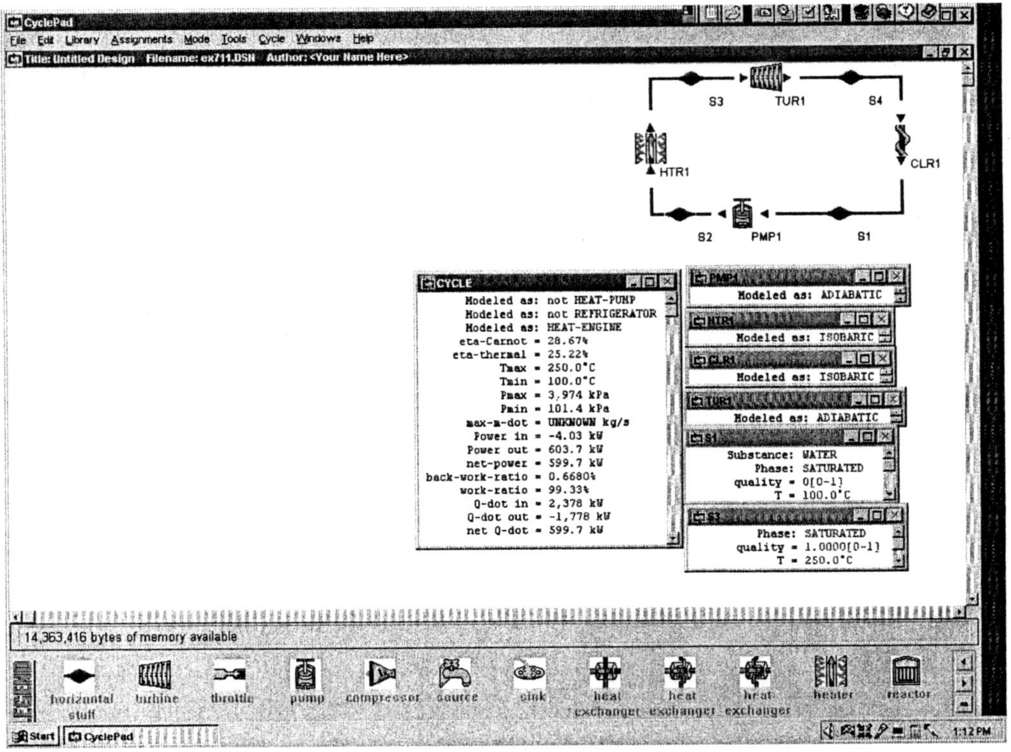

Figure Example 7.1.1. Carnot vapor cycle

Comments: The Carnot vapor cycle as illustrated by Example 7.1.1 is not practical. Difficulties arise in the isentropic processes of the cycle. One difficulty is that the isentropic turbine will have to handle steam with low quality. The impingement of liquid droplets on the turbine blade causes erosion and wear. Another difficult is the isentropic compression of a liquid-vapor mixture. The two phase mixture of the steam causes serious cavitation problem during the compression process. Also, since the specific volume of the saturated mixture is high, the pump power required is also very high. Thus the Carnot vapor cycle is not a realistic model for vapor power cycles.

Homework 7.1 Carnot Cycle

1. Why is excessive moisture in steam undesirable in steam turbine? What is the highest moisture content allowed?
2. Why is the Carnot cycle not a realistic model for steam power plants?
3. A steady flow of 1 kg/s Carnot engine uses water as the working fluid. Water changes phase from saturated liquid to saturated vapor as heat is added from a heat source at 300°C. Heat rejection takes place at a pressure of 10 kPa. Determine (a) the quality at the exit of the turbine, (b) the quality at the inlet of the pump, (c) the heat transfer added in the boiler, (d) the power required for the pump, (e) the power produced by the turbine, (f) the heat transfer rejected in the condenser, and (g) the cycle efficiency.

7.2 Basic Rankine Cycle

The working fluid for vapor cycles is alternately condensed and vaporized. When a working fluid remains in the saturation region at constant pressure, its temperature is also constant. Thus the condensation or the evaporation of a fluid in a heat exchanger is a process that closely approximates the isothermal heat transfer processes of the Carnot cycle. Owing to this fact, vapor cycles are closely approximate the behavior of the Carnot cycle. Thus, in general, they tend to perform efficiently.

The *Rankine cycle* is a modified Carnot cycle to overcome the difficulties with the Carnot cycle when the working fluid is a vapor. In the Rankine cycle, the heating and cooling processes occur at constant pressure. Figure 7.2.1 shows the devices used in a basic Rankine cycle, and Figure 7.2.2 is the T-s diagram of the basic Rankine cycle,

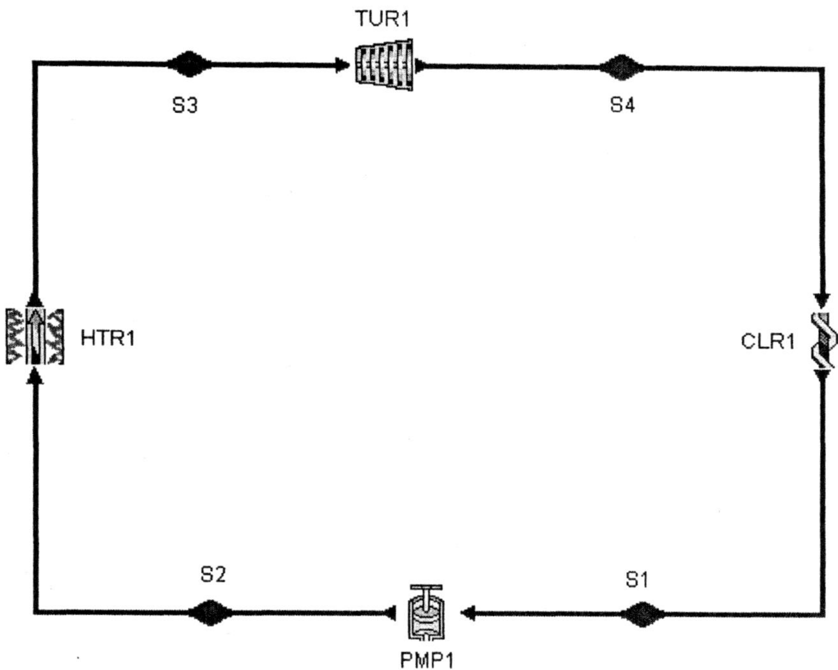

Figure 7.2.1 Basic Rankine cycle

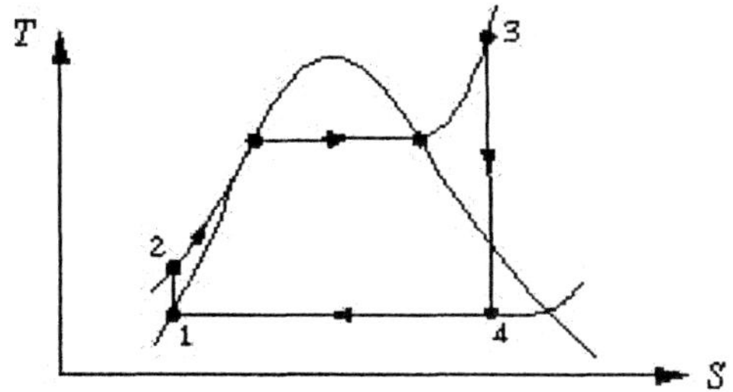

Figure 7.2.2 Basic Rankine cycle T-S diagram

The *basic Rankine cycle* is composed of the following four processes:

1-2 isentropic compression
2-3 isobaric heat addition
3-4 isentropic expansion
4-1 isobaric heat removing

Water enters the pump at state 1 as a low pressure saturated liquid to avoid the cavitation problem and exits at state 2 as a high pressure compressed liquid. The heat supplied in the boiler raises the water from the compressed liquid at state 2 to saturated liquid to saturated vapor and to a much higher temperature superheated vapor at state 3. The superheated vapor at state 3 enters the turbine where it expands to state 4. The superheating moves the isentropic expansion process to the right on the T-s diagram as shown in Figure 7.2.3, thus preventing a high moisture content of the steam as it exits the turbine at state 4 as a saturated mixture. The exhaust steam from the turbine enters the condenser at state 4 and is condensed at constant pressure to state 1 as saturated liquid.

Applying the first law and second law of thermodynamics of the open system to each of the four processes of the Rankine cycle yields:

$$Q_{12} = 0, \tag{7.2.1}$$

$$W_{12} = m(h_1 - h_2) = mv_1 (p_1 - p_2), \tag{7.2.2}$$

$$W_{23} = 0, \tag{7.2.3}$$

$$Q_{23} - 0 = m(h_3 - h_2), \tag{7.2.4}$$

$$Q_{34} = 0, \tag{7.2.5}$$

$$W_{34} = m(h_3 - h_4), \tag{7.2.6}$$

$$W_{41} = 0, \tag{7.2.7}$$

and

$$Q_{41} - 0 = m(h_1 - h_4), \tag{7.2.8}$$

The net work (W_{net}), which is also equal to net heat (Q_{net}), is

$$W_{net} = Q_{net} = Q_{23} + Q_{41} \tag{7.2.9}$$

The thermal efficiency of the cycle is

$$\eta = W_{net} / Q_{23} = Q_{net} / Q_{23} = 1 - Q_{41} / Q_{23} = 1 - (h_4 - h_1)/(h_3 - h_2) \tag{7.2.10}$$

Example 7.2.1 Determine the efficiency and power output of a basic Rankine cycle using steam as the working fluid in which the condenser pressure is 80 kPa. The boiler pressure is 3 Mpa. The steam leaves the boiler as saturated vapor. The mass rate of steam flow is 1 kg/s. Show the cycle on T-s diagram. Plot the sensitivity diagram of cycle efficiency vs boiler pressure.

To solve this problem by CyclePad, we take the following steps:

1. Build
   (A) Take a pump, a boiler, a turbine and a condenser from the inventory shop and connect the four devices to form the basic Rankine cycle.
   (B) Switch to analysis mode.
2. Analysis
   (A) Assume a process each for the four devices: (a) pump as isentropic, (b) boiler as isobaric, (c) turbine as isentropic, and (d) condenser as isobaric.
   (B) Input the given information: (a) working fluid is water, (b) the inlet pressure and quality of the pump are 80 kPa and 0, (c) the inlet pressure and quality of the turbine are 3 Mpa and 1, and (d) the mass flow rate is 1 kg/s.
3. Display results
   (A) Display the T-s diagram and cycle properties results. The cycle is a heat engine. The answers are η=24.61%, $Wdot_{pump}$=-3.07 kW, $Qdot_{boiler}$=2409 kW, $Wdot_{turbine}$=595.7 kW, $Qdot_{condenser}$=-1816 kW, and Net power output=592.7 kW, (B) Display the sensitivity diagram of cycle efficiency vs boiler pressure, and (C) Display the sensitivity diagram of cycle efficiency vs condenser pressure.

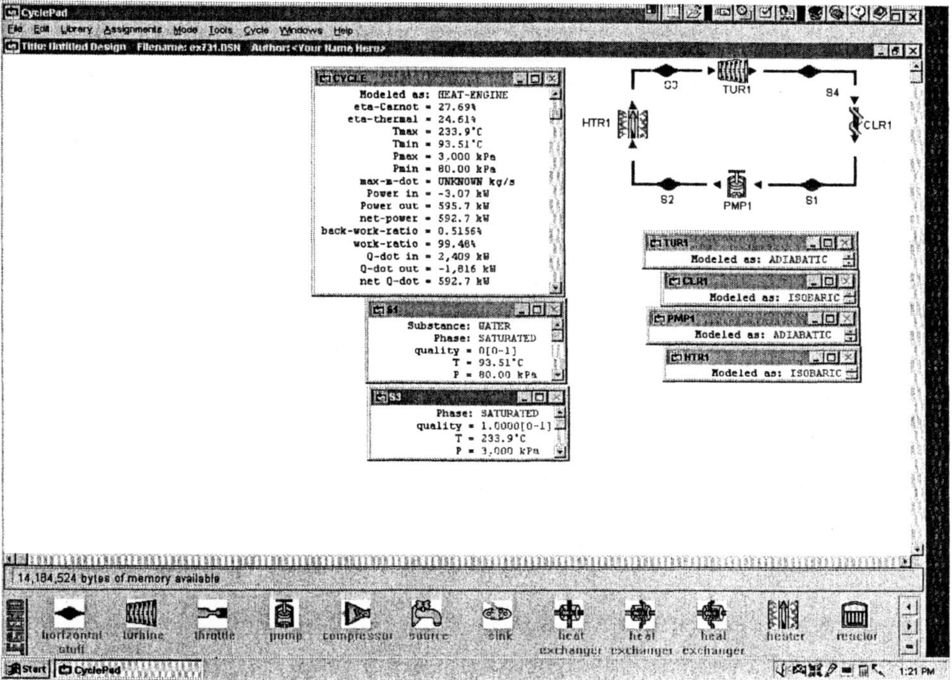

Figure Example 7.2.1a Rankine cycle

Comments:

(1) The sensitivity diagram of cycle efficiency vs boiler pressure demonstrates that increasing the boiler pressure increases the boiler temperature. This raises the average temperature at which heat is added to the steam and thus raises the cycle efficiency. Operating pressures of boilers have increased over the years up to 30 MPa (4500 psia) today.

(2) The sensitivity diagram of cycle efficiency vs condenser pressure demonstrates that decreasing the condenser pressure is decreases the condenser temperature. This dropes the average temperature at which heat is removed to the surroundings and thus raises the cycle efficiency. Operating pressures of condensers have decreased over the years to 5 kPa (0.75 psia) today.

The Rankine efficiency could be increased by increasing the boiler pressure, since the area enclosed by the cycle in the T-s diagram will be increased.

Example 7.2.2. A simple Rankine cycle using water as the working fluid operates between a boiler pressure of 500 psia and a condenser pressure of 20 psia. The mass flow rate of the water is 3 lbm/s. Determine (a) the quality of the steam at the exit of the turbine, (b) the net power of the cycle, and (c) the cycle efficiency. Then change the boiler pressure to 600 psia, and determine (d) the quality of the steam at the exit of the turbine and (e) the net power of the cycle.

To solve this problem by CyclePad, we take the following steps:

1. Build
   (A) Take a pump, a boiler, a turbine and a condenser from the inventory shop and connect the four devices to form the basic Rankine cycle.
   (B) Switch to analysis mode.
2. Analysis
   (A) Assume a process each for the four devices: (a) pump as isentropic, (b) boiler as isobaric, (c) turbine as isentropic, and (d) condenser as isobaric.
   (B) Input the given information: (a) working fluid is water, (b) the inlet pressure and quality of the pump are 10 psia and 0, and (c) the inlet pressure and quality of the turbine are 500 psia and 1.
3. Display results
   (A) Display cycle properties results. The cycle is a heat engine. The answers are $x_{turbine\ outlet}$=0.8082, $Wdot_{net}$=981.8 hp, and η=22.97%.
   (B) Change the boiler pressure to 600 psia and display cycle properties results again. The answers are $x_{turbine\ outlet}$=0.7953, $Wdot_{net}$=1028 hp, and η=24.08%.

Comments: The effect of increasing the boiler pressure on the quality of the steam at the exit of the turbine can be seen by comparing the two cases. The higher the boiler pressure, the higher the moisture content (or the lower the quality) at the exit of the turbine. Steam with qualities less than 90 percent, at the exit of the turbine, cannot be tolerated in the operation of actual Rankine steam power plants. To increase steam quality at the exit of the turbine, superheating and reheating are used.

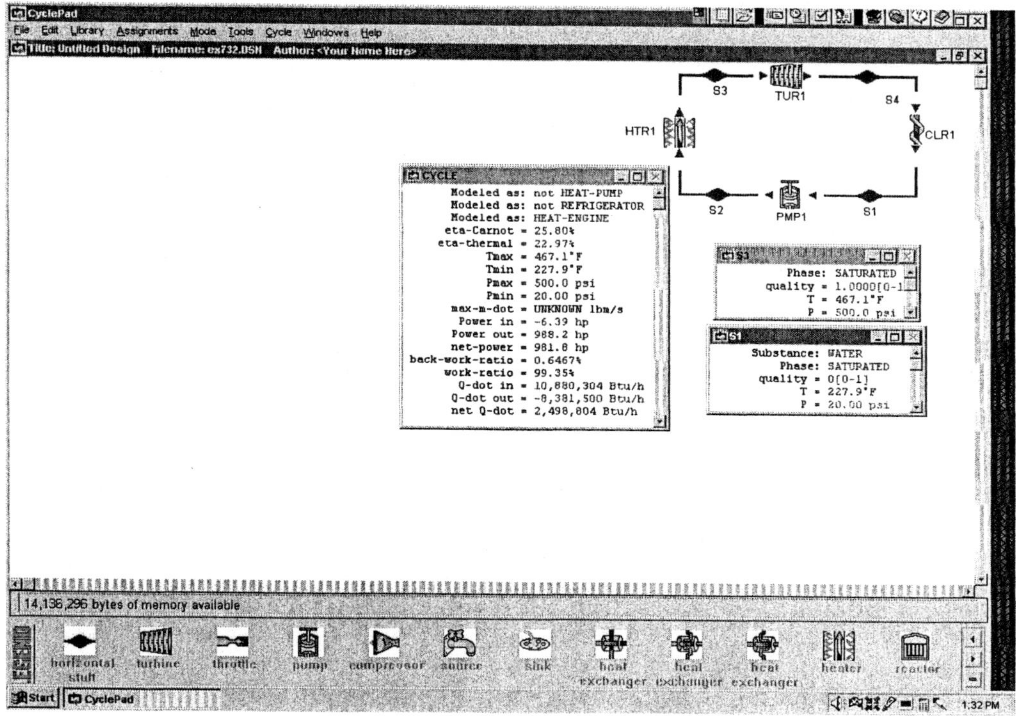

Figure Example 7.2.2a. Rankine cycle

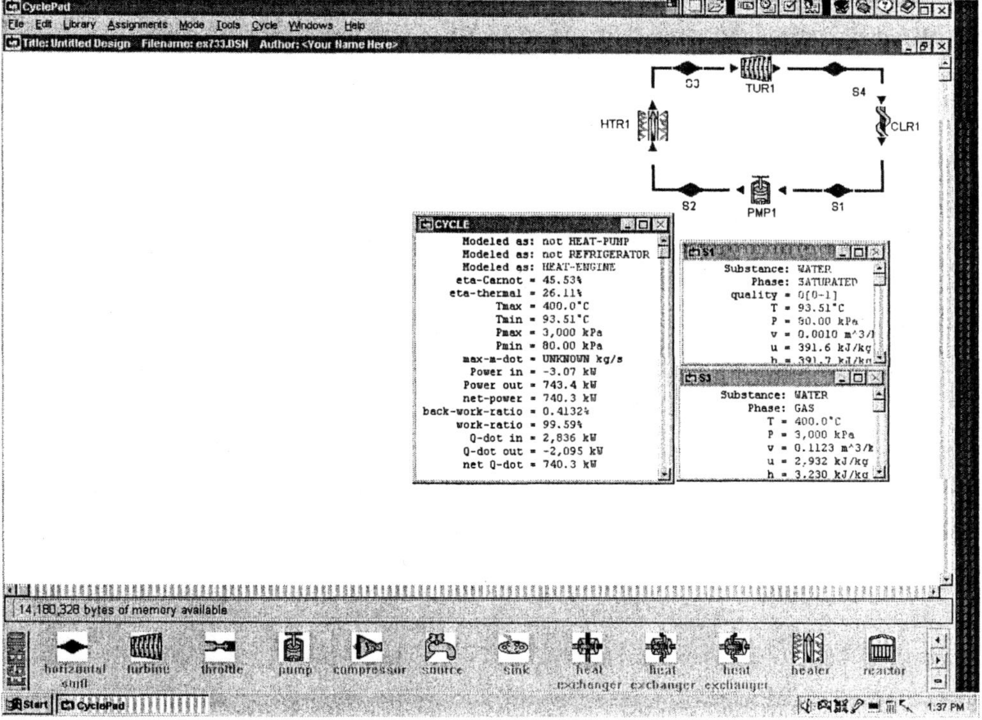

Figure Example 7.2.2b. Rankine cycle

Example 7.2.3. Determine the efficiency and power output of a superheat Rankine cycle using steam as the working fluid in which the condenser pressure is 80 kPa. The boiler pressure is 3 MPa. The steam leaves the boiler at 400°C. The mass rate of steam flow is 1 kg/s. Show the cycle on T-s diagram. Plot the sensitivity diagram of cycle efficiency vs boiler superheat temperature.

Convert the SI unit system to British system and find the answer.

To solve this problem by CyclePad, we take the following steps:

1. Build
   (A) Take a pump, a boiler, a turbine and a condenser from the inventory shop and connect the devices to form the basic Rankine cycle.
   (B) Switch to analysis mode.
2. Analysis
   (A) Assume a process each for the four devices: (a) pump as isentropic, (b) boiler as isobaric, (c) turbine as isentropic, and (d) condenser as isobaric.
   (B) Input the given information: (a) working fluid is water, (b) the inlet pressure and quality of the pump are 80 kPa and 0, (c) the inlet pressure and temperature of the turbine are 3 MPa and 400°C, and (d) the mass flow rate is 1 kg/s.
3. Display results
   (A) Display the T-s diagram and cycle properties results. The cycle is a heat engine. The answers are η=26.11% and Net power output=740.3 kW, and (B) Display the sensitivity diagram of cycle efficiency vs superheat temperature.

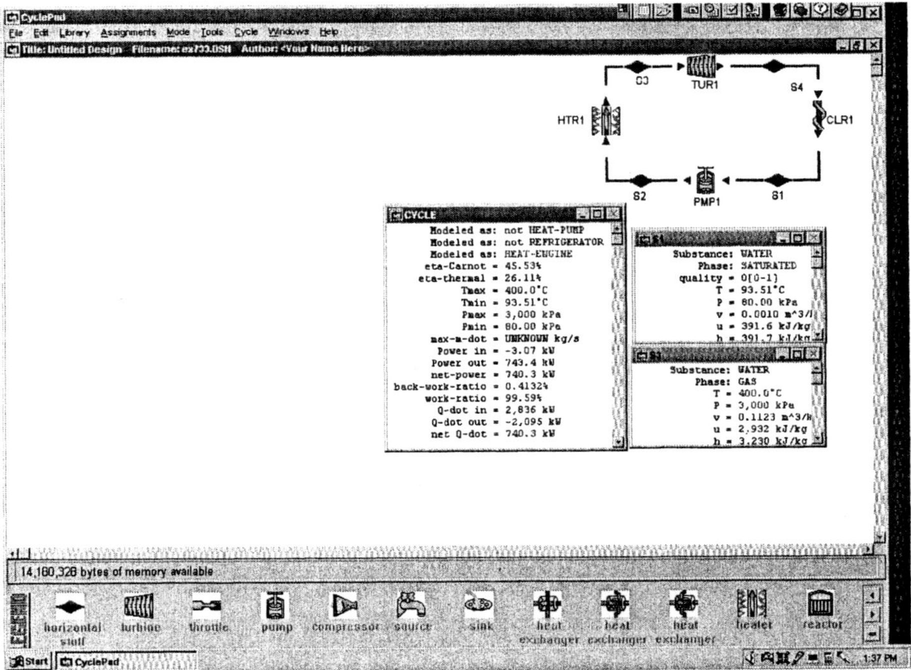

Figure Example 7.2.3a. Superheated Rankine cycle

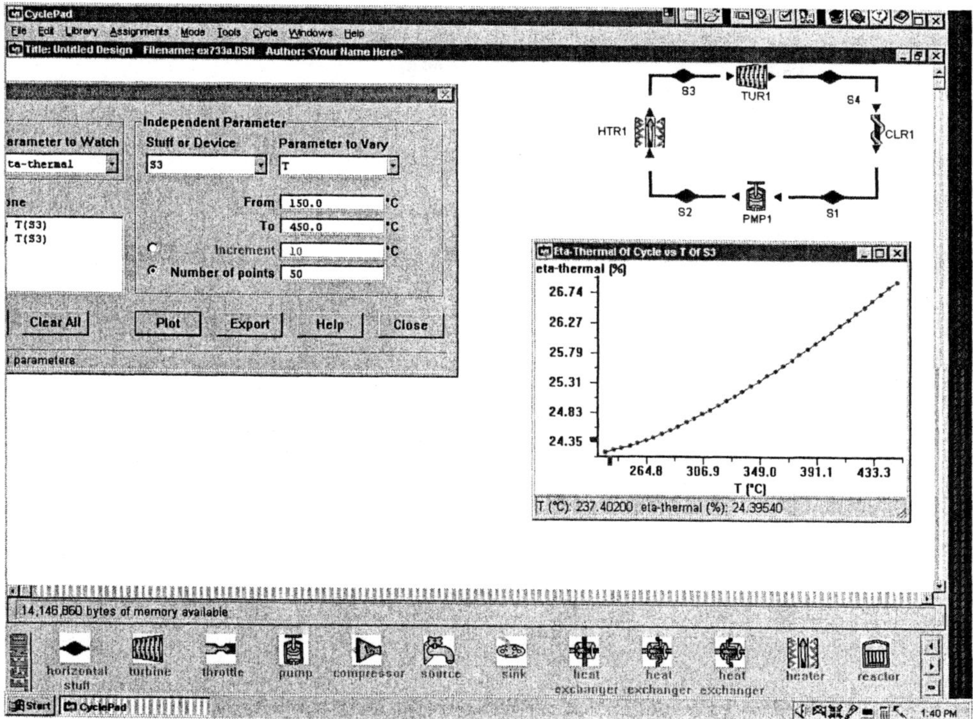

Figure Example 7.2.3b. Superheated Rankine cycle *sensitivity analysis*

Comments: From the sensitivity diagram of cycle efficiency vs superheat temperature, it is seen that the higher the superheat temperature, the higher the cycle efficiency. The superheat temperature is limited, however, due to metallurgical considerations. Presently, the maximum operating superheat temperature is 620°C (1150°F).

7.2 Homework Basic Rankine Cycle

1. What are the processes that make up the simple ideal Rankine cycle?
2. What is the quality of vapor at the inlet of the pump in a simple ideal Rankine cycle? Why?
3. What is the minimum quality of vapor required at the exit of the turbine in a Rankine cycle? Why?
4. Steam in an ideal Rankine cycle flows at a mass rate of flow of 14 lbm/s. It leaves the boiler at 1250 psia and 1000°F, and enters a turbine where it is expanded and then exhausted to the main condenser, which is operating at a pressure of 1 psia. The fluid leaves the condenser as a saturated liquid, where it is pumped by a pump back into the boiler. Determine for the cycle:
   (a) Pump power.
   (b) The rate of heat added by the boiler.
   (c) The ideal turbine power.
   (d) The rate of heat rejected by the condenser.
   (e) The thermal efficiency of the cycle.

5. An ideal Rankine cycle uses water as a working fluid, which circulates at a rate of 80 kg/s. The boiler pressure is 6 MPa, and the condenser pressure is 10 kPa. Determine (a) the power required to operate the pump, (b) the heat transfer added to the water in the boiler, (c) the power developed by the turbine, (d) the heat transfer removed from the condenser, (e) the quality of steam at the exit of the turbine, and (f) the thermal efficiency of the cycle.
6. An ideal Rankine cycle uses water as a working fluid, which circulates at a rate of 80 kg/s. The boiler pressure is 6 MPa, and the condenser pressure is 10 kPa. The steam is superheated and enters the turbine at 600°C and leaves the condenser as a saturated liquid. Determine (a) the power required to operate the pump, (b) the heat transfer added to the water in the boiler, (c) the power developed by the turbine, (d) the heat transfer removed from the condenser, (e) the quality of steam at the exit of the turbine, and (f) the thermal efficiency of the cycle.
7. A Rankine cycle using 1 kg/s of water as the working fluid operates between a condenser pressure of 7.5 kPa and a boiler pressure of 17 MPa. The superheater temperature is 550°C. Determine (a) the pump power, (b) the turbine power, (c) the heat transfer added in the boiler, and (d) the cycle thermal efficiency.
8. In a Rankine power plant, the steam temperature and pressure at the turbine inlet are 1000°F and 2000 psia. The temperature of the condensing steam in the condenser is maintained at 60°F. The power generated by the turbine is 30000 hp. Assuming all processes to be ideal, determine: (a) the pump power required (hp), (b) the mass flow rate, (c) the heat transfer added in the boiler (Btu/hr), (d) the heat transfer removed in the condenser (Btu/hr), and (e) the cycle thermal efficiency (%).
9. Water circulates at a rate of 80 kg/s in an ideal Rankine power plant. The boiler pressure is 6 Mpa and the condenser pressure is 10 kPa. The steam enters the turbine at 600°C and water leaves the condenser as a saturated liquid. Find: (a) the power required to operate the pump, (b) the heat transfer added in the boiler, (c) the power developed by the turbine, (d) the thermal efficiency of the cycle.
10. For an ideal Rankine cycle, steam enters the turbine at 5 MPa and 400°C, and exhausts to the condenser at 10 kPa. The turbine produces 20,000 kW of power.
    (a) Draw a T-s diagram for this cycle with respect to the saturation curve.
    (b) What is the mass flow rate of the steam? (kg/s)
    (c) What is the rate of heat rejection from the condenser, and the rate of heat added in the boiler?
    (d) Find the thermal efficiency for this cycle.
11. Steam is generated in the boiler of a steam power plant operating on an ideal Rankine cycle at 10 MPa and 500 °C at a steady rate of 80 kg/s. The steam expands in the turbine to a pressure of 7.5 kPa. Determine (a) the quality of the steam at the turbine exit, (b) rate of heat rejection in the condenser, (c) the power delivered by the turbine, and (d) the cycle thermal efficiency (%).
12. A steam power plant operating on an ideal Rankine cycle has a boiler pressure of 800 psia and a condenser pressure of 2 psia. The quality at the turbine exit is 95% and the power generated by the turbine is 10,000 hp. Determine (a) the mass flow rate of steam (lbm/s), (b) the turbine inlet temperature (°F), (c) the rate of heat addition in the boiler (Btu/h), and (d) the cycle thermal efficiency (%).

13. Water is the working fluid in an ideal Rankine cycle. The condenser pressure is 8 kPa, and saturated vapor enters the turbine at: (a) 15 MPa, (b) 10 MPa, (c) 7 MPa, and (d) 4 MPa. The net power output of the cycle is 100 MW. Determine for each case the mass flow rate of water, heat transfer in the boiler and the condenser, and the thermal efficiency.
14. A steam power plant operates on the Rankine cycle. The steam enters the turbine at 7 MPa and 550°C. It discharges to the condenser at 20 kPa. Determine the quality of the steam at the exit of the turbine, pump work, turbine work, heat added in the boiler, and thermal cycle efficiency.
15. A steam power plant operates on the Rankine cycle. The steam with a mass rate flow of 10 kg/s enters the turbine at 6 MPa and 600°C. It discharges to the condenser at 10 kPa. Determine the quality of the steam at the exit of the turbine, pump power, turbine power, rate of heat added in the boiler, and thermal cycle efficiency.

7.3 Improvements to Rankine Cycle

The thermal efficiency of the Rankine cycle can be improved by increase the average temperature at which heat is transferred to the working fluid in the heating process, or decrease the average temperature at which heat is transferred to the surroundings from the working fluid in the cooling process. Several modifications to increase the thermal efficiency of the basic Rankine cycle include increasing boiler pressure, decreasing condenser pressure, *superheater*, *reheater*, *regenerator*, pre-heater, etc.

Increasing the average temperature during the heat addition process increases the boiler pressure. The maximum boiler pressure is limited by the tube metallurgical material problem in the boiler. Increasing the boiler pressure increases the moisture content of the steam at the turbine exit which is not desirable.

Increasing the average temperature during the heat addition process without increasing boiler pressure can be done by superheating the steam to high temperature with a superheater. Superheating the steam to a higher temperature also decreases the moisture content of the steam at the turbine exit which is very desirable.

Increasing the average temperature during the heat addition process can be accomplished with a superheater. Moisture content of steam at the turbine exhaust can be decreased by reheating the steam between the stages of a multi-stage turbine.

An increase in the average temperature during the heat addition process can also can be accomplished by regenerating the steam. A portion of the partially expanded steam between the turbine stages of a multi-stage turbine is drawn off to preheat the condensed liquid before it is returned to the boiler. In this way, the amount of heat added at the low temperature is reduced. So the average temperature during the heat addition process is increased.

Decreasing the average temperature during the heat rejection process decreases the condenser pressure and increases cycle efficiency. The minimum condenser pressure is limited by the sealing and leakage problem in the condenser.

7.4 ACTUAL RANKINE CYCLE

7For actual Rankine cycles, many irreversibilities are present in various components. Fluid friction causes pressure drops in the boiler and condenser. These drops in the boiler and condenser are usually small. The major irreversibilities occur within the turbine and pump. To account for these irreversibility effects, turbine efficiency and pump efficiency must be used in computing the actual work produced or consumed. The T-s diagram of the actual Rankine cycle is shown in Figure 7.4.1. The effect of irreversibilities on the thermal efficiency of a Rankine cycle is illustrated in the following example.

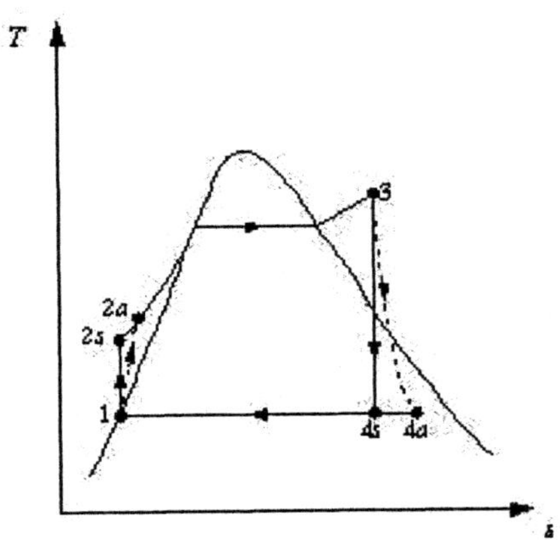

Figure 7.4.1 Actual Rankine cycle T-s diagram

Example 7.4.1 Determine the efficiency and power output of an actual Rankine cycle using steam as the working fluid and having a condenser pressure is 80 kPa. The boiler pressure is 3 MPa. The steam leaves the boiler at 400°C. The mass rate of steam flow is 1 kg/s. The pump efficiency is 85% and the turbine efficiency is 88%. Show the cycle on T-s diagram. Plot the sensitivity diagram of cycle efficiency vs pump efficiency, and turbine efficiency.

Convert the SI unit system to British system and find the answer.

To solve this problem by CyclePad, we take the following steps:

1. Build
   (A) Take a pump, a boiler, a turbine and a condenser from the inventory shop and connect the devices to form the actual Rankine cycle.
   (B) Switch to analysis mode.
2. Analysis
   (A) Assume a process each for the four devices: (a) pump as adiabatic, (b) boiler as isobaric, (c) turbine as adiabatic, and (d) condenser as isobaric.
   (B) Input the given information: (a) working fluid is water, (b) the inlet pressure and quality of the pump are 80 kPa and 0, (c) the inlet pressure and temperature of the turbine are 3 Mpa and 400°C, (d) the mass flow rate is 1 kg/s, (e) the phase at the

exit of turbine is saturated, and (f) The pump efficiency is 85% and the turbine efficiency is 88%.

3. Display results

   (A) Display the T-s diagram and cycle properties results. The cycle is a heat engine. The answers are η=22.95% and Net power output=650.5 kW, (B) Display the sensitivity diagram of cycle efficiency vs pump efficiency, and (C) Display the sensitivity diagram of cycle efficiency vs turbine efficiency.

Comments:

(1) The pump work is quite small compared to the turbine work. Therefore, it is seen from the sensitivity diagram of cycle efficiency vs pump efficiency that the cycle efficiency is not sensitive to the pump efficiency.

(2) The sensitivity diagram of cycle efficiency vs turbine efficiency demonstrates that the cycle efficiency is sensitive to the turbine efficiency.

The power output of the Rankine cycle can be controlled by a throttling valve. The inlet steam pressure and temperature may be throttled down to a lower pressure and temperature if desired. Adding a throttling valve to the Rankine cycle decreases the cycle efficiency. The throttling Rankine cycle is shown in Figure 7.4.2. An example illustrated the throttling Rankine cycle is given in Example 7.4.2.

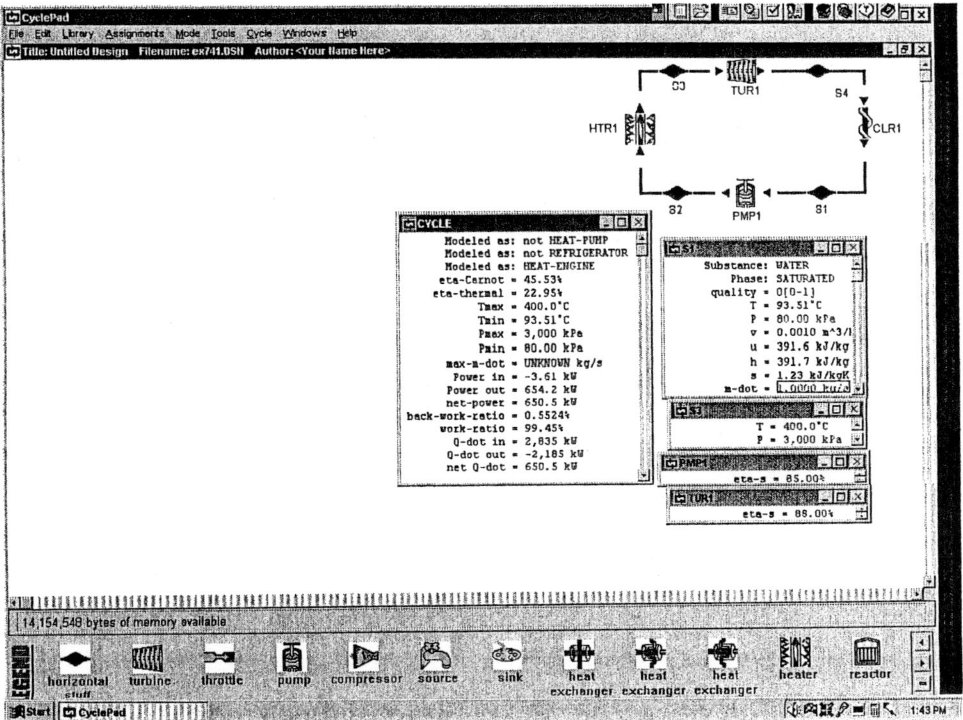

Figure Example 7.4.1a Rankine cycle analysis

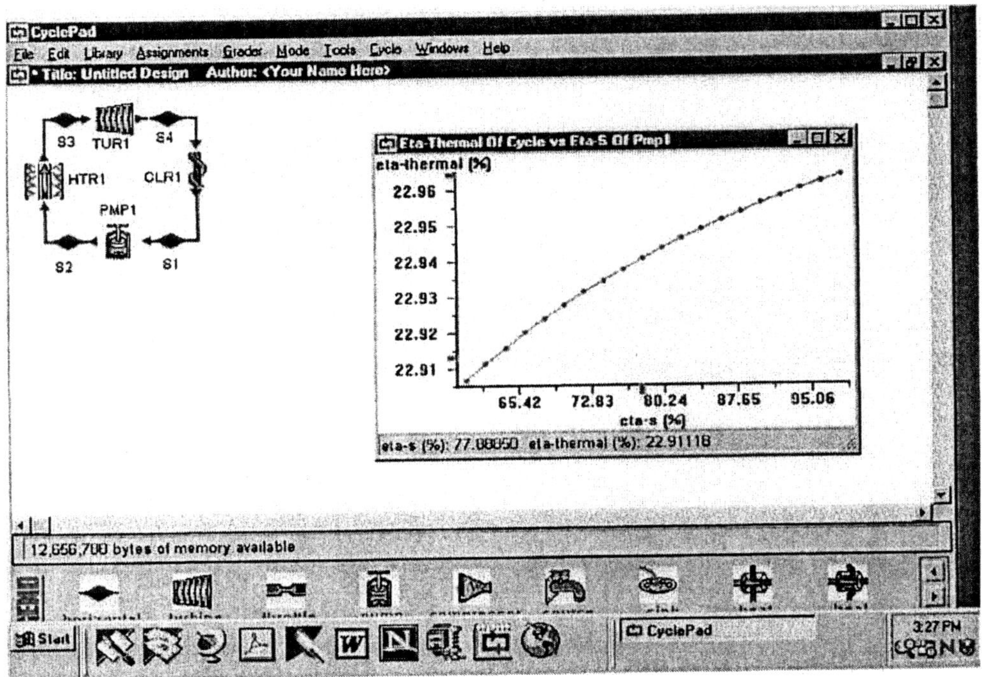

Figure Example 7.4.1b Rankine cycle sensitivity analysis

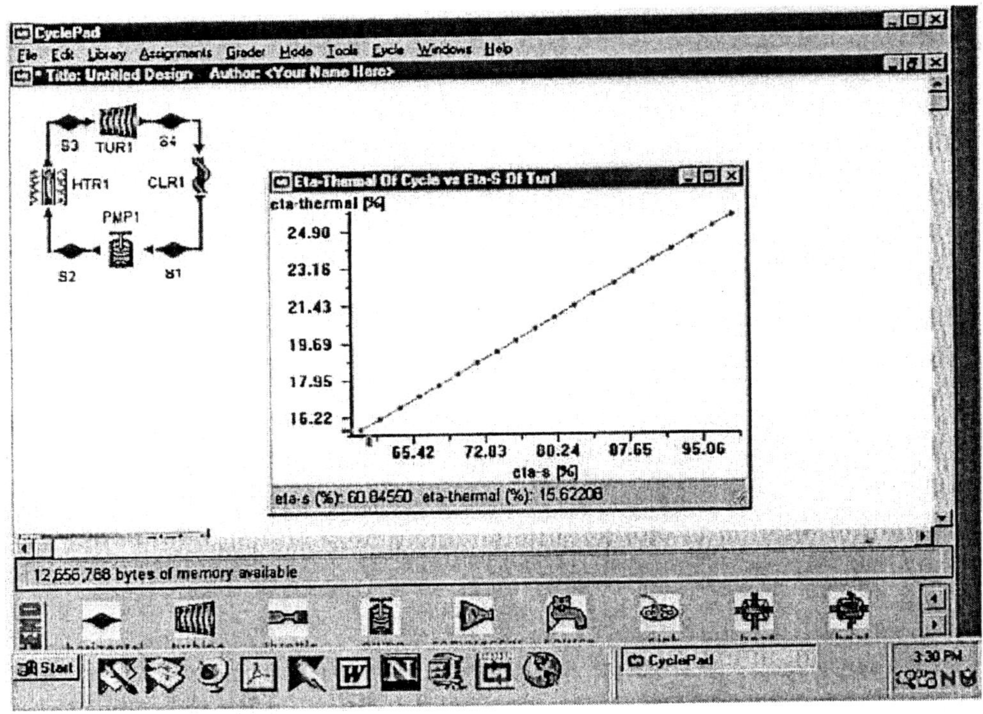

Figure Example 7.4.1c Rankine cycle sensitivity analysis

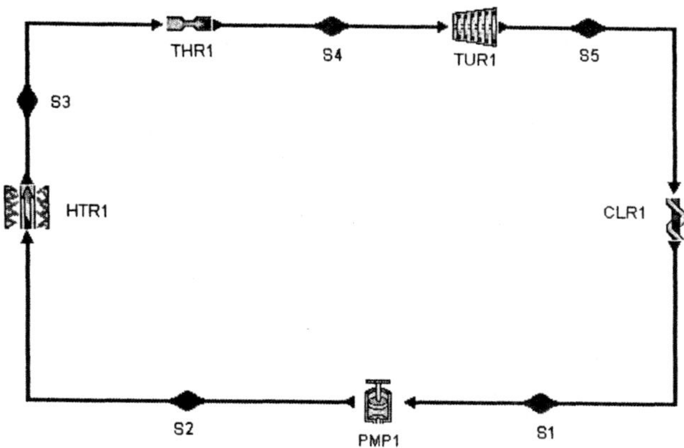

Figure 7.4.2 Throttling Rankine cycle

Example 7.4.2. An actual steam Rankine cycle operates between a condenser pressure of 1 psia and a boiler pressure of 600 psia. The outlet temperature of the superheater is 600°F and the turbine efficiency is 80%. The rate of mass flow in the cycle is 1 lbm/s.

(A) Find the pump power required, turbine power produced, rate of heat added in the boiler, and the cycle efficiency.

(B) If the pressure is throttled down to 400 psia at the inlet of the turbine, find the pump power required, turbine power produced, rate of heat added in the boiler, rate of heat removed in the condenser, and the cycle efficiency. Draw the T-s diagram.

To solve part (A) of this problem, let us make use of Figure 7.4.2.

(A) 1. Assume the pump is isentropic, boiler is isobaric, turbine is adiabatic with 80% efficiency, and condenser is isobaric; 2. Input p_1=1 psia, x_1=0, p_2=600 psia, T_3=600°F, mdot=1lbm/s, and p_4=600 psia;

The results are: $Wdot_{pump}$=-2.55 hp, $Wdot_{turbine}$=491.0 hp, $Qdot_{boiler}$=1218 Btu/s, and η=28.36% as shown in the following diagram.

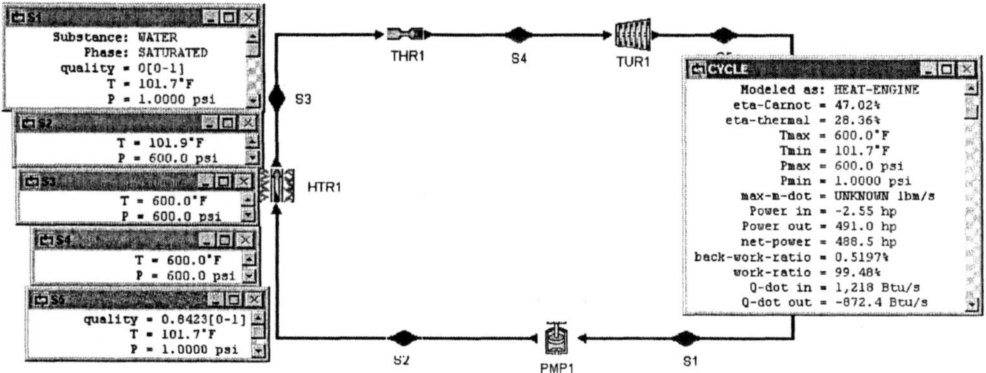

Figure example 7.4.2a Rankine cycle with throttling valve off

To solve part (B) of this problem, let us make use of Figure 7.4.2.

(B) 1. Assume the pump is isentropic, boiler is isobaric, turbine is adiabatic with 80% efficiency, and condenser is isobaric; 2. Input p_1=1 psia, x_1=0, p_2=600 psia, T_3=600°F, mdot=1lbm/s, and p_4=400 psia;

The results are: $Wdot_{pump}$=-2.55 hp, $Wdot_{turbine}$=461.2 hp, $Qdot_{boiler}$=1218 Btu/s, and η=26.62% as shown in the following figure, Figure example 7.4.2b.

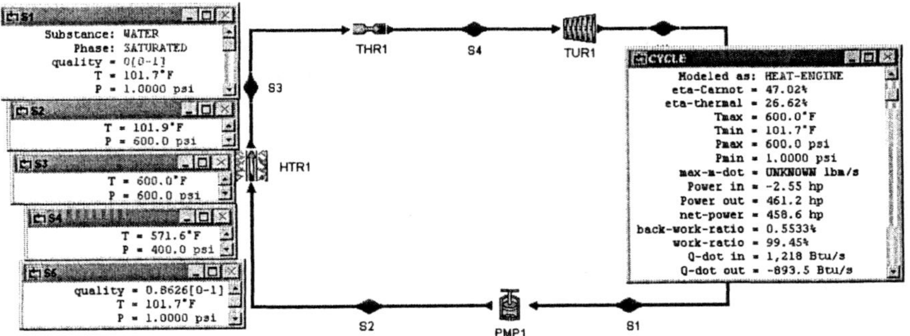

Figure example 7.4.2b Rankine cycle with throttling valve on

The T-s diagram is shown in Figure example 7.4.2c.

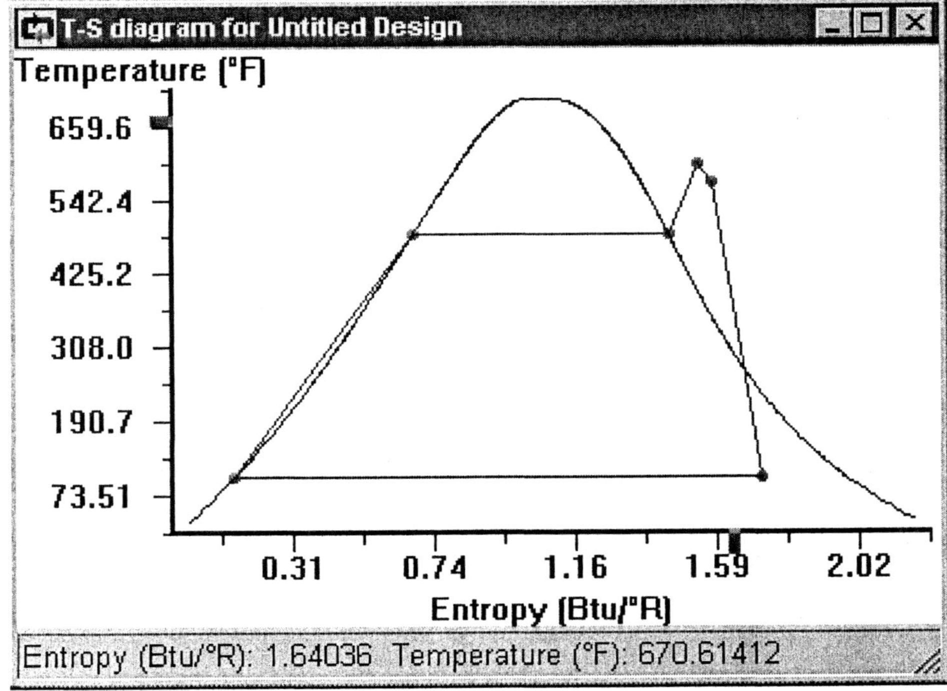

Figure example 7.4.2c Rankine throttling cycle T-s diagram

Homework7.4 Actual Rankine Cycle

1. Is Rankine cycle efficiency sensitive to the pump inefficiency? Why?
2. Is Rankine cycle efficiency sensitive to the turbine inefficiency? Why?
3. What is the purpose of the throttling valve in the Rankine throttling cycle? Does it improve the cycle efficiency?
4. Steam enters a turbine of a Rankine power plant at 5 MPa and 400°C, and exhausts to the condenser at 10 kPa. The turbine produces a power output of 20,000 kW. Given a turbine isentropic efficiency of 90% and a pump isentropic efficiency of 100% :
   (a) What is the mass flow rate of the steam around the cycle?
   (b) What is the rate of heat rejection from the condenser?
   (c) Find the thermal efficiency of the power plant.
5. 12.7 kg/s of superheated steam flow at 2 MPa and 320°C enters the turbine of a Rankine power plant and expands to a condenser pressure of 10 kPa. Assuming the isentropic efficiencies of the turbine and pump are 85 and 100 percent, respectively, find: (a) the actual pump power required, (b) the quality of steam at the exit of the turbine, (c) the actual turbine power produced, (d) the rate of heat supplied in the boiler, (e) the rate of heat removed in the condenser, and (f) the cycle efficiency.
6. Water circulates at a rate of 80 kg/s in a Rankine power plant. The boiler pressure is 6 MPa and the condenser pressure is 10 kPa. The steam enters the turbine at 700°C and water leaves the condenser as a saturated liquid. The actual turbine efficiency is 90%. Find: (a) the power required to operate the pump, (b) the heat transfer added in the boiler, (c) the power developed by the turbine, (d) the thermal efficiency of the cycle.
7. A Rankine cycle using water as the working fluid operates between a condenser pressure of 7.5 kPa and a boiler pressure of 17 MPa. The superheater temperature is 550°C. The rate of mass flow in the cycle is 2.3 lbm/s. The turbine efficiency is 85%. Determine (a) the pump power, (b) the turbine power, (c) the rate of heat added in the boiler, and (d) the cycle thermal efficiency.
8. A throttling Rankine cycle using water as the working fluid operates between a condenser pressure of 7.5 kPa and a boiler pressure of 17 MPa. The superheater temperature is 550°C. The rate of mass flow in the cycle is 2.3 lbm/s. The turbine efficiency is 85%. If the pressure at the exit of the throttling valve is 12 MPa, determine (a) the pump power, (b) the turbine power, (c) the rate of heat added in the boiler, and (d) the cycle thermal efficiency.

7.5 Reheat Rankine Cycle

The thermal efficiency of the Rankine cycle can be significantly increased by using higher boiler pressure. But this requires ever-increasing superheats. Since the maximum temperature in the superheater is limited by the temperature the boiler tubes can stand, superheater temperatures are usually restricted. Since the major fraction of the heat supplied to Rankine cycle is supplied in the boiler, the boiler temperatures (and hence pressures) must be increased if cycle efficiency improvements are to be obtained.

The problem of excessive superheater temperatures may be solved while avoiding two-phase saturated mixtures in the expansion by reheating the expanding steam part way through the expansion as shown in Figure 7.5.1. The steam leaving the boiler section as saturated vapor is superheated to an acceptable temperature and then expanded (while produce work) until it intersects with the maximum moisture (complement of quality, or minimum quality) curve. The steam is then reheated in a second superheater section and expanded in a second turbine (while produce more work) until it intersects with the maximum moisture (complement of quality, or minimum quality) curve again. The steam is then condensed and pump back into the boiler.

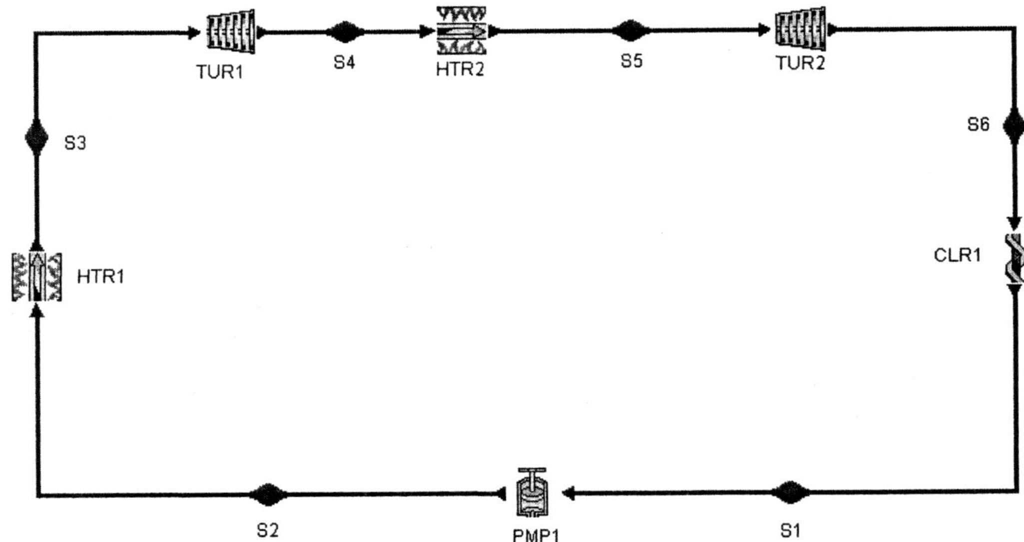

Figure 7.5.1 Reheat Rankine cycle

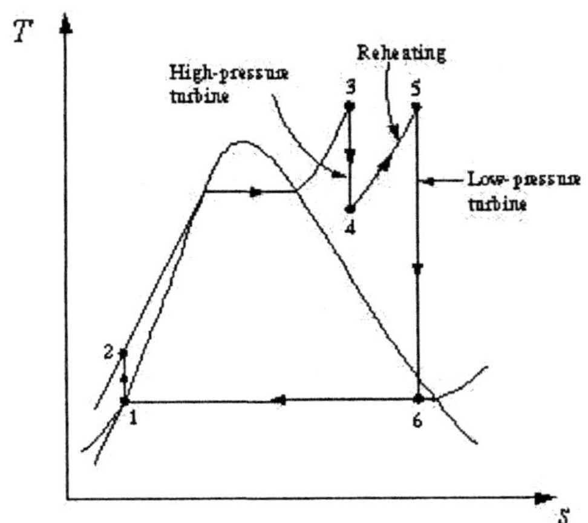

Figure 7.5.2 Reheat Rankine T-S diagram

Applying the First law of thermodynamics of the open system to each of the six processes of the reheat Rankine cycle yields:

$$Q_{12} = 0, \tag{7.5.1}$$

$$W_{12} = m(h_1 - h_2) = mv_1 (p_1 - p_2), \tag{7.5.2}$$

$$W_{23} = 0, \tag{7.5.3}$$

$$Q_{23} - 0 = m(h_3 - h_2), \tag{7.5.4}$$

$$Q_{34} = 0, \tag{7.5.5}$$

$$W_{34} = m(h_3 - h_4), \tag{7.5.6}$$

$$W_{45} = 0, \tag{7.5.7}$$

$$Q_{45} - 0 = m(h_5 - h_4), \tag{7.5.8}$$

$$Q_{56} = 0, \tag{7.5.9}$$

$$W_{56} = m(h_5 - h_6), \tag{7.5.10}$$

$$W_{61} = 0, \tag{7.5.11}$$

and

$$Q_{61} - 0 = m(h_1 - h_6), \tag{7.5.12}$$

The net work (W_{net}), which is also equal to net heat (Q_{net}), is

$$W_{net} = Q_{net} = Q_{23} + Q_{45} + Q_{61} \tag{7.5.13}$$

The thermal efficiency of the cycle is

$$\eta = W_{net} /(Q_{23} + Q_{45}) = Q_{net} /(Q_{23} + Q_{45}) = 1 - Q_{61} /(Q_{23} + Q_{45})$$
$$= 1 - (h_6 - h_1)/[(h_3 - h_2) + (h_5 - h_4)] \tag{7.5.14}$$

Example 7.5.1 A steam reheat Rankine cycle operates between the pressure limits of 5 psia and 1600 psia. Steam is superheated to 600°F before it is expanded to the reheat pressure of 500 psia. Steam is reheated to 600°F. The steam flow rate is 800 lbm/h. Determine the quality of steam at the exit of the turbine, the cycle efficiency, and the power produced by the cycle.

To solve this problem by CyclePad, we take the following steps:

1. Build
   (A) Take a pump, a boiler, a turbine, a reheater, another turbine and a condenser from the inventory shop and connect the devices to form the reheat Rankine cycle.
   (B) Switch to analysis mode.
2. Analysis

(A) Assume a process each for the six devices: (a) pump as adiabatic, (b) boiler and reheater as isobaric, (c) turbines as adiabatic, and (d) condenser as isobaric.

(B) Input the given information: (a) working fluid is water, (b) the inlet pressure and quality of the pump are 5 psia and 0, (c) the inlet pressure and temperature of the first turbine are 1600 psia and 600°F, (d) the mass flow rate is 800 lbm/h, and (e) the inlet pressure and temperature of the first turbine are 500 psia and 600°F.

3. Display results

(A) Display the T-s diagram and cycle properties results. The cycle is a heat engine. The answers are x=82.52 %, η=30.04 % and Net power output=111.4 hp.

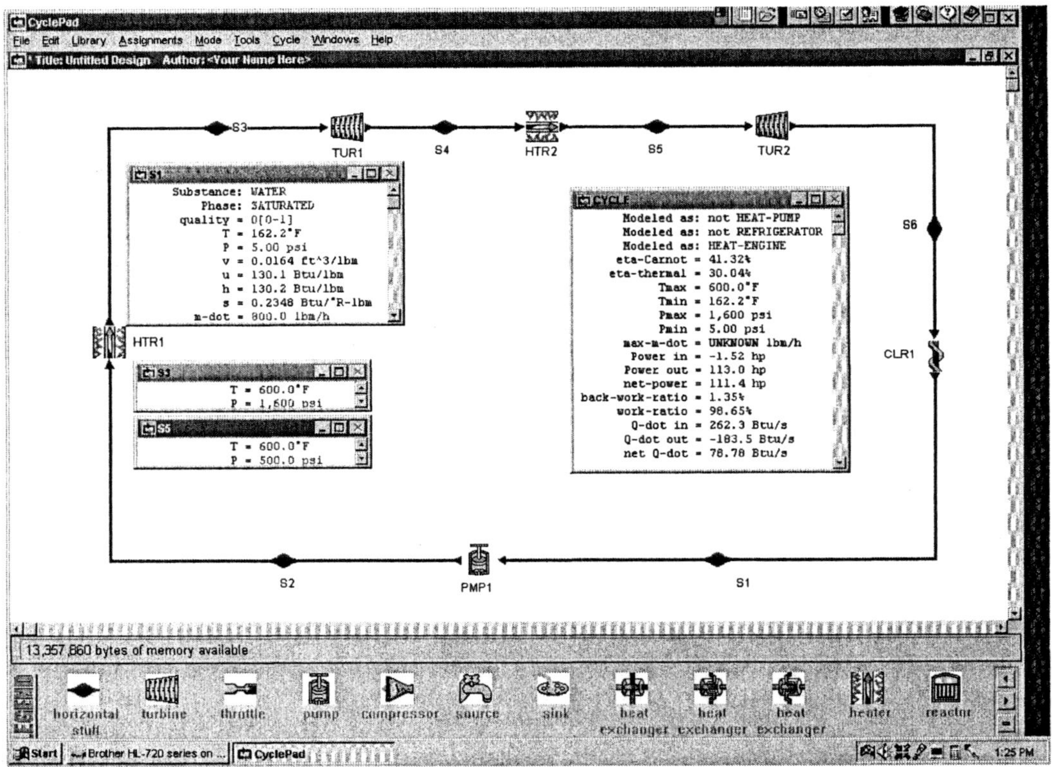

Figure Example 7.5.1 Reheat Rankine cycle

Comments: *The sole purpose of the reheat cycle is to reduce the moisture content of the steam at the final stages of the turbine expansion process. The more reheating process, the higher the quality of the steam at the exit of the last stage turbine. The reheat temperature is often very close or equal to the turbine inlet temperature. The optimum reheat pressure is about one-fourth of the maximum cycle pressure.*

Example 7.5.2 Determine the efficiency and power output of a reheat Rankine cycle using steam as the working fluid in which the condenser pressure is 80 kPa. The boiler pressure is 3 MPa. The steam leaves the boiler at 400°C. The mass flow rate of steam is 1 kg/s. The pump efficiency is 85% and the turbine efficiency is 88%. After expansion in the high-pressure turbine to 800 kPa, the steam is reheated to 400°C and then expanded in the low-pressure turbine to the condenser. Show the cycle on T-s diagram. Plot the sensitivity diagram of cycle efficiency vs reheat pressure. Optimize the reheat pressure for maximum cycle efficiency. Convert the SI unit system to British system and find the answer.

To solve this problem by CyclePad, we take the following steps:

1. Build
   (A) Take a pump, a boiler, a turbine, a reheater, another turbine and a condenser from the inventory shop and connect the devices to form the reheat Rankine cycle.
   (B) Switch to analysis mode.
2. Analysis
   (A) Assume a process each for the six devices: (a) pump as adiabatic, (b) boiler and reheater as isobaric, (c) turbines as adiabatic, and (d) condenser as isobaric.
   (B) Input the given information: (a) working fluid is water, (b) the inlet pressure and quality of the pump are 80 kPa and 0, (c) the inlet pressure and temperature of the turbine are 3 MPa and 400°C, (d) the mass flow rate is 1 kg/s, and (e) The pump efficiency is 85% and the turbines efficiency are 88%.
3. Display results
   (A) Display the T-s diagram and cycle properties results. The cycle is a heat engine. The answers are η=24.57% and Net power output=779.9 kW,(B) Display the sensitivity diagram of cycle efficiency vs reheater pressure, (C) Locate the optimum reheat pressure from the sensitivity diagram, and (D) Change the input value of reheat pressure.

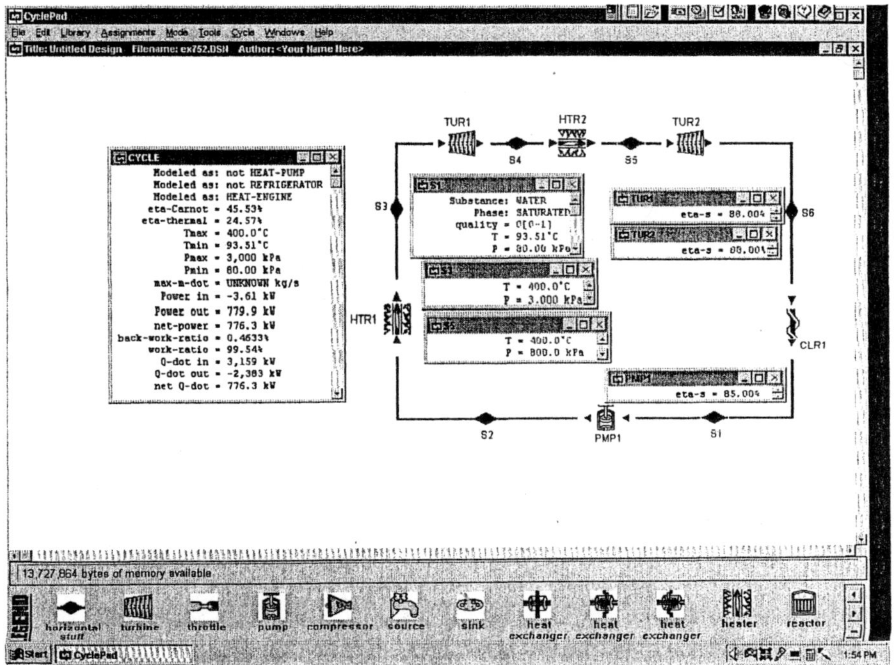

Figure Example 7.5.2a Reheat Rankine cycle

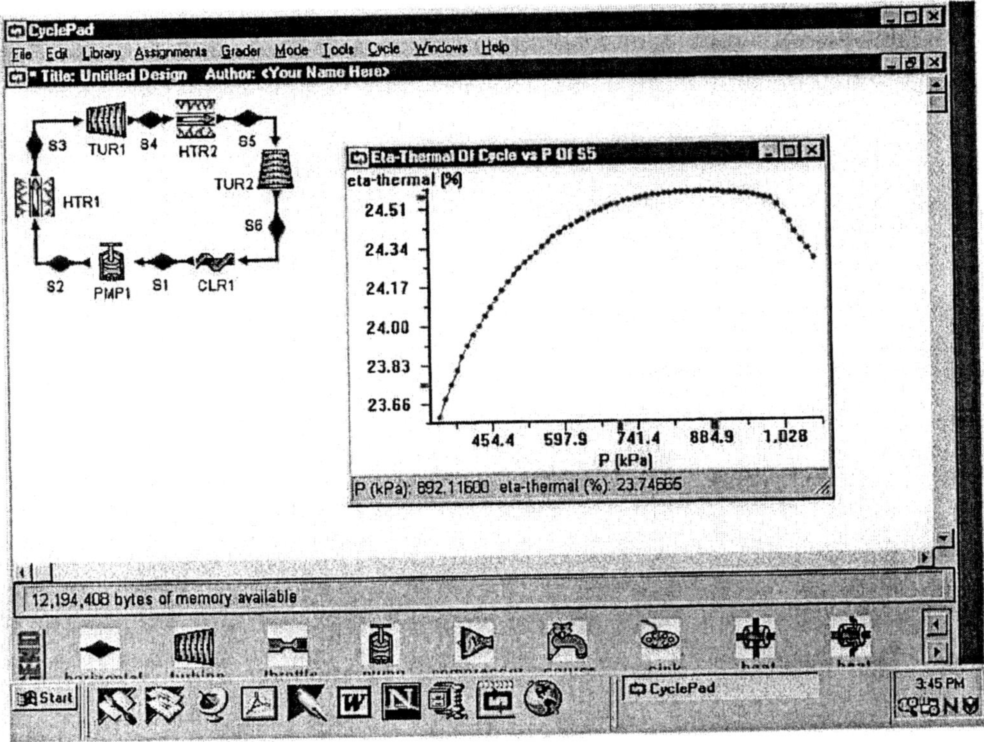

Figure Example 7.5.2a Reheat Rankine cycle *sensitivity analysis*

Comment: The advantage of using reheat is to reduce the moisture content at the exit of the low pressure turbine and increase the net power of the Rankine power plant. The one reheat Rankine basic cycle shown in Figure 7.5.1 can be expanded into more than one reheat if desired. In this fashion it is possible to use higher boiler pressure without having to increase the maximum superheater temperature above the limit of the superheater tubes.

Homework 7.5 Reheat Rankine Cycle

1. What is the purpose of reheat in a reheat Rankine cycle?
2. Is the efficiency of a reheat Rankine cycle always higher than the efficiency of a simple Rankine cycle operating between the same boiler pressure and condenser pressure?
3. Consider a steam power plant operating on the ideal reheat Rankine cycle. 1 kg/s of steam flow enters the high-pressure turbine at 15 MPa and 600°C and leaves at 5 MPa. Steam is reheated to 600°C and enters the low-pressure turbine. Exhaust steam from the turbine is condensed in the condenser at 10 kPa. Determine (a) the pump power required, (b) rate of heat added in the boiler, (c) rate of heat added in the reheater, (d) power produced by the high-pressure turbine, (e) power produced by the low-pressure turbine, (f) rate of heat removed from the condenser, (g) quality of steam at the exit of the low-pressure turbine, and (h) the thermal cycle efficiency.

4. Consider a steam power plant operating on the ideal reheat Rankine cycle. 1 kg/s of steam flow enters the high-pressure turbine at 15 MPa and 600°C and leaves at 5 MPa. Steam is reheated to 600°C and enters the low-pressure turbine. Exhaust steam from the turbine is condensed in the condenser at 10 kPa. Both turbines have 90% efficiency. Determine (a) the pump power required, (b) rate of heat added in the boiler, (c) rate of heat added in the reheater, (d) power produced by the high-pressure turbine, (e) power produced by the low-pressure turbine, (f) rate of heat removed from the condenser, (g) quality of steam at the exit of the low-pressure turbine, and (h) the thermal cycle efficiency.

7.6 Regenerative Rankine Cycle

The thermal efficiency of the Rankine cycle can be increased by the use of regenerative heat exchange as shown in Figure 7.6.1. In the regenerative cycle, a portion of the partially expanded steam is drawn off between the high- and low-pressure turbines. The steam is used to preheat the condensed liquid before it returned to the boiler. In this way, the amount of heat added at the low temperatures is reduced. Therefore the mean effective temperature of heat addition is increased, and cycle efficiency is increased. In the case of an ideal regenerative Rankine cycle, the best result is obtained by heating the feedwater to a temperature equal to the saturation temperature corresponding to the boiler pressure. To carry out the ideal regenerative process, the regenerative heat exchanger is called feedwater heater. The T-s diagram of the ideal *regenerative Rankine cycle* is shown in Figure 7.6.2.

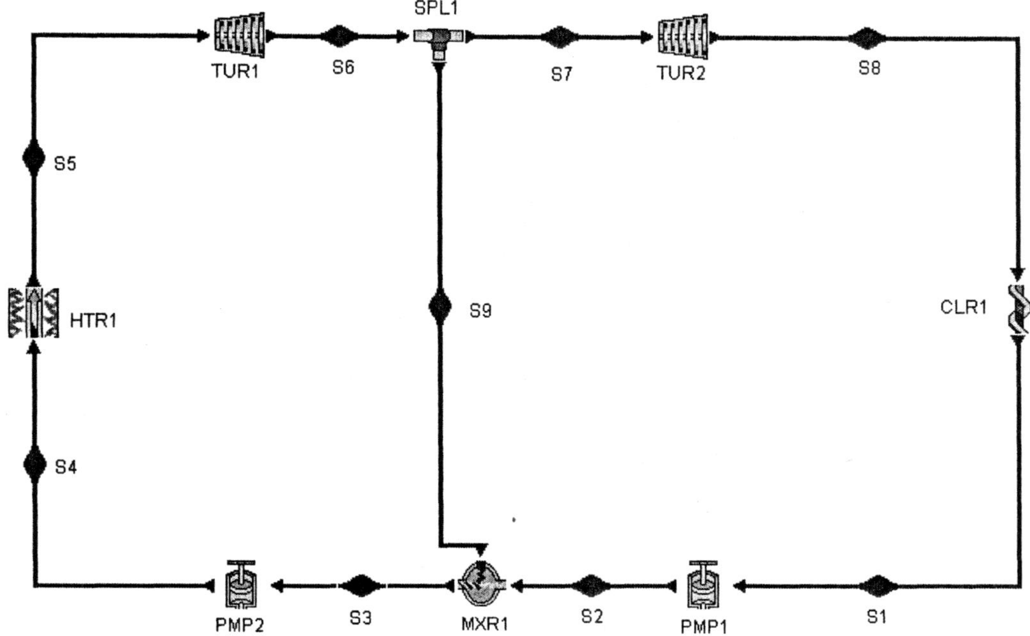

Figure 7.6.1. Regenerative Rankine cycle

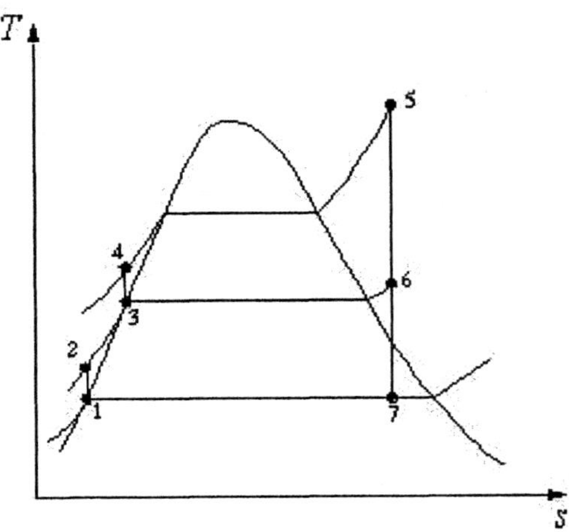

Figure 7.6.2 Regenerative Rankine cycle T-S diagram

The one regenerative Rankine basic cycle is composed of the following six processes:

1-2 isentropic compression
2-3 isobaric heat addition
3-4 isentropic compression
4-5 isobaric heat addition
5-6 isentropic expansion
6-7 isentropic expansion
7-1 isobaric heat removing

Applying the mass balance and the First law of thermodynamics of the open system to each of the six processes of the regenerative Rankine cycle yields:

$$m_1=m_2=m_7, \tag{7.6.1}$$

$$m_4=m_5, \tag{7.6.2}$$

$$m_4 = m_2 + m_6, \tag{7.6.3}$$

$$Q_{12} = 0, \tag{7.6.4}$$

$$W_{12} = m_1(h_1 - h_2) = m_1v_1 (p_1 - p_2), \tag{7.6.5}$$

$$m_4(h_4 - h_3) + m_2(h_2 - h_3)+m_6(h_6 - h_3) = 0, \tag{7.6.6}$$

$$Q_{34} = 0, \tag{7.6.7}$$

$$W_{34} = m_4(h_3 - h_4) = m_4v_3 (p_3 - p_4), \tag{7.6.8}$$

$$W_{45} = 0, \tag{7.6.9}$$

$$Q_{45} - 0 = m_4(h_5 - h_4), \tag{7.6.10}$$

$$Q_{56} = 0, \tag{7.6.11}$$

$$W_{56} = m_4(h_3 - h_4), \tag{7.6.12}$$

$$Q_{67} = 0, \tag{7.6.13}$$

$$W_{67} = m_1(h_6 - h_7), \tag{7.6.14}$$

$$W_{71} = 0, \tag{7.6.15}$$

and

$$Q_{71} - 0 = m_1(h_1 - h_6), \tag{7.6.16}$$

The net work (W_{net}), which is also equal to net heat (Q_{net}), is

$$W_{net} = W_{56} + W_{67} + W_{12} + W_{34} \tag{7.6.17}$$

The thermal efficiency of the cycle is

$$\eta = W_{net} / Q_{45} \tag{7.6.18}$$

Example 7.6.1 Determine the efficiency and power output of a regenerative Rankine cycle using steam as the working fluid and the condenser pressure is 80 kPa. The boiler pressure is 3 MPa. The steam leaves the boiler at 400°C. The mass rate of steam flow is 1 kg/s. The pump efficiency is 85% and the turbine efficiency is 88%. After expansion in the high-pressure turbine to 400 kPa, some of the steam is extracted from the turbine exit for the purpose of heating the feed-water in an open feed-water heater, the rest of the steam is reheated to 400°C and then expanded in the low-pressure turbine to the condenser. The water leaves the open feed-water heater at 400 kPa as saturated liquid. Determine the steam fraction extracted from the turbine exit, cycle efficiency, and net power output of the cycle.

To solve this problem by CyclePad, we take the following steps:

1. Build
   (A) Take two pumps, a boiler, a turbine, a reheater, another turbine, a splitter, a mixing chamber (open feed-water heater) and a condenser from the inventory shop and connect the devices to form the regenerating Rankine cycle.
   (B) Switch to analysis mode.
2. Analysis
   (A) Assume a process for each of the devices: (a) pumps as adiabatic, (b) boiler and reheater as isobaric, (c) turbines as adiabatic, (d) splitter as iso-parametric, and (e) condenser as isobaric.

(B) Input the given information: (a) working fluid is water, (b) the inlet pressure and quality of the pump are 10 kPa and 0, (c) the inlet pressure and temperature of the turbine are 4 MPa and 400°C, (d) the mass flow rate is 1 kg/s through the boiler, and (e) The pump efficiency is 85% and the turbines efficiency is 88%.

3. Display results

(A) Display the T-s diagram and cycle properties results. The cycle is a heat engine. The answers are fraction extraction=0.0877, η=24.28% and Net power output=636.8 kW, (B) Display the sensitivity diagram of cycle efficiency vs reheater pressure, and (C) Locate the optimum regeneration pressure (about 730 kPa) from the sensitivity diagram, (D) Redo the problem with 730 kPa: fraction extraction=0.1233, η=24.42% and Net power output=616.1 kW

Comments: (1) From the example, it is seen that the efficiency of the regenerative Rankine cycle is better than the efficiency of the Rankine cycle without regenerator. (2) Suppose infinitive number of regenerators are used, the regenerative cycle would have the same cycle efficiency as that of a Carnot cycle operating between the same temperature limits. This is physically not practical. Large number of regenerators may not be economically justified. Consequently, the finite number of regenerators is a design decision. In practice, six or seven regenerators is the maximum number employed for huge Rankine commercial power plants.

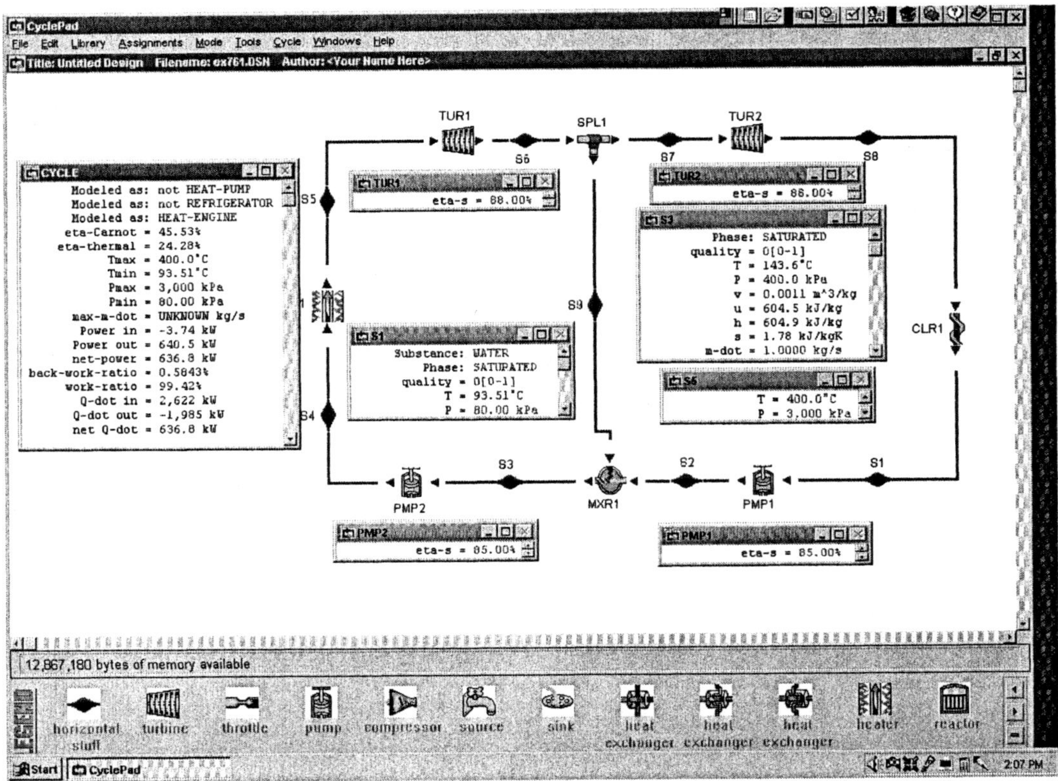

Figure Example 7.6.1a Regenerative Rankine cycle

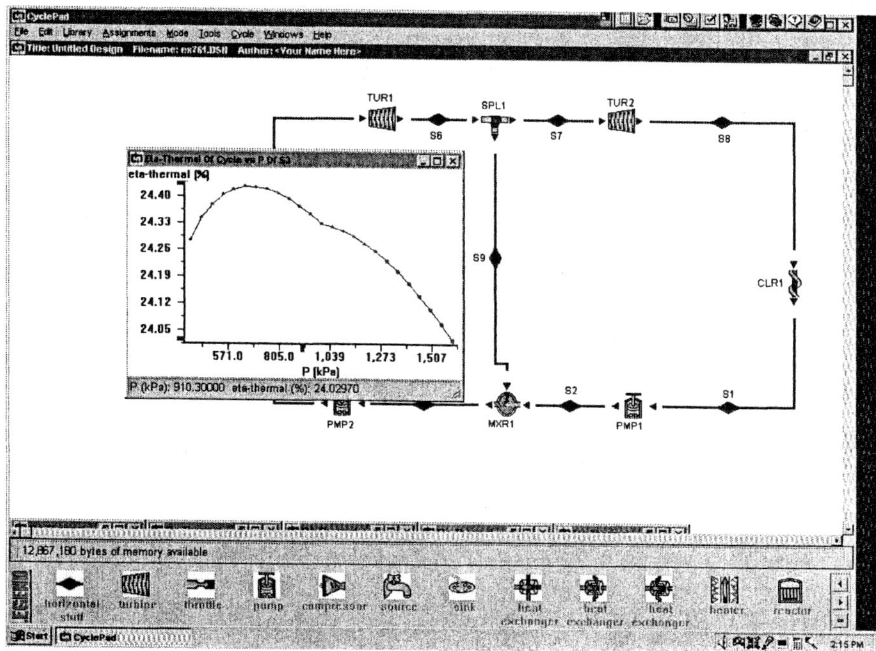

Figure Example 7.6.1b Regenerative Rankine cycle

Example 7.6.2 Determine the efficiency and power output of a regenerative Rankine (without superheater or reheater) cycle using steam as the working fluid in which the condenser temperature is 50°C. The boiler temperature is 350°C. The steam leaves the boiler as saturated vapor. The mass rate of steam flow is 1 kg/s. After expansion in the high-pressure turbine to 100°C, some of the steam is extracted from the turbine exit for the purpose of heating the feed-water in an open feed-water heater, the rest of the steam is then expanded in the low-pressure turbine to the condenser. The water leaves the open feed-water heater at 100°C as saturated liquid. (A) Determine the steam fraction extracted from the turbine exit, cycle efficiency, and net power output of the cycle, (B) Plot the sensitivity diagram of cycle efficiency vs open feed-water heater temperature, and (C) Determine the steam fraction extracted from the turbine exit, cycle efficiency, and net power output of the cycle at the optimal cycle efficiency.

To solve this problem by CyclePad, we take the following steps:

1. Build
   (A) Take two pumps, a boiler, a turbine, a reheater, another turbine, a splitter, a mixing chamber (open feed-water heater) and a condenser from the inventory shop and connect the devices to form the regenerating Rankine cycle.
   (B) Switch to analysis mode.
2. Analysis
   (A) Assume a process each for the devices: (a) pumps as adiabatic, (b) boiler as isobaric, (c) turbines as adiabatic, (d) splitter as iso-parametric, and (e) condenser as isobaric.
   (B) Input the given information: (a) working fluid is water, (b) the inlet temperature and quality of the pump are 50°C and 0, (c) the turbine inlet steam quality and

temperature are 1 and 350°C, (d) the mass flow rate is 1 kg/s through the boiler, and (e) the regenerator exit steam quality and temperature are 0 and 100°C.

3. Display results

(A) Display the cycle properties results. The cycle is a heat engine. The answers are fraction extraction=0.1258, η=40.16% and net power output=871.4 kW.

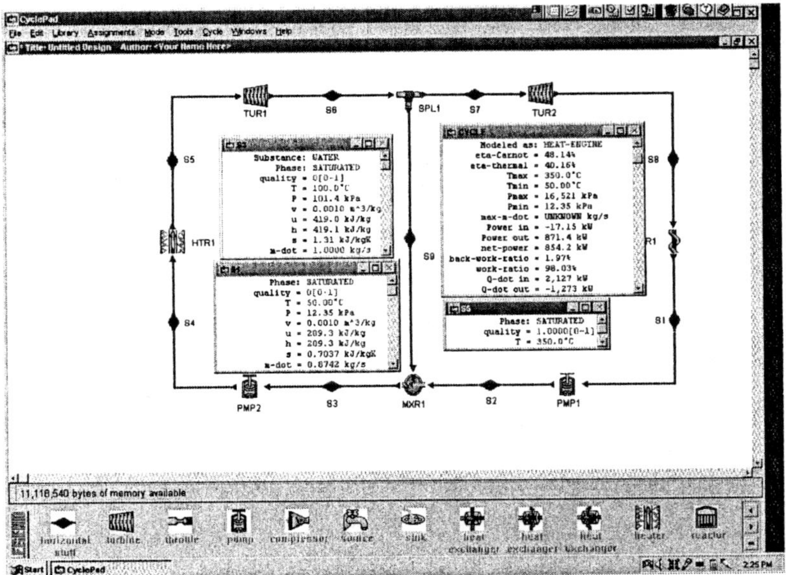

Figure Example 7.6.2a Regenerative Rankine cycle

(B) The sensitivity diagram of cycle efficiency vs open feed-water heater temperature (η vs T_3) is

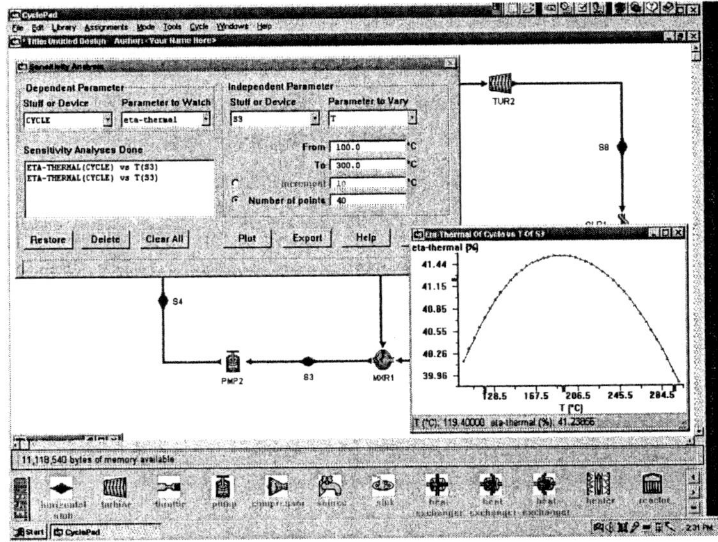

Figure Example 7.6.2b Regenerative Rankine cycle sensitivity diagram

Comment: The temperature difference between condenser and regenerator is approximately equal to the temperature difference between boiler and regenerator. From the diagram, the temperature is approximately 187°C.

(C) Change the temperature of water leaving the open feed-water heater to 187°C as saturated liquid, the cycle properties give fraction extraction=0.2970, η=41.55% and net power output=727.9 kW.

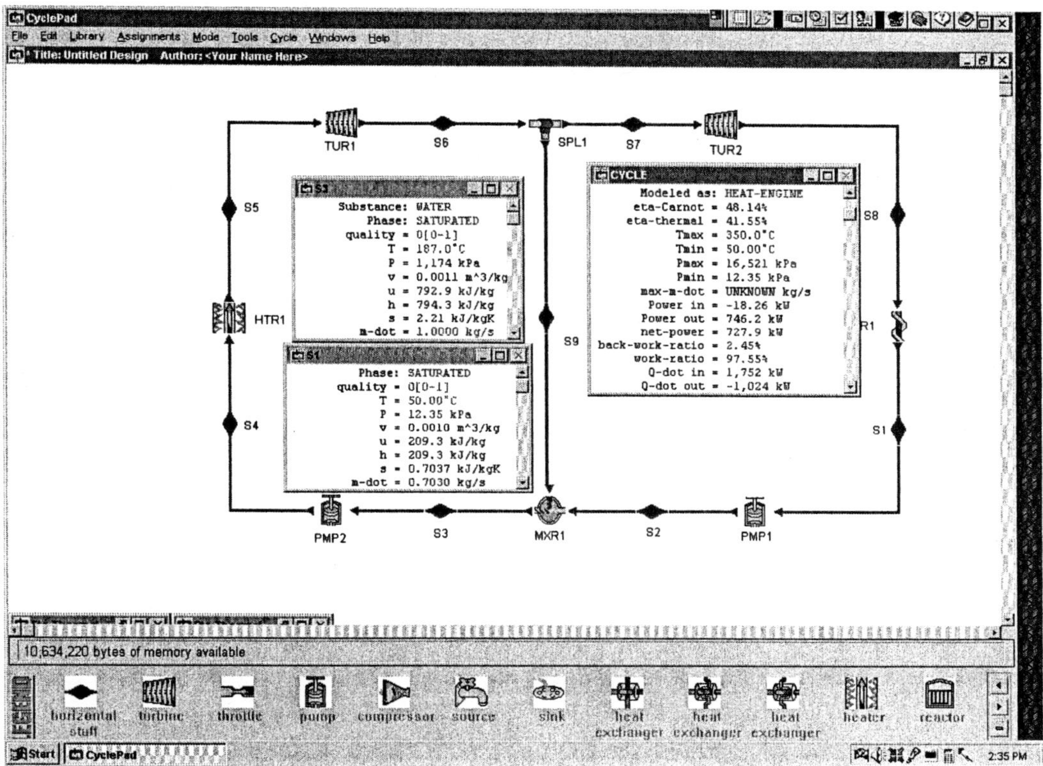

Figure Example 7.6.2c Regenerative Rankine cycle optimization

Example 7.6.3 Determine the efficiency and power output of a 4-stage steam regenerative Rankine cycle (Figure Example 7.6.3a). The following information is provided:

$p_1=p_{24}=10$ kPa, $p_2=p_3=p_{21}=p_{22}=p_{23}=p_{28}=3000$ kPa, $p_4=p_5=p_{18}=p_{19}=p_{20}=p_{27}=7000$ kPa, $p_6=p_7=p_{15}=p_{16}=p_{17}=p_{26}=10000$ kPa, $p_8=p_9=p_{12}=p_{13}=p_{14}=p_{25}=13000$ kPa, $p_{10}=p_{11}=16000$ kPa, $T_{11}=T_{14}=T_{17}=T_{20}=T_{23}=400$°C, $x_1=x_3=x_5=x_7=x_9=0$, $mdot_9=1$ kg/s, $\eta_{tur1}=\eta_{tur2}=\eta_{tur3}=\eta_{tur4}=\eta_{tur5}=85\%$, and $\eta_{pmp1}=\eta_{pmp2}=\eta_{pmp3}=\eta_{pmp4}=\eta_{pmp5}=85\%$.

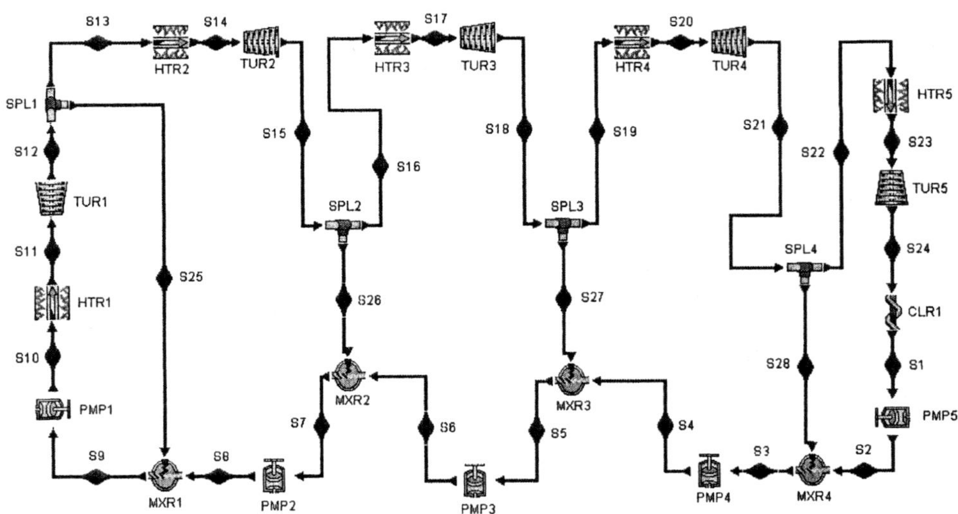

Figure Example 7.6.3a 4-stage steam regenerative Rankine cycle

To solve this problem by CyclePad, we take the following steps:

1. Build
   (A) Take five pumps, five heaters (1 boiler and 4 reheaters), five turbines, four splitters, four mixing chambers (open feed-water heaters) and a condenser from the inventory shop and connect the devices to form the regenerating Rankine cycle.
   (B) Switch to analysis mode.
2. Analysis
   (A) Assume a process each for the devices: (a) pumps as adiabatic, (b) boiler, reheater and regenerators as isobaric, (c) turbines as adiabatic, (d) splitters as iso-parametric, and (e) condenser as isobaric.
   (B) Input the given information: working fluid is water, $p_1=p_{24}=10$ kPa, $p_2=p_3=p_{21}=p_{22}=p_{23}=p_{28}=3000$ kPa, $p_4=p_5=p_{18}=p_{19}=p_{20}=p_{27}=7000$ kPa, $p_6=p_7=p_{15}=p_{16}=p_{17}=p_{26}=10000$ kPa, $p_8=p_9=p_{12}=p_{13}=p_{14}=p_{25}=13000$ kPa, $p_{10}=p_{11}=16000$ kPa, $T_{11}=T_{14}=T_{17}=T_{20}=T_{23}=400°C$, $x_1=x_3=x_5=x_7=x_9=0$, $mdot_9=1$ kg/s, $\eta_{tur1}=\eta_{tur2}=\eta_{tur3}=\eta_{tur4}=\eta_{tur5}=85\%$, and $\eta_{pmp1}=\eta_{pmp2}=\eta_{pmp3}=\eta_{pmp4}=\eta_{pmp5}=85\%$ as shown in Figure 7.6.3b.
3. Display results
   (A) Display the cycle properties results. The cycle is a heat engine. The results are: $\eta_{cycle}=36.92\%$, $Wdot_{input}=-97.09$ kW, $Wdot_{output}=-777.3$ kW, $Wdot_{net\ output}=680.2$ kW, $Qdot_{add}=1842$ kW, $Qdot_{remove}=-1162$ kW, $Wdot_{pmp1}=-44.23$ kW, $Wdot_{pmp2}=-42.45$ kW, $Wdot_{pmp3}=-4.14$ kW, $Wdot_{pmp4}=-4.33$ kW, $Wdot_{pmp5}=-1.93$ kW, $Wdot_{tur1}=39.74$ kW, $Wdot_{tur2}=50.45$ kW, $Wdot_{tur3}=69.28$ kW, $Wdot_{tur4}=147.7$ kW, $Wdot_{tur5}=470.1$ kW, $Qdot_{htr1}=1372$ kW, $Qdot_{htr2}=112.8$ kW, $Qdot_{htr3}=107.2$ kW, $Qdot_{htr4}=107.2$ kW, $Qdot_{htr5}=143.5$ kW, $Qdot_{clr1}=-1162$ kW, $mdot_2=mdot_{22}=0.5372$ kg/s, $mdot_3=mdot_{19}=0.7605$ kg/s, $mdot_5=mdot_{16}=0.8705$ kg/s, $mdot_8=mdot_{12}=0.9459$ kg/s, $mdot_{25}=0.0541$ kg/s, $mdot_{26}=0.0754$ kg/s, $mdot_{27}=0.1100$ kg/s, and $mdot_{28}=0.2234$ kg/s.

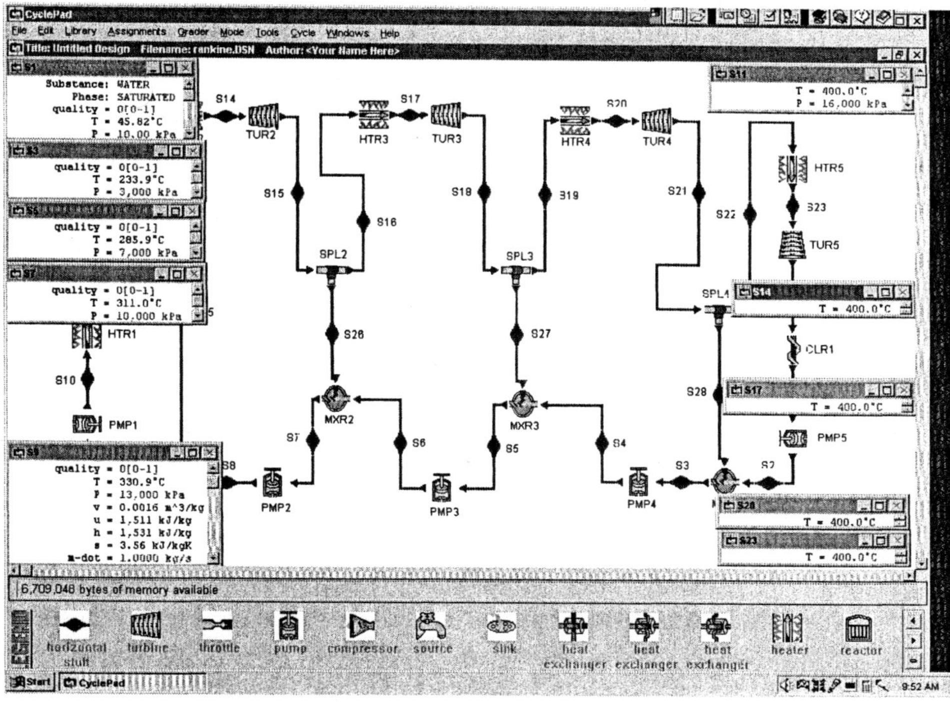

Figure Example 7.6.3b 4-stage steam regenerative Rankine cycle input

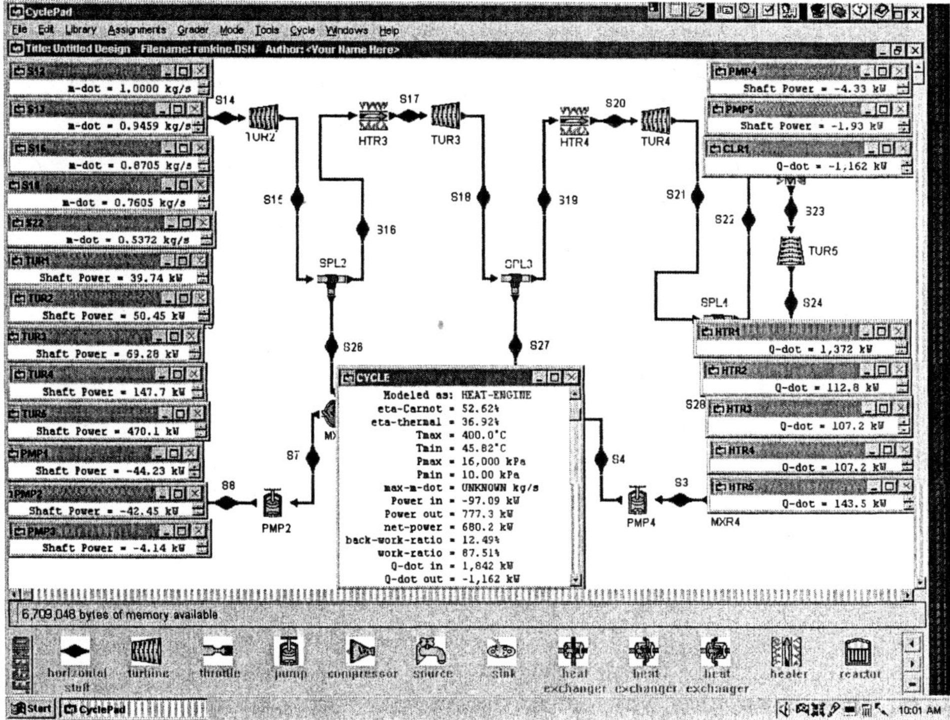

Figure Example 7.6.3b 4-stage steam regenerative Rankine cycle output

Homework 7.6 Regenerative Rankine Cycle

1. What is the purpose of regeneration in a regenerative Rankine cycle?
2. Consider a steam power plant operating on the ideal regenerating Rankine cycle. 1 kg/s of steam flow enters the turbine at 15 MPa and 600°C and is condensed in the condenser at 10 kPa. Some steam leaves the high pressure turbine at 1.2 MPa and enters the open feed-water heater. If the steam at the exit of the open feed-water heater is saturated liquid, determine (a) the fraction of steam not extracted from the high pressure turbine, (b) the rate of heat added in the boiler, (c) the rate of heat removed from the condenser, (d) the turbine power produced by the high pressure turbine, (e) the turbine power produced by the low pressure turbine, (f) the power required by the low-pressure pump, (g) the power required by the high-pressure pump, and (h) the thermal cycle efficiency.
3. Consider a steam power plant operating on the ideal regenerating Rankine cycle. 1 kg/s of steam flow enters the turbine at 15 MPa and 600°C and is condensed in the condenser at 10 kPa. Some steam leaves the high pressure turbine at 1.2 Mpa and enters the open feed-water heater. The turbine efficiency is 90%. If the steam at the exit of the open feed-water heater is saturated liquid, determine (a) the fraction of steam not extracted from the high pressure turbine, (b) the rate of heat added in the boiler, (c) the rate of heat removed from the condenser, (d) the turbine power produced by the high pressure turbine, (e) the turbine power produced by the low pressure turbine, (f) the power required by the low-pressure pump, (g) the power required by the high-pressure pump, and (h) the thermal cycle efficiency.
4. Determine the efficiency and power output of a regenerative Rankine (without superheater or reheater) cycle using steam as the working fluid in which the condenser temperature is 50°C. The boiler temperature is 350°C. The steam leaves the boiler as saturated vapor. The mass rate of steam flow is 1 kg/s. After expansion in the high-pressure turbine to 100°C, some of the steam is extracted from the turbine exit for the purpose of heating the feed-water in an open feed-water heater, the rest of the steam is then expanded in the low-pressure turbine to the condenser. The water leaving the open feed-water heater at 120°C as saturated liquid. Both turbine efficiency are 85%. (A) Determine the steam fraction extracted from the turbine exit, cycle efficiency, and net power output of the cycle, (B) Plot the sensitivity diagram of cycle efficiency vs open feed-water heater temperature, and (C) Determine the steam fraction extracted from the turbine exit, cycle efficiency, and net power output of the cycle at the optimal cycle efficiency.

   ANSWER: (A) y=0.1605, η=35.17%, $Wdot_{net}$=718.4 kW; (B) T =187°C; (C) y=0.2884, η=35.97%, $Wdot_{net}$=630.2 kW

7.7 Low-Temperature Rankine Cycles

Much of this chapter has been concerned with various modifications to the simple Rankine cycle at high-temperature. In this section, the Rankine cycle which makes the possible use of energy sources at low-temperature such as solar, geothermal, ocean thermal, solar pond, and waste heat will be discussed. Because of the small temperature range available, only a simple Rankine cycle can be used and the cycle efficiency will be low. This is not critical economically, because the fuel is free.

The working fluids such as ammonia and freons used in refrigerators and heat pumps are more desirable than steam for the low-temperature Rankine cycles. The reason is that the specific volume of such working fluids at low temperature is much less than that of steam, resulting turbine sizes can be much smaller and less expensive.

The sun provides a direct flux of solar thermal radiation about 1350 kW/m^2 outside the atmospheric air, and about 630 kW/m^2 on the average on the earth's surface. If flat-plate collector, rather than a concentrate collector used to focus the power of the sun is used, the *solar thermal power system* is a low-temperature heat engine.

There are regions where geothermal energy in dry-steam form, hot-water form, or hot-rock form could be tapped and used for power generation. Dry-steam is the most desirable form of geothermal energy. Figure 7.7.1 shows the basic Rankine cycle for a dry-steam geothermal source. More commonly than it puts out dry steam, a geothermal well puts out a mixture of steam and water. A separator is needed to separate the flashing steam from the hot-water. The steam is then used to drive the turbine as illustrated in Figure 7.7.2. In a hot-rock (no steam nor hot water) geothermal well, water will need to be injected into the well to tap the thermal energy. As with the hot-water geothermal energy, a secondary closed, simple Rankine cycle will be required to produce *geothermal power* as illustrated in Figure 7.8.3.

Incident solar energy is absorbed by the surface water of the oceans. The ocean surface temperatures in excess of 26°C occur near the equator. Pure water has a maximum density at a temperature of 4°C. The chilled water tends to settle to the depths of the ocean. The combination of the warmed ocean surface water and cold deep ocean water provides the thermodynamic condition needed to operate a heat engine called *ocean thermal energy conversion* (OTEC). A typical closed-cycle OTEC Rankine cycle using a working fluid such as ammonia or a freon is suggested as illustrated in Figure 7.7.3.

Temperature differences have been found in nature ponds having high concentration gradient of dissolved salt. Solar radiation is absorbed in the lower water levels and at the bottom of the pond. The water near the bottom (70 to 80°C) is at a higher temperature than that of the top surface (30°C), with the density of the hot concentrated lower level water higher than the density of the more dilute and cooler top levels. A typical closed-cycle *solar pond* Rankine cycle using a working fluid such as ammonia or a freon is also suggested as illustrated in Figure 7.7.3.

Waste heat from factory processes and exhaust from gas turbine engines and Diesel engines could also be used for power generation. A typical closed-cycle *waste heat* Rankine cycle using a working fluid such as ammonia or a freon is illustrated in Figure 7.7.3.

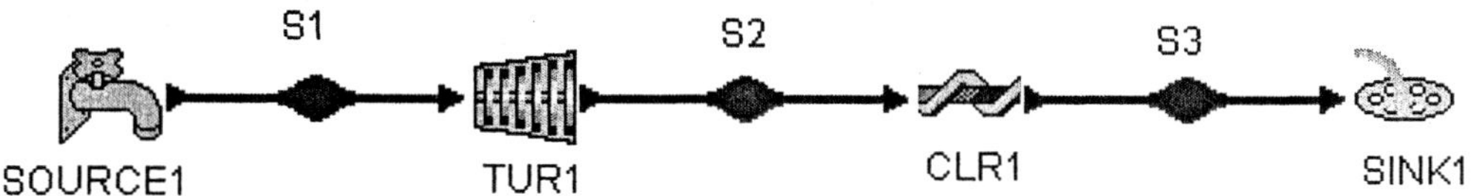

Figure 7.7.1. Dry-steam geothermal power plant

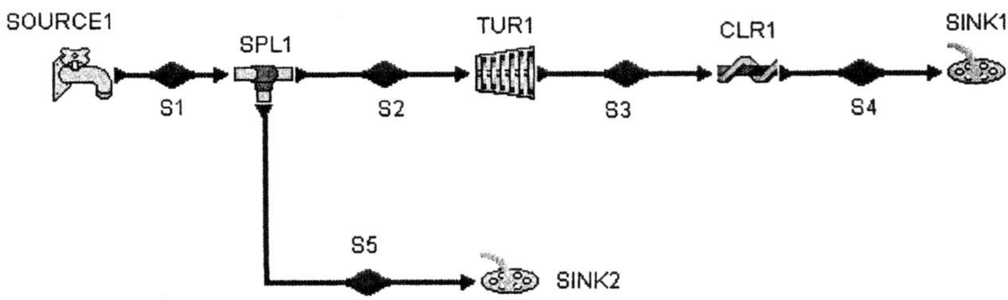

Figure 7.7.2. Hot water-steam mixture geothermal power plant

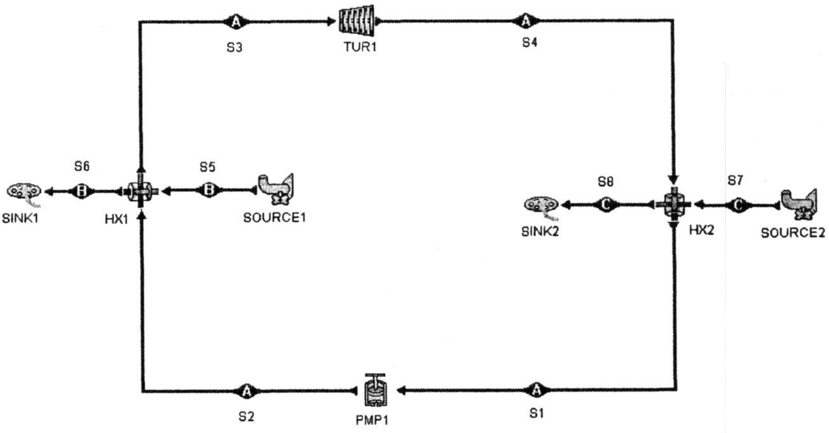

Figure 7.7.3. A closed-cycle low-temperature Rankine cycle

Example 7.7.1. A typical closed-cycle OTEC Rankine cycle using ammonia is suggested as illustrated in Figure 7.7.3 with the following information:

Condenser temperature	12°C
Boiler temperature	24°C
Mass flow rate of ammonia	1 kg/s
Surface ocean warm water entering heat exchanger	28°C
Surface ocean warm water leaving heat exchanger	26°C
Deep ocean cooling water entering heat exchanger	5°C
Deep ocean cooling water leaving heat exchanger	9°C
Turbine efficiency	100%
Pump efficiency	100%

(A) Determine the pump power, turbine power, net power output, rate of heat added in the heat exchanger by surface ocean warm water, rate of heat removed in the heat exchanger by deep ocean cooling water, cycle efficiency, specific volume of ammonia entering the turbine, boiler pressure, condenser pressure, mass flow rate of surface ocean warm water, and mass flow rate of deep ocean cooling water.

(B) Change the working fluid to R-134a. Determine the pump power, turbine power, net power output, rate of heat added in the heat exchanger by surface ocean warm water, rate of heat removed in the heat exchanger by deep ocean cooling water, cycle efficiency, specific volume of ammonia entering the turbine, boiler pressure, condenser pressure, mass flow rate of surface ocean warm water, and mass flow rate of deep ocean cooling water.

To solve this problem with CyclePad, we take the following steps:

1. Build as shown in Figure 7.7.3
2. Analysis
   (A) Assume a process each for the four devices: (a) pump as adiabatic with 100% efficiency, (b) turbine as adiabatic with 100% efficiency, (c) heat exchanger 1 (boiler) as isobaric on both cold-side and hot-side, and (d) heat exchanger 2 (condenser) as isobaric on both cold-side and hot-side.
   (B) Input the given information: (a) working fluid of cycle is ammonia, working fluid of cycle is water, and working fluid of cycle is water, (b) the inlet temperature and quality of the pump are 12°C and 0, (c) the inlet temperature and quality of the turbine are 24°C and 1, and (d) the mass flow rate is 1 kg/s.
3. Display result
   The answers are: (A) $Wdot_{pump}$=-6.58 kW, $Wdot_{turbine}$=48.91 kW, $Wdot_{net}$=42.33 kW, $Qdot_{boiler}$=1220 kW, $Qdot_{condenser}$=-1178 kW, η=3.47%, v=, p_{boiler}=972.4 kPa, $p_{condenser}$=658.5 kPa, $mdot_{warm\ water}$=145.8 kg/s, and $mdot_{cold\ water}$=70.23 kg/s.
4. (A) Retract the working fluid: working fluid of cycle is R-134a, and (B) Display result as shown in Figure Example 7.8.1.
   The answers are: (B) $Wdot_{pump}$=-1.38 kW, $Wdot_{turbine}$=7.74 kW, $Wdot_{net}$=6.36 kW, $Qdot_{boiler}$=194.2 kW, $Qdot_{condenser}$=-187.9 kW, η=3.27%, v=, p_{boiler}=445.3 kPa, $p_{condenser}$=647.6 kPa, $mdot_{warm\ water}$=23.21 kg/s, and $mdot_{cold\ water}$=11.21 kg/s.

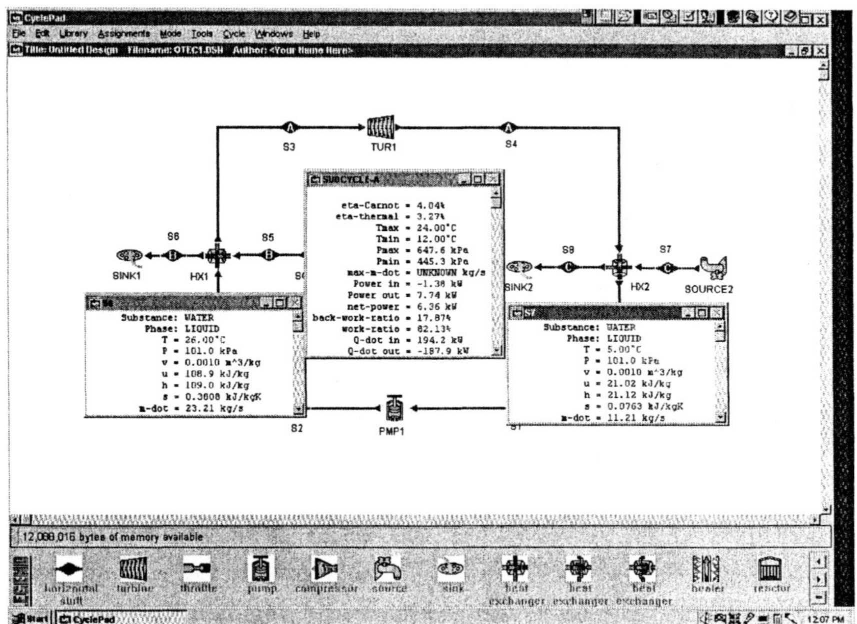

Figure Example 7.7.1 OTEC Rankine cycle

Homework 7.7 Low-Temperature Rankine Cycles

1. Is the cycle efficiency of the low-temperature heat engine higher than that of the high-temperature heat engine?
2. What is the fuel cost of the low-temperature heat engine?
3. List at least three low-temperature energy resources.
4. Why working fluids such as ammonia and freons used in refrigerators and heat pumps are more desirable than steam for the low-temperature Rankine cycles?
5. A typical closed-cycle OTEC Rankine cycle using ammonia is suggested as illustrated in Figure 7.8.3 with the following information:

Condenser temperature	10°C
Boiler temperature	22°C
Mass flow rate of ammonia	1 kg/s
Surface ocean warm water entering heat exchanger	28°C
Surface ocean warm water leaving heat exchanger	22°C
Deep ocean cooling water entering heat exchanger	4°C
Deep ocean cooling water leaving heat exchanger	10°C
Turbine efficiency	100%
Pump efficiency	100%

   Determine the pump power, turbine power, net power output, rate of heat added in the heat exchanger by surface ocean warm water, rate of heat removed in the heat exchanger by deep ocean cooling water, cycle efficiency, boiler pressure, condenser pressure, mass flow rate of surface ocean warm water, and mass flow rate of deep ocean cooling water.
   ANSWERS: $Wdot_{pump}$=-6.36 kW, $Wdot_{turbine}$=49.55 kW, $Wdot_{net}$=43.18 kW, $Qdot_{boiler}$=1228 kW, $Qdot_{condenser}$=-1185 kW, η=3.52%, p_{boiler}=913.4 kPa, $p_{condenser}$=614.9 kPa, $mdot_{warm\ water}$=47.08 kg/s, and $mdot_{cold\ water}$=36.71 kg/s.
6. A typical closed-cycle OTEC Rankine cycle using R-12 is suggested as illustrated in Figure 7.8.3 with the following information:

Condenser temperature	10°C
Boiler temperature	22°C
Mass flow rate of ammonia	1 kg/s
Surface ocean warm water entering heat exchanger	28°C
Surface ocean warm water leaving heat exchanger	22°C
Deep ocean cooling water entering heat exchanger	4°C
Deep ocean cooling water leaving heat exchanger	10°C
Turbine efficiency	100%
Pump efficiency	100%

   Determine the pump power, turbine power, net power output, rate of heat added in the heat exchanger by surface ocean warm water, rate of heat removed in the heat exchanger by deep ocean cooling water, cycle efficiency, boiler pressure, condenser pressure, mass flow rate of surface ocean warm water, and mass flow rate of deep ocean cooling water.
   ANSWERS: $Wdot_{pump}$=-1.06 kW, $Wdot_{turbine}$=6.06 kW, $Wdot_{net}$=5.00 kW, $Qdot_{boiler}$=150.1 kW, $Qdot_{condenser}$=-145.1 kW, η=3.33%, p_{boiler}=600.7 kPa, $p_{condenser}$=423.0 kPa, $mdot_{warm\ water}$=4.49 kg/s, and $mdot_{cold\ water}$=5.77 kg/s.

7. A typical closed-cycle OTEC Rankine cycle using R-22 is suggested as illustrated in Figure 7.8.3 with the following information:

Condenser temperature	10°C
Boiler temperature	22°C
Mass flow rate of ammonia	1 kg/s
Surface ocean warm water entering heat exchanger	28°C
Surface ocean warm water leaving heat exchanger	22°C
Deep ocean cooling water entering heat exchanger	4°C
Deep ocean cooling water leaving heat exchanger	10°C
Turbine efficiency	100%
Pump efficiency	100%

Determine the pump power, turbine power, net power output, rate of heat added in the heat exchanger by surface ocean warm water, rate of heat removed in the heat exchanger by deep ocean cooling water, cycle efficiency, boiler pressure, condenser pressure, mass flow rate of surface ocean warm water, and mass flow rate of deep ocean cooling water.
ANSWERS: $Wdot_{pump}$=-2.23 kW, $Wdot_{turbine}$=8.09 kW, $Wdot_{net}$=5.86 kW, $Qdot_{boiler}$=198.4 kW, $Qdot_{condenser}$=-192.5 kW, η=2.95%, p_{boiler}=963.5 kPa, $p_{condenser}$=680.7 kPa, $mdot_{warm\ water}$=5.93 kg/s, and $mdot_{cold\ water}$=7.65 kg/s.

7.8 Cascaded and Combined Cycle

There are situations where it is desirable to combine several cycles in series in order to take advantage of a very wide temperature range or to utilize what would otherwise be waste heat to improve efficiency. Such a cycle is called *cascaded cycle.* A cascaded cycle made of three Rankine cycles in series is shown schematically in Figure 7.8.1. The cascaded cycle is made of three sub-cycles. Sub-cycle A, 1-2-3-4-1, is the topping cycle; Sub-cycle B, 5-6-7-8-5, is the middle cycle; and Sub-cycle C, 9-10-11-12-9, is the bottom cycle. The waste heat of the upstream cycle is the heat input of the downstream cycle. Since the net work output is equal to the sum of the three outputs and the heat input is that of the topping cycle alone, a substantial efficiency increase is possible.

A cascaded cycle made of two Rankine cycles in series called *combined cycle* is shown schematically in Figure 7.8.2. The combined cycle is made of two sub-cycles. Sub-cycle A, 1-2-3-4-1, is the upstream topping cycle; and Sub-cycle B, 5-6-7-8-5, is the downstream bottom cycle. The waste heat of the upstream topping cycle is the heat added to the downstream bottom cycle. The power output is the sum of the output of the upstream topping cycle and the output of the downstream bottom cycle. The energy flow of the combined cycle is shown in Figure 7.8.3.

A numerical example is given in the following to illustrate the cycle analysis of the combined cycle.

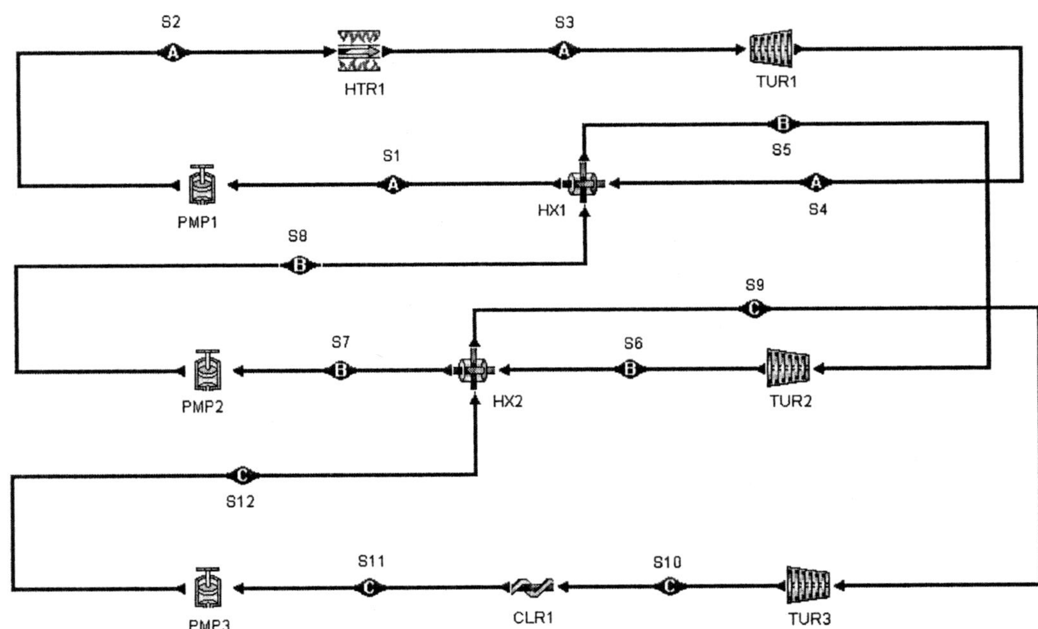

Figure 7.8.1 Cascaded cycle

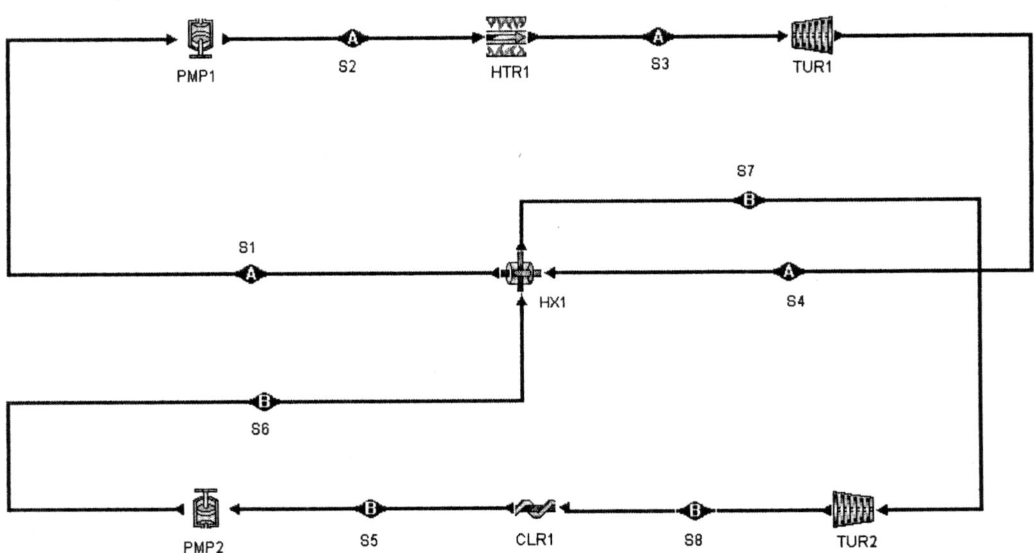

Figure 7.8.2 Combined cycle

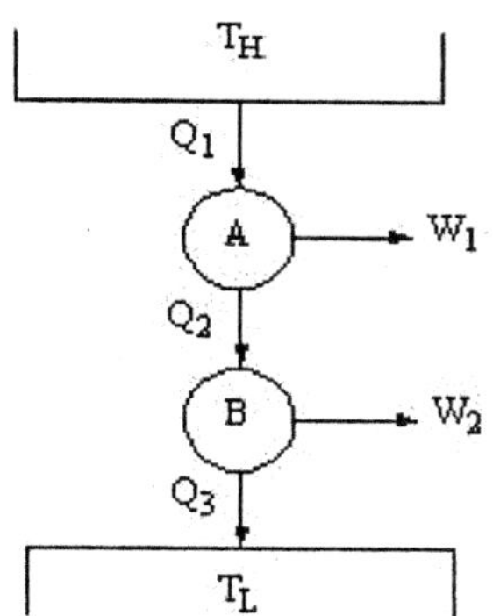

Figure 7.8.3. Combined cycle energy flow diagram

Example 7.8.1 A combined cycle made of two cycles is shown in Figure 7.8.2. The upstream topping cycle is a steam Rankine cycle and the downstream bottom cycle is an ammonia Rankine cycle. The following information is provided:

> steam boiler pressure=2 MPa, steam superheater temperature=400°C, steam condenser (heat exchanger) pressure=20 kPa, ammonia boiler (heat exchanger) pressure=1200 kPa, ammonia condenser pressure=800 kPa, and mass flow rate of steam is 1 kg/s.

Determine the total pump power input, total turbine power output, rate of heat added, rate of heat removed, cycle efficiency, and mass flow rate of ammonia.

To solve this problem with CyclePad, we take the following steps:

1. Build as shown in Figure 7.8.2
2. Analysis
   (A) Assume a process each for the seven devices: (a) pumps as adiabatic with 100% efficiency, (b) turbines as adiabatic with 100% efficiency, (c) heat exchanger as isobaric on both cold-side and hot-side, and (d) ammonia condenser as isobaric, and (d) steam boiler as isobaric.
   (B) Input the given information: (a) working fluid of cycle B is ammonia, and working fluid of cycle A is water, (b) the inlet pressure and quality of the ammonia pump are 800 kPa and 0, (c) the inlet temperature and pressure of the steam turbine are 400°C and 2000 kPa, (d) the inlet quality and pressure of the ammonia turbine are 1 and 1200 kPa, (e) the inlet pressure and quality of the water pump are 20 kPa and 0, and (f) the steam mass flow rate is 1 kg/s.
3. Display result
   The answers are:
   Combined cycle–Power input=-13.28 kW, Power output=988.0 kW, Net Power output=974.7 kW, Rate of heat added=2994 kW, Rate of heat removed=-2019 kW, η=32.56%.

Topping steam cycle–Power input=-2.02 kW, Power output=898.5 kW, Net Power output=896.5 kW, Rate of heat added=2994 kW, Rate of heat removed=-2098 kW, η=29.94%.

Bottom ammonia cycle–Power input=-11.26 kW, Power output=89.53 kW, Net Power output=78.27 kW, Rate of heat added=2098 kW, Rate of heat removed=-2019 kW, η=3.73%, Mass rate flow=1.75 kg/s.

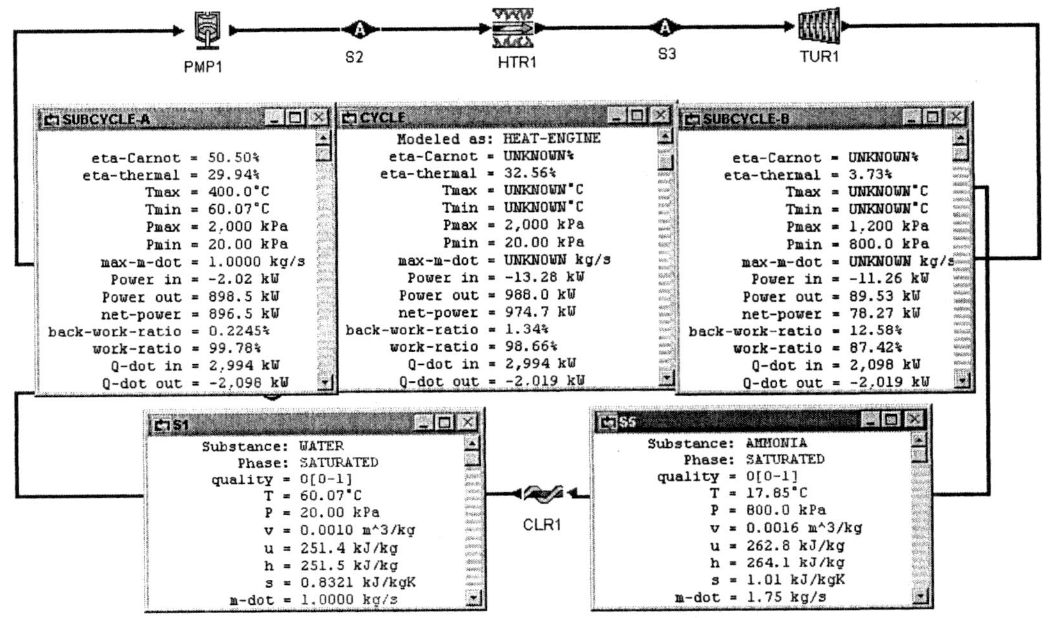

Figure Example 7.8.1 Combined Rankine cycle

Example 7.8.2. Figure Example 7.8.2a depicts a combined plant in which a closed Brayton helium nuclear plant releases heat to a recovery steam generator, which supplies heat to a Rankine steam plant. The generator is provided with a gas burner for supplementary additional heat when the demand of steam power is high. The Rankine plant is a regenerative cycle.

The data given below correspond approximately to the design conditions for the combined plant.

p_1=100 kPa, T_1=30°C, p_2=800 kPa, T_3=1400°C, T_5=200°C, p_6=5 kPa, x_6=0, p_8=1000 kPa, x_8=0, p_9=6000 kPa, T_{10}=400°C, $mdot_{10}$=1 kg/s, T_{11}=500°C, and $\eta_{compressor}=\eta_{turbine}=\eta_{pump}$=100%.

Determine the power required by the compressor, power required by pump #1 and #2, power produced by turbine #1, power produced by turbine #2, produced by turbine #3, rate of heat added by the nuclear reactor, net power produced by the Brayton gas-turbine plant, net power produced by the Rankine plant, rate of heat removed by cooler #1, rate of heat removed by cooler #2, rate of heat exchanged in the heat exchanger, rate of heat added in the gas burner, mass rate flow of helium in the Brayton cycle, mass rate flow of steam extracted to the feed water heater (mixing chamber), cycle efficiency of the Brayton plant, cycle efficiency of the Rankine plant, and cycle efficiency of the combined Brayton-Rankine plant.

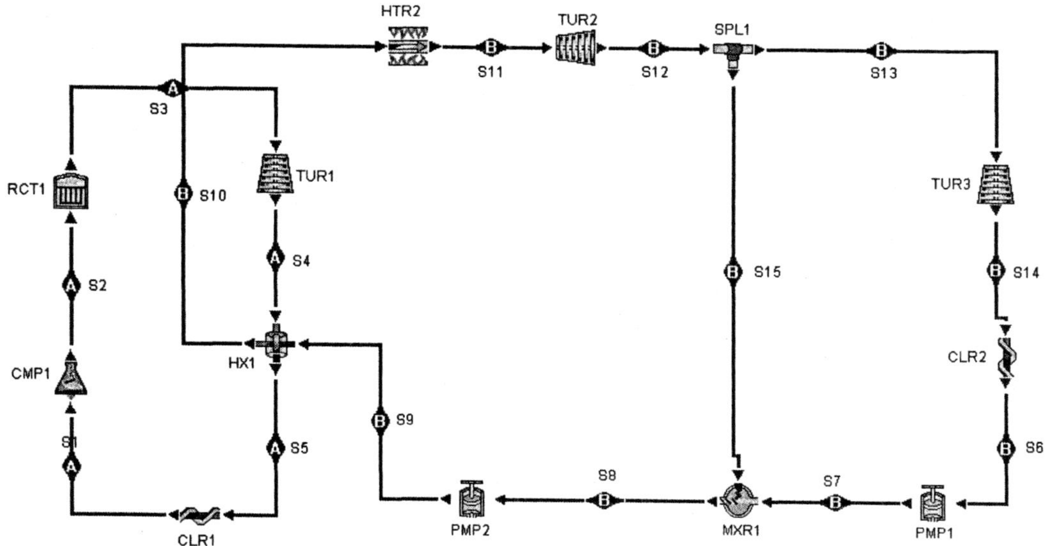

Figure Example 7.8.2a Combined Rankine cycle

To solve this problem with CyclePad, we take the following steps:

1. Build as shown in Figure Example 7.8.2a
2. Analysis
   (A) Assume a process each for the twelve devices: (a) pumps as adiabatic with 100% efficiency, (b) turbines as adiabatic with 100% efficiency, (c) heat exchanger as isobaric on both cold-side and hot-side, (d) nuclear reactor, mixing chamber, heater, and coolers as isobaric, and (e) splitter as iso-parametric.
   (B) Input the given information: (a) working fluid of cycle A is helium, and working fluid of cycle B is water, (b) p_1=100 kPa, T_1=30°C, p_2=800 kPa, T_3=1400°C, T_5=200°C, p_6=5 kPa, x_6=0, p_8=1000 kPa, x_8=0, p_9=6000 kPa, T_{10}=400°C, $mdot_{10}$=1 kg/s, and T_{11}=500°C.
3. Display result The answers are: $Wdot_{compressor}$=-3756 kW, $Wdot_{pump\ \#1\ and\ \#2}$=-6.46 kW, $Wdot_{turbine\#1}$=9001 kW, $Wdot_{turbine\#2}$=512.7 kW, $Wdot_{turbine\#3}$=637.5 kW, $Qdot_{reactor}$=9270 kW, $Wdot_{nee\ Brayton}$=5245 kW, $Wdot_{nee\ Rankine}$=1134 kW, $Qdot_{cooler\#1}$=-1616 kW, $Qdot_{cooler\#2}$=-1520 kW, $Qdot_{HX}$=2408 kW, $Qdot_{gas\ burner}$=245.3 kW, $mdot_{helium}$=1.84 kg/s, $mdot_{15}$=0.2245 kg/s, $\eta_{Brayton}$=5245/9270=56.58 %, $\eta_{Rankine}$=1134/2654=42.73 %, and $\eta_{combined}$=(5245+1134)/(9270+245.3)=67.04 %.

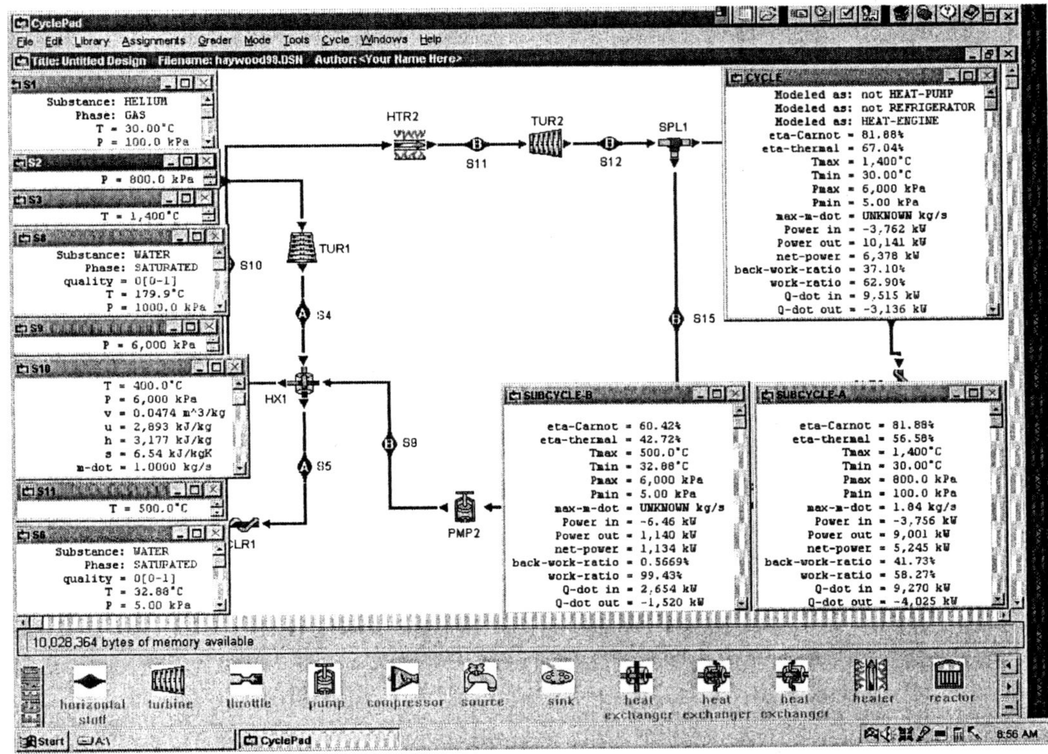

Figure Example 7.8.2b Combined Rankine cycle

Homework 7.8 Cascaded Cycle

1. What is a cascaded cycle?
2. What is the heat input to the whole cascaded cycle?
3. What is the total work output of the whole cascaded cycle?
4. Is the efficiency of the cascaded cycle better than any of the individual efficiency of the cycles which made the cascaded cycle?

7.9 COGENERATION

The cycles considered so far in this chapter are power cycles. However, there are applications in which Rankine cycles are used for the combined supply of power and process heat. The heat may be used as process steam for industrial processes, or steam to heat water for central or district heating. This type of combined heat and power plant is called *cogeneration*. A schematic cogeneration plant is illustrated in Figure 7.9.1. A different schematic cogeneration plant is illustrated in Figure 7.9.2.

The one regenerative Rankine basic cycle is composed of the following six processes:

1-2 isentropic compression

3-4 isobaric heat addition

5-6 isentropic expansion
6-1 isobaric heat removing
7-8 constant enthalpy throttling
8-9 isobaric heat removing
9-10 isentropic compression

Applying the mass balance and the First law of thermodynamics of the open system to the mixing chamber and the splitter of the cogeneration Rankine cycle yields:

$$m_3=m_2+m_{10}, \tag{7.9.1}$$

$$m_4 = m_7 + m_5, \tag{7.9.2}$$

$$Q_{89} = m_9(h_9 - h_8), \tag{7.9.3}$$

and

$$W_{56} = m_5(h_5 - h_6), \tag{7.9.4}$$

The net work (W_{net}) is

$$W_{net} = W_{56} + W_{9\text{-}10} + W_{12} \tag{7.9.5}$$

The thermal efficiency of the cycle is

$$\eta = W_{net} / Q_{34} \tag{7.9.6}$$

To take account of the desired heat output from process 8-9 (Q_{89}), the cogeneration ratio λ is

$$\lambda = Q_{89} / Q_{34} \tag{7.9.7}$$

Therefore, the combined power and heat cogeneration energy utility factor (EUF) is

$$EUF=\eta+\lambda. \tag{7.9.8}$$

Example 7.9.1. A cogeneration cycle as shown in Figure 7.9.1 is to be designed according to the following specifications:

Boiler temperature=500°C, Boiler pressure=7000 kPa, Condenser pressure=5 kPa, Process steam (cooler #2) pressure=500 kPa, Mass rate flow through the boiler=15 kg/s, and Mass rate flow through the turbine=14 kg/s.

Determine the rate of heat supply, net power output, process heat output, cycle efficiency, cogeneration ratio, and energy utility factor of the cycle.

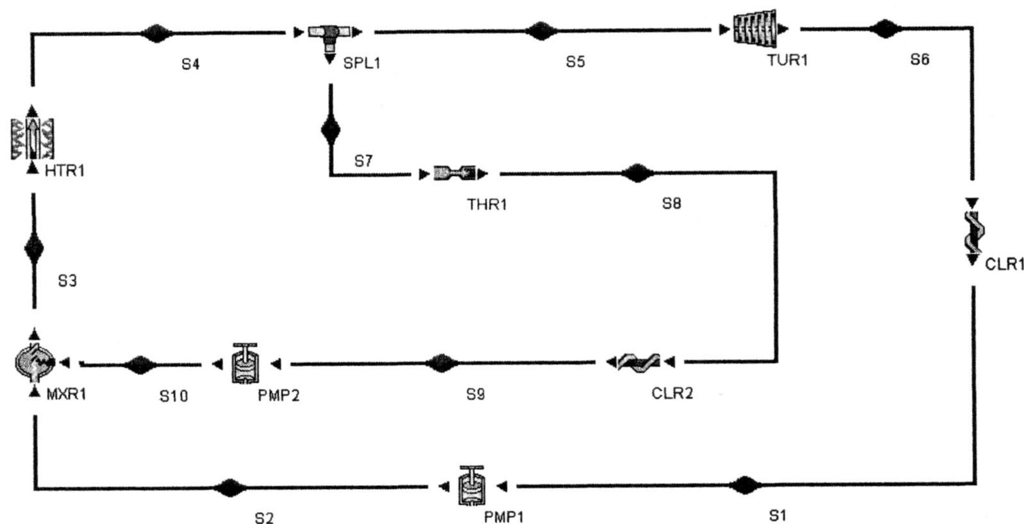

Figure 7.9.1.Cogeneration plant

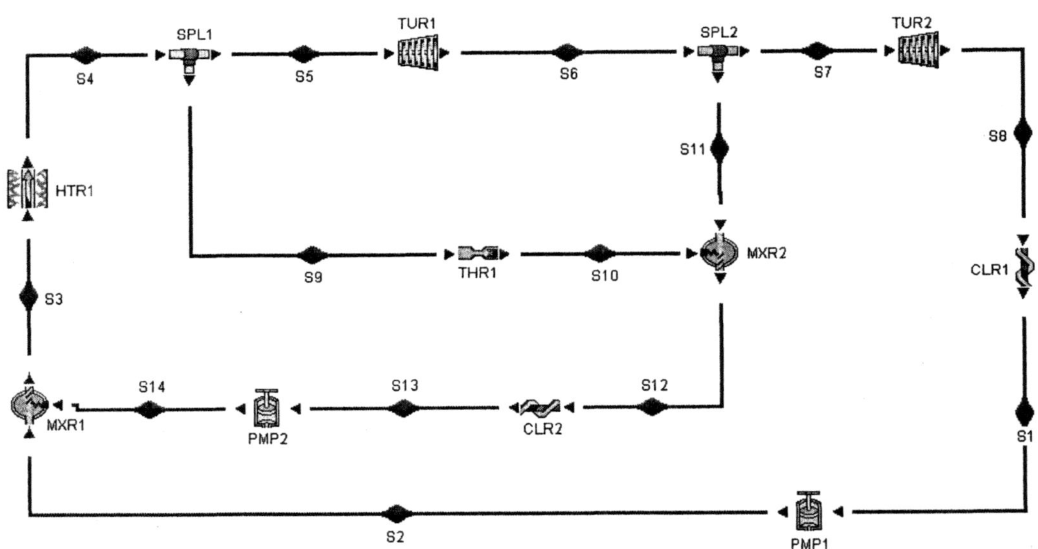

Figure 7.9.2. Cogeneration plant

To solve this problem with CyclePad, we take the following steps:

1. Build as shown in Figure 7.10.1
2. Analysis
   (A) Assume a process each for the following devices: (a) pumps as adiabatic with 100% efficiency, (b) turbines as adiabatic with 100% efficiency, (c) splitter as isoparametric, (d) mixing chamber as isobaric, (d) condenser and cooler as isobaric, and (e) boiler as isobaric.

(B) Input the given information: (a) working fluid is water, (b) the inlet pressure and quality of pump #1 are 5 kPa and 0, (c) the inlet temperature and pressure of the turbine are 500°C and 7000 kPa, (d) the inlet quality and pressure of pump #2 are 0 and 500 kPa, (e) the steam mass flow rate is 15 kg/s at state 4, and (f) the steam mass flow rate is 14 kg/s at state 5.

3. Display result

The answers are: rate of heat supply=48482 kW, net power output=18607 kW, process heat output=2770 kW,

cycle efficiency=0.3838, cogeneration ratio=2770/48482=0.0571, and energy utility factor of the cycle=0.3838+0.0571=0.4409.

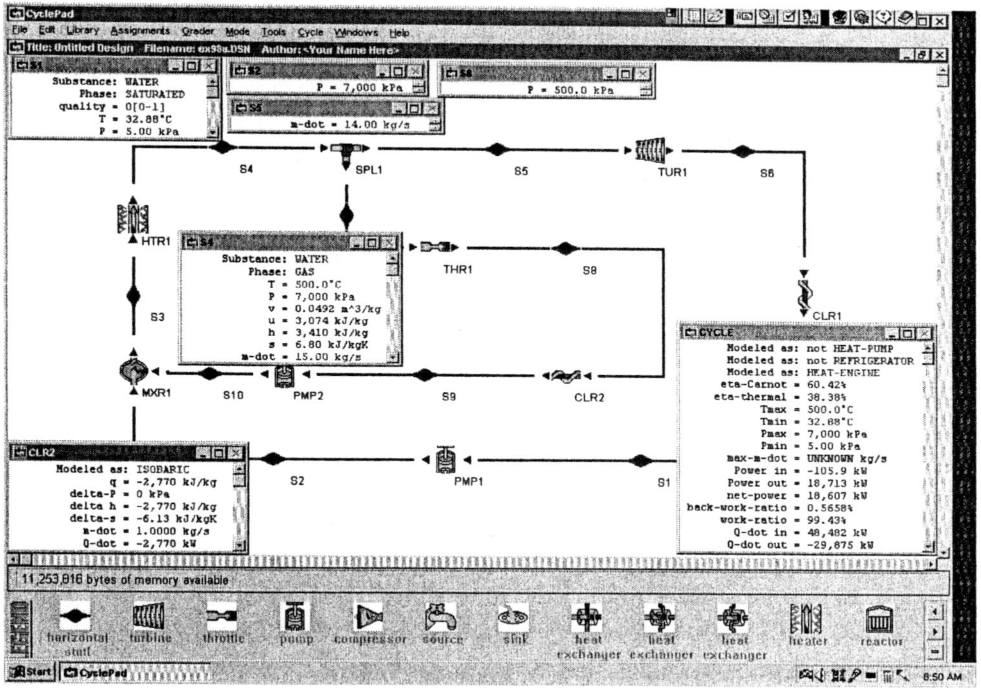

Figure Example 7.9.1. Cogeneration

Example 7.9.2. If the cogeneration cycle as shown in Figure 7.9.1 is to produce power only according to the following specifications:

Boiler temperature=500°C, Boiler pressure=7000 kPa, Condenser pressure=5 kPa, Process steam (cooler #2) pressure=500 kPa, Mass rate flow through the boiler=15 kg/s, and Mass rate flow through the turbine=15 kg/s.

Determine the net power output, rate of heat supply, and cycle efficiency of the cycle.

To solve this problem with CyclePad, we take the following steps:

1. Build as shown in Figure 7.10.1
2. Analysis

(A) Assume a process each for the following devices: (a) pumps as adiabatic with 100% efficiency, (b) turbines as adiabatic with 100% efficiency, (c) splitter as isoparametric, (d) mixing chamber as isobaric, (d) condenser and cooler as isobaric, and (e) boiler as isobaric.

(B) Input the given information: (a) working fluid is water, (b) the inlet pressure and quality of pump #1 are 5 kPa and 0, (c) the inlet temperature and pressure of the turbine are 500°C and 7000 kPa, (d) the inlet quality and pressure of pump #2 are 0 and 500 kPa, (e) the steam mass flow rate is 15 kg/s at state 4, and (f) the steam mass flow rate is 15 kg/s at state 5.

3. Display result

The answers are: rate of heat supply=48985 kW, net power output=19944 kW, and cycle efficiency=0.4071.

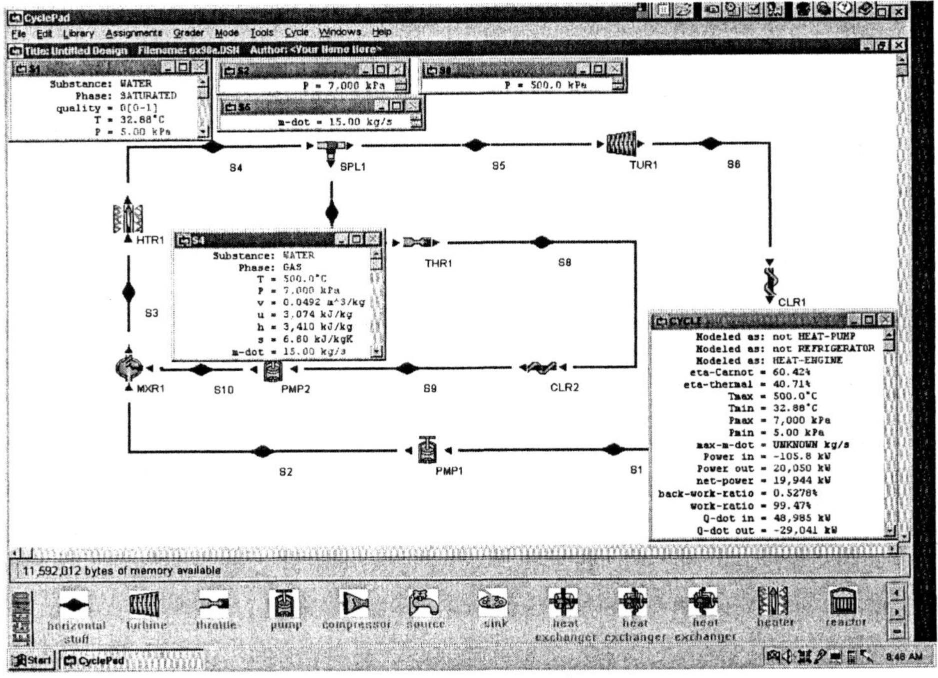

Figure Example 7.9.2. Cogeneration without heat output

Example 7.9.3. A cogeneration cycle as shown in Figure 7.9.2 is to be designed according to the following specifications:

Boiler temperature=400°C, Boiler pressure=40 bar, Inlet pressure of low-pressure turbine=10 bar, Condenser pressure=0.1 bar, Process steam (cooler #2) pressure=10 bar, Mass rate flow through the boiler=1 kg/s, Mass rate flow through the turbine #1=0.98 kg/s, and Mass rate flow through the turbine #1=0.95 kg/s.

Determine the rate of heat supply, net power output, process heat output, cycle efficiency, cogeneration ratio, and energy utility factor of the cycle.

To solve this problem with CyclePad, we take the following steps:

1. Build as shown in Figure 7.9.2
2. Analysis
   (A) Assume a process each for the following devices: (a) pumps as adiabatic with 100% efficiency, (b) turbines as adiabatic with 100% efficiency, (c) splitter as isoparametric, (d) mixing chamber as isobaric, (d) condenser and cooler as isobaric, and (e) boiler as isobaric.
   (B) Input the given information: (a) working fluid is water, (b) p_1=0.1 bar, x_1=0, p_2=40 bar, p_6=10 bar, $mdot_4$=1 kg/s, $mdot_9$=0.02 kg/s, $mdot_{11}$=0.03 kg/s, T_4=400°C, and x_{13}=0.
3. Display result
   The answers are: rate of heat supply=2989 kW, net power output=1023 kW, process heat output=112 kW, cycle efficiency=0.3421, cogeneration ratio=112/2989=0.03747, and energy utility factor of the cycle=0.3421+0.03747=0.3796.

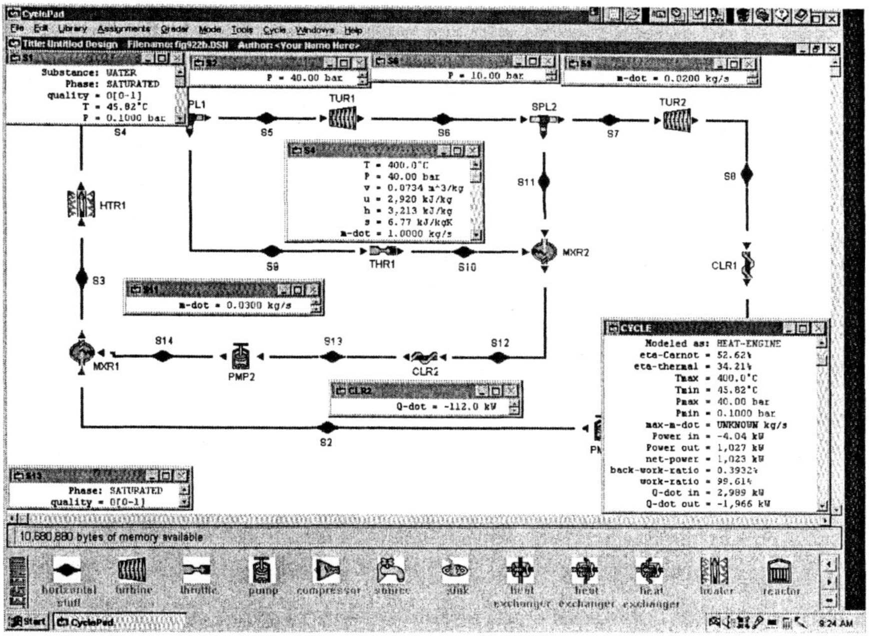

Figure Example 7.9.3. Cogeneration

Homework 7.9 Cogeneration

1. What is cogeneration?
2. How is cogeneration ratio is defined?
3. How is cogeneration energy utility factor is defined?
4. A cogeneration cycle as shown in Figure 7.10.1 is to be designed according to the following specifications:

Turbine efficiency=89%, Boiler temperature=500°C, Boiler pressure=7000 kPa, Condenser pressure=5 kPa, Process steam (cooler #2) pressure=500 kPa, Mass rate flow through the boiler=15 kg/s, and Mass rate flow through the turbine=13 kg/s.
Determine the rate of heat supply, net power output, process heat output, cycle efficiency, cogeneration ratio, and energy utility factor of the cycle.
ANSWER: cycle efficiency=32.01%, rate of heat supply=47980 kW, rate of cogeneration heat=-5540 kW.

5. A cogeneration cycle as shown in Figure 7.10.1is to be designed according to the following specifications:
Turbine efficiency=89%, Boiler temperature=500°C, Boiler pressure=6000 kPa, Condenser pressure=5 kPa, Process steam (cooler #2) pressure=500 kPa, Mass rate flow through the boiler=15 kg/s, and Mass rate flow through the turbine=13 kg/s.
Determine the rate of heat supply, net power output, process heat output, cycle efficiency, cogeneration ratio, and energy utility factor of the cycle.
ANSWER: cycle efficiency=31.63%, rate of heat supply=48173 kW, rate of cogeneration heat=-5564 kW.
6. A cogeneration cycle as shown in Figure 7.10.1 is to be designed according to the following specifications:
Turbine efficiency=89%, Boiler temperature=500°C, Boiler pressure=7000 kPa, Condenser pressure=5 kPa, Process steam (cooler #2) pressure=500 kPa, Mass rate flow through the boiler=15 kg/s, and Mass rate flow through the turbine=14 kg/s.
Determine the rate of heat supply, net power output, process heat output, cycle efficiency, cogeneration ratio, and energy utility factor of the cycle.
ANSWER: cycle efficiency=38.38%, rate of heat supply=48482 kW, rate of cogeneration heat=-2770 kW.
7. A cogeneration cycle as shown in Figure 7.10.2 is to be designed according to the following specifications:
High-pressure Turbine efficiency=84%, Low-pressure Turbine efficiency=100%, Boiler temperature=400°C, Boiler pressure=40 bar, Inlet pressure of low-pressure turbine=10 bar, Condenser pressure=0.1 bar, Process steam (cooler #2) pressure=10 bar, Mass rate flow through the boiler=1 kg/s, Mass rate flow through the turbine #1=0.98 kg/s, and Mass rate flow through the turbine #1=0.95 kg/s.
Determine the rate of heat supply, net power output, process heat output, cycle efficiency, cogeneration ratio, and energy utility factor of the cycle.
ANSWER: cycle efficiency=33.02%, rate of heat supply=2989 kW, rate of cogeneration heat=-113.7 kW.

7.10 VAPOR CYCLE WORKING FLUIDS

Water has been used mainly as the working fluid in the vapor power examples of this chapter. In fact, water is the most common fluid in large central power plants, though by no means is it the only working fluid used in vapor power cycles. The desirable properties of the vapor cycle working fluid include the following eight important characters.

1. High critical temperature– to permit evaporation at high temperature.

2. Low saturation (boiling) pressure at high temperature– to minimize the pressure vessel and piping costs.
3. Pressure around ambient pressure at condenser temperature– to eliminate serious air leakage and sealing problems.
4. Rapidly diverging pressure lines on the h-s diagram– to minimize the back-work ratio and to make reheat modification most effective.
5. Large enthalpy of evaporation– to minimize the mass flow rate for given power output.
6. No degrading aspects– noncorrosive, nonclogging, etc.
7. No hazardous features– nontoxic, inflammable, etc.
8. Low cost readily available.

There are six vapor working fluids listed on the menu of CyclePad. The fluids are ammonia, methane, refrigerant 12, refrigerant 22, refrigerant 134a, and water. Water has the characteristics of items 4, 5, 7 and 8 above and water remains a top choice for industrial central vapor power plant. Hence steam power engineering remains the most important area of applied thermodynamics.

Homework 7.10

1. Why water is the most popular working fluid choice in central vapor power plants?
2. The ocean surface water is warm (27°C at equator) and deep ocean water is cold (5°C at 2000 m depth). If a vapor cycle operates between these two thermal reservoirs, is water or refrigerant a better choice as the working fluid for this power plant?

7.11 DESIGN EXAMPLES

CyclePad is to a power engineer what a word processor is to a journalist. The benefits of using this software are numerous. The first benefit is that significantly less time is spent doing numerical analysis. As an engineer, this is much appreciated because design computation work that would have taken hours before can now be done in seconds. Second, CyclePad is capable of analysing cycles with various working fluids. Third, due to its computer assisted modeling capabilities, the software allows users to immediately view the effects of varying input parameters, either through calculated results or in the form of graphs and diagrams, giving the user a greater appreciation of how a system actually works. More specifically, there is the feature that provides the designer the opportunity to optimize a specific power cycle parameter. The designer is able to gain extensive design experience in a short time. Last, and most important, is the built-in coaching facility that provides definitions of terms and descriptions of calculations. CyclePad goes a step further by informing the inexperience designer if a contradiction or an incompatability exists within a cycle and why.

When viewing the applicability of CyclePad, users can the benefits at all stages of an engineering career. For the young engineer, just beginning the design learning process, less time is spent doing iterations resulting in more time dedicated to reinforcing the fundamentals and gaining valuable experience. In the case of the seasoned engineer, who is well indoctrinated in the principles and has gained an engineer's intuition, they can augment their abilities by becoming more computer literate. The following examples illustrate the design of vapor power cycles using CyclePad.

Example 7.11.1 A 4-stage turbine with reheat and 3-stage regenerative steam Rankine cycle as shown in Figure Example 7.11.1a was designed by a junior engineer. The following design information are provided:

p_1=103 kPa, T_1=15°C, T_2=25°C, p_4=16000 kPa, T_4=570°C, $mdot_4$=1000 kg/s, p_5=8000 kPa, T_6=540°C, p_8=4000 kPa, p_{10}=2000 kPa, p_{12}=15 kPa, x_{13}=0, x_{15}=0, x_{19}=0, $\eta_{turbine\#1}$=0.85, $\eta_{turbine\#2}$=0.85, $\eta_{turbine\#3}$=0.85, $\eta_{turbine\#4}$=0.85, $\eta_{pump\#1}$=0.9, $\eta_{pump\#2}$=0.9, $\eta_{pump\#3}$=0.9, and $\eta_{pump\#4}$=0.9.

Find η_{cycle}, $Wdot_{input}$, $Wdot_{output}$, $Wdot_{net\ output}$, $Qdot_{add}$, $Qdot_{remove}$, $Wdot_{turbine\#1}$, $Wdot_{turbine\#2}$, $Wdot_{turbine\#3}$, $Wdot_{turbine\#4}$, $Wdot_{pump\#1}$, $Wdot_{pump\#2}$, $Wdot_{pump\#3}$, $Wdot_{pump\#4}$, $Qdot_{htr\#1}$, $Qdot_{htr\#2}$, $Qdot_{HX1}$, $mdot_{20}$, $mdot_{21}$, $mdot_{22}$, $mdot_{12}$, $mdot_{15}$, $mdot_{17}$, and $mdot_{19}$.

Based on the preliminary design results, try to improve his design. Use η_{cycle} as the objective function and p_5, p_8 and p_{10} as design parameters.

Draw the η_{cycle} vs p_5 sensitivity diagram, the η_{cycle} vs p_8 sensitivity diagram, and the η_{cycle} vs p_{10} sensitivity diagram.

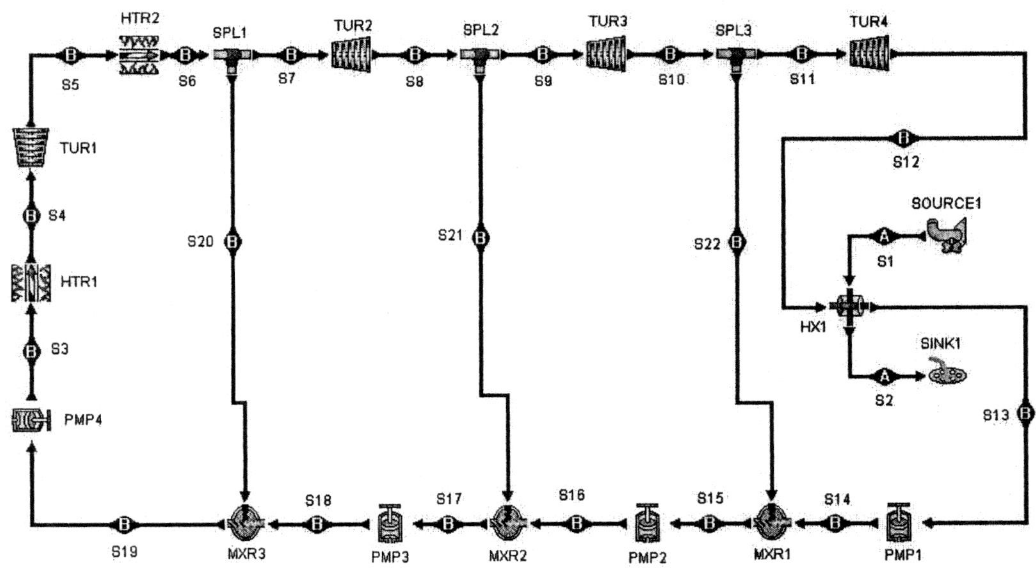

Figure Example 7.11.1a 4-stage turbine with reheat and 3-stage regenerative Rankine cycle

To solve this problem by CyclePad, we take the following steps:

1. Build
   (A) Take a source, a sink, four pumps, a boiler (HTR1), four turbines, a reheater (HTR2), three splitters, three mixing chambers (open feed-water heaters) and a heat exchanger

(condenser) from the inventory shop and connect the devices to form the 4-stage turbine with reheat and 3-stage regenerative Rankine cycle.

(B) Switch to analysis mode.

2. Analysis

(A) Assume a process for each of the devices: (a) pumps as adiabatic, (b) boiler and reheater as isobaric, (c) turbines as adiabatic, (d) splitters as iso-parametric, (e) mixing chambers and (f) heat exchanger as isobaric.

(B) Input the given information: working fluid is water in Cycle A and Cycle B, p_1=103 kPa, T_1=15°C, T_2=25°C, p_4=16000 kPa, T_4=570°C, $mdot_4$=1000 kg/s, p_5=8000 kPa, T_6=540°C, p_8=4000 kPa, p_{10}=2000 kPa, p_{12}=15 kPa, x_{13}=0, x_{15}=0, x_{19}=0, $\eta_{turbine\#1}$=0.85, $\eta_{turbine\#2}$=0.85, $\eta_{turbine\#3}$=0.85, $\eta_{turbine\#4}$=0.85, $\eta_{pump\#1}$=0.9, $\eta_{pump\#2}$=0.9, $\eta_{pump\#3}$=0.9, and $\eta_{pump\#4}$=0.9 as shown in Figure Example 7.11.1b.

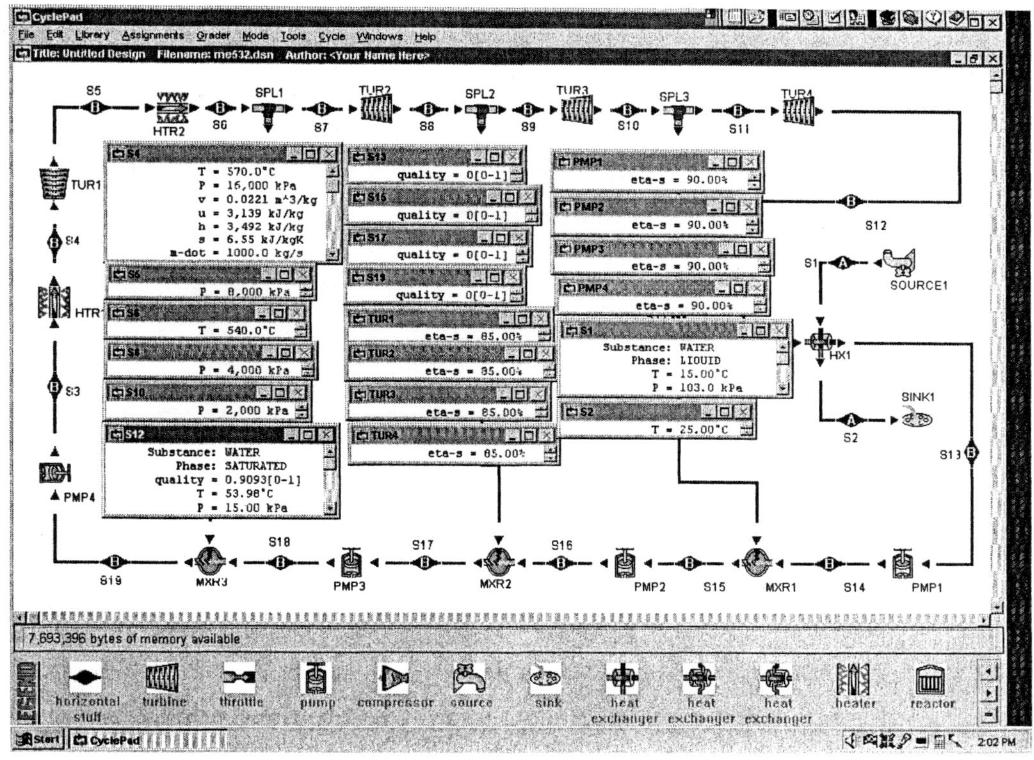

Figure Example 7.11.1b. 4-stage turbine with reheat and 3-stage regenerative Rankine cycle input

3. Display the preliminary design results as shown in Figure Example 7.11.1c and Figure Example 7.11.1d. The results are: η_{cycle}=41.24%, $Wdot_{input}$=-20903 kW, $Wdot_{output}$=994618 kW, $Wdot_{net\ output}$=973716 kW, $Qdot_{add}$=2361193 kW, $Qdot_{remove}$=-1387477 kW, $Wdot_{turbine\#1}$=193305 kW, $Wdot_{turbine\#2}$=176531 kW, $Wdot_{turbine\#3}$=144004 kW, $Wdot_{turbine\#4}$=480778 kW, $Wdot_{pump\#1}$=-1464 kW, $Wdot_{pump\#2}$=-2201 kW, $Wdot_{pump\#3}$=-5034 kW, $Wdot_{pump\#4}$=-12203 kW, $Qdot_{htr\#1}$=2163587 kW, $Qdot_{htr\#2}$=197605 kW, $Qdot_{HX1}$=-1387477 kW, $mdot_{20}$=93.11 kg/s, $mdot_{21}$=66.74 kg/s, $mdot_{22}$=197.0 kg/s, $mdot_{12}$=643.2 kg/s, $mdot_{15}$=840.2 kg/s, $mdot_{17}$=906.9 kg/s and $mdot_{19}$=1000 kg/s.

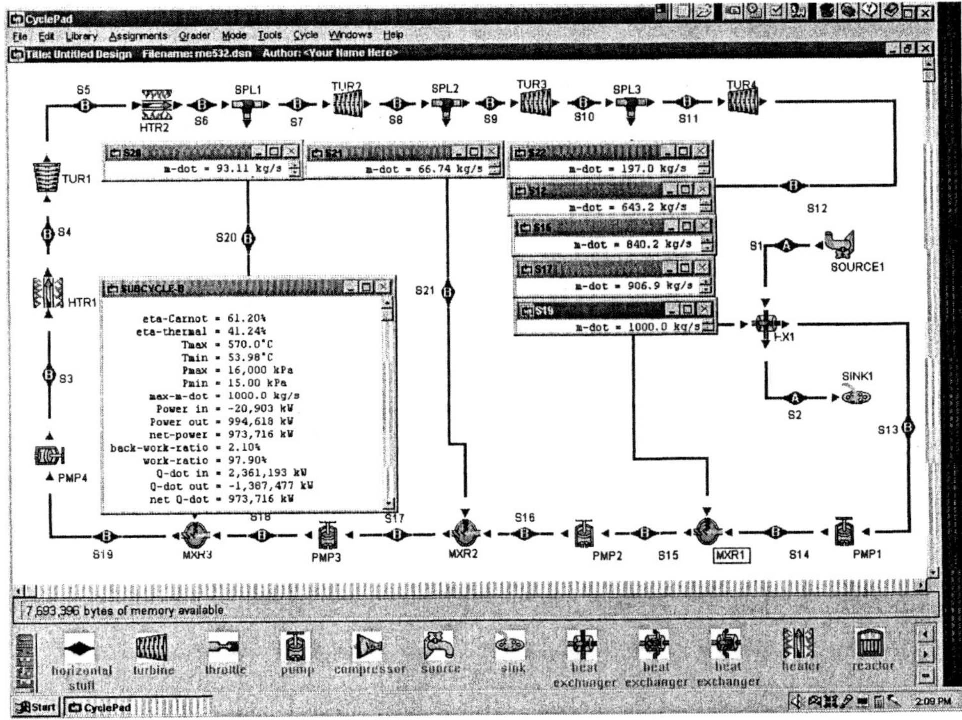

Figure Example 7.11.1c. 4-stage turbine with reheat and 3-stage regenerative Rankine cycle results

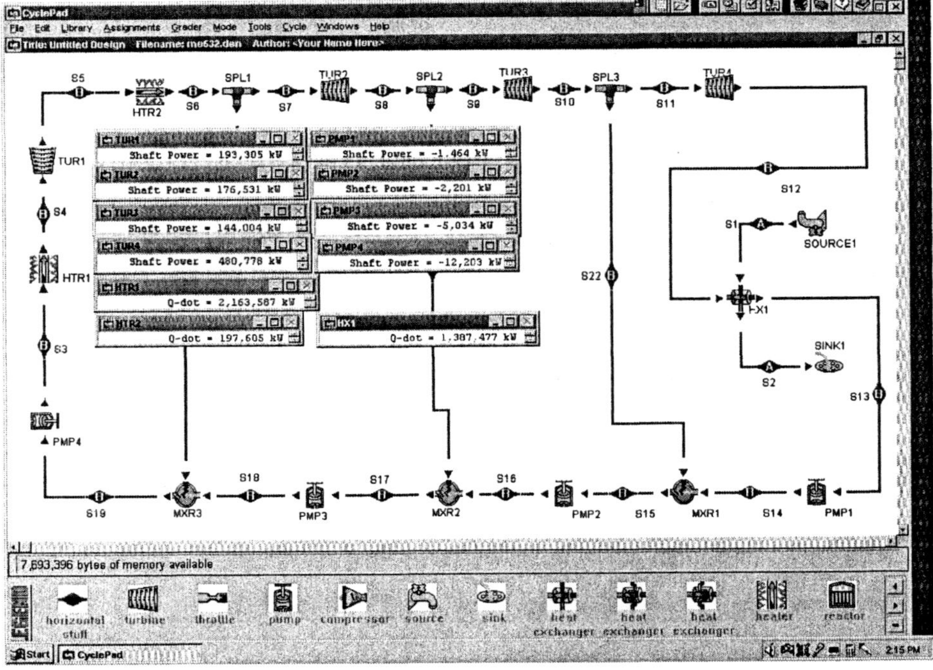

Figure Example 7.11.1d. 4-stage turbine with reheat and 3-stage regenerative Rankine cycle results

4. The η_{cycle} vs p_5 sensitivity diagram, the η_{cycle} vs p_8 sensitivity diagram, and the η_{cycle} vs p_{10} sensitivity diagram are drawn as shown in Figure Example 7.11.1e, Figure Example 7.11.1f, and Figure Example 7.11.1g, respectively. Based on these sensitivity diagrams, η_{cycle} can be optimized.

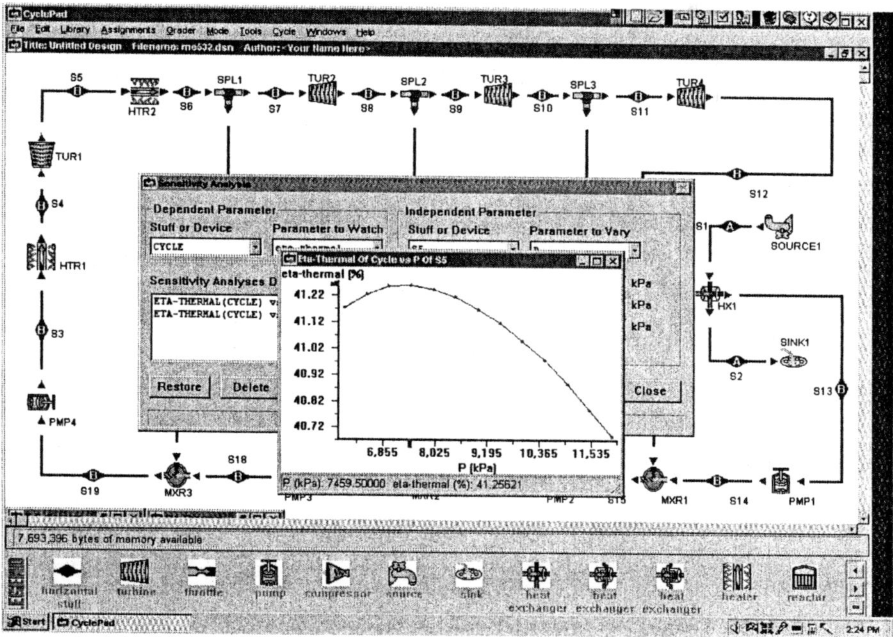

Figure Example 7.11.1e. Rankine cycle sensitivity diagram

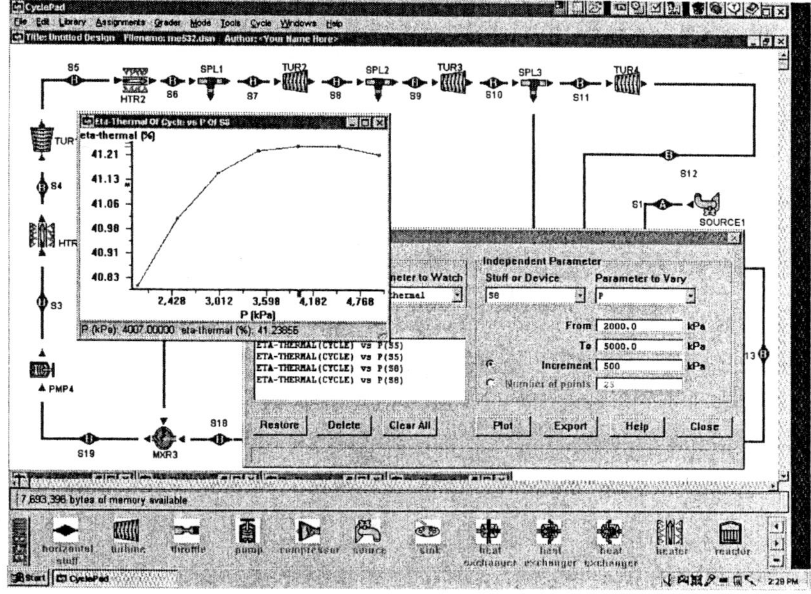

Figure Example 7.11.1f. 4-stage turbine with reheat and 3-stage regenerative Rankine cycle sensitivity diagram

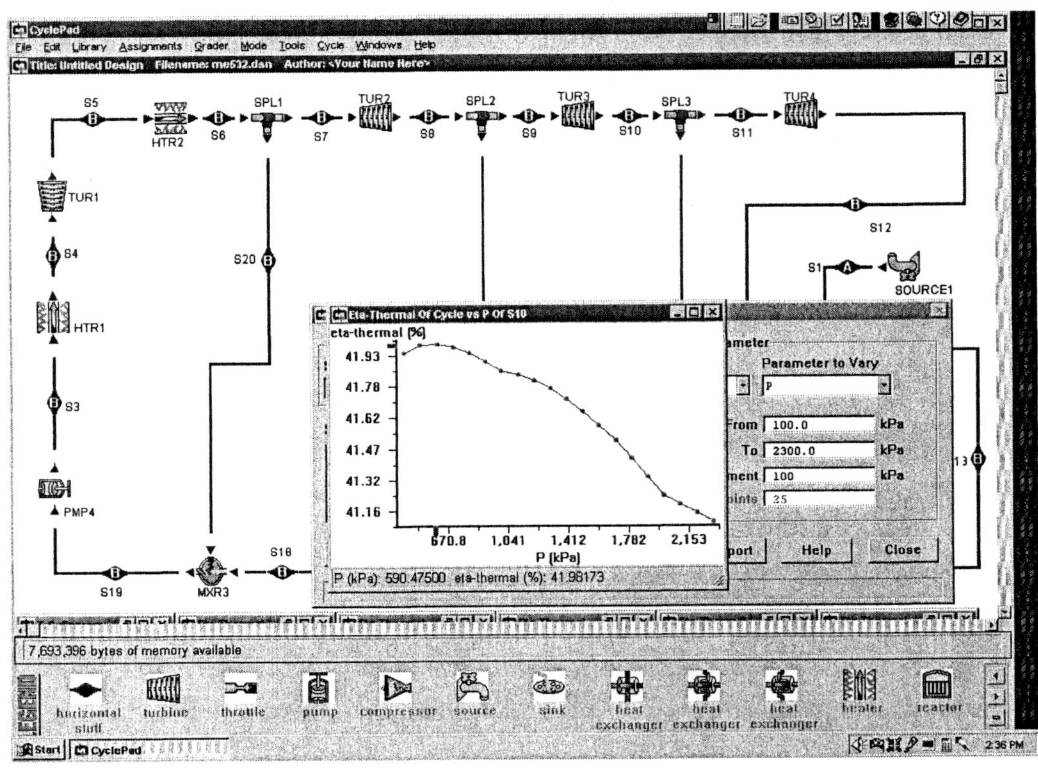

Figure Example 7.11.1g. 4-stage turbine with reheat and 3-stage regenerative Rankine cycle sensitivity diagram

Example 7.11.2. A closed-cycle steam Rankine cycle without superheating is designed by a junior engineer as illustrated in Figure Example 7.11.2a with the following preliminary design information (Figure Example 7.11.2b):

Condenser pressure	5 psia
Boiler pressure	3000 psia
Mass flow rate of steam	1 lbm/s
Flue gas temperature entering high-temperature side heat exchanger	3500°F
Flue gas pressure entering high-temperature side heat exchanger	14.7 psia
Flue gas leaving high-temperature side heat exchanger	1500°F
Cooling water temperature entering low-temperature side heat exchanger	60°F
Cooling water pressure entering low-temperature side heat exchanger	14.7 psia
Cooling water leaving low-temperature side heat exchanger	80°F
Turbine efficiency	88%
Pump efficiency	88%

Use net power output as the objective function and boiler pressure as the independent design parameter. Try to improve the preliminary design.

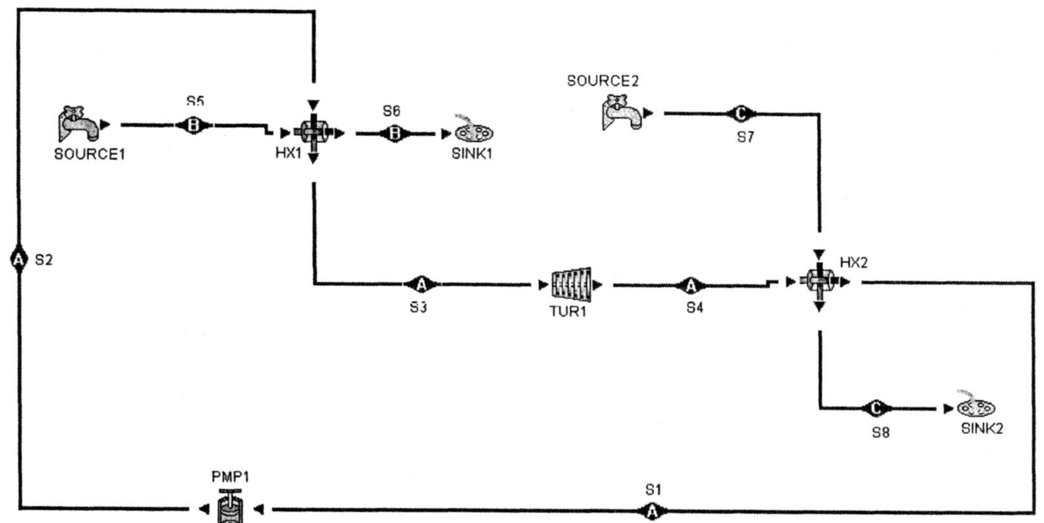

Figure Example 7.11.2a Rankine cycle preliminary design

(A) To improve the design with CyclePad, we take the following steps:

1. Build as shown in Figure Example 7.11.2a
2. Analysis

   Assume a process for each of the four devices: (a) pump as adiabatic with 88% efficiency, (b) turbine as adiabatic with 88% efficiency, (c) heat exchanger 1 (boiler) as isobaric on both cold-side and hot-side, and (d) heat exchanger 2 (condenser) as isobaric on both cold-side and hot-side.

   Input the given information: (a) working fluid of heat source is air (flue gas), p_5=14.7 psia, T_5=3500°F and T_6=1500°F, (b) working fluid of Rankine cycle is water, p_1=5 psia, x_1=0, p_3=3000 psia, and x_3=1, and (c) working fluid of heat sink is water, p_7=14.7 psia, T_7=60°F and T_6=80°F as shown in Figure Example 7.11.2b.

(B) Determine the preliminary design results.

Display result: The preliminary design results are given in Figure Example 7.11.2c as follows:

$Wdot_{pump}$=-14.58 hp, $Wdot_{turbine}$=389.3 hp, $Wdot_{net}$=374.7 hp, $Qdot_{boiler}$=877.5 Btu/s, $Qdot_{condenser}$=-612.6 Btu/s, η=30.18%, $mdot_{flue\ gas}$=1.83 lbm/s, and $mdot_{cold\ water}$=30.66 lbm/s.

(C) Draw the net power vs p_3 sensitivity diagram as shown in Figure Example 7.11.2d. It is shown that the maximum net power is about 430 hp at about 1500 psia.

(D) Change design input information

Change p_3 from 3000 psia to 1500 psia and displace results. The results are shown in Figure Example 7.11.2e.

Display result: The optimized design results are:

$Wdot_{pump}$=-7.31 hp, $Wdot_{turbine}$=440.5 hp, $Wdot_{net}$=433.2 hp, $Qdot_{boiler}$=1033 Btu/s, $Qdot_{condenser}$=-727.7 Btu/s, η=29.63%, $mdot_{flue\ gas}$=2.16 lbm/s, and $mdot_{cold\ water}$=36.39 lbm/s.

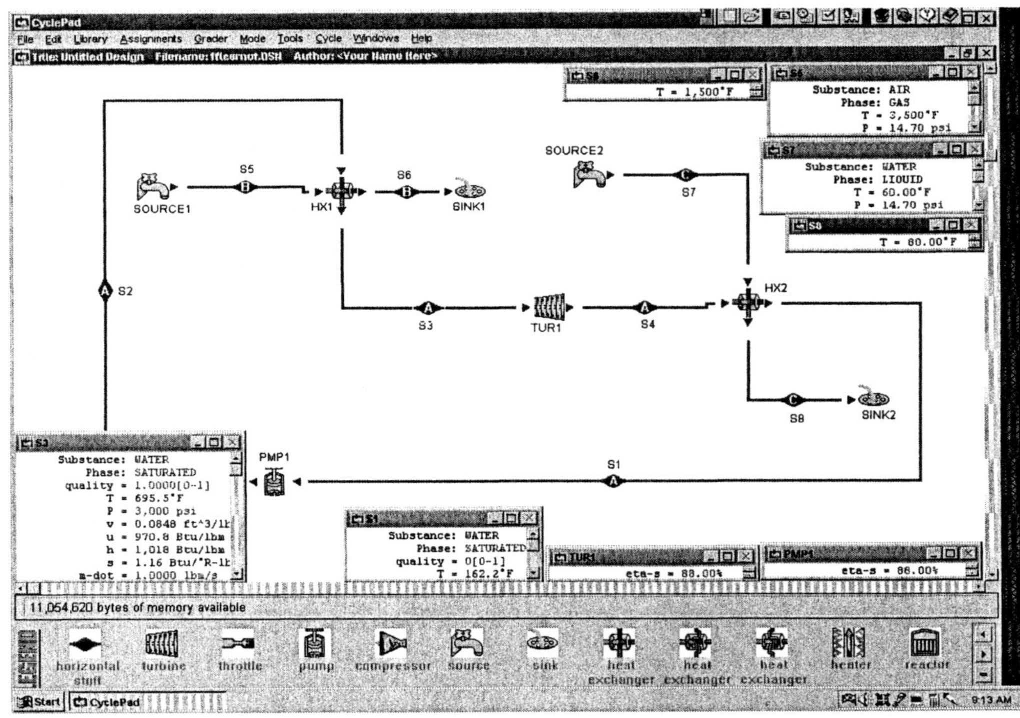

Figure Example 7.11.2b Rankine cycle preliminary design input data

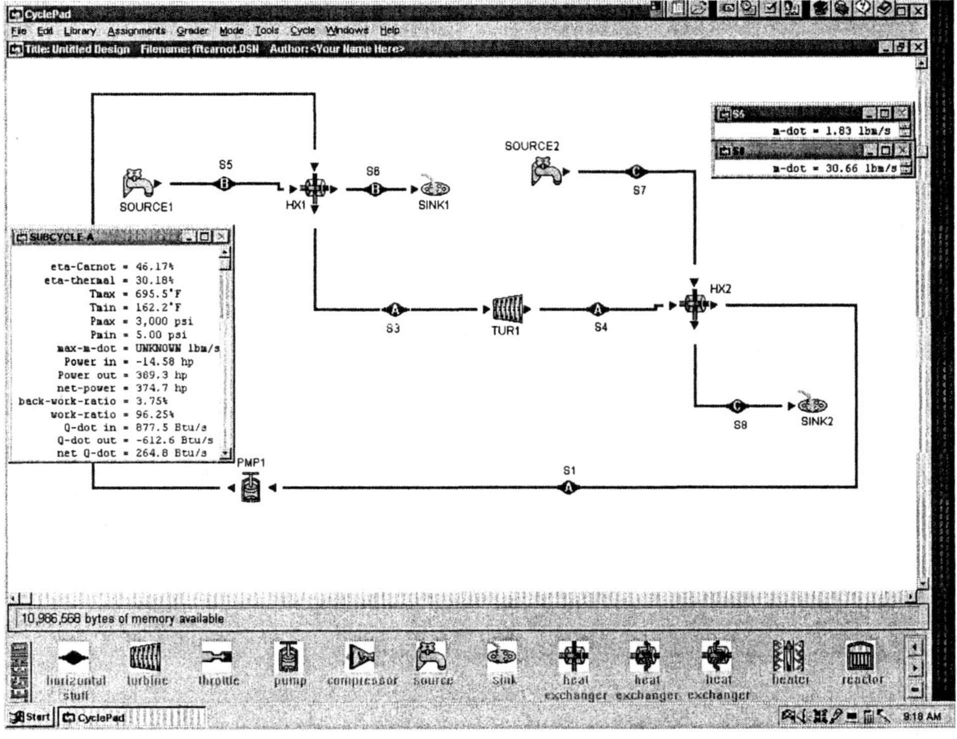

Figure Example 7.11.2b Rankine cycle preliminary design output data

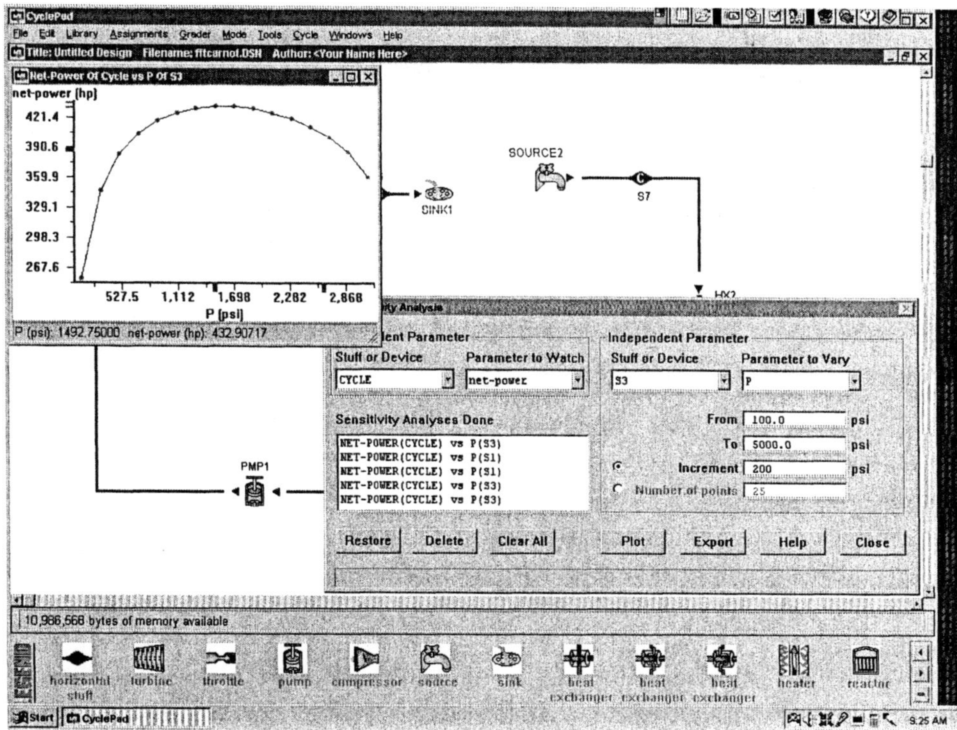

Figure Example 7.11.2d Rankine cycle sensitivity diagram

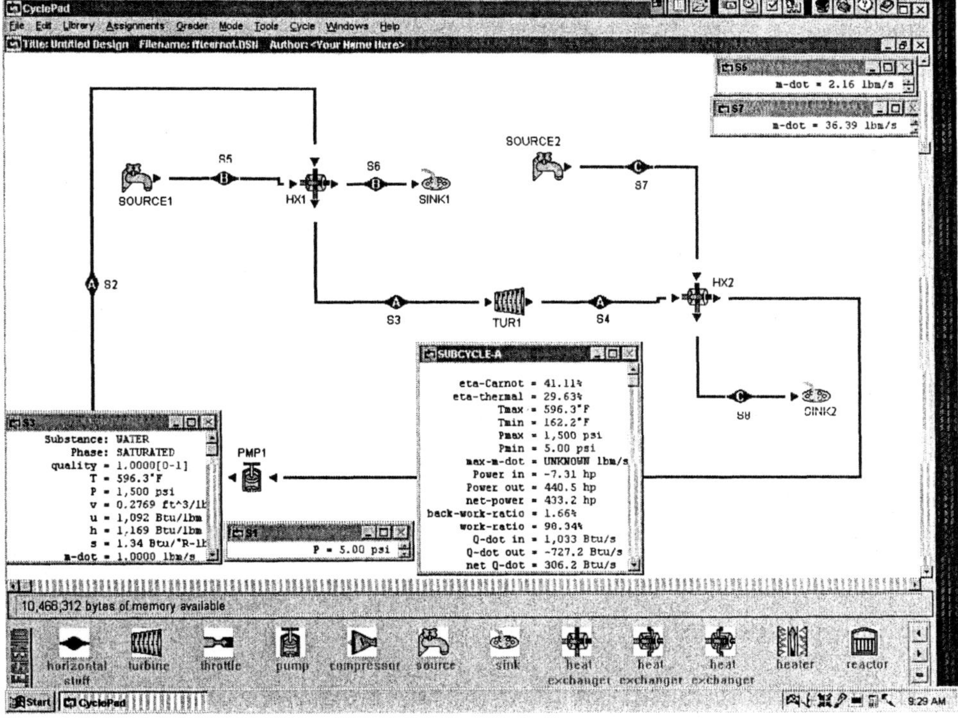

Figure Example 7.11.2e Rankine cycle optimized design output data

Homework 7.11 Design

1. A 4-stage turbine with reheat and 3-stage regenerative Rankine cycle as shown in Figure Example 7.6.3a using steam as the working fluid. The following information are provided:
   p_1=103 kPa, T_1=15°C, T_2=25°C, p_4=16000 kPa, T_4=600°C, $mdot_4$=1000 kg/s, p_5=8000 kPa, T_6=540°C, p_8=4000 kPa, p_{10}=2000 kPa, p_{12}=15 kPa, x_{13}=0, x_{15}=0, x_{19}=0, $\eta_{turbine\#1}$=0.85, $\eta_{turbine\#2}$=0.85, $\eta_{turbine\#3}$=0.85, $\eta_{turbine\#4}$=0.85, $\eta_{pump\#1}$=0.85, $\eta_{pump\#2}$=0.85, $\eta_{pump\#3}$=0.85, and $\eta_{pump\#4}$=0.85.
   Find η_{cycle}, $Wdot_{input}$, $Wdot_{output}$, $Wdot_{net\ output}$, $Qdot_{add}$, $Qdot_{remove}$, $Wdot_{turbine\#1}$, $Wdot_{turbine\#2}$, $Wdot_{turbine\#3}$, $Wdot_{turbine\#4}$, $Qdot_{htr\#1}$, $Qdot_{htr\#2}$, $Qdot_{HX1}$, $mdot_{20}$, $mdot_{21}$, $mdot_{22}$, $mdot_{12}$, $mdot_{15}$, $mdot_{17}$, and $mdot_{19}$.
   Draw the η_{cycle} vs p_5 sensitivity diagram, the η_{cycle} vs p_8 sensitivity diagram, and the η_{cycle} vs p_{10} sensitivity diagram.
   ANSWER: η_{cycle}=41.75%, $Wdot_{input}$=-20985 kW, $Wdot_{output}$=1067259 kW, $Wdot_{net\ output}$=1046274 kW, $Qdot_{add}$=2505993 kW, $Qdot_{remove}$=-1459718 kW, $Wdot_{turbine\#1}$=193305 kW, $Wdot_{turbine\#2}$=193484 kW, $Wdot_{turbine\#3}$=158701 kW, $Wdot_{turbine\#4}$=521769 kW, $Qdot_{htr\#1}$=2163587 kW, $Qdot_{htr\#2}$=342405 kW, $Qdot_{HX1}$=-1459718 kW, $mdot_{20}$=87.82 kg/s, $mdot_{21}$=63.73 kg/s, $mdot_{22}$=191.5 kg/s, $mdot_{12}$=656.9 kg/s, $mdot_{15}$=848.5 kg/s, $mdot_{17}$=912.2 kg/s and $mdot_{19}$=1000 kg/s.
2. A 4-stage turbine with reheat and 3-stage regenerative Rankine cycle as shown in Figure Example 7.6.3a using steam as the working fluid. The following information are provided:
   p_1=103 kPa, T_1=15°C, T_2=25°C, p_4=16000 kPa, T_4=600°C, $mdot_4$=1000 kg/s, p_5=7000 kPa, T_6=540°C, p_8=4000 kPa, p_{10}=2000 kPa, p_{12}=15 kPa, x_{13}=0, x_{15}=0, x_{19}=0, $\eta_{turbine\#1}$=0.85, $\eta_{turbine\#2}$=0.85, $\eta_{turbine\#3}$=0.85, $\eta_{turbine\#4}$=0.85, $\eta_{pump\#1}$=0.9, $\eta_{pump\#2}$=0.9, $\eta_{pump\#3}$=0.9, and $\eta_{pump\#4}$=0.9.
   Find η_{cycle}, $Wdot_{input}$, $Wdot_{output}$, $Wdot_{net\ output}$, $Qdot_{add}$, $Qdot_{remove}$, $Wdot_{turbine\#1}$, $Wdot_{turbine\#2}$, $Wdot_{turbine\#3}$, $Wdot_{turbine\#4}$, $Qdot_{htr\#1}$, $Qdot_{htr\#2}$, $Qdot_{HX1}$, $mdot_{20}$, $mdot_{21}$, $mdot_{22}$, $mdot_{12}$, $mdot_{15}$, $mdot_{17}$, and $mdot_{19}$.
   Draw the η_{cycle} vs p_5 sensitivity diagram, the η_{cycle} vs p_8 sensitivity diagram, and the η_{cycle} vs p_{10} sensitivity diagram.
3. A closed-cycle steam Rankine cycle without superheating is designed by a junior engineer as illustrated in Figure Example 7.11.2a with the following preliminary design information:

Condenser pressure	5 psia
Boiler pressure	2000 psia
Mass flow rate of steam	1 lbm/s
Flue gas temperature entering high-temperature side heat exchanger	3000°F
Flue gas pressure entering high-temperature side heat exchanger	14.7 psia
Flue gas leaving high-temperature side heat exchanger	1000°F
Cooling water temperature entering low-temperature side heat exchanger	50°F
Cooling water pressure entering low-temperature side heat exchanger	14.7 psia
Cooling water leaving low-temperature side heat exchanger	70°F
Turbine efficiency	85%

Pump efficiency 85%

Use net power output as the objective function and boiler pressure as the independent design parameter. Try to improve the preliminary design.

4. A closed-cycle steam Rankine cycle without superheating is designed by a junior engineer as illustrated in Figure Example 7.11.2a with the following preliminary design information:

Condenser pressure	10 kPa
Boiler pressure	16000 kPa
Mass flow rate of steam	1 kg/s
Flue gas temperature entering high-temperature side heat exchanger	2000°C
Flue gas pressure entering high-temperature side heat exchanger	101 psia
Flue gas leaving high-temperature side heat exchanger	1000°C
Cooling water temperature entering low-temperature side heat exchanger	14°C
Cooling water pressure entering low-temperature side heat exchanger	101 kPa
Cooling water leaving low-temperature side heat exchanger	20°C
Turbine efficiency	85%
Pump efficiency	85%

Use net power output as the objective function and boiler pressure as the independent design parameter. Try to improve the preliminary design.

7.12 Summary

The Carnot cycle is not a practical model for vapor power cycles because of cavitation and corrosion problems. The modified Carnot model for vapor power cycles is the basic Rankine cycle, which is made of two isobaric and two isentropic processes. The basic elements of the basic Rankine cycle are pump, boiler, turbine and condenser. Rankine cycle is the most popular heat engine to produce commercial power. The thermal cycle efficiency of the basic Rankine cycle can be improved by adding superheater, regenerating and reheater, among other means.

More Homework Problems

1. Consider a steam power plant operating on the ideal Rankine cycle. The steam enters the turbine at 3 MPa and 350°C and is condensed in the condenser at 10 kPa. Determine (a) the pump work required, (b) the heat added in the boiler, (c) the turbine work produced, (d) the heat removed from the condenser, and (e) the thermal cycle efficiency.
2. Consider a steam power plant operating on the ideal Rankine cycle with superheater. The steam enters the turbine at 3 MPa and 350°C and is condensed in the condenser at 10 kPa. Determine (a) the pump work required, (b) the heat added in the boiler, (c) the turbine work produced, (d) the heat removed from the condenser, and (e) the thermal cycle efficiency, if the steam is superheated to 600°C instead of 350°C.
3. Consider a steam power plant operating on the ideal Rankine cycle. The steam enters the turbine at 3 MPa and 350°C and is condensed in the condenser at 10 kPa. Determine (a) the pump work required, (b) the heat added in the boiler, (c) the turbine work produced,

(d) the heat removed from the condenser, and (e) the thermal cycle efficiency, if the pressure of the steam is raised to 15 Mpa while the turbine inlet temperature is maintained at 600°C.

4. Consider a steam power plant operating on the ideal Rankine cycle. The steam enters the turbine at 1250 psia as saturated vapor and is condensed in the condenser at 5 psia. The mass flow rate of the steam through the cycle is 75 lbm/s. Determine (a) the pump power required, (b) the rate of heat added in the boiler, (c) the turbine power produced, (d) the rate of heat removed from the condenser, and (e) the thermal cycle efficiency.
5. Consider a steam power plant operating on the ideal Rankine cycle with superheater. The steam enters the turbine at 1250 psia and 600°F and is condensed in the condenser at 5 psia. The mass flow rate of the steam through the cycle is 75 lbm/s. Determine (a) the pump power required, (b) the rate of heat added in the boiler, (c) the turbine power produced, (d) the rate of heat removed from the condenser, and (e) the thermal cycle efficiency.
6. Consider a steam power plant operating on the ideal Rankine cycle with superheater. The steam enters the turbine at 1250 psia and 600°F and is condensed in the condenser at 5 psia. The mass flow rate of the steam through the cycle is 75 lbm/s. If the condenser pressure is lowered to 2 psia, determine (a) the pump power required, (b) the rate of heat added in the boiler, (c) the turbine power produced, (d) the rate of heat removed from the condenser, and (e) the thermal cycle efficiency.
7. Consider a steam power plant operating on the ideal Rankine cycle with a superheater. The steam enters the turbine at 1250 psia and 600°F and is condensed in the condenser at 5 psia. The mass flow rate of the steam through the cycle is 75 lbm/s. If the boiler pressure is raised to 1500 psia, determine (a) the pump power required, (b) the rate of heat added in the boiler, (c) the turbine power produced, (d) the rate of heat removed from the condenser, and (e) the thermal cycle efficiency.
8. 12.5 lbm/s of superheated steam at a temperature of 900°F and a pressure of 1200 psia is used to power a naval propulsion turbine. The steam exhausts to a condenser which operates at a pressure of 1 psia. Instrumentation installed in the turbine exhaust trunk indicates that the exhaust steam has a quality of 92%. Find (a) actual turbine work, (b) pump work, (c) rate of heat added in the boiler, (d) turbine efficiency, and (e) cycle efficiency.
9. A saturated Rankine propulsion plant operates between a boiler pressure of 4 MPa and a condense pressure of 0.01 MPa. Determine (a) the pump work, (b) the heat added, (c) the net work of the cycle, and (d) the thermal efficiency of the plant.
10. A Rankine propulsion plant has a boiler pressure of 1200 psia and a temperature of 900°F. The turbine expands isentropically to a condenser operating at 1 psia. The turbine delivers 35,000 horsepower to the load. Determine (a) the pump power required, (b) the rate of heat removed from the condenser, (c) the rate of heat added in the boiler, (d) mass rate of flow, and (e) the cycle efficiency.
11. An ideal Rankine saturated steam cycle operates between pressure limits of 1 and 800 psia. The mass flow rate of the steam through the cycle is 1 lbm/s. Determine (a) the pump power, (b) the rate of heat added in the boiler, (c) the turbine power produced, (d) the quality and temperature of steam leaving the turbine, (e) the rate of heat rejected in the condenser, and (f) the cycle thermal efficiency.

12. An ideal Rankine superheated steam cycle operates between pressure limits of 1 and 1300 psia. The superheater outlet temperature is 900°F. The mass flow rate of the steam through the cycle is 1 lbm/s. Determine (a) the pump power, (b) the rate of heat added in the boiler, (c) the turbine power produced, (d) the quality of steam leaving the turbine, (e) the net power, (f) the rate of heat rejected in the condenser, and (g) the cycle thermal efficiency.
13. An ideal Rankine superheated steam cycle operates between pressure limits of 0.01 and 8 MPa. The superheater outlet temperature is 527°C. The mass flow rate of the steam through the cycle is 1 kg/s. Determine (a) the pump power, (b) the rate of heat added in the boiler, (c) the turbine power produced, (d) the quality of steam leaving the turbine, (e) the net power, (f) the rate of heat rejected in the condenser, and (g) the cycle thermal efficiency.
14. A regenerative Rankine cycle operates between pressure limits of 1 and 1200 psia. The superheater outlet temperature is 900°F. The cycle employs one stage of steam extraction between the HP and LP turbine for feedwater heating. The steam extraction pressure is 80 psia. Determine (a) mass flow rate of extraction steam per lbm of throttle steam, (b) main condensate pump power, (c) main feed pump power, (d) the rate of heat added in the boiler, (e) the HP turbine power produced, (f) the LP turbine power produced, (g) the net power produced by the plant if a steam flow rate is 1 lbm/s, (h) the rate of heat rejected in the condenser, and (i) the cycle thermal efficiency.
15. A regenerative Rankine cycle operates between pressure limits of 10 and 4000 kPa. The superheater outlet temperature is 400°C. The cycle employs one stage of steam extraction between the HP and LP turbine for feedwater heating. The steam extraction pressure is 500 kPa. Determine (a) mass flow rate of extraction steam per kg of throttle steam, (b) pump power, (c) the rate of heat added in the boiler, (d) the turbine power produced by the plant if a steam flow rate is 10 kg/s, (e) the rate of heat rejected in the condenser, and (f) the cycle thermal efficiency.
16. A regenerative Rankine cycle operates between pressure limits of 2 and 1200 psia. The superheater outlet temperature is 900°F. The cycle employs one stage of steam extraction between the HP and LP turbine for feed water heating. The steam extraction pressure is to be determined. Water leaving the feed water heater is a saturated liquid. Determine the optimal extraction pressure and the cycle thermal efficiency.
17. Consider a steam power plant operating on the ideal reheat Rankine cycle. The steam enters the turbine at 15 MPa and 600°C and is condensed in the condenser at 10 kPa. The mass flow rate of the steam through the cycle is 100 kg/s. If the moisture content of the steam at the exit of the low pressure turbine is not to exceed 9 percent, determine (a) the pressure at which the steam should be reheated, (b) the rate of heat added in the boiler, (c) the rate of heat added in the reheater, (d) the rate of heat removed from the condenser, (e) the HP turbine power produced, (f) the LP turbine power produced, and (e) the thermal cycle efficiency.
18. Consider a steam power plant operating on the ideal reheat Rankine cycle. The steam enters the turbine at 15 MPa and 600°C and is condensed in the condenser at 10 kPa. The mass flow rate of the steam through the cycle is 100 kg/s. If the moisture content of the steam at the exit of the low pressure turbine is not to exceed 10 percent, determine (a) the pressure at which the steam should be reheated, (b) the rate of heat added in the boiler, (c)

the rate of heat added in the reheater, (d) the rate of heat removed from the condenser, (e) the HP turbine power produced, (f) the LP turbine power produced, (g) pump power required, and (h) the thermal cycle efficiency.

19. A binary vapor power cycle consists of two ideal Rankine cycles with steam and ammonia as the working fluids. In the steam cycle, superheated vapor enters the turbine at 6 MPa and 900 K, and saturated liquid exits the condenser at 330 K. The heat rejected from the steam cycle is provided to the ammonia cycle, producing saturated vapor at 323 K, which enters the ammonia turbine. Saturated liquid leaves the ammonia condenser at 1 MPa. For a net power of 20 MW from the binary cycle, determine (a) the power output of the steam and ammonia turbines, (b) the rate of heat addition to the binary cycle, and (c) the thermal efficiency.

Chapter 8

GAS CLOSED SYSTEM CYCLES

OBJECTIVES

After reading and studying the material in this chapter, you should be able to:

1. Differentiate between internal-combustion and external-combustion gas heat engines.
2. Describe the sequence of processes in the Otto and Diesel cycles.
3. Sketch the Otto and Diesel cycles on both p-v and T-s diagrams.
4. Perform volume compression ratio sensitivity analysis on the Otto cycle efficiency using CyclePad.
5. Complete a pressure compression ratio sensitivity analysis on the Diesel cycle efficiency using CyclePad.
6. Know that the Otto cycle efficiency is better than the Diesel cycle efficiency with the same volume compression ratio.
7. Understand why the Diesel cycle can be operated with higher volume compression ratios than the Otto cycle.
8. Understand why the Dual cycle is proposed as the prototype for actual internal-combustion gas heat engines.
9. Analyze the Otto, Diesel and Dual cycles using CyclePad.
10. Analyze the Lenoir and Stirling cycles using CyclePad.
11. Design gas closed system cycles using CyclePad.

8.1 OTTO CYCLE

A four stroke internal combustion engine was built by a German engineer, Nicholas Otto, in 1876. The cycle patterned after his design is called the *Otto cycle*. It is the most widely used internal combustion heat engine in automobiles.

The piston in a four stroke internal combustion engine executes four complete strokes as the crankshaft completes two revolution per cycle as shown in Figure 8.1.1. On the intake stroke, the intake valve is open and the piston moves downward in the cylinder, drawing in a premixed charge of gasoline and air until the piston reaches its lowest point of the stroke called *bottom dead center* (BDC). During the compression stroke the intake valve closes and the piston moves toward the top of the cylinder, compressing the gas-air mixture. As the piston approaches the top of the cylinder called *top dead center* (TDC), the spark plug is energized and the mixture ignites, creating an increase in the temperature and pressure of the gas. During the expansion stroke the piston is forced down by the high pressure gas, producing a useful work output. The cycle is then completed when the exhaust valve opens and the piston moves toward the top of the cylinder, expelling the products of combustion. The schematic diagram of the Otto cycle is shown in Figure 8.1.2.

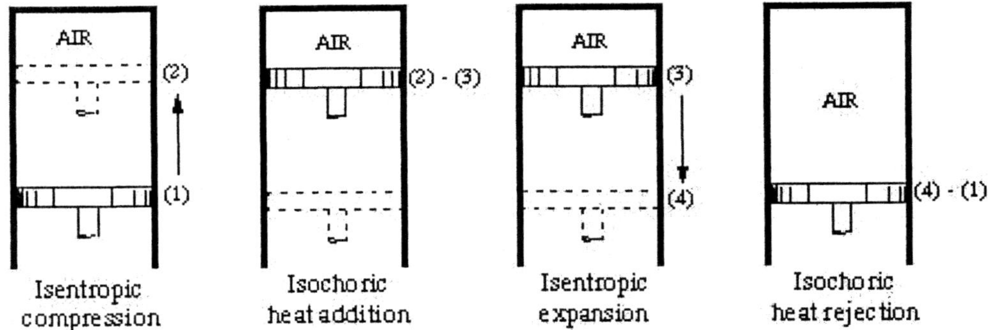

Figure 8.1.1 Otto cycle

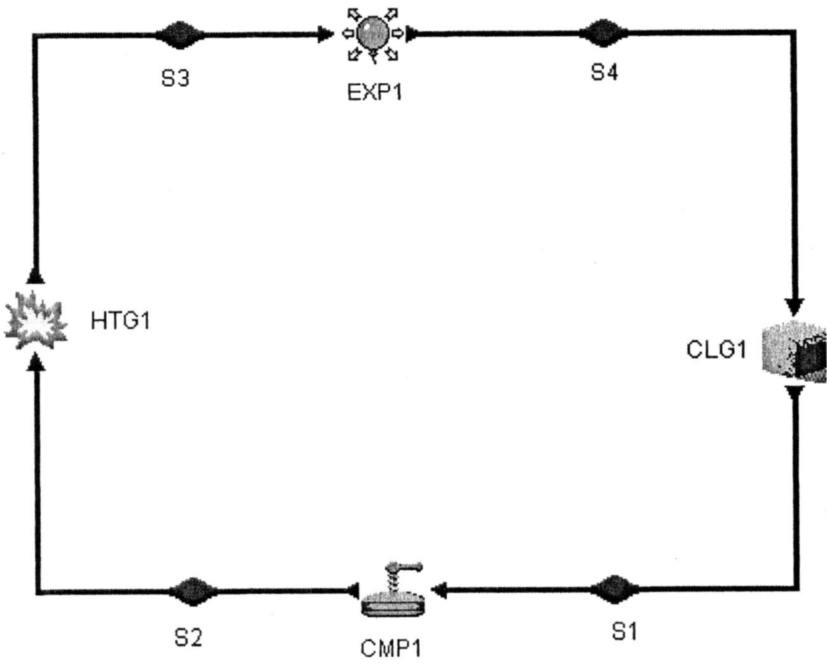

Figure 8.1.2 Otto cycle

The thermodynamic analysis of an actual Otto cycle is complicated. To simplify the analysis, we consider an ideal Otto cycle composed entirely of internally reversible processes. In the Otto cycle analysis, a closed piston-cylinder assembly is used as a control mass system.

The cycle is made of the following four processes:

1-2 isentropic compression
2-3 constant volume heat addition
3-4 isentropic expansion
4-1 constant volume heat removing

The p-v and T-s process diagrams for the ideal Otto cycle are illustrated in Figure 8.1.3.

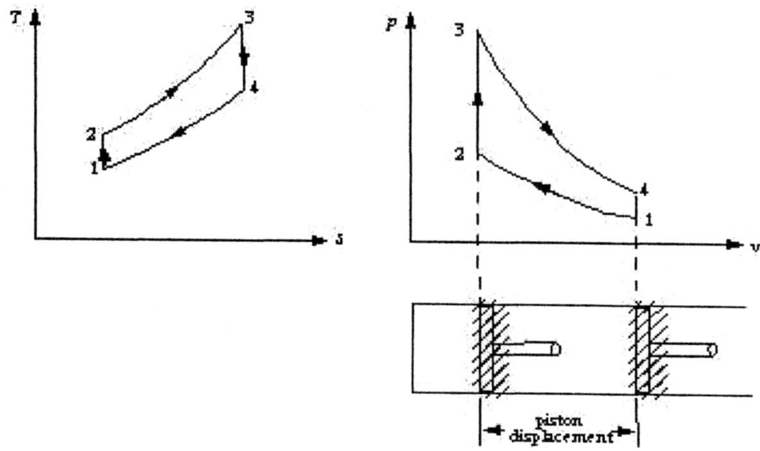

Figure 8.1.3 Otto cycle p-v and T-s diagrams

Applying the first law and second law of thermodynamics of the closed system to each of the four processes of the cycle yields:

$$W_{12} = \int pdV, \tag{8.1.1}$$

$$Q_{12} - W_{12} = m(u_2 - u_1),\ Q_{12}=0 \tag{8.1.2}$$

$$W_{23} = \int pdV = 0, \tag{8.1.3}$$

$$Q_{23} - 0 = m(u_3 - u_2), \tag{8.1.4}$$

$$W_{34} = \int pdV, \tag{8.1.5}$$

$$Q_{34} - W_{34} = m(u_4 - u_3),\ Q_{34} = 0 \tag{8.1.6}$$

$$W_{41} = \int pdV = 0, \tag{8.1.7}$$

and

$$Q_{41} - 0 = m(u_1 - u_4). \tag{8.1.8}$$

The net work (W_{net}), which is also equal to net heat (Q_{net}), is

$$W_{net} = W_{12} + W_{34} = Q_{net} = Q_{23} + Q_{41} \tag{8.1.9}$$

The thermal efficiency of the cycle is

$$\eta = W_{net} / Q_{23} = Q_{net} / Q_{23} = 1 - Q_{41} / Q_{23} = 1 - (u_4 - u_1)/(u_3 - u_2) \tag{8.1.10}$$

This expression for thermal efficiency of an ideal Otto cycle can be simplified if air is assumed to be the working fluid with constant specific heats. Equation (8.1.3) is reduced to:

$$\eta = 1 - (T_4 - T_1)/(T_3 - T_2) = 1 - (r)^{1-k} \tag{8.1.11}$$

where r is the *compression ratio* for the engine defined by the equation

$$r = V_1 / V_2 \tag{8.1.12}$$

The compression ratio is the ratio of the cylinder volume at the beginning of the compression process (BDC) to the cylinder volume at the end of the compression process (TDC).

Equation (8.1.11) shows that the thermal efficiency of the Otto cycle is only a function of the compression ratio of the engine. Therefore, any engine design that increases the compression ratio should result in an increased engine efficiency. The compression ratio cannot be increased indefinitely. As the compression ratio increases, the temperature of the working fluid also increases during the compression process. Eventually, a temperature is reached that is sufficiently high to ignite the air-fuel mixture prematurely without the presence of a spark. This condition causes the engine to produce a noise called knock. The presence of *engine knock* places a barrier on the upper limit of Otto engine compression ratios. To reduce engine knock problem of a high compression ratio Otto cycle, one must use gasoline with higher octane rating. In general, the higher the octane rating number of gasoline, the higher the resistance of engine knock.

One way to simplify the calculation of the net work of the cycle and to provide a comparative measure of the performance of an Otto heat engine is to introduce the concept of the mean effective pressure. The *mean effective pressure* (MEP) is the average pressure of the cycle. The net work of the cycle is equal to the mean effective pressure multiplied by the displacement volume of the cylinder. That is

$$MEP = (\text{cycle net work})/(\text{cylinder displacement volume}) = W_{net}/(V_1 - V_2)$$

The engine with the larger MEP value of two engines of equal cylinder displacement volume would be the better one, because it would produce a greater net work output.

Example 8.1.1. An engine operates on the Otto cycle and has a compression ratio of 8. Fresh air enters the engine at 27°C and 100 kPa. The amount of heat addition is 700 kJ/kg. The amount of air mass in the cylinder is 0.01 kg. Determine the pressure and temperature at the end of the combustion, the pressure and temperature at the end of the expansion, MEP, efficiency and work output per kilogram of air. Show the cycle on T-s diagram. Plot the sensitivity diagram of cycle efficiency vs compression ratio.

To solve this problem by CyclePad, we take the following steps:

1. Build
   (A) Take a compression device, a combustion chamber, an expander and a cooler from the closed system inventory shop and connect the four devices to form the Otto cycle as shown in Figure 8.1.2.
   (B) Switch to analysis mode.
2. Analysis
   (A) Assume a process for each of the four processes: (a) compression device as adiabatic and isentropic, (b) combuster as isochoric, (c) expander as adiabatic and isentropic, and (d) cooler as isochoric.
   (B) Input the given information: (a) working fluid is air, (b) the inlet pressure and temperature of the compression device are 100 kPa and 27°C, (c) the compression ratio of the compression device is 8, (d) the heat addition is 700 kJ/kg in the combustion chamber, and (e) m=0.01 kg.
3. Display results
   (A) Display the T-s diagram and cycle properties results. The cycle is a heat engine. The answers are: p=4441 kPa and T=1393°C (the pressure and temperature at the end of the combustion), p=241.6 kPa and T=452.1°C (the pressure and temperature at the end of the expansion), MEP=525.0 kPa, η=56.47% and Wnet=3.95 kJ, and (B) Display the sensitivity diagram of cycle efficiency vs compression ratio.

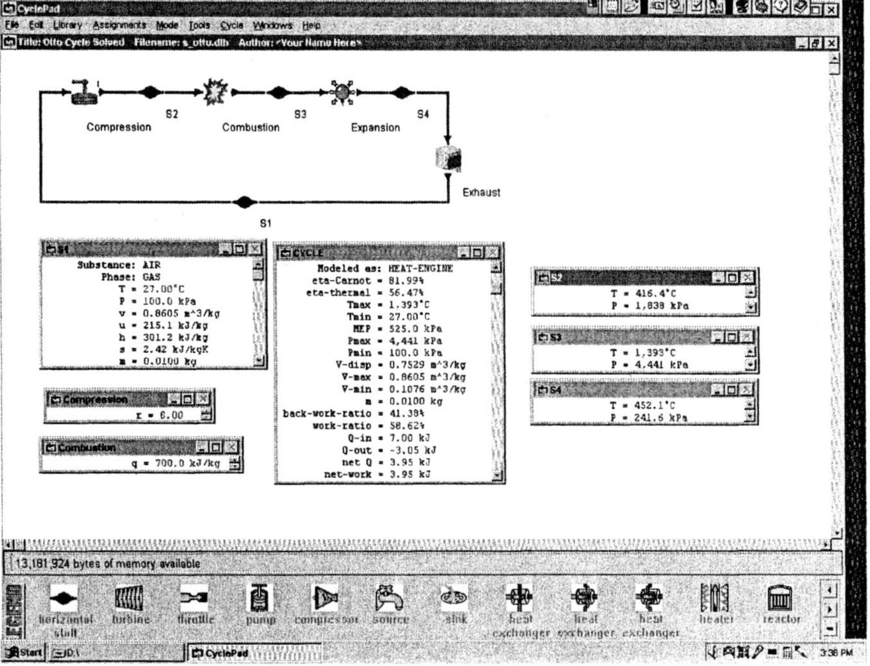

Figure Example 8.1.1a. Otto cycle

Comment: Efficiency increases as compression ratio increases.

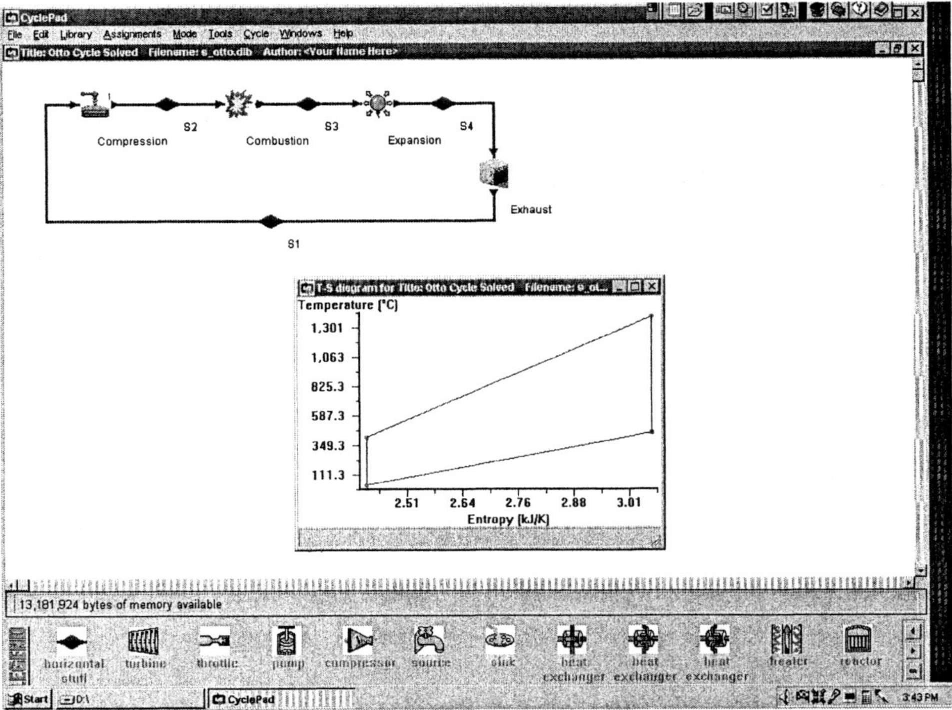

Figure Example 8.1.1b. Otto cycle T-s diagram

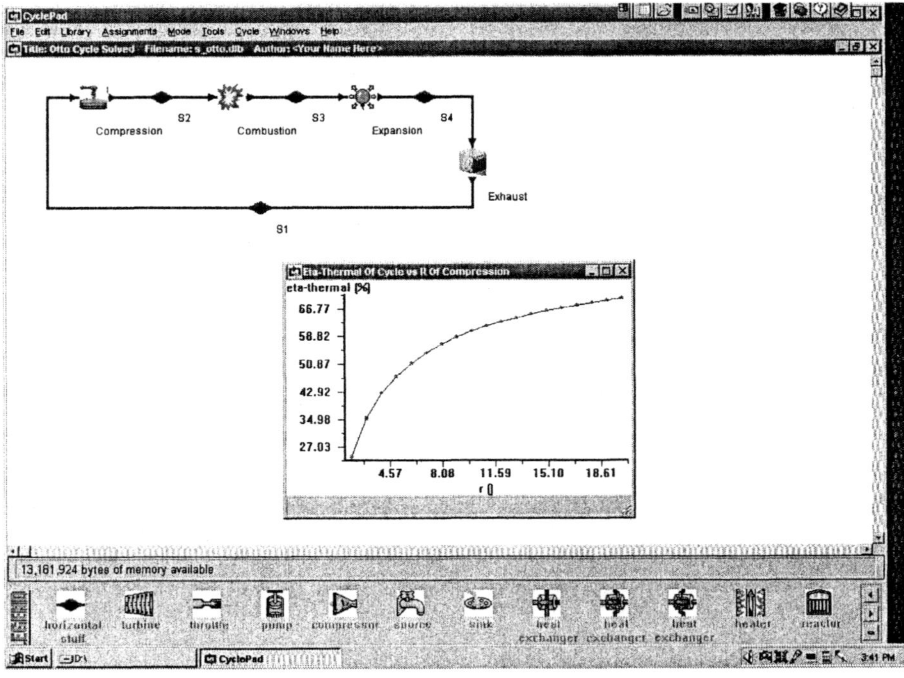

Figure Example 8.1.1a. Otto cycle sensitivity analysis

Example 8.1.2. The compression ratio in an Otto cycle is 8. If the air before compression (state 1) is at 60°F and 14.7 psia and 800 Btu/lbm is added to the cycle and the mass of air contained in the cylinder is 0.025 lbm, Calculate (a) temperature and pressure at each point of the cycle, (b) the heat must be removed, (c) the thermal efficiency, and (d) the MEP of the cycle.

To solve this problem by CyclePad, we take the following steps:

1. Build
   (A) Take a compression device, a combustion chamber, an expander and a cooler from the closed system inventory shop and connect the four devices to form the Otto cycle.
   (B) Switch to analysis mode.
2. Analysis
   (A) Assume a process for each of the four processes: (a) compression device as isentropic, (b) combuster as isochoric, (c) expander as isentropic, and (d) cooler as isochoric.
   (B) Input the given information: (a) working fluid is air, (b) the inlet pressure and temperature of the compression device are 60°F and 14.7 psia, (c) the compression ratio of the compression device is 8, and (d) the heat addition is 800 Btu/lbm in the combustion chamber, and (e) the mass of air is 0.025 lbm.
3. Display results
   (A) Display the T-s diagram and cycle properties results. The cycle is a heat engine. The answers are T_2=734.2 °F, p_2=270.2 psia, T_3=5,407 °F, p_3=1,328 psia, T_4=2,094 °F, p_4=72.24 psia, η=56.47%, Q_{41}=-8.71 Btu, and MEP=213.3 psia.
   (B) Display the T-s diagram.

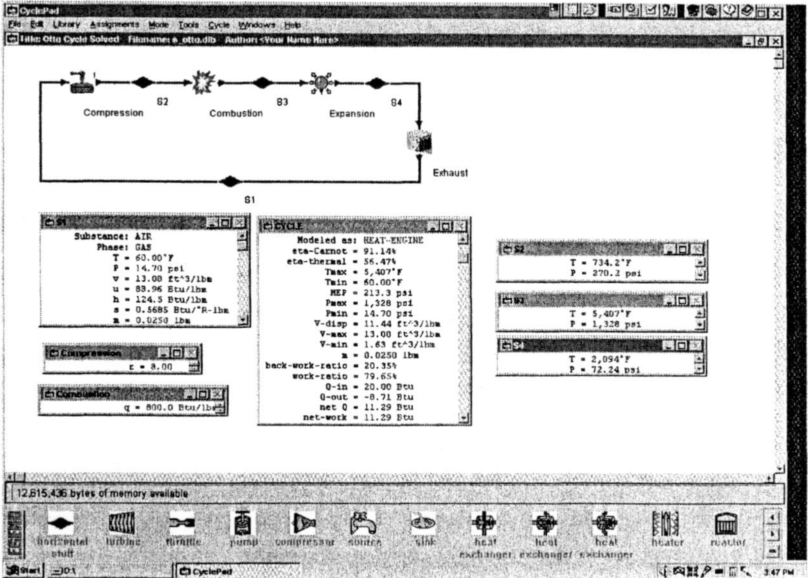

Figure Example 8.1.2a Otto cycle

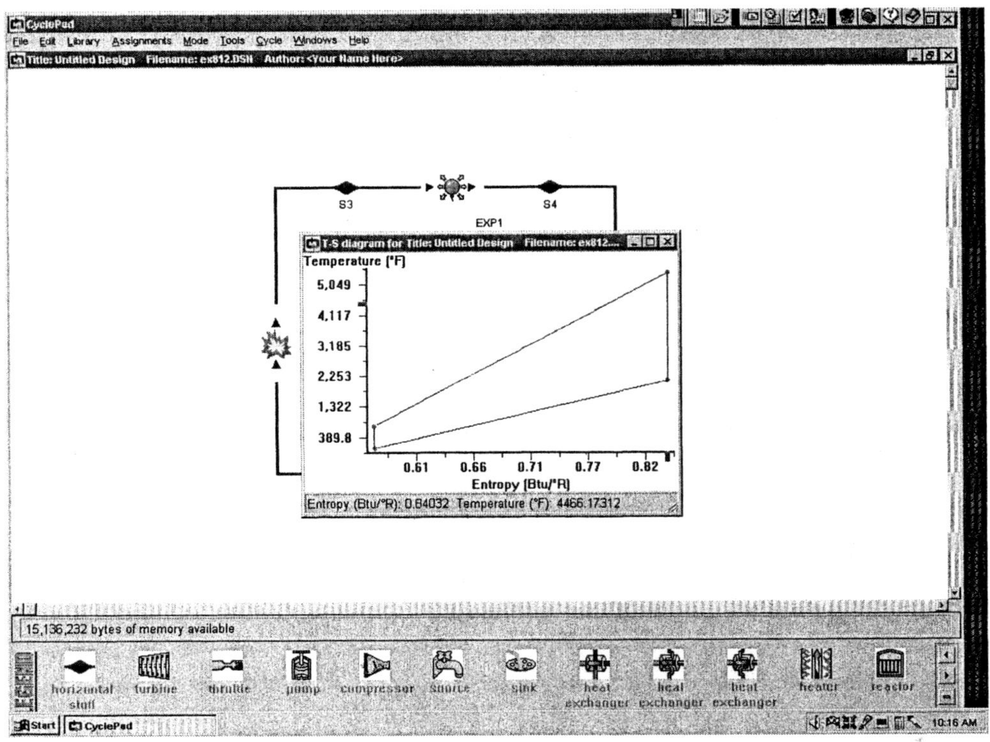

Figure Example 8.1.2b Otto cycle T-s diagram plot

The power output of the Otto cycle can be increased by *turbo-charging* the air before it enters the cylinder in the Otto engine. Since the inlet air density is increased due to higher inlet air pressure, the mass flow rate of air is increased. Turbo-charging raises the inlet air pressure of the engine above atmospheric pressure and raise the power output of the engine, but it may not improve the efficiency of the cycle. The schematic diagram of the Otto cycle with turbo-charging is illustrated in Figure 8.1.4. Example 8.1.3 and Example 8.1.4 show the power increase due to turbo-charging.

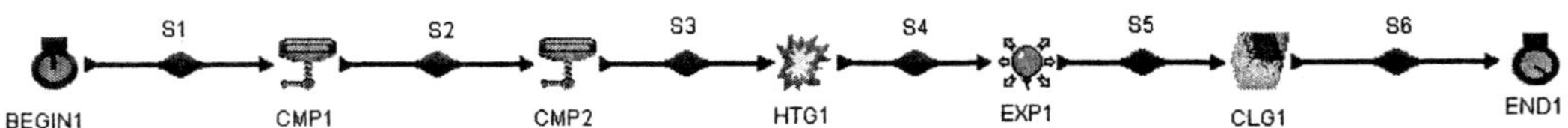

Figure 8.1.4 Otto engine with turbo-charging

Example 8.1.3 Determine the heat supplied, work output, MEP, and thermal efficiency of an ideal Otto cycle with a compression ratio of 10. The highest temperature of the cycle is 3000°F. The volume of the cylinder before compression is 0.1 ft^3. What is the mass of air in the cylinder? The atmosphere conditions are 14.7 psia and 70°F.

To solve this problem, we build the cycle as shown in Figure 8.1.4. Then (A) Assume isentropic for compression process 1-2, isentropic for compression process 2-3, isochoric for the heating process 3-4, isentropic for expansion process 4-5, and isochoric for the cooling process 5-6; (B) input p_1 =14.7 psia, T_1=70°F, V_1=0.1 ft^3; p_2=14.7 psia, T_2=70°F, V_2=0.1 ft^3 (no turbo-charger); compression ratio=10, T_4=3000°F, p_6=14.7 psia, and T_6=70°F; and (C) display results. The results are: W_{12}=-0 Btu, W_{23}=-1.03 Btu, Q_{34}=2.73 Btu, W_{45}=2.67 Btu, W_{net}=1.65 Btu, Q_{56}=-1.09 Btu, MEP=98.80 psia, η=60.19%, and m=0.0075 lbm.

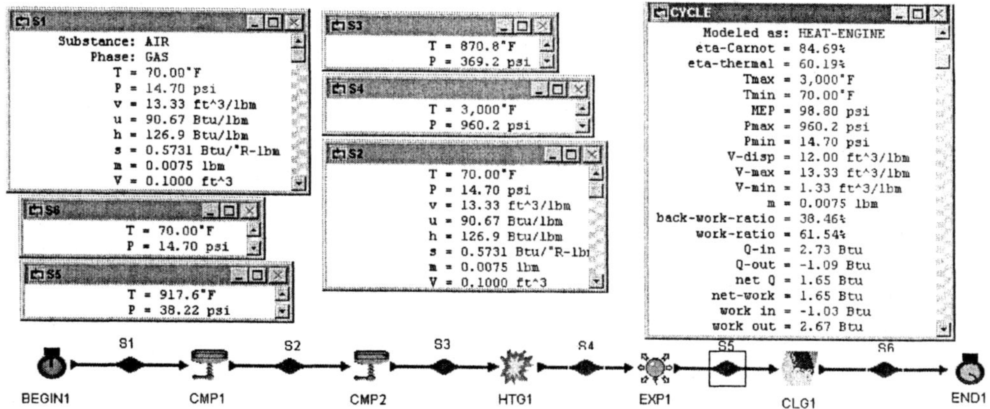

Figure Example 8.1.3a Otto engine without turbo-charging

Example 8.1.4 Determine the heat supplied, work output, MEP, and thermal efficiency of an ideal Otto cycle with a turbo-charger which compresses air to 20 psia and compression ratio of 10. The highest temperature of the cycle is 3000°F. The volume of the cylinder before compression is 0.1 ft^3. What is the mass of air in the cylinder? The atmosphere conditions are 14.7 psia and 70°F.

To solve this problem, we build the cycle as shown in Figure 8.2.1. Then (A) Assume isentropic for compression process 1-2, isentropic for compression process 2-3, isochoric for the heating process 3-4, isentropic for expansion process 4-5, and isochoric for the cooling process 5-6; (B) input p_1 =14.7 psia, T_1=70°F; p_2=20 psia, V_2=0.1 ft^3 (with turbo-charger); compression ratio=10, T_4=3000°F, p_6=14.7 psia, and T_6=70°F; and (C) display results. The results are: W_{13}=-1.48 Btu, Q_{34}=3.21 Btu, W_{45}=3.52 Btu, W_{net}=2.04 Btu, Q_{56}=-1.17 Btu, MEP=96.20 psia, η=63.54%, and m=0.0093 lbm.

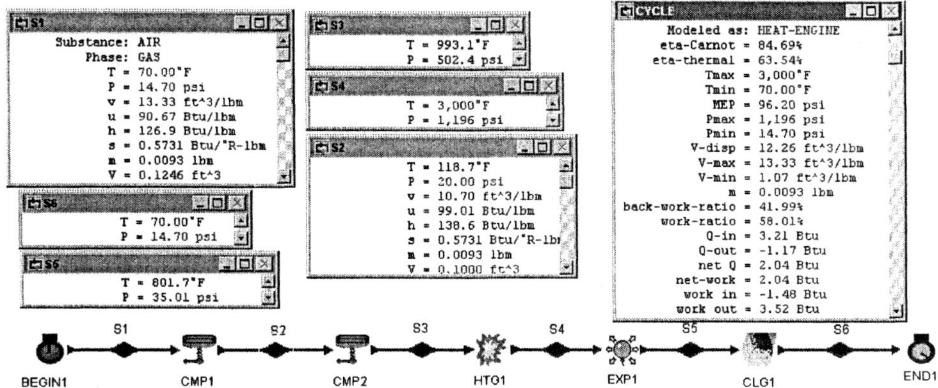

Figure Example 8.1.4 Otto engine with turbo-charging

Modern Otto engine designs are affected by environmental constrains as well as desires to increase gas mileage. Recent design improvements include the use of four valves per cylinder to reduce the restriction to air flow into and out of the cylinder, turbocharges to increase the air and fuel flow to each cylinder, catalytic converters to aid the combustion of unburned hydrocarbons that are expelled by the engine, among others.

Homework 8.1 Otto Cycle

1. Do Otto heat engines operate on a closed system or an open system? Why?
2. What is compression ratio of an Otto cycle? How does it affect the thermal efficiency of the cycle?
3. Do you know the compression ratio of your car? Is there any limit to an Otto cycle? Why?
4. Which area represents cycle net work of an Otto cycle plotted on a T-s diagram and p-v diagram?
5. How do you define MEP (mean effect pressure)?
6. What is engine knock? What cause the engine knock problem?
7. Do you get a better performance using premier gasoline (Octane number 93) for your compact car?
8. An engine operates on an Otto cycle with a compression ratio of 8. At the beginning of the isentropic compression process, the volume, pressure and the temperature of the air are 0.01 m^3, 110 kPa and 50°C. At the end of the combustion process, the temperature is 900°C. Find (a) the temperature at the remaining two states of the Otto cycle, (b) the pressure of the gas at the end of the combustion process, (c) heat added per unit mass to the engine in the combustion chamber, (d) heat removed per unit mass from the engine to the environment, (e) the compression work per unit mass added, (f) the expansion work per unit mass done, (g) MEP, and (h) thermal cycle efficiency.
9. An ideal Otto Cycle with air as the working fluid has a compression ratio of 9. At the beginning of the compression process, the air is at 290 K and 90 kPa. The peak temperature in the cycle is 1800 K. determine: (a) the pressure and temperature at the end of the expansion process (power stroke), (b) the heat per unit mass added in kJ/kg during the combustion process, (c) net work, (d) thermal efficiency of the cycle, and (e) mean effective pressure in kPa.
10. An ideal Otto engine receives air at 15 psia, 0.01 ft^3 and 65°F. The maximum cycle temperature is 3465°F and the compression ratio of the engine is 7.5. Determine (a) the work added during the compression process, (b) the heat added to the air during the heating process, (c) the work done during the expansion process, (d) the heat removed from the air during the cooling process, and (e) the thermal efficiency of the cycle.
11. An ideal Otto engine receives air at 14.6 psia and 55°F. The maximum cycle temperature is 3460°F and the compression ratio of the engine is 10. Determine (a) the work done per unit mass during the compression process, (b) the heat added per unit mass to the air during the heating process, (c) the work done per unit mass during the expansion process, (d) the heat removed per unit mass from the air during the cooling process, and (e) the thermal efficiency of the cycle.

12. An ideal Otto engine receives air at 100 kPa and 25°C. Work is performed on the air in order to raise the pressure at the end of the compression process to 1378 kPa. 400 kJ/kg of heat is added to the air during the heating process. Determine (a) the work done during the compression process, (b) the compression ratio, (c) the work done during the expansion process, (d) the heat removed from the air during the cooling process, (e) the MEP (mean effective pressure), and (f) the thermal efficiency of the cycle.
13. At the beginning of the compression process of an air-standard Otto cycle, p=100 kPa, T=290 K and V=0.04 m^3. The maximum temperature in the cycle is 2200 K and the compression ratio is 8. Determine (a) the heat addition, (b) the net work, (c) the thermal efficiency, and (d) the MEP.
    ANSWER: (a) 52.89 kJ, (b) 29.87 kJ, (c) 56.47%, (d) 853.3 kPa.
14. An Otto engine operates with a compression ratio of 8.5. The following information is known:
    Temperature prior to the compression process: 70°F.
    Volume prior to the compression process: 0.05 ft^3
    Pressure prior to the compression process: 14.7 psia.
    Heat added during the combustion process: 345 Btu/lbm.
    (a) Determine the mass of air in the cylinder
    (b) Determine the temperature and pressure at each process endpoint.
    (c) Find the compression work and expansion work in Btu/lbm.
    (d) the thermal efficiency.
15. The compression ratio in an Otto cycle is 8. If the air before compression (state 1) is at 80°F and 14.7 psia and 800 Btu/lbm is added to the cycle and the mass of air contained in the cylinder is 0.02 lbm, Find the heat added, heat removed, work added, work produced, net work produced, MEP and efficiency of the cycle.
    ANSWER: heat added=16 Btu, heat removed=-6.96 Btu, work added=-2.4 Btu, work produced=11.43 Btu, net work produced=9.04 Btu, MEP=205.4 psia, and efficiency of the cycle=56.47%.
16. The compression ratio in an Otto cycle is 10. If the air before compression (state 1) is at 60°F and 14.7 psia and 800 Btu/lbm is added to the cycle and the mass of air contained in the cylinder is 0.02 lbm, Find the heat added, heat removed, work added, work produced, net work produced, MEP and efficiency of the cycle.
    ANSWER: heat added=16 Btu, heat removed=-6.37 Btu, work added=-2.69 Btu, work produced=12.32 Btu, net work produced=9.63 Btu, MEP=221.0 psia, and efficiency of the cycle=60.19%.
17. The compression ratio in an Otto cycle is 16. If the air before compression (state 1) is at 60°F and 14.7 psia and 800 Btu/lbm is added to the cycle and the mass of air contained in the cylinder is 0.02 lbm, Find the heat added, heat removed, work added, work produced, net work produced, MEP and efficiency of the cycle.
    ANSWER: heat added=16 Btu, heat removed=-5.28 Btu, work added=-3.61 Btu, work produced=14.34 Btu, net work produced=10.72 Btu, MEP=236.2 psia, and efficiency of the cycle=67.01%.
18. An Otto engine with a turbo-charger operates with a compression ratio of 8.5. The following information is known:
    Temperature prior to the turbo-charging compression process: 70°F.
    Pressure prior to the turbo-charging compression process: 14.7 psia.

Pressure after the turbo-charging compression process: 20 psia.
Heat added during the combustion process: 345 Btu/lbm.
Volume after the compression process: 0.05 ft^3

(a) Determine the mass of air in the cylinder
(b) Determine the temperature and pressure at each process endpoint.
(c) Find the compression work and expansion work in Btu/lbm.
(d) the thermal efficiency.

19. An ideal Otto Cycle with a turbo-charger using air as the working fluid has a compression ratio of 9. The volume of the cylinder is 0.01 m^3. At the beginning of the turbo-charging compression process, the air is at 290 K and 90 kPa. The air pressure is 150 kPa after the turbo-charging compression process. The peak temperature in the cycle is 1800 K. determine: (a) the pressure and temperature at the end of the expansion process (power stroke), (b) the heat per unit mass added in kJ/kg during the combustion process, (c) net work, (d) thermal efficiency of the cycle, and (e) mean effective pressure in kPa.

8.2 DIESEL CYCLE

The *Diesel cycle* was proposed by Rudolf Diesel in the 1890s. The Diesel cycle as shown in Figure 8.2.1 is somewhat similar to the Otto cycle, except that ignition of the fuel-air mixture is caused by spontaneous combustion owing to the high temperature that results from compressing the mixture to a very high pressure. The basic components of the Diesel cycle are the same as the Otto cycle, except that the spark plug is replaced by a fuel injector and the stroke of the piston is lengthened to provide a larger compression ratio.

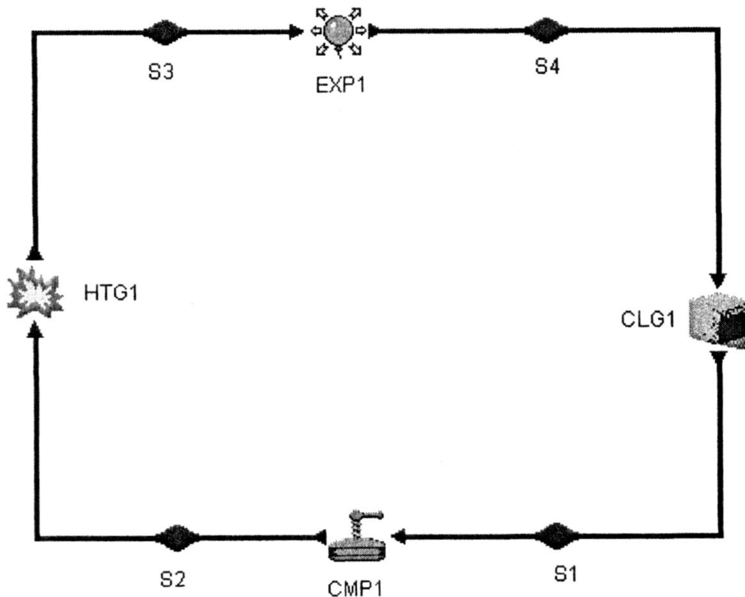

Figure 8.2.1 Diesel cycle

The Diesel cycle consists of the following four processes:

1-2 isentropic compression
2-3 constant pressure heat addition
3-4 isentropic expansion
4-1 constant volume heat removing

Since the duration of the heat addition process is extended, this process is modeled by a constant pressure process. The p-v and T-s diagram for the Diesel cycle are illustrated in Figure 8.2.2.

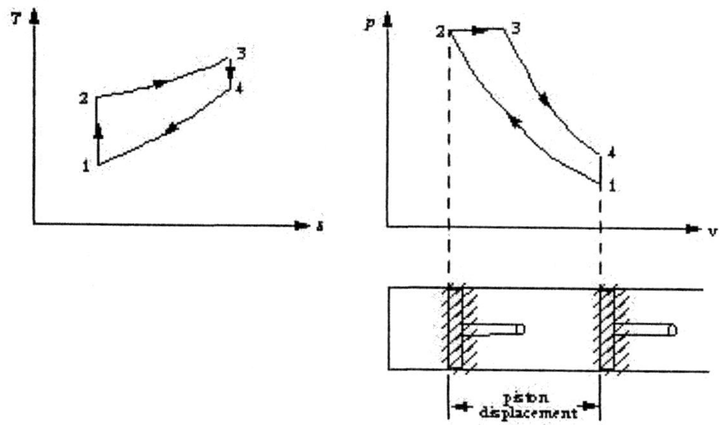

Figure 8.2.2 Diesel cycle p-v and T-s diagrams

Applying the First law of thermodynamics of the closed system to each of the four processes of the cycle yields:

$$W_{12} = \int pdV, \quad (8.2.1)$$

$$Q_{12} - W_{12} = m(u_2 - u_1), \; Q_{12} = 0, \quad (8.2.2)$$

$$W_{23} = \int pdV = m(p_3v_3 - p_2v_2), \quad (8.2.3)$$

$$Q_{23} = m(u_3 - u_2) + W_{23} = m(h_3 - h_2), \quad (8.2.4)$$

$$W_{34} = \int pdV, \quad (8.2.5)$$

$$Q_{34} - W_{34} = m(u_4 - u_3), Q_{34} = 0, \quad (8.2.6)$$

$$W_{41} = \int pdV = 0, \quad (8.2.7)$$

and

$$Q_{41} - 0 = m(u_1 - u_4). \quad (8.2.8)$$

The net work (W_{net}), which is also equal to net heat (Q_{net}), is

$$W_{net} = W_{12}+W_{23}+W_{34} = Q_{net} = Q_{23} + Q_{41} \quad (8.2.9)$$

The thermal efficiency of the cycle is

$$\eta = W_{net} /Q_{23} = Q_{net} /Q_{23} = 1- Q_{41} /Q_{23} = 1- (u_4 - u_1)/(h_3 - h_2) \quad (8.2.10)$$

This expression for thermal efficiency of an ideal Otto cycle can be simplified if air is assumed to be the working fluid with constant specific heats. Equation (8.2.10) is reduced to:

$$\eta = 1- (T_4 - T_1)/[k(T_3 - T_2)] = 1-(r)^{1-k} \{[(r_c)^k -1]/k(r_c -1)\} \quad (8.2.11)$$

where r is the *compression ratio*, and r_c is the *cut-off ratio* for the engine defined by the equation

$$r = V_1 /V_2 \quad (8.2.12)$$

and

$$r_c = V_3 /V_2 \quad (8.2.13)$$

A comparison of the thermal efficiency of the Diesel cycle and the thermal efficiency of the Otto cycle

shows that the two thermal cycle efficiencies differ by the quantity in the brackets of equation (8.2.11). This bracket factor is always larger than one, hence the Diesel cycle efficiency is always less than the Otto cycle efficiency operating at the same compression ratio.

Since the fuel is not injected into the cylinder until after the air has been completely compressed in the Diesel cycle, there is no engine knock problem. Therefore the Diesel engine can be designed to operate at much higher compression ratios and less refined fuel than those of the Otto cycle. As a result of the higher compression ratio, Diesel engines are slightly more efficient than Otto engines.

Example 8.2.1. A Diesel engine receives air at 27°C and 100 kPa. The compression ratio is 18. The amount of heat addition is 500 kJ/kg. The mass of air contained in the cylinder is 0.0113 kg. Determine (a) the maximum cycle pressure and maximum cycle temperature, (b) the efficiency and work output, and (c) the MEP. Plot the sensitivity diagram of cycle efficiency vs compression ratio.

To solve this problem by CyclePad, we take the following steps:

1. Build
   (A) Take a compression device, a combustion chamber, an expander and a cooler from the closed system inventory shop and connect the four devices to form the Diesel cycle.

(B) Switch to analysis mode.

2. Analysis

   (A) Assume a process for each of the four processes: (a) compression device as isentropic, (b) combustion as isobaric, (c) expander as isentropic, and (d) cooler as isochoric.

   (B) Input the given information: (a) working fluid is air, (b) the inlet pressure and temperature of the compression device are 100 kPa and 27°C, (c) the compression ratio of the compression device is 18, and (d) the heat addition is 500 kJ/kg in the combustion chamber,and (e) the mass of air is 0.0113 kg.

3. Display results

   (A) Display cycle properties results. The cycle is a heat engine. The answers are T_{max}=1179 °C, p_{max}=5720 kPa, η=65.53%, MEP=403.2 kPa and Wnet=3.72 kJ, and

   (B) Display the sensitivity diagram of cycle efficiency vs compression ratio.

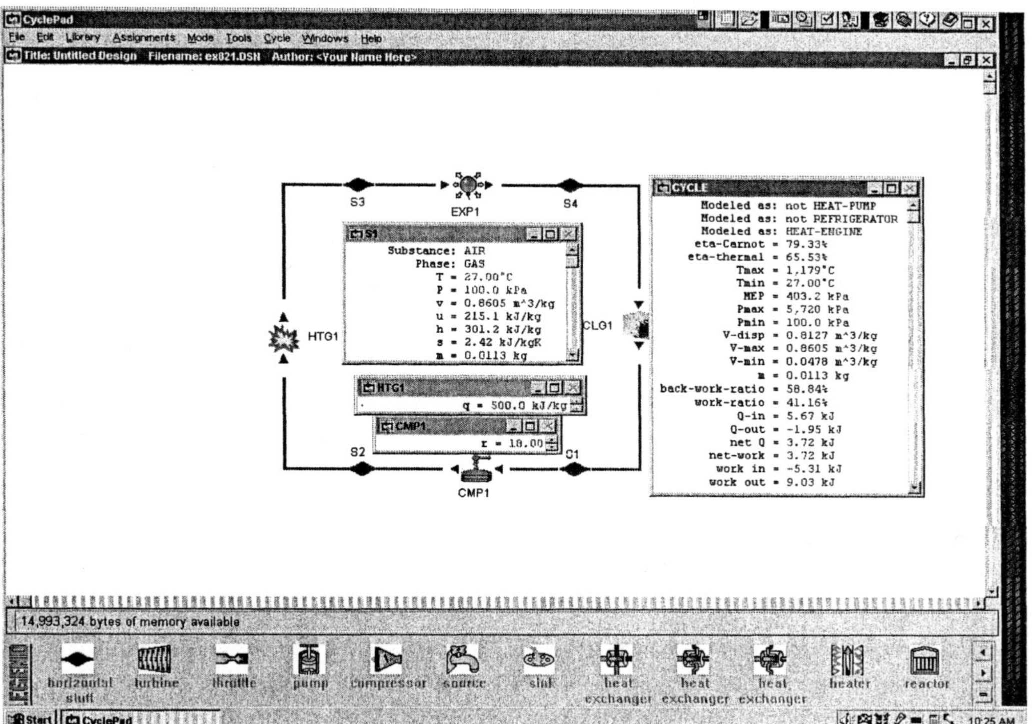

Figure Example 8.2.1a Diesel cycle

Comment: Efficiency increases as compression ratio increases.

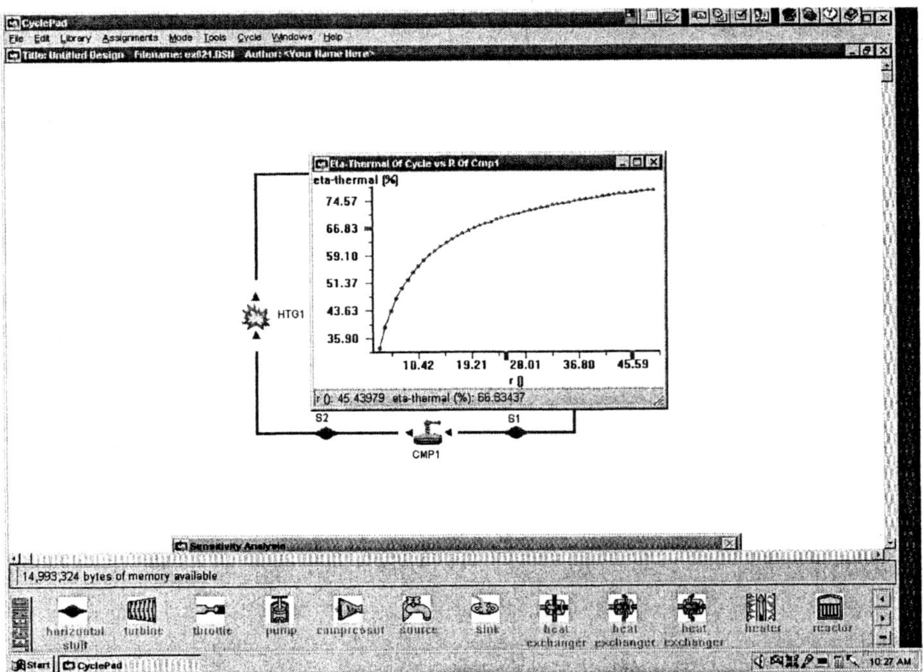

Figure Example 8.2.1b Diesel cycle sensitivity analysis

Example 8.2.2. A Diesel engine receives air at 60°F and 14.7 psia. The compression ratio is 16. The amount of heat addition is 800 Btu/lbm. The mass of air contained in the cylinder is 0.02 lbm. Determine the maximum cycle temperature, heat added, heat removed, work added, work produced, net work produced, MEP and efficiency of the cycle.

To solve this problem by CyclePad, we take the following steps:

1. Build
   (A) Take a compression device, a combustion chamber, an expander and a cooler from the closed system inventory shop and connect the four devices to form the Diesel cycle.
   (B) Switch to analysis mode.
2. Analysis
   (A) Assume a process for each of the four processes: (a) compression device as isentropic, (b) combustion as isobaric, (c) expander as isentropic, and (d) cooler as isochoric.
   (B) Input the given information: (a) working fluid is air, (b) the inlet temperature and pressure of the compression device are 60°F and 14.7 psia, (c) the compression ratio of the compression device is 16, and (d) the heat addition is 800 Btu/lbm in the combustion chamber, and (e) the mass of air is 0.02 lbm.
3. Display results
   (A) Display cycle properties results. The cycle is a heat engine. The answers are T_{max}=4454°F, Q_{add}=16 Btu,Q_{remove}=-6.97 Btu, W_{add}=-3.61 Btu, $W_{expansion}$=12.65 Btu, Wnet=9.03 Btu, MEP=199 psia, and η=56.45%.

The power output of the Diesel cycle can be increased by super-charging, turbo-charging and pre-cooling the air before it enters the cylinder in the Otto engine. The difference between a super-charger and a turbo-charger is the manner in which they are powered. Since the inlet air density is increased due to higher inlet air pressure or lower air temperature, the mass rate flow of air is increased. Turbo-charging raises the inlet air pressure of the engine above atmospheric pressure and raise the power output of the engine, but it may not improve the efficiency of the cycle. The schematic diagram of the Diesel cycle with turbo-charging or super-charging is illustrated in Figure 8.2.3. The bottom schematic diagram of Figure 8.2.4 illustrates the Diesel cycle with turbo-charging and pre-cooling. The following three examples (Example 8.2.2, Example 8.2.3, and Example 8.2.4) show the power increase due to super-charging, and pre-cooling and super-charging.

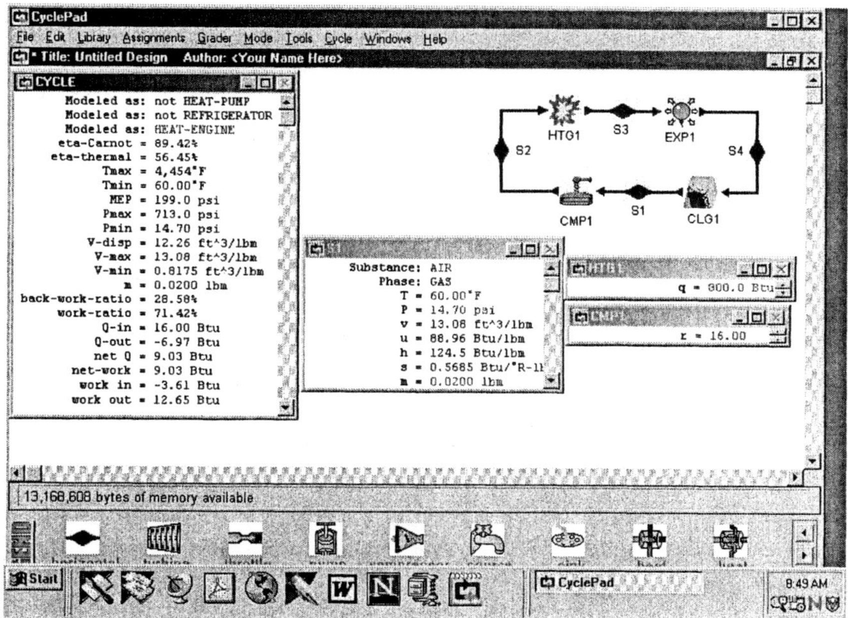

Figure Example 8.2.2 Diesel cycle

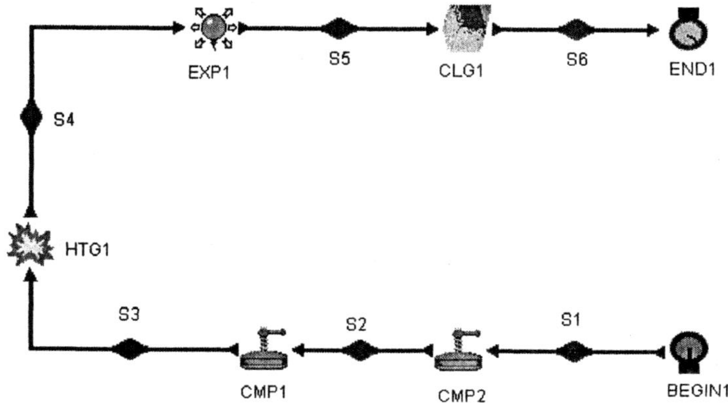

Figure 8.2.3. Diesel cycle with super-charging

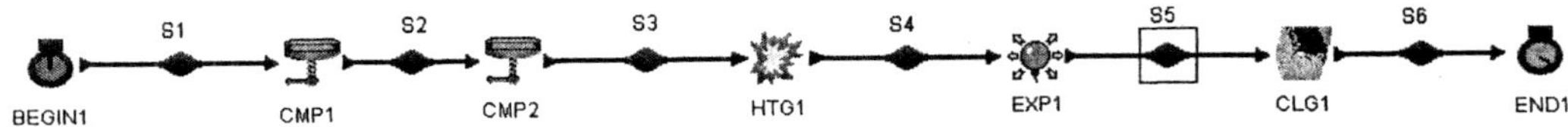

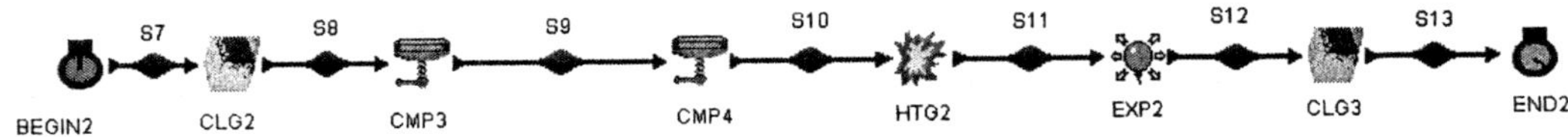

Figure 8.2.4. Diesel cycle with super-charging and pre-cooling

Example 8.2.3. Find the pressure and temperature of each state of an ideal Diesel cycle with a compression ratio of 15 and a cut-off ratio of 2. The cylinder volume before compression is 0.16 ft^3. The atmosphere conditions are 14.7 psia and 70°F. Also determine the mass of air in the cylinder, heat supplied, net work produced, MEP, and cycle efficiency.

To solve this problem, we build the cycle as shown in Figure 8.2.4. Then (A) Assume isobaric for the pre-cooling process 7-8, isentropic for compression process 8-9, isentropic for compression process 9-10, isobaric for the heating process 10-11, isentropic for expansion process 11-12, and isochoric for the cooling process 12-13; (B) input p_7=14.7 psia, T_7=70°F; p_{13}=14.7 psia, T_{13}=70°F; p_8=14.7 psia, T_8=70°F; p_9=14.7 psia, V_9=0.16 ft^3 (no turbo-charger and no pre-cooler); compression ratio=15, and cut-off ratio=2; and (C) display results. The results are: T_8=70°F, T_9=96.87°F, T_{10}=1105°F, T_{11}=2670°F, T_{12}=938.1°F, Q_{in}=4.5 Btu, W_{net}=2.72 Btu, MEP=98.31 psia, η=60.37%, and m=0.012 lbm.

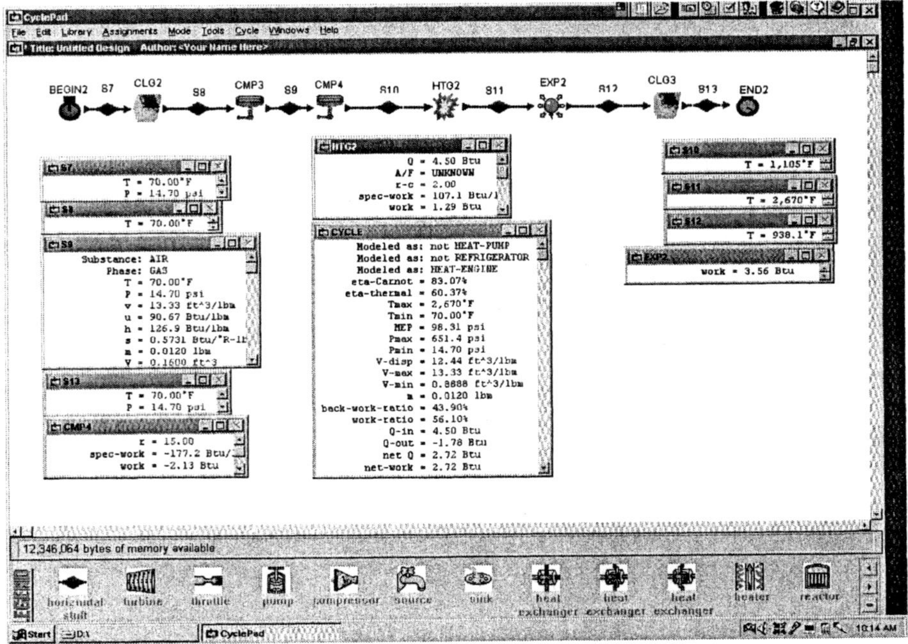

Figure Example 8.2.3. Diesel cycle without pre-cooler and without turbo-charger

Example 8.2.4. Find the pressure and temperature of each state of an ideal Diesel cycle with a compression ratio of 15 and a cut-off ratio of 2, and a super-charger which compresses fresh air to 20 psia before it enters the cylinder of the engine. The cylinder volume before compression is 0.16 ft^3. The atmosphere conditions are 14.7 psia and 70°F. Also determine the mass of air in the cylinder, heat supplied, net work produced, MEP, and cycle efficiency.

To solve this problem, we build the cycle as shown in Figure 8.2.4. Then (A) Assume isobaric for the pre-cooling process 7-8, isentropic for compression process 8-9, isentropic for compression process 9-10, isobaric for the heating process 10-11, isentropic for expansion process 11-12, and isochoric for the cooling process 12-13; (B) input p_7=14.7 psia, T_7=70°F; p_{13}=14.7 psia, T_{13}=70°F; T_8=70°F; p_9=20 psia, V_9=0.16 ft^3 (with turbo-charger and no pre-cooler); compression ratio=15, and cut-off ratio=2; and (C) display results. The results are: T_8=70°F, T_9=96.87°F, T_{10}=1249°F, T_{11}=2958°F, T_{12}=938.1°F, Q_{in}=6.12 Btu, W_{net}=3.90 Btu, MEP=106.5 psia, η=64.51%, and m=0.015 lbm.

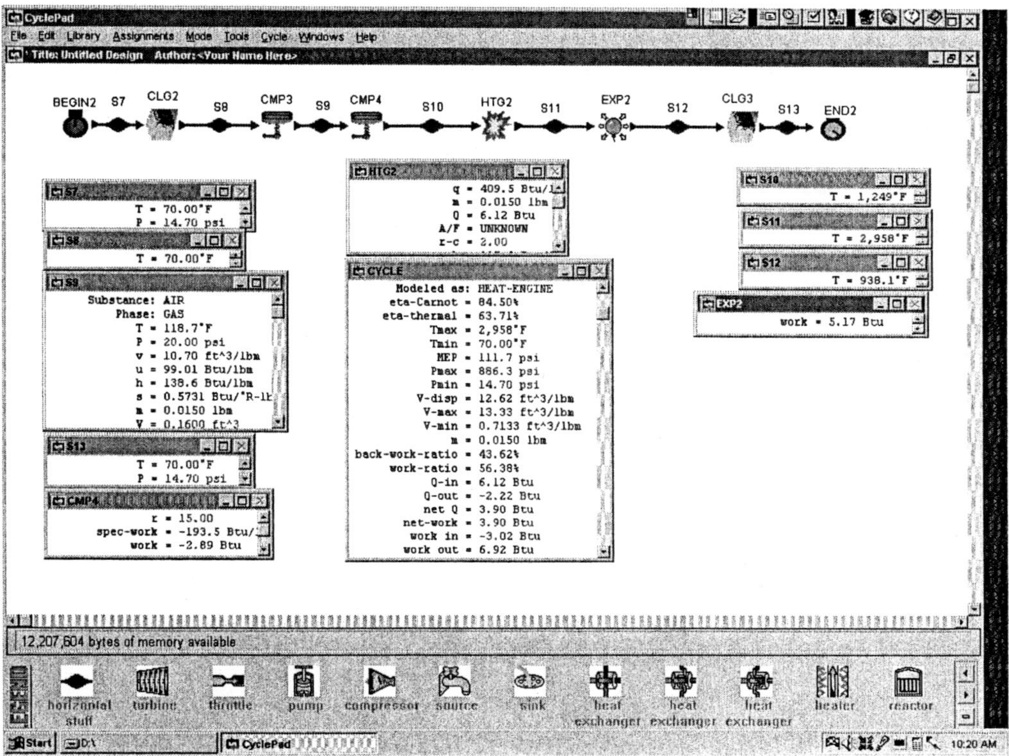

Figure Example 8.2.4. Diesel cycle with turbo-charger

Example 8.2.5. Find the pressure and temperature of each state of an ideal Diesel cycle with a compression ratio of 15 and a cut-off ratio of 2. A pre-cooler which cools the atmospheric air from 70°F to 50°F, and a super-charger which compresses fresh air to 20 psia before it enters the cylinder of the engine are added to the engine. The cylinder volume before compression is 0.16 ft^3. The atmosphere conditions are 14.7 psia and 70°F. Also determine the mass of air in the cylinder, heat supplied, net work produced, MEP, and cycle efficiency.

To solve this problem, we build the cycle as shown in Figure 8.2.4. Then (A) Assume isobaric for the pre-cooling process 7-8, isentropic for compression process 8-9, isentropic for compression process 9-10, isobaric for the heating process 10-11, isentropic for expansion process 11-12, and isochoric for the cooling process 12-13; (B) input p_7=14.7 psia, T_7=70°F; p_8=14.7 psia, p_{13}=14.7 psia, T_{13}=70°F; T_8=50°F; p_9=20 psia, V_9=0.16 ft^3 (with turbo-charger and pre-cooler); compression ratio=15, and cut-off ratio=2; and (C) display results. The results are: T_8=50°F, p_8 =14.7 psia, T_9=96.87°F, p_9=20 psia, T_{10}=1184°F, p_{10}=886.3 psia, T_{11}=2829°F, p_{11}=886.3 psia, T_{12}=864.8°F,p_{12}=36.76 psia; Q_{78}=-0.0745 Btu, W_{89}=-0.1247 Btu, W_{910}=-2.89 Btu, W_{1011}=1.75 Btu, Q_{1011}=6.12 Btu, W_{1112}=5.2 Btu,W_{net}=3.93 Btu, MEP=108.2 psia, η=64.25%, and m=0.0155 lbm.

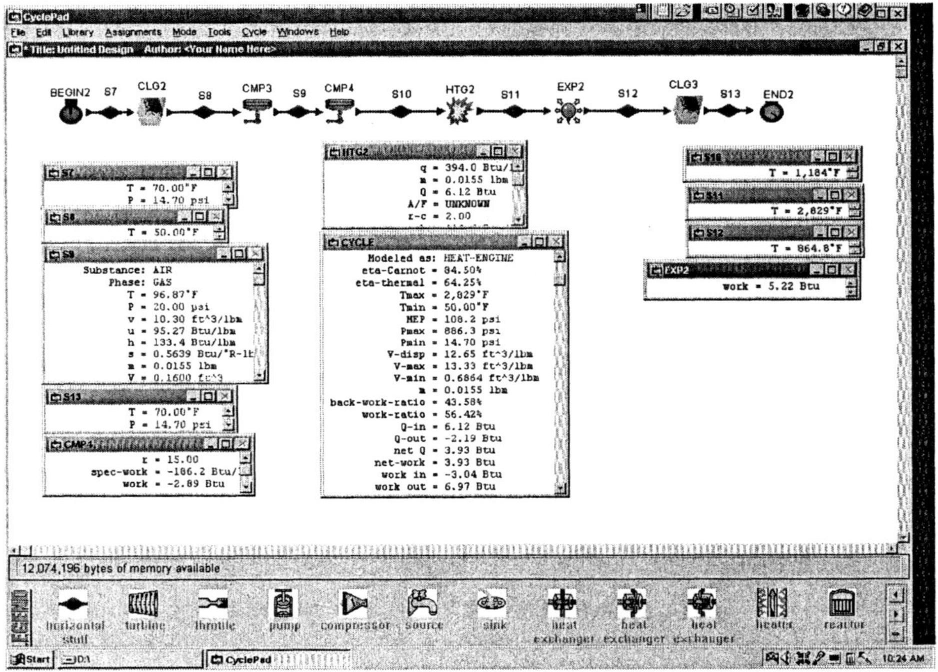

Figure Example 8.2.5.Diesel cycle with pre-cooler and turbo-charger

Homework 8.2 Diesel Cycle Analysis and Optimization

1. What is the difference between the compression ratio and cut-off ratio?
2. What is the difference between the Otto and Diesel engine?
3. Does the Diesel engine have sparkling plugs? If yes, for what reason?
4. Does the Diesel engine have engine knock problem? Why?
5. Is the Otto cycle more efficient than a Diesel cycle with the same compression ratio?
6. Why is the Diesel engine usually used for big trucks and the Otto engine usually used for compact cars?
7. Can the Diesel engine afford to have a large compression ratio? Why?
8. Suppose a large amount of power is required. Which engine would you choose between Otto and Diesel? Why?

9. The compression ratio of an air-standard Diesel cycle is 15. At the beginning of the compression stroke, the pressure is 14.7 psia and the temperature is 80°F. The maximum temperature of the cycle is 4040°F. Find the temperature at the end of the compression stroke, the temperature at the beginning of the exhaust process, the heat addition to the cycle, the net work produced by the cycle, the thermal efficiency, and the MEP of the cycle.
10. An ideal Diesel cycle with a compression ratio of 17 and a cutoff ratio of 2 has an air temperature of 105°F and a pressure of 15 psia at the beginning of the isentropic compression process. Determine (a) the temperature and pressure of the air at the end of the isentropic compression process, (b)the temperature and pressure of the air at the end of the combustion process, and (c) the thermal efficiency of the cycle.
11. An ideal Diesel cycle with a compression ratio of 20 and a cutoff ratio of 2 has a temperature of 105°F and a pressure of 15 psia at the beginning of the compression process. Determine (a) the temperature and pressure of the gas at the end of the compression process, (b) the temperature and pressure of the gas at the end of the combustion process, (c) heat added to the engine in the combustion chamber, (d) heat removed from the engine to the environment, and (e) thermal cycle efficiency
12. The pressure and temperature at the start of compression in an air Diesel cycle are 101 kPa and 300 K. The compression ratio is 15. The amount of heat addition is 2000 kJ/kg of air. Determine (a) the maximum cycle pressure and maximum temperature of the cycle, and (b) the cycle thermal efficiency.
13. An ideal Diesel engine receives air at 103.4 kPa and 27°C. Heat added to the air is 1016.6 kJ/kg, and the compression ratio of the engine is 13. Determine (a) the work added during the compression process, (b) the cut-off ratio, (c) the work done during the expansion process, (d) the heat removed from the air during the cooling process, (e) the MEP (mean effective pressure), and (f) the thermal efficiency of the cycle.
14. An ideal Diesel engine receives air at 15 psia and 65°F. Heat added to the air is 160 Btu/lbm, and the compression ratio of the engine is 6. Determine (a) the work added during the compression process, (b) the cut-off ratio, (c) the work done during the expansion process, (d) the heat removed from the air during the cooling process, (e) the MEP (mean effective pressure), and (f) the thermal efficiency of the cycle.
15. An ideal Diesel engine receives air at 100 kPa and 25°C. The maximum cycle temperature is 1460°C and the compression ratio of the engine is 16. Determine (a) the work done during the compression process, (b) the heat added to the air during the heating process, (c) the work done during the expansion process, (d) the heat removed from the air during the cooling process, and (e) the thermal efficiency of the cycle.
16. A Diesel engine receives air at 60°F and 14.7 psia. The compression ratio is 20. The amount of heat addition is 800 Btu/lbm. The mass of air contained in the cylinder is 0.02 lbm. Determine the maximum cycle temperature, heat added, heat removed, work added, work produced, net work produced, MEP and efficiency of the cycle.

    Answer: T_{max}=4601°F, Q_{add}=16 Btu,Q_{remove}=-6.27 Btu, W_{add}=-4.12 Btu, $W_{expansion}$=13.85 Btu, Wnet=9.73 Btu, MEP=211.7 psia, and η=60.84%.
17. A Diesel engine receives air at 80°F and 14.7 psia. The compression ratio is 20. The amount of heat addition is 800 Btu/lbm. The mass of air contained in the cylinder is 0.02

lbm. Determine the maximum cycle temperature, heat added, heat removed, work added, work produced, net work produced, MEP and efficiency of the cycle.
Answer: T_{max}=4667°F, Q_{add}=16 Btu,Q_{remove}=-6.22 Btu, W_{add}=-4.28 Btu, $W_{expansion}$=14.05 Btu, Wnet=9.78 Btu, MEP=204.7 psia, and η=61.11%.

18. An ideal Diesel engine receives air at 15 psia, 70°F. The air volume is 7 ft^3 before compression. Heat added to the air is 200 Btu/lbm, and the compression ratio of the engine is 11. Determine (a) the work added during the compression process, (b) the maximum temperature of the cycle, (c) the work done during the expansion process, (d) the heat removed from the air during the cooling process, (e) the MEP (mean effective pressure), and (f) the thermal efficiency of the cycle.
19. A Diesel cycle has a compression ratio of 18. Air intake conditions (prior to compression) are 72°F and 14.7 psia, and the highest temperature in the cycle is limited to 2500°F to avoid damaging the engine block. Calculate:
(a) thermal efficiency, (b) net work, and (c) mean effective pressure. (d) Compare engine efficiency to that of a Carnot cycle engine operating between the same temperatures.
20. A Diesel engine is modeled with an ideal Diesel cycle with a compression ratio of 17. The following information is known:
Temperature prior to the compression process: 70°F.
Pressure prior to the compression process: 14.7 psia.
Heat added during the combustion process: 245 Btu/lbm.
(a) Determine the temperature and pressure at each process endpoint.
(b) Solve for the net cycle work (Btu/lbm).
(c) Solve for the thermal efficiency.
21. An ideal Diesel cycle with a compression ratio of 17 and a cutoff ratio of 2 has a temperature of 313 K and a pressure of 100 kPa at the beginning of the isentropic compression process. Use the cold air-standard assumptions, assume that k=1.4, determine (a) the temperature and pressure of the air at the end of the isentropic compression process and at the end of the combustion process, and (b) the thermal efficiency of the cycle.
22. Find the pressure and temperature of each state of an ideal Diesel cycle with a compression ratio of 15 and a cut-off ratio of 2. A pre-cooler which cools the atmospheric air from 80°F to 50°F, and a super-charger which compresses fresh air to 20 psia before it enters the cylinder of the engine are added to the engine. The cylinder volume before compression is 0.1 ft^3. The atmosphere conditions are 14.7 psia and 80°F. Also determine the mass of air in the cylinder, heat supplied, net work produced, MEP, and cycle efficiency.
ANSWER: MEP=106.5 psia, η=64.51%
23. Find the pressure and temperature of each state of an ideal Diesel cycle with a compression ratio of 15 and a cut-off ratio of 2. A pre-cooler which cools the atmospheric air from 80°F to 50°F, and a super-charger which compresses fresh air to 25 psia before it enters the cylinder of the engine are added to the engine. The cylinder volume before compression is 0.1 ft^3. The atmosphere conditions are 14.7 psia and 80°F. Also determine the mass of air in the cylinder, heat supplied, net work produced, MEP, and cycle efficiency.
ANSWER: MEP=116.4 psia, η=66.7%

8.3 Atkinson Cycle

A cycle called Atkinson cycle is similar to the Otto cycle except that the isochoric exhaust and intake process at the end of the Otto cycle power stroke is replaced by an isobaric process. The schematic diagram of the cycle is shown in Figure 8.3.1. The cycle is made of the following four processes:

1-2	isentropic compression
2-3	isochoric heat addition
3-4	isentropic expansion
4-1	isobaric heat removing

Applying the first law and second law of thermodynamics of the closed system to each of the four processes of the cycle yields:

$$W_{12} = \int pdV, \tag{8.3.1}$$

$$Q_{12} - W_{12} = m(u_2 - u_1),\ Q_{12}=0 \tag{8.3.2}$$

$$W_{23} = \int pdV = 0, \tag{8.3.3}$$

$$Q_{23} - 0 = m(u_3 - u_2), \tag{8.3.4}$$

$$W_{34} = \int pdV, \tag{8.3.5}$$

$$Q_{34} - W_{34} = m(u_4 - u_3),\ Q_{34} = 0 \tag{8.3.6}$$

$$W_{41} = \int pdV = p\ m(v_1 - v_4), \tag{8.3.7}$$

and

$$Q_{41} - W_{41} = m(u_1 - u_4). \tag{8.3.8}$$

The net work (W_{net}), which is also equal to net heat (Q_{net}), is

$$W_{net} = W_{12} + W_{34} + W_{41} = Q_{net} = Q_{23} + Q_{41} \tag{8.3.9}$$

The thermal efficiency of the cycle is

$$\eta = W_{net}/Q_{23} = Q_{net}/Q_{23} = 1 - Q_{41}/Q_{23} = 1 - (h_4 - h_1)/(u_3 - u_2) \tag{8.3.10}$$

This expression for thermal efficiency of an ideal Otto cycle can be simplified if air is assumed to be the working fluid with constant specific heats. Equation (8.1.3) is reduced to:

$$\eta = 1 - k(T_4 - T_1)/(T_3 - T_2) \tag{8.3.11}$$

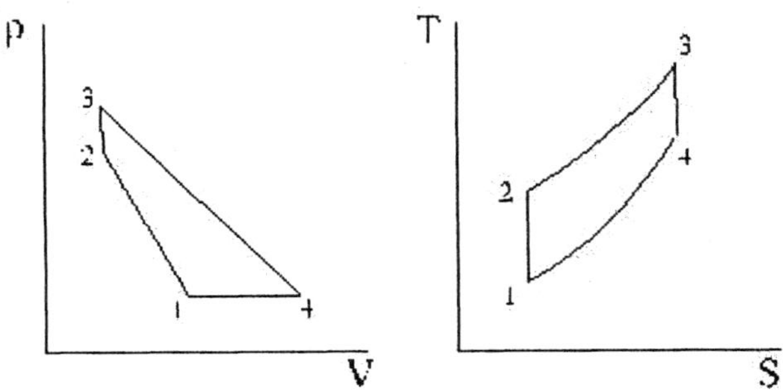

Figure 8.3.1. Atkinson cycle

Example 8.3.1. Find the pressure and temperature of each state of an ideal Atkinson cycle with a compression ratio of 8. The heat addition in the combustion chamber is 800 Btu/lbm. The atmospheric air is at 14.7 psia and 60°F. The cylinder contains 0.02 lbm of air. Determine the maximum temperature, maximum pressure, heat supplied, heat removed, work added during the compression processes, work produced during the expansion process, net work produced, MEP, and cycle efficiency. Draw the T-s diagram of the cycle.

To solve this problem, we build the cycle as shown in Figure 8.1.2. Then (A) Assume isentropic for the compression process 1-2 and the expansion process 3-4, isochoric for the heating process 2-3, and isobaric for the cooling process 4-1; (B) input p_1=14.7 psia, T_1=60°F, mdot=0.02 lbm; r=8 for the compression process 1-2, and q=800 Btu/lbm for the heating process 2-3. and (C) display results. The results are: T_{max}=5407°F, p_{max}=1328 psia, Q_{add}=16 Btu, Q_{remove}=-5.28 Btu, W_{comp}=-3.82 Btu, W_{expan}=14.54 Btu, W_{net}=10.72 Btu, MEP=74.00 psia, and η=67.02%.

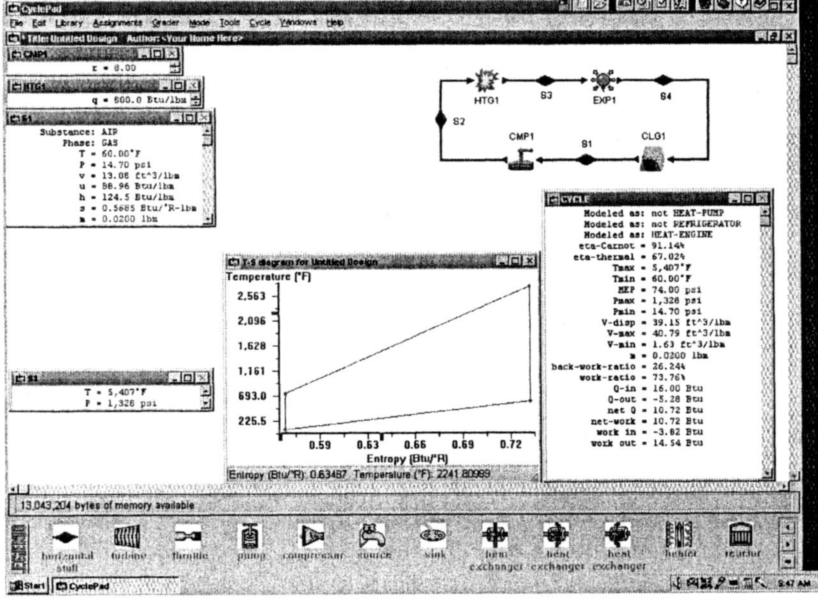

Figure Example 8.3.1. Atkinson cycle

Homework 8.3.1. Atkinson Cycle

1. What are the four processes of the Atkinson cycle?
2. What is the difference between the Otto cycle and the Atkinson cycle?
3. Find the pressure and temperature of each state of an ideal Atkinson cycle with a compression ratio of 16. The heat addition in the combustion chamber is 800 Btu/lbm. The atmospheric air is at 14.7 psia and 60°F. The cylinder contains 0.02 lbm of air. Determine the maximum temperature, maximum pressure, heat supplied, heat removed, work added during the compression processes, work produced during the expansion process, net work produced, MEP, and cycle efficiency.
   ANSWER: T_{max}=5789°F, p_{max}=2828 psia, Q_{add}=16 Btu, Q_{remove}=-4.17 Btu, W_{comp}=-4.81 Btu, W_{expan}=16.63 Btu, W_{net}=11.83 Btu, MEP=73.91 psia, and η=73.91%.
4. Find the pressure and temperature of each state of an ideal Atkinson cycle with a compression ratio of 16. The heat addition in the combustion chamber is 800 Btu/lbm. The atmospheric air is at 101.4 kPa and 18°C. The cylinder contains 0.01 kg of air. Determine the maximum temperature, maximum pressure, heat supplied, heat removed, work added during the compression processes, work produced during the expansion process, net work produced, MEP, and cycle efficiency.
   ANSWER: T_{max}=3121°C, p_{max}=18904 kPa, Q_{add}=18 kJ, Q_{remove}=-4.72 kJ, W_{comp}=-5.59 kJ, W_{expan}=18.86 kJ, W_{net}=13.28 kJ, MEP=631 kPa, and η=73.75%.
5. Find the pressure and temperature of each state of an ideal Atkinson cycle with a compression ratio of 10. The heat addition in the combustion chamber is 800 Btu/lbm. The atmospheric air is at 101.4 kPa and 18°C. The cylinder contains 0.01 kg of air. Determine the maximum temperature, maximum pressure, heat supplied, heat removed, work added during the compression processes, work produced during the expansion process, net work produced, MEP, and cycle efficiency.
   ANSWER: T_{max}=2970°C, p_{max}=11289 kPa, Q_{add}=18 kJ, Q_{remove}=-5.54 kJ, W_{comp}=-4.74 kJ, W_{expan}=17.20 kJ, W_{net}=12.46 kJ, MEP=540.7 kPa, and η=69.21%.

8.4 DUAL CYCLE

Combustion in the Otto cycle is based on a constant volume process; in the Diesel cycle, it is based on a constant pressure process. But combustion in actual spark-ignition engine requires a finite amount of time if the process is to be complete. For this reason, combustion in Otto cycle does not actually occur under the constant volume condition. Similarly, in compression-ignition engines, combustion in Diesel cycle does not actually occur under the constant pressure condition, because of the rapid and uncontrolled combustion process.

The operation of the reciprocating internal combustion engines represents a compromise between the Otto and the Diesel cycle, and can be described as a Dual combustion cycle. Heat transfer to the system may be considered to occur first at constant volume and then at constant pressure. Such a cycle is called Dual cycle.

The Dual cycle is composed of the following five processes:

1-2 isentropic compression
2-3 constant volume heat addition

3-4 constant pressure heat addition
4-5 isentropic expansion
5-1 constant volume heat removing

Figure 8.4.1 shows the Dual cycle on p-v and T-s diagrams.

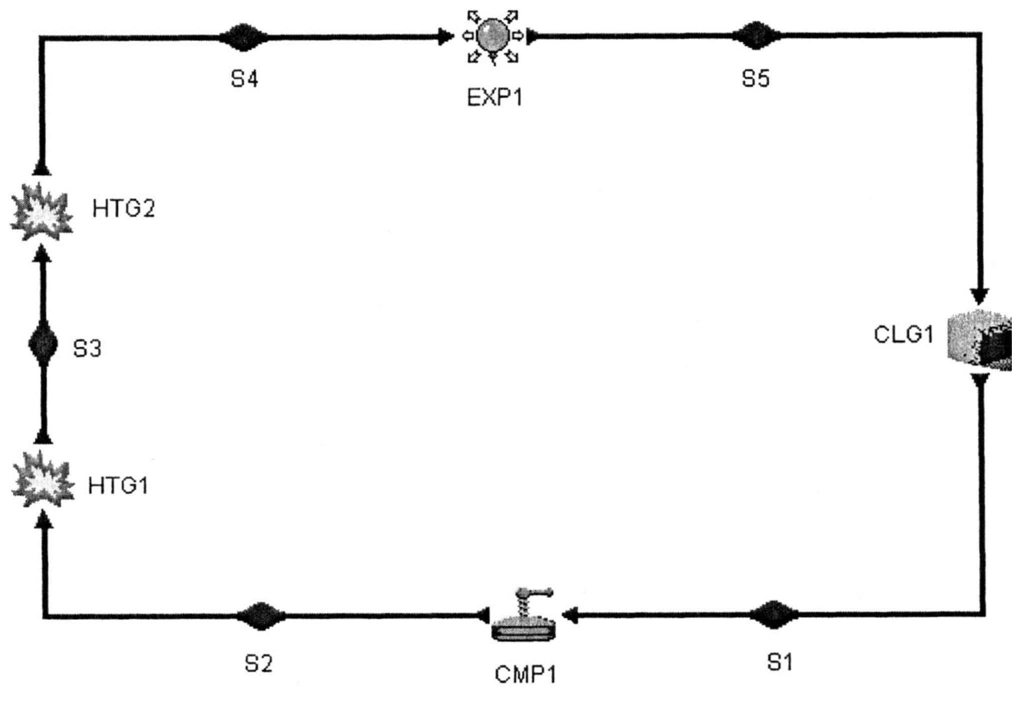

Figure 8.4.1 Dual cycle

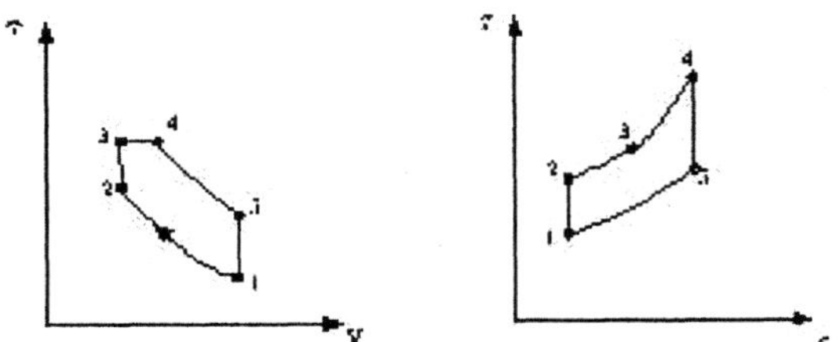

Figure 8.4.2 Dual cycle on p-v and T-s diagrams

Applying the First law of thermodynamics of the closed system to each of the five processes of the cycle yields:

$$W_{12} = \int pdV, \tag{8.4.1}$$

$$Q_{12} - W_{12} = m(u_2 - u_1), Q_{12} = 0, \tag{8.4.2}$$

$$W_{23} = \int pdV = 0, \tag{8.4.3}$$

$$Q_{23} - 0 = m\,(u_3 - u_2), \tag{8.4.4}$$

$$W_{34} = \int pdV = m(p_4v_4 - p_3v_3), \tag{8.4.5}$$

$$Q_{34} = m(u_4 - u_3) + W_{34} = m(h_4 - h_3), \tag{8.4.6}$$

$$W_{45} = \int pdV, \tag{8.4.7}$$

$$Q_{45} - W_{45} = m(u_5 - u_4),\ Q_{45} = 0, \tag{8.4.8}$$

$$W_{51} = \int pdV = 0, \tag{8.4.9}$$

and

$$Q_{51} - W_{51} = m(u_1 - u_5). \tag{8.4.10}$$

The net work (W_{net}), which is also equal to net heat (Q_{net}), is

$$W_{net} = W_{12}+W_{34}+W_{45} = Q_{net} = Q_{23}+Q_{34}+Q_{51} \tag{8.4.11}$$

The thermal efficiency of the cycle is

$$\eta = W_{net}/(Q_{23} + Q_{34}) = Q_{net}/Q_{23} = 1 - Q_{51}/Q_{23} \qquad (8.4.11)$$

This expression for thermal efficiency of an ideal Otto cycle can be simplified if air is assumed to be the working fluid with constant specific heats. Equation (8.3.11) is reduced to:

$$\eta = 1 - (T_5 - T_1)/[\,(T_3 - T_2) + k\,(T_4 - T_3)\,] \tag{8.4.12}$$

Example 8.4.1. Pressure and temperature at the start of compression in a Dual cycle are 14.7 psia and 540°R. The compression ratio is 15. Heat addition at constant volume is 300 Btu/lbm of air, while heat addition at constant pressure is 500 Btu/lbm of air. The mass of air contained in the cylinder is 0.03 lbm. Determine (a) the maximum cycle pressure and maximum cycle temperature, (b) the efficiency and work output per kilogram of air, and (c) the mep. Show the cycle on T-s diagram. Plot the sensitivity diagram of cycle efficiency vs compression ratio.

To solve this problem by CyclePad, we take the following steps:

1. Build
   (A) Take a compression device, two combustion chambers, an expander and a cooler from the closed system inventory shop and connect the four devices to form the Dual cycle.

(B) Switch to analysis mode.

2. Analysis

(A) Assume a process for each the five processes: (a) compression device as isentropic, (b) first combuster as isocbaric and second combuster as isobaric, (c) expander as isentropic, and (d) cooler as isochoric.

(B) Input the given information: (a) working fluid is air, (b) the inlet pressure and temperature of the compression device are 14.7 psia and 540°R, (c) the compression ratio of the compression device is 15, (d) the heat addition is 300 Btu/lbm in the isocbaric combustion chamber, (e) the heat addition is 500 Btu/lbm in the isobaric combustion chamber, and (f)The mass of air contained in the cylinder is 0.03 lbm.

3. Display results

(A) Display the T-s diagram and cycle properties results. The cycle is a heat engine. The answers are T_{max}=5434 R, p_{max}=1367 psia, η=63.78%, MEP=217.3 psia and Wnet=15.31 Btu, and (B) Display the sensitivity diagram of cycle efficiency vs compression ratio.

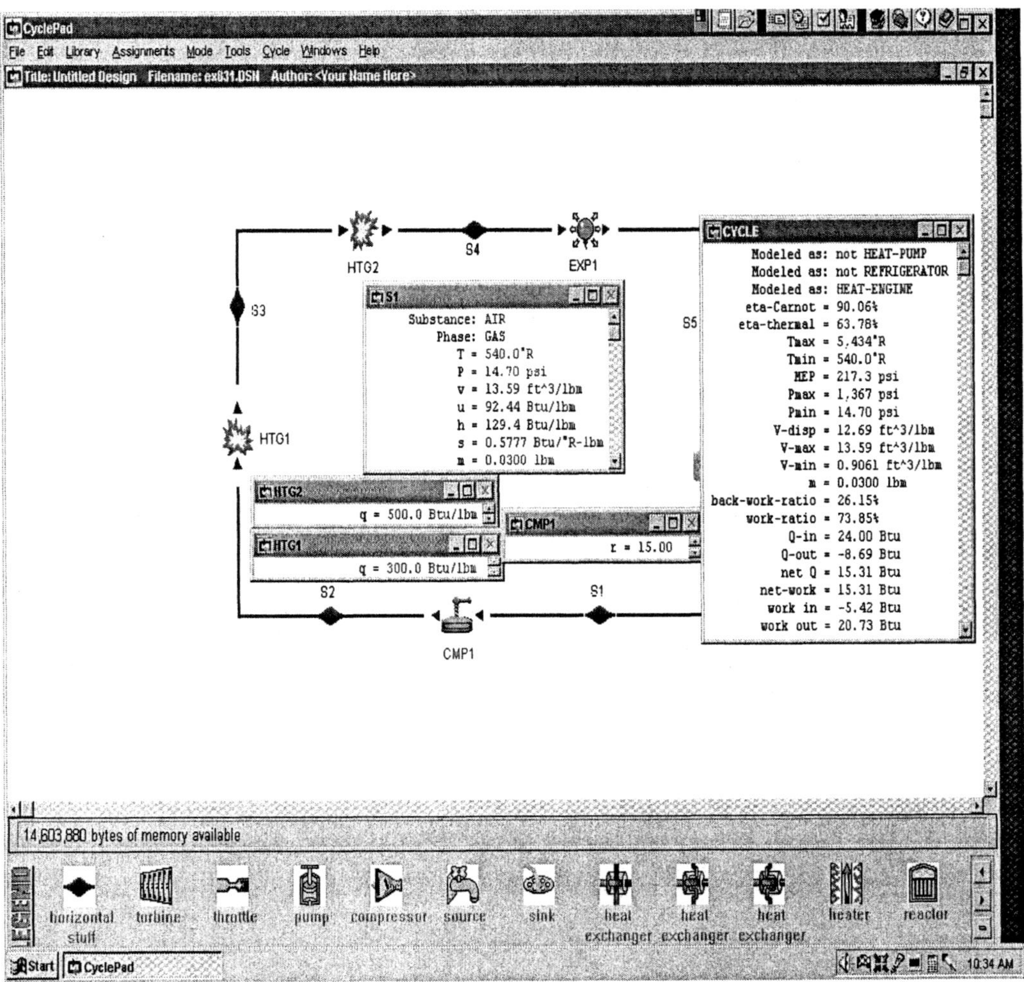

Figure Example 8.4.1a Dual cycle

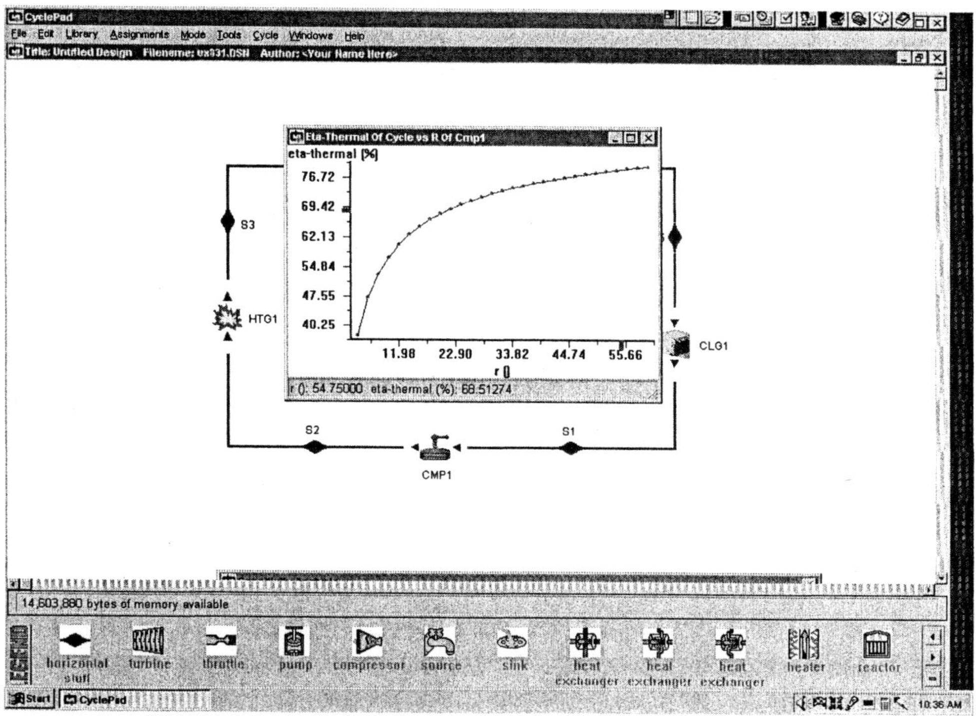

Figure Example 8.4.1b Dual cycle sensitivity analysis

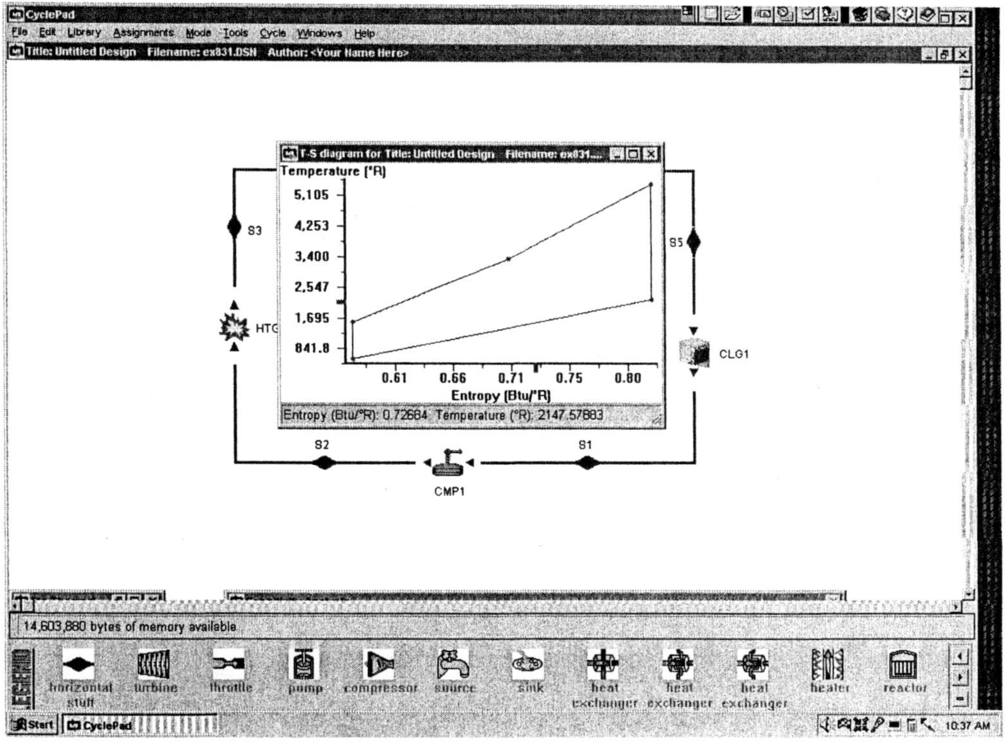

Figure Example 8.4.1c Dual cycle T-s diagram

Homework 8.4 Dual cycle

1. What five processes make up the Dual cycle?
2. The combustion process in internal combustion engines as an isobaric or isometric heat addition process is over simplistic and not realistic. A real cycle p-v diagram of the Otto or Diesel cycle looks like a curve (combination of isobaric and isometric) rather than a linear line. Are the combustion processes in the Dual cycle more realistic?
3. Can we consider the Otto or Diesel cycle to be special cases of the Dual cycle?
4. Pressure and temperature at the start of compression in a Dual cycle are 101 kPa and 15°C. The compression ratio is 8. Heat addition at constant volume is 100 kJ/kg of air, while the maximum temperature of the cycle is limited to 2000°C. The mass of air contained in the cylinder is 0.01 kg. Determine (a) the maximum cycle pressure, the MEP, Heat added, heat removed, compression work added, expansion work produced, net work produced and efficiency of the cycle.
   ANSWER: p_{max}=2248 kPa, MEP=988.1 kPa, Q_{add}=15.77 kJ, Q_{remove}=-8.7 kJ, W_{comp}=-2.68 kJ, $W_{expansion}$=9.75 kJ, Wnet=7.07 kJ, and η=44.85%.
5. Pressure and temperature at the start of compression in a Dual cycle are 101 kPa and 15°C. The compression ratio is 12. Heat addition at constant volume is 100 kJ/kg of air, while the maximum temperature of the cycle is limited to 2000°C. The mass of air contained in the cylinder is 0.01 kg. Determine (a) the maximum cycle pressure, the MEP, Heat added, heat removed, compression work added, expansion work produced, net work produced and efficiency of the cycle.
   ANSWER: p_{max}=3862 kPa, MEP=1067 kPa, Q_{add}=14.60 kJ, Q_{remove}=-6.60 kJ, W_{comp}=-3.51 kJ, $W_{expansion}$=11.51 kJ, Wnet=8.00 kJ, and η=57.48%.
6. Pressure and temperature at the start of compression in a Dual cycle are 101 kPa and 15°C. The compression ratio is 12. Heat addition at constant volume is 100 kJ/kg of air, while the maximum temperature of the cycle is limited to 2200°C. The mass of air contained in the cylinder is 0.01 kg. Determine (a) the maximum cycle pressure, the MEP, Heat added, heat removed, compression work added, expansion work produced, net work produced and efficiency of the cycle.
   ANSWER: p_{max}=3862 kPa, MEP=1189 kPa, Q_{add}=16.60 kJ, Q_{remove}=-7.69 kJ, W_{comp}=-3.51 kJ, W_{comp}=12.43 kJ, Wnet=8.92 kJ, and η=53.71%.

8.5 LENOIR CYCLE

The first commercially successful internal combustion engine was made by the French engineer Lenoir in 1860. He converted a reciprocating steam engine to admit a mixture of air and methane during the first half of the piston's outward suction stroke, at which point it was ignited with an electric spark and resulting combustion pressure acted on the piston for the remainder of the outward expansion stroke. The following inward stroke of the piston was used to expel the exhaust gases, and then the cycle began over again. The *Lenoir cycle* as shown in Figure 8.4.1 is composed of the following three effective processes:

1-2 isochoric combustion process
2-3 isentropic power expansion process

3-1 isobaric exhaust process

The p-v and T-s diagrams of the cycle is shown in Figure 8.4.2.

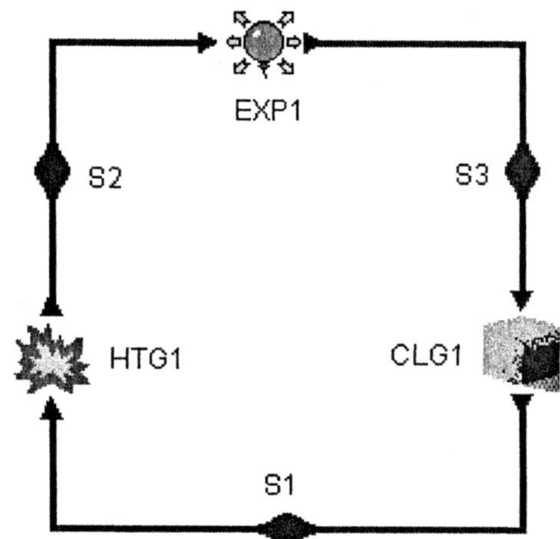

Figure 8.5.1 Lenoir cycle

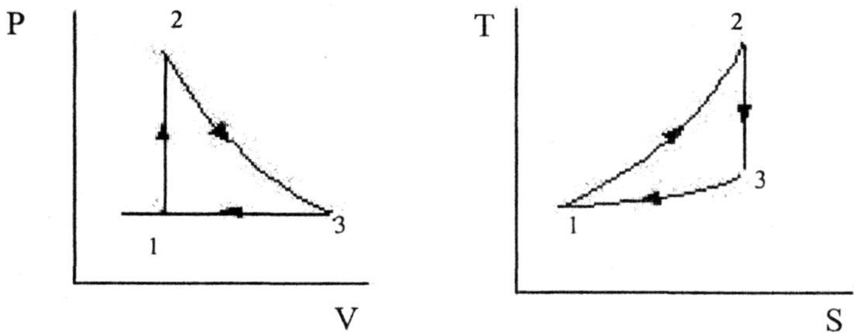

Figure 8.5.2 Lenoir cycle p-v diagram and T-s diagram

Applying the first law and second law of thermodynamics of the closed system to each of the three processes of the cycle yields:

$$W_{12} = 0, \tag{8.5.1}$$

$$Q_{12} - 0 = m(u_2 - u_1), \tag{8.5.2}$$

$$Q_{23} = 0, \tag{8.5.3}$$

$$0 - W_{23} = m\,(u_3 - u_2), \tag{8.5.4}$$

$$W_{31} = \int pdV = m(p_1v_1 - p_3v_3), \tag{8.5.5}$$

and

$$Q_{31} = m(u_1 - u_3) + W_{31} = m(h_1 - h_3). \quad (8.5.6)$$

The net work (W_{net}), which is also equal to net heat (Q_{net}), is

$$W_{net} = W_{23}+W_{31}= Q_{net} = Q_{12}+Q_{31} \quad (8.5.7)$$

The thermal efficiency of the cycle is

$$\eta = W_{net} /Q_{12} = Q_{net} /Q_{12} = 1+ Q_{31} /Q_{12} \quad (8.5.8)$$

This expression for thermal efficiency of an ideal Otto cycle can be simplified if air is assumed to be the working fluid with constant specific heats. Equation (8.3.11) is reduced to:

$$\eta = 1- (h_3 - h_1)/(u_2 - u_1) = 1 - k\, T_2 (r_s -1)/(T_2 -T_1) \quad (8.5.9)$$

where r_s is the isentropic volume compression ratio, $r_s=v_3/v_1$.

Because the air-fuel mixture was not compressed before ignition, the engine efficiency was very low and fuel consumption was very high. The fuel-air mixture was ignited by an electric spark inside the cylinder.

Example 8.5.1. The isochoric heating process of a Lenoir engine receives air at 15°C and 101 kPa. The air is heated to 2000°C. The mass of air contained in the cylinder is 0.01 kg. Determine the pressure at the end of the isochoric heating process, the temperature at the end of the isentropic expansion process, heat added, heat removed, work added, work produced, net work produced, and efficiency of the cycle. Draw the T-s diagram of the cycle.

To solve this problem by CyclePad, we take the following steps:

1. Build
   (A) Take a combustion chamber, an expander and a cooler from the closed system inventory shop and connect the three devices to form the Lenoir cycle.
   (B) Switch to analysis mode.
2. Analysis
   (A) Assume a process for each the three processes: (a) combustion as isochoric, (b) expander as isentropic, and (c) cooler as isobaric.
   (B) Input the given information: (a) working fluid is air, (b) the inlet pressure and temperature of the combustion device are 101 kPa and 15°C, (c) the temperature at the end of combustion device is 2000°C, and (d) the mass of air is 0.01 kg.
3. Display results
   (A) Display cycle properties results. The cycle is a heat engine. The answers are T_3=986.8 °C, p_2=796.8 kPa, Q_{add}=14.23 kJ, Q_{remove}=-9.75 kJ, W_{comp}=-2.79 kJ, W_{expan}=7.26 kJ, W_{net}=4.48 kJ, and η=31.46%; and (B) Display the T-s diagram.

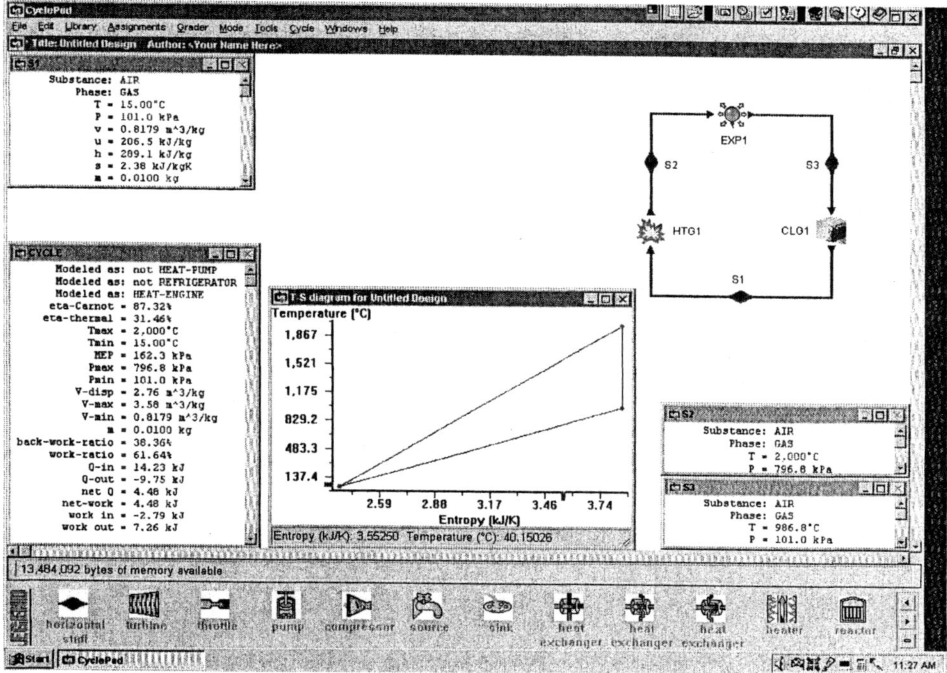

Figure Example 8.5.1 Lenoir cycle

Homework 8.5 Lenoir Cycle

1. What are the five processes that make up the Lenoir cycle?
2. The isochoric heating process of a Lenoir engine receives air at 15°C and 101 kPa. The air is heated to 2200°C. The mass of air contained in the cylinder is 0.01 kg. Determine the pressure at the end of the isochoric heating process, the temperature at the end of the isentropic expansion process, heat added, heat removed, work added, work produced, net work produced, and efficiency of the cycle.
   ANSWER: T_3=1065°C, p_2=866.9 kPa, Q_{add}=15.66 kJ, Q_{remove}=-10.54 kJ, W_{comp}=-3.01 kJ, W_{expan}=8.13 kJ, W_{net}=5.12 kJ, and η=32.72%.
3. The isochoric heating process of a Lenoir engine receives air at 60°F and 14.7 psia. The air is heated to 4000°F. The mass of air contained in the cylinder is 0.02 lbm. Determine the pressure at the end of the isochoric heating process, the temperature at the end of the isentropic expansion process, heat added, heat removed, work added, work produced, net work produced, and efficiency of the cycle.
   ANSWER: T_3=1953°F, p_2=126.2 psia, Q_{add}=13.49 Btu, Q_{remove}=-9.08 Btu, W_{comp}=-2.59 Btu, W_{expan}=7.01 Btu, W_{net}=4.41 Btu, and η=32.72%.
4. The isochoric heating process of a Lenoir engine receives air at 80°F and 14.7 psia. The air is heated to 4500°F. The mass of air contained in the cylinder is 0.02 lbm. Determine the pressure at the end of the isochoric heating process, the temperature at the end of the isentropic expansion process, heat added, heat removed, work added, work produced, net work produced, and efficiency of the cycle.

ANSWER: T_3=2172°F, p_2=135.1 psia, Q_{add}=15.13 Btu, Q_{remove}=-10.03 Btu, W_{comp}=-2.86 Btu, W_{expan}=7.97 Btu, W_{net}=5.11 Btu, and η=33.74%.

8.6 Stirling Cycle

The *Stirling cycle* is composed of the following four processes:

1-2 isothermal compression
2-3 constant volume heat addition
3-4 isothermal expansion
4-1 constant volume heat removing

Stirling-cycle engine is an external-combustion engine. The schematic diagram of the Stirling cycle is illustrated in Figure 8.6.1. Figure 8.6.2 shows the Stirling cycle on p-v and T-s diagrams.

During the isothermal compression process 1-2, heat is rejected to maintain a constant temperature T_L. During the isothermal expansion process 3-4, heat is added to maintain a constant temperature T_H. There are also heat interactions along the constant volume heat addition process 2-3 and the constant volume heat removing process 4-1. The quantities of heat in these two constant volume processes are equal but opposite in direction.

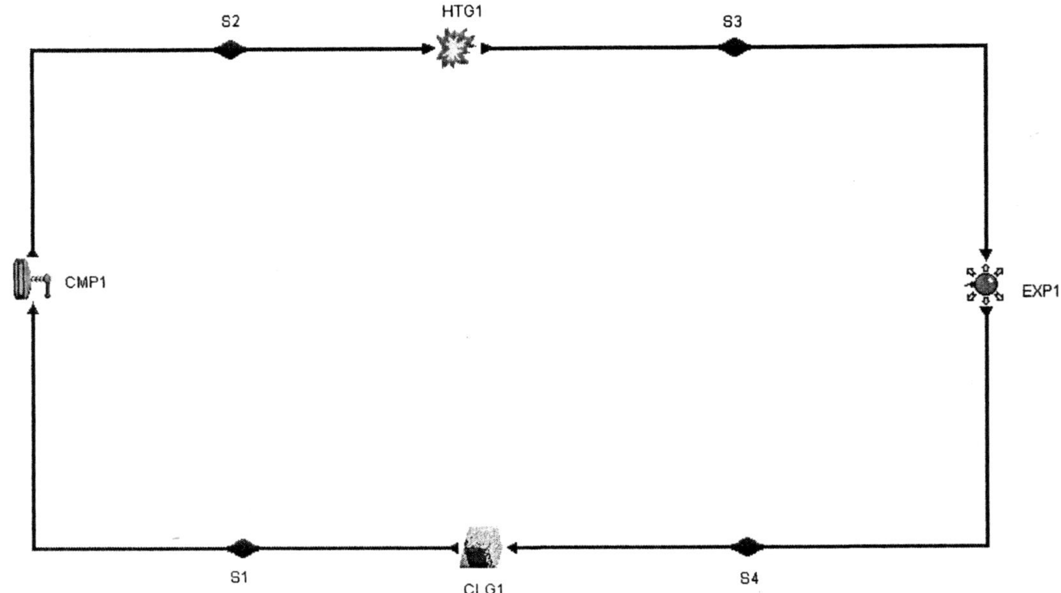

Figure 8.6.1. Stirling cycle schematic diagram

The operation of the Stirling-cycle engine is shown in Figure 8.6.3. There are two pistons in the cylinder. One is a power piston (P), and the other is the displace piston (D). The purpose of the displace piston is to move the working fluid around from one space to another space through the regenerator. At state 1, the power piston is at BDC (bottom dead center), with the displacer at its TDC (top dead center). The power piston moves from its BDC to TDC to compress the working fluid during the compression process 1-2. From 1-2, the working fluid in the cylinder is in contact with the low temperature reservoir, so the temperature remains constant (T_1=T_2) and heat is removed. During the heating process 2-3, the displacer moves downward, pushing the working fluid through the regenerator where it picks up heat to reach T_3. During the expansion process 3-4, the working fluid in the cylinder is in contact with the high temperature reservoir, so the temperature remains constant (T_3=T_4) and heat is added. During the cooling process 4-1, the displacer moves upward, pushing the working fluid through the regenerator where it removes heat to reach T_1.

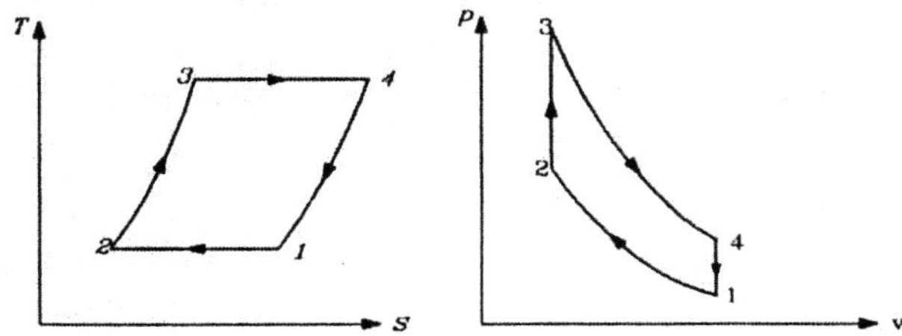

Figure 8.6.2 Stirling cycle on p-v and T-s diagrams.

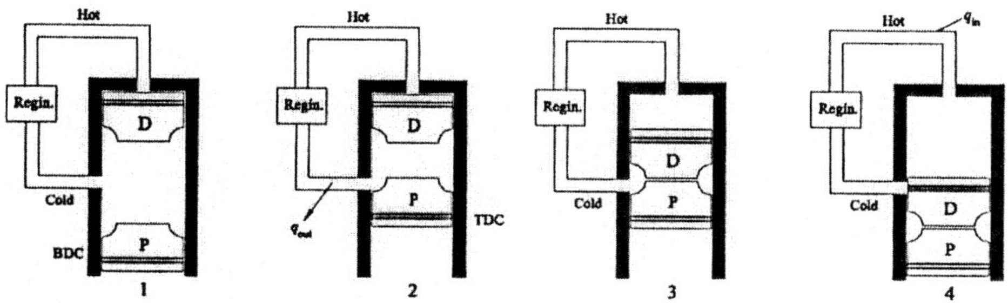

Figure 8.6.3. Stirling cycle operation

Applying the first law and second law of thermodynamics of the closed system to each of the four processes of the cycle yields:

$$W_{12} = \int pdV,\ Q_{12}=\int TdS= T_1(S_2-S_1), \tag{8.6.1}$$

$$Q_{12} - W_{12} = m(u_2 - u_1) =0, \tag{8.6.2}$$

$$W_{23} = \int pdV=0, \tag{8.6.3}$$

$$Q_{23} - 0 = m(u_3 - u_2), \quad (8.6.4)$$

$$W_{34} = \int pdV, \; Q_{34}=\int TdS= T_3(S_4-S_3), \quad (8.6.5)$$

$$Q_{34} - W_{34} = m(u_4 - u_3) =0, \quad (8.6.6)$$

$$W_{41} = \int pdV = 0, \quad (8.6.7)$$

and

$$Q_{41} - 0 = m(u_1 - u_4). \quad (8.6.8)$$

The net work (W_{net}), which is also equal to net heat (Q_{net}), is

$$W_{net} =W_{12} +W_{34} = Q_{net} = Q_{12}+ Q_{23}+ Q_{34} + Q_{41} \quad (8.6.9)$$

The thermal efficiency of the cycle is

$$\eta = W_{net} /(Q_{12}+Q_{41}) \quad (8.6.10)$$

Example 8.6.1 A Stirling cycle operates with 0.1 kg of hydrogen as a working fluid between 1000°C and 30°C. The highest pressure and the lowest pressure during the cycle are 3000 kPa and 500 kPa. Determine the heat and work added in each of the four processes, net work, and cycle efficiency.

To solve this problem by CyclePad, we take the following steps:

1. Build
   (A) Take a compression device, a combustion chamber, an expander and a cooler from the closed system inventory shop and connect the four devices to form the Stirling cycle as shown in Figure 8.4.1.
   (B) Switch to analysis mode.
2. Analysis
   (A) Assume a process for each of the four processes: (a) compression device as isothermal, (b) combuster as isochoric, (c) expander as isothermal, and (d) cooler as isochoric.
   (B) Input the given information: (a) working fluid is helium, (b) the inlet pressure and temperature of the compression device are 500 kPa and 30°C and m=0.1 kg, (c) the inlet pressure and temperature of the expander are 3000 kPa and 1000°C.
3. Display results
   (A) Display the cycle properties results. The cycle is a heat engine. The answers are: $Q_{12}=W_{12}=-22.46$ kJ, $Q_{23}=300.7$ kJ, $Q_{34}=W_{34}=94.33$ kJ, $Q_{41}=-300.7$ kJ, $W_{net}=71.87$ kJ, $Q_{in}=395.0$ kJ, and $\eta=18.19\%$.

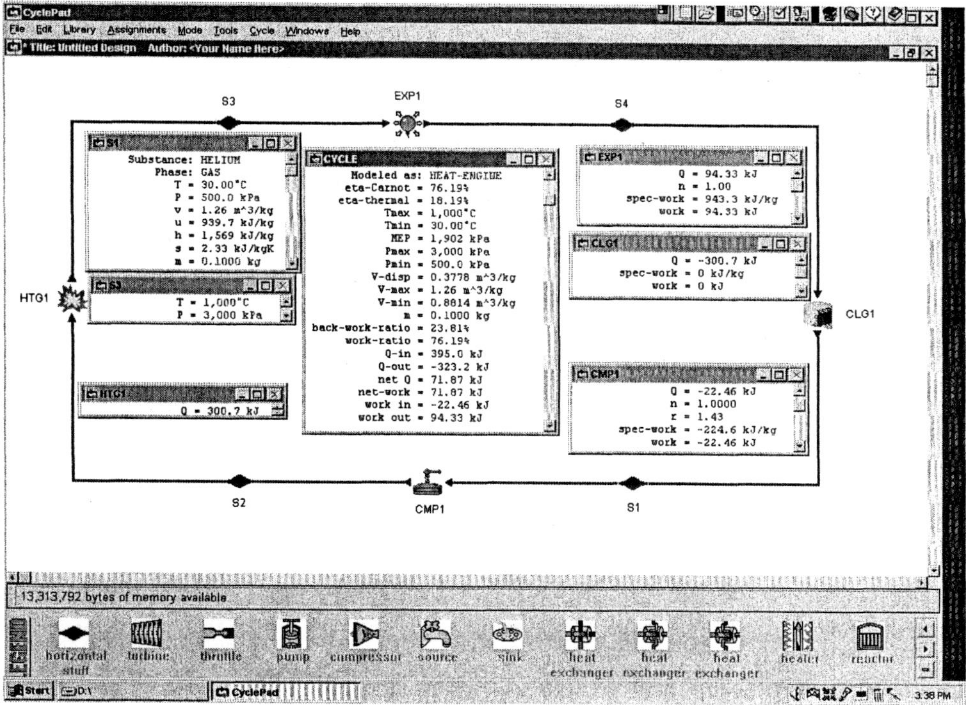

Figure Example 8.6.1 Stirling cycle

The Stirling cycle is an attempt to achieve Carnot efficiency by the use of an ideal regenerator.

A device called regenerator can be used to absorb heat during process 4-1 (Q_{41}) and ideally delivering the same quantity of heat during process 2-3 (Q_{23}). These two quantities of heat are represented by the areas underneath of the process 4-1 and process 2-3 of the T-s diagram in Figure 8.4.2. Using the ideal regenerator, Q_{41} is not counted as a part of the heat input. The efficiency of the Stirling cycle can be reduced from Equation (8.6.10) to

$$\eta = W_{net} / Q_{12} = 1 - T_3 / T_1 \tag{8.6.11}$$

In this respect, the Stirling cycle has the same efficiency as the Carnot cycle.

The regenerative Stirling cycle is illustrated in Figure 8.6.4. In this figure, the combination of heater #1 and cooler #1 is equivalent to the regenerator. Heat removed from the cooler #1 is added to the heater #1. Since this energy transfer occurs within the cycle internally, the amount of heat added to the heater #1 from the cooler #1 is not a part of heat added to the cycle from its surrounding heat reservoirs. Therefore

$$Q_{in} = Q_{12} \tag{8.6.12}$$

Example 8.6.2 illustrates the analysis of the regenerative Stirling cycle.

Practical attempts to follow the Stirling cycle present difficulties primarily due to the difficulty of achieving isothermal compression and isothermal expansion in a machine operating at a reasonable speed.

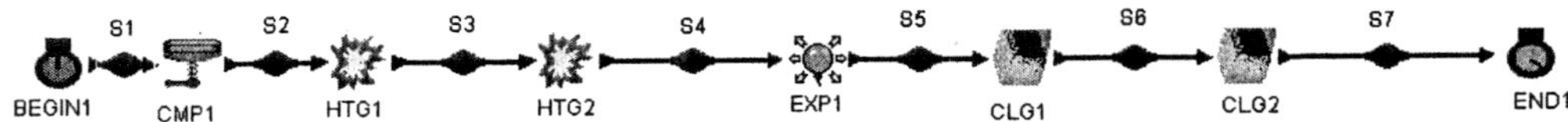

Figure 8.6.4 Regenerative Stirling cycle

Example 8.6.2 A regenerative Stirling cycle operates with 0.1 kg of hydrogen as a working fluid between 1000°C and 30°C. The highest pressure and the lowest pressure during the cycle are 3000 kPa and 500 kPa. The temperature at the exit of the regenerator (heater #1) and inlet to the heater #2 is 990°C and the temperature at the exit of the regenerator (cooler #1) and inlet to the cooler #2 is 40°C. Determine the heat and work added in each of the four processes, net work, and cycle efficiency.

To solve this problem by CyclePad, we take the following steps:

1. Build
   (A) Take a compression device, a heater (cold-side regenerator) combustion chamber, an expander and two coolers (cooler #1 is the hot-side regenerator) from the closed system inventory shop and connect the four devices to form the regenerative Stirling cycle as shown in Figure 8.6.2.
   (B) Switch to analysis mode.
2. Analysis
   (A) Assume a process for each of the six processes: (a) compression device as isothermal, (b) both heaters as isochoric, (c) expander as isothermal, and (d) both coolers as isochoric.
   (B) Input the given information: (a) working fluid is helium, (b) the inlet pressure and temperature of the compression device are 500 kPa and 30°C and m=0.1 kg, (c) the inlet pressure and temperature of the expander are 3000 kPa and 1000°C, (d) temperature at the exit of the regenerator (heater #1)= 990°C, (e) temperature at the exit of the regenerator (cooler #1)=40°C.
3. Display results
   (A) Display the cycle properties results. The cycle is a heat engine. The answers are: $Q_{12}=W_{12}=-22.46$ kJ, $Q_{23}=Q_{htr\#1}=-Q_{clr\#1}=Q_{regenerator}=297.6$ kJ, $Q_{34}=3.1$ kJ, $Q_{45}=W_{45}=94.33$ kJ, $Q_{56}=-3.1$ kJ, $W_{net}=71.87$ kJ, $Q_{in}=94.33+3.1=97.43$ kJ, and $\eta=71.87/97.43=73.77\%$.

Comment: The regenerator used in this example is not ideal. Yet, the regenerator raises the cycle efficiency almost to the Carnot efficiency.

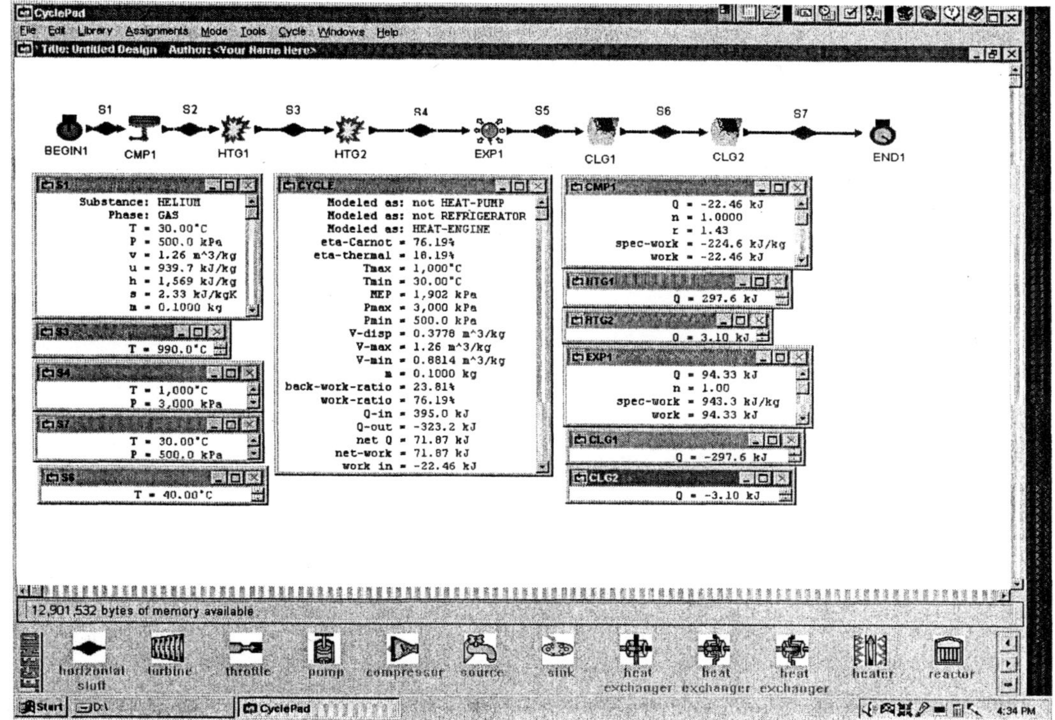

Figure Example 8.6.2 Regenerative Stirling cycle

Homework 8.6 Stirling Cycle

1. What are the four processes of the basic Stirling cycle?
2. What is a regenerator?
3. What is a regenerative Stirling cycle?
4. What would be the cycle efficiency of the Stirling cycle with an ideal regenerator?
5. A cycle is executed in a closed system with 0.004 kg of air and is composed of the following three processes:

   1-2 constant volume heating from 100 kPa and 20°C to 400 kPa

   2-3 isentropic expansion to 100 kPa

   3-1 constant pressure cooling to 100 kPa and 20°C

   Determine the net work per cycle, the thermal cycle efficiency, and MEP. Show the cycle on a T-s diagram.
6. A proposed air standard piston - cylinder arrangement cycle consists of an isentropic compression process, a constant volume heat addition process, an isentropic expansion process and a constant pressure heat rejection process. The compression ratio (v_1/v_2) during the isentropic compression process is 8.5. At the beginning of the compression process, P = 100 kPa and T = 300 K. The constant volume specific heat addition is 1,400 kJ/kg. Assume constant specific heats at 25°C.
   (a) Draw the cycle (with labels) on a T-s diagram.
   (b) Find the specific heat rejection during the constant pressure process.

(c) Find the specific net work for the cycle.
(d) Find the thermal efficiency for the cycle.
(e) Find the mean effective pressure for the cycle.

7. A Stirling cycle operates with 1 lbm of helium as a working fluid between 1800°R and 540°R. The highest pressure and the lowest pressure during the cycle are 450 psia and 75 psia. Determine the heat added, net work, and cycle efficiency.
   ANSWER: W_{net}=367.4 Btu, Q_{in}=1458 Btu, and η=25.20%.
8. A Stirling cycle operates with 0.1 kg of air as a working fluid between 1000°C and 30°C. The highest pressure and the lowest pressure during the cycle are 3000 kPa and 500 kPa. Determine the heat and work added in each of the four processes, net work, and cycle efficiency.
   ANSWER: $Q_{12}=W_{12}$=-3.10 kJ, Q_{23}=69.52 kJ, $Q_{34}=W_{34}$=13.02 kJ, Q_{41}=-69.52 kJ, W_{net}=9.92 kJ, Q_{in}=82.54 kJ, and η=12.02%.
9. A regenerative Stirling cycle operates with 0.1 kg of air as a working fluid between 1000°C and 30°C. The highest pressure and the lowest pressure during the cycle are 3000 kPa and 500 kPa. The temperature at the exit of the regenerator (heater #1) and inlet to the heater #2 is 990°C and the temperature at the exit of the regenerator (cooler #1) and inlet to the cooler #2 is 40°C. Determine the heat and work added in each of the four processes, net work, and cycle efficiency.

8.7 Design Examples

CyclePad is a powerful tool for cycle design and analysis. Due to its capabilities, the software allows users to view the cycle effects of varying design input parameters at once. The following examples illustrate the design of several closed-system power cycles.

Example 8.7.1. A six-process internal combustion engine as shown in Figure Example 8.7.1a is proposed by a junior engineer. Air mass contained in the cylinder is 0.01 kg. The six processes are:

Process 1-2 isentropic compression
Process 2-3 isochoric heating
Process 3-4 isobaric heating
Process 4-5 isentropic expansion
Process 5-6 isochoric cooling
Process 6-1 isobaric cooling

The following information is given as shown in Figure Example 8.7.1b:

p_1=100 kPa, T_1=20°C, $V_1=10V_2$, q_{23}=600 kJ/kg, q_{34}=400 kJ/kg, and T_5=400°C.

Determine the pressure and temperature of each state of the cycle, work and heat of each process, work input, work output, net work output, heat added, heat removed, MEP and cycle efficiency.

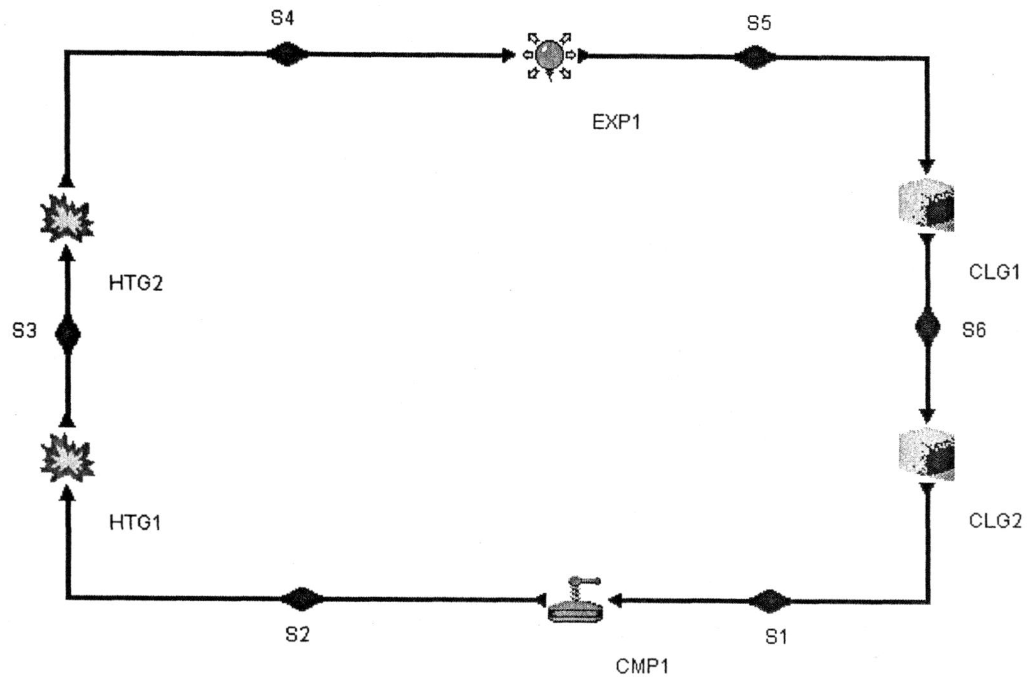

Figure Example 8.7.1a Six-process internal combustion engine design

To evaluate this design by CyclePad, we take the following steps:

1. Build
   (A) Take a compression device, two heaters, an expander and two coolers from the closed system inventory shop and connect the four devices to form the cycle as shown in Figure 8.7.1a.
   (B) Switch to analysis mode.
2. Analysis
   (A) Assume a process for each of the six processes: (a) compression device as isentropic, (b) one heater as isochoric and the other as isobaric, (c) expander as isentropic, and (d) one cooler as isochoric and the other as isobaric.
   (B) Input the given information: working fluid is air, m=0.01 kg, p_1=100 kPa, T_1=20°C, $V_1=10V_2$, q_{23}=600 kJ/kg, q_{34}=400 kJ/kg, and T_5=400°C.
3. Display results
   (A) Display the cycle properties results. The cycle is a heat engine. The results are: p_1=100 kPa, T_1=20°C, p_2=2512 kPa, T_2=463.2°C, p_3=5368 kPa, T_3=1300°C, p_4=5368 kPa, T_4=1699°C, p_5=124.7 kPa, T_5=400°C, p_6=100 kPa, T_6=266.6°C, Q_{12}=0, W_{12}=-3.18 kJ, Q_{23}=6 kJ, W_{23}=0, Q_{34}=4 kJ, W_{34}=1.14 kJ, Q_{45}=0, W_{45}=9.31 kJ, Q_{56}=-0.9558 kJ, W_{56}=0, Q_{61}=-2.47 kJ, W_{61}=-0.7071 kJ, W_{add}=-3.88 kJ, W_{out}=10.45 kJ, W_{net}=6.57 kJ, Q_{in}=10 kJ, Q_{out}=-3.43 kJ, MEP=448.9 kPa, and η=65.69% as shown in Figure 8.7.1b.

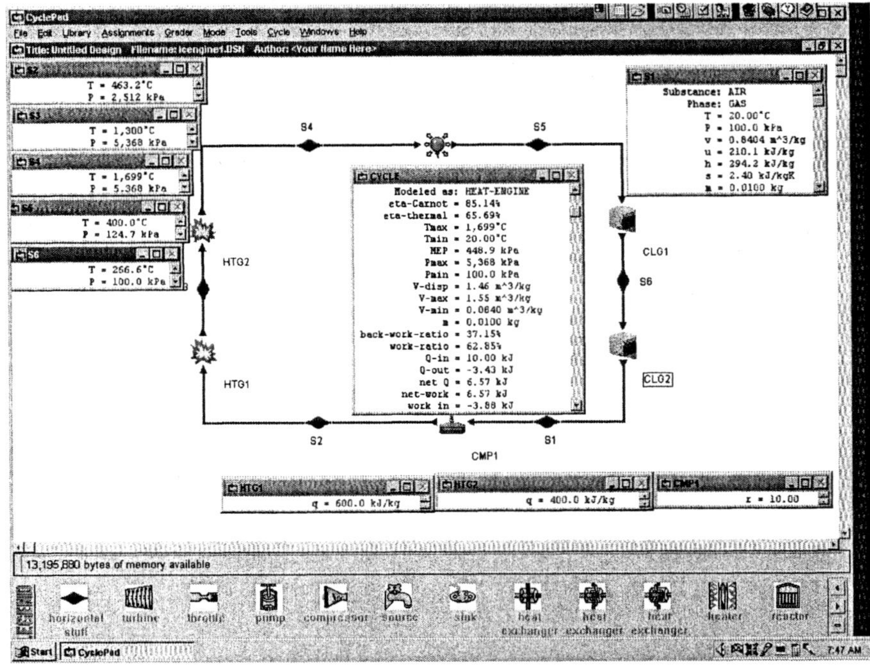

Figure Example 8.7.1b Six-process internal combustion engine design results

The T-s diagram of the cycle is shown in Figure Example 8.7.1c.

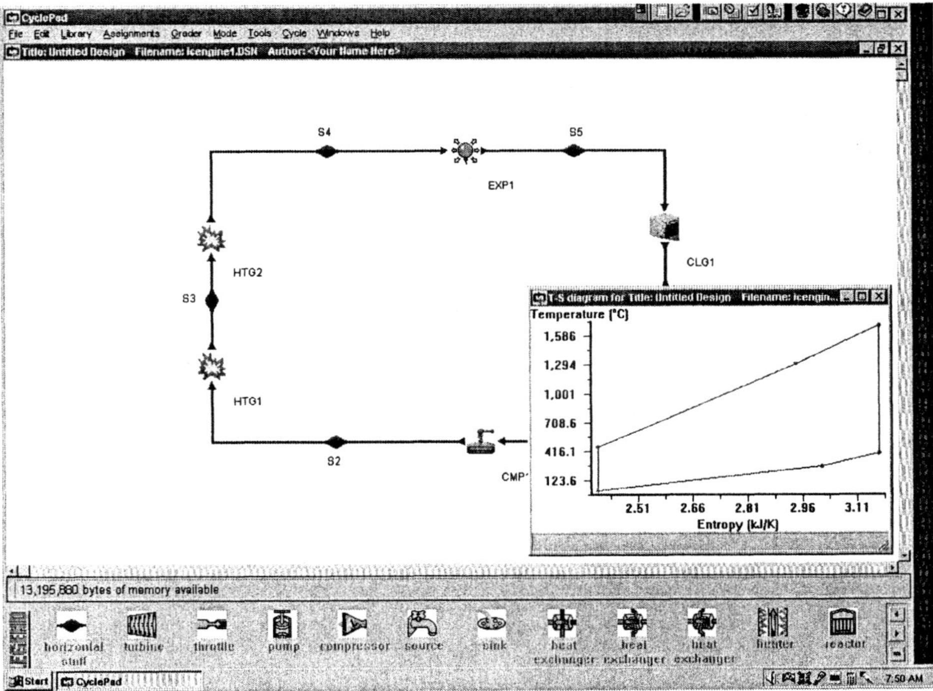

Figure Example 8.7.1c T-s diagram

The sensitivity diagram of η (cycle efficiency) vs r (compression ratio) is plotted in Figure Example 8.7.1d. The figure shows that the larger the compression ratio, the better the cycle efficiency. To improve the proposed engine, a larger compression ratio could be used.

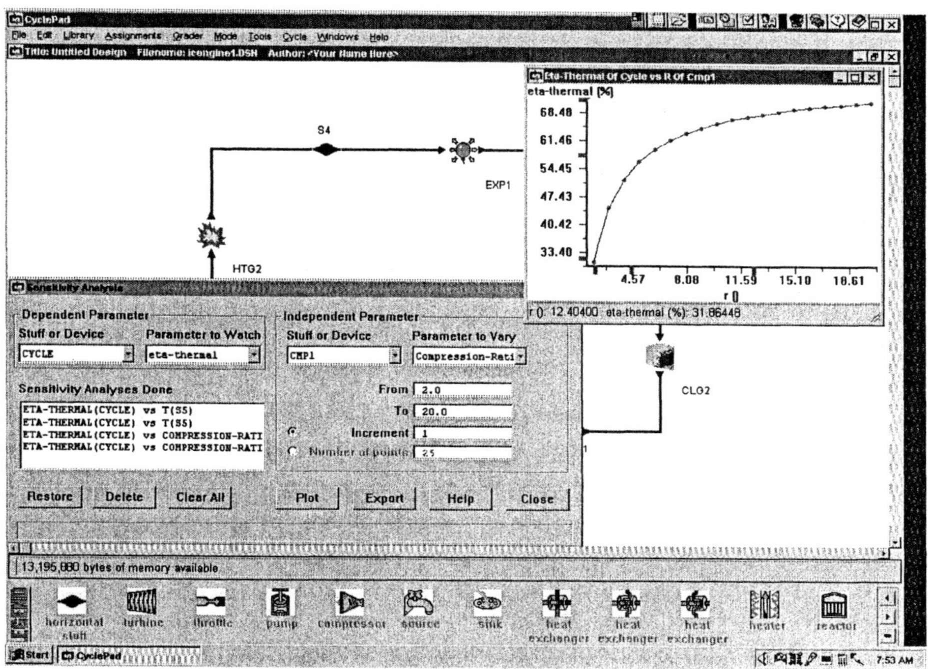

Figure Example 8.7.1d Sensitivity diagram

Example 8.7.2. A six-process internal combustion engine as shown in Figure Example 8.7.2a is proposed by a junior engineer. Air mass contained in the cylinder is 0.01 kg. The six processes are:

Process 1-2 isentropic compression
Process 2-3 isochoric heating
Process 3-4 isobaric heating
Process 4-5 isentropic expansion
Process 5-6 isobaric cooling
Process 6-1 isochoric cooling

The following information is given as shown in Figure Example 8.7.2b:

p_1=100 kPa, T_1=20°C, V_1=10V_2, q_{23}=600 kJ/kg, q_{34}=400 kJ/kg, and T_5=400°C.

Determine the pressure and temperature of each state of the cycle, work and heat of each process, work input, work output, net work output, heat added, heat removed, MEP and cycle efficiency.

Notice that Process 5-6 and Process 6-1of the cycle are different from Process 5-6 and Process 6-1of the cycle proposed in Example 8.7.1.

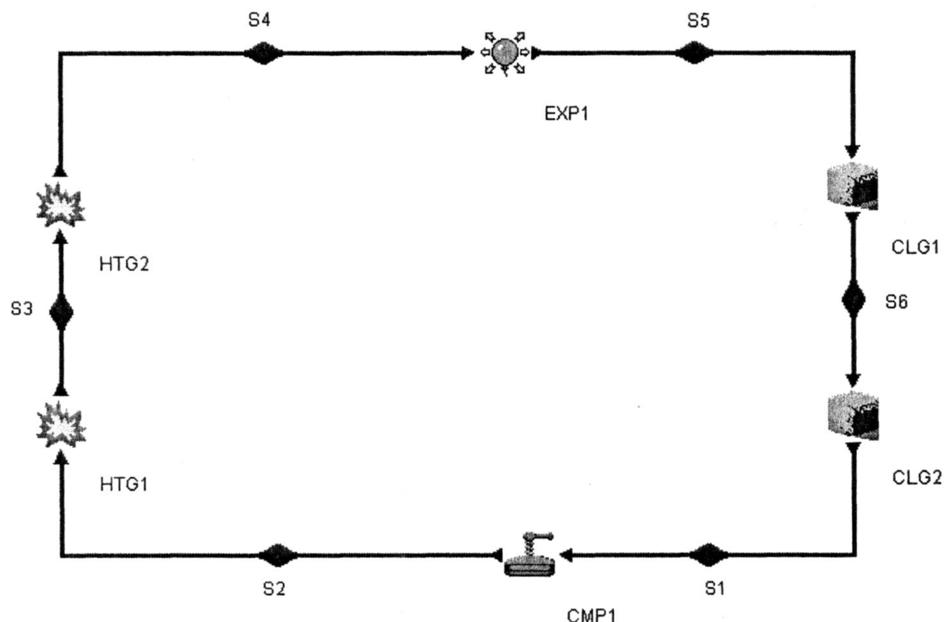

Figure Example 8.7.2a Six process internal combustion engine

To evaluate this design by CyclePad, we take the following steps:

1. Build
   (A) Take a compression device, two heaters, an expander and two coolers from the closed system inventory shop and connect the four devices to form the cycle as shown in Figure 8.7.2a.
   (B) Switch to analysis mode.
2. Analysis
   (A) Assume a process for each of the six processes: (a) compression device as isentropic, (b) one heater as isochoric and the other as isobaric, (c) expander as isentropic, and (d) one cooler as isochoric and the other as isobaric.
   (B) Input the given information: working fluid is air, m=0.01 kg, p_1=100 kPa, T_1=20°C, $V_1=10V_2$, q_{23}=600 kJ/kg, q_{34}=400 kJ/kg, and T_5=400°C.
3. Display results
   (A) Display the cycle properties results. The cycle is a heat engine. The results are: p_1=100 kPa, T_1=20°C, p_2=2512 kPa, T_2=463.2°C, p_3=5368 kPa, T_3=1300°C, p_4=5368 kPa, T_4=1699°C, p_5=124.7 kPa, T_5=400°C, p_6=124.7 kPa, T_6=92.42°C, Q_{12}=0, W_{12}=-3.18 kJ, Q_{23}=6 kJ, W_{23}=0, Q_{34}=4 kJ, W_{34}=1.14 kJ, Q_{45}=0, W_{45}=9.31 kJ, Q_{56}=-3.09 kJ, W_{56}=-0.8818, Q_{61}=-0.5191 kJ, W_{61}=0 kJ, W_{add}=-4.06 kJ, W_{out}=10.45 kJ, W_{net}=6.39 kJ, Q_{in}=10 kJ, Q_{out}=-3.61 kJ, MEP=436.9 kPa, and η=63.95%

It is observed that both the cycle efficiency and MEP of the proposed cycle are less than those of the proposed cycle given by example 8.7.1.

The T-s diagram and sensitivity diagram of η (cycle efficiency) vs r (compression ratio) is plotted in Figure Example 8.7.2c.

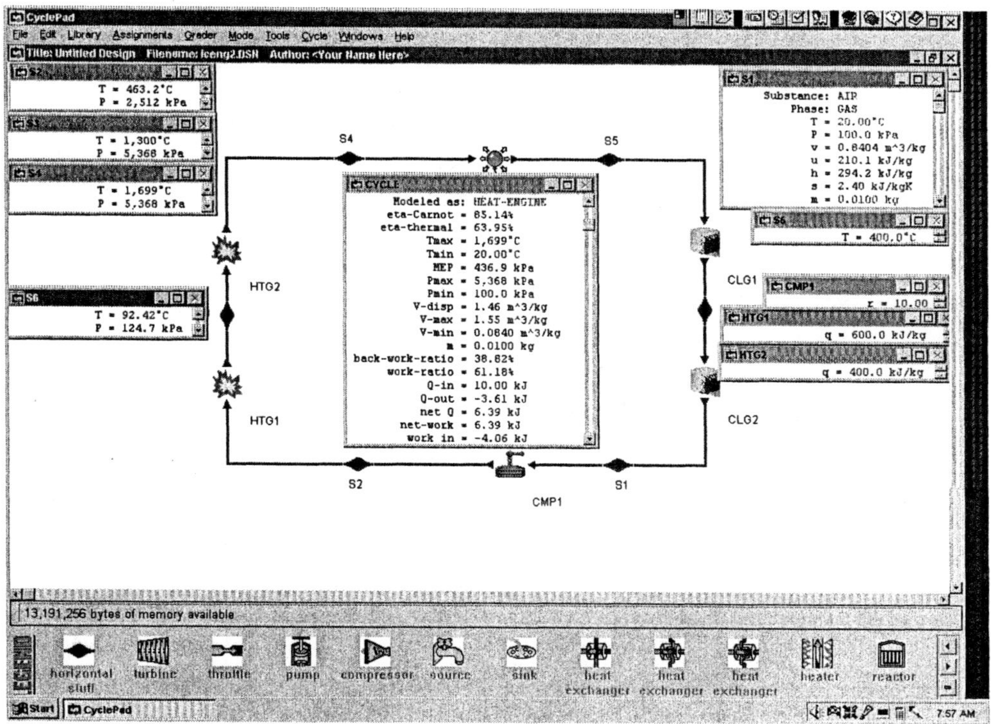

Figure Example 8.7.2b Six-process internal combustion engine design results

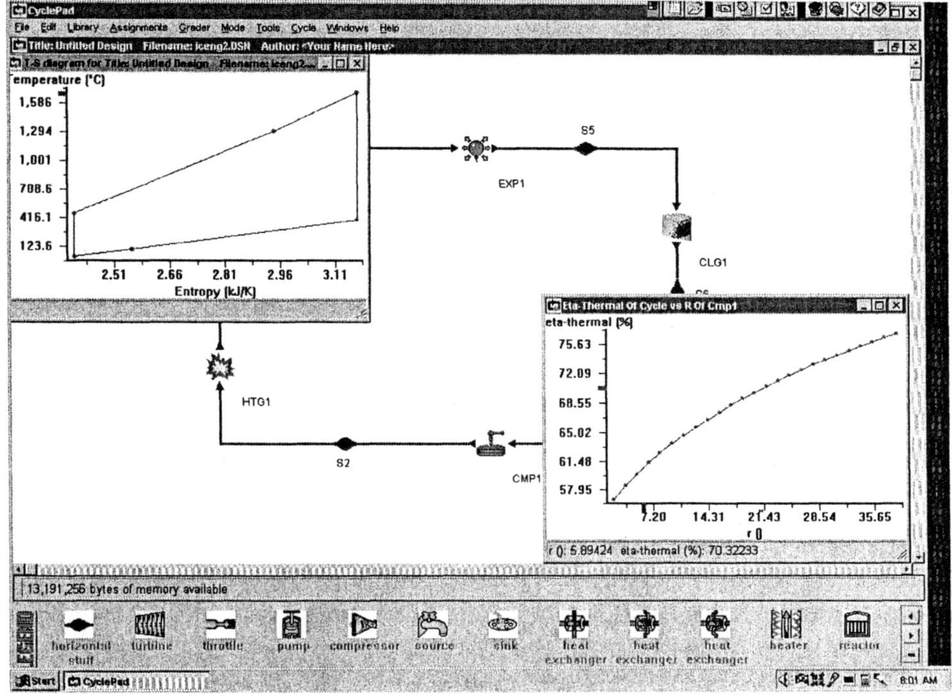

Figure Example 8.7.2c. T-s diagram and sensitivity diagram of η (cycle efficiency) vs r (compression ratio)

Example 8.7.3. Adding a turbo-charger and a pre-cooler to a Dual cycle is proposed as shown in Figure Example 8.7.3a. The cylinder volume of the engine is 0.01 m^3. Evaluate the proposed cycle. The basic Dual cycle and the proposed turbo-charger and pre-cooler Dual cycle information is:

Basic Dual cycle:

$p_1=p_2=p_3=p_8=101$ kPa, $T_1=T_2=T_3=T_8=15°C$, $V_3=0.01$ m^3, r(compression ratio)=10, and $q_{45}=q_{56}=300$ kJ/kg.

Turbo-charger and pre-cooler Dual cycle

$p_1=p_8=101$ kPa, $T_1=T_8=15°C$, $p_2=150$ kPa, $T_3=15°C$, $V_3=0.01$ m^3, r(compression ratio)=10, and $q_{45}=q_{56}=300$ kJ/kg.

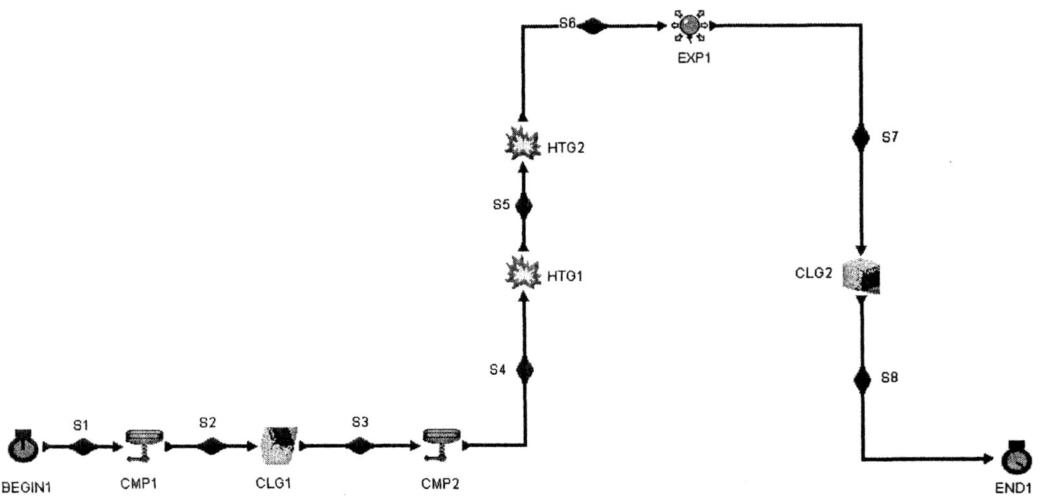

Figure Example 8.7.3a Turbo-charger and pre-cooler Dual cycle

To evaluate this proposed cycle by CyclePad, we take the following steps:

1. Build
   (A) Take two compression devices, two heaters, an expander and two coolers from the closed system inventory shop and connect the devices to form the cycle as shown in Figure 8.7.3a.
   (B) Switch to analysis mode.
2. Analysis
   (A) Assume a process for each of the seven processes: (a) compression devices as isentropic, (b) one heater as isochoric and the other as isobaric, (c) expander as isentropic, and (d) one cooler as isochoric and the other as isobaric.
   (B) Input the given information: working fluid is air, $p_1=p_8=101$ kPa, $T_1=T_8=15°C$, $p_2=150$ kPa, $T_3=15°C$, $V_3=0.01$ m^3, r(compression ratio)=10, and $q_{45}=q_{56}=300$ kJ/kg as shown in Figure Example 8.7.3b.

3. Display results

   (A) Display the cycle properties results. The cycle is a heat engine. The results are: p_1=100 kPa, T_1=15°C, p_2=150 kPa, T_2=49.47°C, p_3=150 kPa, T_3=15°C, p_4=3768 kPa, T_4=450.7°C, p_5=5947 kPa, T_5=869.2°C, p_6=5947 kPa, T_6=1168°C, p_7=188.4 kPa, T_7=264.4°C, Q_{12}=0, W_{12}=-0.4486 kJ, Q_{23}=-0.6281 kJ, W_{23}=-0.1795, W_{34}=-5.67 kJ, Q_{45}=0, W_{45}=11.76 kJ, Q_{56}=-3.25 kJ, W_{add}=-6.30 kJ, W_{out}=13.32 kJ, W_{net}=7.02 kJ, Q_{in}=10.89 kJ, Q_{out}=-3.87 kJ, MEP=506.8 kPa, and η=64.44% as shown in Figure Example 8.7.3c.

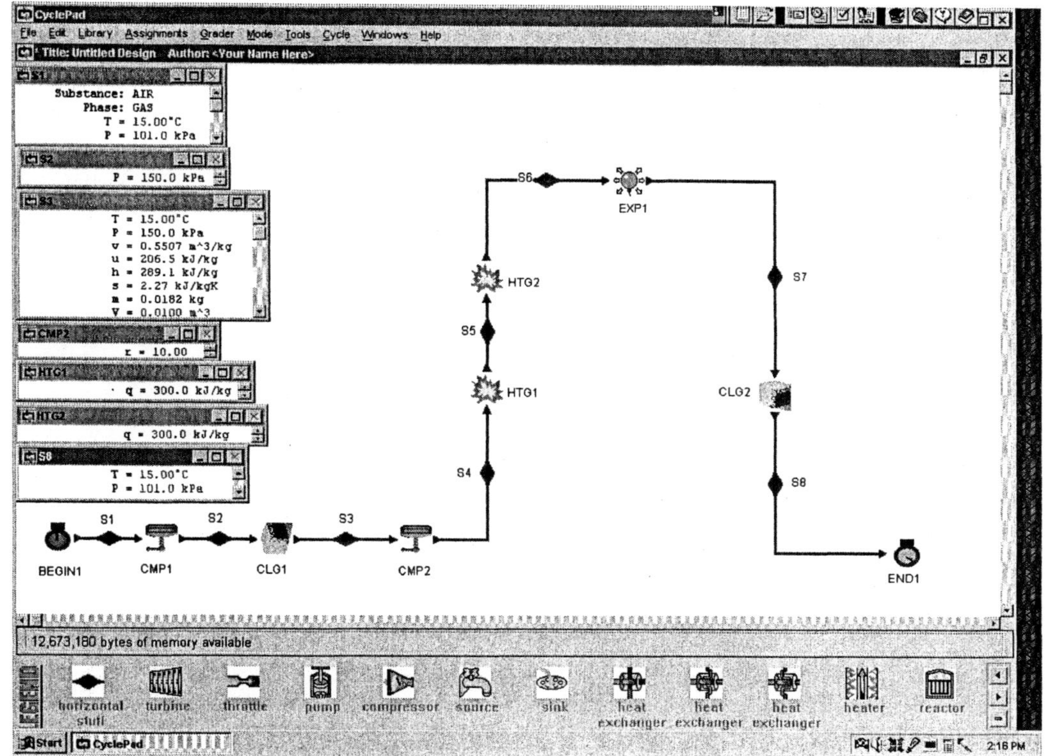

Figure Example 8.7.3b Turbo-charger and pre-cooler Dual cycle input

For the Dual cycle without turbo-charger and pre-cooler, the input are: p_1=p_2=p_3=p_8=101 kPa, T_1=T_2=T_3=T_8=15°C, V_3=0.01 m^3, r(compression ratio)=10, and q_{45}=q_{56}=300 kJ/kg as shown in Figure Example 8.7.3d.

The output results are: p_1=p_2=p_3=101 kPa, T_1=T_2=T_3=15°C, p_4=2537 kPa, T_4=450.7°C, p_5=4004 kPa, T_5=869.2°C, p_6=4004 kPa, T_6=1168°C, p_7=220.7 kPa, T_7=355.6°C, Q_{12}=0, W_{12}=-0 kJ, Q_{23}=-0 kJ, W_{23}=-0 kJ, W_{34}=-3.82 kJ, Q_{45}=0, W_{45}=7.11 kJ, Q_{56}=-2.99 kJ, W_{add}=-3.82 kJ, W_{out}=8.16 kJ, W_{net}=4.34 kJ, Q_{in}=7.34 kJ, Q_{out}=-2.94 kJ, MEP=482.5 kPa, and η=59.20% as shown in Figure Example 8.7.3d.

It is observed that both the cycle efficiency and MEP of the proposed cycle are better than those of the Dual cycle without turbo-charger and pre-cooler.

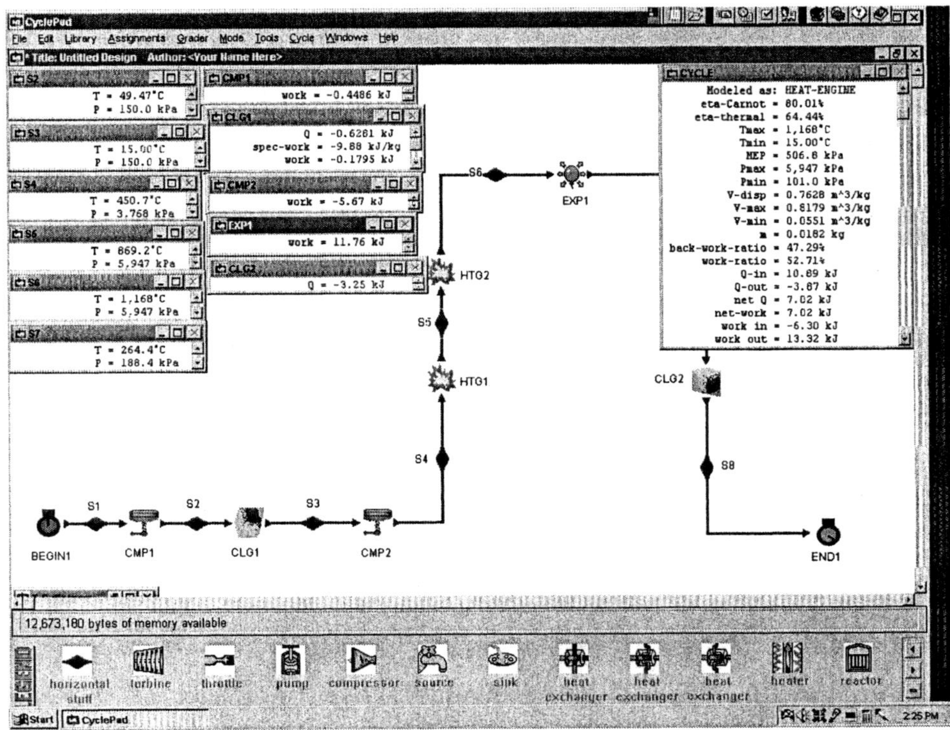

Figure Example 8.7.3c Turbo-charger and pre-cooler Dual cycle result

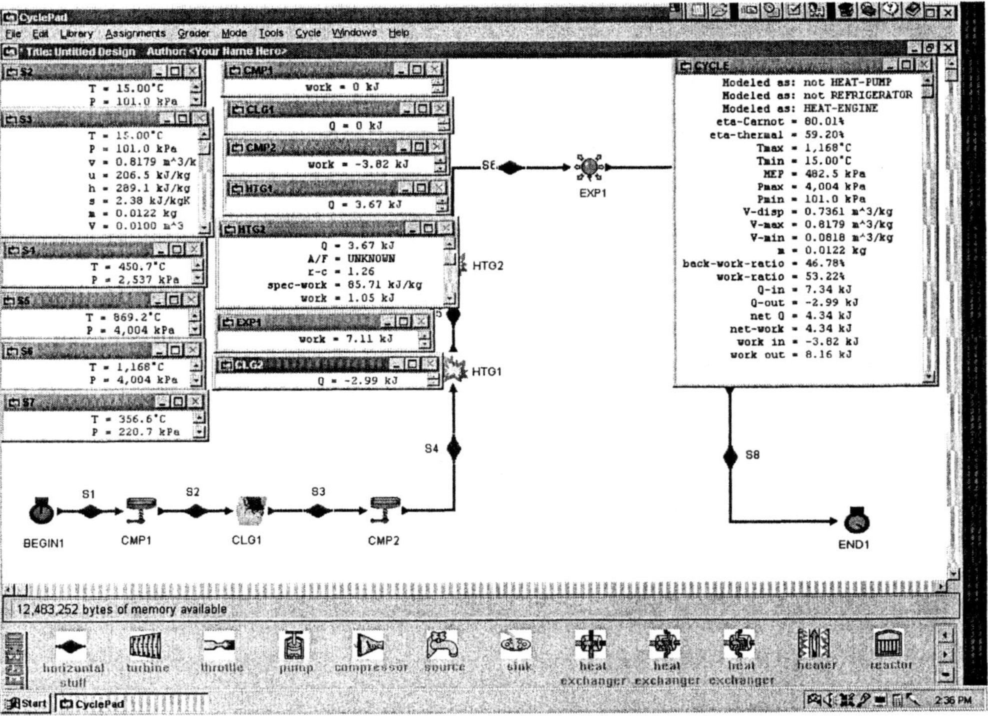

Figure Example 8.7.3c Dual cycle without result turbo-charger and pre-cooler

Homework 8.7 Design

1. A six-process internal combustion engine as shown in Figure Example 8.7.2a is proposed by a junior engineer. Air mass contained in the cylinder is 0.01 kg. The six processes are:
   Process 1-2 isentropic compression
   Process 2-3 isochoric heating
   Process 3-4 isobaric heating
   Process 4-5 isentropic expansion
   Process 5-6 isobaric cooling
   Process 6-1 isochoric cooling
   The following information is given as shown in Figure Example 8.7.2b:
   p_1=100 kPa, T_1=15°C, $V_1=8V_2$, q_{23}=300 kJ/kg, q_{34}=400 kJ/kg, and T_5=400°C.
   Determine the pressure and temperature of each state of the cycle, work and heat of each process, work input, work output, net work output, heat added, heat removed, MEP and cycle efficiency.
2. Adding a turbo-charger and a pre-cooler to a Dual cycle is proposed as shown in Figure Example 8.7.3a. The cylinder volume of the engine is 0.01 m^3. Evaluate the proposed cycle. The basic Dual cycle and the proposed turbo-charger and pre-cooler Dual cycle information are:
   Basic Dual cycle:
   $p_1=p_2=p_3=p_8$=101 kPa, $T_1=T_2=T_3=T_8$=15°C, V_3=0.01 m^3, r(compression ratio)=8, and $q_{45}=q_{56}$=300 kJ/kg.
   Turbo-charger and pre-cooler Dual cycle
   $p_1=p_8$=101 kPa, $T_1=T_8$=15°C, p_2=120 kPa, T_3=15°C, V_3=0.01 m^3, r(compression ratio)=10, and $q_{45}=q_{56}$=300 kJ/kg.

8.8 SUMMARY

Heat engines that use gases as the working fluid in a closed system model were discussed in this chapter. Otto cycle, Diesel, and Dual cycle are internal combustion engines. Stirling cycle is an external combustion engine.

The Otto cycle is a spark-ignition reciprocating engine made of an isentropic compression process, a constant volume combustion process, an isentropic expansion process, and a constant volume cooling process. The thermal efficiency of the Otto cycle depends on its compression ratio. The compression ratio is defined as $r=V_{max}/V_{min}$. The Otto cycle efficiency is limited by the compression ratio because of the engine knock problem.

The Diesel cycle is a compression-ignition reciprocating engine made of an isentropic compression process, a constant pressure combustion process, an isentropic expansion process, and a constant volume cooling process. The thermal efficiency of the Otto cycle depends on its compression ratio and cut-off ratio. The compression ratio is defined as $r=V_{max}/V_{min}$. The cut-off ratio is defined as $r_{cutoff}=V_{combustion\ off}/V_{min}$.

The Dual cycle involves two heat addition processes, one at constant volume and one at constant pressure. It behaves more like an actual cycle than either Otto or Diesel cycle.

The Lenoir cycle was the first commercially successful internal combustion engine.

The Stirling cycle is an attempt to achieve the Carnot efficiency by using of regeneration.

Chapter 9

Gas Open System Cycles

Objectives

After reading and studying the material in this chapter, you should be able to:

1. Describe the sequence of processes in the Brayton cycle.
2. Sketch the Brayton cycle on both p-v and T-s diagrams.
3. Know that the efficiency of the Brayton cycle is a function of the compression ratio only.
4. Understand the purpose of adding reheat, inter-cooler and regenerator to the Brayton cycle.
5. Understand the difference between a closed Brayton cycle and an open Brayton cycle.
6. Understand the split-shaft gas turbine cycle.
7. Analyze the Brayton cycle with all kinds of modifications using CyclePad.
8. Understand the bleed-air gas turbine cycle.
9. Understand the Feher cycle.
10. Understand the Ericsson cycle.
11. Understand the combined, cascaded, Braysson, and Field cycles.
12. Understand the cogeneration cycle.
13. Analyze and design various gas open system cycles using CyclePad.

9.1 Brayton Cycle

The *ideal Brayton gas turbine cycle* is named after an American engineer, George Brayton, who proposed the cycle in the 1870s. The gas turbine cycle consists of four processes: an isentropic compression process 1-2, a constant pressure combustion process 2-3, an isentropic expansion process 3-4, and a constant pressure cooling process 4-1. The p-v and T-s diagrams for an ideal Brayton cycle are illustrated in Figure 9.1.1.

The gas turbine cycle may be either closed or open. The more common cycle is the open one, in which atmospheric air is continuously drawn into the compressor, heat is added to the air by the combustion of fuel in the combustion chamber, and the working fluid expands through the turbine and exhausts to the atmosphere. A schematic diagram of an *open Brayton cycle*, which is assumed to operate steadily as an open system, is shown in Figure 9.1.2.

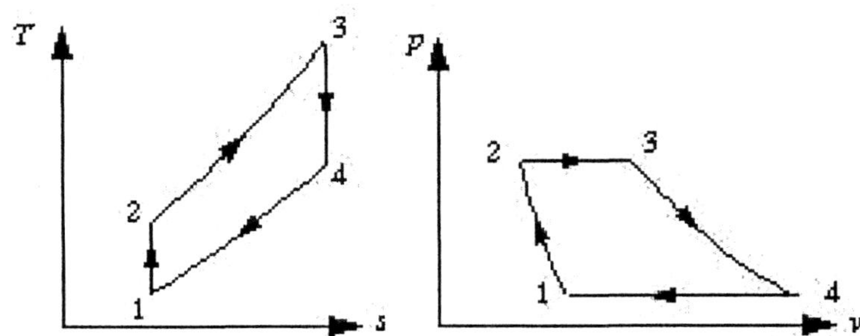

Figure 9.1.1. Brayton cycle p-v and T-s diagrams

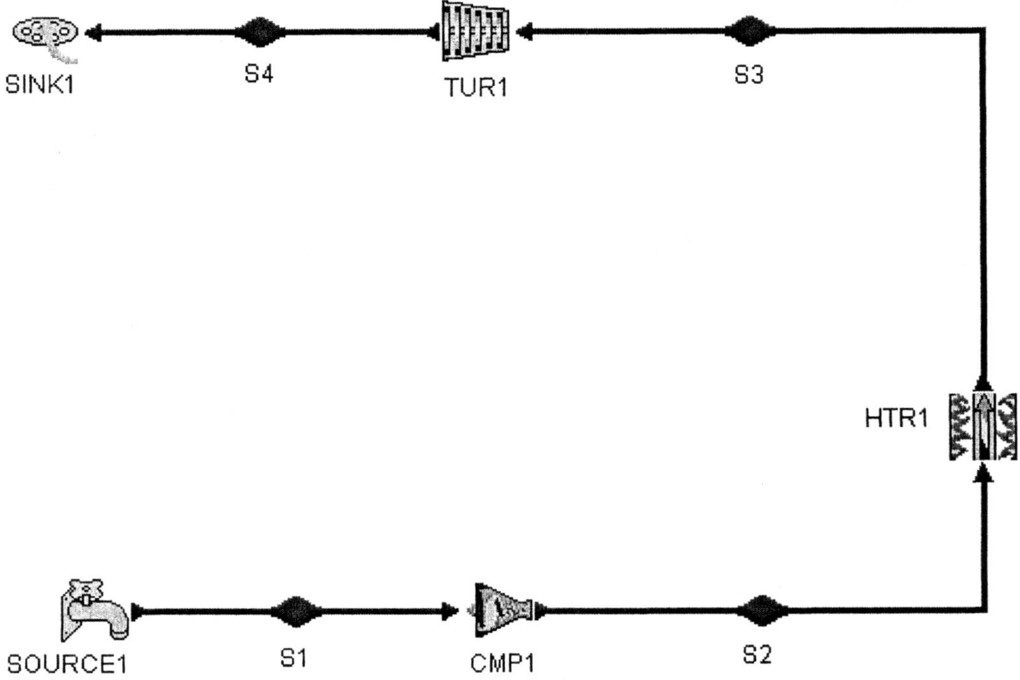

Figure 9.1.2. Open Brayton cycle

In the closed cycle, the heat is added to the fluid in a heat exchanger from an external heat source, such as nuclear power plant, and the fluid is cooled in another heat exchanger after it leaves the turbine and before it enters the compressor. A schematic diagram of a *closed Brayton cycle* is shown in Figure 9.1.3.

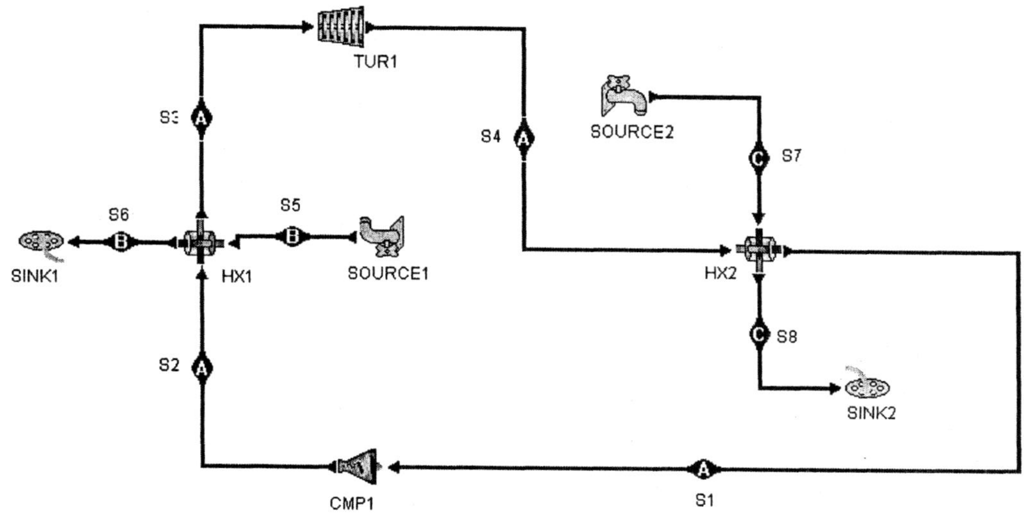

Figure 9.1.3. Closed Brayton cycle

Applying the first law of thermodynamics for an open system to each of the four processes of the Brayton cycle yields:

$$Q_{12} = 0, \tag{9.1.1}$$

$$W_{12} = m(h_1 - h_2), \tag{9.1.2}$$

$$W_{23} = 0, \tag{9.1.3}$$

$$Q_{23} - 0 = m(h_3 - h_2), \tag{9.1.4}$$

$$Q_{34} = 0, \tag{9.1.5}$$

$$W_{34} = m(h_3 - h_4), \tag{9.1.6}$$

$$W_{41} = 0, \tag{9.1.7}$$

and

$$Q_{41} - 0 = m(h_1 - h_4). \tag{9.1.8}$$

The net work (W_{net}), which is also equal to net heat (Q_{net}), is

$$W_{net} = W_{12}+W_{34} = Q_{net} = Q_{23} + Q_{41} \tag{9.1.9}$$

The thermal efficiency of the cycle is

$$\eta = W_{net} /Q_{23} =Q_{net} /Q_{23} = 1- Q_{41} /Q_{23} =1- (h_4 - h_1)/(h_3 - h_2) \tag{9.1.10}$$

This expression for thermal efficiency of an ideal Brayton cycle can be simplified if air is assumed to be the working fluid with constant specific heats. Equation (8.1.3) is reduced to:

$$\eta = 1 - (T_4 - T_1)/(T_3 - T_2) = 1-(r_p)^{(k-1)/k} \tag{9.1.11}$$

where r_p is the *pressure compression ratio* for the compressor defined by the equation

$$r_p = p_2 / p_1 \tag{9.1.12}$$

The highest temperature in the cycle occurs at the end of the combustion process (state 3), and it is limited by the maximum temperature that the turbine blade can withstand. The maximum temperature does have an effect on the optimal performance of the gas turbine cycle.

In the gas turbine cycle, the ratio of the compressor work to the turbine work is called *back work ratio*. The back work ratio is very high, usually more than 40 percent.

Example 9.1.1. An engine operates on the open Brayton cycle and has a compression ratio of eight. Air, at a mass flow rate of 0.1 kg/s, enters the engine at 27°C and 100 kPa. The amount of heat addition is 1000 kJ/kg. Determine the efficiency, compressor power input, turbine power output, back-work-ratio, and net power of the cycle. Show the cycle on T-s diagram. Plot the sensitivity diagram of cycle efficiency vs compression ratio.
Convert the SI unit system to British system and find the answer.

To solve this problem by CyclePad, we take the following steps:

1. Build
   (A) Take a source, a compressor, a combustion chamber (heater), a turbine and a sink from the open system inventory shop and connect the five devices to form the open Brayton cycle.
   (B) Switch to analysis mode.
2. Analysis
   (A) Assume a process for each of the three processes: (a) compressor device as isentropic, (b) combustion chamber as isobaric, and (c) turbine as isentropic.
   (B) Input the given information: (a) working fluid is air, (b) the inlet pressure and temperature of the compression device are 100 kPa and 27°C, (c) the compression ratio of the compressor is 8, (d) the heat addition is 1000 kJ/kg in the combustion chamber, (e) mass flow rate of air is 0.1 kg/s, and (f) the exit pressure of the turbine is 100 kPa.
3. Display results
   (A) Display the T-s diagram and cycle properties results. The cycle is a heat engine. The answers are η=44.80%, Compressor power=-24.44 kW, Turbine power=69.23 kW, back-work-ratio=35.30%, and Net power=69.23 kW, and (B) Display the sensitivity diagram of cycle efficiency vs compression ratio.

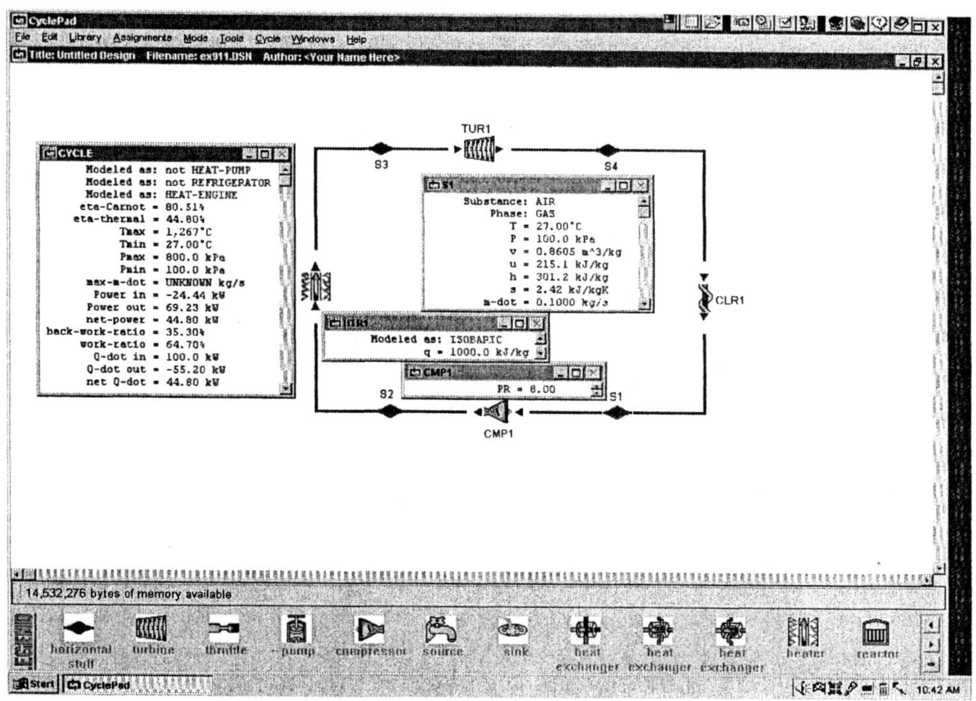

Figure Example 9.1.1a Open Brayton cycle

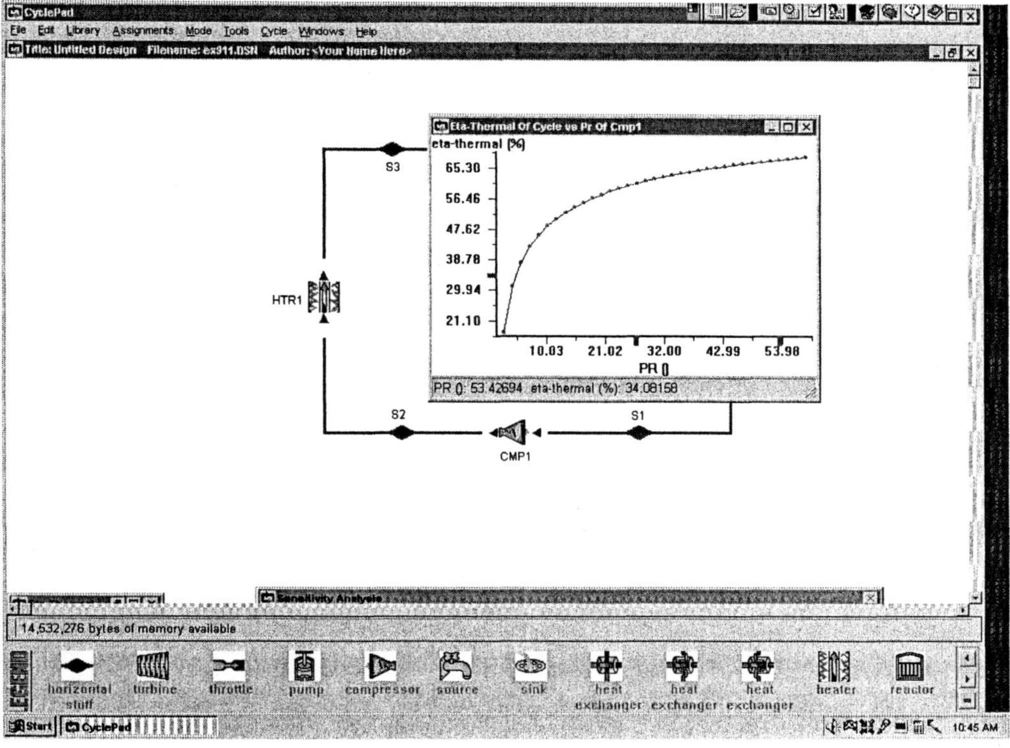

Figure Example 9.1.1a Open Brayton cycle Sensitivity analysis

Comment: The cycle efficiency increases as compressor ratio increases

For *actual Brayton cycles*, many irreversibilities in various components are present. The T-s diagram of an actual Brayton cycle is shown in Figure 9.1.4. The major irreversibilities occurs within the turbine and pump. To account for these irreversibility effects, turbine efficiency and pump efficiency must be used in computing the actual work produced or consumed. The effect of irreversibilities on the thermal efficiency of a Brayton cycle is illustrated in the following example.

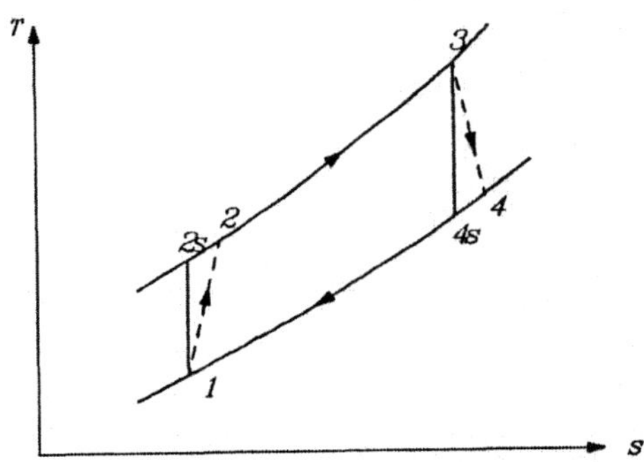

Figure 9.1.4 Actual Brayton cycle T-s diagram

Example 9.1.2. An engine operates on an open actual Brayton cycle and has a compression ratio of 8. The air enters the engine at 27°C and 100 kPa. The mass flow rate of air is 0.1 kg/s. The amount of heat addition is 1000 kJ/kg. The compressor efficiency is 86% and the turbine efficiency is 89%. Determine the efficiency and work output per kilogram of air.

To solve this problem by CyclePad, we take the following steps:

1. Build
   (A) Take a source, a compressor, a combustion chamber (heater), a turbine and a sink from the open system inventory shop and connect the five devices to form the open actual Brayton cycle.
   (B) Switch to analysis mode.
2. Analysis
   (A) Assume a process for each of the three processes: (a) compressor device as adiabatic, (b) combustion chamber as isobaric, and (c) turbine as adiabatic.
   (B) Input the given information: (a) working fluid is air, (b) the inlet pressure and temperature of the compression device are 100 kPa and 27°C, (c) the compression ratio of the compressor is 8, (d) the heat addition is 1000 kJ/kg in the combustion chamber, (e) the compressor efficiency is 86% and the turbine efficiency is 89%, (f) mass flow rate of air is 0.1 kg/s, and (g) the exit pressure of the turbine is 100 kPa.
3. Display results

(A) Display the cycle properties results. The cycle is a heat engine. The answers are η=34.79% and net-power=34.79 kW.

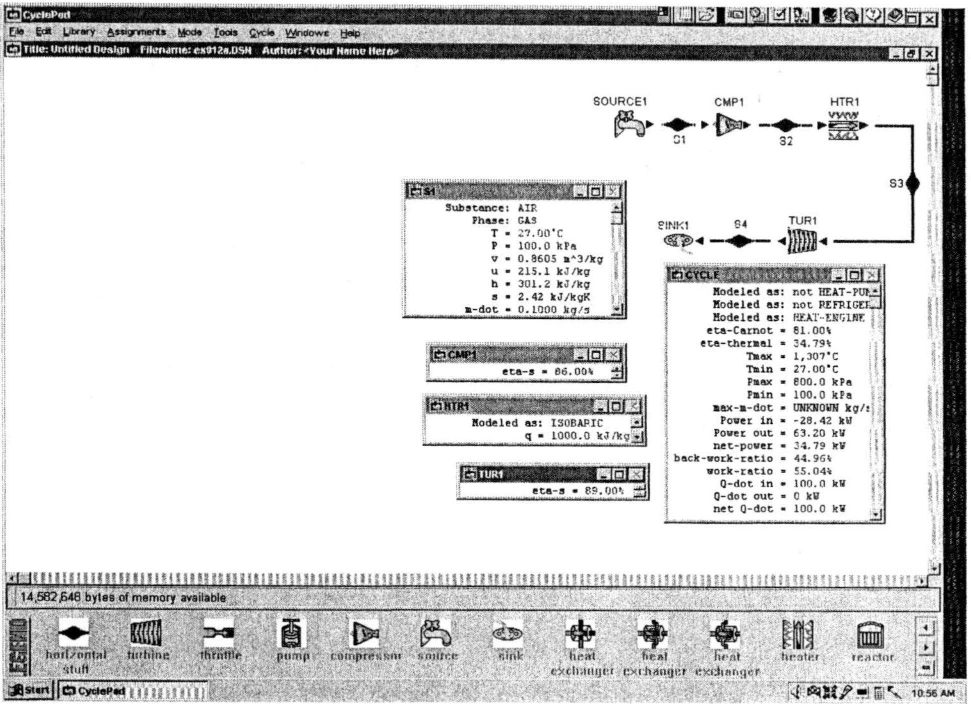

Figure Example 9.1.2. Open actual Brayton cycle

Example 9.1.3. An engine operates on the closed Brayton cycle and has a compression ratio of 8. Helium enters the engine at 47°C and 200 kPa. The mass flow rate of helium is 1.2 kg/s. The amount of heat addition is 1000 kJ/kg. Determine the highest temperature of the cycle, the turbine power produced, the compressor power required, the back work ratio, the rate of heat added, and the cycle efficiency.

To solve this problem by CyclePad, we take the following steps:

1. Build
   (A) Take a compressor, a combustion chamber (heater), a turbine and a cooler from the open system inventory shop and connect the four devices to form the closed Brayton cycle.
   (B) Switch to analysis mode.
2. Analysis
   (A) Assume a process for each of the four processes: (a) compressor device as adiabatic and isentropic, (b) combustion chamber as isobaric, (c) turbine as adiabatic and isentropic, and (d) cooler as isobaric.
   (B) Input the given information: (a) working fluid is helium, (b) the inlet pressure and temperature of the compression device are 200 kPa and 47°C, (c) the compressor exit

pressure is 1600 kPa, (d) the mass flow rate of helium is 1.2 kg/s, and (d) the heat addition is 1000 kJ/kg in the combustion chamber.

3. Display results

(A) Display the cycle properties results. The cycle is a heat engine. The answers are Tmax=657.4°C, Turbine power=3271 kW, Compressor power=-2592 kW, back-work-ratio=79.24%, Qdot in=1200 kW, and η=56.58%.

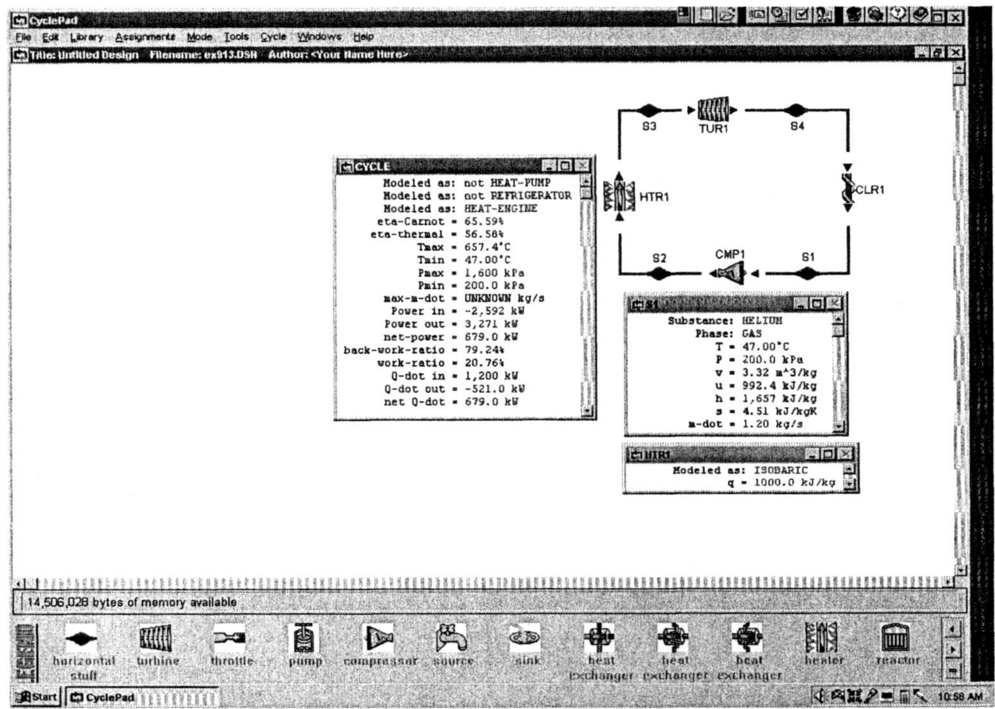

Figure Example 9.1.3. Closed Brayton cycle

Homework 9.1 Brayton Cycle

1. The maximum and minimum temperatures and pressures of a 40 MW turbine shaft output power ideal air Brayton power plant are 1200K (T_3), 0.38 MPa (P_3)and 290K (T_1), 0.095 MPa (P_1), respectively. Determine the temperature at the exit of the compressor (T_2), the temperature at the exit of the turbine (T_4), the compressor work, the turbine work, the heat added, the mass rate of flow of air, the back work ratio (the ratio of compressor work to the turbine work), and the thermal efficiency of the cycle.
2. An ideal Brayton cycle uses air as a working fluid. The air enters the compressor at 100 kPa and 37°C. The pressure ratio of the compressor is 12:1, and the temperature of the air as it leaves the turbine is 497°C. Assuming variable specific heats, determine (a) the specific work required to operate the compressor, (b) the specific work produced by the turbine, (c) the heat transfer added to the air in the combustion chamber, and (d) the thermal efficiency of the cycle.

3. An ideal Brayton engine receives 1 lbm/s of air at 15 psia and 70°F. The maximum cycle temperature is 1300°F and the compressor pressure ratio of the engine is 10. Determine (a) the power added during the compression process, (b) the rate of heat added to the air during the heating process, (c) the power done during the expansion process, and (d) the thermal efficiency of the cycle.
4. A Brayton engine receives air at 15 psia and 70°F. The air mass rate of flow is 4.08 lbm/s. The discharge pressure of the compressor is 78 psia. The maximum cycle temperature is 1740°F and the air turbine discharge temperature is 1161°F. Determine (a) the power added during the compression process, (b) the rate of heat added to the air during the heating process, (c) the power produced during the expansion process, (d) the turbine efficiency, and (e) the thermal efficiency of the cycle.
5. A Brayton engine receives air at 103 kPa and 27°C. The maximum cycle temperature is 1050°C and the compressor discharge pressure is 1120 kPa. The compressor efficiency is 85% and the turbine efficiency is 82%. Determine (a) the work added during the compression process, (b) the heat added to the air during the heating process, (c) the work done during the expansion process, and (d) the thermal efficiency of the cycle.
6. An ideal Brayton engine receives air at 15 psia and 80°F. The maximum cycle temperature is 1800°F and the compressor discharge pressure is 225 psia. The mass rate flow of air is 135 lbm/s. Determine (a) the power added during the compression process, (b) the rate of heat added to the air during the heating process, (c) the power produced during the expansion process, (d) the net power produced by the engine, (e) the back work ratio, and (f) the thermal efficiency of the cycle.
7. An ideal Brayton engine receives air at 14.7 psia and 60°F. The maximum cycle temperature is 1750°F and the compressor discharge pressure is 147 psia. The mass rate flow of air is 15 lbm/s. Determine (a) the power added during the compression process, (b) the heat added to the air during the heating process, (c) the power produced during the expansion process, (d) the net power produced by the engine, (e) the back work ratio, and (f) the thermal efficiency of the cycle.
8. A Brayton engine receives air at 14 psia and 70°F. The air mass rate of flow is 16 lbm/s. The maximum cycle temperature is 1500°F. The compressor efficiency is 79% and the turbine efficiency is 90%. Determine (a) the work done during the compression process, (b) the heat added to the air during the heating process, (c) the work done during the expansion process, (d) the turbine efficiency, and (e) the thermal efficiency of the cycle.
9. A Brayton engine receives air at 100 kPa and 25°C. The maximum cycle temperature is 1082°C and the compressor discharge pressure is 1300 kPa. The compressor efficiency is 87%. The air temperature at the turbine exit is 536°C. Determine (a) the work done during the compression process, (b) the heat added to the air during the heating process, (c) the work done during the expansion process, (d) the turbine efficiency, and (e) the thermal efficiency of the cycle.
10. An ideal air Brayton cycle has air enters the compressor at a temperature of 310 K and a pressure of 100 kPa. The pressure ratio across the compressor is 12, and the temperature of the air leaves the turbine at 780 K. Assume variable specific heats, determine (a) the compressor work required, (b) the turbine work produced, (c) the heat added during the combustion, and (d) the thermal efficiency of the cycle.

11. Air enters the compressor of an ideal Brayton cycle at 100 kPa and 300 K with a volumetric flow rate of 5 m^3/s. The compressor pressure ratio is 10. The turbine inlet temperature is 1400 K. Determine (a) the thermal efficiency of the cycle, (b) the back work ratio, and (c) the net power developed.
12. Air enters the compressor of a simple gas turbine at 14.7 psia and 520 R, and a volumetric flow rate of 1,000 ft^3/s. The compressor discharge pressure is 260 psia. The turbine inlet temperature is 2000 F. The turbine efficiency is 87% and the compressor efficiency is 83%. Determine (a) the thermal efficiency of the cycle, (b) the back work ratio, and (c) the net power developed.

9.2 Split-Shaft Gas Turbine Cycle

Gas turbines can be arranged either in single-shaft or split shaft types. The single-shaft arrangement requires the turbine to provide power to drive both the compressor and the load. This means that the compressor is influenced by the load. The compressor efficiency is a function of the speed. When the load is increased, the compressor speed is slowed down which is not desirable. It is very desirable to make the gas turbine a reliable shaft driven propulsion system, therefore the compressor speed must be held constant. Hence the *split-shaft gas turbines* are developed. In this arrangement, there are two turbines each with its own independent shaft. The sole function of the first turbine is to drive the compressor at a steady speed without being influenced by the load. The net power of the gas turbine is produced by the second turbine as shown by Figure 9.2.1.

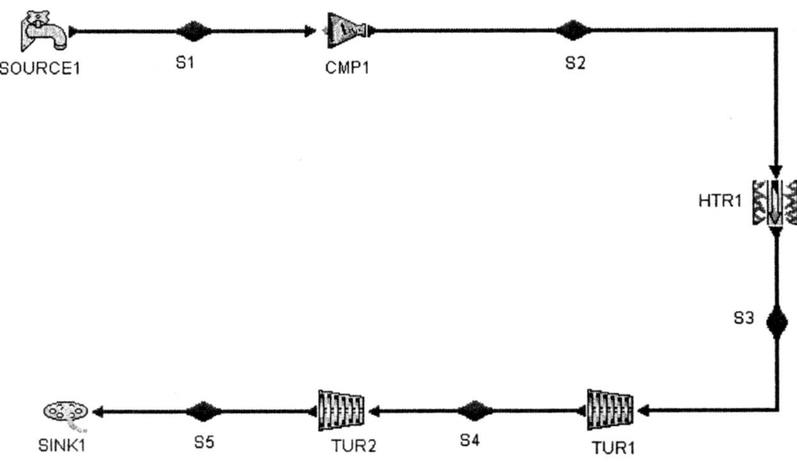

Figure 9.2.1. Split-shaft gas turbine

Example 9.2.1. An engine operates on the split-shaft actual open Brayton cycle and has a compression ratio of 8. The air enters the engine at 27°C and 100 kPa. The mass flow rate of air is 0.1 kg/s. The amount of heat addition is 1000 kJ/kg. The compressor efficiency is 86% and the efficiency is 89% for both turbines. Determine the highest temperature of the cycle, the turbine power produced, the compressor power required, the back work ratio, the rate of heat added, the pressure and temperature between the two turbines, and the cycle efficiency.

To solve this problem by CyclePad, we take the following steps:

1. Build
   (A) Take a source, a compressor, a combustion chamber (heater), two turbines and a sink from the open system inventory shop and connect the five devices to form the open actual Brayton cycle.
   (B) Switch to analysis mode.
2. Analysis
   (A) Assume a process for each of the three processes: (a) compressor device as adiabatic and η=86%, (b) combustion chamber as isobaric, and (c) turbines as adiabatic and η=89%.
   (B) Input the given information: (a) working fluid is air, (b) the inlet pressure and temperature of the compression device are 100 kPa and 27°C, (c) the compressor exit pressure is 800 kPa, (d) the heat addition is 1000 kJ/kg in the combustion chamber, (e) mass flow rate of air is 0.1 kg/s, (f) Display the compressor and find the power required to run the compressor, the finding is -28.42 kW, (g) input the first turbine power (which is used to operate the compressor) 28.42 kW, and (h) the exit pressure of the turbine is 100 kPa.
3. Display results
   (A) Display the cycle properties results. The cycle is a heat engine. The answers are Tmax=1307°C, First turbine power=28.42 kW, Second turbine power=35.75 kW, Compressor power=-28.42 kW, back-work-ratio=44.28%, Qdot in=100 kW, the pressure and temperature between the two turbines are 364.1 kPa and 1024°C, and η=35.75%.

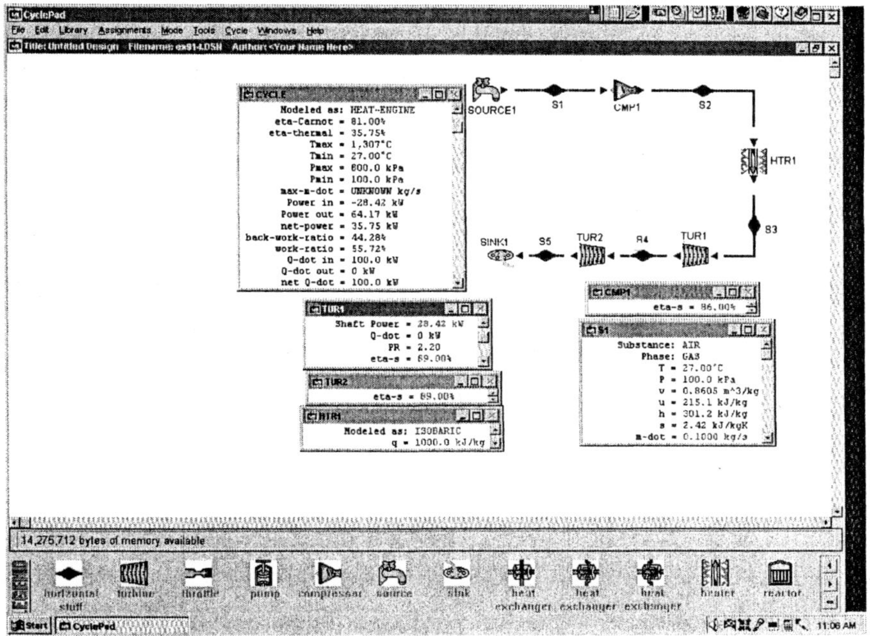

Figure Example 9.2.1 Split-shaft open Brayton cycle

Homework 9.2 Split-Shaft Gas Turbine Cycle

1. Why do we need a split shaft gas turbine engine?
2. What is the function of each turbine in a split shaft gas turbine engine?
3. An ideal split shaft Brayton cycle receives air at 14.7 psia and 70°F. The upper pressure and temperature limits of the cycle are 60 psia and 1500°F, respectively. Find the temperature and pressure of all states of the cycle. Calculate the input compressor work, the output power turbine work, heat supplied in the combustion chamber, and the thermal efficiency of the cycle, based on variable specific heats.
4. An actual split shaft Brayton cycle receives air at 14.7 psia and 70°F. The upper pressure and temperature limits of the cycle are 60 psia and 1500°F, respectively. The turbine efficiency is 85% for both turbines. The compressor efficiency is 80%. Find the temperature and pressure of all states of the cycle. The mass flow rate of air is 1 lbm/s. Calculate the input compressor power, the output power turbine power, rate of heat supplied in the combustion chamber, and the thermal efficiency of the cycle, based on variable specific heats.
   ANSWER: p_2=60 psia, T_2=397.5°F, p_3=60 psia, T_3=1500°F, p_5=14.7 psia, T_5=940.6°F, $Wdot_c$=-111.0 hp, $Wdot_{t1}$=111.0 hp, $Wdot_{t2}$=78.7 hp, $Qdot_s$=264.3 Btu/s, η=21.04%.
5. The following are operating characteristics of a split-shaft gas turbine:
   Atmospheric conditions– p=14.7 psia, T=60°F
   Compressor– inlet pressure=14.5 psia, inlet temperature=60°F, mdot=1 lbm/s, η=0.8, exit pressure=101.5 psia
   Combustion chamber-- exit temperature=1800°F, exit pressure=99 psia
   Turbine #1--η=0.85
   Power Turbine #2--η=0.85, exit pressure=14.9 psia
   Find the temperatures of all states of the cycle, power required by the compressor, power produced by the Turbine #1, power produced by the Power Turbine #2, rate of heat transfer supplied in the combustion chamber, cycle efficiency, and IHP.
   ANSWER: $T_{compressor\ exit}$=543.1°F, $T_{turbine\ \#1\ exit}$=1317°F, $T_{power\ turbine\ \#2\ exit}$=981.3°F, $Power_{compressor}$=-163.8 hp, $Power_{turbine\ \#1}$=163.8 hp, $Power_{power\ turbine\ \#2}$=-113.8 hp, $Qdot_{combustion\ chamber}$=301.2 Btu/s, η=26.71%, IHP=113.8 hp.
6. A split shaft Brayton gas turbine operates with the following information:
   inlet compressor temperature=70°F=T_1
   inlet compressor pressure=14.5 psia=p_1
   inlet combustion chamber pressure=145 psia=p_2
   inlet ggt (gas generator turbine) temperature=2000°F=T_3
   exit pt (power turbine) pressure=14.8 psia=p_{5a}
   air mass rate of flow=2 lbm/s
   power turbine efficiency=85%
   gas generator turbine efficiency=100%
   compressor efficiency=100%
   Draw the T-s diagram of the cycle.
   Find the temperature at the exit of the compressor (T_2), the temperature at the exit of the gas generator turbine (T_4), the temperature at the exit of the power turbine (T_{5a}), compressor work required per unit mass, gas generator turbine work produced per unit

mass, power turbine work produced per unit mass, heat supplied per unit mass, net work produced per unit mass, total power produced by the cycle, and cycle efficiency.
ANSWER: T_2=563.0°F, T_4=1516°F, T_{4a}=1507°F, w_c=-118.1 Btu/lbm, q_s=344.0 Btu/lbm, w_{t1}=118.1 Btu/lbm, w_{t2}=143.7 Btu/lbm, w_{net}=143.7 Btu/lbm, P_{net}=287.4 Btu/s=740.9 hp, η=41.72%.

7. A split shaft Brayton gas turbine operates with the following information:
inlet compressor temperature=60°F=T_1
inlet compressor pressure=14.5 psia=p_1
inlet combustion chamber pressure=145 psia=p_2
inlet ggt (gas generator turbine) temperature=2000°F=T_3
exit pt (power turbine) pressure=14.8 psia=p_{5a}
air mass rate of flow=2 lbm/s
power turbine efficiency=85%
gas generator turbine efficiency=100%
compressor efficiency=100%
Draw the T-s diagram of the cycle.
Find the temperature at the exit of the compressor (T_2), the temperature at the exit of the gas generator turbine (T_4), the temperature at the exit of the power turbine (T_{5a}), compressor work required per unit mass, gas generator turbine work produced per unit mass, power turbine work produced per unit mass, heat supplied per unit mass, net work produced per unit mass, total power produced by the cycle, and cycle efficiency.
ANSWER: T_2=543.7°F, T_4=1516°F, T_{5a}=926°F, w_c=-115.9 Btu/lbm, q_s=349.0 Btu/lbm, w_{t1}=115.9 Btu/lbm, w_{t2}=141.5 Btu/lbm, w_{net}=141.5 Btu/lbm, P_{net}=283.0 Btu/s=400.4 hp, η=40.54%.

9.3 IMPROVEMENTS TO BRAYTON CYCLE

The thermal efficiency or net work of the Brayton cycle can be improved by several modifications to the basic cycle. These modifications include increasing the turbine inlet temperature, reheating, intercooling, regeneration, etc.

Increasing the turbine inlet temperature increases the thermal efficiency of the Brayton cycle. It is limited by the metallurgical material problem in the turbine blade.

Increasing the average temperature during the heat addition process with a reheater without increasing the compressor pressure ratio increases the net work of the Brayton cycle. A multi-stage turbine is used. Gas is reheated between stages.

Using an intercooler without increasing the compressor pressure ratio increases the net work of the Brayton cycle. A multi-stage compressor is used. Gas is cooled between stages.

Increasing the average temperature during the heat addition process can also be done by regenerating the gas. A multi-stage turbine is used. The exhaust gas is used to preheat the air before it is heated in the combustion chamber. In this way, the amount of heat added at the low temperature is reduced. So the average temperature during the heat addition process is increased.

9.4 Reheat and Inter-Cool Brayton Cycle

Methods for improving the gas turbine cycle performance are available.

Two ways to improve the cycle net work are to reduce the compressor work and to increase the turbine work. The inter-cool may be accomplished by compressing in stages with an inter-cooler, cooling the air as it passes from one stage to another. Similarly, the reheat may be accomplished by the expansion in stages with a reheater. Since there is more than sufficient air for combustion, some more can be injected. The reheated products of combustion return to the turbine. The products of combustion reentering the turbine are usually at the same temperature as those entering the turbine. The schematic diagram of a reheat open Brayton cycle is illustrated in Figure 9.4.1. The schematic diagram and T-s diagram for a *reheat and inter-cool gas turbine* cycle is illustrated in Figure 9.4.2 and Figure 9.4.3, respectively.

Notice that reheat and inter-cool increase the net work of the gas turbine cycle, but not necessarily the efficiency, unless a regenerator is also added.

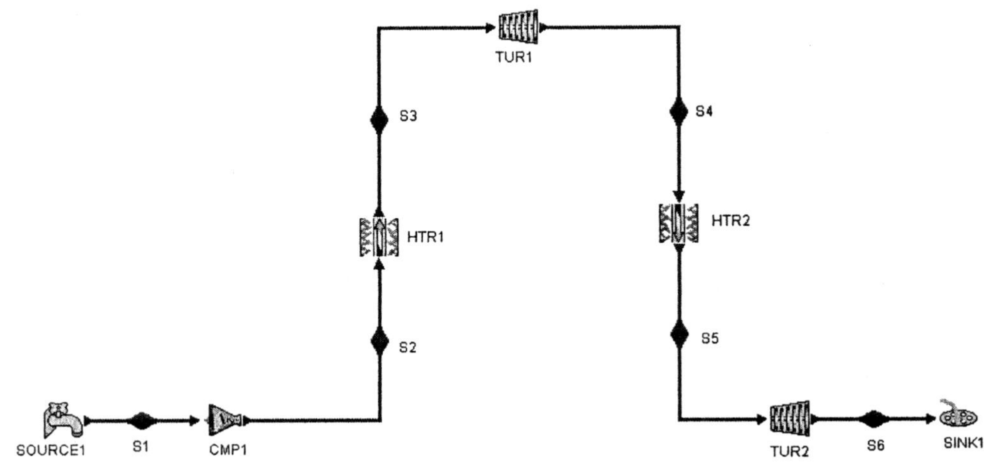

Figure 9.4.1. Reheat Brayton cycle

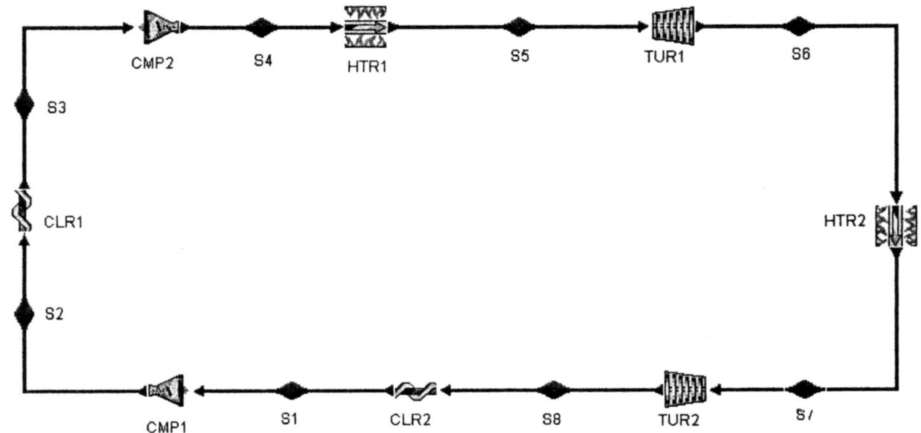

Figure 9.4.2. Reheat and inter-cool Brayton cycle

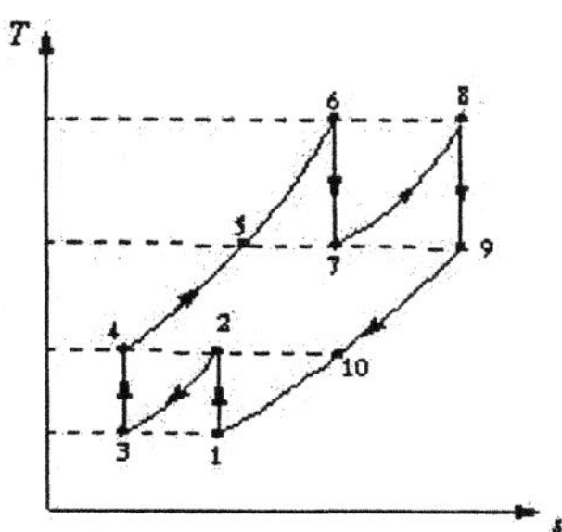

Figure 9.4.3 Reheat and inter-cool Brayton cycle T-s diagram

Example 9.4.1. An engine operates on an ideal reheat and inter-cooling Brayton cycle. The low-pressure compressor has a compression ratio of 2, and the high-pressure compressor has a compression ratio of 4. The air enters the engine at 27°C and 100 kPa. The air is cooled to 27°C at the inlet of the high-pressure compressor. The heat added in the combustion chamber is 1000 kJ/kg and air is heated to the maximum temperature of the cycle. The mass flow rate of air is 0.1 kg/s. Air expands to 200 kPa through the first turbine. Air is heated again by the reheater to the maximum temperature of the cycle and then expanded through the second turbine to 100 kPa. Determine the power required for the first compressor, the power required for the second compressor, the maximum temperature of the cycle (at the exit of the combustion chamber), the power produced by the first turbine, the rate of heat added in the reheater, the power produced by the second turbine, the net power produced, back work ratio, and the efficiency of the cycle. Show the cycle on T-s diagram.

To solve this problem by CyclePad, we take the following steps:

1. Build
   (A) Take two compressors, two coolers (one is the inter-cooler), two combustion chambers (one is the reheater), and two turbines from the open system inventory shop and connect the devices to form the reheat and inter-cooling Brayton cycle.
   (B) Switch to analysis mode.
2. Analysis
   (A) Assume a process for each of the seven processes: (a) compressors as adiabatic and compressor efficiency is 86%, (b) combustion chamber as isobaric, (c) turbines as adiabatic, and both turbine efficiency is 89%, (d) inter-cooler as isobaric, (e) reheater as isobaric.
   (B) Input the given information: (a) working fluid is air, (b) the inlet temperature and pressure of the compressor are 27°C and 100 kPa, (c) the inlet temperature and pressure of the high-pressure compressor are 27°C and 200 kPa, (d) the heat added in the combustion chamber is 1000 kJ/kg, (e) the inlet pressure of the reheater is 200 kPa, (f) display the exit temperature of the combustion chamber (maximum temperature of cycle), (g) input the exit temperature of the reheater (same as the exit temperature of the combustion chamber as found in part f), (h) the mass flow rate of air is 0.1 kg/s, and (i) the exit pressure of the low-pressure turbine is 100 kPa.
3. Display results

(A) Display the T-s diagram and cycle properties results. The cycle is a heat engine. The answers are power required for the first compressor=-6.60 kW, the power required for the second compressor=-14.64 kW, the maximum temperature of the cycle =1169°C, the power produced by the first turbine=47.34 kW, the rate of heat added in the reheater=47.29 kW, the power produced by the second turbine=26.00 kW, the net power produced=52.11 kW, back work ratio=28.95%, and the efficiency of the cycle η=35.38%.

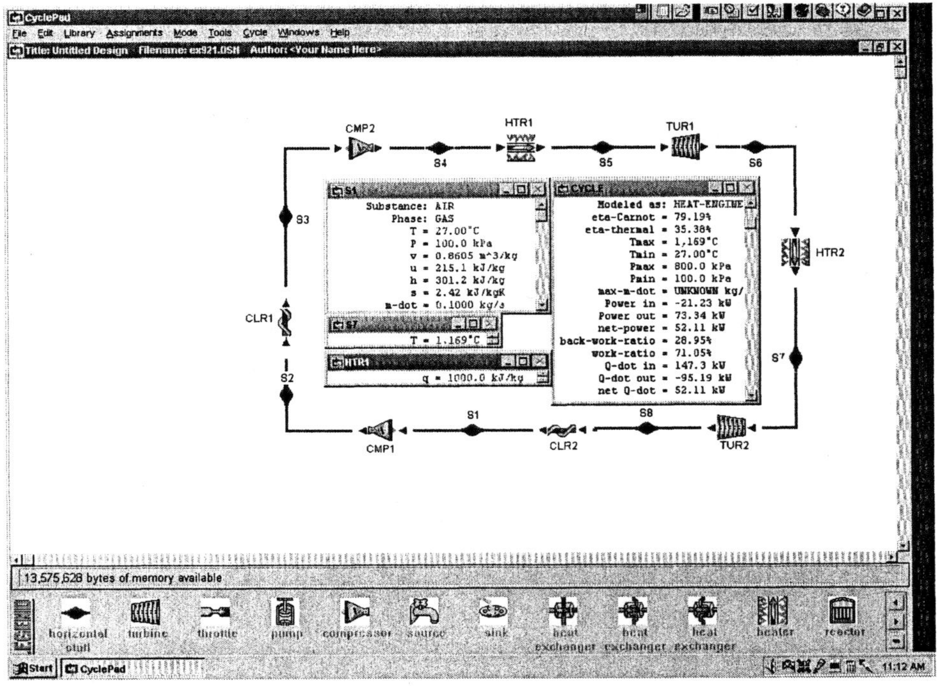

Figure Example 9.4.1. Ideal reheat and inter-cool Brayton cycle

Comment: Comparing with Example 9.1.1, we see that: (1) the efficiency of the reheat and intercooler cycle does not increase, and (2) the net power of the reheat and intercooler cycle does increase.

Example 9.4.2. An engine operates on an actual reheat open Brayton cycle. The air enters the compressor at 60°F and 14.7 psia, and leaves at 120 psia. The maximum cycle temperature (at the exit of the combustion chamber) allowed due to material limitation is 2000°F. The exit pressure of the high-pressure turbine is 50 psia. The air is reheated to 2000°F. The mass flow rate of air is 1 lbm/s. The exit pressure of the low-pressure turbine is 14.7 psia. The compressor efficiency is 86% and the turbine efficiency is 89%. Determine the power required for the compressor, the power produced by the first turbine, the rate of heat added in the reheater, the power produced by the second turbine, the net power produced, back work ratio, and the efficiency of the cycle. Show the cycle on T-s diagram. Plot the sensitivity diagram of cycle efficiency vs inlet pressure of the low-pressure turbine.

To solve this problem by CyclePad, we take the following steps:

1. Build
   (A) Take a source, a compressors, a combustion chamber (heater), a reheater, two turbines and a sink from the open system inventory shop and connect the nine devices to form the actual reheat open Brayton cycle.
   (B) Switch to analysis mode.
2. Analysis
   (A) Assume a process for each of the seven processes: (a) compressor as adiabatic with 85% efficiency, (b) combustion chamber as isobaric, (c) turbines as adiabatic with 89% efficiency, and (d) reheater as isobaric.
   (B) Input the given information: (a) working fluid is air, (b) the inlet pressure and temperature of the compression device are 15 psia and 60°F, (c) the exit pressure of the compressor is 120 psia, (d) the exit temperature of the combustion chamber is 2000°F, (e) the exit pressure of the high-pressure turbine is 50 psia, (f) the inlet temperature of the low-pressure turbine is 2000°F, (g) the mass flow rate of air is 1 lbm/s, and (h) the exit pressure of the turbine is 15 psia.
3. Display results
   (A) Display the cycle properties results. The cycle is a heat engine. The answers are power required for the compressor=-170.4 hp, the power produced by the first turbine=164.3 hp, the rate of heat added in the combustion chamber=334.5 Btu/s, the rate of heat added in the reheater=116.1 Btu/s, the power produced by the second turbine=219.1 hp, the net power produced=213.0 hp, back work ratio=44.45%, and the efficiency of the cycle η=32.68% and
   (B) Display the sensitivity diagram of cycle efficiency vs inlet pressure of the low-pressure turbine, it is seen that the optimal pressure is about 77 psia which gives the maximum cycle efficiency of about 32%.

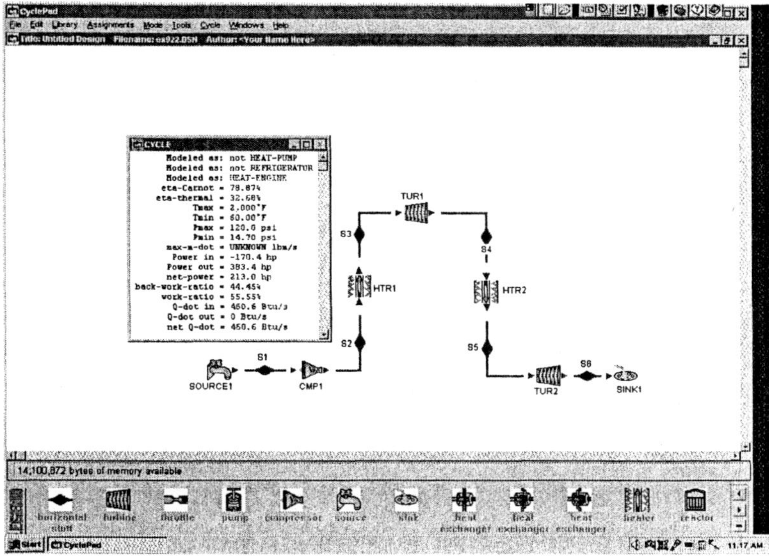

Figure Example 9.4.2a Actual reheat open Brayton cycle

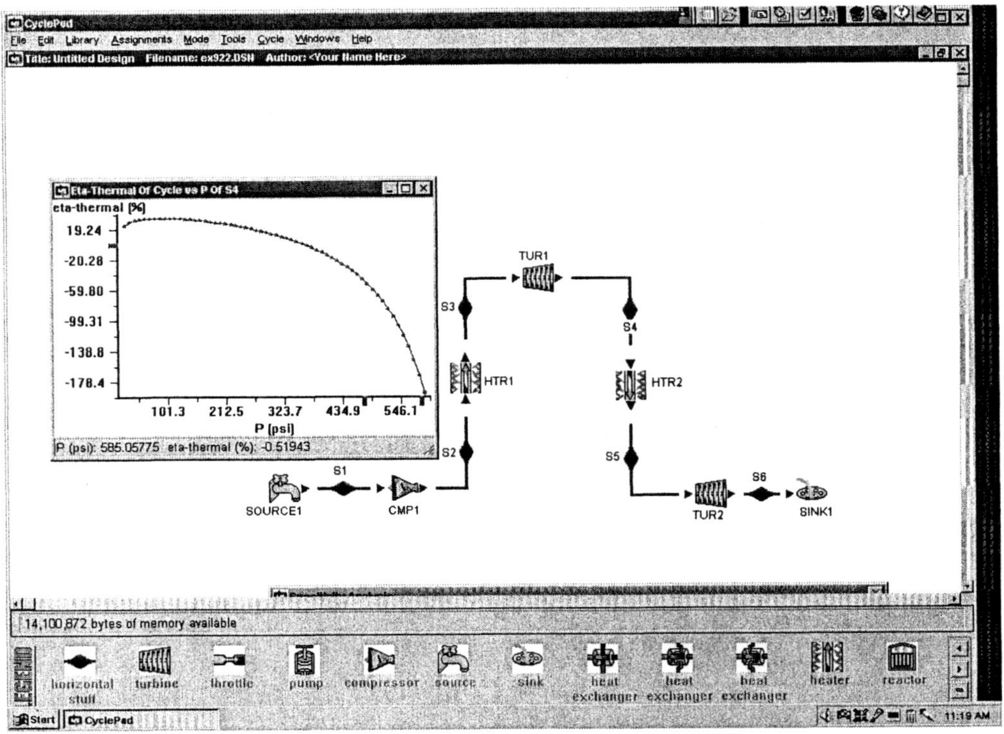

Figure 9.4.2b. Actual reheat open Brayton cycle sensitivity analysis

Homework 9.4 Reheat and Inter-Cool Brayton Cycle

1. Does reheat or inter-cooling increase the net work of the Brayton cycle?
2. Does reheat or inter-cooling increase the efficiency of the Brayton cycle?
3. An ideal Brayton cycle is modified to incorporate multi-stage compression with intercooling, and multi-stage expansion with reheating. As a result of these modifications, does the efficiency increase?
4. An ideal Brayton cycle is modified to incorporate multi-stage compression with intercooling, and multi-stage expansion with reheating. As a result of these modifications, does the net work increase?
5. Atmospheric air is at 100 kPa and 300 K. 1 kg/s of air at 800 kPa and 1200 K enters an actual two-stage (high-pressure stage and low-pressure stage) adiabatic turbine at steady state and exits to 100 kPa. Air is reheated to 1200 K and enters the low-pressure turbine. Air pressure at 300.1 kPa is measured at the exit of the high-pressure stage turbine. The high-pressure stage turbine, low-pressure stage turbine, and the compressor all are known to have an isentropic efficiency of 85%. Determine (a) the actual temperature at the exit of the high-pressure stage turbine, (b) the actual temperature at the exit of the low-pressure stage turbine, (c) the cycle efficiency, (d) net power produced by the cycle, (e) power required by the compressor, (f) power produced by the high-pressure turbine, (g) power produced by the low-pressure turbine, (h) the back work ratio, (i) the rate of heat added in the combustion chamber, and (j) the rate of heat added in the reheater.

6. An ideal Brayton cycle with a one-stage compressor and a two-stage turbine has an overall pressure ratio of 10. Atmospheric air is at 101 kPa and 292 K. The high-pressure stage turbine, low-pressure stage turbine, and the compressor all are known to have an isentropic efficiency of 85%. Air enters each stage of the turbine at 1350 K. The mass rate of air flow is 0.56 kg/s. The air pressure at the inlet of the second stage turbine is 307 kPa. Determine (a) the power required by the compressor, (b) power produced by the turbine, (c) rate of heat added, (d) back work ratio, (e) net power produced, and (f) the cycle efficiency.
7. An ideal Brayton cycle with a one-stage compressor and a two-stage turbine has an overall pressure ratio of 10. Atmospheric air is at 101 kPa and 292 K. The high-pressure stage turbine, low-pressure stage turbine, and the compressor all are known to have an isentropic efficiency of 85%. Air enters each stage of the turbine at 1350 K. The mass rate of air flow is 0.56 kg/s. Use air pressure at the inlet of the second stage turbine as a parameter (from 400 to 800 kPa), determine (a) the maximum cycle efficiency by showing the sensitivity diagram, (b) the optimum air pressure at the inlet of the second stage turbine at the maximum cycle efficiency condition, (c) power required by the compressor, (d) power produced by the turbine, (e) rate of heat added, (f) back work ratio, and (g) net power produced.
8. An ideal Brayton cycle with a one-stage compressor and a two-stage turbine has an overall pressure ratio of 10. Atmospheric air is at 14.6 psia and 65 F. The high-pressure stage turbine, low-pressure stage turbine, and the compressor all are known to have an isentropic efficiency of 85%. Air enters each stage of the turbine at 2000 F. The mass rate of air flow is 1.5 lbm/s. The air pressure at the inlet of the second stage turbine is 50 psia. Determine (a) the power required by the compressor, (b) power produced by the turbine, (c) rate of heat added, (d) back work ratio, (e) net power produced, and (f) the cycle efficiency.

9.5 REGENERATIVE BRAYTON CYCLE

The thermal efficiency of the gas turbine cycle is not high. It is observed that the exhaust temperature of the turbine is quite high, indicating that a large portion of available energy is wasted. One way to put this high-temperature available energy to use is to preheat the combustion air before it enters the combustion chamber. This increases the overall efficiency by decreasing the fuel required, hence heat added. The schematic diagram and T-s diagram for an ideal *regenerative gas turbine cycle* is illustrated in Figure 9.5.1 and Figure 9.5.2, respectively.

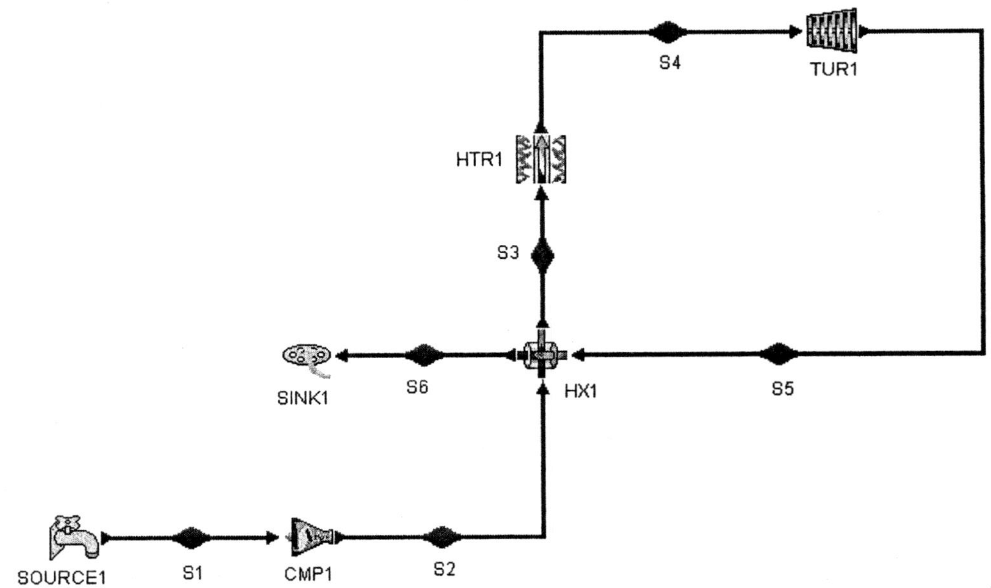

Figure 9.5.1. Regenerative Brayton cycle

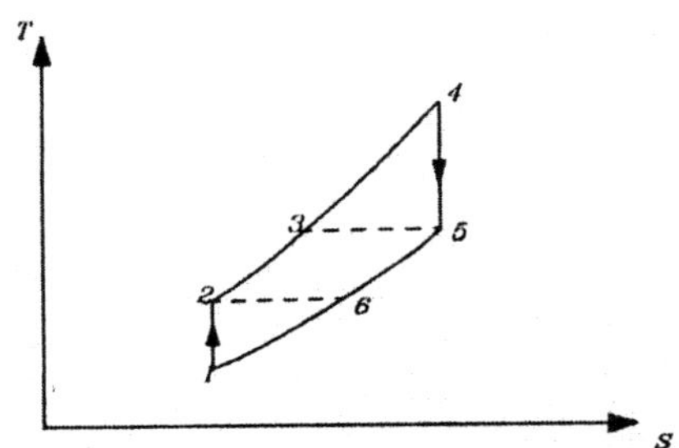

Figure 9.5.2. Regenerative Brayton cycle T-s diagram

Example 9.5.1. An engine operates on the actual regenerative Brayton cycle. Air enters the engine at 60°F and 14.7 psia. The maximum cycle temperature and the maximum pressure are 2000°F and 120 psia. The compressor efficiency is 85% and the turbine efficiency is 89%. The mass flow rate of air is 1 lbm/s. Determine the power required for the compressor, the power produced by the turbine, the rate of heat added in the combustion chamber, the net power produced, back work ratio, and the efficiency of the cycle. Show the cycle on T-s diagram. Plot the sensitivity diagram of cycle efficiency vs inlet pressure of the low-pressure turbine.

Show the cycle on T-s diagram. Plot the sensitivity diagram of cycle efficiency vs exit temperature of the turbine exhaust stream in the heat exchanger.

Convert the British unit system to SI system and find the answer.

To solve this problem by CyclePad, we take the following steps:

1. Build
   (A) Take a source, a compressor, a combustion chamber (heater), a heat exchanger, a turbine and a sink from the open system inventory shop and connect the six devices to form the actual regenerative Brayton cycle.
   (B) Switch to analysis mode.
2. Analysis
   (A) Assume a process for each of the four devices: (a) compressor as adiabatic with efficiency of 85%, (b) combustion chamber as isobaric, (c) turbine as adiabatic with efficiency of 89%, and (d) heat exchanger as isobaric on both hot and cold sides.
   (B) Input the given information: (a) working fluid is air, (b) the inlet pressure and temperature of the compression device are 14.7 psia and 60°F, (c) the inlet pressure and temperature of the turbine are 120 psia and 2000°F, (d) the mass flow rate of air is 1 lbm/s, (e) the exit pressure of the turbine is 14.7 psia, (f) display the exit temperature of the compressor, it is 562.5°F, and (g) input the exit temperature of the exhaust turbine gas be the same as the compressor exit temperature, 562.5°F, by assuming perfect regeneration.
3. Display results
   (A) Display the T-s diagram and cycle properties results. The cycle is a heat engine. The answers are power required for the compressor=-170.4 hp, the power produced by the turbine=334.9 hp, the rate of heat added in the combustion chamber=236.7 Btu/s, the net power produced=164.5 hp, back work ratio=50.88%, and the efficiency of the cycle η=49.12%.

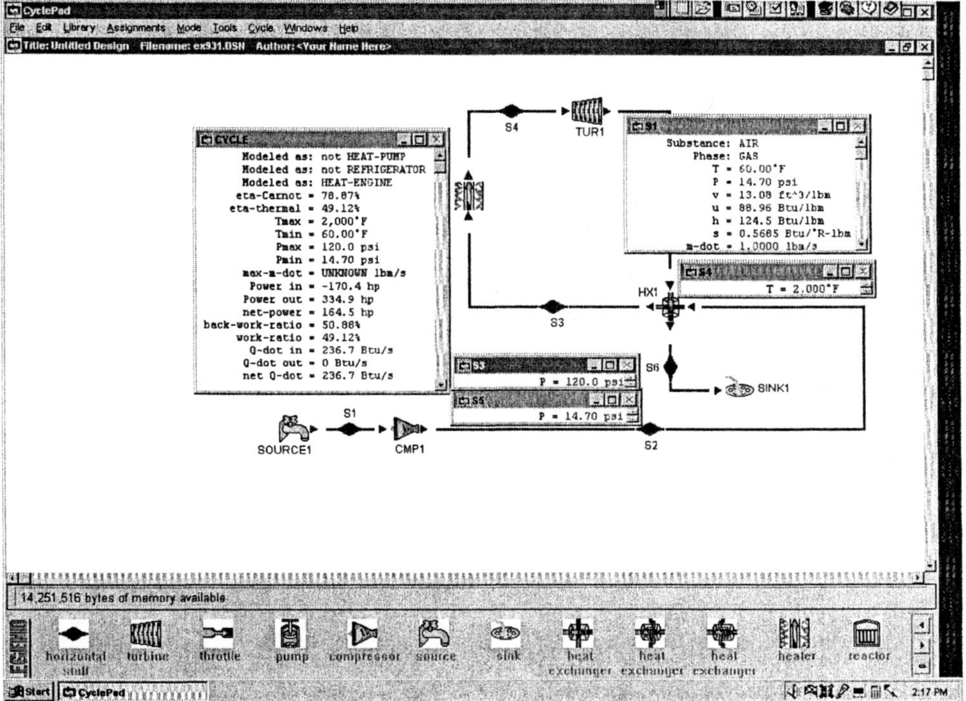

Figure Example 9.5.1a. Actual regenerative Brayton cycle

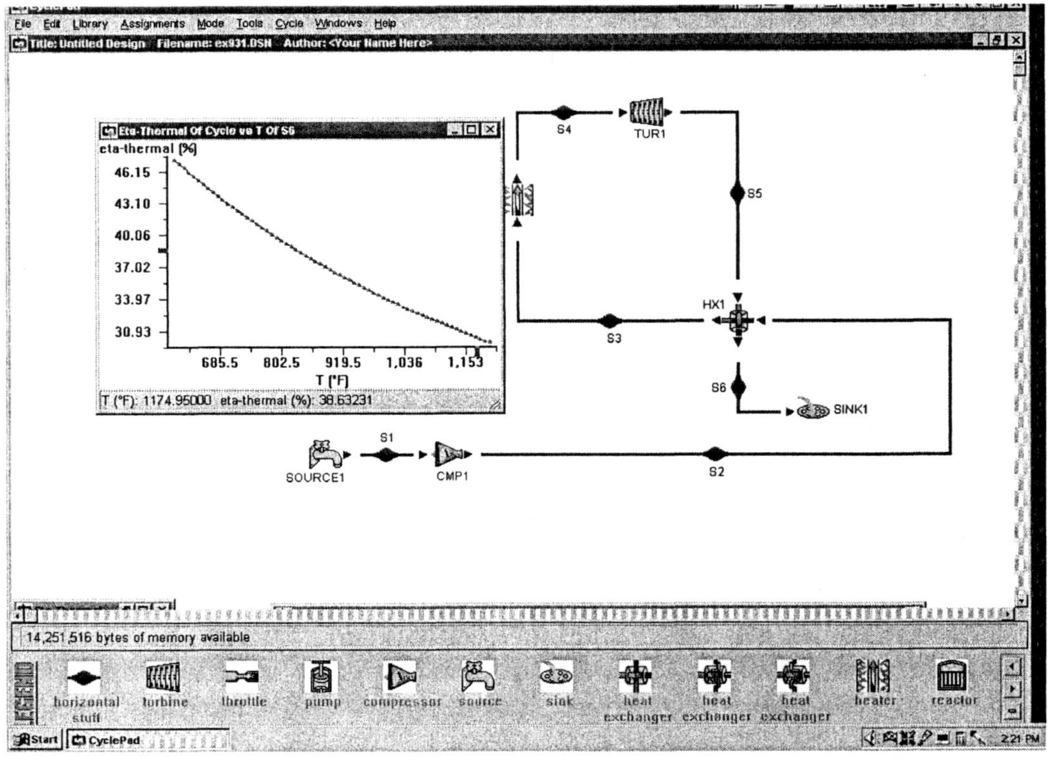

Figure Example 9.5.1b Actual Regenerative Brayton cycle sensitive analysis

Comment: The cycle efficiency is increased because the rate of heat added is decreased by making use of the waste energy. The higher the exit temperature of the heat exchanger in the exhaust turbine gas stream, the less waste energy is used, and therefore the less the efficiency as shown by the sensitivity diagram.

Homework 9.5 Regenerative Brayton Cycle

1. What is regeneration?
2. Why is regeneration added to a Brayton cycle?
3. How does regeneration affect the efficiency of the Brayton cycle?
4. An ideal Brayton cycle is modified to incorporate multi-stage compression with intercooling, multi-stage expansion with reheating, and regeneration. As a result of these modifications, does the efficiency increase?
5. An ideal Brayton cycle with regeneration has a pressure ratio of 10. Air enters the compressor at 101 kPa and 290 K. Air leaves the regenerator and enters the combustion chamber at 580 K. Air enters the turbine at 1220 K. The air mass flow rate is 0.46 kg/s. Determine (a) the power required by the compressor, (b) power produced by the turbine, (c) rate of heat added, (d) back work ratio, (e) net power produced, and (f) the cycle efficiency.
6. An ideal Brayton cycle with regeneration has a pressure ratio of 10. Air enters the compressor at 14.7 psia and 29°F. Air enters the combustion chamber at 610°F. Air

enters the turbine at 1520°F. The turbine exit air pressure is 15.0 psia. The air mass flow rate is 0.41 lbm/s. The turbine efficiency is 85%, and the compressor efficiency is 82%. Determine (a) the exit air temperature of the compressor, (b) the inlet air temperature of the combustion chamber, (c) the power required by the compressor, (d) power produced by the turbine, (e) rate of heat added, (f) back work ratio, (g) net power produced, and (h) the cycle efficiency.

7. A regenerative gas-turbine power plant is to be designed according to the following specifications:
   Maximum cycle temperature=1200 K
   Turbine efficiency=85%
   Compressor efficiency=82%
   Inlet air temperature of the combustion chamber is 20 K higher than the exit air temperature of the compressor.
   Inlet air pressure to the compressor=100 kPa
   Turbine exit air pressure=110 kPa
   Exit air pressure of the compressor=800 kPa
   Mass flow rate of air=1.4 kg/s
   Determine (a) the exit air temperature of the compressor, (b) the inlet air temperature of the combustion chamber, (c) the power required by the compressor, (d) power produced by the turbine, (e) rate of heat added, (f) back work ratio, (g) net power produced, and (h) the cycle efficiency.

9.6 Bleed Air Brayton Cycle

Real gas turbine engines send a portion of the air supplied by the compressor through alternative flow paths to provide cooling to the outside of the engine, to protect nearby components, to be remixed with the combustion products, and to drive ancillary equipments such as air conditioning and ventilation. The rate of the bleed air can be controlled. The schematic diagram for the *bleed air Brayton cycle* is illustrated in Figure 9.6.1. It can be seen from the diagram that the engine's entire flow rate passes through the compressor, but only a fraction of the flow passes through the combustion chamber and the turbine.

Bleed air is a necessary feature of practical gas turbine engines. For example, when one enters a commercial aircraft, the cabin temperature is normally pleasant and the small vents above the seats are providing plenty of fresh air. The ventilation reduces dramatically as the aircraft prepares to take off, because the pilot has temporarily reduced the bleed air so that more air will flow through the combustion chamber to give the pilot more power available to get airborne. Bleed air does not improve the efficiency of the Brayton cycle.

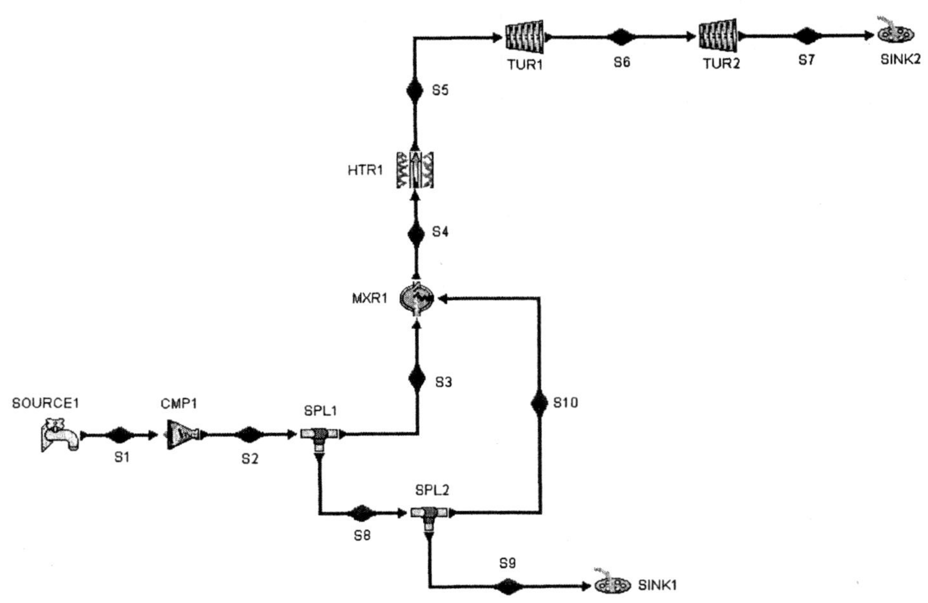

Figure 9.6.1. Bleed air Brayton cycle

Example 9.6.1. An open, split-shaft, air bleed Brayton cycle illustrated in Figure 9.5.1 has the following information:

Compressor efficiency=80%, turbine efficiency=80%, compressor inlet pressure=14.5 psia, compressor inlet temperature=60°F, compressor exit pressure=145 psia, combustion chamber exit temperature=1800°F, power turbine exit pressure=14.9 psia, air mass flow rate through compressor=1 lbm/s, and air mass flow rate through combustion chamber=0.9 lbm/s.

Find the temperature of all states, power required by the compressor, power produced by turbine #1 which drives the compressor, power produced by the power turbine, rate of heat supplied by the combustion chamber, and cycle efficiency.

To solve this problem by CyclePad, we take the following steps:

(A) Build the cycle as shown by Figure 9.6.1
(B) Assume compressor as adiabatic and 80% efficient, combustion chamber and mixing chamber as isobaric, and turbines as adiabatic and 80% efficient.
(C) Input working fluid is air, compressor inlet pressure=14.5 psia, compressor inlet temperature=60°F, compressor exit pressure=145 psia, combustion chamber exit temperature=1800°F, power turbine exit pressure=14.9 psia, air mass flow rate through compressor=1 lbm/s, air mass flow rate through state 9=0 lbm/s, and air mass flow rate through combustion chamber=0.9 lbm/s.
(D) Display compressor power (-205 hp), input turbine #1 power=205 hp.
(E) Display the cycle properties results and state results. The answers are T_2=664.6°F, T_3=T_4=T_8=T_9=T_{10}= 664.6°F, T_6=1128°F, T_7=913.1°F, compressor power=-205 hp, turbine #1 power=205 hp, turbine #2 power=65.66 hp, rate of heat added in the combustion chamber=244.9 Btu/s, net cycle power=65.66 hp, and cycle efficiency=18.95%.

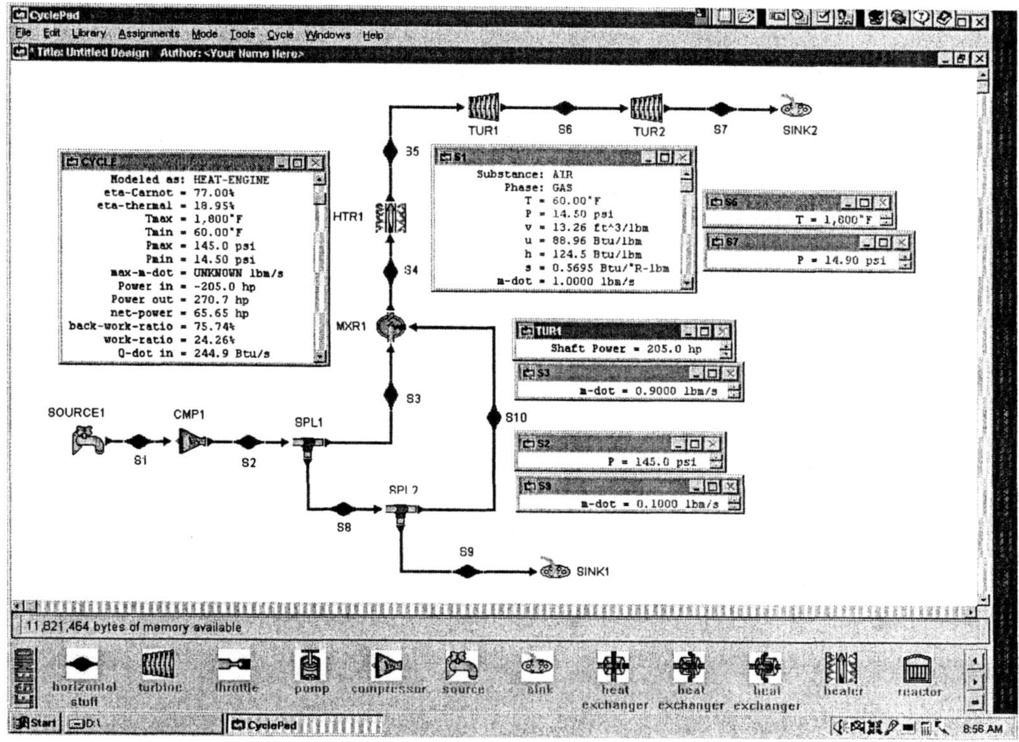

Figure Example 9.6.1. Bleed air Brayton cycle

Example 9.6.2. An open, split-shaft, air bleed Brayton cycle illustrated in Figure 9.5.1 has the following information:

Compressor efficiency=80%, turbine efficiency=80%, compressor inlet pressure=14.5 psia, compressor inlet temperature=60°F, compressor exit pressure=145 psia, combustion chamber exit temperature=1800°F, power turbine exit pressure=14.9 psia, air mass flow rate through compressor=1 lbm/s, and air mass flow rate through combustion chamber=0.9 lbm/s.

Find the temperature of all states, power required by the compressor, power produced by turbine #1 which drives the compressor, power produced by the power turbine, rate of heat supplied by the combustion chamber, and cycle efficiency.

To solve this problem by CyclePad, we take the following steps:

(A) Build the cycle as shown by Figure 9.6.1
(B) Assume compressor as adiabatic and 80% efficient, combustion chamber and mixing chamber as isobaric, and turbines as adiabatic and 80% efficient.
(C) Input working fluid is air, compressor inlet pressure=14.5 psia, compressor inlet temperature=60°F, compressor exit pressure=145 psia, combustion chamber exit temperature=1800°F, power turbine exit pressure=14.9 psia, air mass flow rate through compressor=1 lbm/s, air mass flow rate through state 9=0.05 lbm/s, and air mass flow rate through combustion chamber=0.9 lbm/s.
(D) Display compressor power (-205 hp), input turbine #1 power=205 hp.

(E) Display the cycle properties results and state results. The answers are T_2=664.6°F, T_3=T_4=T_8=T_9=T_{10}= 664.6°F, T_6=1164°F, T_7=911.1°F, compressor power=-205 hp, turbine #1 power=205 hp, turbine #2 power=81.33 hp, rate of heat added in the combustion chamber=258.5 Btu/s, net cycle power=81.33 hp, and cycle efficiency=22.24%.

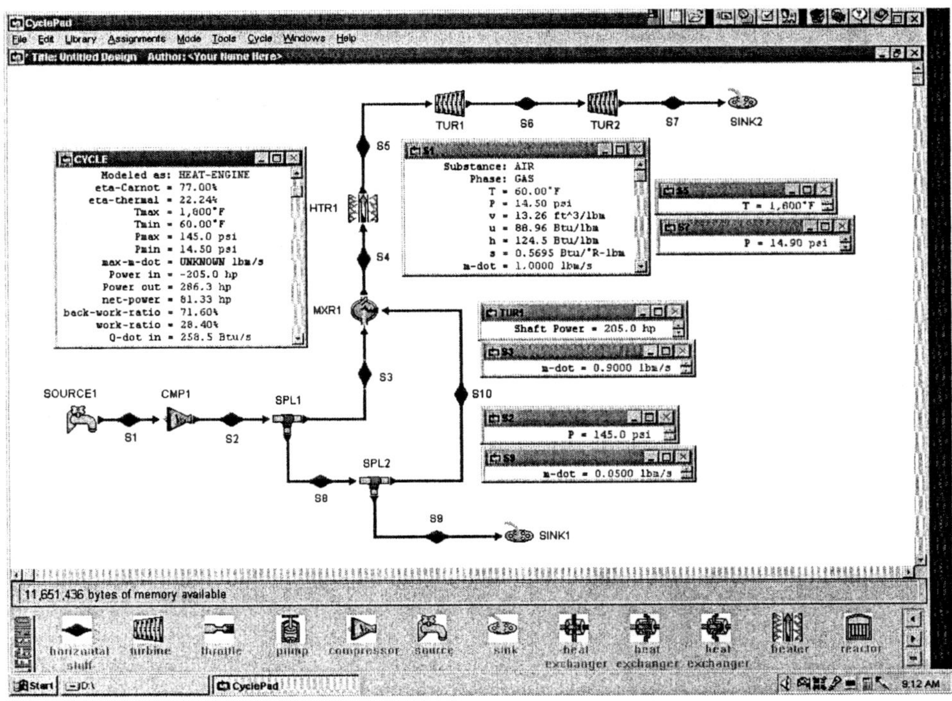

Figure Example 9.6.2. Bleed air Brayton cycle

Example 9.6.3. An open, split-shaft, air bleed Brayton cycle with two compressors (cmp1 and cmp2), three turbines (tur1, tur2, and tur3), one inter-cooler (clr1), one combustion chamber (htr1), one reheater (htr2), and one regenerator (hx1) is illustrated in Figure Example 9.5.3a has the following information:

Compressor efficiency=80%, turbine efficiency=80%, compressor inlet pressure=14.5 psia, compressor inlet temperature=60°F, compressor #1 exit pressure=40 psia, inter-cooler exit temperature=100°F, compressor #2 exit pressure=140 psia, combustion chamber exit temperature=1800°F, reheater exit temperature=1800°F, regenerator hot-side exit temperature=500°F, power turbine exit pressure=14.9 psia, air mass flow rate through compressor #2=1 lbm/s, and air mass flow rate through combustion chamber=0.9 lbm/s, air mass flow rate through state 15=0.1 lbm/s, air mass flow rate through state 17=0.01 lbm/s, and air mass flow rate through state 13=0 lbm/s (regenerator is off).

Find the temperature of all states, power required by the compressors, power produced by turbine #1 which drives the compressor #1, power produced by turbine #2 which drives the compressor #2, power produced by the power turbine #3, rate of heat supplied by the combustion chamber, rate of heat supplied by the reheater, rate of heat removed from the inter-cooler, and cycle efficiency.

To solve this problem by CyclePad, we take the following steps:

(A) Build the cycle as shown by Figure Example 9.6.3a.

(B) Assume all compressors as adiabatic and 80% efficient, combustion chamber, reheater, inter-cooler, heat exchanger (both hot- and cold-side) and mixing chamber as isobaric, and all turbines as adiabatic and 80% efficient.

(C) Input working fluid is air, compressor inlet pressure=14.5 psia, compressor inlet temperature=60°F, compressor #1 exit pressure=40 psia, inter-cooler exit temperature=100°F, compressor #2 exit pressure=140 psia, combustion chamber exit temperature=1800°F, reheater exit temperature=1800°F, regenerator hot-side exit temperature=500°F, power turbine exit pressure=14.9 psia, air mass flow rate through compressor #2=1 lbm/s, and air mass flow rate through combustion chamber=0.9 lbm/s, air mass flow rate through state 15=0.1 lbm/s, air mass flow rate through state 17=0.01 lbm/s, and air mass flow rate through state 13=0 lbm/s (regenerator is off).

(D) Display compressor power #1 (-74.08 hp), input turbine #1 power=74.08 hp; Display compressor power #2 (-102.1 hp), input turbine #2 power=102.1 hp;

(E) Display the cycle properties results and state results. The answers are T_2=278.5°F, T_3=100°F, T_4=401.1°F= T_5=T_{15}=T_{17}=401.1°F, T_7=T_{11}=1800°F, T_8=1673°F, T_9=1452°F, T_{10}=1148°F, T_{12}=1359°F, compressor #1power=-74.08 hp, turbine #1 power=74.08 hp, compressor #2 power=-102.1 hp, turbine #2 power=102.1 hp, turbine #3 power=148.1 hp, rate of heat added in the combustion chamber=301.8 Btu/s, rate of heat added in the reheater=154.7 Btu/s, rate of heat removed in the inter-cooler=-42.78 Btu/s, net cycle power=148.1 hp, and cycle efficiency=22.94%.

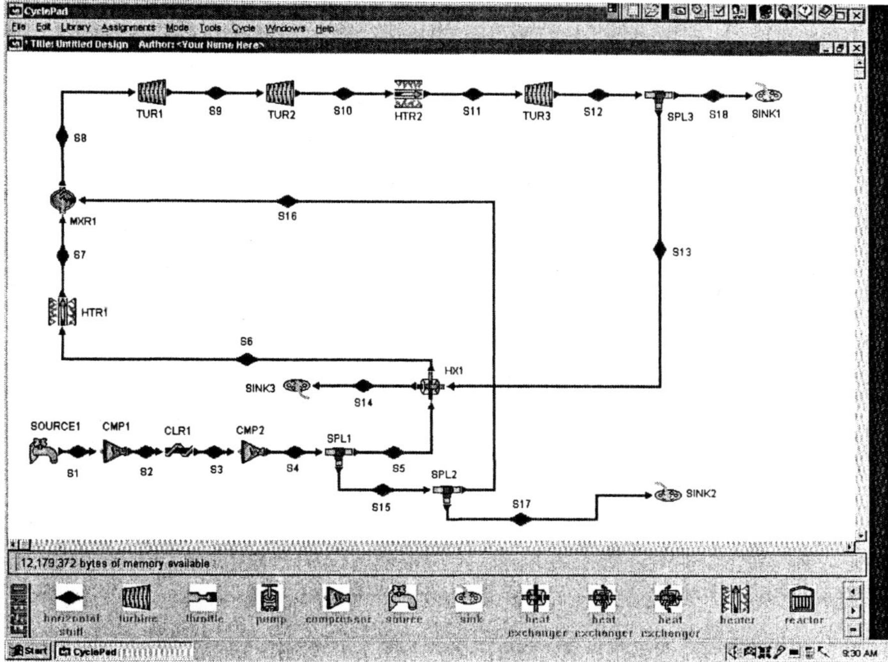

Figure Example 9.6.3a. Bleed air Brayton cycle

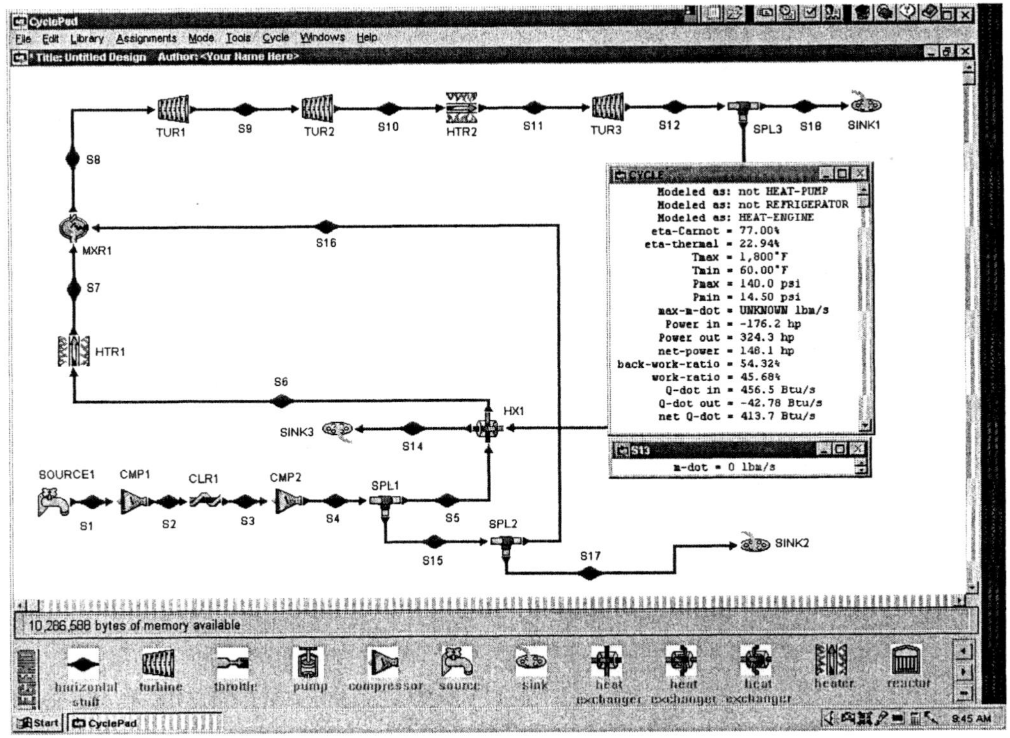

Figure Example 9.6.3b. Bleed air Brayton cycle (regenerator off)

Example 9.6.4. An open, split-shaft, air bleed Brayton cycle with two compressors (cmp1 and cmp2), three turbines (tur1, tur2, and tur3), one inter-cooler (clr1), one combustion chamber (htr1), one reheater (htr2), and one regenerator (hx1) is illustrated in Figure Example 9.5.3a has the following information:

Compressor efficiency=80%, turbine efficiency=80%, compressor inlet pressure=14.5 psia, compressor inlet temperature=60°F, compressor #1 exit pressure=40 psia, inter-cooler exit temperature=100°F, compressor #2 exit pressure=140 psia, combustion chamber exit temperature=1800°F, reheater exit temperature=1800°F, regenerator hot-side exit temperature=500°F, power turbine exit pressure=14.9 psia, air mass flow rate through compressor #2=1 lbm/s, and air mass flow rate through combustion chamber=0.9 lbm/s, air mass flow rate through state 15=0.1 lbm/s, air mass flow rate through state 17=0.01 lbm/s, and air mass flow rate through state 13=0.99 lbm/s (regenerator is on).

Find the temperature of all states, power required by the compressors, power produced by turbine #1 which drives the compressor #1, power produced by turbine #2 which drives the compressor #2, power produced by the power turbine #3, rate of heat supplied by the combustion chamber, rate of heat supplied by the reheater, rate of heat removed from the inter-cooler, and cycle efficiency.

To solve this problem by CyclePad, we take the following steps:

(A) Build the cycle as shown by Figure Example 9.6.3a.

(B) Assume all compressors as adiabatic and 80% efficient, combustion chamber, reheater, inter-cooler, heat exchanger (both hot- and cold-side) and mixing chamber as isobaric, and all turbines as adiabatic and 80% efficient.

(C) Input working fluid is air, compressor inlet pressure=14.5 psia, compressor inlet temperature=60°F, compressor #1 exit pressure=40 psia, inter-cooler exit temperature=100°F, compressor #2 exit pressure=140 psia, combustion chamber exit temperature=1800°F, reheater exit temperature=1800°F, regenerator hot-side exit temperature=500°F, power turbine exit pressure=14.9 psia, air mass flow rate through compressor #2=1 lbm/s, and air mass flow rate through combustion chamber=0.9 lbm/s, air mass flow rate through state 15=0.1 lbm/s, air mass flow rate through state 17=0.01 lbm/s, and air mass flow rate through state 13=0.99 lbm/s (regenerator is on).

(D) Display compressor power #1 (-74.08 hp), input turbine #1 power=74.08 hp; Display compressor power #2 (-102.1 hp), input turbine #2 power=102.1 hp;

(E) Display the cycle properties results and state results. The answers are T_2=278.5°F, T_3=100°F, T_4=401.1°F= T_5=T_{15}=T_{17}=401.1°F, T_7=T_{11}=1800°F, T_8=1673°F, T_9=1452°F, T_{10}=1148°F, T_{12}=1359°F, compressor #1power=-74.08 hp, turbine #1 power=74.08 hp, compressor #2 power=-102.1 hp, turbine #2 power=102.1 hp, turbine #3 power=148.1 hp, rate of heat added in the combustion chamber=301.8 Btu/s, rate of heat added in the reheater=154.7 Btu/s, rate of heat removed in the inter-cooler=-42.78 Btu/s, net cycle power=148.1 hp, and cycle efficiency=22.94%.

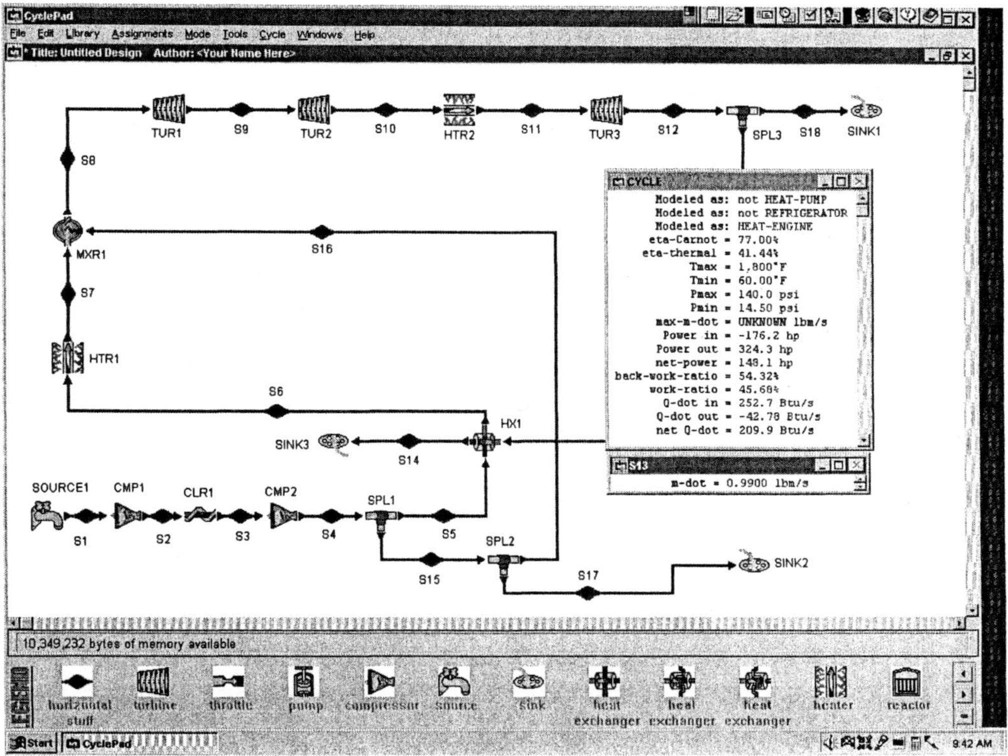

Figure Example 9.6.4. Bleed air Brayton cycle (regenerator on)

Homework 9.6 Bleed Air Brayton Cycle

1. Why do we need bleed air in practical gas turbine cycles?
2. How do you control the power output of a real gas turbine cycle using bleed air?
3. Does the efficiency of a real gas turbine cycle increase using bleed air?
4. Bleed air is used in a split-shaft gas turbine cycle. The following information is provided:
   Compressor–air inlet temperature and pressure are 60°F and14.5 psia, efficiency=85%, exit pressure is 145 psia, and mdot=1 lbm/s
   Combustion chamber–q_{supply}=400 Btu/lbm, and mdot=1 lbm/s (no bleed air)
   Turbine #1 – efficiency=92.3%
   Turbine #2 – efficiency=92.3%, air exit pressure=15 psia
   Determine temperature of all states of the cycle, compressor power required, power produced by turbine #1, power produced by turbine #2, and cycle efficiency.
   ANSWER: T_2=629°F, T_3=629°F, T_4=1881°F, T_5=1881°F, T_6=1312°F, T_7=837.6°F, $Wdot_{compressor}$=-192.9 hp, $Wdot_{turbine \#1}$=192.9 hp, $Wdot_{turbine \#2}$=160.8 hp, η=37.88%.
5. Bleed air is used in a split-shaft gas turbine cycle. The following information is provided:
   Compressor–air inlet temperature and pressure are 60°F and14.5 psia, efficiency=85%, exit pressure is 145 psia, and mdot=1 lbm/s
   Combustion chamber–q_{supply}=400 Btu/lbm, and mdot=0.95 lbm/s
   Turbine #1 – efficiency=92.3%
   Turbine #2 – efficiency=92.3%, air exit pressure=15 psia
   Determine temperature of all states of the cycle, compressor power required, power produced by turbine #1, power produced by turbine #2, and cycle efficiency.
   ANSWER: T_2=629°F, T_3=629°F, T_4=1881°F, T_5=1881°F, T_6=1282°F, T_7=837.6°F, $Wdot_{compressor}$=-192.9 hp, $Wdot_{turbine \#1}$=192.9 hp, $Wdot_{turbine \#2}$=143.2 hp, η=35.49%.
6. Bleed air is used in a split-shaft gas turbine cycle. The following information is provided:
   Compressor–air inlet temperature and pressure are 60°F and14.5 psia, efficiency=85%, exit pressure is 145 psia, and mdot=1 lbm/s
   Combustion chamber–q_{supply}=400 Btu/lbm, and mdot=0.9 lbm/s
   Turbine #1 – efficiency=92.3%
   Turbine #2 – efficiency=92.3%, air exit pressure=15 psia
   Determine temperature of all states of the cycle, compressor power required, power produced by turbine #1, power produced by turbine #2, and cycle efficiency.
   ANSWER: T_2=629°F, T_3=629°F, T_4=1881°F, T_5=1881°F, T_6=1249°F, T_7=837.6°F, $Wdot_{compressor}$=-192.9 hp, $Wdot_{turbine \#1}$=192.9 hp, $Wdot_{turbine \#2}$=125.4 hp, η=32.83%.
7. Bleed air is used in a split-shaft gas turbine cycle. The following information is provided:
   Compressor–air inlet temperature and pressure are 60°F and14.5 psia, efficiency=85%, exit pressure is 145 psia, and mdot=1 lbm/s
   Combustion chamber–q_{supply}=400 Btu/lbm, and mdot=0.9 lbm/s
   Turbine #1 – efficiency=92.3%, mdot=1 lbm/s
   Turbine #2 – efficiency=92.3%, air exit pressure=15 psia, mdot=1lbm/s
   Determine temperature of all states of the cycle, compressor power required, power produced by turbine #1, power produced by turbine #2, and cycle efficiency.
   ANSWER: T_2=629°F, T_3=629°F, T_4=1881°F, T_5=1756°F, T_6=1187°F, T_7=768.2°F, $Wdot_{compressor}$=-192.9 hp, $Wdot_{turbine \#1}$=192.9 hp, $Wdot_{turbine \#2}$=141.9 hp, η=37.14%.

9.7 FEHER CYCLE

The Feher cycle as shown in Figure 9.7.1 is a single-phase cycle operating above the critical point of the working fluid. It incorporates the efficient pumping of the Rankine cycle with the regenerative heating features of the Brayton cycle to achieve higher theoretical efficiencies. The components of the cycle are identical to those of Brayton cycle except that the higher pressures permit a substantial reduction in the size of all components. The T-s diagram of the cycle is shown in Figure 9.7.2. The four processes of the Feher cycle are isentropic compression process 1-2, isobaric heat addition process 2-3, isentropic expansion process 3-4, and isobaric heat removal process 4-1. Carbon dioxide with a critical pressure of 73.9 bars (1072 psia) and critical temperature of 304 K (88°F) appears to be the most suitable working fluid for such a cycle.

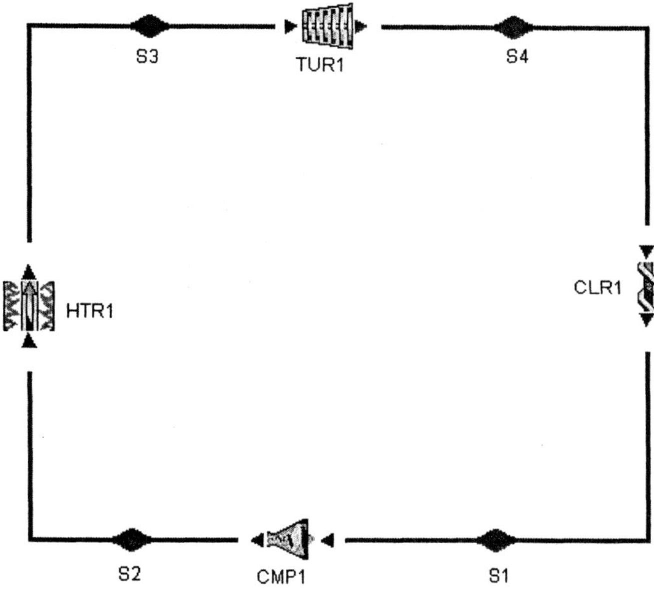

Figure 9.7.1 Feher cycle

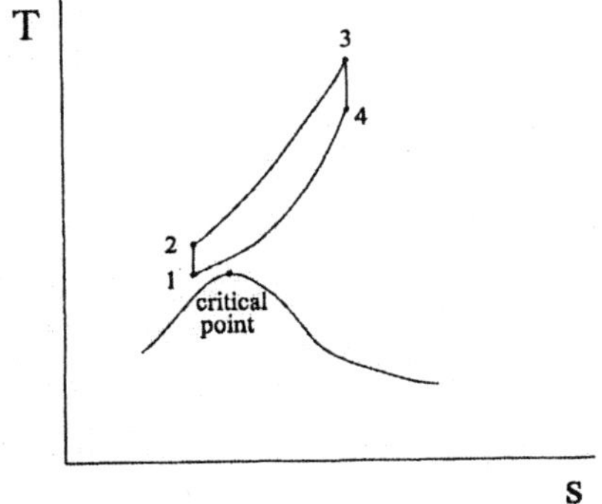

Figure 9.7.2 Feher cycle T-s diagram

Applying the first law and second law of thermodynamics of the closed system to each of the four processes of the cycle yields:

$$Q_{12} - W_{12} = m(h_2 - h_1),\ Q_{12}=0, \tag{9.7.1}$$

$$Q_{23} - W_{23} = m(h_3 - h_2), W_{23}=0, \tag{9.7.2}$$

$$Q_{34} - W_{34} = m(h_4 - h_3), Q_{34}=0, \tag{9.7.3}$$

and

$$Q_{41} - W_{41} = m(h_1 - h_4),\ W_{41} = 0. \tag{9.7.4}$$

The net work (W_{net}), which is also equal to net heat (Q_{net}), is

$$W_{net} = W_{12} + W_{34} = Q_{net} = Q_{23} + Q_{41} \tag{9.7.5}$$

The thermal efficiency of the cycle is

$$\eta = W_{net} / Q_{23} \tag{9.7.6}$$

Example 9.7.1. A proposed Feher cycle using carbon dioxide has the following design information:

turbine efficiency=0.88, compressor efficiency=0.88, mass rate flow of carbon dioxide=1 lbm/s, compressor inlet pressure=1950 psia, compressor inlet temperature=100°F, turbine inlet pressure=4000 psia, and turbine inlet temperature=1300°F.

Determine the compressor power, turbine power, rate of heat added, rate of heat removed, and cycle efficiency.

The answers are: $Wdot_{compressor}$=-31.66 hp, $Wdot_{turbine}$=-65.59 hp, $Qdot_{heater}$=218.5 Btu/s, $Qdot_{cooler}$=-194.5 Btu/s, and η=10.98%.

If a regenerator is added to the Feher cycle as shown in Figure 9.7.3 and the following example. The cycle would have an efficiency of 39.82%, which is comparable to today's best steam power plant.

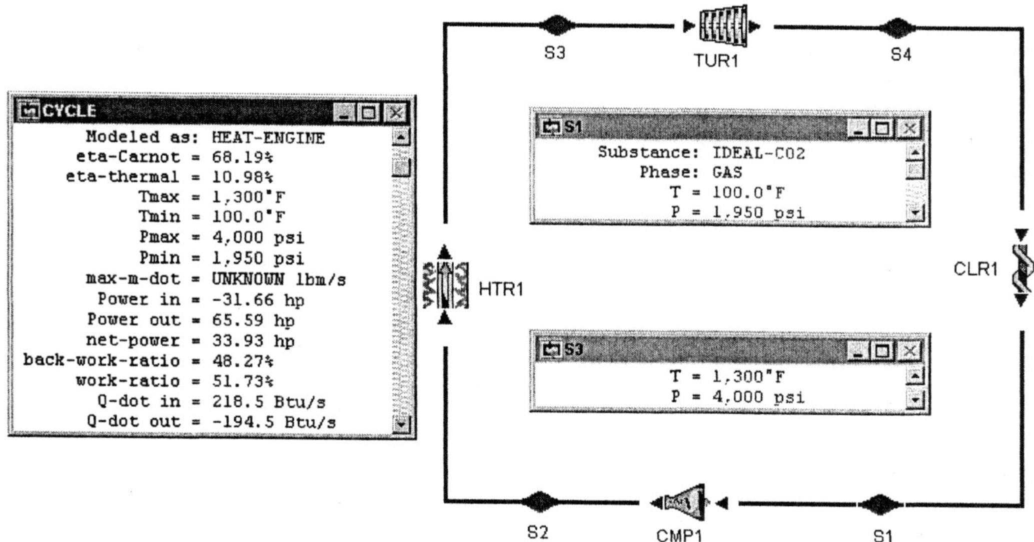

Figure Example 9.7.1 Feher cycle

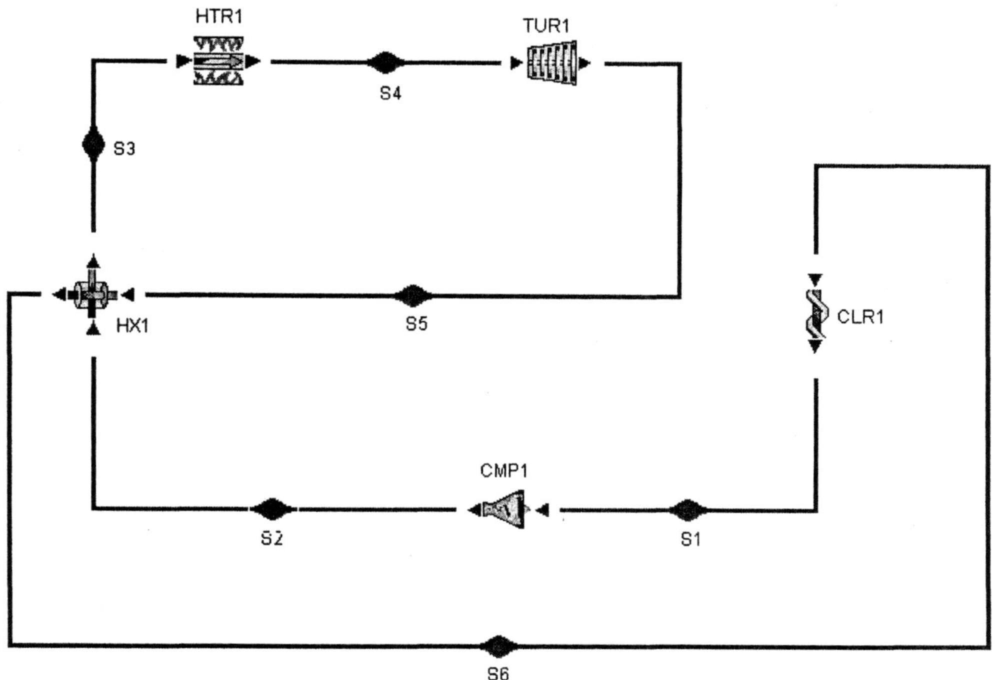

Figure 9.7.3 Feher cycle with regenerator

Example 9.7.2. A proposed Feher cycle with a regenerator using carbon dioxide has the following design information:

turbine efficiency=0.88, compressor efficiency=0.88, mass rate flow of carbon dioxide=1 lbm/s, compressor inlet pressure=1950 psia, compressor inlet temperature=100°F, combustion chamber inlet temperature=1000°F, turbine inlet pressure=4000 psia, and turbine inlet temperature=1300°F.

Determine the compressor power, turbine power, rate of heat added, rate of heat removed, and cycle efficiency.

The answers are: $Wdot_{compressor}$=-31.66 hp, $Wdot_{turbine}$=-65.59 hp, $Qdot_{heater}$=218.5 Btu/s, $Qdot_{cooler}$=-194.5 Btu/s, η=39.82%.

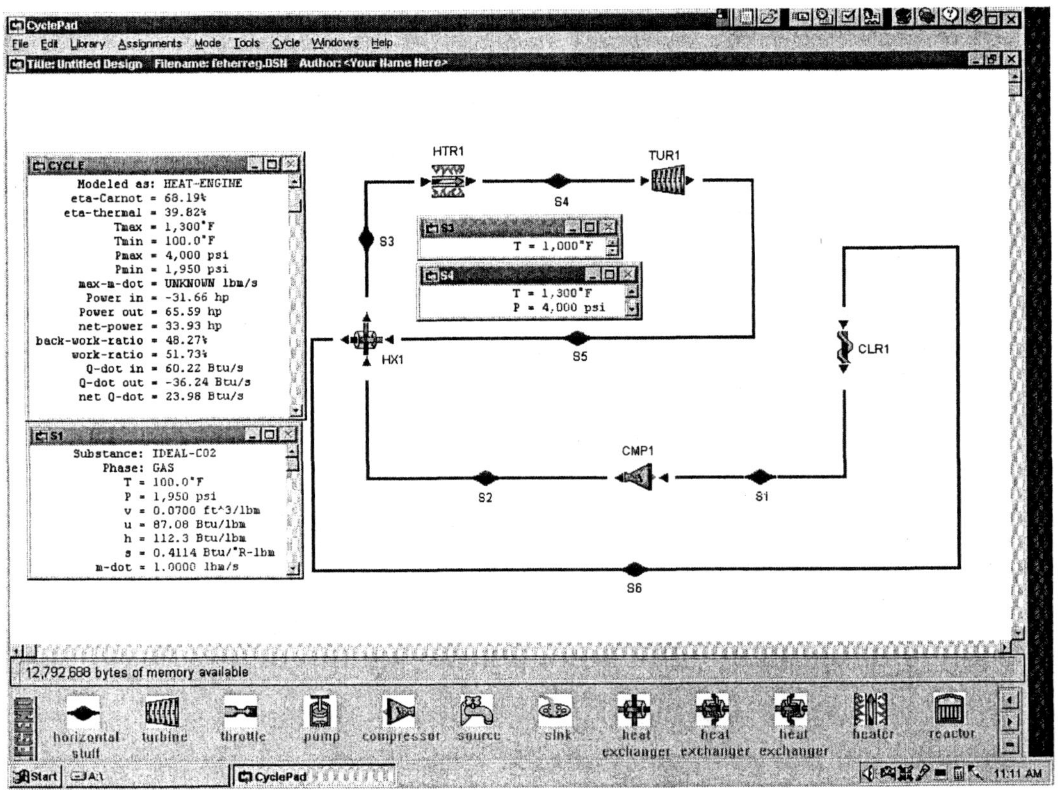

Figure Example 9.7.2 Feher cycle with regenerator

Homework 9.7 Feher Cycle

1. What are the four processes of the Feher cycle?
2. What is the difference between the Feher and the Brayton cycle?
3. A proposed Feher cycle using carbon dioxide has the following design information: turbine efficiency=0.9, compressor efficiency=0.85, mass rate flow of carbon dioxide=1 lbm/s, compressor inlet pressure=1950 psia, compressor inlet temperature=120°F, turbine inlet pressure=4200 psia, and turbine inlet temperature=1400°F.
   Determine the compressor power, turbine power, rate of heat added, rate of heat removed, and cycle efficiency.

4. A proposed Feher cycle with a regenerator using carbon dioxide has the following design information:
turbine efficiency=0.9, compressor efficiency=0.85, mass rate flow of carbon dioxide=1 lbm/s, compressor inlet pressure=1950 psia, compressor inlet temperature=120°F, combustion chamber inlet temperature=1000°F, turbine inlet pressure=4200 psia, and turbine inlet temperature=1300°F.
Determine the compressor power, turbine power, rate of heat added, rate of heat removed, and cycle efficiency.

9.8 Ericsson Cycle

Thermal cycle efficiency of a Brayton cycle can be increased by adding more intercoolers, compressors, reheaters, turbines, and regeneration. However, there is an economic limit to the number of stages of intercoolers and compressors, and reheaters and turbines.

If an infinite number of intercoolers and compressors, and reheaters and turbines are added to a basic ideal Brayton cycle, the intercooling and multi-compression processes approaches an isothermal process. Similarly, the reheat and multi-expansion processes approaches another isothermal process. This limiting Brayton cycle becomes an Ericsson cycle.

The schematic Ericsson cycle is shown in Figure 9.8.1. The p-v and T-s diagrams of the cycle is shown in Figure 9.8.2. The cycle consists of two isothermal processes and two isobaric processes. The four processes of the Ericsson cycle are isothermal compression process 1-2 (compressor), isobaric compression heating process 2-3 (heater), isothermal expansion process 3-4 (turbine), and isobaric expansion cooling process 4-1 (cooler).

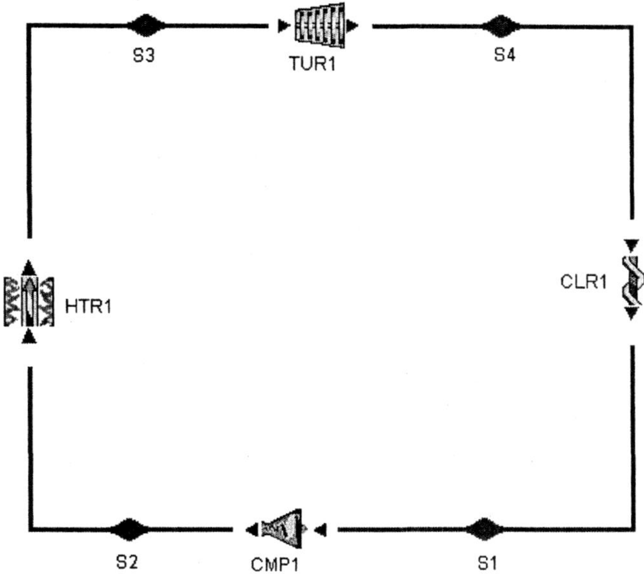

Figure 9.8.1. Schematic Ericsson cycle

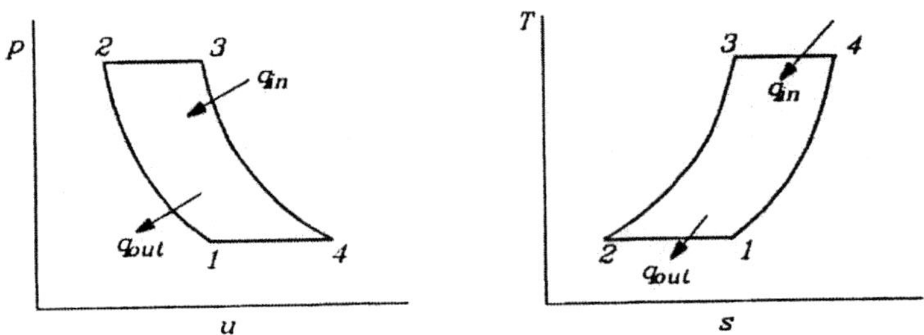

Figure 9.8.1. Ericsson cycle p-v and T-s diagram

Applying the basic laws of thermodynamics, we have

$$q_{12} - w_{12} = h_2 - h_1 \tag{9.8.1}$$

$$q_{23} - w_{23} = h_3 - h_2,\ w_{23} = 0 \tag{9.8.2}$$

$$q_{34} - w_{34} = h_4 - h_3 \tag{9.8.3}$$

$$q_{41} - w_{41} = h_1 - h_4,\ w_{41} = 0 \tag{9.8.4}$$

The net work produced by the cycle is

$$w_{net} = w_{12} + w_{34} \tag{9.8.5}$$

The heat added to the cycle is

$$q_{add} = q_{23} + q_{34}. \tag{9.8.6}$$

The cycle efficiency is

$$\eta = w_{net} / q_{add}. \tag{9.8.7}$$

Example 9.8.1. Air, at a mass flow rate of 1 kg/s, is compressed and heated from 100 kPa and 100°C in an Ericsson cycle to a turbine inlet at 1000 kPa and 1000°C. Determine the pressure and temperature of each of the four states, power and rate of heat added in each of the four devices, and cycle efficiency.

To solve this problem by CyclePad, we do the following steps:

(A) Build the cycle as shown in Figure 9.8.1. Assuming the compressor is isothermal, the heater is isobaric, the turbine is isothermal, and the cooler is isobaric.

(B) Input working fluid=air, mass flow rate=1 kg/s, compressor inlet pressure=100 kPa, compressor inlet temperature=100°C, turbine inlet pressure=1000 kPa, and turbine inlet temperature=1000°C.

(C) Display results. The answers are: T_2=100°C, p_2=1000 kPa, T_4=1000°C, p_2=100 kPa, $Qdot_{htr}$=903.1 kW, $Qdot_{clr}$=-903.1 kW, $Qdot_{cmp}$=-246.3 kW, $Wdot_{cmp}$=-246.3 kW, $Qdot_{tur}$=840.4 kW, $Wdot_{cmp}$=840.4 kW, and η=34.08. Notice that η_{Carnot}=70.69%.

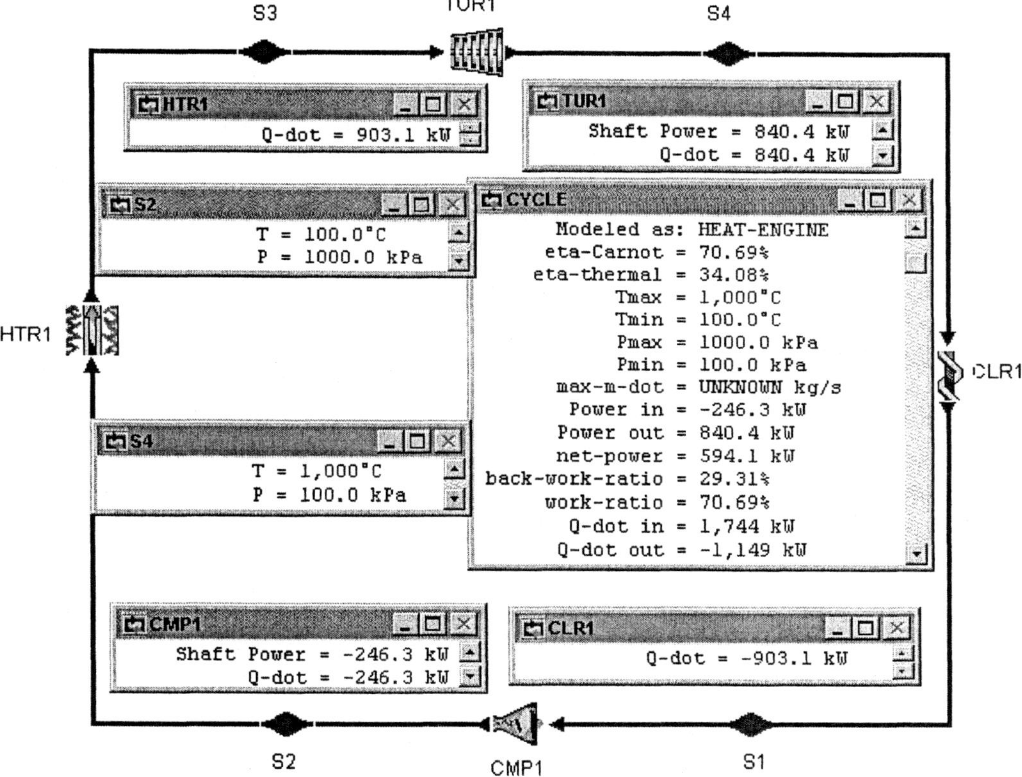

Figure Example 9.8.1 Ericsson cycle

An attempt to achieve Carnot cycle efficiency is made by the Ericsson cycle using an ideal regenerator. Figure 9.8.2 shows a schematic Ericsson cycle with a regenerator. In the regenerator, gas from the compressor enters as a cold-side stream at a low temperature (T_2) and leaves at a high temperature (T_1). The gas from the turbine enters as a hot-side stream at a high temperature (T_5)and leaves at a low temperature (T_6). Suppose there is only a small temperature difference between the two gas streams at any one section of the regenerator, so that the operation of the regenerator is almost ideal. Then the heat loss by the hot-side stream gas (Q_{56}) equals to the heat gain by the cold-side stream gas (Q_{23}). The cycle efficiency is close to that of the Carnot cycle operating between the same two temperatures.

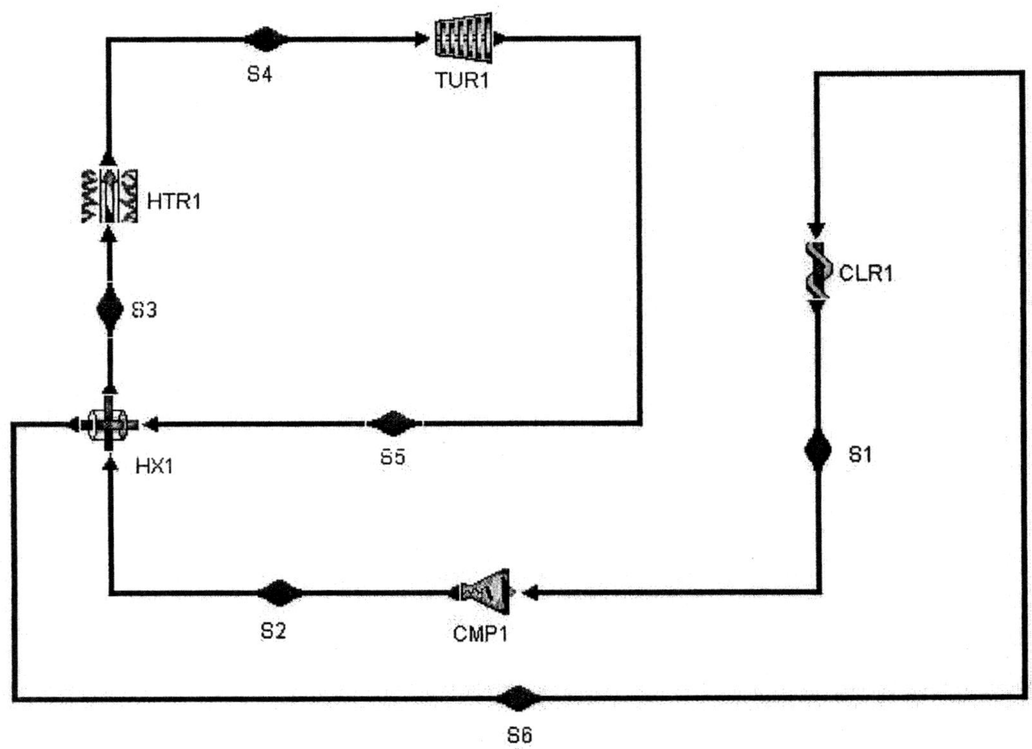

Figure 9.8.2 Schematic Ericsson cycle with a regenerator

Example 9.8.2. Air, at a mass flow rate of 1 kg/s, is compressed and heated from 100 kPa and 100°C in an Ericsson cycle to a turbine inlet at 1000 kPa and 1000°C. A regenerator is added. The inlet temperature of the hot stream in the regenerator is 995°C. Determine the pressure and temperature of each of the four states, power and rate of heat added in each of the four devices, and cycle efficiency.

To solve this problem by CyclePad, we do the following steps:

(A) Build the cycle as shown in Figure 9.8.2. Assuming the compressor is isothermal, the heater is isobaric, the turbine is isothermal, the cooler is isobaric, and the hot-side and cold-side of the regenerator are isobaric.

(B) Input working fluid=air, mass flow rate=1 kg/s, compressor inlet pressure=100 kPa, compressor inlet temperature=100°C, turbine inlet pressure=1000 kPa, turbine inlet temperature=1000°C, and regenerator hot-side inlet temperature=995°C.

(C) Display results. The answers are: T_2=100°C, p_2=1000 kPa, T_3=995°C, T_4=1000°C, p_4=100 kPa,T_6=105°C, $Qdot_{htr}$=903.1 kW, $Qdot_{clr}$=-903.1 kW, $Qdot_{cmp}$=-246.3 kW, $Wdot_{cmp}$=-246.3 kW, $Qdot_{tur}$=840.4 kW, $Wdot_{cmp}$=840.4 kW, and η=70.52. Notice that η=70.52 is very close to η_{Carnot}=70.69%.

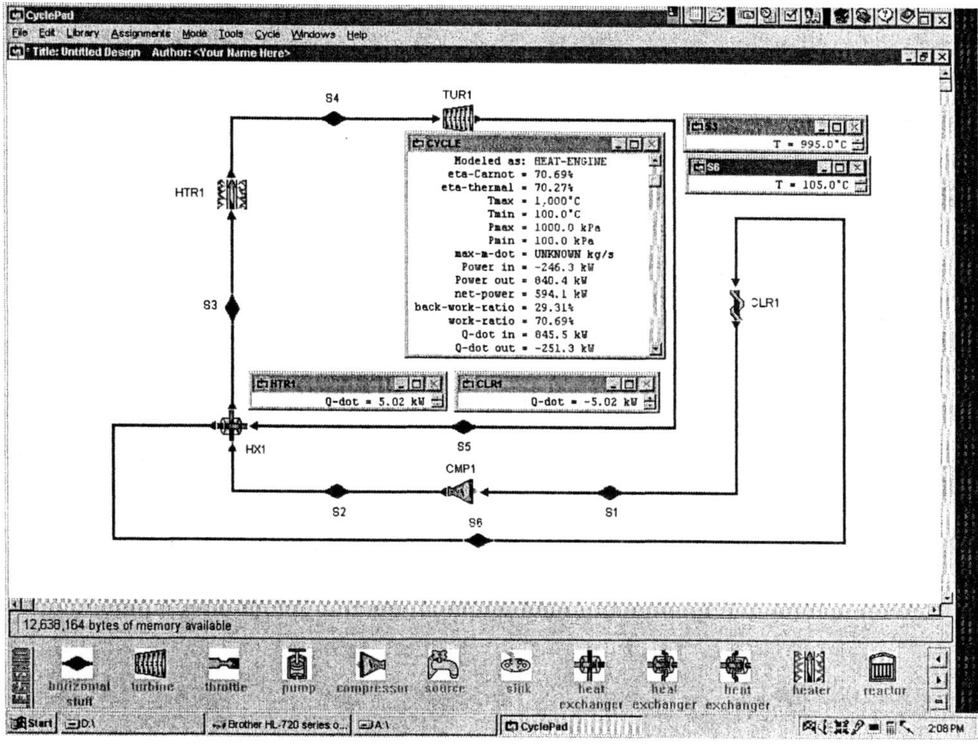

Figure Example 9.8.2Ericsson cycle with regenerator

Homework 9.8 Ericsson cycle

1. What are the four processes of the Ericsson cycle?
2. What is the function of a regenerator?
3. Does the regenerator improve the efficiency of the Ericsson cycle?
4. Suppose an ideal regenerator is added to an Ericsson cycle. The regenerator would absorb heat from the system during part of the cycle and return exactly the same amount of heat to the system during another part of the cycle. What would be the difference between the Ericsson cycle efficiency and the Carnot cycle efficiency?
5. Air, at a mass flow rate of 1.2 kg/s, is compressed and heated from 100 kPa and 300 K in an Ericsson cycle to a turbine inlet at 1200 kPa and 1500 K. The turbine efficiency is 85% and the compressor efficiency is 88%. Determine the pressure and temperature of each of the four states, power and rate of heat added in each of the four devices, and cycle efficiency.
6. Air, at a mass flow rate of 1.2 kg/s, is compressed and heated from 100 kPa and 300 K in an Ericsson cycle to a turbine inlet at 1200 kPa and 1500 K. The turbine efficiency is 85% and the compressor efficiency is 88%. A regenerator is added. The inlet temperature of the hot stream in the regenerator is 1495 K. Determine the pressure and temperature of each of the states, power and rate of heat added in each of the four devices, and cycle efficiency. What would be the efficiency of the Carnot cycle operating between 300 K and 1500 K?

9.9 BRAYSSON CYCLE

A Braysson cycle proposed by Frost, Anderson and Agnew [**Reference**: Frost, T.H., A. Anderson, and B. Agnew, A hybrid gas turbine cycle (Brayton/Ericsson): an alternative to conventional combined gas and steam turbine power plant, *Proceedings of the Institution of Mechanical Engineers, Part A, Journal of Power and Energy*, vol.211, n.A2, pp121-131, 1997] is an alternative to the Brayton/Rankine combined gas and steam turbine power plant. The Braysson cycle is a combination of a single Brayton cycle and an Ericsson cycle. The cycle takes advantage of the high-temperature heat addition process of the Brayton cycle and the low-temperature heat rejection process of the Ericsson cycle. It employs one working fluid in the two cycles in such a way that the full waste heat from the top Brayton cycle serves as the heat source for the bottom Ericsson cycle. The total power output of the Braysson cycle is the summation of the power produced by the top and bottom cycles.

A design of such a novel Braysson cycle [**Reference**: Wu, C., Intelligent computer aided optimization of power and energy systems, *Proceedings of the Institution of Mechanical Engineers, Part A, Journal of Power and Energy*, vol.213, n.A1, pp1-6, 1999] consisting of four compressors, one combustion chamber, two turbines, and two coolers is shown in Figure 9.9.1. The T-s diagram of the Braysson cycle is shown in Figure 9.9.2. Another arrangement of the Braysson cycle consisting of four compressors, one combustion chamber, two turbines, and two coolers is shown in Figure 9.9.3.

Neglecting kinetic and potential energy changes, a steady state and steady flow mass and energy balance on the components of the Braysson cycle have the general forms

$$\sum \text{mdot}_e = \sum \text{mdot}_i, \tag{9.9.1}$$

and

$$\text{Qdot-Wdot}=\sum \text{mdot}_e h_e - \sum \text{mdot}_i h_i. \tag{9.9.2}$$

The energy input of the cycle is the heat added in the heater. The net energy output of the cycle is the sum of the work added to the indiviual compressors and work produced by the turbines

$$\text{Wdot}=\sum \text{Wdot}_{compressor}+\sum \text{Wdot}_{turbine} \tag{9.9.3}$$

and the efficiency of the cycle is

$$\eta=\text{Wdot}/\text{Qdot}_{\text{lowest T evaporator}} \tag{9.9.4}$$

The following examples illustrate the analysis of the Braysson cycle.

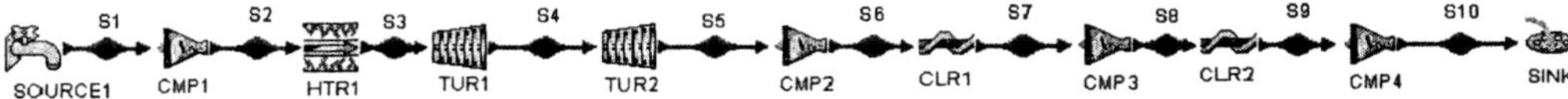

Figure 9.9.1. Braysson cycle

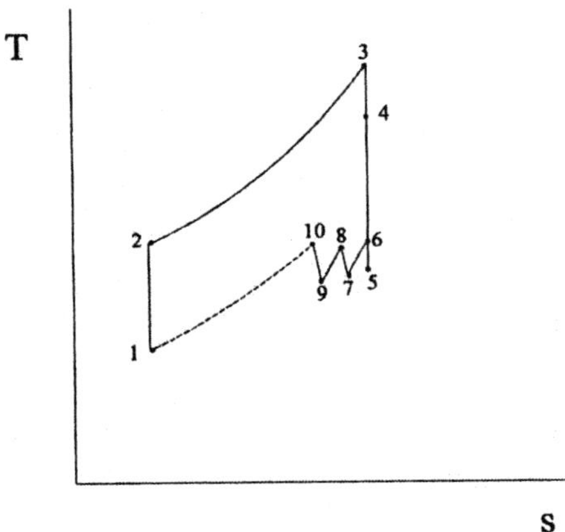

Figure 9.9.2. Braysson cycle T-s diagram

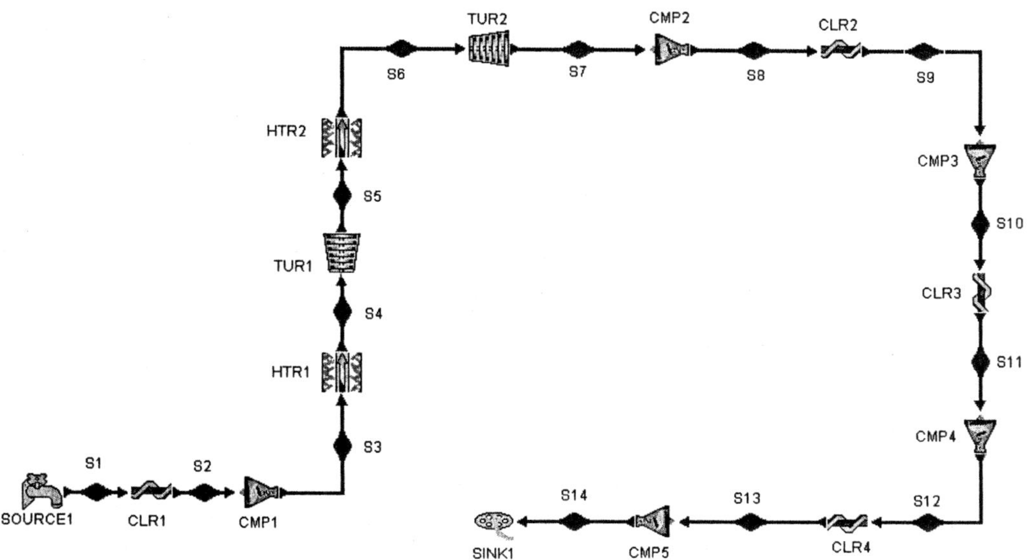

Figure 9.9.3. Braysson cycle

Example 9.9.1. A Braysson cycle (Fig. 9.9.1) uses air as the working fluid with 1 kg/s of mass flow rate through the cycle. In the Brayton cycle, air enters from the atmospheric source to an isentropic compressor at 20°C and 1 bar (state 1) and leaves at 8 bar (state 2); air enters an isobaric heater (combustion chamber) and leaves at 1100°C (state 3); air enters a high pressure isentropic turbine and leaves at 1 bar (state 4). In the Ericsson cycle, air enters a low pressure isentropic turbine and leaves at 0.04 bar(state 5); air enters a first-stage isentropic compressor and leaves at 0.2 bar (state 6); air enters an isobaric inter-cooler and leaves at 20°C (state 7); air enters a second-stage isentropic compressor and leaves at 1 bar (state 8); and air is discharged to the atmospheric sink.

Determine the thermodynamic efficiency and the net power output of the Brayson combined plant. Plot the sensitivity diagram of η (cycle efficiency) vs p_6 (pressure at state 6) and sensitivity diagram of η (cycle efficiency) vs p_8 (pressure at state 8).

To solve this problem by CyclePad, we do the following steps:

(A) Build the cycle as shown in Figure 9.10.1. Assuming the compressors are adiabatic and isentropic, the heater is isobaric, the turbines are adiabatic and isentropic, and the coolers are isobaric.

(B) Input working fluid=air, p_1=1bar, T_1=20°C, mdot=1 kg/s, p_2=8 bar, T_3=1100°C, p_4=1bar, p_5=0.04 bar, p_6=0.2 bar, compressor inlet temperature=100°C, turbine inlet pressure=1000 kPa, turbine inlet temperature=1000°C, T_7=20°C, p_8=0.6 bar, T_9=20°C, and p_{10}=1bar.

(C) Display results. The answers are η=59.67%, and power input=-570.4 kW, power output=1075 kW, net power output=504.2 kW, Qdot in=845.0 kW.

(D) Display sensitivity diagram of η (cycle efficiency) vs p_6 (pressure at state 6) and sensitivity diagram of η (cycle efficiency) vs p_8 (pressure at state 8).

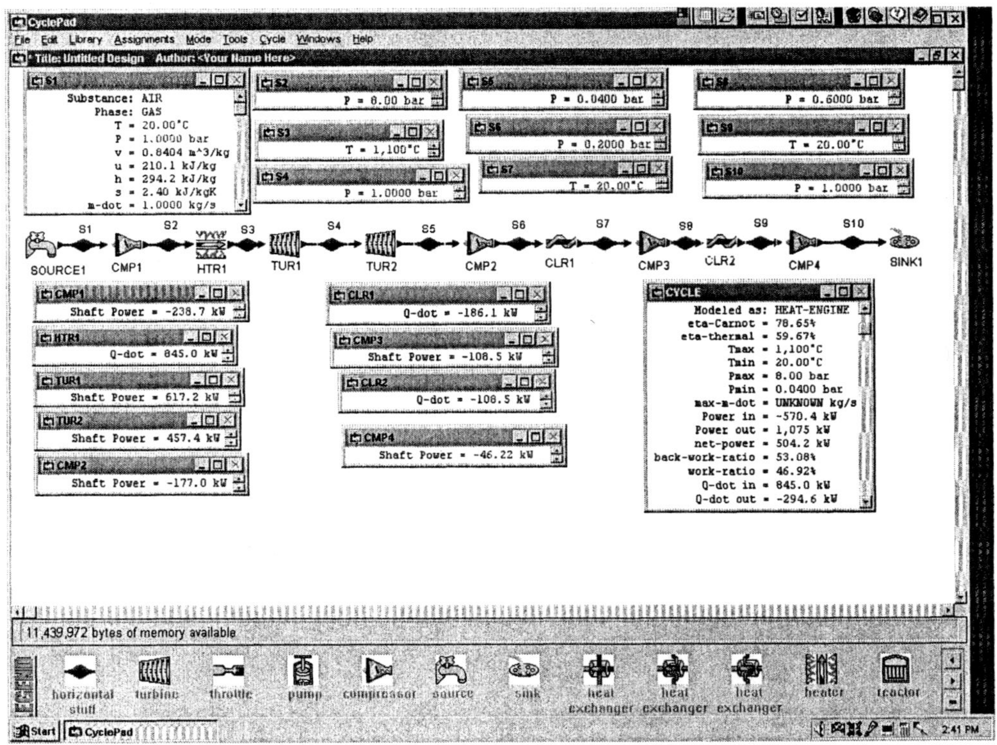

Figure Example 9.9.1a. Braysson cycle

Design of differently arranged Braysson cycles such as those shown in Figures 6 and 7 can be easily performed by the usage of the computer software. In Figures 6 and 7 a simple Brayson cycle (two-stage compressor) and a Brayson cycle (four-stage compressor) are shown. The four-stage compressor Braysson cycle is a refinement of the three-stage compressor Brayson cycle. Other improvement include (1) a reheater added in the Brayton cycle of the combined cycle, (2) more reheaters and more turbines added in the Brayton cycle, (3) more compressors and more inter-coolers added in the Ericsson cycle, etc.

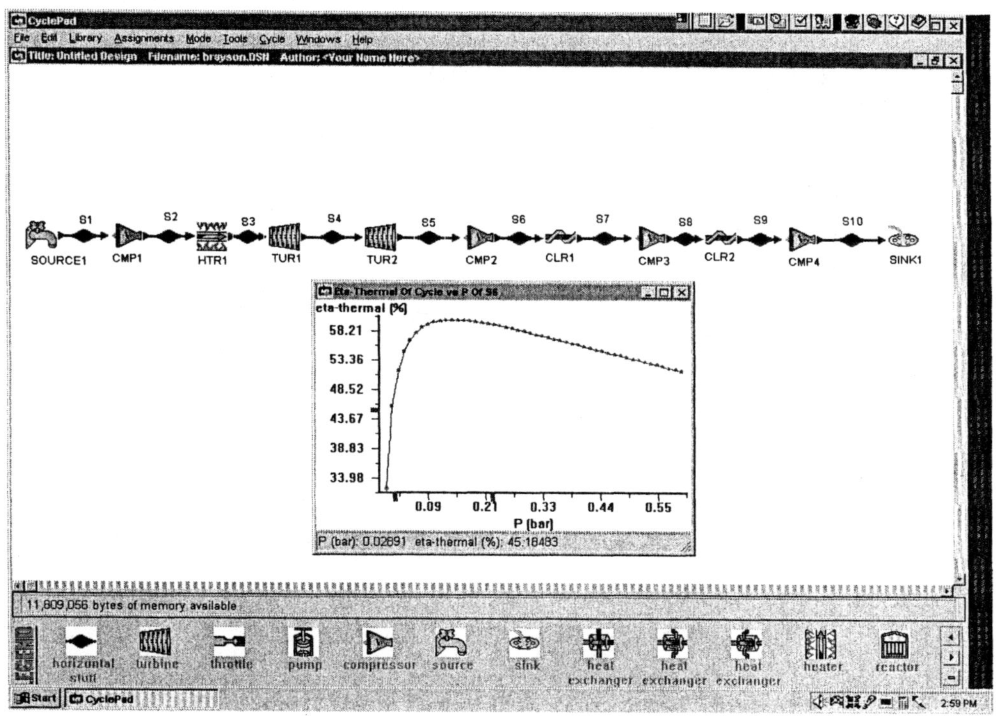

Figure Example 9.9.1b. Braysson cycle sensitivity diagram

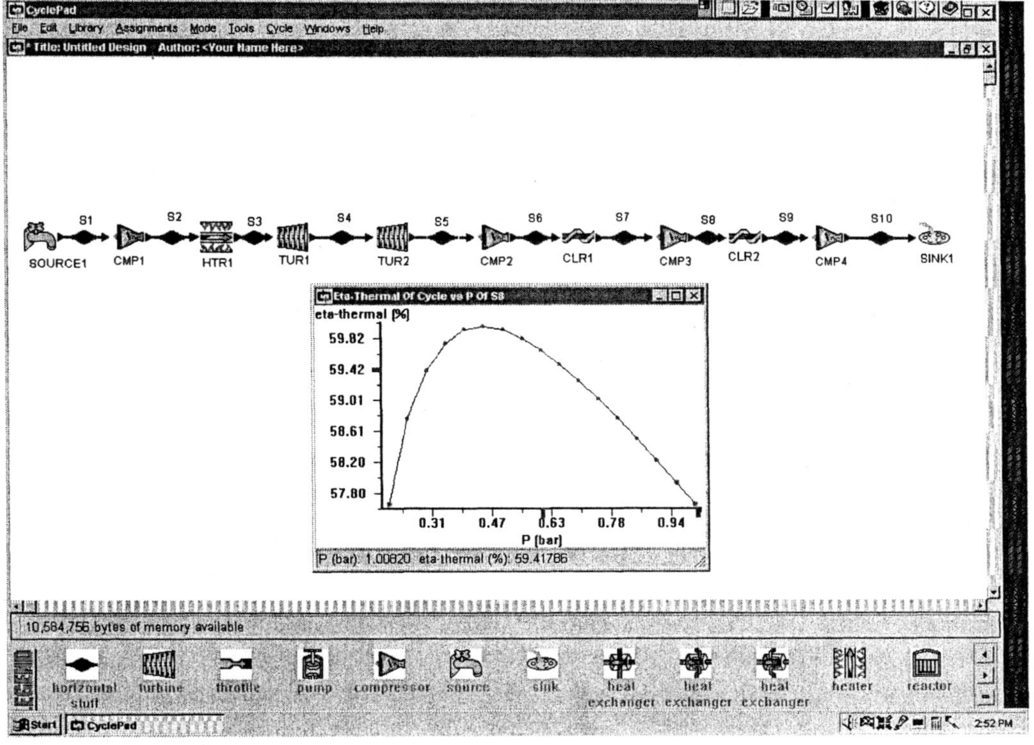

Figure Example 9.9.1c. Braysson cycle sensitivity diagram

Homework 9.9 Braysson Cycle

1. What is a Braysson cycle?
2. Why is the efficiency of the Braysson cycle high?
3. A Braysson cycle (Fig. 9.10.1) uses air as the working fluid with 1 kg/s of mass flow rate through the cycle. In the Brayton cycle, air enters from the atmospheric source to a compressor at 20°C and 1 bar (state 1) and leaves at 8 bar (state 2); air enters an isobaric heater (combustion chamber) and leaves at 1100°C (state 3); air enters a high pressure turbine and leaves at 1 bar (state 4). In the Ericsson cycle, air enters a low pressure turbine and leaves at 0.04 bar(state 5); air enters a first-stage compressor and leaves at 0.2 bar (state 6); air enters an isobaric inter-cooler and leaves at 20°C (state 7); air enters a second-stage compressor and leaves at 1 bar (state 8); and air is discharged to the atmospheric sink. Assume all turbines and compressors have 85% efficiency.
   Determine the thermodynamic efficiency and the net power output of the Braysson combined plant. Plot the sensitivity diagram of η (cycle efficiency) vs p_6 (pressure at state 6) and sensitivity diagram of η (cycle efficiency) vs p_8 (pressure at state 8).
   ANSWER: η=26.34%, Power input=-749.3 kW, Power output=960.7 kW, Qdot in=802.9 kW.
4. A Braysson cycle (Fig. 9.10.1) uses air as the working fluid with 1 kg/s mass flow rate through the cycle. In the Brayton cycle, air enters from the atmospheric source to a compressor at 20°C and 1 bar (state 1) and leaves at 8 bar (state 2); air enters an isobaric heater (combustion chamber) and leaves at 1100°C (state 3); air enters a high pressure isentropic turbine and leaves at 1 bar (state 4). In the Ericsson cycle, air enters a low pressure isentropic turbine and leaves at 0.04 bar(state 5); air enters a first-stage compressor and leaves at 0.2 bar (state 6); air enters an isobaric inter-cooler and leaves at 20°C (state 7); air enters a second-stage compressor and leaves at 1 bar (state 8); and air is discharged to the atmospheric sink. Assume all compressors have 85% efficiency.
   Determine the thermodynamic efficiency and the net power output of the Braysson combined plant. Plot the sensitivity diagram of η (cycle efficiency) vs p_6 (pressure at state 6) and sensitivity diagram of η (cycle efficiency) vs p_8 (pressure at state 8).
   ANSWER: η=50.26%, Power input=-671.1 kW, Power output=1075 kW, Qdot in=802.9 kW.

9.10 CASCADED AND COMBINED CYCLE

There are applications when the temperature difference between the heat source and the heat sink is quite large. A single power cycle usually can not be used to utilize the full range of the available temperature difference. A cascade cycle must be used. A cascade cycle is several power cycles connecting in series. A cascade cycle of three cycles in series is shown in Figure 9.10.1. The energy flow diagram of the cascade cycle is illustrated in

Figure 9.10.2. The cooler of the highest-temperature cycle (cycle A) provides the heat input to the heater of the mid-temperature cycle (cycle B). The cooler of the mid-temperature cycle provides the heat input to the heater of the lowest-temperature cycle (cycle C).

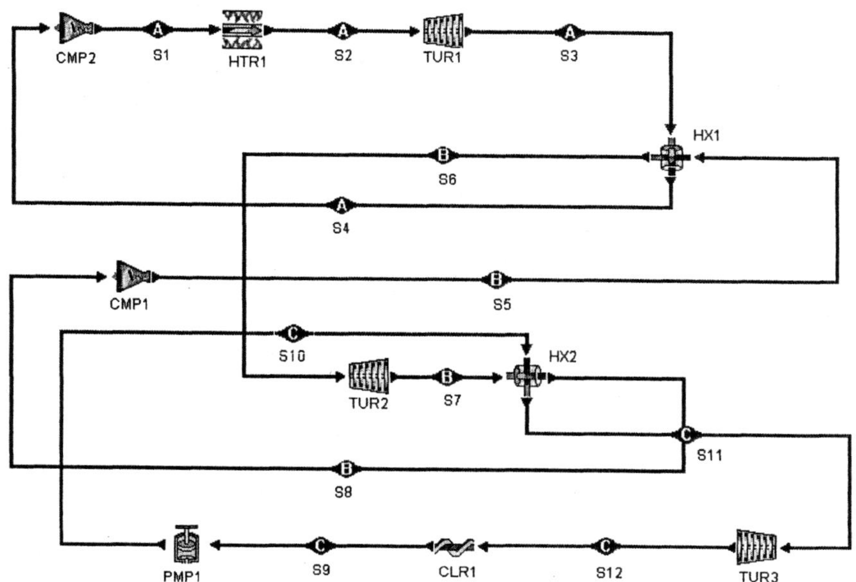

Figure 9.10.1. Cascade cycle

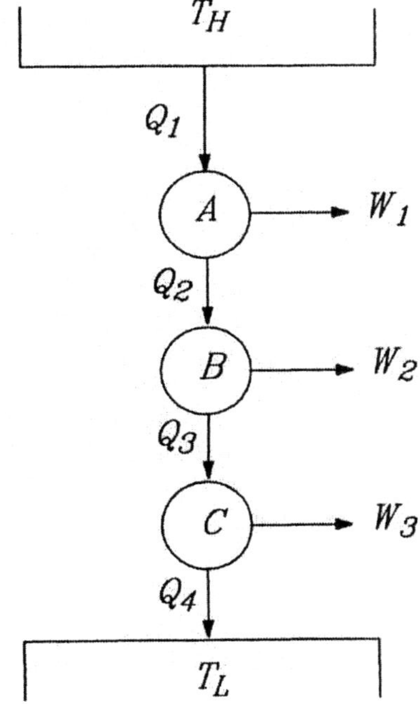

Figure 9.10.2. Cascade cycle energy flow diagram

The overall efficiency of the cascaded cycle is the total output work ($W_1+W_2+W_3$) divided by the heat input, Q_1. Referring to Figure 9.10.2, we have

$$\eta = (W_1+W_2+W_{Co})/Q_1, \tag{9.10.1}$$

$$W_1=\eta_A Q_1, \tag{9.10.2}$$

$$W_2=\eta_B Q_2, \tag{9.10.3}$$

and

$$W_3=\eta_C Q_3. \tag{9.10.4}$$

Substituting W_1,W_2 and W_3 into Equation (9.10.1), the following efficiency expression is obtained.

$$\eta=1-(1-\eta_A)(1-\eta_B)(1-\eta_c) \tag{9.10.5}$$

The combined cycle efficiency therefore may be substantially greater than the cycle efficiency of any of its components operating alone.

Eq. (9.10.5) can be extended to a cascaded cycle with n-component cycles.

$$\eta=1-(1-\eta_1)(1-\eta_2)...(1-\eta_n) \tag{9.10.6}$$

9.10.1 Brayton/Rankine Cycle

Further improvement on cycle efficiency can be achieved by using the hot exhaust waste heat of a high-temperature cycle to either partially or totally power a low-temperature cycle. For example, since the boiler temperature of the basic Rankine cycle is about 500°C, the exhaust gases of the gas turbine cycle with a temperature of 500°C could be used for the boiler heat the boiler heat input. One arrangement of the Brayton/Rankine cycle which is a combination of a two-stage reheat Brayton cycle and a two-stage reheat Rankine cycle is shown in Figure 9.10.1.1. In general, modifications of both the Brayton and Rankine cycles could also be included. Since the net work output is equal to the sum of the two outputs and the heat input is that of the topping cycle alone, a substantial cycle efficiency increase is possible. Another arrangement of the Brayton/Rankine cycle which is a combination of a two-stage reheat Brayton cycle and a two-stage reheat Rankine cycle is shown in Figure 9.10.1.2.

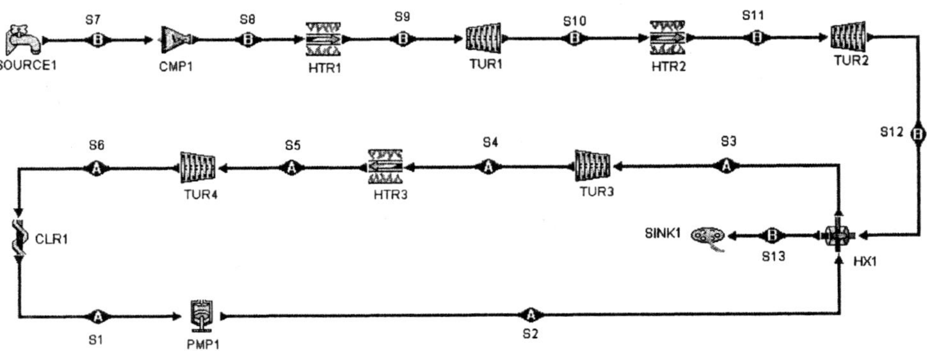

Figure 9.10.1.1.Brayton/Rankine cycle

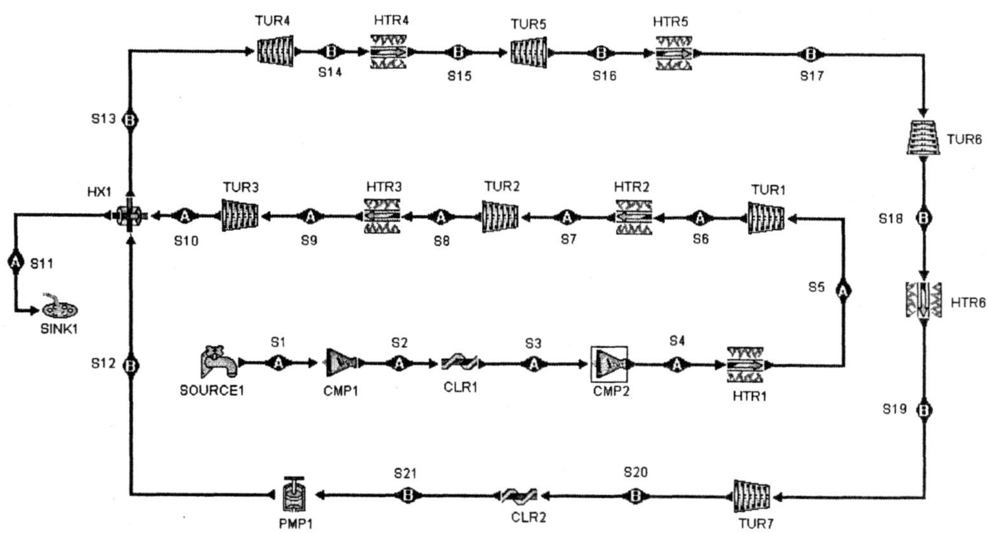

Figure 9.10.1.2. Brayton/Rankine cycle

Example 9.10.1.1 A Brayton/Rankine cycle (Fig. 9.10.1.1) uses water as the working fluid with 1 kg/s mass flow rate through the Rankine cycle, and air as the working fluid in the Brayton cycle. In the Rankine cycle, the condenser pressure is 15 kPa (p_1); the boiler pressure is 8000 kPa (p_2); the reheater pressure is 5000 kPa (p_4); the superheater and reheater temperature (T_3 and T_5) are both 400°C. In the Brayton cycle, air enters from the atmospheric source to an isentropic compressor at 20°C and 100 kPa (T_7 and p_7), and leaves at 1000 kPa (p_8); air enters an isobaric heater (combustion chamber) and leaves at 1800°C (T_9); air enters a high pressure isentropic turbine and leaves at 600 kPa (p_{11}). Air enters a low pressure isentropic turbine and leaves at 100 kPa (p_{12}); air enters an isobaric regenerator and leaves at 500°C (T_{13}); and air is discharged to the atmospheric sink.

Determine the mass rate flow of air through the Brayton cycle, thermodynamic efficiency and the net power output of the Brayton/Rankine combined plant. Plot the sensitivity diagram of η (cycle efficiency) vs p_{11} (pressure at state 11).

To solve this problem by CyclePad, we do the following steps:

(A) Build the cycle as shown in Figure 9.10.1.1. Assuming the compressor, turbines and pump are adiabatic and isentropic, and the heater, cooler and regenerator are isobaric.

(B) Input working fluid=air, p_1=15 kPa, x_1=0, mdot=1 kg/s, p_3=8000 kPa, T_3=400°C, p_5=5000 kPa, T_5=400°C, p_7=100 kPa, T_7=20°C, p_9=1000 kPa, T_9=1800°C, p_{11}=600 kPa, T_{11}=1600°C, p_{12}=100 kPa, and T_{13}=500°C.

(C) Display results. The answers are: (1) Cycle A: η=37.52%, power input=-8.12 kW, power output=1165 kW, net power output=1157 kW, Qdot in=3084 kW; (2) Cycle B: η=47.79%, power input=-2267 kW, power output=8575 kW, net power output=6308 kW, Qdot in=13200 kW, mdot=8.28 kg/s; and (3) combined Cycle: η=55.79%, power input=-2275 kW, power output=9740 kW, net power output=7465 kW, Qdot in=13380 kW.

(D) Display sensitivity diagram of η (cycle efficiency) vs p_{11}(pressure at state 11).

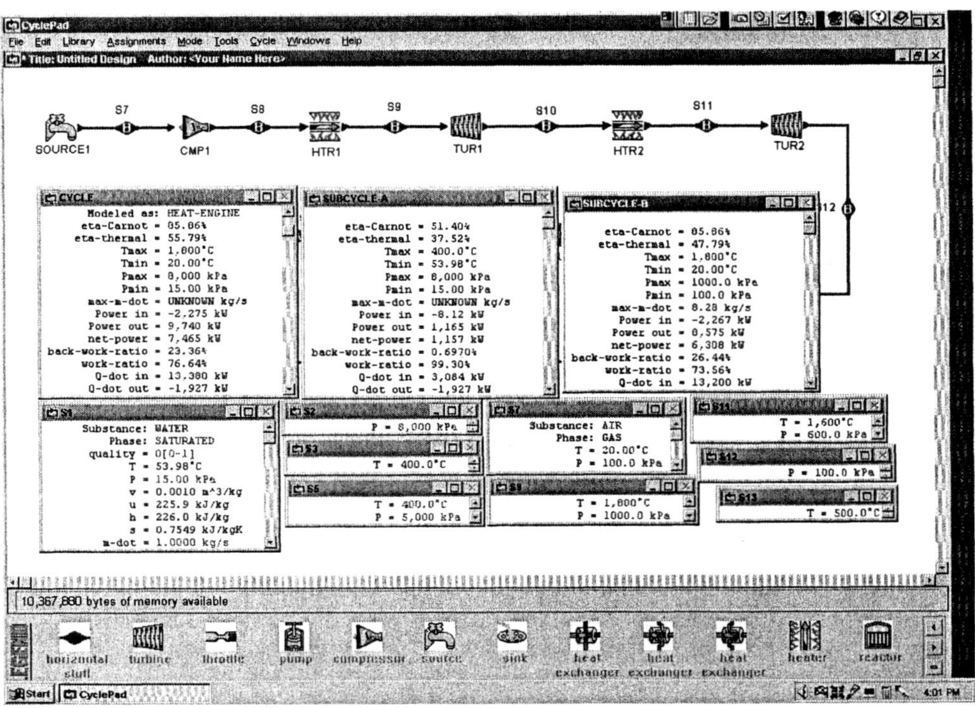

Figure Example 9.10.1.1 aBrayton/Rankine cycle

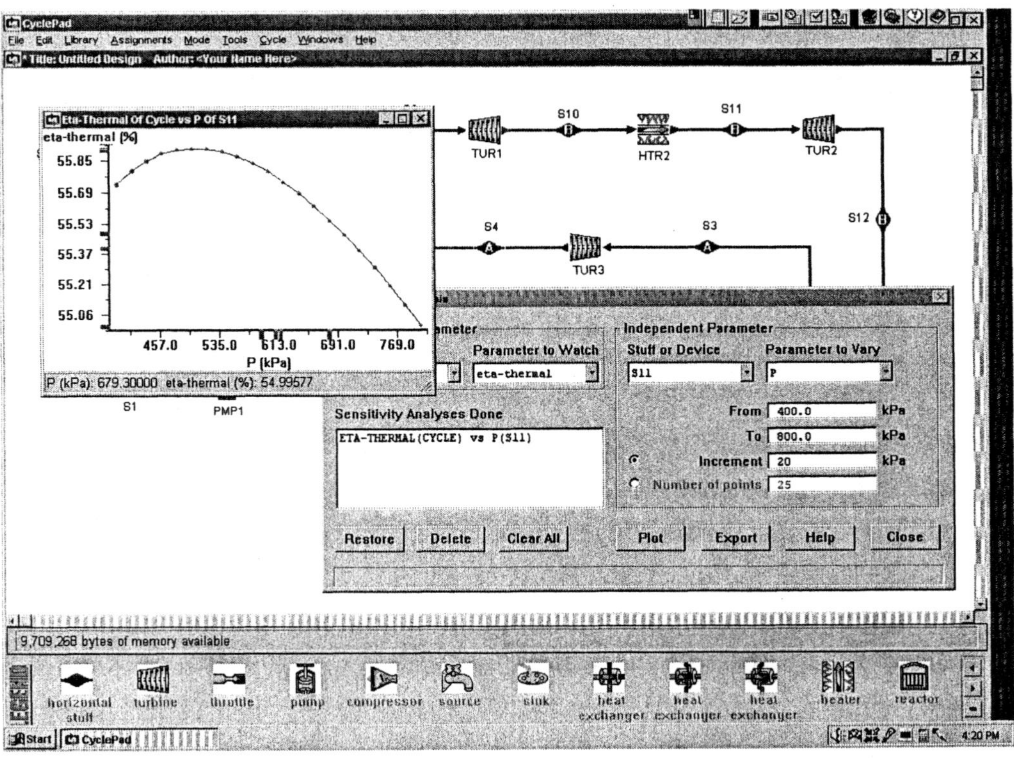

Figure Example 9.10.1.1 b Brayton/Rankine cycle

Homework 9.10.1

1. What is a combined Brayton/Rankine cycle? What is its purpose?
2. What is the heat source for the Rankine cycle in the combined Brayton/Rankine cycle?
3. Why is the combined Brayton/Rankine cycle more efficient than either of the cycles operating alone?
4. A Brayton/Rankine cycle (Fig. 9.10.1.1) uses water as the working fluid with 1 kg/s mass flow rate through the Rankine cycle, and air as the working fluid in the Brayton cycle. In the Rankine cycle, the condenser pressure is 15 kPa (p_1); the boiler pressure is 8000 kPa (p_2); the reheater pressure is 5000 kPa (p_4); the superheater and reheater temperature (T_3 and T_5) are both 400°C. In the Brayton cycle, air enters from the atmospheric source to an isentropic compressor at 20°C and 100 kPa (T_7 and p_7), and leaves at 1000 kPa (p_8); air enters an isobaric heater (combustion chamber) and leaves at 1800°C (T_9); air enters a high pressure isentropic turbine and leaves at 600 kPa (p_{11}). Air enters a low pressure isentropic turbine and leaves at 100 kPa (p_{12}); air enters an isobaric regenerator and leaves at 500°C (T_{13}); and air is discharged to the atmospheric sink. Assume the compressor efficiency is 85%.
   Determine the mass flow rate of air through the Brayton cycle, the thermodynamic efficiency and the net power output of the Brayton/Rankine combined plant.
   ANSWER: (1) Cycle A: η=37.52%, power input=-8.12 kW, power output=1165 kW, net power output=1157 kW, Qdot in=3084 kW; (2) Cycle B: η=46.16%, power input=-2667 kW, power output=8575 kW, net power output=5908 kW, Qdot in=12800 kW, mdot=8.28 kg/s; and (3) combined Cycle: η=54.43%, power input=-2675 kW, power output=9740 kW, net power output=7065 kW, Qdot in=12980 kW.
5. A Brayton/Rankine cycle (Fig. 9.10.1.1) uses water as the working fluid with 1 kg/s of mass flow rate through the Rankine cycle, and air as the working fluid in the Brayton cycle. In the Rankine cycle, the condenser pressure is 15 kPa (p_1); the boiler pressure is 8000 kPa (p_2); the reheater pressure is 5000 kPa (p_4); the superheater and reheater temperature (T_3 and T_5) are both 400°C. In the Brayton cycle, air enters from the atmospheric source to an isentropic compressor at 20°C and 100 kPa (T_7 and p_7), and leaves at 1000 kPa (p_8); air enters an isobaric heater (combustion chamber) and leaves at 1800°C (T_9); air enters a high pressure isentropic turbine and leaves at 600 kPa (p_{11}). Air enters a low pressure isentropic turbine and leaves at 100 kPa (p_{12}); air enters an isobaric regenerator and leaves at 500°C (T_{13}); and air is discharged to the atmospheric sink. Assume the gas turbines and compressor efficiency are 85%.
   Determine the mass rate flow of air through the Brayton cycle, thermodynamic efficiency and the net power output of the Brayton/Rankine combined plant.
   ANSWER: (1) Cycle A: η=37.52%, power input=-8.12 kW, power output=1165 kW, net power output=1157 kW, Qdot in=3084 kW; (2) Cycle B: η=37.12%, power input=-2017 kW, power output=5513 kW, net power output=3496 kW, Qdot in=9416 kW, mdot=6.26 kg/s; and (3) combined Cycle: η=48.48%, power input=-2025 kW, power output=6678 kW, net power output=4653 kW, Qdot in=9596 kW.
6. A Brayton/Rankine cycle (Fig. 9.10.1.1) uses water as the working fluid with 1 kg/s mass rate of flow through the Rankine cycle, and air as the working fluid in the Brayton cycle. In the Rankine cycle, the condenser pressure is 15 kPa (p_1); the boiler pressure is 8000

kPa (p_2); the reheater pressure is 5000 kPa (p_4); the superheater and reheater temperature (T_3 and T_5) are both 400°C. In the Brayton cycle, air enters from the atmospheric source to an isentropic compressor at 20°C and 100 kPa (T_7 and p_7), and leaves at 1000 kPa (p_8); air enters an isobaric heater (combustion chamber) and leaves at 1800°C (T_9); air enters a high pressure isentropic turbine and leaves at 600 kPa (p_{11}). Air enters a low pressure isentropic turbine and leaves at 100 kPa (p_{12}); air enters an isobaric regenerator and leaves at 500°C (T_{13}); and air is discharged to the atmospheric sink. Assume all turbine and compressor efficiencies are 85%.

Determine the mass flow rate of air through the Brayton cycle, thermodynamic efficiency and the net power output of the Brayton/Rankine combined plant.

ANSWER: (1) Cycle A: η=32.04%, power input=-8.12 kW, power output=990.3 kW, net power output=982.2 kW, Qdot in=3066 kW; (2) Cycle B: η=37.12%, power input=-2017 kW, power output=5513 kW, net power output=3496 kW, Qdot in=9416 kW, mdot=6.26 kg/s; and (3) combined Cycle: η=46.75%, power input=-2025 kW, power output=6503 kW, net power output=4478 kW, Qdot in=9578 kW.

9.11 Field Cycle

The field cycle is a super-generative cycle which makes use of the high-temperature heat addition of the Brayton cycle and the low-temperature heat removal of Rankine cycle. Therefore, it is able to achieve a high mean temperature of heat addition. The gain due to high-temperature heat addition, however, is offset by the reduction in cycle efficiency resulting from the irreversibility of the mixing process. The schematic diagram of the Field cycle is shown in Figure 9.11.1. The arrangement includes one compressor, five turbines, three pumps, one boiler and one reheater (heaters), one regenerator (heat exchanger), one condenser (cooler), three mixing chambers, and two splitters. Process 1-2, 2-3, 3-4, 4-5, 5-6, and 6-7 takes advantage of the high-temperature heat addition of Brayton cycle, and the rest of the processes takes advantage of the low-temperature heat removing and regenerative condensing of Rankine cycle.

Example 9.11.1. An ideal field cycle with perfect regeneration as shown in Figure 9.11.1 is designed according to the following data:

p_{16}=10 kPa, x_{16}=0, p_{19}=200 kPa, x_{19}=0, p_{22}=1000 kPa, x_{22}=0, p_{23}=2000 kPa, p_2=6000 kPa, T_4=500 °C, $mdot_4$=1 kg/s, p_5=4000 kPa, T_6=500 °C, T_8=300 °C, $mdot_{10}$=0.9 kg/s, and $mdot_{17}$=0.1 kg/s.

Determine (1) the pressure and temperature of each state of the cycle, (2) power produced by each of the five turbines, rate of heat added by each of the two heaters, power reuired by the compressor and each of the three pumps, rate of heat removed by the condenser, (3) net power produced by the cycle, and cycle efficiency.

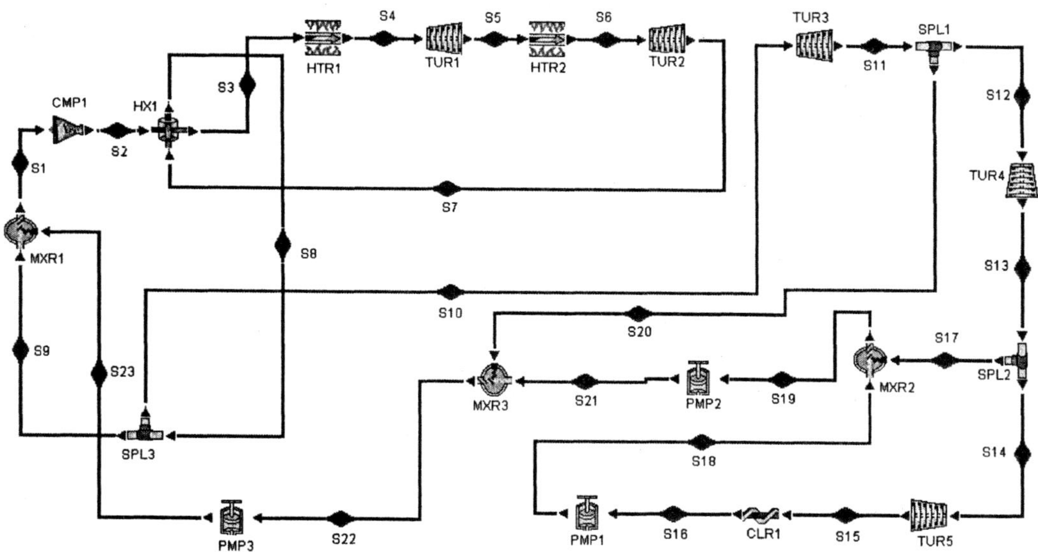

Figure 9.11.1. Field cycle schematic diagram

To solve this problem by CyclePad, we do the following steps:

(A) Build the cycle as shown in Figure 9.11.1. Assuming the compressor, turbines and pump are adiabatic and isentropic, the heaters, mixing chambers, cooler and regenerator are isobaric, and the splitters are iso-parametric.

(B) Input working fluid=water, p_{16}=10 kPa, x_{16}=0, p_{19}=200 kPa, x_{19}=0, p_{22}=1000 kPa, x_{22}=0, p_{23}=2000 kPa, p_2=6000 kPa, T_4=500 °C, $mdot_4$=1 kg/s, p_5=4000 kPa, T_6=500 °C, T_8=300 °C, $mdot_{10}$=0.9 kg/s, and $mdot_{17}$=0.9 kg/s.

(C) Display results. The answers are: (1) p_1=2000 kPa, T_1=212.4 °C, p_2=6000 kPa, T_2=237.3 °C, p_3=6000 kPa, T_3=275.6 °C, p_4=6000 kPa, T_4=500 °C, p_5=4000 kPa, T_5=432.8 °C, p_6=4000 kPa, T_6=500°C, p_7=2000 kPa, T_7=388.9 °C, p_8=2000 kPa, T_8=300°C, p_9=2000 kPa, T_9=300 °C, p_{10}=2000 kPa, T_{10}=300°C, p_{11}=1000 kPa, T_{11}=214.6°C, p_{12}=1000 kPa, T_{12}=214.6°C, p_{13}=200 kPa, T_{13}=120.2°C, p_{14}=200 kPa, T_{14}=120.2°C, p_{15}=10 kPa, T_{15}=45.82°C, p_{16}=10 kPa, T_{16}=45.82°C, p_{17}=200 kPa, T_{17}=120.2°C, p_{18}=200 kPa, T_{18}=45.83°C, p_{19}=200 kPa, T_{19}=120.2°C, p_{20}=1000 kPa, T_{20}=214.6°C, p_{21}=1000 kPa, T_{21}=120.3°C, p_{22}=2000 kPa, T_{17}=179.9°C, and p_{23}=2000 kPa, T_{17}=185.7°C, (2) $Wdot_{T\#1}$=131.9 kW, $Wdot_{T\#2}$=222.4 kW, $Wdot_{T\#3}$=144.7 kW, $Wdot_{T\#4}$=238.9 kW, $Wdot_{T\#5}$=293.2 kW, $Wdot_{P\#1}$=-0.1598 kW, $Wdot_{P\#2}$=-0.6973 kW, $Wdot_{P\#3}$=-28.45 kW, $Wdot_{Compressor}$=-7.57 kW, $Qdot_{Htr\#1}$=2197 kW, $Qdot_{Htr\#2}$=155 kW, and $Qdot_{Condenser}$=-1358 kW, (3) $Wdot_{net}$=994.0 kW, and η=42.26%.

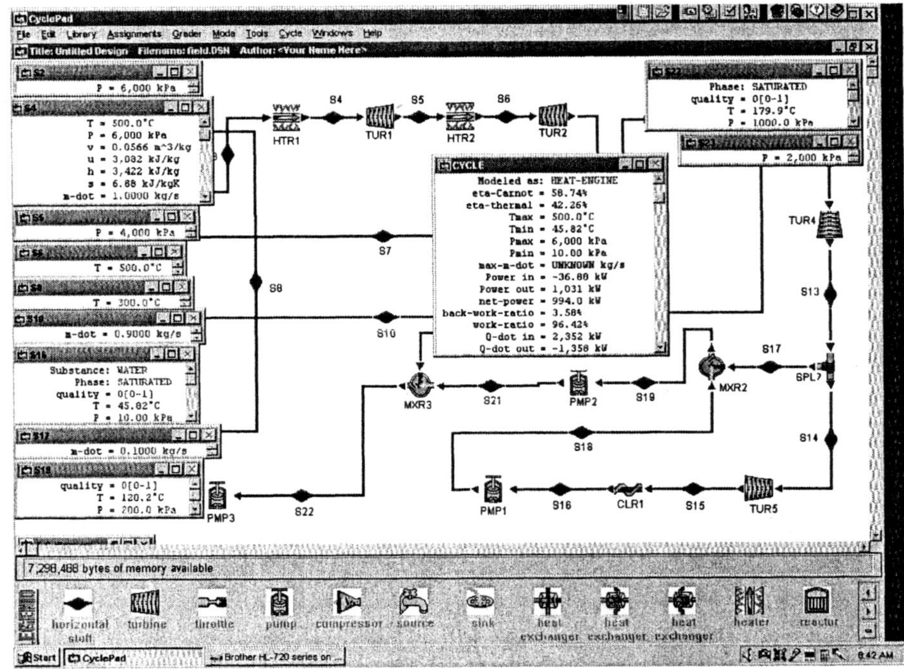

Figure Example 9.11.1. Field cycle

Homework 9.11.Field Cycle

1. What is the concept of the Field cycle?
2. An ideal field cycle with perfect regeneration as shown in Figure 9.11.1 is designed according to the following data:
   p_{16}=10 kPa, x_{16}=0, p_{19}=200 kPa, x_{19}=0, p_{22}=1000 kPa, x_{22}=0, p_{23}=2000 kPa, p_2=7000 kPa, T_4=500 °C, $mdot_4$=1 kg/s, p_5=4000 kPa, T_6=500 °C, T_8=300 °C, $mdot_{10}$=0.9 kg/s, and $mdot_{17}$=0.1 kg/s.
   Determine rate of heat added by the heaters, total power produced by the turbines, total power required by the pumps and compressor, net power produced by the cycle, and cycle efficiency.
   ANSWER: $Qdot_{add}$=2393 kW, $Wdot_{Turbines}$=1073 kW, $Wdot_{Pumps\ and\ Compressor}$=-38.1 kW, $Wdot_{net}$=1035 kW, and η=43.26%.
3. An ideal field cycle with perfect regeneration as shown in Figure 9.11.1 is designed according to the following data:
   p_{16}=10 kPa, x_{16}=0, p_{19}=200 kPa, x_{19}=0, p_{22}=1000 kPa, x_{22}=0, p_{23}=2000 kPa, p_2=7000 kPa, T_4=500 °C, $mdot_4$=1 kg/s, p_5=4000 kPa, T_6=500 °C, T_8=300 °C, $mdot_{10}$=0.9 kg/s, and $mdot_{17}$=0.12 kg/s.
   Determine rate of heat added by the heaters, total power produced by the turbines, total power required by the pumps and compressor, net power produced by the cycle, and cycle efficiency.
   ANSWER: $Qdot_{add}$=2393 kW, $Wdot_{Turbines}$=1074 kW, $Wdot_{Pumps\ and\ Compressor}$=-38.1 kW, $Wdot_{net}$=1036 kW, and η=43.28%.

9.12 COGENERATION

There are applications in which gas cycles are used to supply both gas power and process heat. The heat may be used as process steam for industrial processes, or steam to heat water for central or district heating. This type of combined heat and gas power plant is called *cogeneration*. A schematic cogeneration plant is illustrated in Figure 9.12.1. Figure 9.12.1 depicts a cogeneration plant in which an open Brayton gas-turbine plant exhausts to a heat recovery steam generator (heat exchanger), which supplies process steam to a dairy factory. The generator is provided with a gas burner (heater #2) for supplementary heat when the demand of process steam is high. The open Brayton gas-turbine is a split-shaft plant.

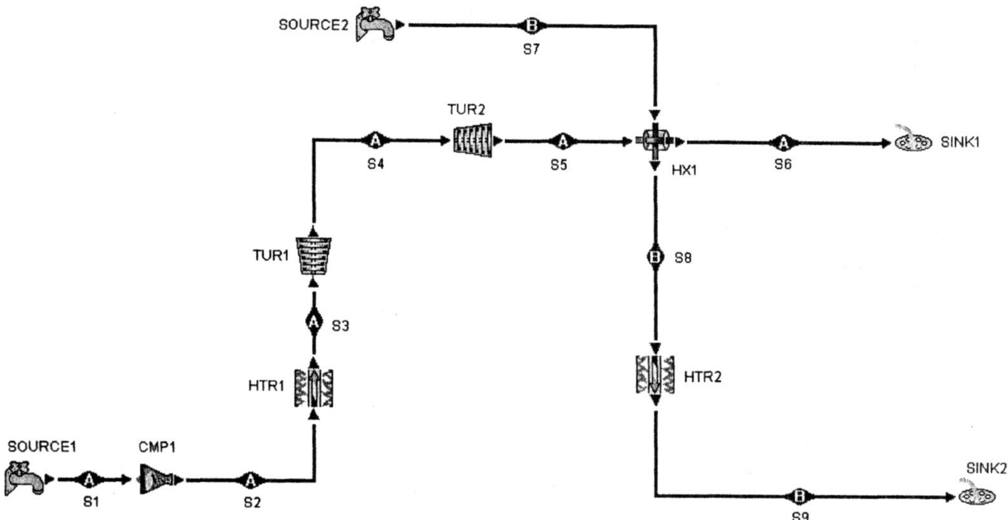

Figure 9.12.1 Cogeneration

The cogeneration cycle is composed of the following six processes:

1-2	isentropic compression
2-3	isobaric heat addition
3-4	isentropic expansion
4-5	isentropic expansion
5-6	isobaric heat removing
7-8	isobaric heat addition
8-9	isobaric heat addition

Applying the First law of thermodynamics of the open system to the cogeneration cycle yields:

$$Wdot_{12}=Wdot_{34}, \tag{9.12.1}$$

$$mdot_5(h_5 - h_6) = mdot_7 (h_8 - h_7), \tag{9.12.2}$$

The net work (W_{net}) is

$$W_{net} = W_{12} + W_{34} + W_{45} = W_{45} \quad (9.12.3)$$

and

The combined power and heat cogeneration energy utility factor (EUF) is

$$EUF= [W_{net} + mdot_5(h_5 - h_6)] / (Qdot_{23}+Qdot_{89}). \quad (9.12.4)$$

Example 9.12.1. The data given below correspond approximately to the design conditions for the dairy factory cogeneration plant:

$mdot_1$=20.45 kg/s, p_1=1 bar, T_1=25°C, p_2=7 bar, T_3=850°C, p_5=1 bar, T_6=138°C, $\eta_{compressor}=\eta_{turbine\#1}=\eta_{turbine\#2}$=85%, T_7=90°C, p_7=13 bar, and x_9=1.

Determine the power required by the compressor, power produced by turbine #1, power produced by turbine #2, rate of heat added to the combustion chamber, net power produced by the open Brayton gas-turbine plant, cycle efficiency of the open Brayton gas-turbine plant, rate of heat added to the process steam, rate of process steam, and energy utility of the cogeneration plant.

To solve this problem by CyclePad, we do the following steps:

(A) Build the cycle as shown in Figure 9.12.1. Assuming the compressor and turbines are adiabatic and 85% efficient, and the heaters and heat exchanger are isobaric.
(B) Input Cycle A working fluid=air, p_1=1 bar, T_1=25°C, $mdot_1$=20.45 kg/s, p_2=7 bar, T_3=850°C, $-Wdot_{12}=Wdot_{34}$, T_6=138°C, Cycle B working fluid=water, T_7=90°C, p_8=13 bar, and x_8=1.
(C) Display results. The answers are: $Wdot_{compressor}$=-5352 kW, $Wdot_{turbine\#1}$=5352 kW, $Wdot_{turbine\#2}$=3172 kW, $Wdot_{net}$=3172 kW, $Qdot_{comb\ chamber}$=11576 kW, η=3172/11576=27.40 %, $Qdot_{HX}$=-6086 kW, $mdot_{steam}$=2.53 kg/s, and EUF=(3172+6086)/11576=0.7998.

Example 9.12.2. Referring to the dairy factory cogeneration design conditions where 5 kg/s of process steam is needed, determine the rate of heat provided by the gas burner.

To solve this problem by CyclePad, we do the same thing as Example 9.12.1, delete x_8=1, and let x_9=1and $mdot_8$=5 kg/s.

The answer are Qdot=5960 kW and and EUF=(3172+6086)/(11576+5960)=0.5279.

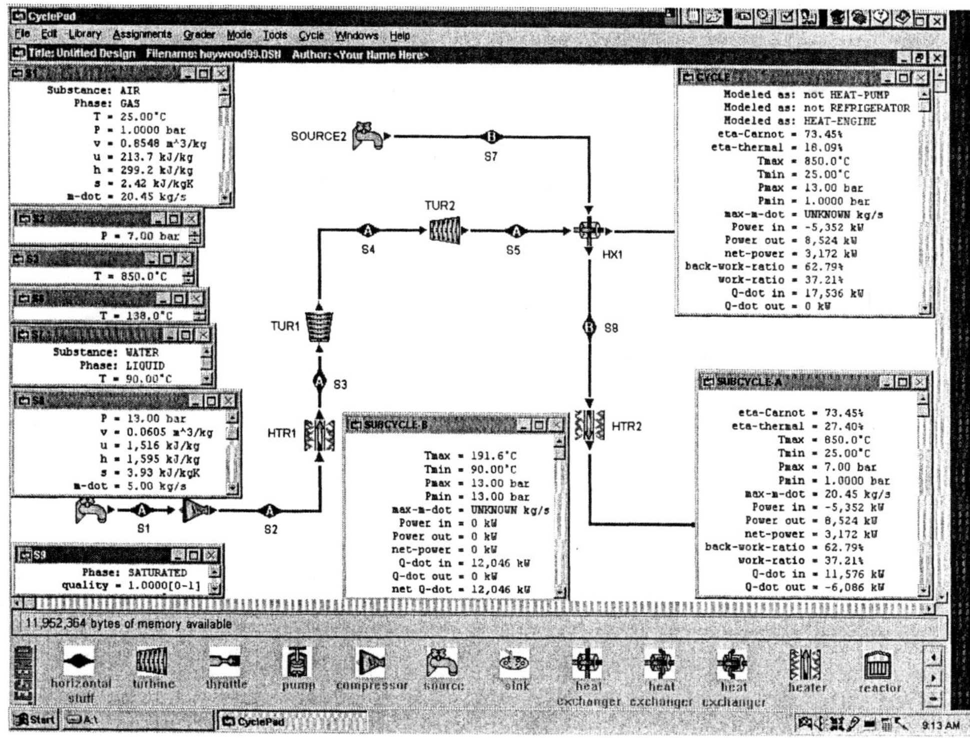

Figure Example 9.12.1 Cogeneration

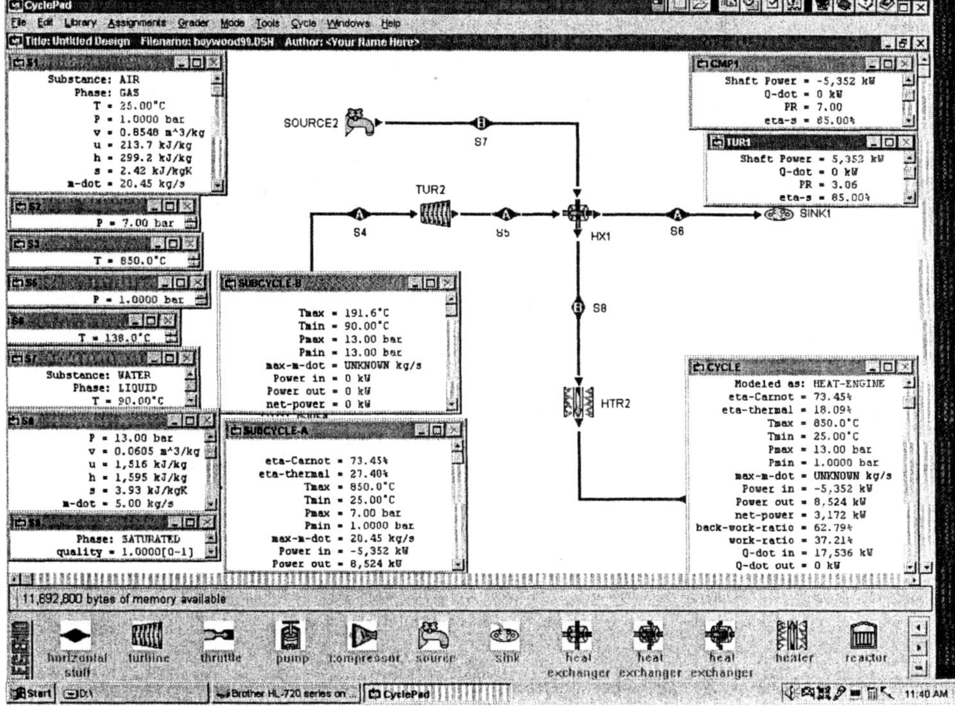

Figure Example 9.12.2 Cogeneration

Homework 9.12 Cogeneration

1. Referring to the dairy factory design conditions, except that the compressor efficiency is 80% and the mass flow rate of process steam is 4 kg/s, determine the power required by the compressor, power produced by turbine #1, power produced by turbine #2, rate of heat added to the combustion chamber, net power produced by the open Brayton gas-turbine plant, cycle efficiency of the open Brayton gas-turbine plant, rate of heat added to the process steam, rate of heat added in the gas burner, and energy utility of the cogeneration plant.
ANSWER: $Wdot_{compressor}$=-5352 kW, $Wdot_{turbine\#1}$=5352 kW, $Wdot_{turbine\#2}$=3172 kW, $Wdot_{net}$=3172 kW, $Qdot_{comb\ chamber}$=11576 kW, η=3172/11576=27.40 %, $Qdot_{HX}$=-6086 kW, $Qdot_{gas\ burner}$=2.53 kg/s, and EUF=(3172+6086)/11576=0.7998.

9.13 Design Examples

Typically, CyclePad's "build" mode allows the designer to select parts from one of its two inventory shops (closed system and open system) and connect them. Each component is clearly labeled with its input and output denoted by arrows. After connecting the parts in a complete cycle, the software will prompt the designer to stay in the "build" mode or move on to "analysis" mode. In the "analysis" mode, CyclePad combines user input and thermodynamic principles to numerically sove cycles. Here the designer selects the working fluid, component properties, and boundary conditions. CyclePad solves all possible variables as the designer adds conditions to the cycle. When all necessary entries have been made, the software will have solved for properties such as total heat input, total heat output, net power, thermal efficiency, etc. In the event that the designer enters conflicting conditions, CyclePad enters the "contradiction" mode. In this mode, the designer is told of a conflict and the program will not proceed until the contradiction is resolved. This is done in a pop-up window which shows all inputs that contributes to the error. In this way, the designer does not necessarily have to remove the entry that forced the contradiction, a very useful function indeed. If the designer wishes, CyclePad will explain what the contradiction is and how the current assumptions cannot be correct. This editing technique is very similar to the senior design engineer monitoring juniors who are not yet experienced enough to see the error in advance.

The intelligent computer software CyclePad is a very effective tool in design cycles. Any complicated gas cycle can be easily designed and analyzed using CyclePad. Optimization of design parameters of the cycle is demonstrated by the following examples.

Example 9.13.1. A 4-stage reheat and 4-stage inter-cool Brayton air cycle as shown in Figure 9.13.1a has been designed by a junior engineer with the following design input information as shown in Figure 9.13.1b:

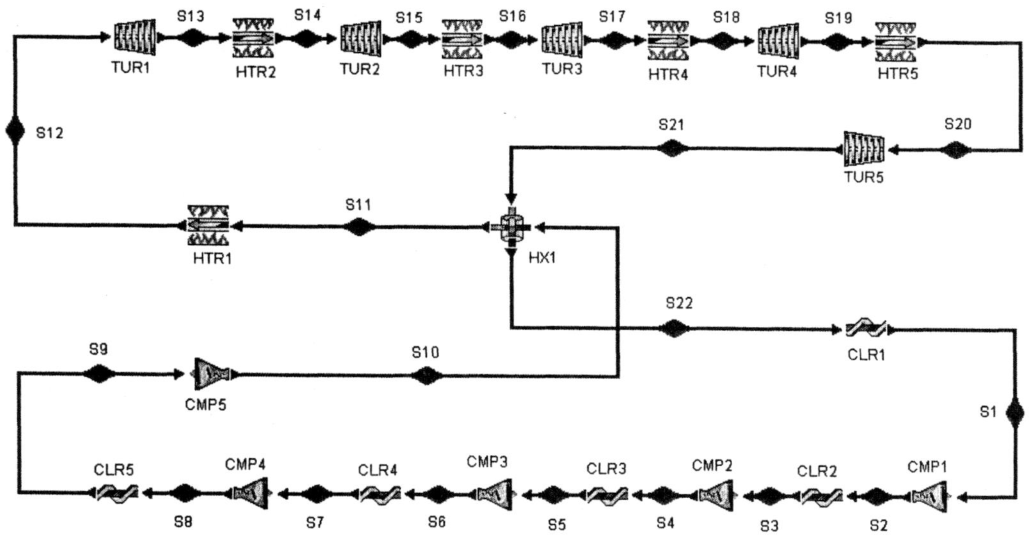

Figure Example 9.13.1a 4-stage reheat and 4-stage inter-cool Brayton air cycle

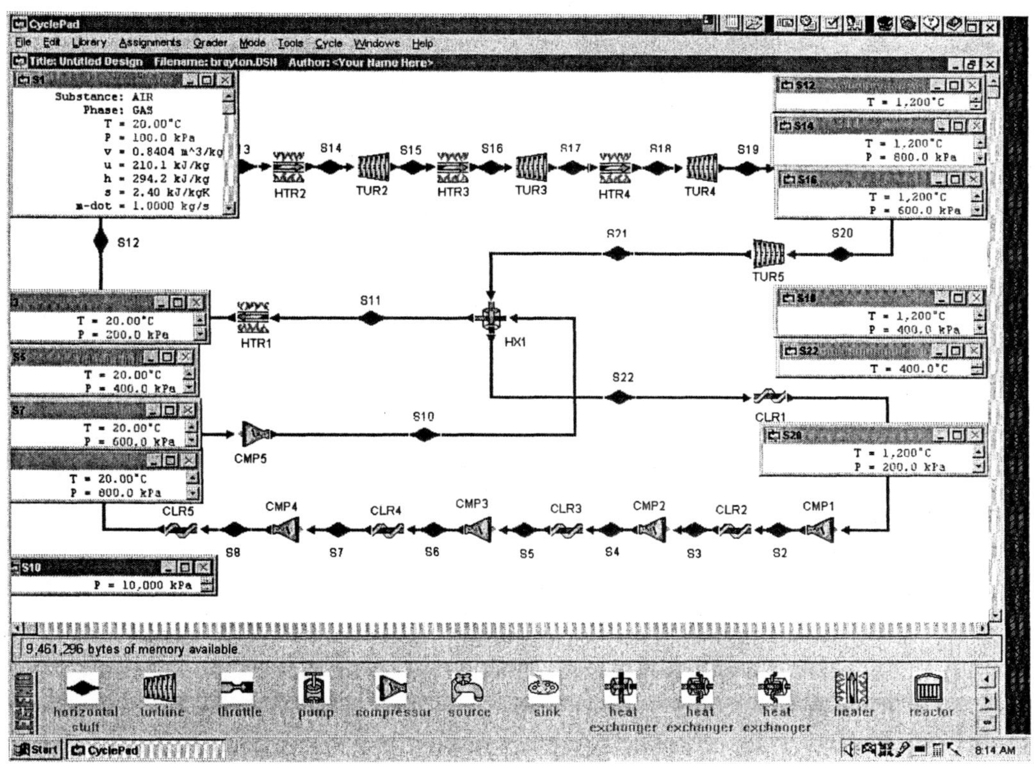

Figure Example 9.13.1b Brayton air cycle design input

Design input information:

$p_1=p_{21}=p_{22}=100$ kPa, $p_2=p_3=p_{20}=p_{19}=200$ kPa, $p_4=p_5=p_{17}=p_{18}=400$ kPa, $p_6=p_7=p_{15}=p_{16}=600$ kPa, $p_8=p_9=p_{13}=p_{14}=800$ kPa, $p_{10}=p_{11}=p_{12}=1000$ kPa, $T_1=T_3=T_5=T_7=T_9=20°C$, $T_{12}=T_{14}=T_{16}=T_{18}=T_{20}=1200°C$, $T_{22}=400°C$, $mdot_1=1$ kg/s, $\eta_{tur1}=\eta_{tur2}=\eta_{tur3}=\eta_{tur4}=\eta_{tur5}=85\%$, and $\eta_{cmpr1}=\eta_{cmpr2}=\eta_{cmpr3}=\eta_{cmpr4}=\eta_{cmpr5}=85\%$.

The following output results as shown in Figure 9.13.1c are obtained from his design:

$\eta_{cycle}=55.16\%$, $Wdot_{input}=-589.8$ kW, $Wdot_{output}=-1334$ kW, $Wdot_{net\ output}=744.1$ kW, $Qdot_{add}=1349$ kW, $Qdot_{remove}=-605.0$ kW, $Wdot_{cmp1}=-75.79$ kW, $Wdot_{cmp2}=-75.79$ kW, $Wdot_{cmp3}=-42.50$ kW, $Wdot_{cmp4}=-29.65$ kW, $Wdot_{cmp5}=-366.1$ kW, $Wdot_{tur1}=645.9$ kW, $Wdot_{tur2}=99.14$ kW, $Wdot_{tur3}=137.4$ kW, $Wdot_{tur4}=225.7$ kW, $Wdot_{tur5}=225.7$ kW, $Qdot_{htr1}=241.0$ kW, $Qdot_{htr2}=645.9$ kW, $Qdot_{htr3}=99.14$ kW, $Qdot_{htr4}=137.4$ kW, $Qdot_{htr5}=225.7$ kW, $Qdot_{clr1}=-381.4$ kW, $Qdot_{clr2}=-75.79$ kW, $Qdot_{clr3}=-75.79$ kW, $Qdot_{clr4}=-42.50$ kW, $Qdot_{clr5}=-29.65$ kW, $T_{10}=384.8°C$, $T_{11}=959.8°C$, and $T_{21}=400°C$.

The T-s diagram of the cycle is shown in Figure Example 9.13.1d.

Try to modify his design (use p_{16}, p_{18}, p_6 and p_8 as design parameters only) to get a better cycle thermal efficiency than his $\eta_{cycle}=55.16\%$.

The sensitivity analyses of η_{cycle} vs p_{16}, and η_{cycle} vs p_{18} are shown in the following diagrams. The optimization design values of p_{16} and p_{18} can be easily identified.

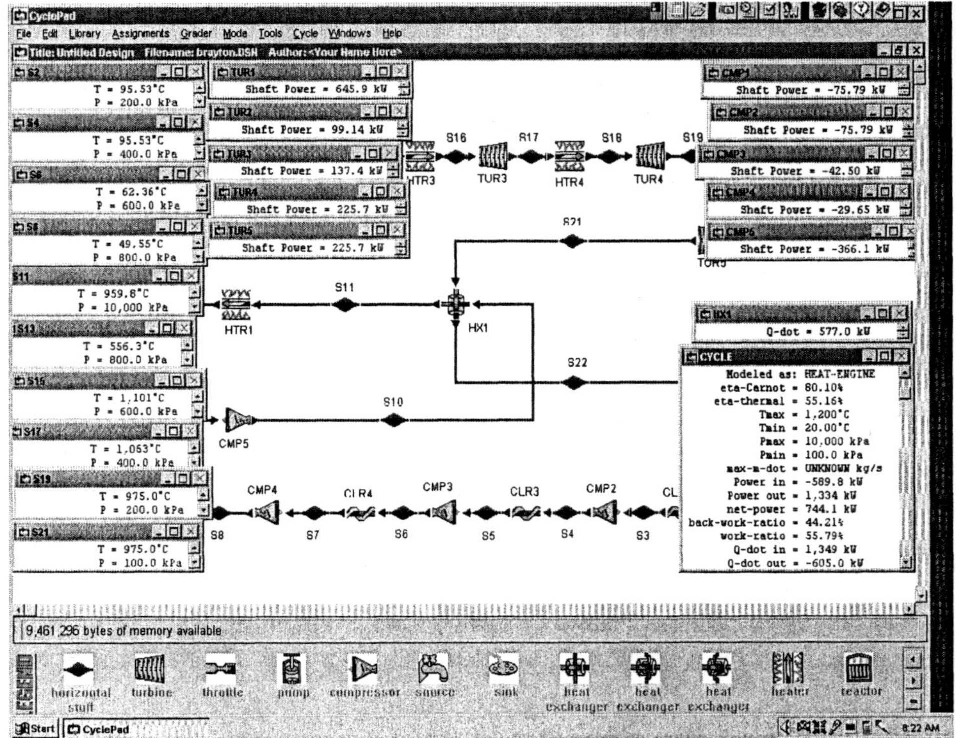

Figure Example 9.13.1c Brayton air cycle design output

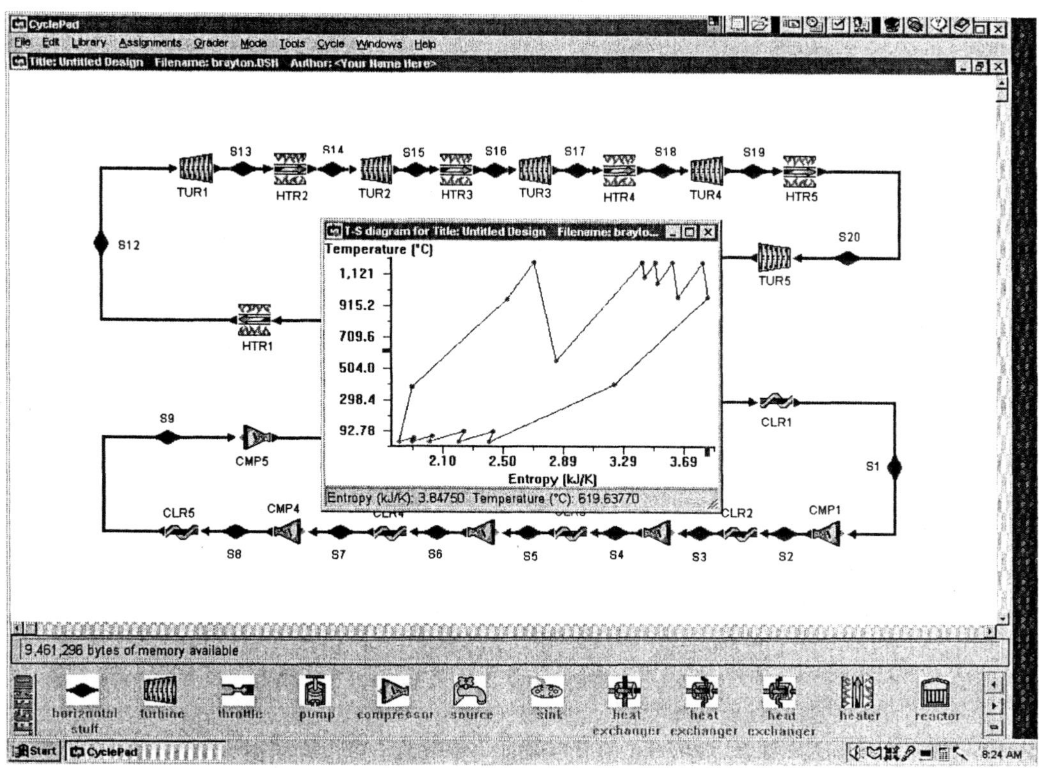

Figure Example 9.13.1d Brayton air cycle T-s diagram

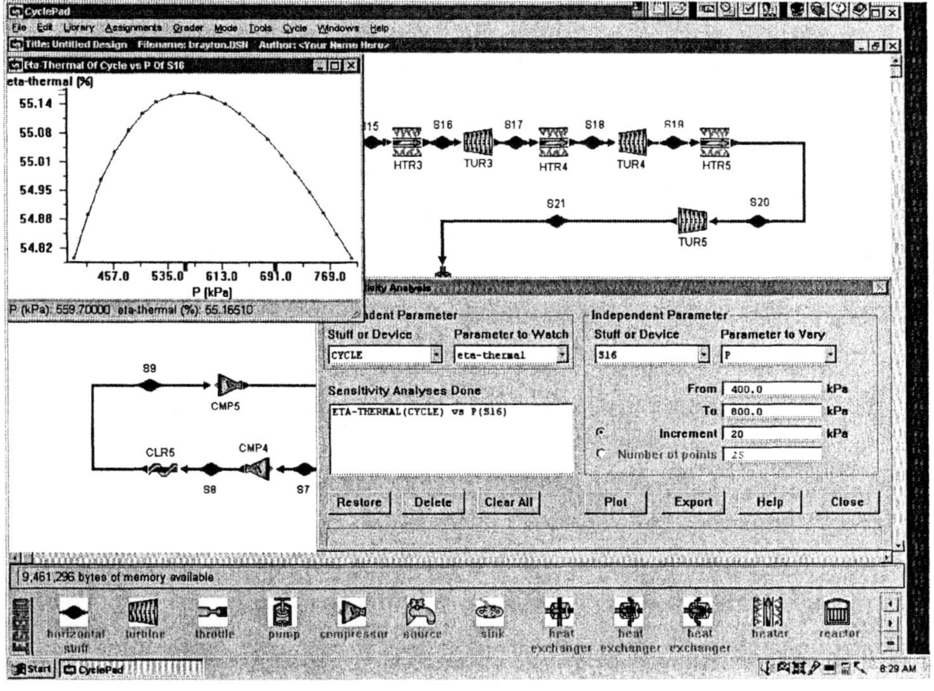

Figure Example 9.13.1e Brayton air cycle design parameter optimization

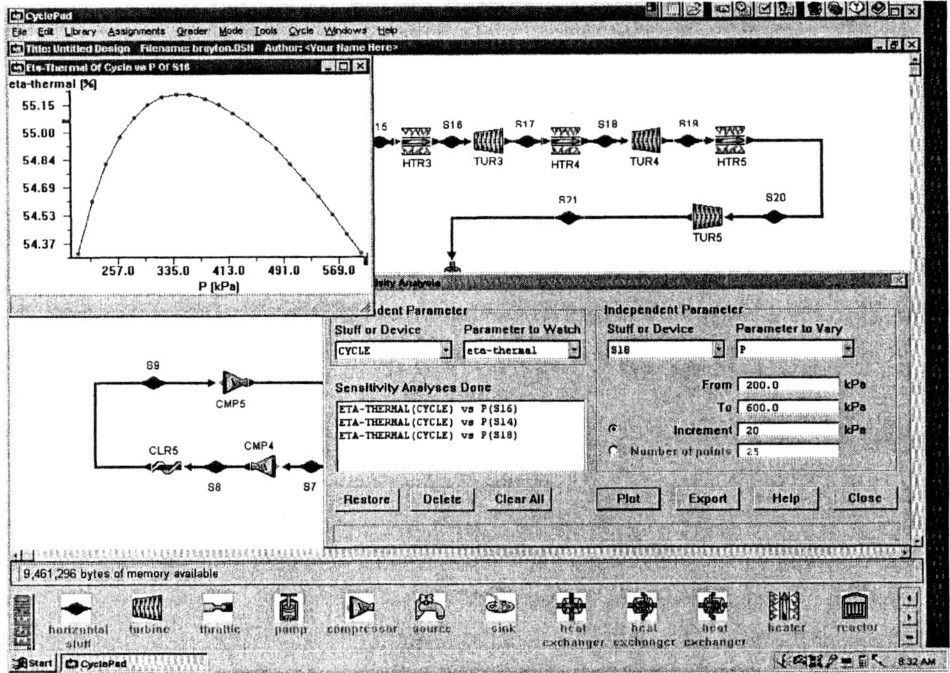

Figure Example 9.13.1f Brayton air cycle design parameter optimization

Example 9.13.2. A 3-stage regenerative steam Rankine cycle and a 4-stage intercool and 4-stage reheat air Brayton cycle is combined by a heat exchanger as shown in Figure 9.13.2a has been designed by a junior engineer with the following design input information as shown in Figure Example 9.13.2b:

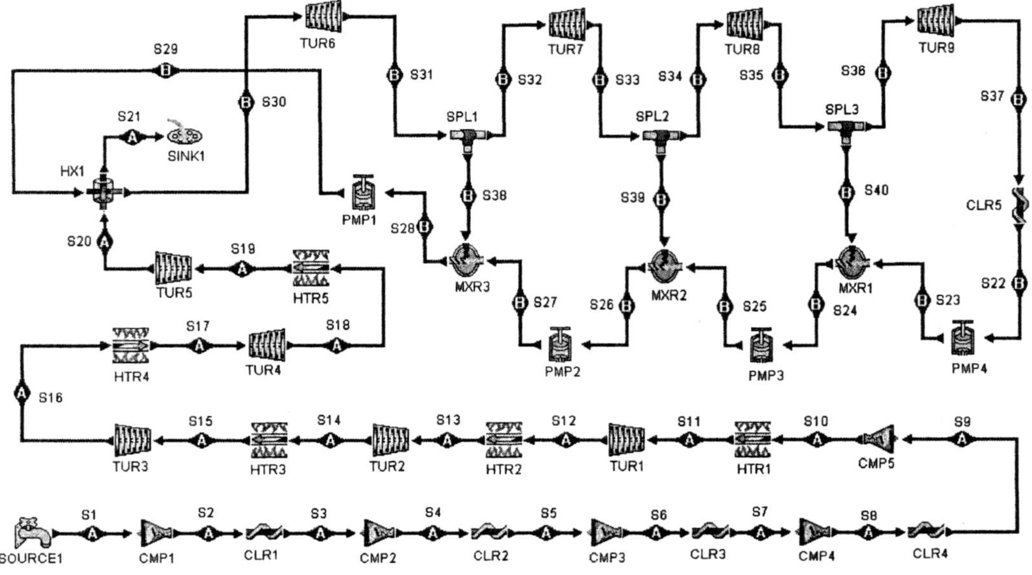

Figure Example 9.13.2a Combined Brayton-Rankine cycle

The preliminary design information is:

Brayton cycle

p_1=100 kPa, T_1=20°C, p_3=200 kPa, T_3=20°C, p_5=300 kPa, T_5=20°C, p_7=500 kPa, T_7=20°C, p_9=800 kPa, T_9=20°C, p_{11}=1200 kPa, T_{11}=1200°C, p_{13}=800 kPa, T_{13}=1200°C, p_{15}=500 kPa, T_{15}=1200°C, p_{17}=300 kPa, T_{17}=1200°C, p_{19}=200 kPa, T_{19}=1200°C, p_{20}=100 kPa, T_{21}=550°C, $mdot_1$=1 kg/s, $\eta_{tur1}=\eta_{tur2}=\eta_{tur3}=\eta_{tur4}=\eta_{tur5}$=85%, and $\eta_{cmp1}=\eta_{cmp2}=\eta_{cmp3}=\eta_{cmp4}=\eta_{cmp5}$=85%.

Rankine cycle

p_{22}=7 kPa, x_{22}=0, p_{24}=2000 kPa, x_{24}=0, p_{26}=4000 kPa, x_{26}=0, p_{28}=8000 kPa, x_{28}=0, p_{30}=12000 kPa, T_{30}=500°C, $\eta_{tur6}=\eta_{tur7}=\eta_{tur8}=\eta_{tur9}$=85%, and $\eta_{pmp1}=\eta_{pmp2}=\eta_{pmp3}=\eta_{pmp4}$=85%.

The following output results as shown in Figure 9.13.2c are obtained from his design:

Combined cycle

η_{cycle}=41.67%, $Wdot_{input}$=-280.1 kW, $Wdot_{output}$=1007 kW, $Wdot_{net\ output}$=727.1 kW, $Qdot_{add}$=1745 kW, $Qdot_{remove}$=-486.0 kW.

Brayton cycle

η_{cycle}=32.33%, $Wdot_{input}$=-264.9 kW, $Wdot_{output}$=829.1 kW, $Wdot_{net\ output}$=564.2 kW, $Qdot_{add}$=1745 kW, $Qdot_{remove}$=-648.9 kW, $Wdot_{cmp1}$=-75.79 kW, $Wdot_{cmp2}$=-42.50 kW, $Wdot_{cmp3}$=-54.38 kW, $Wdot_{cmp4}$=-49.74 kW, $Wdot_{cmp5}$=-42.50 kW, $Wdot_{tur1}$=137.4 kW, $Wdot_{tur2}$=157.9 kW, $Wdot_{tur3}$=170.6 kW, $Wdot_{tur4}$=137.4 kW, $Wdot_{tur5}$=225.7 kW, $Qdot_{clr1}$=-75.79 kW, $Qdot_{clr2}$=-42.50 kW, $Qdot_{clr3}$=-54.38 kW, $Qdot_{clr4}$=-49.74 kW, $Qdot_{htr1}$=1142 kW, $Qdot_{htr2}$=137.4 kW, $Qdot_{htr3}$=157.9 kW, $Qdot_{htr4}$=170.6 kW, $Qdot_{htr5}$=137.4 kW, $Qdot_{heat\ exch}$=-426.5 kW.

Rankine cycle

η_{cycle}=38.20%, $Wdot_{input}$=-15.20 kW, $Wdot_{output}$=178.1 kW, $Wdot_{net\ output}$=162.9 kW, $Qdot_{add}$=426.5 kW, $Qdot_{remove}$=-263.6 kW, $Wdot_{pmp1}$=-13.26 kW, $Wdot_{pmp2}$=-1.14 kW, $Wdot_{pmp3}$=-0.4943 kW, $Wdot_{pmp4}$=-0.3131 kW, $Wdot_{tur6}$=22.91 kW, $Wdot_{tur7}$=31.64 kW, $Wdot_{tur8}$=25.56 kW, $Wdot_{tur9}$=98.01 kW, $Qdot_{clr5}$=-263.6 kW, $Qdot_{heat\ exch}$=426.5 kW, $mdot_{24}$=0.1782 kg/s, $mdot_{26}$=0.1939 kg/s, $mdot_{28}$=0.2164 kg/s, $mdot_{31}$=0.2164 kg/s, $mdot_{32}$=0.1939 kg/s, $mdot_{34}$=0.1782 kg/s, $mdot_{36}$=0.1304 kg/s, $mdot_{38}$=0.0225 kg/s, $mdot_{39}$=0.0157 kg/s, and $mdot_{40}$=0.0478 kg/s.

Let us try to modify his design (use p_5, p_7, p_{15}, and p_{17} as design parameters only) to get a better cycle thermal efficiency than his η_{cycle}=41.67%.

The sensitivity analysis of η_{cycle} vs p_5, η_{cycle} vs p_7, η_{cycle} vs p_{15}, and η_{cycle} vs p_{17} are shown in the following diagrams. The optimization design values of p_5, p_7, p_{15} and p_{18} can be easily identified.

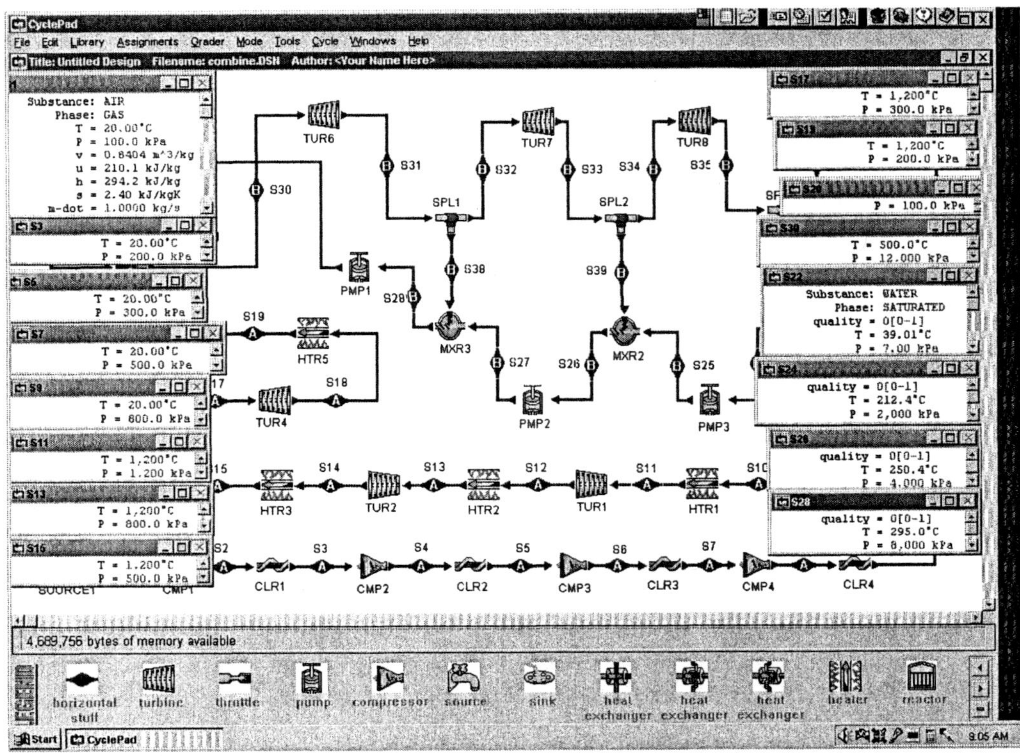

Figure Example 9.13.2b Combined Brayton-Rankine cycle input

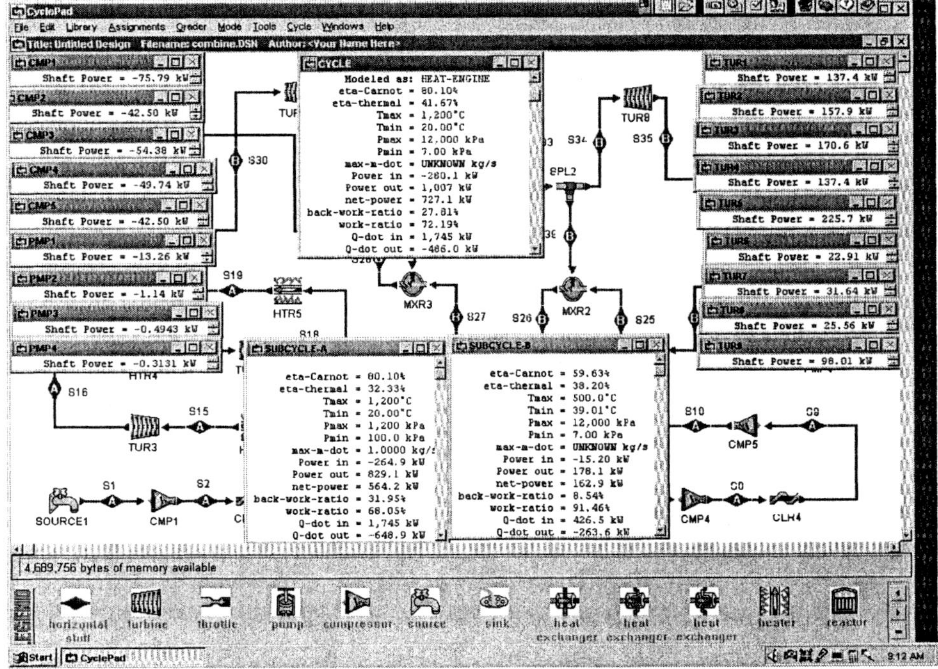

Figure Example 9.13.2c Combined Brayton-Rankine cycle output

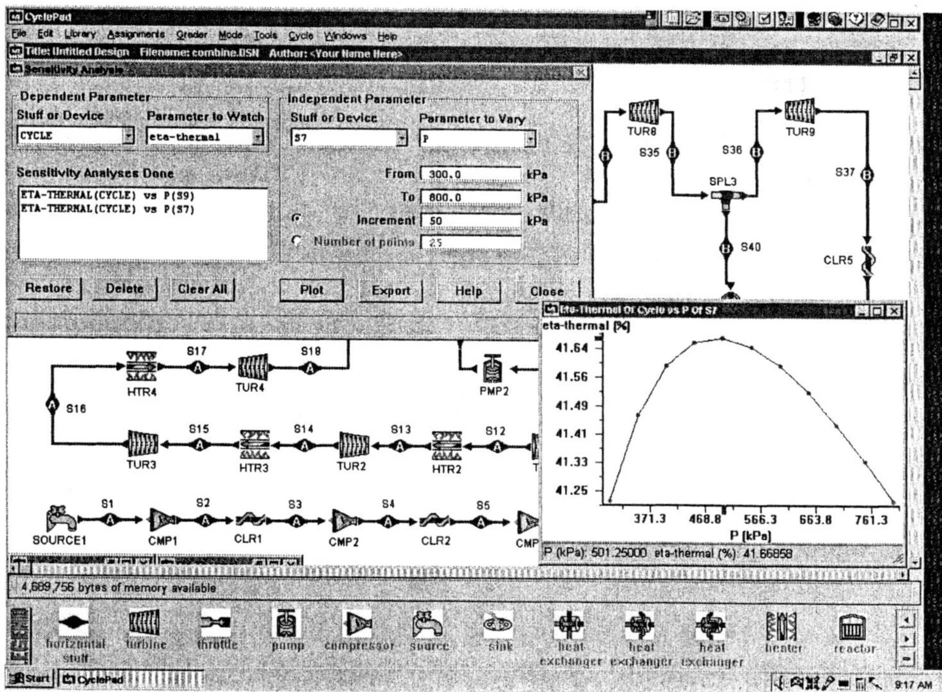

Figure Example 9.13.2c Combined Brayton-Rankine cycle sensitivity diagram

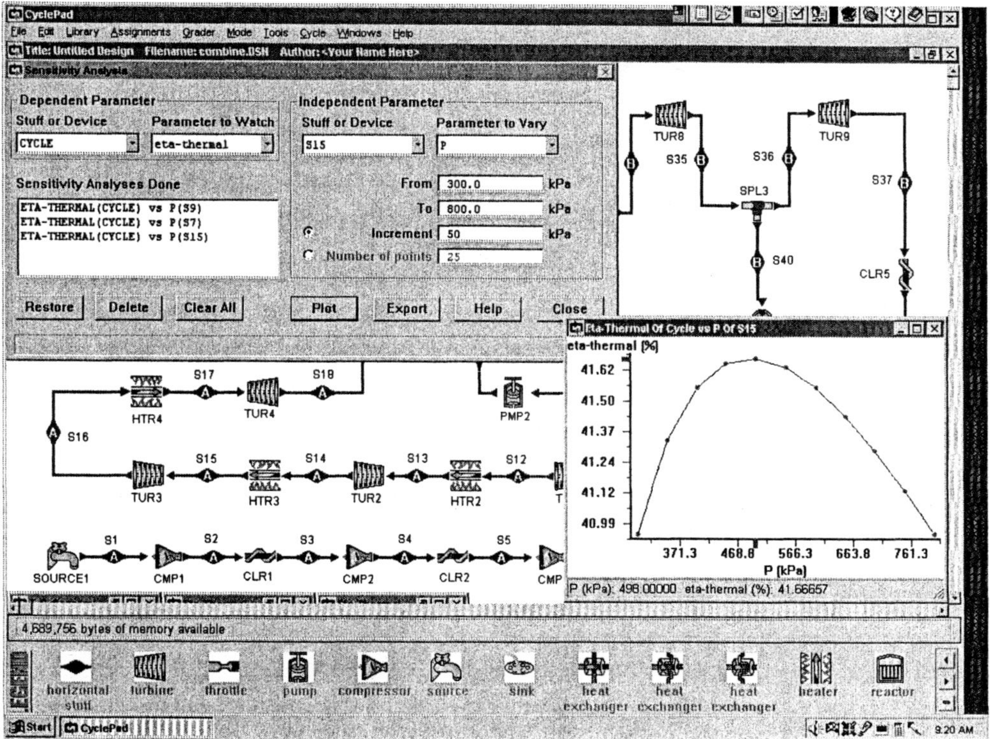

Figure Example 9.13.2d Combined Brayton-Rankine cycle sensitivity diagram

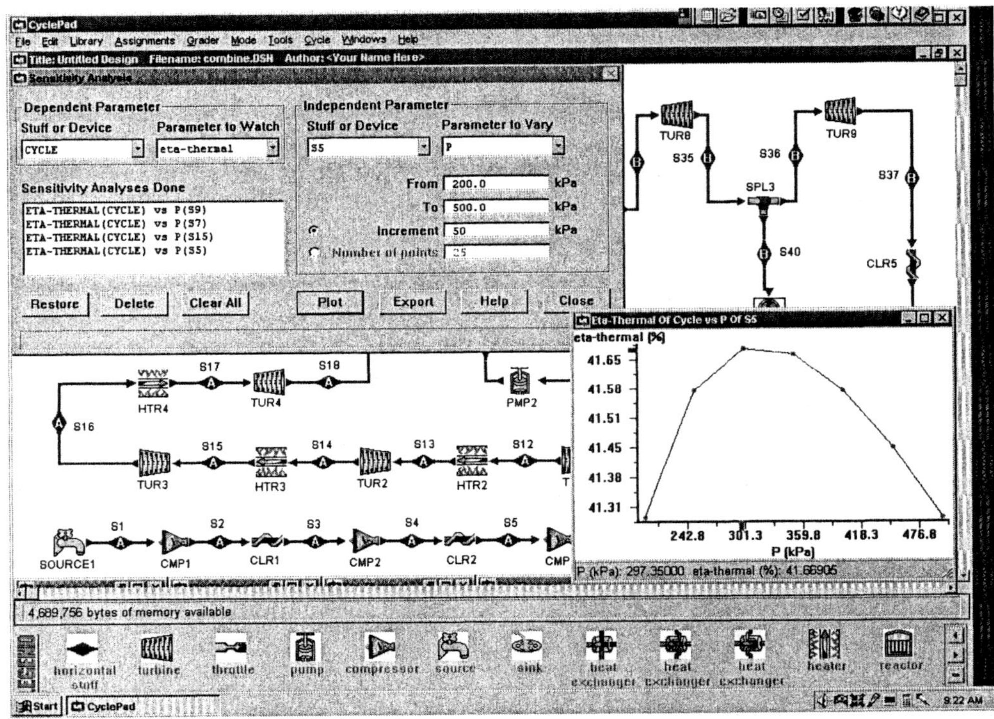

Figure Example 9.13.2e Combined Brayton-Rankine cycle sensitivity diagram

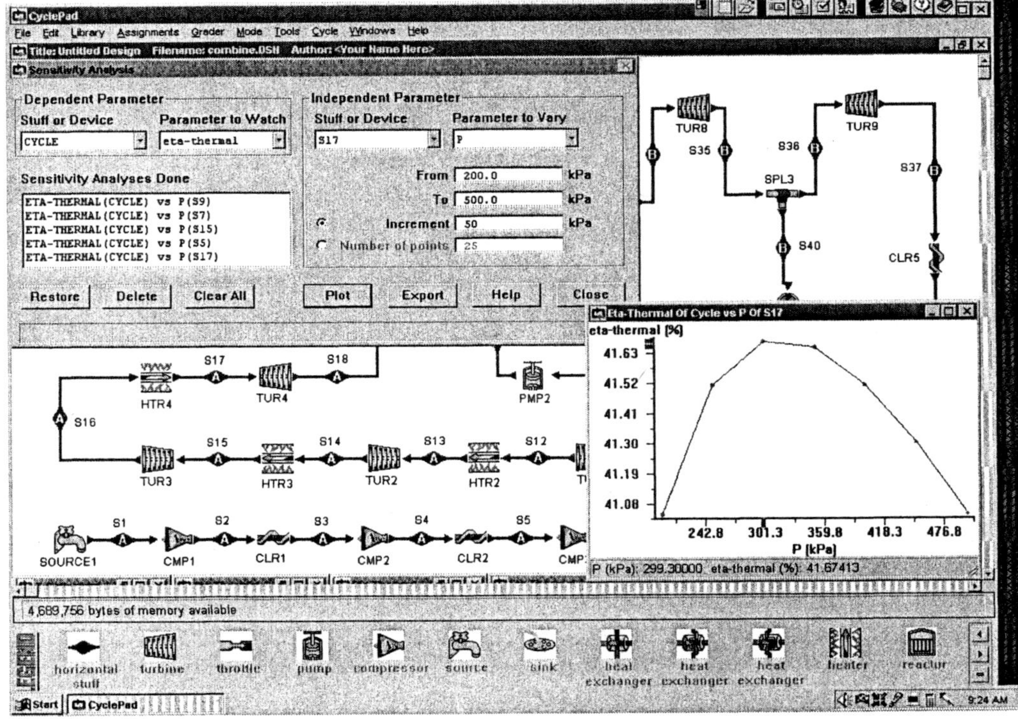

Figure Example 9.13.2f Combined Brayton-Rankine cycle sensitivity diagram

Homework 9.13 Design

1. A 3-stage regenerative steam Rankine cycle and a 4-stage intercool and 4-stage reheat air Brayton cycle combined with a heat exchanger, as shown in Figure 9.13.2a, has been designed by a junior engineer with the following design input information:
   Brayton cycle
   p_1=100 kPa, T_1=20°C, p_3=200 kPa, T_3=20°C, p_5=300 kPa, T_5=20°C, p_7=500 kPa, T_7=20°C, p_9=800 kPa, T_9=20°C, p_{11}=1200 kPa, T_{11}=1300°C, p_{13}=800 kPa, T_{13}=1300°C, p_{15}=500 kPa, T_{15}=1300°C, p_{17}=300 kPa, T_{17}=1300°C, p_{19}=200 kPa, T_{19}=1300°C, p_{20}=100 kPa, T_{21}=550°C, $mdot_1$=1 kg/s, $\eta_{tur1}=\eta_{tur2}=\eta_{tur3}=\eta_{tur4}=\eta_{tur5}$=85%, and $\eta_{cmp1}=\eta_{cmp2}=\eta_{cmp3}=\eta_{cmp4}=\eta_{cmp5}$=85%.
   Rankine cycle
   p_{22}=7 kPa, x_{22}=0, p_{24}=2000 kPa, x_{24}=0, p_{26}=4000 kPa, x_{26}=0, p_{28}=8000 kPa, x_{28}=0, p_{30}=12000 kPa, T_{30}=500°C, $\eta_{tur6}=\eta_{tur7}=\eta_{tur8}=\eta_{tur9}$=85%, and $\eta_{pmp1}=\eta_{pmp2}=\eta_{pmp3}=\eta_{pmp4}$= 85%.
   The following output results are obtained from his design:
   Combined cycle
   η_{cycle}=43.26%, $Wdot_{input}$=-283.2 kW, $Wdot_{output}$=1099 kW, $Wdot_{net\ output}$=815.9 kW, $Qdot_{add}$=1886 kW, $Qdot_{remove}$=-538.5 kW.
   Try to improve his design (use p_3, p_5, p_7, p_9, p_{24}, p_{26}, and p_{28} as design parameters only) and get a better cycle thermal efficiency than his η_{cycle}=43.26%.
2. A 3-stage regenerative steam Rankine cycle and a 4-stage intercool and 4-stage reheat air Brayton cycle combined with a heat exchanger, as shown in Figure 9.13.2a, has been designed by a junior engineer with the following design input information:
   Brayton cycle
   p_1=100 kPa, T_1=20°C, p_3=200 kPa, T_3=20°C, p_5=300 kPa, T_5=20°C, p_7=500 kPa, T_7=20°C, p_9=800 kPa, T_9=20°C, p_{11}=1200 kPa, T_{11}=1300°C, p_{13}=800 kPa, T_{13}=1300°C, p_{15}=500 kPa, T_{15}=1300°C, p_{17}=300 kPa, T_{17}=1300°C, p_{19}=200 kPa, T_{19}=1300°C, p_{20}=100 kPa, T_{21}=550°C, $mdot_1$=1 kg/s, $\eta_{tur1}=\eta_{tur2}=\eta_{tur3}=\eta_{tur4}=\eta_{tur5}$=85%, and $\eta_{cmp1}=\eta_{cmp2}=\eta_{cmp3}=\eta_{cmp4}=\eta_{cmp5}$=88%.
   Rankine cycle
   p_{22}=7 kPa, x_{22}=0, p_{24}=2000 kPa, x_{24}=0, p_{26}=4000 kPa, x_{26}=0, p_{28}=8000 kPa, x_{28}=0, p_{30}=12000 kPa, T_{30}=500°C, $\eta_{tur6}=\eta_{tur7}=\eta_{tur8}=\eta_{tur9}$=85%, and $\eta_{pmp1}=\eta_{pmp2}=\eta_{pmp3}=\eta_{pmp4}$ =85%.
   The following output results are obtained from his design:
   Combined cycle
   η_{cycle}=43.70%, $Wdot_{input}$=-274.1 kW, $Wdot_{output}$=1099 kW, $Wdot_{net\ output}$=824.9 kW, $Qdot_{add}$=1888 kW, $Qdot_{remove}$=-530.9 kW.
   Try to improve his design (use p_3, p_5, p_7, p_9, p_{24}, p_{26}, and p_{28} as design parameters only) and get a better cycle thermal efficiency than his η_{cycle}=43.70%.

9.14 SUMMARY

Heat engines that use gases as the working fluid in an open system model are treated in this chapter. The modern gas-turbine engine operates on the Brayton cycle. The basic Brayton cycle consists of an isentropic compression process, an isobaric combustion process, an isentropic expansion process, and an isobaric cooling process. The thermal efficiency of the basic Brayton cycle depends on the compression ratio across the compressor. The compression ratio is defined as $r_p = p_{compressor\ exit}/p_{compressor\ inlet}$. The thermal efficiency of the basic Brayton cycle can be improved by split-shaft, and regeneration. The net work of the basic Brayton cycle can be improved by inter-cooling, and reheating. The power of the basic Brayton cycle can be controlled by air-bleed.

The Ericsson cycle is a modified Brayton cycle with many stages of intercooling and reheat. It has the same efficiency of the Carnot cycle operating between the same temperature limits. The Feher cycle is a cycle operating above the critical point of the working fluid.

A combined cycle couples two cycles in series such that the heat discharged from the top cycle is used for the heat input of the bottom cycle. A cascaded cycle couples several cycles in series such that the heat discharged from the upstream cycle is used for the heat input of the downstream cycle. A Braysson cycle is a combined cycle of a Brayton cycle and an Ericsson cycle.

A Field cycle is a combined cycle of a Brayton cycle and a Rankine cycle. A cogeneration cycle produces both work and heat.

Chapter 10

REFRIGERATION AND HEAT PUMP CYCLES

OBJECTIVES

After reading and studying the material in this chapter, you should be able to:

1. Understand that refrigeration and heat pump are the reverse of a power-producing cycle.
2. Describe the sequence of processes in the basic vapor refrigeration and heat pump cycles.
3. Sketch the basic and actual vapor refrigeration and heat pump cycles on T-s diagrams.
4. Describe the sequence of processes in the absorption refrigeration cycle.
5. Analyze the vapor refrigeration and heat pump cycles using CyclePad.
6. Quantitatively understand the construction and operation of the major pieces of equipment used on heat pump, refrigerator and air conditioning system.
7. Know the desirable properties of refrigerants.
8. Know the domestic refrigerator-freezer and air conditioning-heat pump system.
9. Understand cascade and multi-stage vapor refrigeration and heat pump cycles.
10. Analyze Brayton, Stirling and Ericsson refrigeration and heat pump cycles using CyclePad.
11. Analyze gas liquefaction and solidification cycles using CyclePad.
12. Design various refrigeration and heat pump cycles using CyclePad.

10.1 CARNOT REFRIGERATOR AND HEAT PUMP

A system is called a refrigerator or a heat pump depending on the purpose of the system. If the purpose of the system is to remove heat from a low temperature thermal reservoir, it is a refrigerator. If the purpose of the system is to deliver heat to a high temperature thermal reservoir, it is a heat pump.

Carnot cycle is a reversible cycle. Reversing the cycle will also reverse the directions of heat and work interactions. The reversed Carnot heat engine cycles are Carnot refrigeration and heat pump cycles. Therefore a reversed Carnot vapor heat engine is either a Carnot vapor refrigerator or a Carnot vapor heat pump depending upon the function of the cycle.

A schematic diagram of the *Carnot refrigerator* or *Carnot heat pump* is illustrated in Figure 10.1.1.

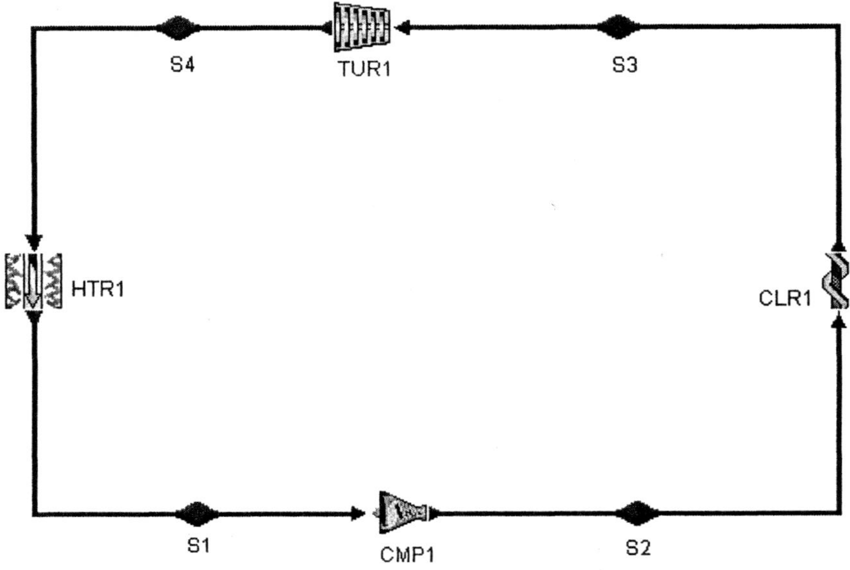

Figure 10.1.1. Carnot refrigerator or Carnot heat pump

1-2	isentropic compression
2-3	isothermal cooling
3-4	isentropic expansion
4-1	isothermal heating

Applying the First law of thermodynamics of the open system to each of the four processes of the basic vapor refrigeration cycle at steady flow and steady state condition yields:

$$Q_{12} = 0, \tag{10.1.1}$$

$$0 - W_{12} = m(h_2 - h_1), \tag{10.1.2}$$

$$W_{23} = 0, \tag{10.1.3}$$

$$Q_{23} - 0 = m(h_3 - h_2), \tag{10.1.4}$$

$$Q_{34} = 0 \tag{10.1.5}$$

$$0 - W_{34} = m(h_3 - h_4), \tag{10.1.6}$$

$$W_{41} = 0, \tag{10.1.7}$$

and

$$Q_{41} - 0 = m(h_1 - h_4), \tag{10.1.8}$$

The desirable energy output of the refrigeration cycle is the heat added to the evaporator (or heat removed from the inner space or the low temperature reservoir of the refrigerator). The energy input to the cycle is the compressor work required. The energy produced is the turbine work. The net work (W_{net}) required to operate the cycle is (W_{12}+W_{34}). Thus the coefficient of performance (COP) of the cycle is

$$\beta_R = Q_{41} / W_{net} \tag{10.1.9}$$

The rate of the heat removed from the inner space of the refrigerator is called cooling load or cooling capacity. The cooling load of a refrigeration system is sometimes giving a unit in tons of refrigeration. A *ton of refrigeration* is the removal of heat from the cold space at a rate of 200 Btu/min (12000 Btu/h) or 211 kJ/min (3.52 kW). A ton of refrigeration is the rate of heat required to make a ton of ice per unit of time.

Example 10.1.1. Determine the COP, horsepower required and cooling load of a Carnot vapor refrigeration cycle using R-12 as the working fluid in which the condenser temperature is 100°F and the evaporation temperature is 20°F. The circulation rate of fluid is 0.1 lbm/s. Determine the compressor power required, turbine power produced, net power required, cooling load, quality at the inlet of the evaporator, quality at the inlet of the compressor, and COP of the refrigerator.

To solve this problem by CyclePad, we take the following steps:

1. Build
   (A) Take a compressor, a cooler (condenser), a turbine, and a heater (evaporator) from the open system inventory shop and connect the four devices to form the basic vapor refrigeration cycle.
   (B) Switch to analysis mode.
2. Analysis
   (A) Assume a process for each of the four devices: (a) compressor as isentropic, (b) condenser as isobaric, (c)turbine as isentropic, and (d) evaporator as isobaric.
   (B) Input the given information: (a) working fluid is R-12, (b) the inlet temperature and quality of the compressor are 20°F and 1, (c) the inlet temperature and quality of the turbine are 100°F and 0, (d) the phase of the exit refrigerant from the turbine is saturated mixture, and (e) the mass flow rate is 0.1 lbm/s.
3. Display results
   (A) Display the cycle properties results. The cycle is a refrigerator. The answers are the compressor power required=-1.36 hp, turbine power produced=0.2234 hp, net power required=-1.13 hp, cooling load=4.79 Btu/s, quality at the inlet of the evaporator=0.2504, and COP=5.99

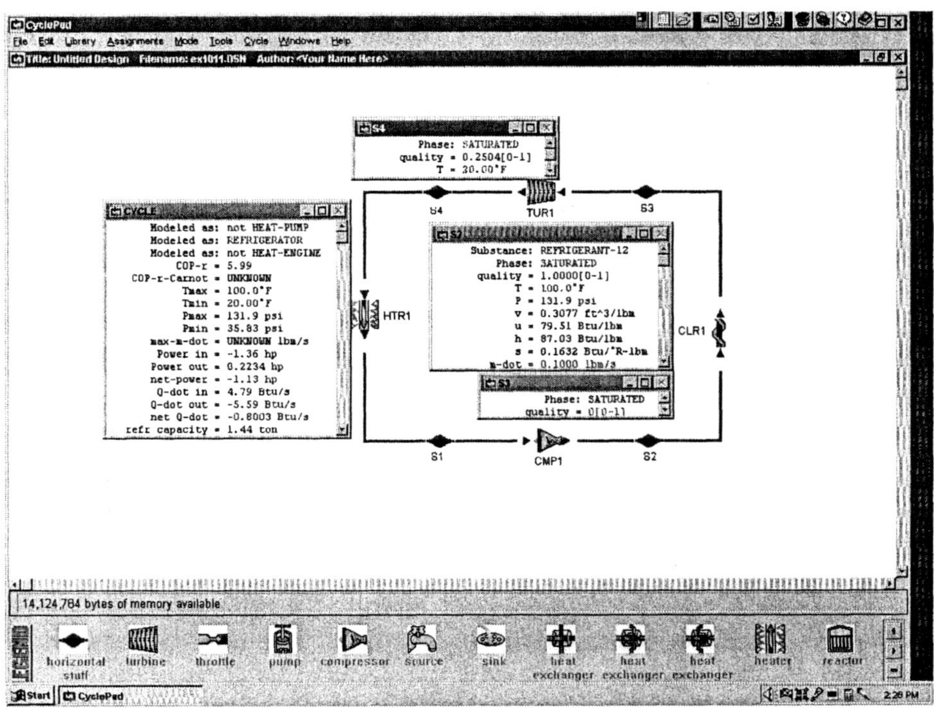

Figure Example 10.1.1 Carnot vapor refrigeration cycle

Comments: 1. The turbine work produced is very small. It does not pay to install an expansive device to produce a small amount of work. The expansion process can be achieved by a simple throttling valve. 2. The compressor handles the refrigerant as a mixture of saturated liquid and saturated vapor. It is not practical. Therefore, the compression process should be move out of the mixture region to the superheated region.

10.2 BASIC VAPOR REFRIGERATION CYCLE

The Carnot refrigerator is not a practical cycle, because the compressor is designed to handle superheated vapor or gas. The turbine in the small temperature and pressure range produces very small amount work. It is not worth of to have an expensive turbine in the cycle to produce a very small amount of work. Therefore the Carnot refrigeration cycle is modified to have the compression process completely in the superheated region and the turbine is replaced by an inexpensive throttling valve termed thermal expansion valve to form a basic vapor refrigeration cycle. The schematic diagram of the basic vapor refrigeration system is shown in Figure 10.2.1. The components of the *basic vapor refrigeration cycle* include a compressor, a condenser, an expansion valve and an evaporator. The T-s diagram of the cycle is shown in Figure 10.2.2. Notice that the throttling process 3-4 is an irreversible process and is indicated by a broken line on the T-s diagram.

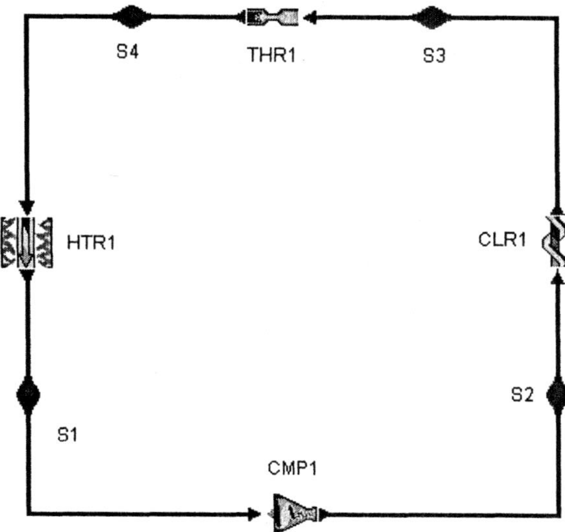

Figure 10.2.1 Basic vapor refrigeration system

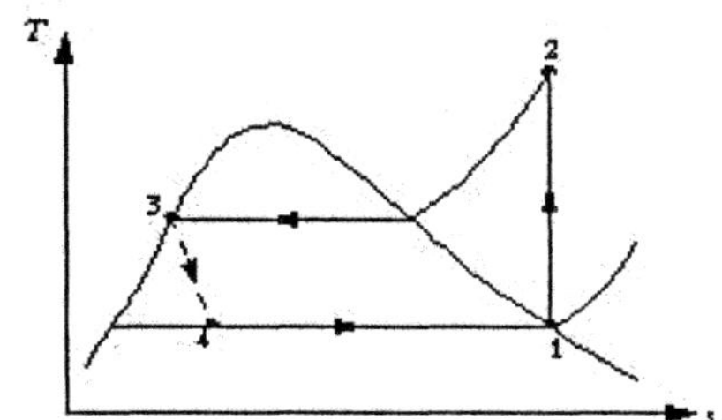

Figure 10.2.2 T-s diagram of the basic vapor refrigeration cycle

The basic vapor refrigeration cycle consists of the following four processes:

1-2 isentropic compression
2-3 isobaric cooling
3-4 throttling
4-1 isobaric heating

Applying the First law of thermodynamics of the open system to each of the four processes of the basic vapor refrigeration cycle yields:

$$Q_{12} = 0, \quad (10.3.1)$$

$$0 - W_{12} = m(h_2 - h_1), \quad (10.1.2)$$

$$W_{23} = 0, \quad (10.1.3)$$

$$Q_{23} - 0 = m(h_3 - h_2), \quad (10.1.4)$$

$Q_{34} = 0$ and $W_{34} = 0$ (10.1.5)

$0 - 0 = m(h_3 - h_4)$, (10.1.6)

$W_{41} = 0$, (10.1.7)

and

$Q_{41} - 0 = m(h_1 - h_4)$, (10.1.8)

The desirable energy output of the basic vapor refrigeration cycle is the heat added to the evaporator (or heat removed from the inner space of the refrigerator, Q_{41}). The energy input to the cycle is the compressor work required (W_{12}). Thus the coefficient of performance (COP) of the basic refrigeration cycle is

$\beta_R = Q_{41} / W_{12} = (h_4 - h_1)/(h_3 - h_2)$ (10.1.9)

The rate of the heat removed from the inner space of the refrigerator is called cooling load or cooling capacity. The cooling load of a refrigeration system is sometimes giving a unit in tons of refrigeration. A ton of refrigeration is the removal of heat from the cold space at a rate of 12000 Btu/h.

Example 10.2.1. Determine the COP, horsepower required and cooling load of a basic vapor refrigeration cycle using R-12 as the working fluid in which the condenser pressure is 130 psia and the evaporation pressure is 35 psia. The circulation rate of fluid is 0.1 lbm/s. Determine the compressor power required, cooling load, quality at the inlet of the evaporator, and COP of the refrigerator.

To solve this problem by CyclePad, we take the following steps:

1. Build
   (A) Take a compressor, a condenser, a valve, and a heater (evaporator) from the open system inventory shop and connect the four devices to form the basic vapor refrigeration cycle.
   (B) Switch to analysis mode.
2. Analysis
   (A) Assume a process for each of the four devices: (a) compressor as isentropic, (b) condenser as isobaric, (c) valve as constant enthalpy, and (d) evaporator as isobaric.
   (B) Input the given information: (a) working fluid is R-12, (b) the inlet pressure and quality of the compressor are 35 psia 20°F and 1, (c) the inlet pressure and quality of the valve are 130 psia and 0, and (d) the mass flow rate is 0.1 lbm/s.
3. Display results
   (A) Display the cycle properties results. The cycle is a refrigerator. The answers are the compressor power required=-1.27 hp, cooling load=4.84 Btu/s=1.45 ton, quality at the inlet of the evaporator=0.2738, and COP=5.38.

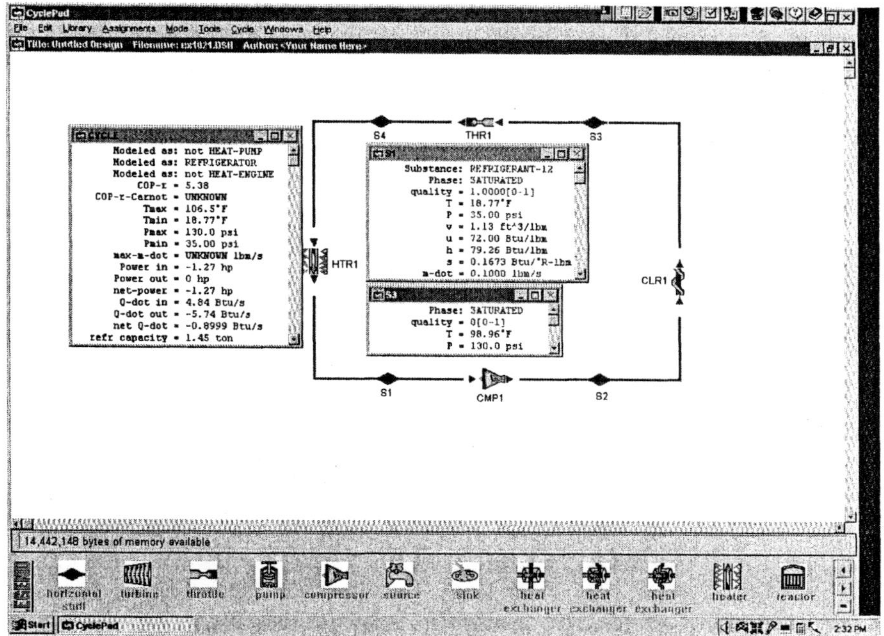

Figure Example 10.2.1 Basic vapor refrigeration cycle

Example 10.2.2. Determine the COP, horsepower required and cooling load of a basic vapor refrigeration cycle using R-12 as the working fluid in which the condenser pressure is 130 psia and the evaporation pressure is 35 psia. The circulation rate of fluid is 0.1 lbm/s. The temperature of the refrigerant at the exit of the compressor is 117°F. Determine the compressor power required, cooling load, quality at the inlet of the evaporator, and COP of the refrigerator.

To solve this problem by CyclePad, we take the following steps:

1. Build
   (A) Take a compressor, a condenser, a valve, and a heater (evaporator) from the open system inventory shop and connect the four devices to form the basic vapor refrigeration cycle.
   (B) Switch to analysis mode.
2. Analysis
   (A) Assume a process for each of the four devices: (a) compressor as adiabatic, (b) condenser as isobaric, (c) valve as constant enthalpy, and (d) evaporator as isobaric.
   (B) Input the given information: (a) working fluid is R-12, (b) the inlet pressure and quality of the compressor are 35 psia and 1, (c) the temperature of the refrigerant at the exit of the compressor is 117°F, (d) the inlet pressure and quality of the valve are 130 psia and 0, and (e) the mass flow rate is 0.1 lbm/s.
3. Display results
   (A) Display the cycle properties results. The cycle is a refrigerator. The answers are the compressor power required=-1.54 hp, cooling load=4.84 Btu/s=1.45 ton, quality at the inlet of the evaporator=0.2738, and COP=4.44.

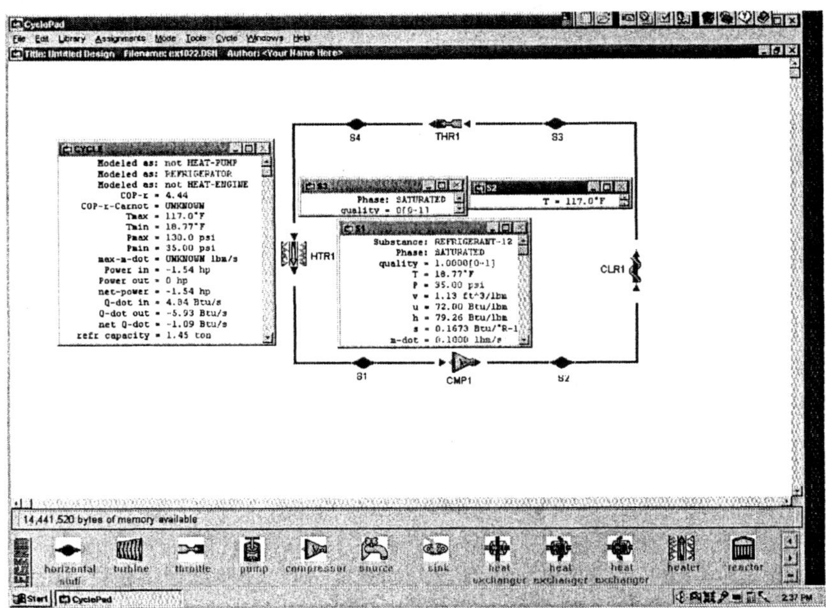

Figure Example 10.2.2 Actual vapor refrigeration cycle

Homework 10.2 Basic Vapor Refrigeration Cycle

1. Why is the Carnot refrigeration cycle executed within the saturation dome not a realistic model for refrigeration cycles?
2. What is the difference between a refrigerator and a heat pump?
3. Why is the throttling valve not replaced by an isentropic turbine in the ideal refrigeration cycle?
4. What is the area enclosed by the refrigeration cycle on a T-s diagram?
5. Does the ideal vapor compression refrigeration cycle involve any internal irreversibilities?
6. A steady flow ideal 0.4 tons refrigerator use refrigerant R134a as the working fluid. The evaporator pressure is 120 kPa. The condenser pressure is 600 kPa. Determine (a) the mass rate flow, (b) the compressor power required, (c) the rate of heat absorbed from the refrigerated space, (d) the rate of heat removed from the condenser, and (e) the COP.
7. An actual vapor compression refrigeration cycle operates at steady state with refrigerant 134a as the working fluid. Saturated vapor enters the compressor at 263 K. Superheated vapor enters the condenser at 311 K. Saturated liquid leaves the condenser at 301 K. The mass flow rate of refrigerant is 0.1 kg/s. Determine (a) the cooling load, (b) the compressor work required, (c) the condenser pressure, (d) the rate of heat removed from the condenser, (e) the compressor efficiency, and (f) the COP.
8. An actual vapor compression refrigeration cycle operates at steady state with refrigerant 134a as the working fluid. The evaporator pressure is 120 kPa. The condenser pressure is 600 kPa. The mass flow rate of refrigerant is 0.1 kg/s. The efficiency of the compressor is 85%. Determine (a) the compressor power, (b) the refrigerating capacity, and (c) the coefficient of performance (COP).

9. Refrigerant R134a enters the compressor of a steady flow vapor compression refrigeration cycle as superheated vapor at 0.14 Mpa and -10°C at a rate of 0.04 kg/s, and it leaves at 0.7 Mpa and 50°C. The refrigerant is cooled in the condenser to 24°C and saturated liquid. Determine (a) the compressor power required, (b) the rate of heat absorbed from the refrigerated space, (c) the compressor efficiency, and (d) the COP.
10. Find the compressor power required, quality of the refrigerant at the end of the throttling process, cooling load and COP for a refrigerator that uses refrigerant-12 as the working fluid and is designed to operate at an evaporator temperature of 5°C and a condenser temperature of 30°C. The compressor efficiency is 68 percent. The mass rate flow of refrigerant 12 is 0.22 kg/s.
11. Consider an ideal refrigerator which uses refrigerant-12 as the working fluid. The temperature of the refrigerant in the evaporator is -10°C and in the condenser it is 38°C. The refrigerant is circulated at the rate of 0.031 kg/s. Determine the compressor power required, quality of the refrigerant at the end of the throttling process, cooling load and COP of the refrigerator.
12. An ideal refrigerator uses ammonia as the working fluid. The temperature of the refrigerant in the evaporator is 20°F and the pressure in the condenser is 140 psia. The refrigerant is circulated at the rate of 0.051 lbm/s. Determine the compressor power required, cooling load and COP of the refrigerator.
13. An ice-making machine operates on an ideal refrigeration cycle using refrigerant-134a. The refrigerant enters the compressor as saturated vapor at 20 psia and leaves the condenser as saturated liquid at 80 psia. For 1 ton (12,000 Btu/h) of refrigeration, determine the compressor power required, mass rate flow of refrigerant-134a, and COP of the refrigerator.
14. An ice-making machine operates on an ideal refrigeration cycle using refrigerant-134a. The refrigerant enters the compressor at 18 psia and 0°F, and leaves the condenser at 125 psia and 90°F. For 1 ton (12,000 Btu/h) of refrigeration, determine the compressor power required, mass rate flow of refrigerant-134a, and COP of the refrigerator.
15. Find the compressor power required, cooling load and COP for a refrigerator that uses refrigerant-12 as the working fluid and is designed to operate at an evaporator temperature of 5°C and a condenser temperature of 30°C. The mass flow rate of refrigerant 12 is 0.22 kg/s.
16. Consider a 2-ton (24,000 Btu/h) air conditioning unit that operates on an ideal refrigeration cycle with refrigerant-134a as the working fluid. The refrigerant enters the compressor as saturated vapor at 140 kPa and is compressed to 800 kPa. Determine the compressor power required, mass rate flow of refrigerant-134a, quality of the refrigerant at the end of the throttling process, and COP of the refrigerator.

10.3 Actual Vapor Refrigeration Cycle

The actual vapor refrigeration cycle deviates from the ideal cycle primarily because of the inefficiency of the compressor as shown in Figure 10.3.1.

In industry, pressure drops associated with fluid flow and heat transfer to or from the surroundings are also considered. The vapor that enters the compressor is usually superheated rather than at saturated vapor state. The degree of superheat of the refrigerant at the inlet of the compressor determines the extent of opening of the expansion valve. This is a principal way to control the refrigeration cycle. The refrigerant that enters the throttling valve is usually compressed rather than at saturated liquid state. The T-s diagram of the actual vapor refrigeration cycle is shown in Figure 10.3.2.

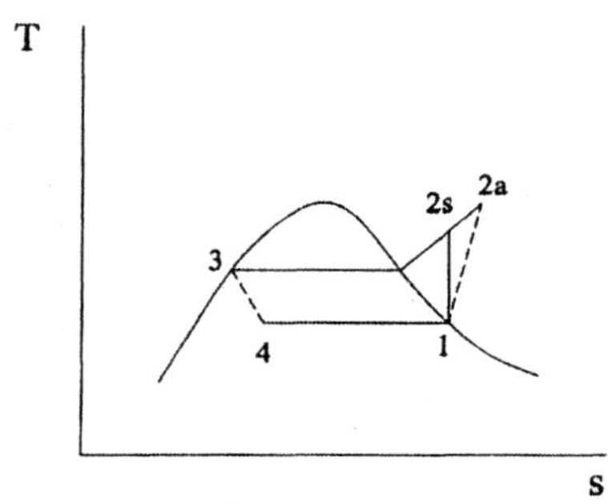

Figure 10.3.1. T-s diagram of actual vapor refrigeration cycle

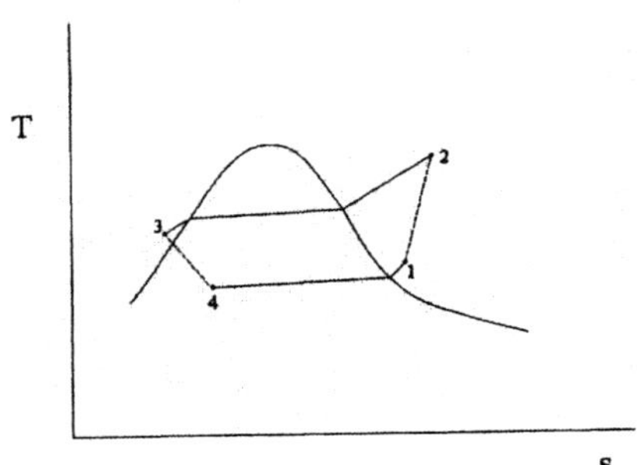

Figure 10.3.2. T-s diagram of actual vapor refrigeration cycle

Example 10.3.1. Determine the COP, horsepower required and cooling load of an actual air conditioning unit using R-12 as the working fluid. The refrigerant enters the compressor at 100 kPa and 5°C. The compressor efficiency is 87%. The refrigerant enters the throttling valve at 1.2 MPa and 45°C. The circulation rate of fluid is 0.05 kg/s. Show the cycle on T-s diagram. Determine the COP and cooling load of the air conditioning unit, and the power required for the compressor. Plot the sensitivity diagram of COP vs condenser pressure.

To solve this problem by CyclePad, we take the following steps:

1. Build
   (A) Take a compressor, a condenser, a valve, and a heater (evaporator) from the open system inventory shop and connect the four devices to form the actual vapor refrigeration cycle.
   (B) Switch to analysis mode.
2. Analysis
   (A) Assume a process for each of the four devices: (a) compressor as adiabatic, (b) condenser as isobaric, (c) valve as constant enthalpy, and (d) evaporator as isobaric.
   (B) Input the given information: (a) working fluid is R-12, (b) the inlet temperature and pressure of the compressor are 5°C and 100 kPa, (c) the inlet pressure and temperature of the valve are 1.2 MPa and 45°C, (d) the mass flow rate is 0.05 kg/s, and (e) The compressor efficiency is 87%.
3. Display results
   (A) Display the T-s diagram and cycle properties results. The cycle is a refrigerator. The answers are COP=2.09, cooling load=5.88 kW=1.67 ton, and Net power input=-2.81 kW, and
   (B) Display the sensitivity diagram of cycle COP vs condenser pressure.

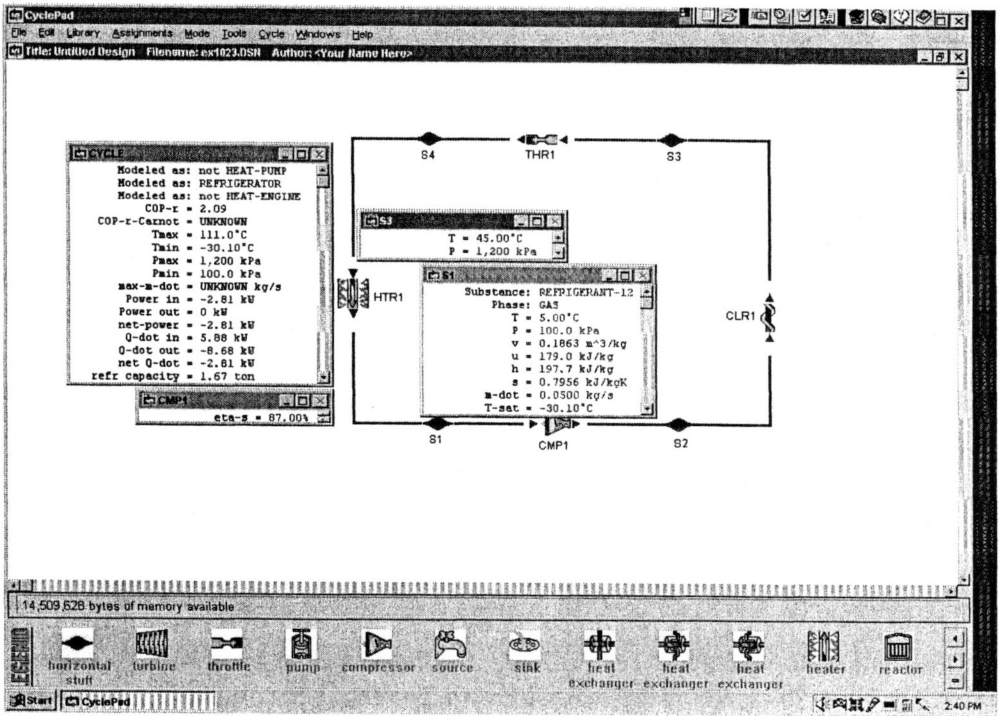

Figure Example 10.3.1a Actual refrigeration cycle

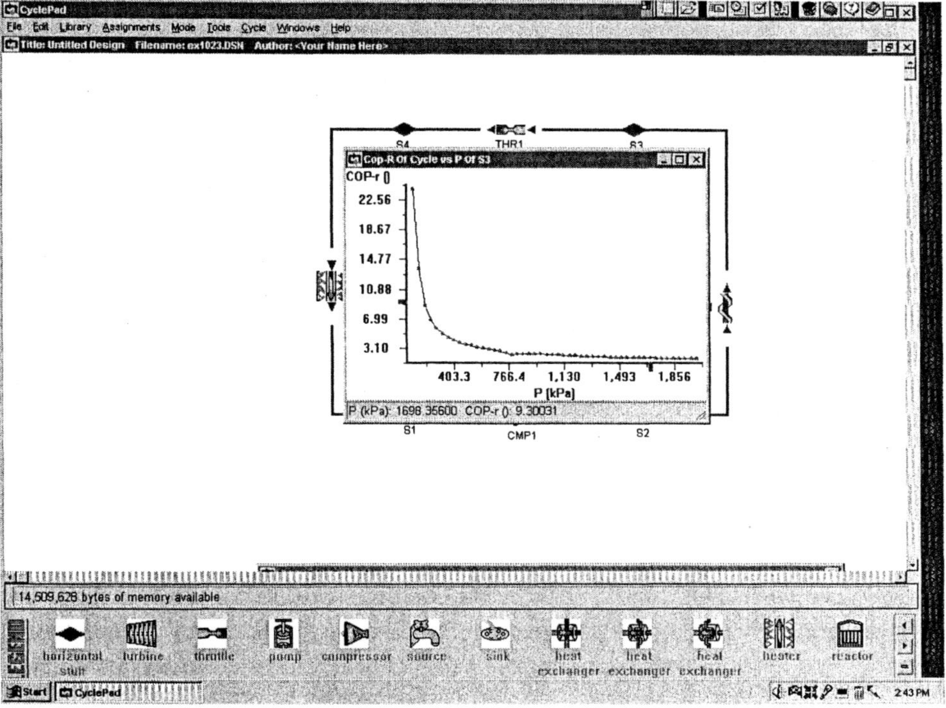

Figure Example 10.3.1b Actual refrigeration cycle sensitivity analysis

Homework 10.3 Actual Vapor Refrigeration Cycle

1. Does the ideal vapor refrigeration cycle involve any internal irreversibilities?
2. Why is the inlet state of the compressor of the actual vapor refrigeration cycle in the superheated vapor region?
3. Why is the inlet state of the throttling process of the actual vapor refrigeration cycle in the compressed liquid region?
4. R-134a enters the compressor of an actual refrigerator at 140 kPa and -10°C at a rate of 0.05 kg/s and leaves at 800 kPa. The compressor efficiency is 80%. The refrigerant is cooled in the condenser to 26°C. Determine the rate of heat added, rate of heat removed, power input, cooling load, and COP of the actual refrigerator.
5. R-22 enters the compressor of an actual refrigerator at 140 kPa and -10°C at a rate of 0.05 kg/s and leaves at 800 kPa. The compressor efficiency is 80%. The refrigerant is cooled in the condenser to 26°C. Determine the rate of heat added, rate of heat removed, power input, cooling load, and COP of the actual refrigerator.
6. Ammonia enters the compressor of an actual refrigerator at 140 kPa and -10°C at a rate of 0.05 kg/s and leaves at 800 kPa. The compressor efficiency is 80%. The refrigerant is cooled in the condenser to 26°C. Determine the rate of heat added, rate of heat removed, power input, cooling load, and COP of the actual refrigerator.

10.4 Basic Vapor Heat Pump Cycle

A system is called a refrigerator or a heat pump depending on the purpose of the system. If the purpose of the system is to remove heat from a low temperature thermal reservoir, it is a refrigerator. If the purpose of the system is to deliver heat to a high temperature thermal reservoir, it is a heat pump.

Consequently, the methodology of analysis for heat pump is identical to that for refrigerator.

The schematic diagram of a basic vapor heat pump is shown in Figure 10.4.1. The components of the basic vapor heat pump include a compressor, a condenser, an expansion valve and an evaporator.

The T-s diagram of the basic vapor heat pump cycle which consists of the following four processes is shown in Figure 10.4.2:

1-2 isentropic compression
2-3 isobaric cooling
3-4 throttling
4-1 isobaric heating

Applying the First law of thermodynamics of the open system to each of the four processes of the basic vapor heat pump yields:

$$Q_{12} = 0, \qquad (10.3.1)$$

$$0 - W_{12} = m(h_2 - h_1), \qquad (10.1.2)$$

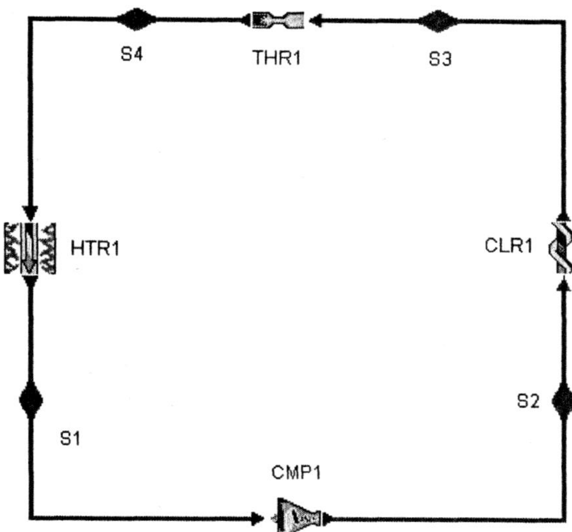

Figure 10.4.1 Basic vapor heat pump

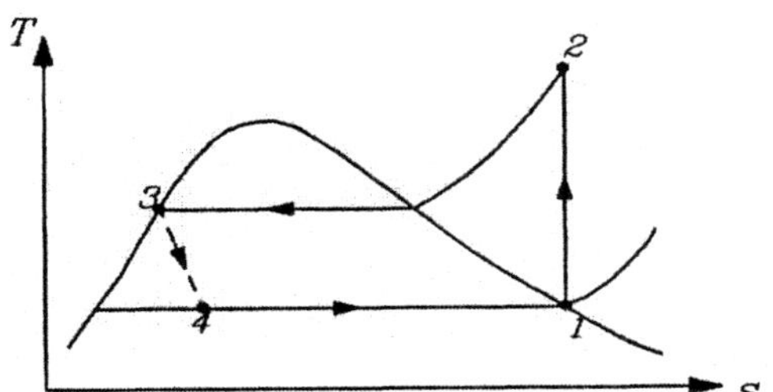

Figure 10.4.2 T-s diagram of the basic vapor heat pump cycle

$$W_{23} = 0, \tag{10.1.3}$$

$$Q_{23} - 0 = m(h_3 - h_2), \tag{10.1.4}$$

$$Q_{34} = 0 \text{ and } W_{34} = 0 \tag{10.1.5}$$

$$0 - 0 = m(h_3 - h_4), \tag{10.1.6}$$

$$W_{41} = 0, \tag{10.1.7}$$

and

$$Q_{41} - 0 = m(h_1 - h_4), \tag{10.1.8}$$

The desirable energy output of the basic vapor heat pump is the heat removed from the condenser (or heat added to the high temperature thermal reservoir). The energy input to the cycle is the compressor work required. Thus the coefficient of performance (COP) of the cycle is

$$\beta_{HP} = Q_{23} / W_{12} = (h_3 - h_2)/(h_3 - h_2) \tag{10.1.9}$$

The rate of the heat removed from the inner space of the heat pump is called heating load or heating capacity.

Example 10.4.1 Determine the COP, horsepower required and heating load of a basic vapor heat pump cycle using R-134a as the working fluid in which the condenser pressure is 900 kPa and the evaporation pressure is 240 kPa. The circulation rate of fluid is 0.1 kg/s. Show the cycle on a T-s diagram. Plot the sensitivity diagram of COP vs condenser pressure. Convert the SI unit system to British system and find the answer.

To solve this problem by CyclePad, we take the following steps:

1. Build
   (A) Take a compressor, a condenser, a valve, and a heater (evaporator) from the open system inventory shop and connect the four devices to form the basic heat pump cycle.
   (B) Switch to analysis mode.
2. Analysis
   (A) Assume a process for each of the four devices: (a) compressor as isentropic, (b) condenser as isobaric, (c) valve as constant enthalpy, and (d) evaporator as isobaric.
   (B) Input the given information: (a) working fluid is R-134a, (b) the inlet pressure and quality of the compressor are 240 kPa and 1, (c) the inlet pressure and quality of the valve are 900 kPa and 0, and (d) the mass flow rate is 0.1kg/s.
3. Display results
   (A) Display the T-s diagram and cycle properties results. The cycle is a heat pump. The answers are COP=6.28, heating load=-17.27 kW, and Net power input=-2.75 kW, and
   (B) Display the sensitivity diagram of cycle COP vs evaporation pressure.

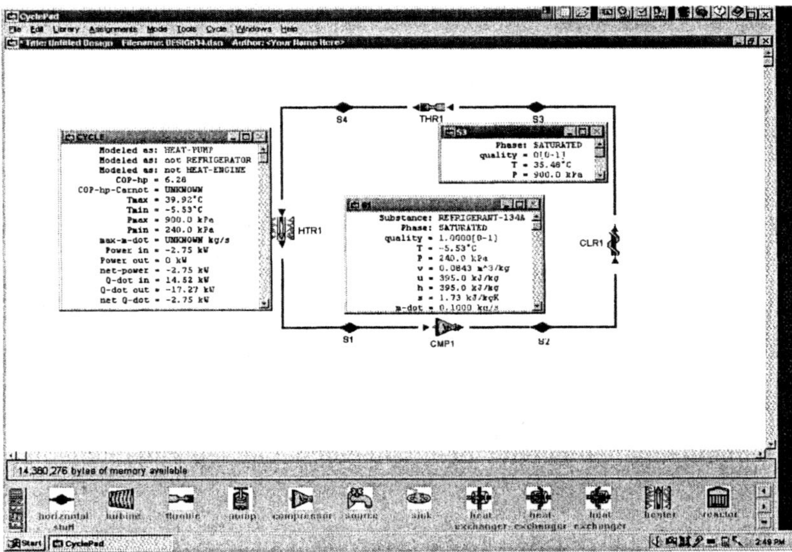

Figure Example 10.4.1a Basic vapor heat pump cycle

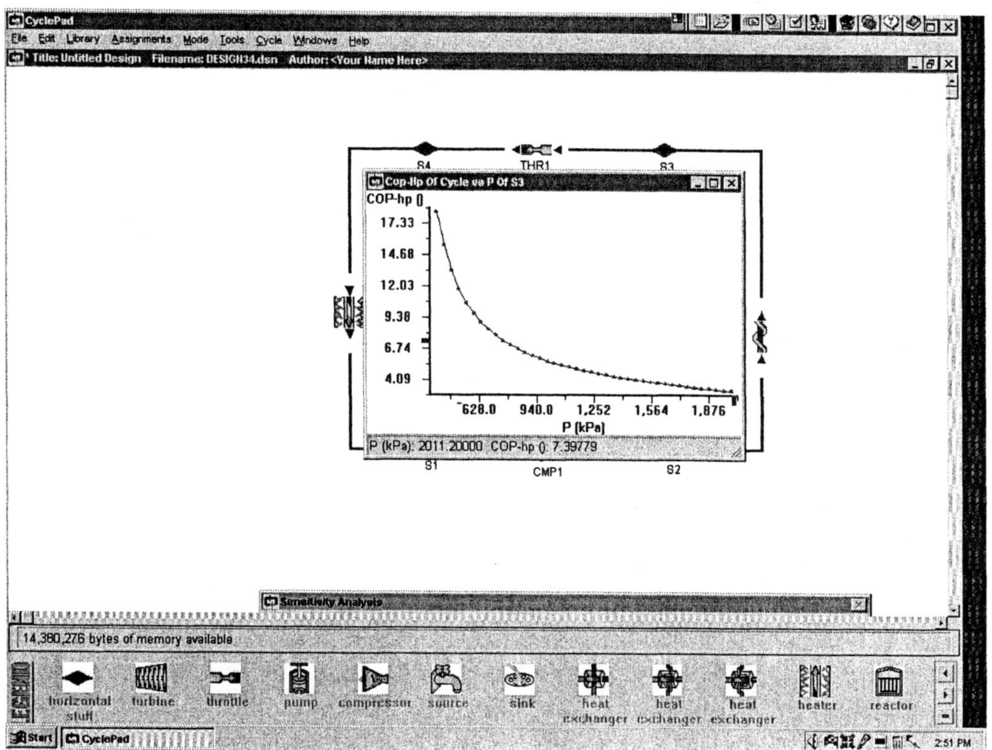

Figure Example 10.4.1b Basic vapor heat pump cycle sensitivity analysis

Homework 10.4 Heat Pump

1. Find the compressor power required, quality of the refrigerant at the end of the throttling process, heating load and COP for a heat pump that uses refrigerant-12 as the working fluid and is designed to operate at an evaporator saturation temperature of 10°C and a condenser saturation temperature of 40°C. The mass rate flow of refrigerant 12 is 0.22 kg/s.
2. Consider an ideal heat pump which uses refrigerant-12 as the working fluid. The saturation temperature of the refrigerant in the evaporator is 6°C and in the condenser it is 58°C. The refrigerant is circulated at the rate of 0.021 kg/s. Determine the compressor power required, quality of the refrigerant at the end of the throttling process, heating load and COP of the heat pump.
3. An ideal heat pump uses ammonia as the working fluid. The saturation temperature of the refrigerant in the evaporator is 22°F and in the condenser it is 98°F. The refrigerant is circulated at the rate of 0.051 lbm/s. Determine the compressor power required, heating load and COP of the heat pump.
4. Find the compressor power required, heating load and COP for a heat pump that uses refrigerant-12 as the working fluid and is designed to operate at an evaporator saturation temperature of 2°C and a condenser temperature of 39°C. The compressor efficiency is 78 percent. The mass rate flow of refrigerant 12 is 0.32 kg/s.

5. Find the compressor power required, turbine power produced, heating load and COP for a heat pump that uses refrigerant-12 as the working fluid and is designed to operate at an evaporator saturation temperature of 2°C and a condenser saturation temperature of 39°C. The mass rate flow of refrigerant 12 is 0.32 kg/s. The throttling valve is replaced by an adiabatic turbine with 74% efficiency.

10.5 Actual Vapor Heat Pump Cycle

The actual vapor heat pump cycle deviates from the ideal cycle primarily because of inefficiency of the compressor, pressure drops associated with fluid flow and heat transfer to or from the surroundings. The vapor entering the compressor must be superheated sightly rather than a saturated vapor. The refrigerant entering the throttling valve is usually compressed liquid rather than a saturated liquid.

Example 10.5.1. Determine the COP, horsepower required and heating load of a basic vapor heat pump cycle using R-134a as the working fluid in which the condenser pressure is 900 kPa and the evaporation pressure is 240 kPa. The circulation rate of fluid is 0.1 kg/s. The compressor efficiency is 88%. Show the cycle on T-s diagram.

To solve this problem by CyclePad, we take the following steps:

1. Build
   (A) Take a compressor, a cooler (condenser), a valve, and a heater (evaporator) from the open system inventory shop and connect the four devices to form the basic heat pump cycle.
   (B) Switch to analysis mode.
2. Analysis
   (A) Assume a process for each of the four devices: (a) compressor as isentropic, (b) condenser as isobaric, (c) valve as constant enthalpy, and (d) evaporator as isobaric.
   (B) Input the given information: (a) working fluid is R-134a, (b) the inlet pressure and quality of the compressor are 240 kPa and 1, (c) the inlet pressure and quality of the valve are 900 kPa and 0, (d) the compressor efficiency is 88%. and (e) the mass flow rate is 0.1kg/s.
3. Display results
   (A) Display the T-s diagram and cycle properties results. The cycle is a heat pump. The answers are COP=5.65, heating load=14.52 kW, and Net power input=-3.12 kW.

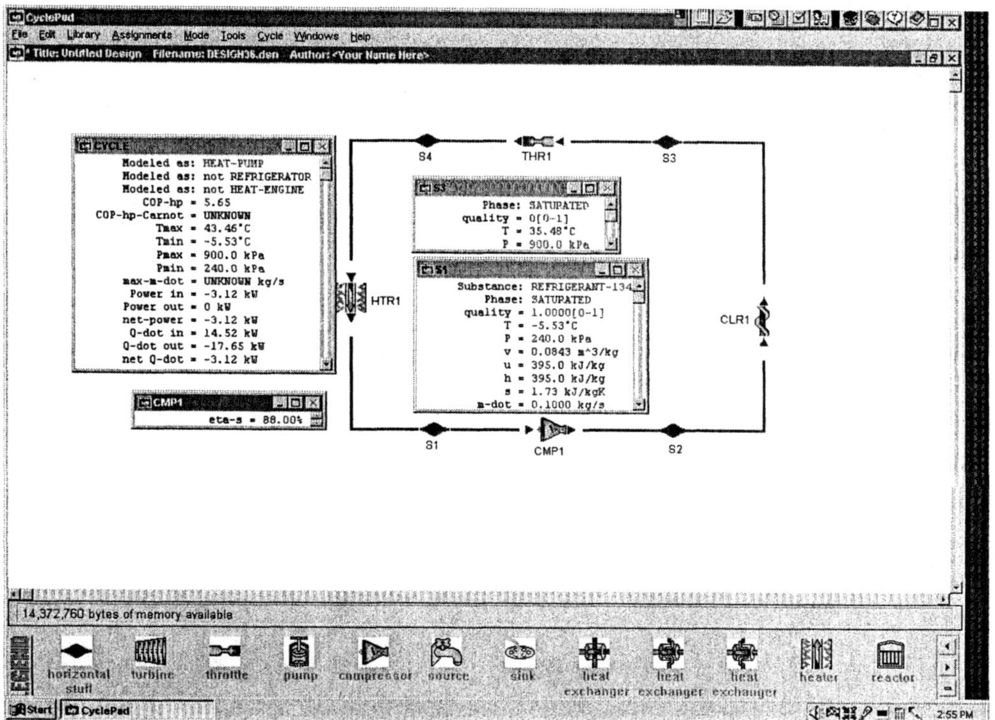

Figure Example 10.5.1 Basic vapor heat pump cycle

Homework 10.5 Actual Vapor Heat Pump

1. Find the compressor power required, quality of the refrigerant at the end of the throttling process, heating load and COP for a heat pump that uses refrigerant-12 as the working fluid and is designed to operate at an evaporator saturation temperature of 10°C and a condenser saturation temperature of 40°C. The compressor efficiency is 68 percent. The mass rate flow of refrigerant 12 is 0.22 kg/s.
2. Consider a heat pump which uses refrigerant-12 as the working fluid. The compressor efficiency is 80%. The saturation temperature of the refrigerant in the evaporator is 6°C and in the condenser it is 58°C. The refrigerant is circulated at the rate of 0.021 kg/s. Determine the compressor power required, quality of the refrigerant at the end of the throttling process, heating load and COP of the heat pump.
3. A heat pump uses ammonia as the working fluid. The compressor efficiency is 80%. The saturation temperature of the refrigerant in the evaporator is 22°F and in the condenser it is 98°F. The refrigerant is circulated at the rate of 0.051 lbm/s. Determine the compressor power required, heating load and COP of the heat pump.
4. Find the compressor power required, heating load and COP for a heat pump that uses refrigerant-12 as the working fluid and is designed to operate at an evaporator saturation temperature of 2°C and a condenser temperature of 39°C. The compressor efficiency is 78 percent. The mass rate flow of refrigerant 12 is 0.32 kg/s.

5. Find the compressor power required, turbine power produced, heating load and COP for a heat pump that uses refrigerant-12 as the working fluid and is designed to operate at an evaporator temperature of 2°C and a condenser temperature of 39°C. The compressor efficiency is 78 percent. The mass rate flow of refrigerant 12 is 0.32 kg/s. The throttling valve is replaced by an adiabatic turbine with 74% efficiency.
6. Determine the COP, horsepower required and heating load of a basic vapor heat pump cycle using R-134a as the working fluid in which the condenser pressure is 900 kPa and the evaporator pressure is 240 kPa. The circulation rate of fluid is 0.1 kg/s. The compressor efficiency is 78%.
7. R-134a enters the compressor of a refrigerator at 0.14 Mpa and -10°C at a rate of 0.1 kg/s and leaves at 0.7 Mpa and 50°C. The refrigerant is cooled in the condenser to 24°C and 0.65 Mpa. The refrigerant is throttled to 0.15 Mpa. Determine the compressor efficiency, power input to the compressor, cooling effect, and COP of the refrigerator.
8. R-134a enters the compressor of a refrigerator at 0.14 Mpa and -10°C at a rate of 0.1 kg/s and leaves at 0.7 Mpa. The compressor efficiency is 85%. The refrigerant is cooled in the condenser to 24°C and 0.65 Mpa. The refrigerant is throttled to 0.15 Mpa. Determine the power input to the compressor, cooling effect, and COP of the refrigerator.
9. R-134a enters the compressor of a refrigerator at 0.14 Mpa and -10°C at a rate of 0.1 kg/s and leaves at 0.8 Mpa. The compressor efficiency is 85%. The refrigerant is cooled in the condenser to 26°C and 0.75 Mpa. The refrigerant is throttled to 0.15 Mpa. Determine the power input to the compressor, cooling effect, and COP of the refrigerator.

10.6 WORKING FLUIDS FOR VAPOR REFRIGERATION AND HEAT PUMP SYSTEMS

Ammonia, carbon dioxide, and sulphur dioxide were used widely in early years of refrigeration in industrial refrigeration applications. For domestic and industrial applications now, the principal refrigerants have been man-made freons. This family of substances are known by an R number of the general form RN, R signifying refrigerant and the number N specifically identifying the chemical compound. The number allocated to the halogenated hydrocarbons (freons) are derived as follows: For refrigerants derived from methane (CH_4), N is a two digit integer. The first digit indicates the number of hydrogen atoms +1 and the second digit indicates the number of fluorine atoms, e.g., CCl_2F_2 is R12. For refrigerants derived from ethane (C_2H_6), N is a three digit integer. The first digit is always 1, the second digit is the number of hydrogen atoms +1 and the third digit is the number of fluorine atoms, e.g., $C_2Cl_2F_4$ is R114.

There are three R number refrigerants on the substance menu of CyclePad. The three refrigerants are R12, R22 and R134a.

The desirable properties of working fluids for vapor refrigeration and heat pump systems include high critical temperature and low pressure, low specific volume, inexpensive, non-flammable, non-explosive, non-toxic, non- corrosive, inert and stable, etc.

In recent years, the effects of freons on the ozone layer have been critically evaluated. Some freons such as R12 having leaked from refrigeration systems into the atmosphere, spend many years slowly diffusing upward into the stratosphere. There it is broken down, releasing chlorine which depletes the protective ozone layer surrounding the Earth's stratosphere. Ozone is a critical component of the atmospheric system both for climate control and for reducing solar radiation. It is therefore important to human beings to ban these freons such as the widely used but life threatening R12. New desirable refrigerants which contains no chlorine atoms are found to be suitable and acceptable replacements.

Homework 10.6 Working Fluids for Vapor Refrigeration and Heat Pump Systems

1. Why are Ammonia, carbon dioxide, and sulphur dioxide no longer used in domestic refrigerators and heat pumps?
2. What devastating consequence result in our environment by refrigerants leaking out from refrigeration systems into the atmosphere?
3. List five desirable properties of working fluids for vapor refrigeration and heat pump systems.
4. What does the refrigerant number mean?

10.7 Cascade and Multi-Staged Vapor Refrigerators

There are several variations of the basic vapor refrigeration cycle. A cascade cycle is used when the temperature difference between the evaporator and the condenser is quite large. The multi-staged cycle is used to reduced the required compressor power input.

10.7.1 Cascade Vapor Refrigerators

There are applications when the temperature difference between the evaporator and the condenser is quite large. A single vapor refrigeration cycle usually can not be used to achieve the large difference. To solve this problem and still using vapor refrigeration cycles, a cascade vapor refrigeration cycle must be used. A cascade cycle is several vapor refrigeration cycles connecting in series. A three cycle in series is illustrated in Figure 10.7.1.1 The condenser of the lowest-temperature cycle (cycle A, 1-2-3-4-1) provides the heat input to the evaporator of the mid-temperature cycle (cycle B, 5-6-7-8-5); The condenser of the mid-temperature cycle (cycle B) provides the heat input to the evaporator of the highest-temperature cycle (cycle C, 9-10-11-12-9). Different working fluids may be used in each of the individual cycle.

Neglecting kinetic and potential energy changes, a steady state and steady flow mass and energy balance on the components of the cascade vapor refrigeration cycle have the general forms

$$\sum mdot_e = \sum mdot_i, \tag{10.7.1.1}$$

and

$$Qdot\text{-}Wdot=\sum mdot_e h_e - \sum mdot_i h_i. \tag{10.7.1.2}$$

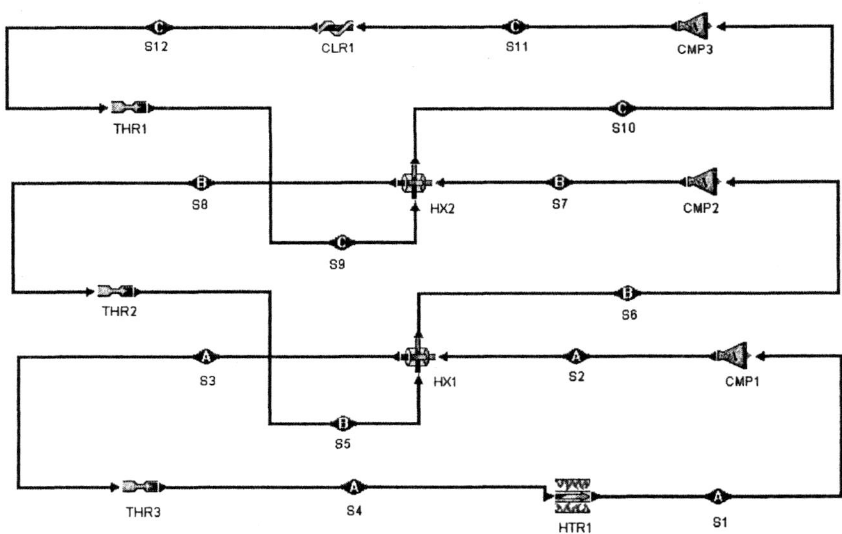

Figure 10.7.1.1 Cascade vapor refrigerator

The cooling load of the cascade vapor refrigeration cycle is the rate of heat added in the evaporator of the lowest temperature cycle. The power added to the cycle and is the sum of the power added to the indiviual compressors

$$Wdot=\sum Wdot_{compressor} \tag{10.7.1.3}$$

and the COP of the cycle is

$$\beta=Qdot_{lowest\ T\ evaporator}/Wdot \tag{10.7.1.4}$$

The following examples illustrate the analysis of the cascade vapor refrigeration cycle.

Example 10.7.1.1 A cascade vapor refrigeration cycle made of two separate vapor refrigeration cycles as shown has the following information:

Topping cycle: working fluid=R134a, p_5=200 kPa, x_5=1, p_7=500 kPa, and x_7=0.
Bottoming cycle: working fluid=R134a, p_1=85 kPa, $mdot_1$=1 kg/s, x_1=1, p_3=250 kPa, and x_3=0.

Determine the mass rate flow of the topping cycle, power required by compressor #1, power required by compressor #2, total power required by the compressors, rate of heat added in the evaporator, cooling load, and COP of the cascade vapor refrigeration cycle.

To solve the problem by CyclePad, we do the following steps:

1. Build the cycle as shown in Figure Example 10.7.1.1a.
2. Assume compressors are adiabatic with 100% efficiency, heater and cooler be isobaric, and both hot-side and cold-side of the heat exchanger are isobaric.

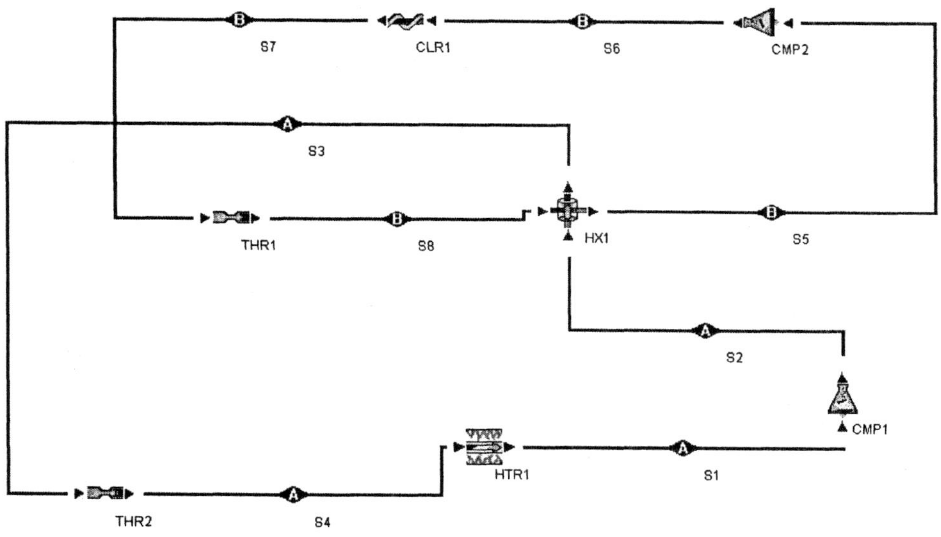

Figure Example 10.7.1.1a Cascade refrigerator

3. Input working fluid is R134a at state 1, p_1=85 kPa, $mdot_1$=1 kg/s, x_1=1, p_3=250 kPa, and x_3=0; working fluid is R134a at state 5, p_5=200 kPa, x_5=1, p_7=500 kPa, and x_7=0.
4. Display the cycle property results: The results are: $mdot_5$=1.21 kg/s, COP=4.17, power input by compressor 1=-21.64 kW, power input by compressor 2=-22.87 kW, power input by compressors =-44.5 kW, rate of heat removed by the condenser=-230.2 kW, rate of heat added to the evaporator=185.7 kW, cooling load=52.81 ton, and COP=4.17.

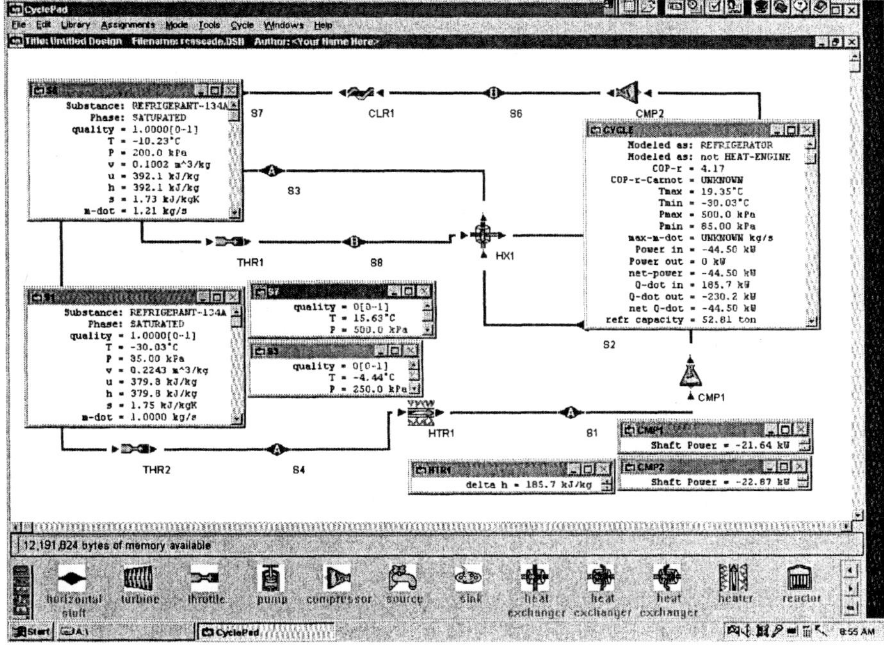

Figure Example 10.7.1.1b Cascade refrigerator

Homework 10.7.1Cascade Vapor Refrigerators

1. What is the purpose of Cascade vapor refrigerators?
2. A cascade vapor refrigeration cycle made of two separate vapor refrigeration cycles as shown in Figure Example 10.7.1.1a has the following information:
   Topping cycle: working fluid=R12, p_5=200 kPa, x_5=1, p_7=500 kPa, and x_7=0.
   Bottoming cycle: working fluid=R12, p_1=85 kPa, $mdot_1$=1 kg/s, x_1=1, p_3=250 kPa, and x_3=0.
   Determine the total power required by the compressors, rate of heat added in the evaporator, cooling load, and COP of the cascade vapor refrigeration cycle.
   ANSWER: power input by compressors =-44.5 kW, rate of heat removed by the condenser=-179.9 kW, rate of heat added to the evaporator=142.2 kW, cooling load=40.42 ton, and COP=3.77.
3. A cascade vapor refrigeration cycle made of two separate vapor refrigeration cycles as shown has the following information:
   Topping cycle: working fluid=ammonia, p_5=200 kPa, x_5=1, p_7=500 kPa, and x_7=0.
   Bottoming cycle: working fluid=ammonia, p_1=85 kPa, $mdot_1$=1 kg/s, x_1=1, p_3=250 kPa, and x_3=0.
   Determine total power required by the compressors, rate of heat added in the evaporator, cooling load, and COP of the cascade vapor refrigeration cycle.
   ANSWER: power input by compressors =-279.7 kW, rate of heat added to the evaporator=1276 kW, cooling load=362.9 ton, and COP=4.56.

10.7.2 Multi-Staged Vapor Refrigerators

A flash chamber may have better heat transfer characteristics than the heat exchanger employed between the upstream cycle and down-stream cycle of the cascaded cycle. It is therefore used to replace the heat exchanger in *multi-staged vapor refrigerators*. In this arrangement, the working fluid flowing throughout the whole system must be the same.

A schematic diagram of a two-stage vapor refrigerator is shown in Figure 10.7.2.1. The liquid leaving the condenser is throttling into a flashing chamber (separator use to separate mixture to vapor and liquid) maintained at a pressure between the evaporator pressure and condenser pressure. Saturated vapor separated from the liquid in the flashing chamber enters a mixing chamber, where it mixes with the vapor leaving the low-pressure compressor at state 2. The saturated liquid is throttled to the evaporator pressure at state 9. By adjusting the mass flow rate flowing in the separator, the cooling load of the refrigeration cycle can be controlled. The cycle analsis of the cycle is illustrated in Example 10.7.2.1.

Neglect kinetic and potential energy changes. A steady state and steady flow mass and energy balance on the components of the cascade vapor refrigeration cycle have the general forms

$$\sum mdot_e = \sum mdot_i, \tag{10.7.2.1}$$

and

Qdot-Wdot=$\sum$mdot$_e$h$_e$ - $\sum$mdot$_i$h$_i$. (10.7.2.2)

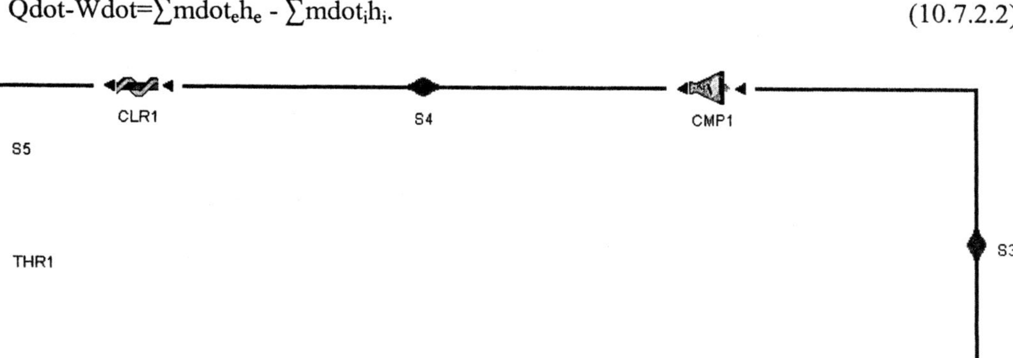

Figure 10.7.2.1 Multi-stage vapor refrigerator

The cooling load of the cascade vapor refrigeration cycle is the rate of heat added in the evaporator of the lowest temperature cycle. The power added to the cycle and is the sum of the power added to the indiviual compressors

Wdot=$\sum$Wdot$_{compressor}$ (10.7.2.3)

and the COP of the cycle is

β=Qdot$_{lowest\ T\ evaporator}$ /Wdot (10.7.2.4)

The following examples illustrate the analysis of the cascade vapor refrigeration cycle.

Example 10.7.2.1 A two-stage vapor refrigeration cycle as shown in Figure 10.7.2.1 has the following information:

working fluid=R134a, p_1=85 kPa, x_1=0, p_2=200 kPa, p_4=500 kPa, m_4=1 kg/s, m_7=0.8 kg/s, and x_5=0.

Determine the power required by compressor #1, power required by compressor #2, total power required by the compressors, rate of heat added in the evaporator, cooling load, and COP of the cycle. Plot the cooling load vs m_7 sensitivity diagram.

To solve the problem by CyclePad, we do the following steps:

1. Build the cycle as shown in Figure 10.7.2.1a.
2. Assume compressors are adiabatic with 100% efficiency, heater and cooler be isobaric, and both hot-side and cold-side of the heat exchanger are isobaric.
3. Input working fluid is R134a at state 1, p_1=85 kPa, x_1=0, p_2=200 kPa, p_4=500 kPa, m_4=1 kg/s, m_7=0.8 kg/s, and x_5=0.
4. Display the cycle property results: The results are: COP=4.84, power input by compressor 1=-18.35 kW, power input by compressor 2=-13.58 kW, power input by compressors =-31.93 kW, rate of heat removed by the condenser=-186.6 kW, rate of heat added to the evaporator=154.7 kW, and cooling load=43.99 ton.

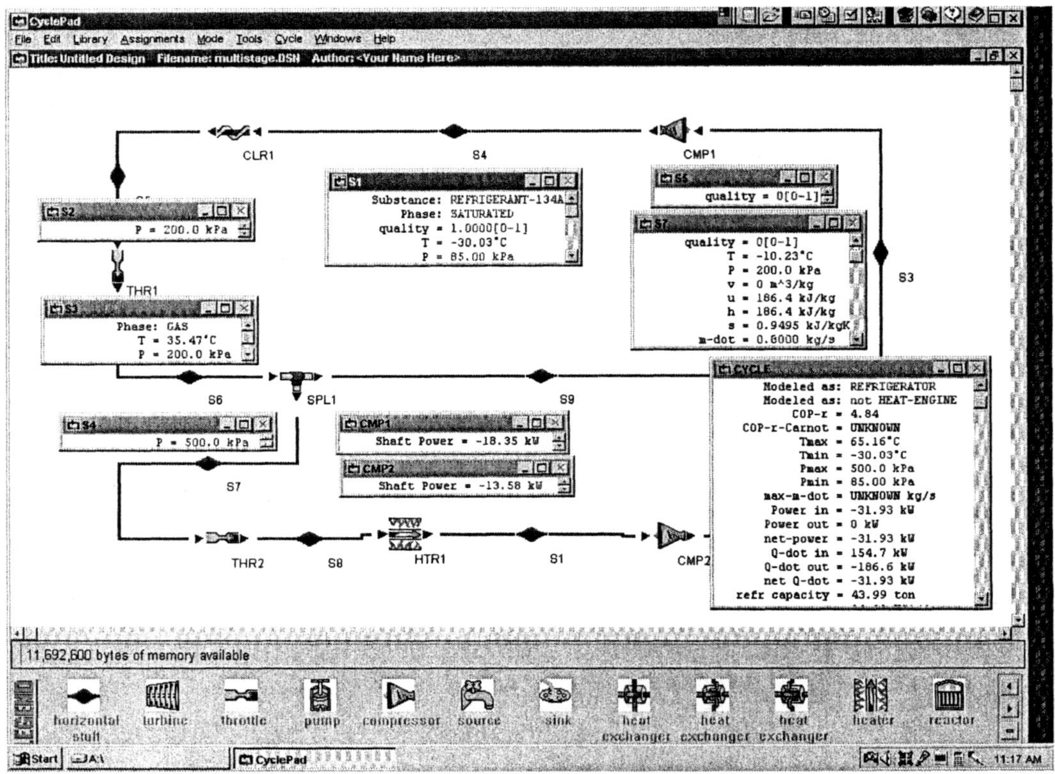

Figure Example 10.7.2.1a Two-stage vapor refrigerator

An arrangement of either a cascaded or multi-staged refrigerator can be made as illustrated in Figure 10.7.2.2. In this arrangement, the system can be either a cascaded refrigerator or a multi-staged refrigerator.

Suppose $mdot_3$=0 and $mdot_8$=0, the working fluids of the top and that of the bottom cycle do not mix. Therefore, it is a cascaded refrigerator.

Suppose $mdot_3$=1, the working fluids of the top and that of the bottom cycle do mix. It becomes a multi-staged refrigerator.

Example 10.7.2.2 illustrates the system as a multi-staged refrigerator if $mdot_3$=1.

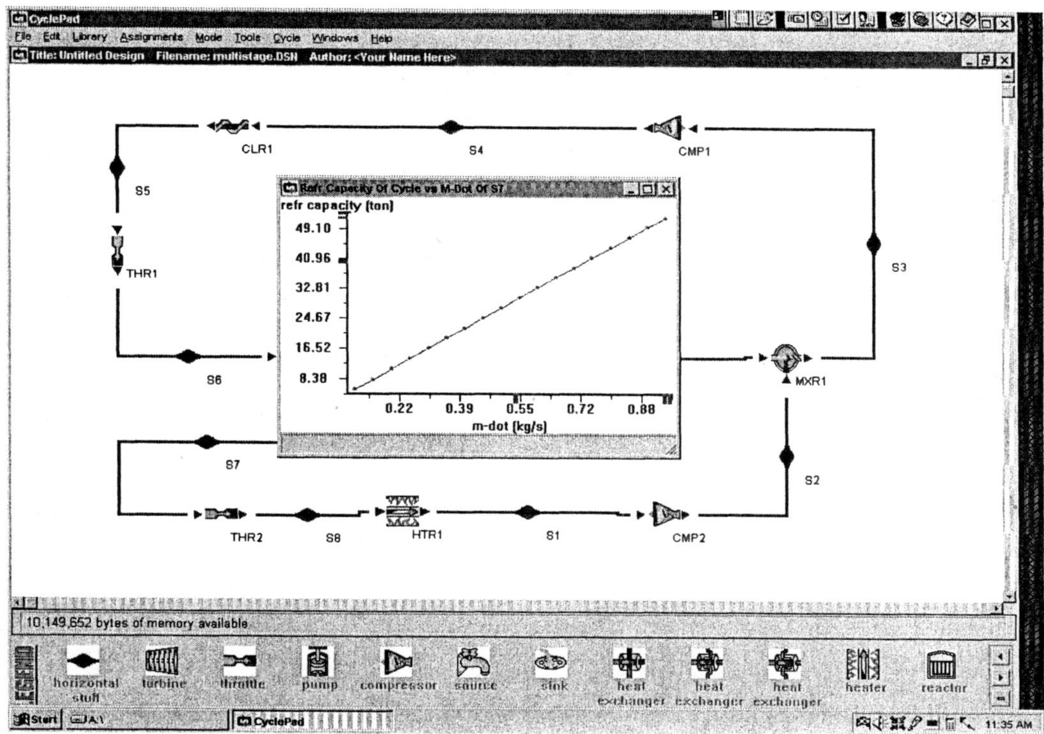

Figure Example 10.7.2.1b **Two-stage vapor refrigerator cooling load sensitivity analysis**

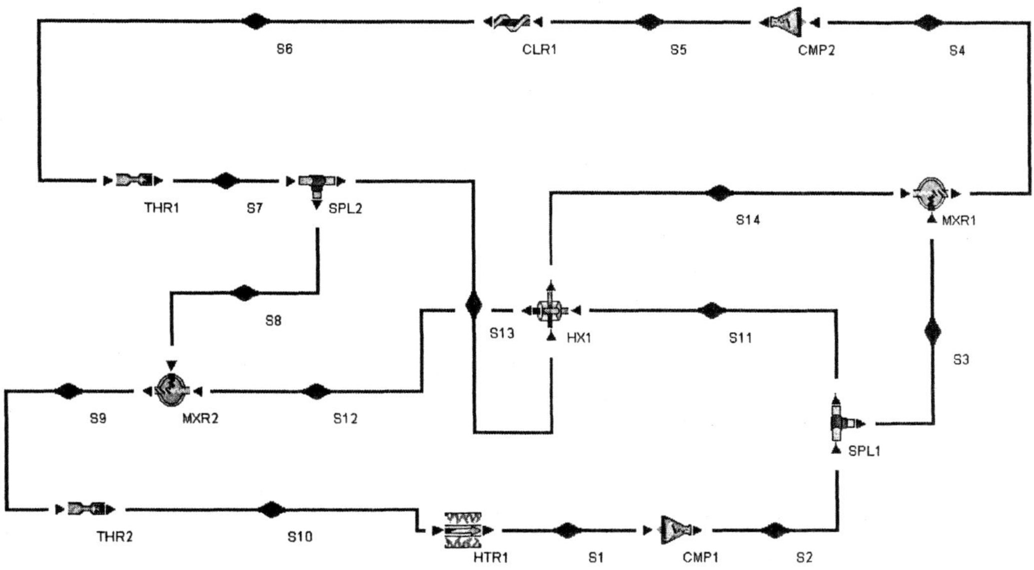

Figure 10.7.2.2. **Cascaded or multi-staged refrigerator**

Example 10.7.2.2 A cycle as shown in Figure 10.7.2.2 has the following information:

> working fluid=R134a, p_1=85 kPa, x_1=0, p_2=200 kPa, p_5=500 kPa, m_4=1 kg/s, m_{11}=0 kg/s, x_6=0, x_8=0 and x_{13}=1.

Determine the power required by compressor #1, power required by compressor #2, total power required by the compressors, rate of heat added in the evaporator, cooling load, and COP of the cycle. Plot the cooling load vs m_7 sensitivity diagram.

To solve the problem by CyclePad, we do the following steps:

1. Build the cycle as shown in Figure 10.6.2.1a.
2. Assume compressors are adiabatic with 100% efficiency, heater and cooler be isobaric, and both hot-side and cold-side of the heat exchanger are isobaric.
3. Input working fluid is R134a at state 1, p_1=85 kPa, x_1=0, p_2=200 kPa, p_5=500 kPa, m_4=1 kg/s, m_{11}=0 kg/s, x_6=0, x_8=0 and x_{13}=1.
4. Display the cycle property results: The results are: COP=4.81, power input by compressor 1=-33.37 kW, rate of heat removed by the condenser=-193.9 kW, rate of heat added to the evaporator=160.5 kW, and cooling load=45.64 ton.

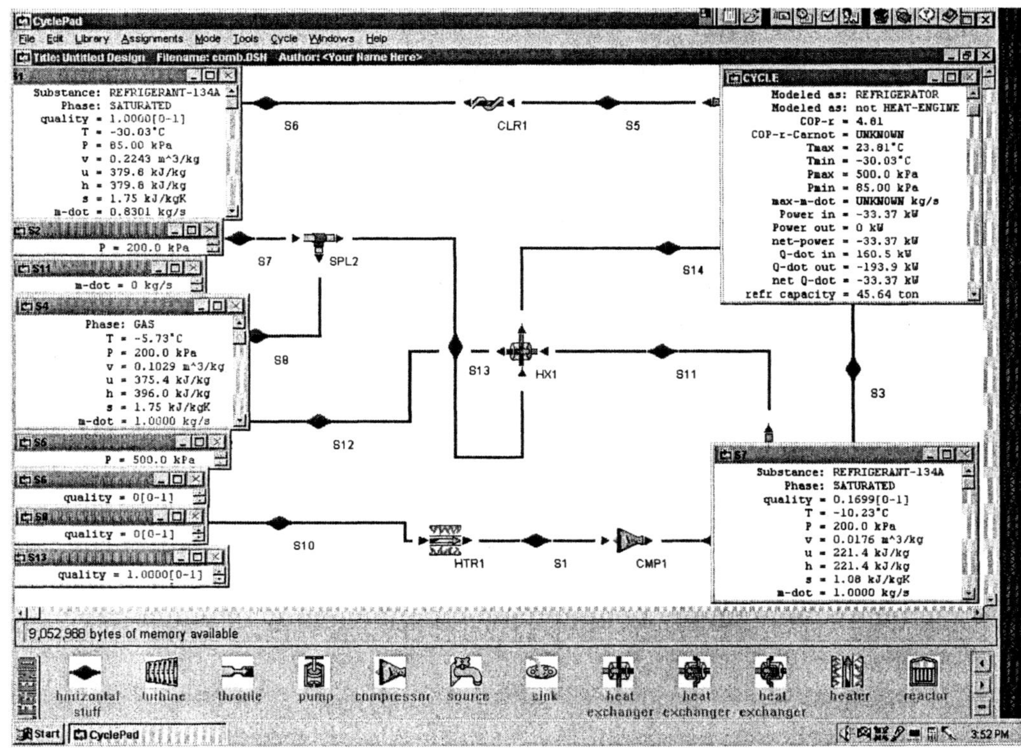

Figure Example 10.7.2.2 Cascaded or multi-staged refrigerator

Homework 10.7.2 Multi-Stage Vapor Refrigerators

1. What is the purpose of multi-staged vapor refrigerators?
2. What is the difference between cascaded and multi-staged vapor refrigerators?
3. A two-stage vapor refrigeration cycle as shown in Figure 10.6.2.1 has the following information:
   working fluid=R22, p_1=85 kPa, x_1=0, p_2=200 kPa, p_4=500 kPa, m_4=1 kg/s, m_7=0.8 kg/s, and x_5=0.
   Determine the power required by compressor #1, power required by compressor #2, total power required by the compressors, rate of heat added in the evaporator, cooling load, and COP of the cycle.
   ANSWER: power input by compressors =-38.12 kW, rate of heat removed by the condenser=-210.2 kW, rate of heat added to the evaporator=172.1 kW, cooling load=48.92 ton, and COP=4.51.
4. A two-stage vapor refrigeration cycle as shown in Figure 10.6.2.1 has the following information:
   working fluid=ammonia, p_1=85 kPa, x_1=0, p_2=200 kPa, p_4=500 kPa, m_4=1 kg/s, m_7=0.8 kg/s, and x_5=0.
   Determine the power required by compressor #1, power required by compressor #2, total power required by the compressors, rate of heat added in the evaporator, cooling load, and COP of the cycle.
   ANSWER: power input by compressors =-217.5 kW, rate of heat removed by the condenser=-1257 kW, rate of heat added to the evaporator=1040 kW, cooling load=295.6 ton, and COP=4.78.
5. A two-stage vapor refrigeration cycle as shown in Figure 10.6.2.1 has the following information:
   working fluid=R134a, p_1=85 kPa, x_1=0, p_2=200 kPa, p_4=500 kPa, m_4=1 kg/s, m_7=0.9 kg/s, and x_5=0.
   Determine the power required by compressor #1, power required by compressor #2, total power required by the compressors, rate of heat added in the evaporator, cooling load, and COP of the cycle.
   ANSWER: power input by compressors =-35.92 kW, rate of heat removed by the condenser=-209.9 kW, rate of heat added to the evaporator=174.0 kW, cooling load=49.49 ton, and COP=4.84.

10.8. Domestic Refrigerator-Freezer System, and Air Conditioning-Heat Pump System

10.8.1 Domestic Refrigerator-Freezer System

The household refrigerator-freezer combination uses one evaporator (heater) in the freezer section to keep that region at the desired temperature (-18°C or 0°F). Cold air from the freezer is transferred into the refrigerator section to keep it at a higher temperature (2°C or 35°F). The COP of the refrigerator-freezer combination suffers because the COP of the combination is equal to the COP of the freezer. It is known that the COP of a refrigeration cycle is inversely proportional to (T_H -T_L). The lower the T_L, the lower the COP.

One method of improving the COP of the refrigerator-freezer combination is to employ an evaporator for both the refrigerator region and the freezer region with a single compressor as illustrated in Figure 10.8.1.1. A numerical example of this arrangement is shown in the following example.

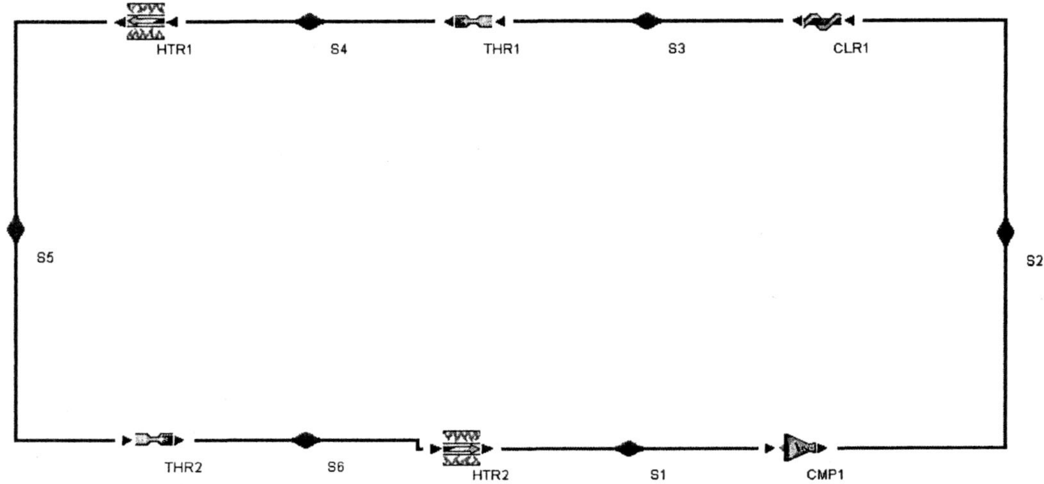

Figure 10.8.1.1. Refrigerator and freezer with dual evaporator

Example 10.8.1.1. A two-region-section refrigerator requires refrigeration at -37°C and -19°C. Using ammonia as the refrigerant, design a dual evaporator refrigerator and find the COP, compressor input power, and cooling load of the refrigerator based on one unit mass flow rate of refrigerant.

To design the refrigerator by CyclePad, we do the following :

(A) Built the two-region-section refrigerator as shown in Figure 10.8.1.1,
(B) Assume the compressor is adiabatic and 100% efficient, and cooler and heaters are isobaric,
(C) Let working fluid be ammonia, T_1=-37°C, x_1=1, mdot =1 kg/s, p_2=800 kPa, x_3=0, T_4=-19°C, and x_6=0.4, and
(D) Display cycle property results. The results are: COP=3.39, $Qdot_{htr\#1}$=301.3 kW, $Qdot_{htr\#2}$=828.2 kW, compressor input power=-333.4 kW, and cooling load=321.2 ton.

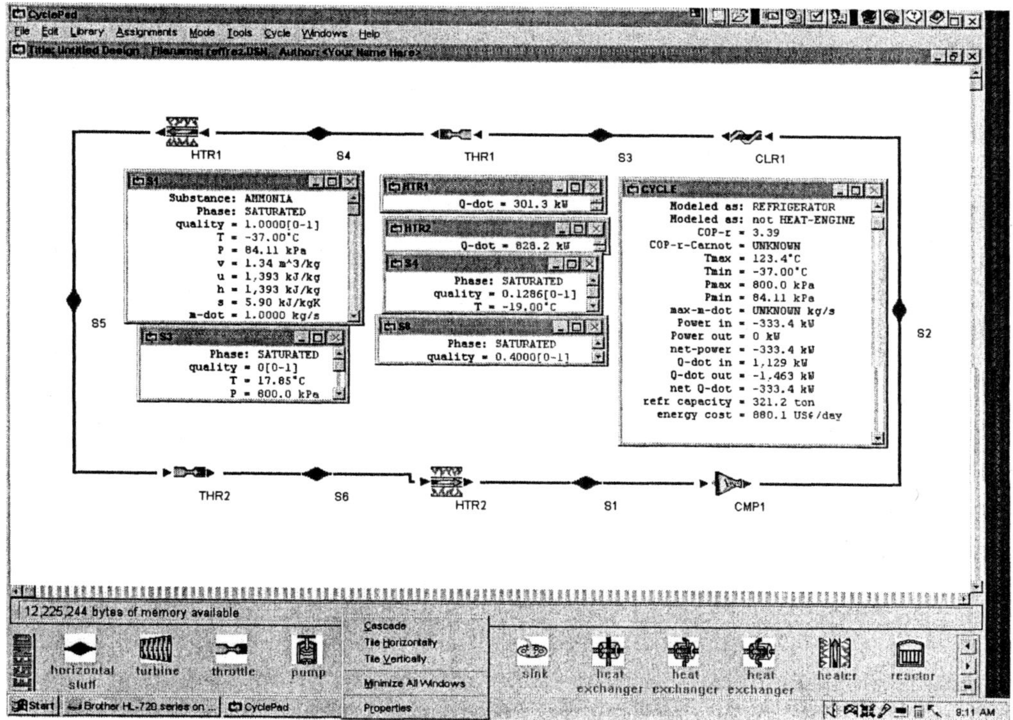

Figure Example 10.8.1.1. Refrigerator and freezer with dual evaporator

Homework 10.8.1.1. Refrigerator and Freezer with Dual Evaporator

1. What is the purpose of the Refrigerator and freezer with dual evaporator?
2. A two-region-section refrigerator requires refrigeration at -17°C and 2°C. Using ammonia as the refrigerant, design a dual evaporator refrigerator and find the COP, compressor input power, and cooling load of the refrigerator based on one unit mass flow rate of refrigerant.

10.8.2 Domestic Air Conditioning-Heat Pump System

Refrigerators and heat pumps have the same energy flow diagram (Figure 5.2.2.1and Figure 5.2.3.1) and have the same components (Figure 5.2.2.2 and Figure 5.2.3.2). A domestic air conditioning and heat pump system as shown in Figure 10.8.2.1 can therefore be used as a heat pump in the winter as well as an air conditioning unit in the summer. Notice that both the domestic air conditioning and heat pump system share the same equipments. Thus the investment in the heat pump can also be used for air conditioning to provide year-round house comfort control.

In the air conditioning mode, the cycle (cycle A) is 1-2-3-4-5-6-7-8-9-10-1. The heat exchanger removes heat from the building replaces the evaporator. Atmospheric hot air entering the dwelling at state 11of cycle B is cooled by the heat exchanger by removing heat from the building to vaporize the refrigerant and leaving the dwelling at state 12 of cycle B.

In the heat pump mode, the cycle (cycle A) is 1-2-15-8-9-16-5-6-13-14-1. The heat exchanger adds heat to the building replacing the condenser. Atmospheric cold air entering the dwelling at state 11of cycle B is heated by the heat exchanger by discharging heat to the building to condense the refrigerant and leaving the dwelling at state 12 of cycle B.

Notice that the heat exchanger is an evaporator in the air conditioning mode, and a condenser In the heat pump mode. Therefore, the refrigerant is on the hot-side in the heat pump mode, and on the cold-side in the air conditioning mode when the system is build using CyclePad.

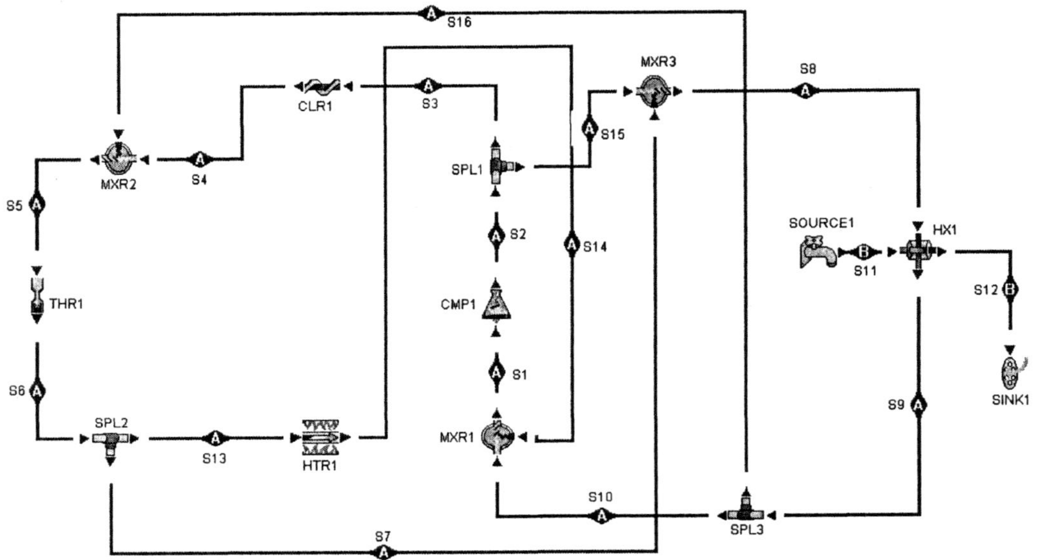

Figure 10.8.2.1 Domestic air conditioning and heat pump system

Example 10.8.2.1. In the air conditioning mode, a domestic air conditioning and heat pump system as shown in Figure 10.6.4.1 uses R-134a as the refrigerant. The refrigerant saturated vapor is compressed from 140 kPa to 700 kPa. Summer ambient air at 30°C is to be cooled down to 17°C. Find the compressor power required, heat removed from the ambient air in the heat exchanger, COP of the system, and mass rate of air flow per unit of mass rate of refrigerant flow.

To solve this problem by CyclePad, we do the following :

(A) Built the system as shown in Figure 10.8.2.1,
(B) Assume the compressor is adiabatic and 100% efficient, and cooler and heaters are isobaric,
(C) Let working fluid be R-134a, p_1=140 kPa, x_1=1, mdot =1 kg/s, p_2=700 kPa, x_3=0, T_{11}=30°C, T_{12}=17°C, $mdot_{13}$=0, $mdot_{15}$=0, and $mdot_{16}$=0,
(D) Display results. The results are: compressor input power=-33.46 kW, $Qdot_{HX1}$=-149.8 kW, and $mdot_{11}$=11.49 kg/s, and
(E) the COP of the system is $Qdot_{HX1}$/compressor input power=149.8/33.46=4.477.

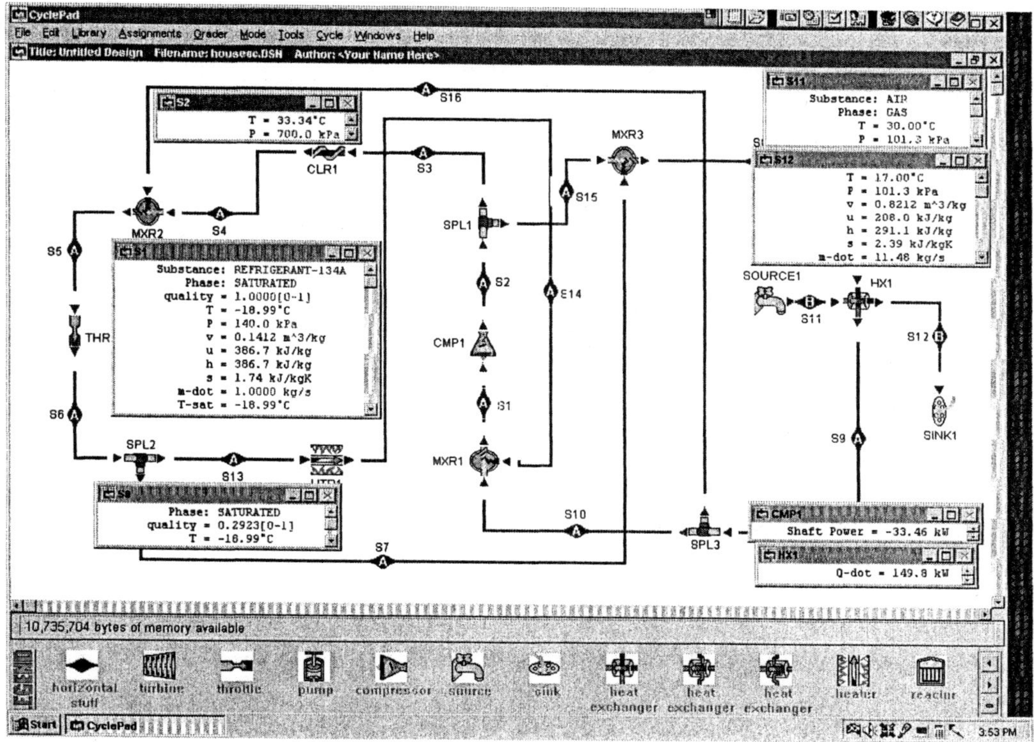

Figure Example 10.8.2.1. Domestic air conditioning system

10.9 ABSORPTION AIR-CONDITIONING

The *absorption air-conditioning or refrigeration system* shown in Figure 10.9.1 is a system in which heat instead of work is employed to produce a refrigeration effect. In a conventional refrigeration system, high-quality and expensive electric work is consumed by the compressor which compresses vapor from a low pressure to a high pressure. Since pumping involves only a liquid, the pump consumes very little electric work. Therefore, the compressor of the basic vapor refrigeration cycle is replaced by a refrigerant generator (heater), an absorber (mixing chamber), a separator (splitter), a valve, and a liquid pump in the absorption air-conditioning system. The major energy input to the absorption system is heat added to the refrigerant generator, which generates refrigerant.

To illustrate the operation of the absorption air-conditioning or refrigeration system, consider the working fluids employed to be ammonia-water. Ammonia is the refrigerant and water the absorber in the absorption air-conditioning or refrigeration system shown in Figure 10.9.1. The vaporized ammonia refrigerant leaving the evaporator at state 9 is absorbed by the weak absorber solution at state 5 and is accompanied by a release of heat. The absorbent solution with a high concentration of refrigerant at state 1 is pumped to an upper pressure at state 2 corresponding to that of the condenser system. The heat input to the refrigerant generator from state 2 to state 3 boils off the refrigerant to state 6, leaving a weak absorber solution at state 4. The vaporized refrigerant that enters the condenser at state 6, the expansion valve at state 7, and then the evaporator at state 9 completes the refrigeration cycle as in the basic vapor refrigeration cycle. The most commonly used working fluid combinations in the absorption air-conditioning or refrigeration systems is water and lithium bromide, where water is the refrigerant and lithium bromide is the absorber.

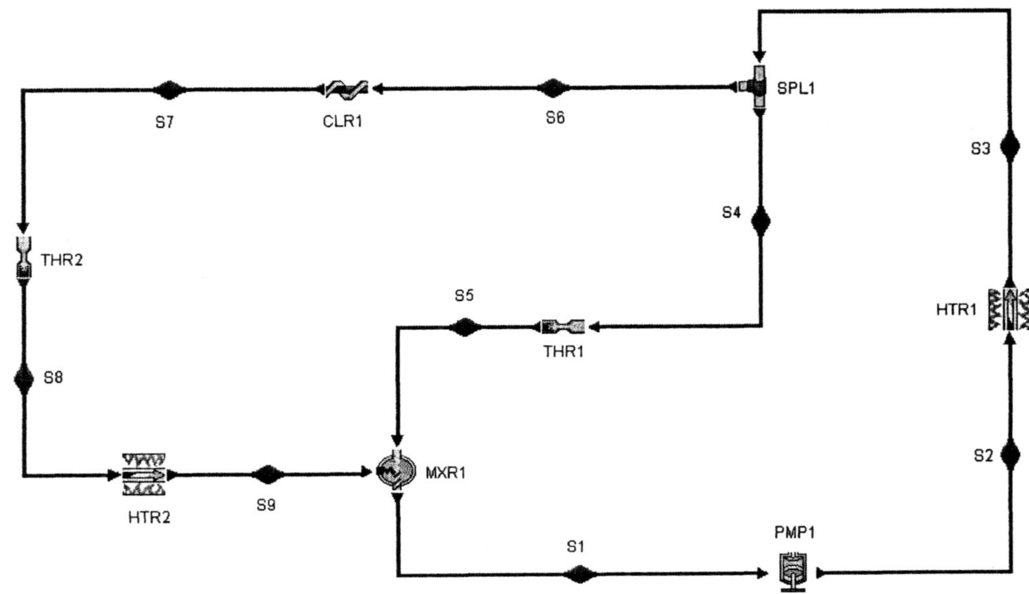

Figure 10.9.1 Absorption air-conditioning

Neglect kinetic and potential energy changes. A steady state and steady flow mass and energy balance on the components of the cycle have the general forms

$$\sum mdot_e = \sum mdot_i, \qquad (10.9.1)$$

and

$$Qdot\text{-}Wdot=\sum mdot_e h_e - \sum mdot_i h_i. \qquad (10.9.2)$$

The cooling load of the cycle is the rate of heat added in the evaporator (heater between state 8 and 9). The rate of energy added to the cycle and is the sum of the pump power and the rate of heat added in the generator (heater between state 2 and 3). Since the pump requires very little power (neglect pump power), the rate of energy added to the cycle is the rate of heat added in the generator

$$Qdot_{in}=Qdot_{23} \qquad (10.9.3)$$

and the COP of the cycle is

$$\beta=Qdot_{89} /Qdot_{23} \qquad (10.9.4)$$

The absorption cycle efficiency can be improved by adding a heat exchanger between the generator and the absorber as a regenerator as shown in Figure 10.9.2.

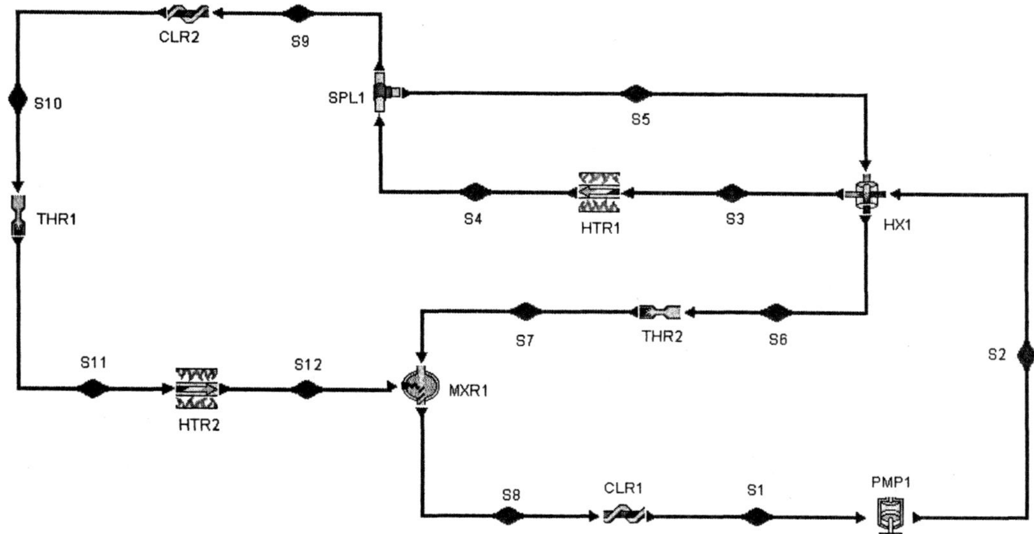

Figure 10.9.2. Absorption air-conditioning with regenerator

Homework 10.9 Absorption Air-Conditioning

1. What is absorption refrigeration?
2. How does an absorption refrigeration system differ from a vapor compression refrigeration system?
3. What are the advantages of absorption refrigeration?

10.10 Brayton Gas Refrigeration Cycle

The *basic gas Brayton refrigeration cycle* is the reversed Brayton gas power cycle. The components of the basic gas refrigeration cycle include a compressor, a cooler, a turbine, and a heater as shown in Figure 10.10.1.

The T-s diagram of the basic gas Brayton refrigeration cycle which consists of the following four processes is shown in Figure 10.10.2:

1-2 isentropic compression
2-3 isobaric cooling
3-4 isentropic expansion
4-1 isobaric heating

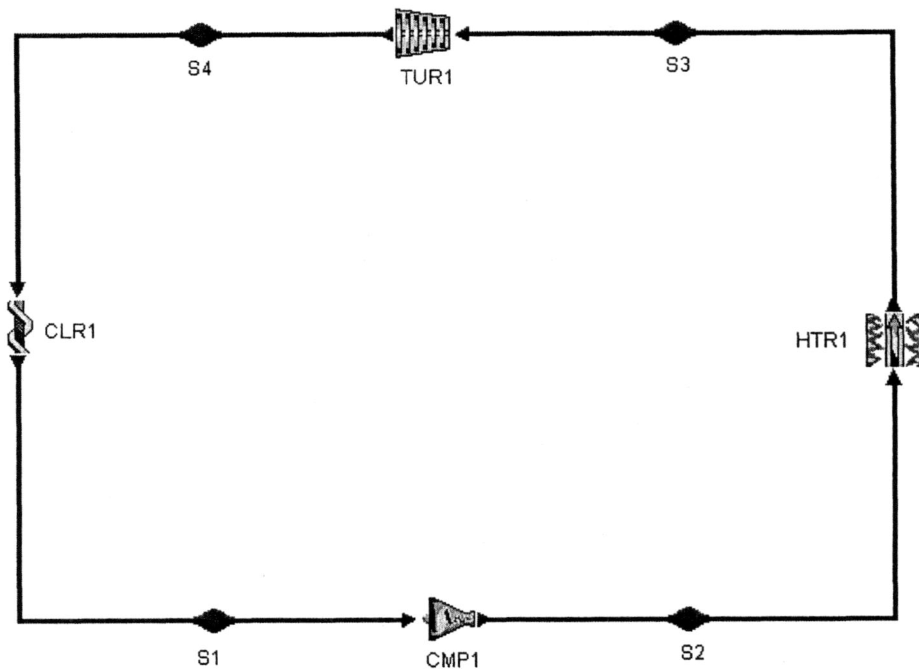

Figure 10.10.1 Basic Brayton refrigeration cycle

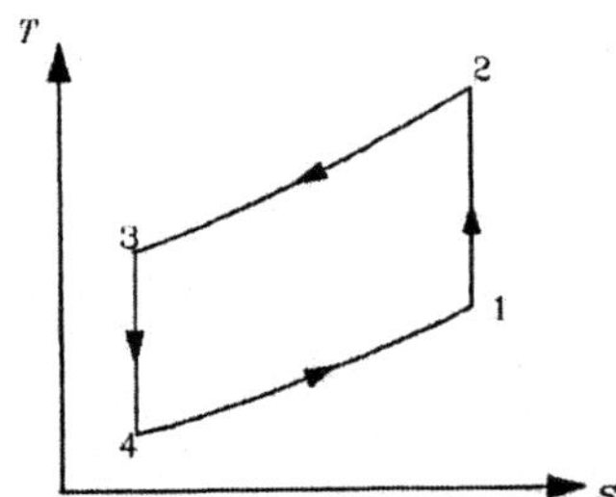

Figure 10.10.2 T-s diagram of the basic Brayton refrigeration cycle

Applying the First law of thermodynamics of the open system to each of the four processes of the basic gas refrigeration yields:

$$Q_{12} = 0, \quad (10.10.1)$$

$$0 - W_{12} = m(h_2 - h_1), \quad (10.10.2)$$

$$W_{23} = 0, \quad (10.10.3)$$

$$Q_{23} - 0 = m(h_3 - h_2), \quad (10.10.4)$$

$$Q_{34} = 0, \tag{10.10.5}$$

$$0 - W_{34} = m(h_3 - h_4), \tag{10.10.6}$$

$$W_{41} = 0, \tag{10.10.7}$$

and

$$Q_{41} - 0 = m(h_1 - h_4), \tag{10.10.8}$$

The net work (W_{net}) added to the cycle is the sum of the compressor work (W_{12}) and the turbine work (W_{34})

$$W_{net} = W_{12} + W_{34} \tag{10.10.9}$$

The desirable energy output of the basic gas refrigeration cycle is the heat added to the heater (or heat removed from the inner space of the refrigerator). The energy input to the cycle is the net work required. Thus the coefficient of performance (COP) of the cycle is

$$\beta_R = Q_{41} / W_{net} = (h_4 - h_1)/[(h_1 - h_2) + (h_4 - h_3)] \tag{10.10.10}$$

Assuming gas has constant specific heats, Equation (10.8.10) can be simplified to

$$\beta_R = 1/\{[r_p]^{(k-1)/k} - 1\} \tag{10.10.11}$$

where $r_p = p_2/p_1 = p_3/p_4$ is the pressure ratio.

The basic gas Brayton refrigeration cycle analysis is given by Example 10.10.1.

Example 10.10.1. Consider the design of an ideal air refrigeration cycle according to the following specifications:

Pressure of air at compressor inlet=15 psia
Pressure of air at turbine inlet=60 psia,
Temperature of air at compressor inlet=20°F
Temperature of air at turbine inlet=80°F
Mass rate of air flow=0.1 lbm/s

Determine the COP, the compressor horsepower required, turbine power produced, net power required, and cooling load for the cycle.

To solve this problem by CyclePad, we take the following steps:

1. Build
   (A) Take a compressor, a cooler, a turbine, and a heater from the open system inventory shop and connect the four devices to form the gas refrigeration cycle.
   (B) Switch to analysis mode.

2. Analysis
   (A) Assume a process for each of the four devices: (a) compressor as isentropic, (b) cooler as isobaric, (c) turbine as isentropic, and (d) heater as isobaric.
   (B) Input the given information: (a) working fluid is air, (b) the inlet temperature and pressure of the compressor are 20°F and 15psia, (c) the inlet temperature and pressure of the are 80°F and 60 psia,and (d) the mass flow rate is 0.1 lbm/s.
3. Display results
   (A) Display the T-s diagram and cycle properties results. The cycle is a refrigerator. The answers are COP=2.06, compressor horsepower required=-7.90 hp, turbine power produced=5.98 hp, net power required=-1.92 hp, cooling load=2.79 Btu/s=0.8377 ton, and Net power input= hp, and (B) Display the sensitivity diagram of cycle COP vs compression pressure ratio.

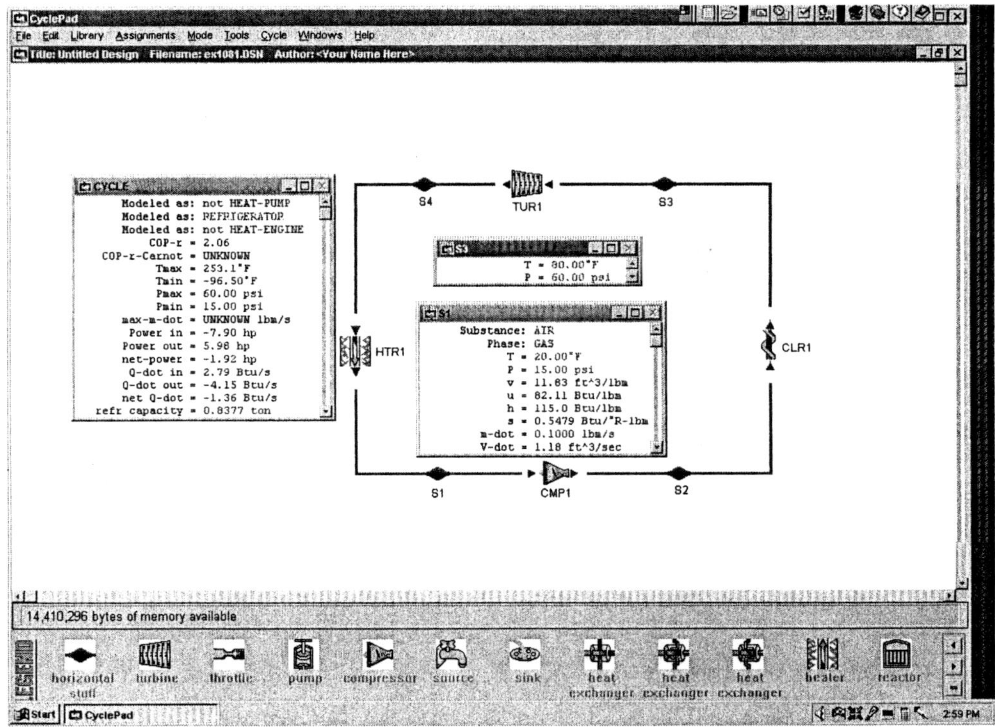

Figure Example 10.10.1a Ideal gas refrigeration cycle

Comment: The sensitivity diagram of cycle COP vs compression ratio indicates that the COP is increased as the compression pressure ratio is decreased. Unfortunately, the volume of the gas also increases when the compression pressure ratio decreases. Thus this type of air refrigeration cycle is very bulky when the compression pressure ratio is too low.

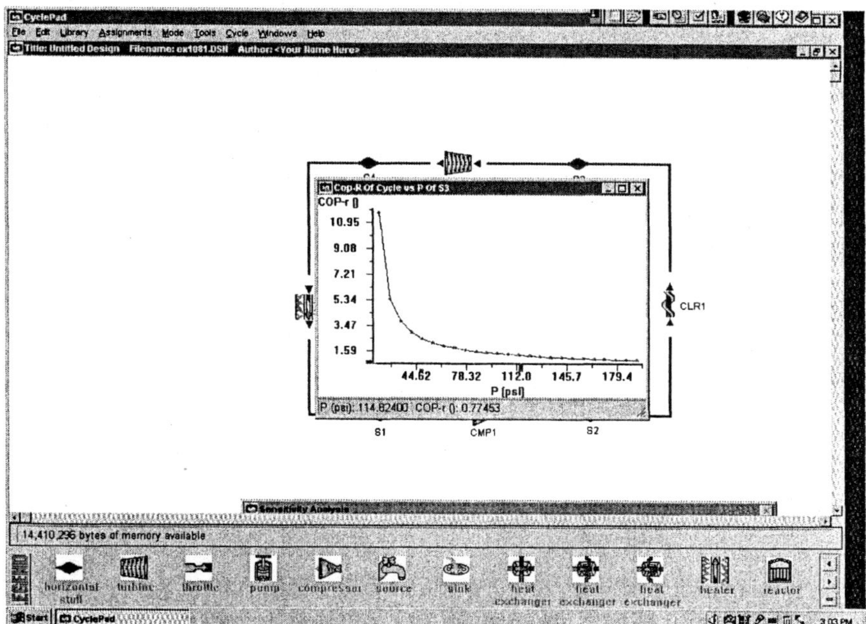

Figure Example 10.10.1b Ideal gas refrigeration cycle sensitivity analysis

Example 10.10.2. The refrigeration cycle shown at the left in Figure Example 10.10.2a is proposed to replace the more conventional cycle at the right. The compressor of the conventional system is isentropic. The low-pressure compressor of the non-conventional system is isentropic, but the high-pressure compressor is isothermal. If the circulating fluid is CO_2 and the temperature entering (100°F) and leaving (10°F) the cooler are to be the same for both systems (same refrigeration cooling load), and the cooling water temperature entering (50°F) and leaving (70°F) the heat exchanger are also to be the same for both systems. The heater pressure (1100 psia) and cooler pressure (150 psia) are also to be the same for both systems. The CO_2 pressure between the two compressors is 420 psia. With the same refrigeration cooling load, find the compressor power required for both systems.

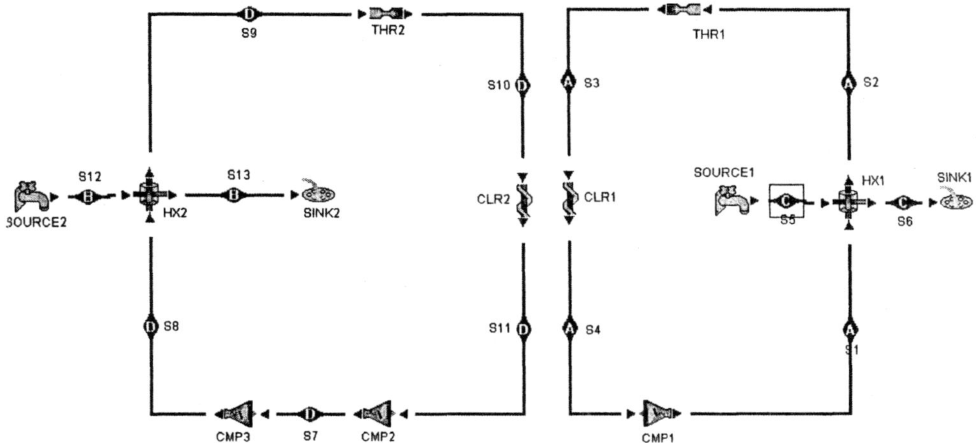

Figure Example 10.10.2a Gas refrigeration systems comparison

To solve this problem by CyclePad, we take the following steps:

1. Build
   (A) Take three compressors, two coolers, two throttling valves, two heat exchangers, two sources, and two sinks from the open system inventory shop and connect the devices to form the two gas refrigeration cycles (cycle A and cycle D).
   (B) Switch to analysis mode.
2. Analysis
   (A) Assume a process for each of the following devices: (a) compressor #1 and #2 as isentropic, (b) coolers and heat exchangers as isobaric, and (c) compressor #3 as isothermal.
   (B) Input the given information: (a) working fluid is CO_2 and the mass flow rate is 1 lbm/s ($mdot_3$=$mdot_{10}$), (b) the inlet temperature and pressure of the coolers are 100°F (T_3=T_{10}) and 10°F (T_4=T_{11}) 150 psia (p_3=p_{10}), (c) the water inlet and outlet temperature and pressure of the heat exchangers are 50°F (T_5=T_{12}) and 70°F (T_6=T_{13}) at atmospheric pressure 14.7 psia (p_5=p_{12}),(d) the working fluid high-pressure of the cycles are 1100 psia (p_9=p_2), (e) the working fluid low-pressure of the cycles are 150 psia (p_4=p_{11}), and (d) the working fluid pressure between the two compressor of the cycle at the left is 420 psia (p_7).
3. Display results: The answers are compressor power required for the system (cycle D) at the left=-71.13 hp, and compressor power required for the system (cycle A) at the right=-75.36 hp.

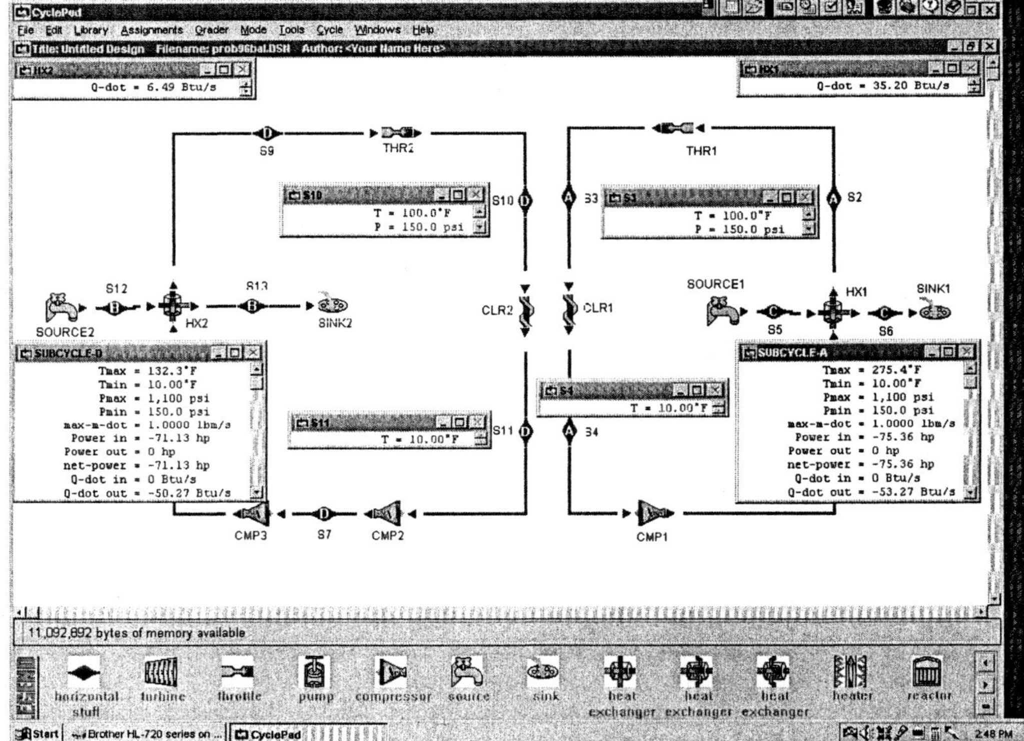

Figure Example 10.10.2b Gas refrigeration systems comparison

Homework 10.10 Gas Refrigeration

1. A 5-ton air ideal Brayton refrigeration system is to be designed according to the following specification:
   Air pressure at compressor inlet: 100 kPa
   Air pressure at turbine inlet: 420 kPa
   Air temperature at compressor inlet: -5 °C
   Air temperature at turbine inlet: 25 °C
   Determine (a) the air mass rate flow, (b) the compressor power required, (c) the turbine power produced, and (d) the cycle COP.
2. A 5-ton air Brayton refrigeration system is to be designed according to the following specification:
   Compressor efficiency: 82%
   Turbine efficiency: 84%
   Air pressure at compressor inlet: 100 kPa
   Air pressure at turbine inlet: 420 kPa
   Air temperature at compressor inlet: -5 °C
   Air temperature at turbine inlet: 25 °C
   Determine (a) the air mass rate flow, (b) the compressor power required, (c) the turbine power produced, and (d) the cycle COP.
3. An ideal Brayton refrigeration system uses air as a refrigerant. The pressure and temperature of air at compressor inlet are 14.7 psia and 100°F. The pressure and temperature of air at turbine inlet are 60 psia and 260° F. The mass rate of air flow is 0.03 lbm/s. Determine (a) the cooling load, (b) the compressor power required, (c) the turbine power produced, and (d) the cycle COP.
4. An ideal air Brayton refrigeration system is operated between -10°F and 120°F. The air pressure at compressor inlet is 14.5 psia. The air pressure at turbine inlet is 75.5 psia. The mass rate of air flow is 0.031 lbm/s. Determine (a) the cooling load, (b) the compressor power required, (c) the turbine power produced, and (d) the cycle COP.

10.11 STIRLING REFRIGERATION CYCLE

An ideal reciprocating Stirling refrigeration cycle is shown in Figure 10.11.1. It is the reversible Stirling heat engine cycle, which is composed of two isothermal processes and two isochoric processes. Working fluid is compressed in an isothermal process 1-2 at T_H. Heat is then removed at a constant volume process 2-3. Working fluid is expanded in an isothermal process 3-4 at T_L. The cycle is completed by a constant volume heat addition process 4-1. The T-s diagram of the cycle is illustrated in Figure 10.11.2. Work has been done recently on developing a practical refrigeration device based on the Stirling refrigeration cycle in extremely low temperatures (less than 200 K or -100°F). In the presence of an ideal regenerator in the cycle, the heat quantity Q_{23} and Q_{41}, which are equal in magnitude but opposite in sign, are exchanged between fluid streams within the device. Hence the only external heat transfer occurs in processes 1-2 at constant temperatures T_H and 3-4 at constant temperatures T_L. Consequently, the coefficient of performance (COP) of the Stirling refrigeration cycle theoretically equals that of the Carnot refrigeration cycle, $T_L/(T_H-T_L)$. The

Stirling refrigeration cycle with regenerator is shown in Figure 10.11.3. In this figure, the heater #1 and cooler #1 are the regenerator. Heat removed from cooler #1 is added to the heater #1.

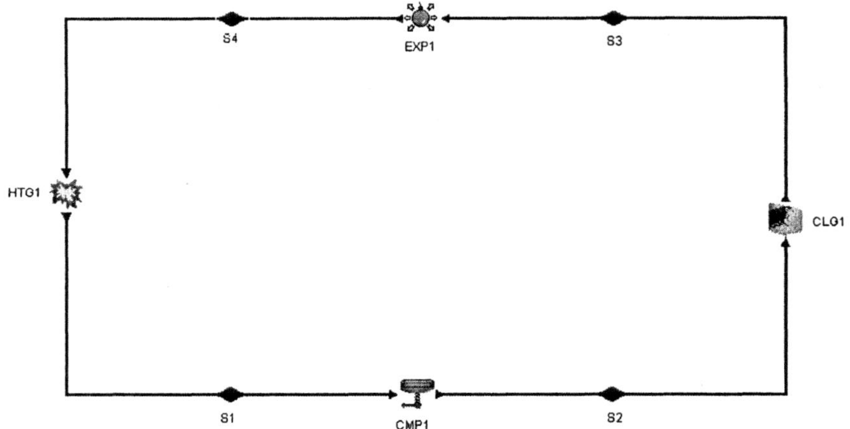

Figure 10.11.1. Stirling refrigeration cycle

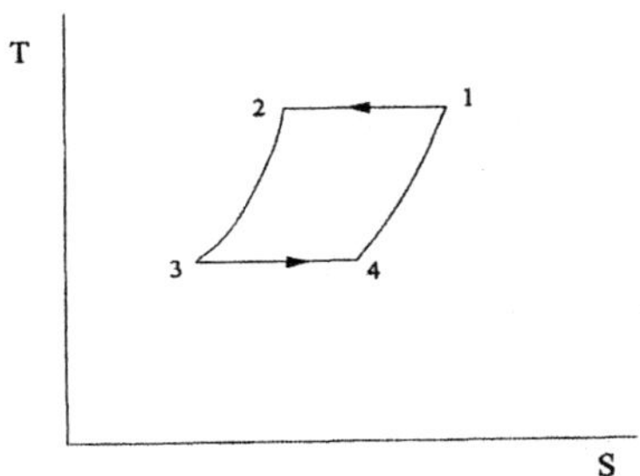

Figure 10.11.2. Stirling refrigeration cycle T-s diagram

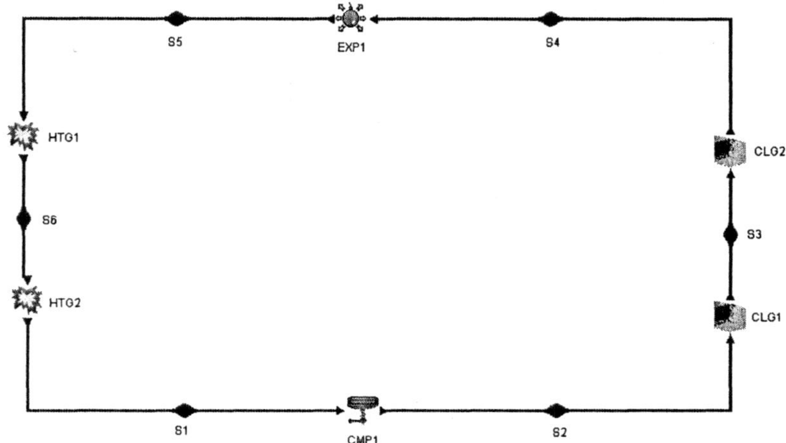

Figure 10.11.3. Stirling refrigeration cycle with regeneration

Example 10.11.1 In an ideal reciprocating Stirling refrigeration cycle, 0.01 kg of air at 235 K and 10 bars is expanded isothermally to 1 bar. It is then heated to 320 K isometrically. Compression at 320 K isothermally follows, and the cycle is completed by isometric heat removal. Determine the heat added, heat removed, work added, work done, and COP of the cycle.

To solve this problem by CyclePad, we take the following steps:

1. Build
   (A) Take a compression device, a cooling device, an expansion device, and a heating device from the closed system inventory shop and connect the four devices to form the Stirling refrigeration cycle.
   (B) Switch to analysis mode.
2. Analysis
   (A) Assume a process for each the four devices: (a) compression and expansion device as isothermal, and (b) cooling and heating device as isochoric.
   (B) Input the given information: (a) working fluid is air, (b) the inlet temperature and pressure of the expansion are 235 K and 1o bar, m=0.01 kg, (c) the inlet temperature of the compression is 320 K, and exit pressure of the expansion is 1 bar.
3. Display results
   (A) Display the cycle properties results. The cycle is a refrigerator. The answers are COP=3.05, Q_{in}=2.16 kJ, Q_{out}=-2.72 kJ, W_{in}=-2.11 kJ, and W_{out}=1.55 kJ.

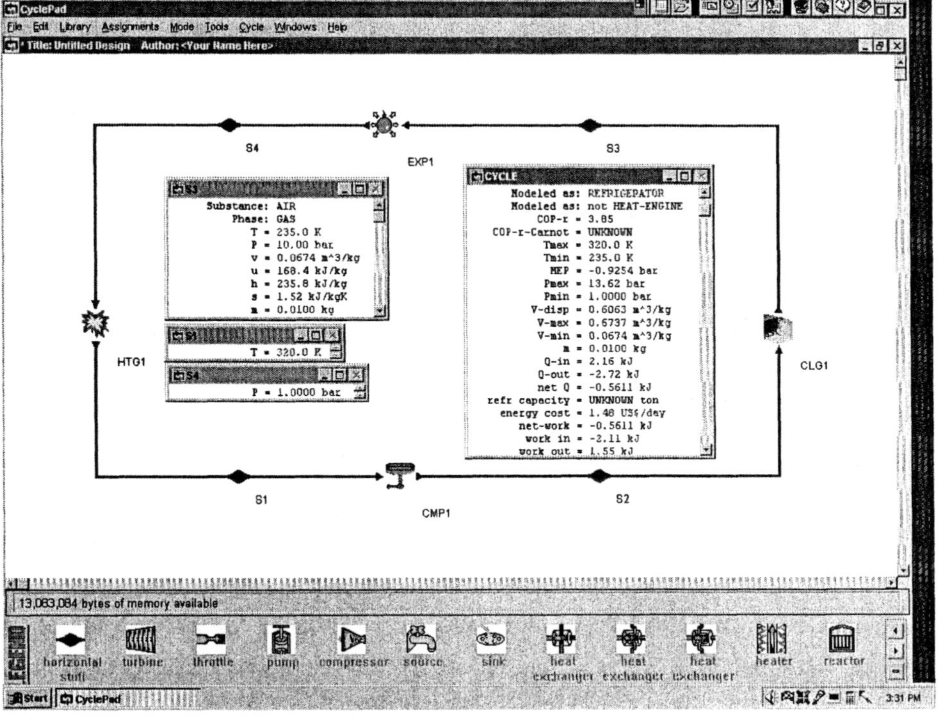

Figure Example 10.11.1. Stirling refrigeration cycle

Homework 10.11 Stirling Refrigeration Cycle

1. What are the four basic processes of the Stirling refrigeration cycle?
2. What temperature range is the main application of the Stirling refrigeration cycle?
3. In an ideal reciprocating Stirling refrigeration cycle, 0.01 kg of carbon dioxide at 235 K and 10 bars is expanded isothermally to 1 bar. It is then heated to 320 K isometrically. Compression at 320 K isothermally follows, and the cycle is completed by isometric heat removal. Determine the heat added, heat removed, work added, work done, and COP of the cycle.
   ANSWER: COP=4.26
4. In an ideal reciprocating Stirling refrigeration cycle, 0.01 kg of helium at 235 K and 10 bars is expanded isothermally to 1 bar. It is then heated to 320 K isometrically. Compression at 320 K isothermally follows, and the cycle is completed by isometric heat removal. Determine the heat added, heat removed, work added, work done, and COP of the cycle.
   ANSWER: COP=3.41

10.12 ERICSSON CYCLE

The Ericsson refrigeration cycle is a reversible Ericsson power cycle. The schematic Ericsson cycle is shown in Figure 10.12.1. The working fluid usually used in the Ericsson refrigerator is helium or hydrogen. The Ericsson refrigerator may be used for very low temperature application. The cycle consists of two isothermal processes and two isobaric processes. The four processes of the Ericsson refrigeration cycle are isothermal compression process 1-2 (compressor), isobaric cooling process 2-3 (cooler), isothermal expansion process 3-4 (turbine), and isobaric heating process 4-1 (heater).

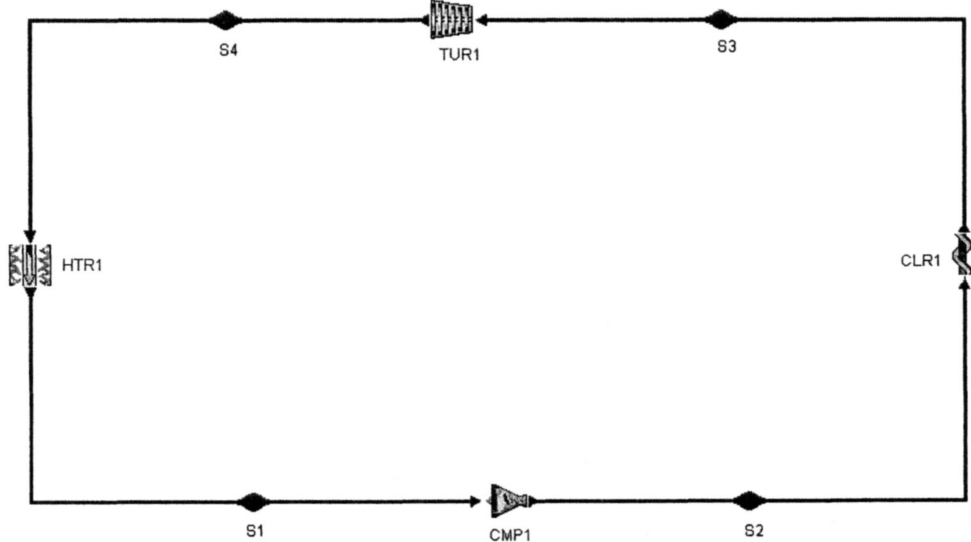

Figure 10.12.1 Ericsson refrigeration cycle

Applying the basic laws of thermodynamics, we have

$$q_{12} - w_{12} = h_2 - h_1 \tag{10.12.1}$$

$$q_{23} - w_{23} = h_3 - h_2,\ w_{23} = 0 \tag{10.12.2}$$

$$q_{34} - w_{34} = h_4 - h_3 \tag{10.12.3}$$

$$q_{41} - w_{41} = h_1 - h_4,\ w_{41} = 0 \tag{10.12.4}$$

The net work produced by the cycle is

$$w_{net} = w_{12} + w_{34} \tag{10.12.5}$$

The heat added to the cycle in the heater is q_{41}, and The cycle COP is

$$\beta_R = q_{41}/w_{net}. \tag{10.12.6}$$

Example 10.12.1. 0.001 kg/s mass flow rate of helium is compressed and heated from 100 kPa and 300 K in an Ericsson refrigeration cycle to a turbine inlet at 800 kPa and 100 K. Determine power required by the compressor, power produced by the turbine, rate of heat removed from the cooler, rate of heat added in the heater, and cycle COP. Draw the T-s diagram of the cycle.

To solve this problem by CyclePad, we do the following steps:

(A) Build the cycle as shown in Figure 10.12.1. Assume the compressor is isothermal, the heater is isobaric, the turbine is isothermal, and the cooler is isobaric.
(B) Input working fluid=helium, mass flow rate=0.001 kg/s, compressor inlet pressure=100 kPa, compressor inlet temperature=300 K, turbine inlet pressure=800 kPa, and turbine inlet temperature=100 K.
(C) Display results. The answers are: $Qdot_{htr}$=1.47 kW, $Qdot_{clr}$=-2.33 kW, $Wdot_{cmp}$=-1.3 kW, $Wdot_{tur}$=0.4319 kW, $Wdot_{net}$=-0.8638 kW, and β_R=1.7.

Homework 10.12.1 Ericsson Refrigeration Cycle

1. 0.001 kg/s mass flow rate of helium is compressed and heated from 100 kPa and 300 K in an Ericsson refrigeration cycle to a turbine inlet at 500 kPa and 100 K. Determine power required by the compressor, power produced by the turbine, rate of heat removed from the cooler, rate of heat added in the heater, and cycle COP.
ANSWER: $Qdot_{htr}$=1.37 kW, $Qdot_{clr}$=-2.04 kW, $Wdot_{cmp}$=-1.0 kW, $Wdot_{tur}$=0.3343 kW, $Wdot_{net}$=-0.6685 kW, and β_R=2.05.

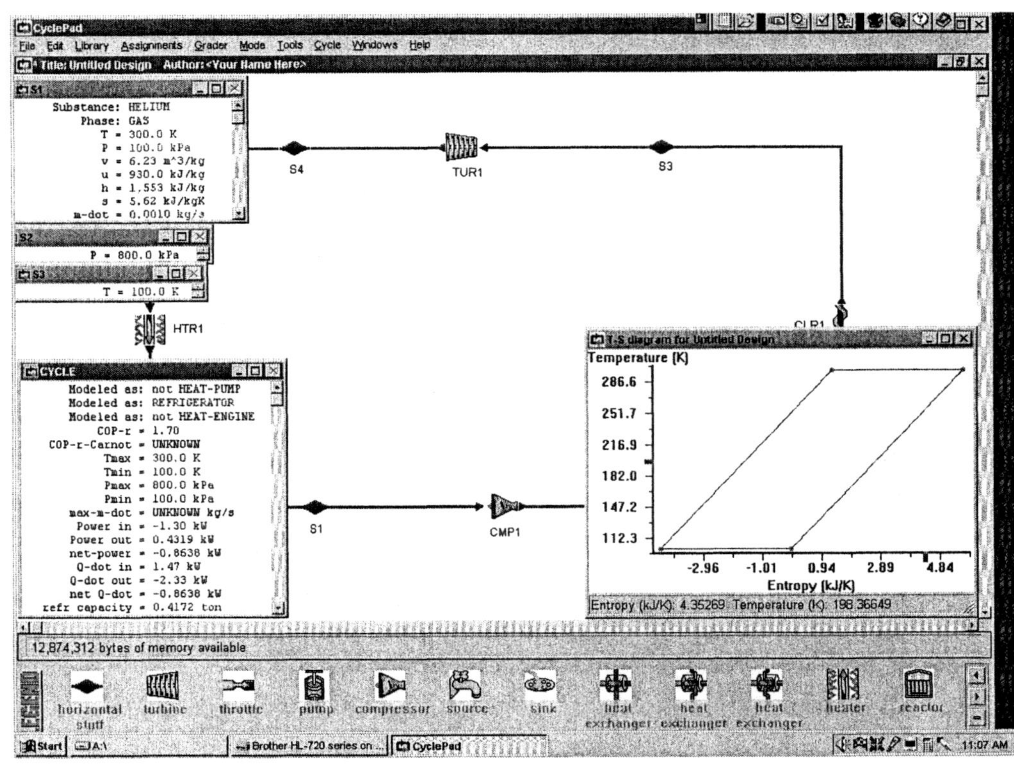

Figure Example 10.12.1 Ericsson refrigeration cycle

10.13 LIQUEFACTION OF GASES

The liquefaction of gases is a very important area in refrigeration at very low temperature. Methods of producing very low temperatures refrigeration, liquefying gases, or solidifaction solids are based on the adiabatic expansion of a high-pressure gas either through a throttling valve or in an expansion turbine. The schematic and T-s diagrams for an ideal Hampson-Linde gas liquefaction system are shown in Figure 10.13.1 and Figure 10.13.2. Makeup gas at state 1 is mixed with the uncondensed portion of the gas at state 15 from the previous cycle, and the mixture at state 2 is compressed by a four-stage compressor with inter-coolers (compressor 1, inter-cooler 1, compressor 2, inter-cooler 2, compressor 3, inter-cooler 3, and compressor 4) to state 9. After the multi-stage compression, the gas is cooled from state 9 to state 10 at constant pressure in a cooler. The gas is further cooled to state 11 in a regenerative heat exchanger. After expansion through a throttle valve, the fluid at state 12 is in the liquid-vapor mixture state and is separated into liquid (state 13) and vapor (state 14) states. The liquid at state 13 is drawn off as the desired product, and the vapor at state 14 flows through the regenerative heat exchanger to cool high-pressure gas flowing toward the throttle valve. The gas at state 15 is finally mixed with fresh makeup gas, and the cycle is repeated. This cycle can also be used for the solidifaction of gases at even lower temperature.

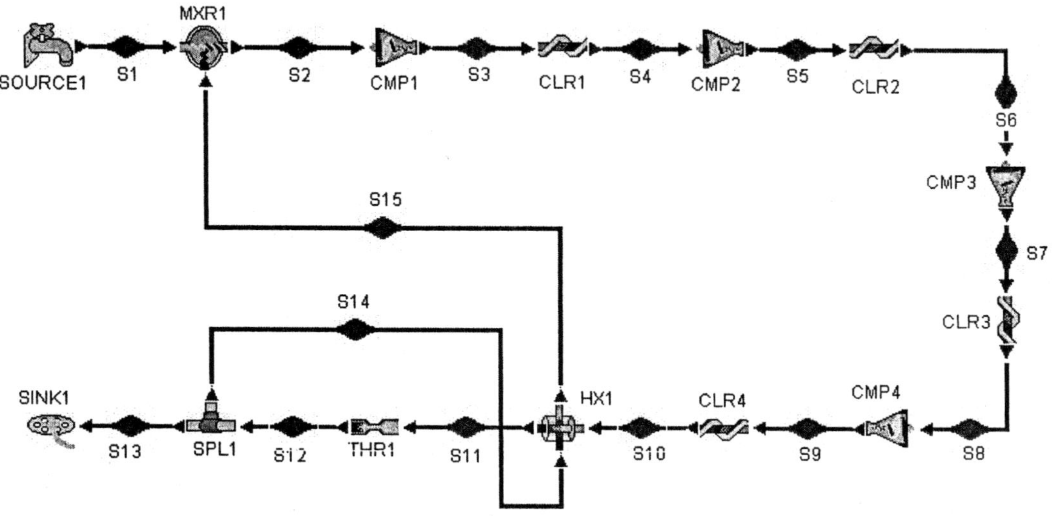

Figure 10.13.1 Hampson-Linde gas liquefaction system schematic diagram

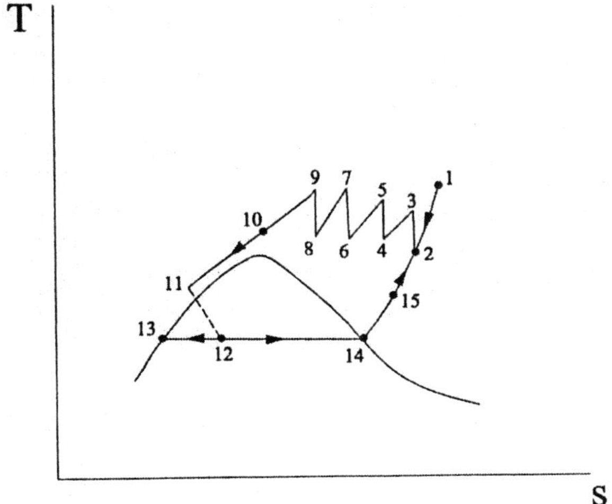

Figure 10.13.2 Hampson-Linde gas liquefaction system T-s diagram

Figure 10.13.3 is the Claude gas liquefaction system, a modification of the Hampson-Linde gas liquefaction system. The Claude system has a turbine in the expansion process to replace a part of the highly irreversible throttling process of the Hampson-Linde system. From state 1 to state 10, the Claude system processes are the same as those of the Hampson-Linde system. After the gas is cooled to state by the regenerative cooler (heat exchanger 1), most of it is expanded through a turbine from state 12 to state 19 and then is mixed with vapor at state 17 from the separator (splitter 2) and flows back toward the compressor through a heat exchanger which precools the small fraction of the flow that is directed toward the throttle valve instead of the turbine.

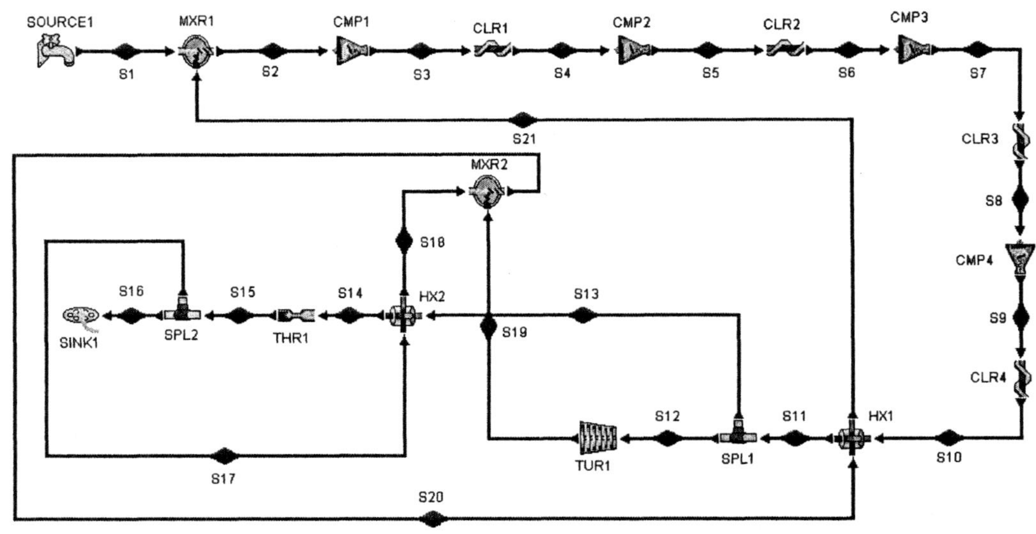

Figure 10.13.3 Claude gas liquefaction system

10.14 Design Examples

Design of refrigeration and heat pump cycles would typically be tedious. With the use of the intelligent computer-aided software CyclePad, however, the design can be completed rapidly and modifications are simple to make. CyclePad makes designing a refrigeration or heat pump plant as simple as point and click. To demonstrate the power CyclePad offers designers in the design of refrigeration and heat pump cycles, the following design examples are considered.

Example 10.14.1. A combined split-shaft gas turbine power plant and gas refrigeration system to be used in an airplane as illustrated in Figure Example 10.14.1a is designed by a junior engineer.

The following design information are provided:

> $mdot_1$=1 kg/s, p_1=101 kPa, T_1=15°C, p_2=1000 kPa, T_5=1200°C, p_7=102 kPa, T_{11}=400°C, p_{12}=102 kPa, T_{13}=15°C, $\eta_{turbine}$=85% and $\eta_{compressor}$=85%.
>
> (A) During the cruise condition, the split-shaft gas turbine power plant requires to produce 240 kW and 10% of the compressed air is used to the gas refrigeration system which requires to removes 7 kW from the cabin.
>
> (B) During the take-off condition, the split-shaft gas turbine power plant requires to produce 300 kW. and (C) During the high-wind condition, the split-shaft gas turbine power plant requires to produce 270 kW while at least 3.5 kW of cabin refrigeration is to be provided.

Check if all these conditions can be met by the design.

To check this design by CyclePad, we do the following steps:

1. Build the cycle as shown in Figure example 10.14.1a. Assume the compressor and turbines are adiabatic with 85% efficiency, the heaters, mixing chambers and cooler are isobaric, and the splitters are iso-parametric.
2. Input working fluid=air, $mdot_1$=1 kg/s, p_1=101 kPa, T_1=15°C, p_2=1000 kPa, T_5=1200°C, p_7=102 kPa, T_{11}=400°C, p_{12}=102 kPa, T_{13}=15°C, $\eta_{turbine}$=85% and $\eta_{compressor}$=85%. Display the compressor power (-314.7 kW) and input $Wdot_{turbine\ \#1}$=314.7 kW as shown in Figure example 10.14.1b.

   (A) Cruise condition: input $mdot_8$=0.1 kg/s, and $mdot_9$=0 kg/s.
   Display results. The answers are: $Wdot_{turbine\ \#2}$=240.2 kW and $Qdot_{heater\#2}$=7.16 kW as shown in Figure example 10.14.1c. The design requirement is met.
   (B) Take-off condition: input $mdot_9$=0.01 kg/s, and $mdot_{10}$=0 kg/s.
   Display results. The answers are: $Wdot_{turbine\ \#2}$=301.7 kW and $Qdot_{heater\#2}$=0 kW as shown in Figure example 10.14.1d. The design requirement is met.
   (C) High-wind condition: input $mdot_9$=0.05 kg/s, and $mdot_{10}$=0.05 kg/s.
   Display results. The answers are: $Wdot_{turbine\ \#2}$=271.0 kW and $Qdot_{heater\#2}$=3.58 kW as shown in Figure example 10.14.1e. The design requirement is met.

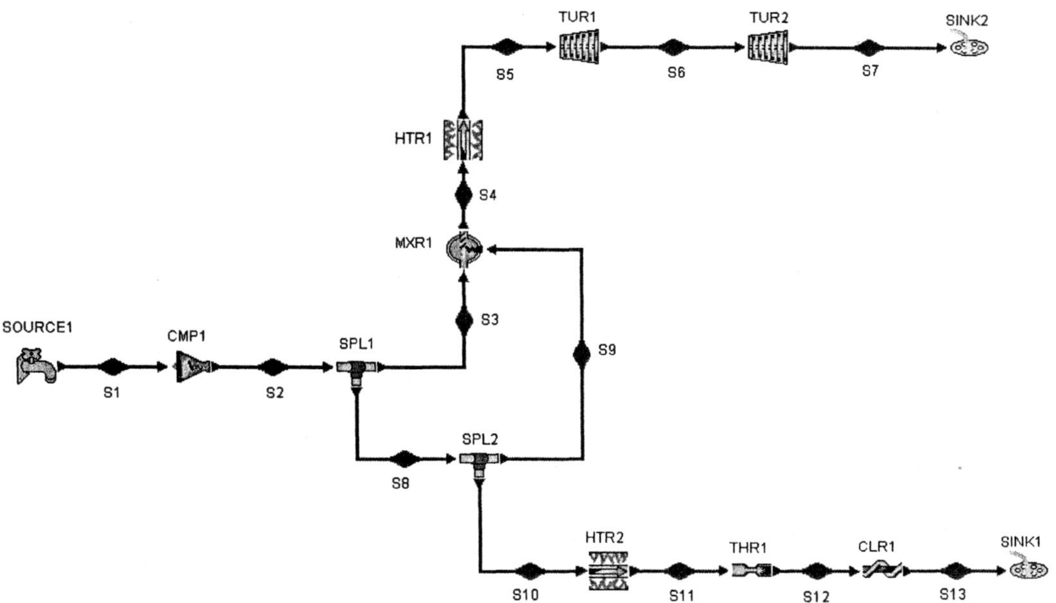

Figure Example 10.14.1a. Combined gas turbine power plant and gas refrigeration system design

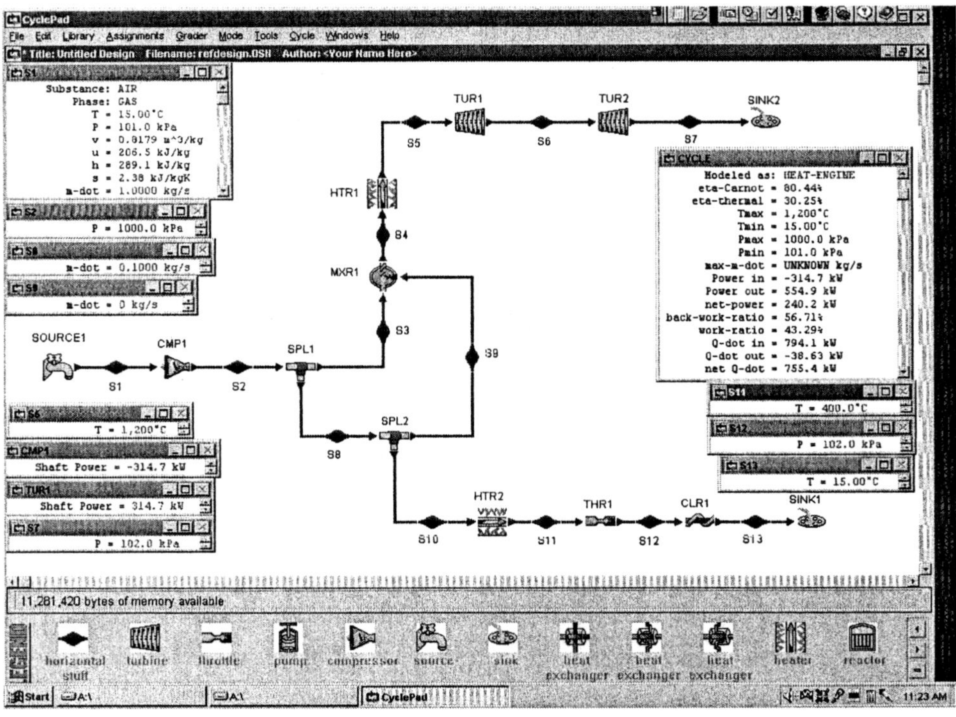

Figure Example 10.14.1b. Combined gas turbine power plant and gas refrigeration system design input

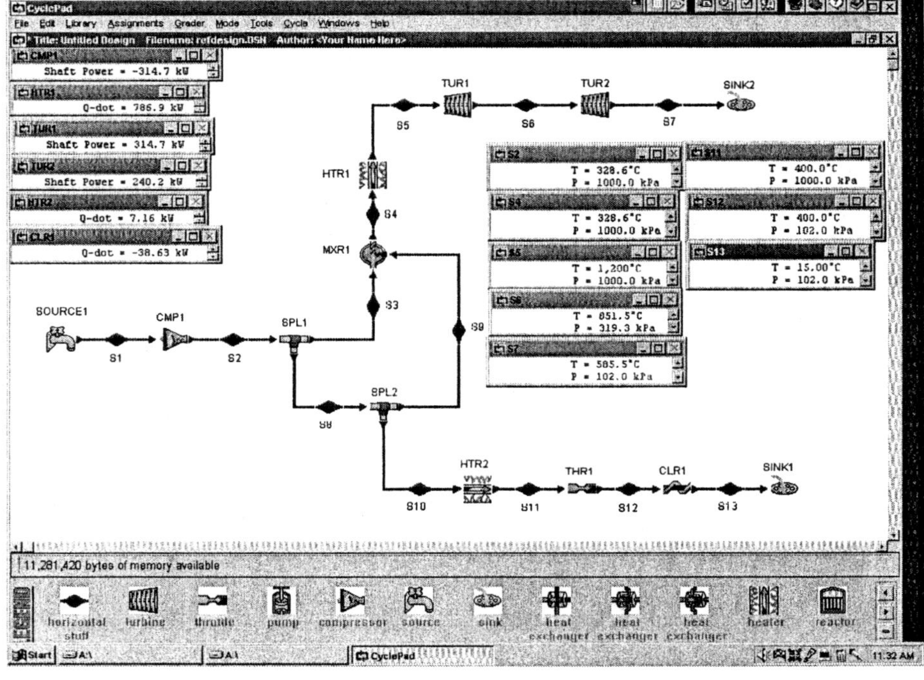

Figure Example 10.14.1c. Combined gas turbine power plant and gas refrigeration system design at cruise condition

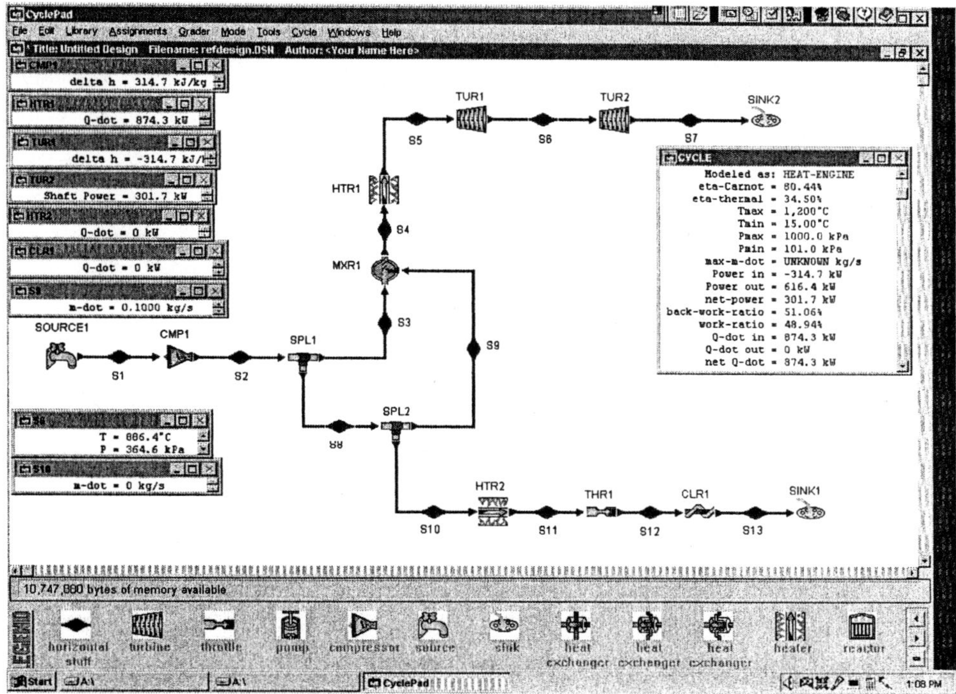

Figure Example 10.14.1d. Combined gas turbine power plant and gas refrigeration system design at take-off condition

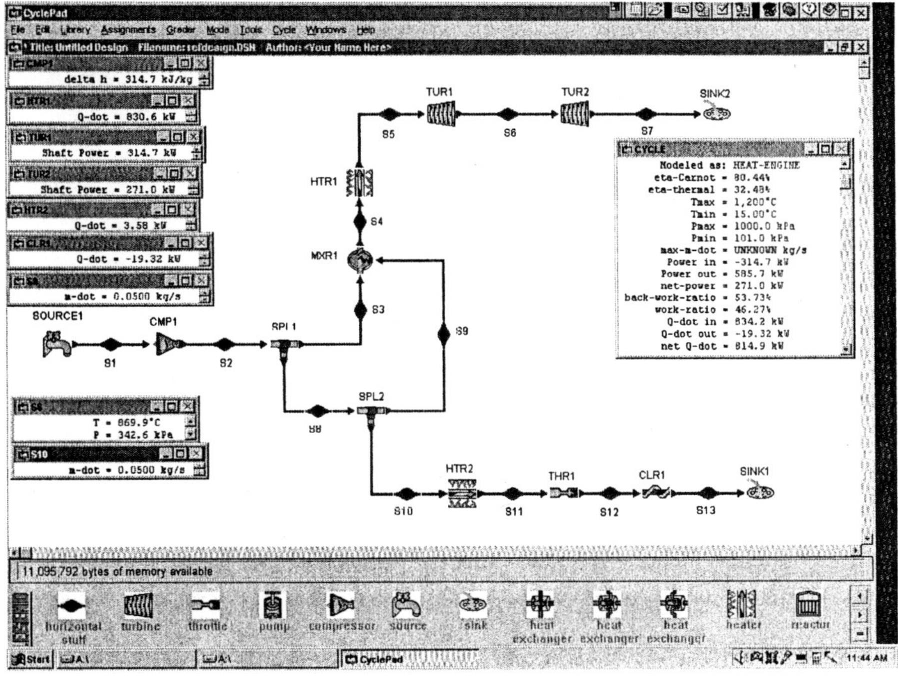

Figure Example 10.14.1e. Combined gas turbine power plant and gas refrigeration system design at high-wind condition

Example 10.14.2. An engineer claims that the performance of a simple refrigeration cycle can be improved by using his three-stage compression process as shown in Figure Example 10.14.2a. Refrigerant enters all compressors as a saturated vapor, and enters the throttling valve as a saturated liquid. The highest and lowest pressure of the cycle are 1200 kPa and 150 kPa, respectively. His design information are:

Refrigerant: R-12, $mdot_1$=1 kg/s, p_1=150 kPa, x_1=1, p_3=300 kPa, x_3=1, p_5=600 kPa, x_5=1, p_7=1200 kPa, x_7=0, and $\eta_{compressor}$=85%.
His design results are: β (COP)=2.16, $Wdot_{in}$=-43.64 kW, $Qdot_{in}$=94.48 kW, $Qdot_{out}$=-138.1 kW, and cooling load=26.87 tons.

(A) Check on his claim, (B) What are the the performance of the cycle if ammonia, R-134a, or R-22 is used instead of R-12, (C) Try to improve the COP by varying p_3 and p_5.

(A) To check this design by CyclePad, we do the following steps:
   (1) Build the three-stage-compressor refrigeration cycle as shown in Figure example 10.14.2a. Assume the compressors are adiabatic with 85% efficiency, and the heater xand cooler are isobaric.
   (2) Input working fluid=R-12, $mdot_1$=1 kg/s, p_1=150 kPa, x_1=1, p_3=300 kPa, x_3=1, p_5=600 kPa, x_5=1, p_7=1200 kPa, x_7=0, and $\eta_{compressor}$=85%.
   (3) Display the COP=2.22, compressor power (-42.56 kW), $Qdot_{in}$=94.48 kW, $Qdot_{out}$=-137.0 kW and cooling capacity=26.87 tons as shown in Figure Example 10.14.2b.

(B) For the one-compressor refrigeration system using R-12, (1) retract p_3=300 kPa, x_3=1, p_5=600 kPa, and x_5=1. (2) let $Wdot_{compressor\#1}$=0, $Wdot_{compressor\#2}$=0, $Qdot_{cooler\#1}$=0, and $Qdot_{cooler\#2}$=0. (3) Display results, the results are: COP=2.16, compressor power (-43.64 kW), $Qdot_{in}$=94.48 kW, $Qdot_{out}$=-138.1 kW and cooling capacity=26.87 tons as shown in Figure Example 10.14.2c. The COP is indeed improved.

(B) For the three-compressor refrigeration system using ammonia, (1) retract working fluid and let the working fluid be ammonia. (2) Display results, the results are: COP=3.36, compressor power (-323.0 kW), $Qdot_{in}$=1084 kW, $Qdot_{out}$=-1407 kW and cooling capacity=308.2 tons as shown in Figure Example 10.14.2d.

(B) For the three-compressor refrigeration system using R-134a, (1) retract working fluid and let the working fluid be R-134a. (2) Display results, the results are: COP=2.41, compressor power (-50.41 kW), $Qdot_{in}$=121.6 kW, $Qdot_{out}$=-172.1 kW and cooling capacity=34.59 tons as shown in Figure Example 10.14.2e.

(B) For the three-compressor refrigeration system using R-22, (1) retract working fluid and let the working fluid be R-22. (2) Display results, the results are: COP=2.64, compressor power (-58.82 kW), $Qdot_{in}$=155.2 kW, $Qdot_{out}$=-214.0 kW and cooling capacity=44.13 tons as shown in Figure Example 10.14.2d.

(C) To improve the COP by varying p_3 and p_5, draw the COP vs p_3 sensitivity diagram and COP vs p_5 sensitivity diagram as shown in Figure Example 10.14.2g. The maximum COP is about 2.225 when p_3 is about 309.7 kPa, and the maximum COP is about 2.384 when p_5 is about 898.8 kPa.

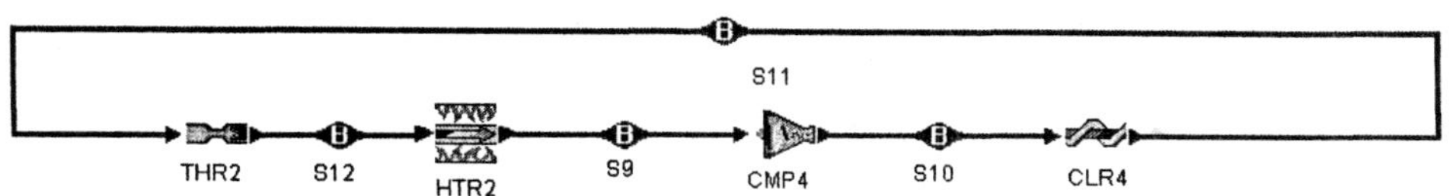

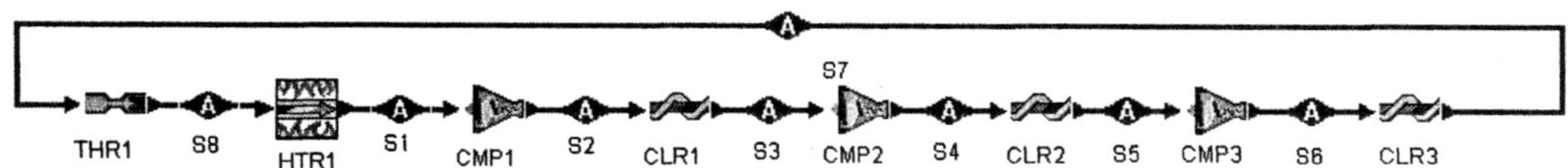

Figure Example 10.14.2a. Single-stage-compressor and Three-stage-compressor refrigeration systems

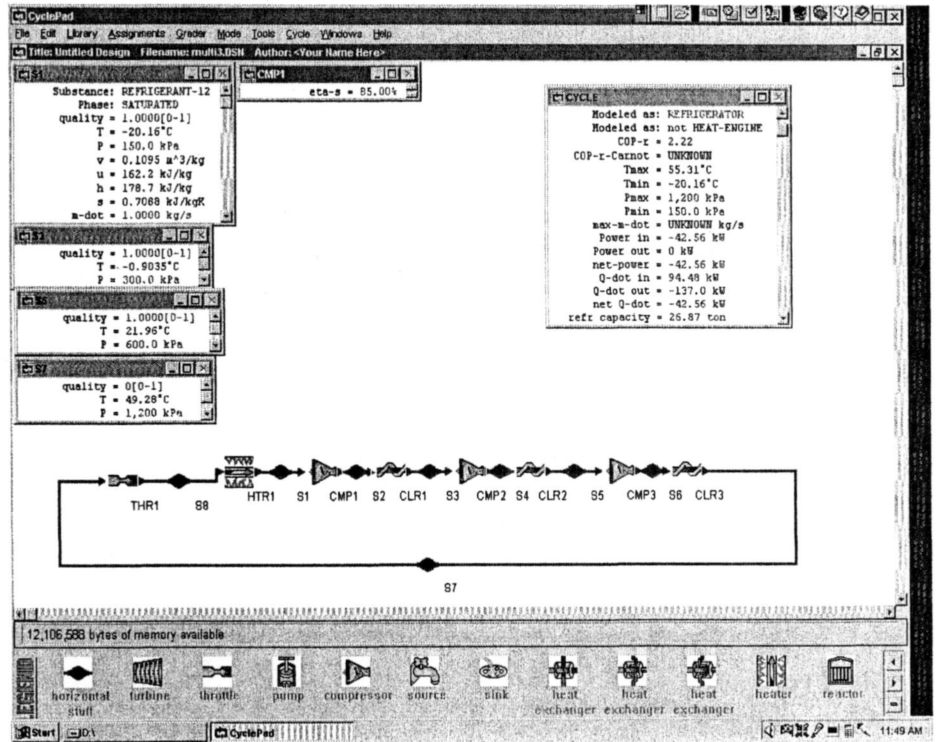

Figure Example 10.14.2b. Three-stage-compressor refrigeration system using R-12

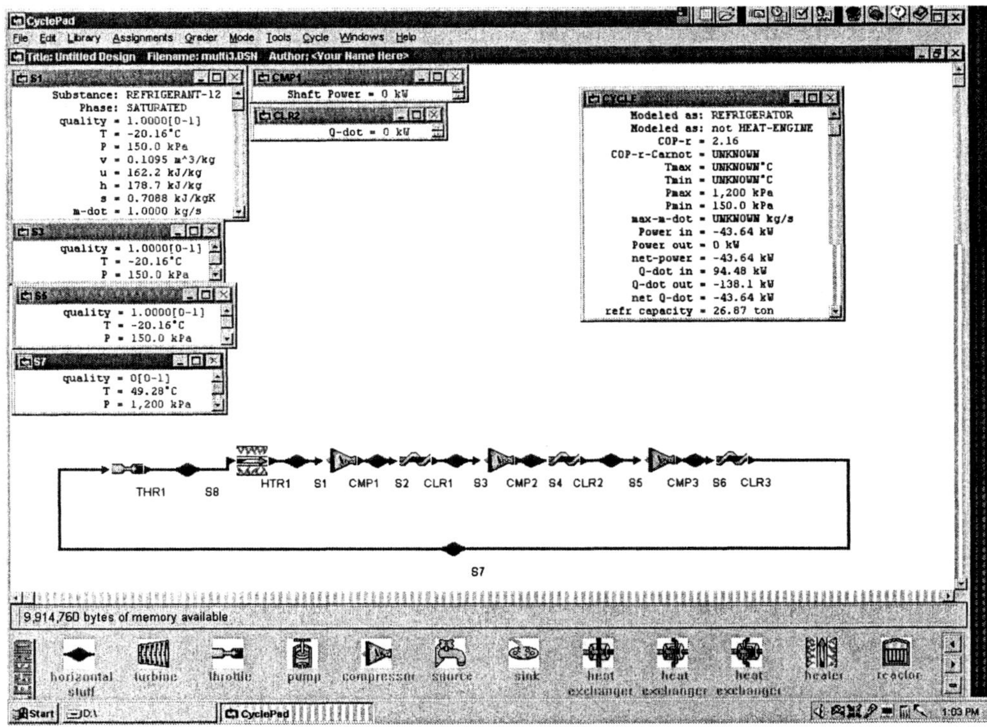

Figure Example 10.14.2c. One-stage-compressor refrigeration system using R-12

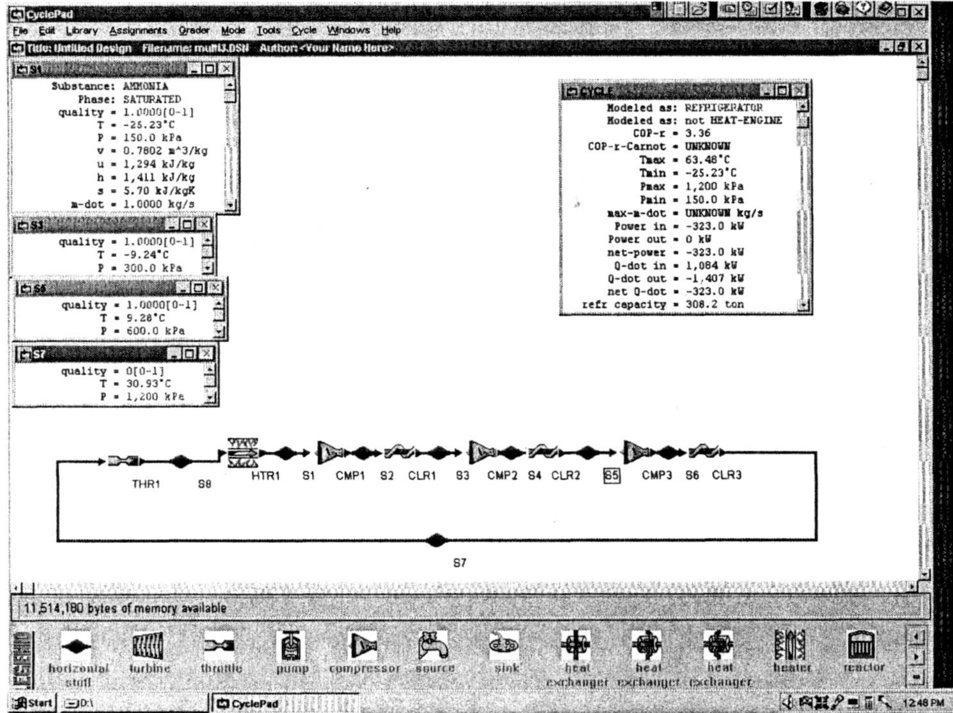

Figure Example 10.14.2d. Three-stage-compressor refrigeration system using ammonia

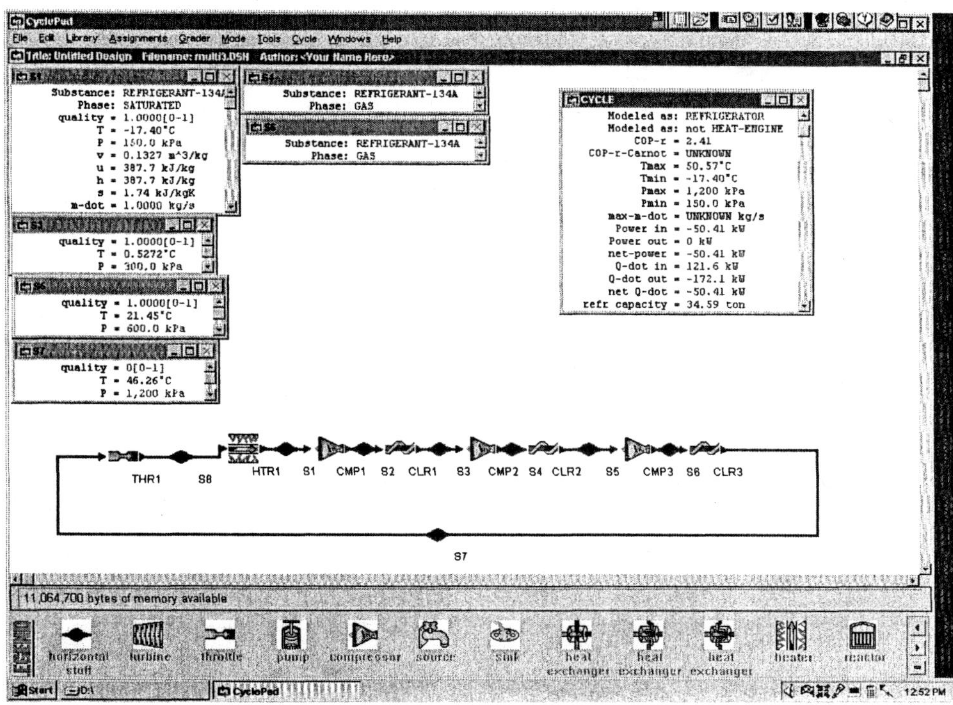

Figure Example 10.14.2e. Three-stage-compressor refrigeration system using R-134a

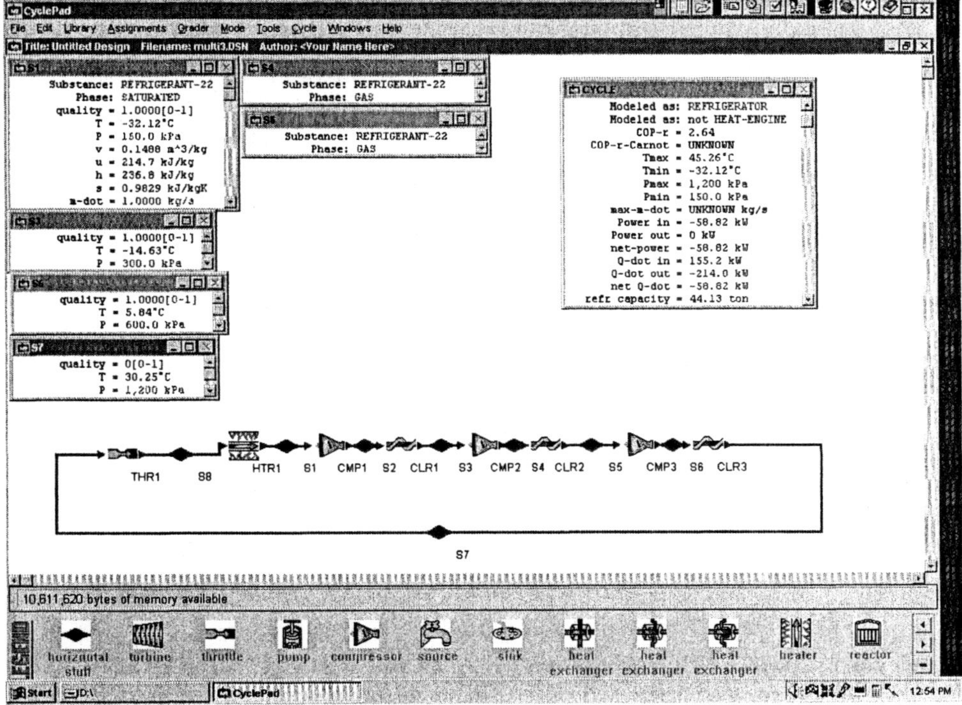

Figure Example 10.14.2f. Three-stage-compressor refrigeration system using R-22

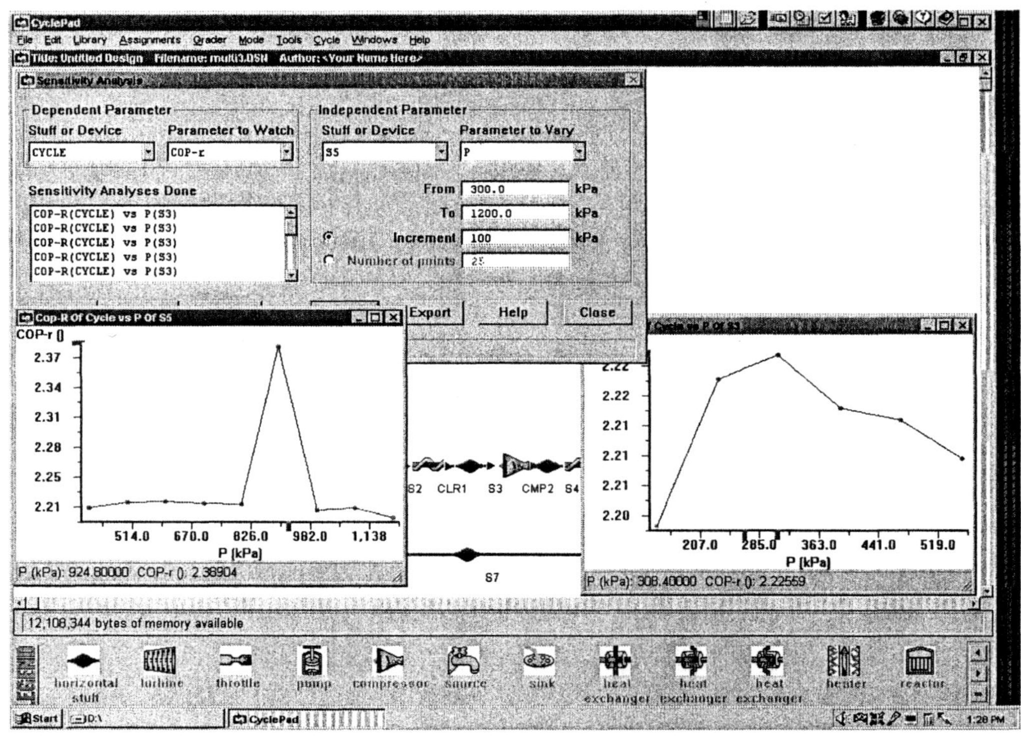

Figure Example 10.14.2g. Three-stage-compressor refrigeration system sensitivity diagrams

Homework 10.14 Design Examples

1. The performance of a simple refrigeration cycle can be improved by using cascaded refrigeration cycle. An engineer claims that he has developed a separate three-loop cascaded refrigeration cycle. The cascaded cycle consists of three separate loops--one at high pressure, one at low pressure and one at mid pressure using R-12 as working fluid in all three loops. The three loops are connected by two heat exchangers. His design information are:
   Low-pressure loop– mdot=1 lbm/s, high pressure=60 psia, low pressure=20 psia, $\eta_{compressor}$=85%, R-12 quality at inlet of compressor=1, and R-12 quality at inlet of throttling valve=0.
   Mid-pressure loop– high pressure=110 psia, low pressure=60 psia, $\eta_{compressor}$=85%, R-12 quality at inlet of compressor=1, and R-12 quality at inlet of throttling valve=0.
   High-pressure loop– high pressure=160 psia, low pressure=110 psia, $\eta_{compressor}$=85%, R-12 quality at inlet of compressor=1, and R-12 quality at inlet of throttling valve=0.
   His design results are: Whole system: β (COP)=2.74, $Wdot_{in}$=-29.59 hp, $Qdot_{in}$=57.21 Btu/s, $Qdot_{out}$=-78.12 Btu/s, and cooling load=17.16 tons; Low-pressure loop: $Wdot_{in}$=-13.53 hp, $Qdot_{in}$=57.21 Btu/s, $Qdot_{out}$=-66.78 Btu/s, and mdot=1 lbm/s; Mid-pressure loop: $Wdot_{in}$=-9.41 hp, $Qdot_{in}$=66.78 Btu/s, $Qdot_{out}$=-73.43 Btu/s, and mdot=1.23 lbm/s;

and High-pressure loop: $Wdot_{in}$=-6.65 hp, $Qdot_{in}$=73.43 Btu/s, $Qdot_{out}$=-78.12 Btu/s, and mdot=1.43 lbm/s.

(A) Check on his claim, (B) What are the the performance of the cycle if ammonia, R-134a, or R-22 is used instead of R-12, (C) Try to improve the COP by varying the two pressures of the mid-pressure loop as design variables.

ANSWERS: (B) Ammonia– Whole system: β (COP)=3.42, $Wdot_{in}$=-219.0 hp, $Qdot_{in}$=529.9 Btu/s, $Qdot_{out}$=-684.7 Btu/s, and cooling load=159.0 tons; Low-pressure loop: $Wdot_{in}$=-104.2 hp, $Qdot_{in}$=529.9 Btu/s, $Qdot_{out}$=-603.6 Btu/s, and mdot=1 lbm/s; Mid-pressure loop: $Wdot_{in}$=-68.13 hp, $Qdot_{in}$=603.6 Btu/s, $Qdot_{out}$=-651.7 Btu/s, and mdot=1.18 lbm/s; and High-pressure loop: $Wdot_{in}$=-46.65 hp, $Qdot_{in}$=651.7 Btu/s, $Qdot_{out}$=-684.7 Btu/s, and mdot=1.33 lbm/s.

(B) R-22– Whole system: β (COP)=2.95, $Wdot_{in}$=-40.72 hp, $Qdot_{in}$=85.05 Btu/s, $Qdot_{out}$=-113.8 Btu/s, and cooling load=25.51 tons; Low-pressure loop: $Wdot_{in}$=-18.78 hp, $Qdot_{in}$=85.05 Btu/s, $Qdot_{out}$=-98.32 Btu/s, and mdot=1 lbm/s; Mid-pressure loop: $Wdot_{in}$=-12.84 hp, $Qdot_{in}$=98.32 Btu/s, $Qdot_{out}$=-107.4 Btu/s, and mdot=1.22 lbm/s; and High-pressure loop: $Wdot_{in}$=-9.10 hp, $Qdot_{in}$=107.4 Btu/s, $Qdot_{out}$=-113.8 Btu/s, and mdot=1.41 lbm/s.

2. The performance of a simple refrigeration cycle can be improved by using cascaded refrigeration cycle. An engineer claims that he has developed a separate four-loop cascaded refrigeration cycle. The cascaded cycle consists of four separate loops--one between 20 psia and 40 psia, one between 40 psia and 80 psia, one between 80 psia and 120 psia, and one between 120 psia and 160 psia. The four loops are connected by three heat exchangers. His design information are:

   Loop A– mdot=1 lbm/s, high pressure=40 psia, low pressure=20 psia, $\eta_{compressor}$=85%, R-12 quality at inlet of compressor=1, and R-12 quality at inlet of throttling valve=0.

   Loop B– high pressure=80 psia, low pressure=40 psia, $\eta_{compressor}$=85%, R-12 quality at inlet of compressor=1, and R-12 quality at inlet of throttling valve=0.

   Loop C– high pressure=120 psia, low pressure=80 psia, $\eta_{compressor}$=85%, R-12 quality at inlet of compressor=1, and R-12 quality at inlet of throttling valve=0.

   Loop D– high pressure=160 psia, low pressure=120 psia, $\eta_{compressor}$=85%, R-12 quality at inlet of compressor=1, and R-12 quality at inlet of throttling valve=0.

   His design results are: Whole system: β (COP)=2.84, $Wdot_{in}$=-31.06 hp, $Qdot_{in}$=62.26 Btu/s, $Qdot_{out}$=-84.21 Btu/s, and cooling load=18.68 tons; Loop A: $Wdot_{in}$=-8.33 hp, $Qdot_{in}$=62.26 Btu/s, $Qdot_{out}$=-68.15 Btu/s, and mdot=1 lbm/s; Loop B: $Wdot_{in}$=-10.37 hp, $Qdot_{in}$=68.15 Btu/s, $Qdot_{out}$=-75.48 Btu/s, and mdot=1.20 lbm/s; Loop C: $Wdot_{in}$=-6.86 hp, $Qdot_{in}$=75.48 Btu/s, $Qdot_{out}$=-80.32 Btu/s, and mdot=1.38 lbm/s; and Loop D: $Wdot_{in}$=-5.50 hp, $Qdot_{in}$=80.32 Btu/s, $Qdot_{out}$=-84.21 Btu/s, and mdot=1.55 lbm/s.

   (A) Check on his claim, (B) What are the the performance of the cycle if ammonia, R-134a, or R-22 is used instead of R-12, (C) Try to improve the COP by varying the three pressures (40 psia, 80 psia and 120 psia) of the loops as design variables.

10.15 Summary

The reversed Carnot cycle is modified for the most widely used vapor heat pump and refrigerator. The basic vapor heat pump and refrigerator cycle is made of an isentropic compression process, an isobaric cooling process, an irreversible throttling process, and an isobaric heating process. The coefficient of performance (COP) of refrigerators is defined as Q_L (desirable heat output or cooling effect)/W_{net}. The coefficient of performance (COP) of heat pumps is defined as Q_H (desirable heat output or heating effect)/W_{net}.

Large temperature difference can be achieved by cascaded refrigerators and heat pumps.

Multi-staged refrigerators and heat pumps reduce the compressor power.

Stirling and Ericsson refrigerators have practical applications in very low temperature.

Domestic refrigerator-freezer and air conditioning-heat pump systems share equipments to reduce cost.

An absorption refrigerator or heat pump is economically attractive because it uses inexpensive heat input rather than the expensive electric work input to produce the refrigeration or heat pump effect.

Brayton gas refrigeration cycle is a reversed Brayton gas power cycle.

Liquefaction and solidification of gases are obtained by compression of gas followed by cooling and throttling leading to a change of phase for part of the fluid.

Chapter 11

FINITE-TIME THERMODYNAMICS

OBJECTIVES

After reading and studying the material in this chapter, you should be able to:

1. know that time is not involved in classical equilibrium thermodynamics.
2. understand what is finite-time thermodynamics.
3. know that the Carnot cycle efficiency is too high, and Carnot cycle does not produce any power.
4. know the purpose of a heat exchanger and several types.
5. understand that power or specific power is more important than work in industrial heat engine design.
6. understand and analyze a finite-time endoreversible Carnot cycle.
7. understand and analyze a finite-time Rankine cycle.
8. understand and analyze a finite-time Brayton cycle.

11.1 INTRODUCTION

Among the important topics in thermodynamics is the formulation of criteria for comparing the performance of real and ideal processes. Carnot showed that any heat engine absorbing heat from a high temperature heat source reservoir to produce work must transfer some heat to a heat sink reservoir of lower temperature. He also showed that no heat engine could be better than the Carnot heat engine. The early tradition was carried on by Clausius, Kelvin and others using thermodynamics as a tool to find limits on work, heat transfer, efficiency, coefficient of performance, energy effectiveness and energy figure of merit of energy conversion devices. The basic laws of thermodynamics were all conceived based on irreversible processes. However, the subsequent development of thermodynamics has turned from the process variable of heat and work toward state variables since Gibbs. The Carnot-Clausius-Kelvin view emphasizes the interaction of a thermodynamic system with its surroundings, while the Gibbs view makes the properties of the system dominant and focuses on equilibrium states. Contemporary classical thermodynamics gives a fairly complete description of equilibrium states and reversible processes. The only fact that it tells about real processes is that these irreversible processes always produce less work and more entropy than the corresponding reversible processes. Reversible processes are defined only in the limit of infinitely slow execution.

In the real engineering world, actual changes in enthalpy and free energy in an irreversible process rarely approach the corresponding ideal enthalpy and free energy changes. No practicing engineer wants to design a heat engine that runs infinitely slowly without producing power. The need to produce power in real energy conversion devices is one reason why the high efficiences of ideal, reversible performance are seldom approached.

Classical equilibrium thermodynamics can be extended to quasi-static processes. Conventional irreversible thermodynamics has become increasingly powerful, but its microscopic view does not lend itself to the macroscopic view preferred by practicing engineers. This is a significant extension, since quasi-static processes happen in finite time, produce entropy and provide a better approximation of real processes than provided by equilibrium thermodynamics. System parameters in equilibrium thermodynamics are the measurable quantities: volume, temperature, pressure, and heat capacity. To rigorously model real time dependent processes, the set of parameters must also include transport properties, relaxation time, etc. In general, irreversible thermodynamic problems are too difficult for practicing engineers to solve exactly.

The literature of finite-time thermodynamics started with Curzon and Ahlborn [**Reference:** Curzon, F.L. and B. Ahlborn, Efficiency of a Carnot engine at maximum power output, *Amer. J. Phys.*, 1975, v.41, n.1, pp22-24] in 1975. They treated an internally reversible but externally irreversible Carnot heat engine with power, rate of work, limited by the rates of heat transfer to and from the working substance. They remarked that the Carnot efficiency [$\eta_{Carnot}=1-(T_L/T_H)$, where T_L and T_H are the temperatures of the heat sink and heat source for the engine] is realized only by a completely reversible heat engine operating at zero speed and hence at zero power. They showed that the efficiency of an engine operating at maximum power is given by a remarkably simple formula [$\eta_{Curzon\text{-}Ahlborn}=1-(T_L/T_H)^{1/2}$], which of course always gives a lower value than the Carnot formula. They also verified that their formula agrees much better with the measured efficiencies of operating installations.

Finite time thermodynamics is an extension to traditional thermodynamics in order to obtain more realistic limits to the performance of real processes, and to deal with processes or devices with finite time characteristics. Finite time thermodynamics is a method for the modeling and optimization of real devices that owe their thermodynamic imperfection to heat transfer, mass transfer and fluid flow irreversibilities.

A literature survey of finite-time thermodynamics is given by Wu, Chen and Chen.

Engineering thermodynamic cycle analysis is based on the concept of equilibrium and does not deal with time. Heat transfer does deal with time but not cycle analysis. Finite-time thermodynamics fills in a gap which has long existed between equilibrium thermodynamics and heat transfer.

Homework 11.1 Introduction

1. What is the basic concept of classical engineering equilibrium thermodynamics? Does engineering thermodynamic cycle analysis deal with time?

2. Are the heat transfers between the Carnot heat engine and its surrounding heat source and heat sink reversible?
3. Why cannot the Carnot cycle efficiency be approached in real world?
4. What is finite time thermodynamics?

11.2 Rate of Heat Transfer

Heat is an amount of microscopic energy transfer across the boundary of a system in an energy interaction with its surroundings. The symbol Q is used to denote heat.

The *rate of heat transfer* is an amount of microscopic energy transfer per unit time across the boundary of a system in an energy interaction with its surroundings. The symbol Qdot is used to denote rate of heat transfer.

The relation between Q and Qdot are:

$$Qdot=\delta Q/dt \qquad (11.2.1)$$

and

$$Q=\int(Qdot)dt \qquad (11.2.2)$$

There are three modes of rate of heat transfer: conduction ($Qdot_k$), convection ($Qdot_c$) and radiation ($Qdot_r$).

Conduction is a rate of heat transfer through a medium without mass transfer. The basic rate of conduction heat transfer equation is Fourier's law.

$$Qdot_k=-kA(dT/dx) \qquad (11.2.3)$$

where k is a heat transfer property called *thermal conductivity*, A is the cross section area normal to the heat transfer,
and dT/dx is the temperature gradient in the direction of the heat transfer, respectively.

For example, in the case of a linear temperature gradient, the rate of conduction heat transfer from a high temperature T_H on one side to a low temperature T_L on the other side through a solid wall with thickness L is

$$Qdot_k=kA(T_H -T_L)/L \qquad (11.2.4)$$

Convection is a rate of heat transfer leaving a surface to a fluid. The basic rate of convection heat transfer equation is Newton's law.

$$dot_c=hA(T_{surface}-T_{fluid}) \qquad (11.2.5)$$

where h is a heat transfer transport property called *convection coefficient*, A is the surface area, and ($T_{surface}$-T_{fluid}) is the temperature difference between the surface and the fluid, respectively.

Radiation is a rate of heat transfer by electromagnetic waves emitted by matters. Unlike conduction and convection, radiation does not require an intervening medium to propagate. The basic rate of radiation heat transfer equation between a high temperature (T_H) black body and a low temperature (T_L) black body is Stefan-Boltzmann's law.

$$Qdot_r=\sigma A\ [(T_H)^4-(T_L)^4] \tag{11.2.6}$$

where σ is the Stefan-Boltzmann's constant, and A is the emitting surface area, respectively.

Radiation heat transfer is usually not important in ordinary heat exchanger design and analysis, unless significant temperature differences are present.

Equation (11.2.4) and Equation (11.2.5) can be rewritten in the Ohm's law forms as

$$Qdot_k=(T_H-T_L)/(L/kA)=(T_H-T_L)/R_k \tag{11.2.7}$$

and

$$Qdot_c=(T_{surface}-T_{fluid})/(1/hA)=(T_{surface}-T_{fluid})/R_c \tag{11.2.8}$$

where R_k is the conduction heat transfer resistance and R_c is the convection heat transfer resistance.

Usually, the rate of heat transfer is a combination of conduction and convection in a heat exchanger system as illustrated in Figure 11.2.1 and only the fluid temperature on either side of the solid surface is known. For steady state, the rate of conduction heat transfer and the rate of convection heat transfer are equal. The total resistance (R) of the combined rate of heat transfer is

$$R=\sum R_k+\sum R_c \tag{11.2.9}$$

The rate of the combined heat transfer (Qdot) is

$$Qdot=(T_{fluid1}-T_{fluid2})/(1/UA)=(T_{fluid1}-T_{fluid2})/R \tag{11.2.10}$$

where U is the overall heat transfer coefficient.

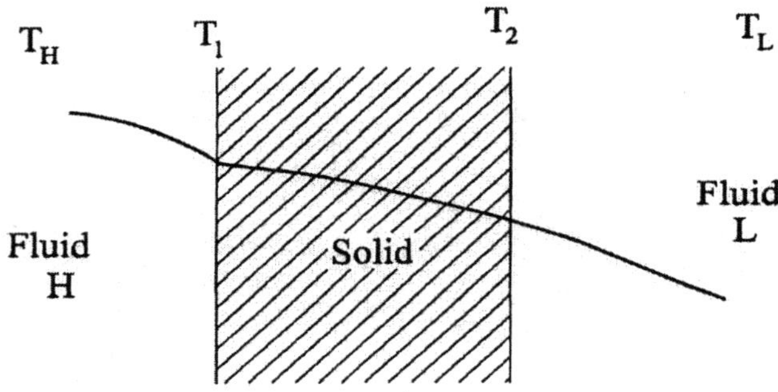

Figure 11.2.1. Rate of heat transfer in a heat exchanger

Homework 11.2. Rate of Heat Transfer

1. What are the three modes of heat transfer?
2. How does conduction differ from convection?
3. What is the mechanism of radiation heat transfer?
4. What is the overall heat transfer coefficient, U?

11.3 HEAT EXCHANGERS

One of the most important thermodynamic devices is the heat exchanger. Heat exchangers can be classified as mixed flow, recuperative and regenerative types.

In the mixed flow type heat exchanger, one fluid being cooled and another fluid being heated are mixed together.

In the recuperative flow type heat exchanger, the fluid being cooled is physically separated from the fluid being heated by some solid boundary.

In the regenerative flow type heat exchanger, There is only one set of flow channels. The hot fluid enters through the channels and heats the material in the heat exchanger surrounding the channels. The hot fluid then exits the

heat exchanger. Next, The cold fluid enters through the channels and is heated by the material in the heat exchanger surrounding the channels. The cold fluid then exits the heat exchanger.

The three commonly used recuperative flow type heat exchanger in power and refrigeration industry are parallel-flow, counter-flow and cross-flow heat exchangers. Consider the case where a fluid is flowing through a pipe and exchanging energy with another fluid flowing around the pipe. When the fluids flow in the same direction, it is a

parallel-flow heat exchanger. When the fluids flow in the opposite directions, it is a counter-flow heat exchanger. When the fluids flow in the normal directions, it is a cross-flow heat exchanger. The operation of parallel-flow and counter-flow heat exchangers and their associated temperature profiles are shown in Figure 11.3.1.

As can be seen from the non-linear temperature profiles, the temperature difference between the fluids varies from one end of the heat exchanger to the other end. To find an effective temperature difference between the two fluids, A logarithmic mean temperature difference (LMTD) is defined as

$$LMTD=[(T_{B1}-T_{A1})-(T_{B2}-T_{A2})]/Ln[(T_{B1}-T_{A1})/(T_{B2}-T_{A2})] \tag{11.3.1}$$

The rate of heat transfer between the two fluids is

$$Qdot=UA(LMTD) \tag{11.3.2}$$

where U is the overall heat transfer coefficient, A is the surface area of the tube(s), respectively.

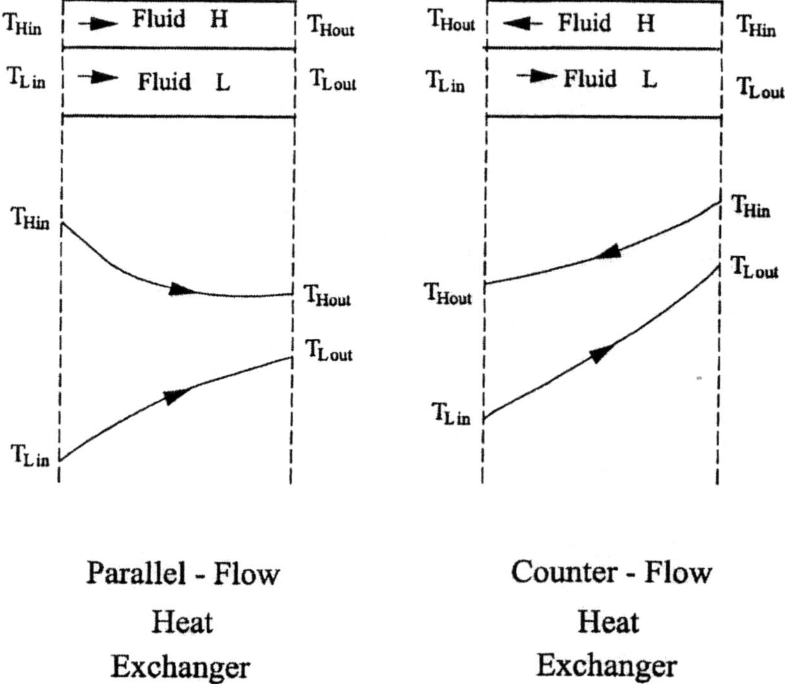

Figure 11.3.1. Operation of parallel-flow and counter-flow heat exchangers and their associated temperature profiles

Usually, the counter-flow heat exchanger is smaller than the parallel-flow heat exchanger because the LMTD of the counter-flow heat exchanger is larger than the LMTD of the parallel-flow heat exchanger when the inlet and outlet temperatures of the hot fluid and the inlet and outlet temperatures of the cold fluid are identical. The following example illustrate this comparison.

Example 11.3.1. A counter-flow lubricating oil cooler with a net heat transfer area of 258 ft^2 cools 60000 lbm of oil per hour from a temperature of 145°F to 120°F. The temperature of the cooling water entering and leaving are 75°F and 90°F. The specific heat of the oil is 0.5 Btu/[lbm(°F)]. Find the LMTD and overall heat transfer coefficient under these operating conditions. Also find the required area for a parallel-flow heat exchanger under these identical operating conditions. The temperature profiles of the counter-flow and parallel-flow heat exchanger are shown in Figure Example 11.3.1

Solution: $Qdot=[mdot(c)(T_{in}-T_{exit})]_{oil}=60000(0.5)(145-120)=750000$ Btu/hr

$LMTD=[(145-90)-(120-75)]/Ln[(145-90)/(120-75)]=49.8°F$

$U=Qdot/[A(LMTD)]=750000/[258(49.8)]=58.4$ Btu/[hr(ft^2)°F]

For the parallel-flow heat exchanger

$LMTD=[(145-75)-(120-90)]/Ln[(145-75)/(120-90)]=47.2°F$

and

A=Qdot/[U(LMTD)]=750000/[58.4(47.2)]=272 ft^2. This area is larger than 258 ft^2.

Example 11.3.2. A counter-flow heater as shown in Figure Example 11.3.2a heats helium at 101 kPa from a temperature of 20°C to 800°C. The temperature of the heating flue gas (air) entering and leaving are 1800°C and 1200°C at 101 kPa. (A) Find the LMTD, rate of helium flow and heat transfer based on unit of heating flue gas. (B) Also find the LMTD, rate of helium flow and heat transfer for a parallel-flow heat exchanger under these identical operating conditions.

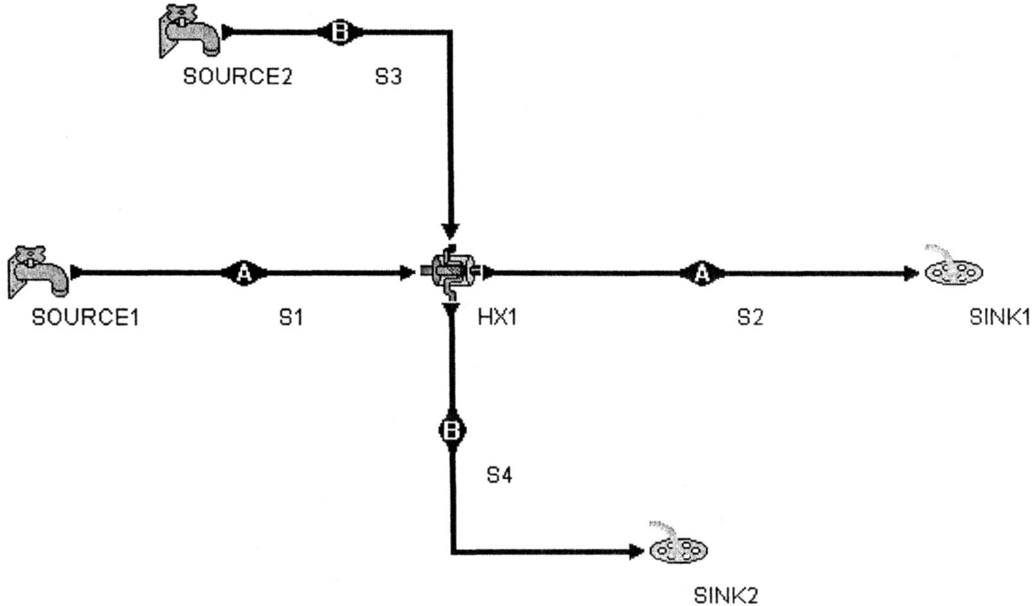

Figure Example 11.3.2a Heat exchanger

(A) To solve this problem by CyclePad, we take the following steps:

(1) Build the heat exchanger as shown in Figure Example 11.3.2a.

(2) Analysis: (a) Assuming both hot- and cold-side of the heat exchanger are isobaric, and type is counter-flow. (b) Input hot-side fluid=air, T_1=1800°C, p_1=101 kPa, $mdot_1$=1 kg/s, and T_2=1200°C; cold-side fluid=helium, T_3=20°C, p_3=101 kPa, and T_4=800°C.

(3) Display results: The results are: LMTD=1088°C, Qdot=602 kW, and $mdot_3$=0.1491 kg/s as shown in Figure Example 11.3.2b.

(B) (1) Analysis: (a) retract the heat exchanger type is counter-flow, and (b) input the heat exchanger type is co-current (parallel)-flow.

(2) Display results: The results are: LMTD=924.4°C, Qdot=602 kW, and $mdot_3$=0.1491 kg/s as shown in Figure Example 11.3.2c.

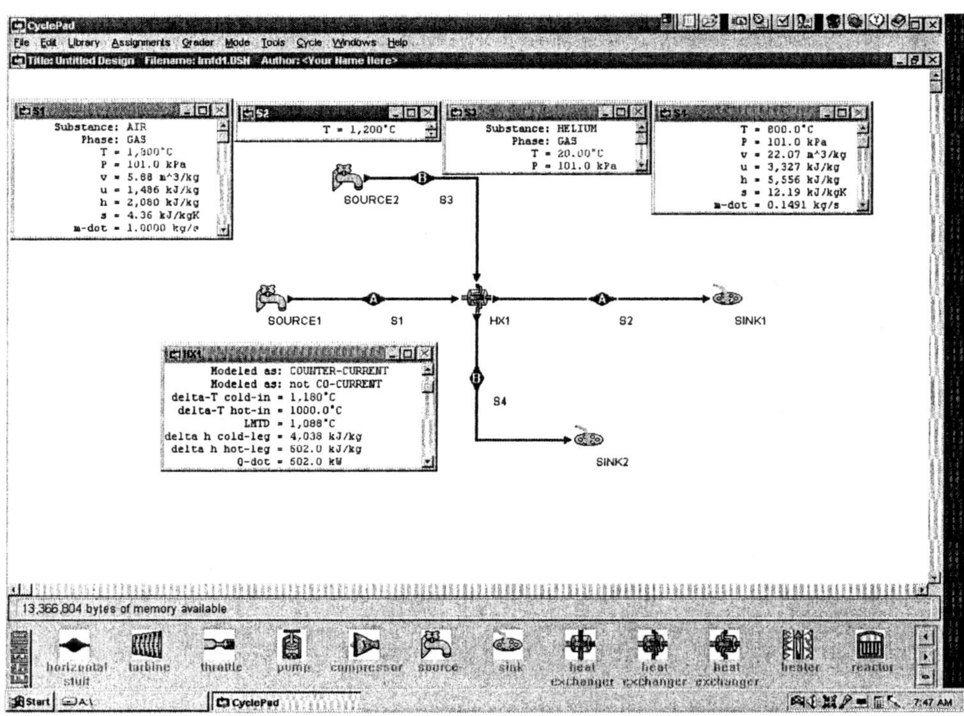

Figure Example 11.3.2b. Heat exchanger input and output results

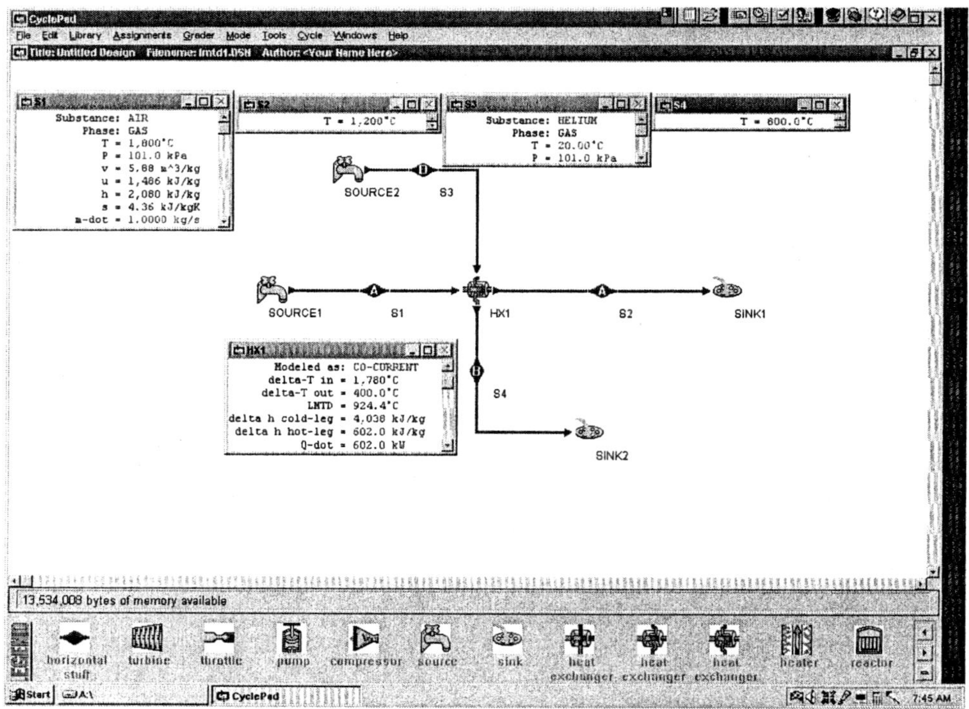

Figure Example 11.3.2c. Heat exchanger input and output results

Example 11.3.3. A counter-flow heat exchanger heats water at 101 kPa from saturated liquid state to saturated vapor state. The temperature of the heating flue gas (air) entering and leaving are 1800°C and 1200°C at 101 kPa. (A) Find the LMTD, rate of water flow and heat transfer based on unit mass of heating flue gas. (B) Also find the LMTD, rate of helium flow and heat transfer for a parallel-flow heat exchanger under these identical operating conditions.

(A) To solve this problem by CyclePad, we take the following steps:

(1) Build the heat exchanger as shown in Figure Example 11.3.2a.

(2) Analysis: (a) Assume both hot- and cold-side of the heat exchanger are isobaric, and type is counter-flow. (b) Input hot-side fluid=air, T_1=1800°C, p_1=101 kPa, $mdot_1$=1 kg/s, and T_2=1200°C; cold-side fluid=water, p_3=101 kPa, x_3=0, and x_4=1.

(3) Display results: The results are: LMTD=1378°C, Qdot=602 kW, and $mdot_3$=0.2668 kg/s as shown in Figure Example 11.3.3a.

(B) (1) Analysis: (a) retract the heat exchanger type is counter-flow, and (b) input the heat exchanger type is co-current (parallel)-flow.

(2) Display results: The results are: LMTD=1378°C, Qdot=602 kW, and $mdot_3$=0.2668 kg/s as shown in Figure Example 11.3.3b.

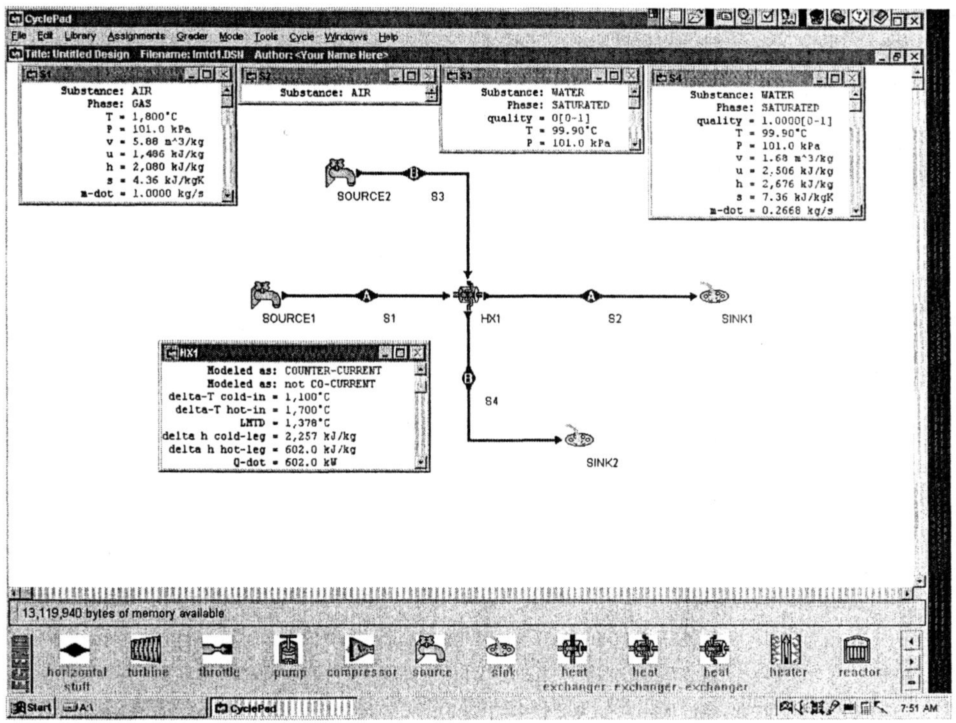

(a)

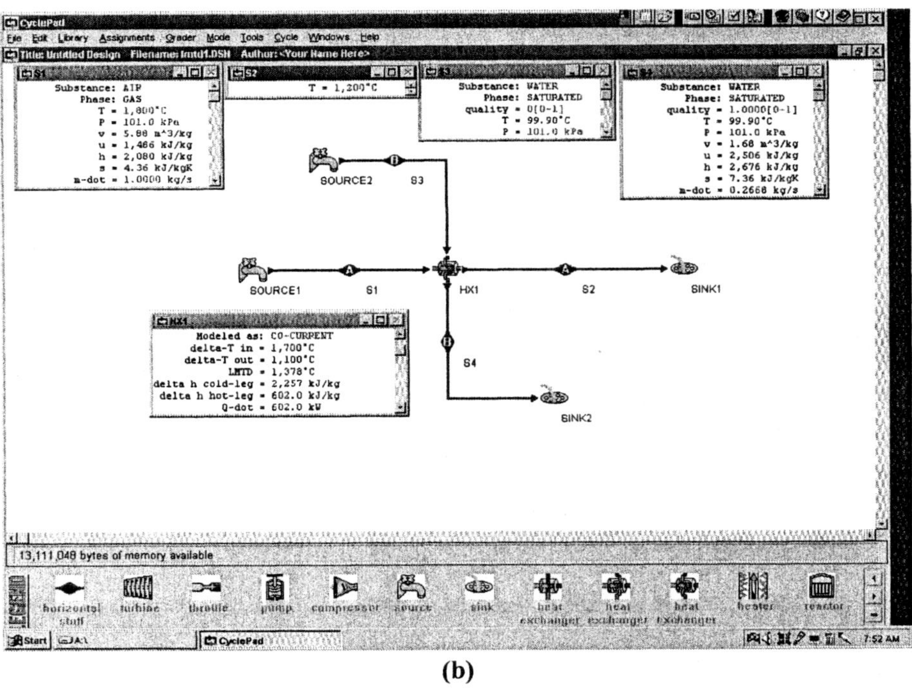

(b)

Figure Example 11.3.3a and b.Heat exchanger input and output results

Homework 11.3 Heat Exchanger

1. A counter-flow heater heats ammonia at 101 kPa from saturated liquid state to saturated vapor state. The temperature of the heating air entering and leaving are 33°C and 17°C at 101 kPa. (A) Find the LMTD, rate of water flow and heat transfer based on unit of heating flue gas. (B) Also find the LMTD, rate of helium flow and heat transfer for a parallel-flow heat exchanger under these identical operating conditions.
ANSWER: (A) LMTD=58.05°C, Qdot=16.05 kW, and $mdot_3$=0.0117 kg/s. (B) LMTD=58.05°C, Qdot=16.05 kW, and $mdot_3$=0.0117 kg/s.

11.4 CURZON AND AHLBORN (ENDOREVERSIBLE CARNOT) CYCLE

The T-s diagram and schematic diagram of the Curzon and Ahlborn (endoreversible Carnot) cycle are shown in Figure 11.4.1 and Figure 11.4.2, respectively. The cycle operates between a heat source at temperature T_H and a heat sink at temperature T_L. The temperatures of the working fluid in the isothermal heat addition and heat rejection processes are T_W and T_C. The finite temperature difference (T_H-T_W) allows a finite heat transfer from the heat source to the working fluid in the heat addition process. Similarly, the finite temperature difference (T_C-T_L) allows a heat transfer from the working fluid to the heat sink in the heat rejection process. The cycle is a modified Carnot cycle. Other than the external irreversibilities due to the two heat transfer processes, the modified cycle is an internal reversible heat engine.

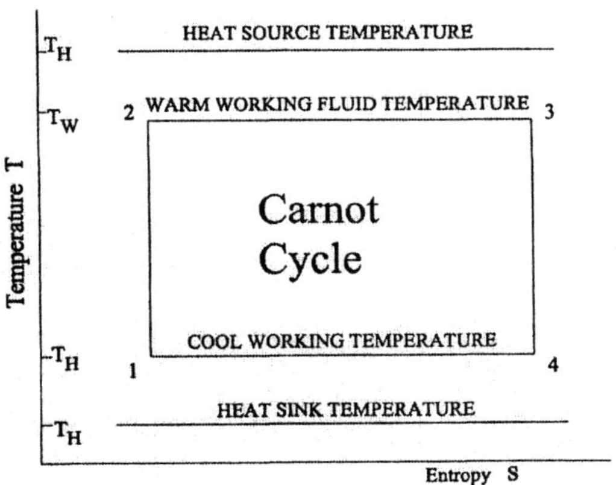

Figure11.4.1 Curzon and Ahlborn cycle T-s diagram

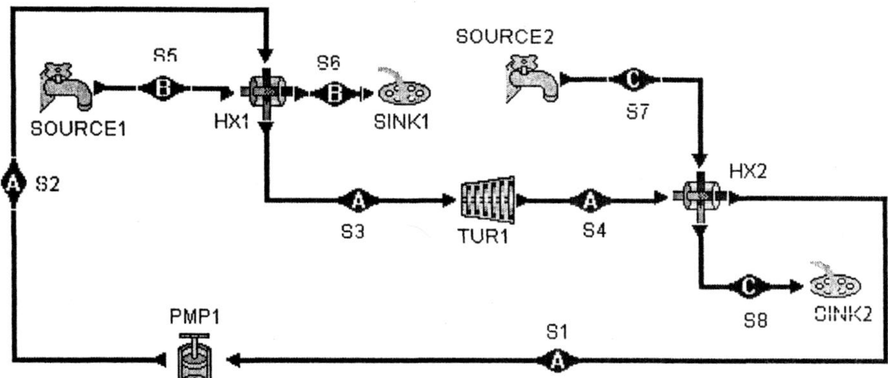

Figure11.4.2 Curzon and Ahlborn cycle schematic diagram

Assume that the working fluid flows through the heat engine in a steady-state fashion. Such a steady-state endoreversible Carnot heat engine is schematically illustrated in Figure 11.4.2. The rates of heat rejection and addition of the heat engine are:

$$Qdot_H = U_H A_H (T_H - T_W) \tag{11.4.1}$$

$$Qdot_L = U_L A_L (T_C - T_L) \tag{11.4.2}$$

where U_H is the heat transfer coefficient and A_H is the heat transfer surface area of the high-temperature side heat exchanger between the heat engine and the heat source; U_L is the heat transfer coefficient and A_L is the heat transfer surface area of the low-temperature side heat exchanger between the heat engine and the heat sink.

The total heat transfer surface area (A) of the two heat exchangers is assumed to be a constant.

$$A = A_H + A_L \tag{11.4.3}$$

The power output (P) of the heat engine according to the first law of thermodynamics is

$$P=Qdot_H-Qdot_L' \quad (11.4.4)$$

The second law of thermodynamics requires that

$$Qdot_H/T_W=Qdot_L'/T_C \quad (11.4.5)$$

The efficiency (η) of the heat engine is

$$\eta=P/Qdot_H \quad (11.4.6)$$

We define a heat transfer surface area ratio (f)

$$f=A_H/A_L \quad (11.4.7)$$

Combining Eqs.(11.4.1)-(11.4.7) gives the optimum value of f at maximum power output (f_a) and the optimal power output for given values of T_H, T_L, U_H, U_L, A and η.

$$f=f_a=(U_L/U_H)^{1/2} \quad (11.4.8)$$

$$P=[T_H-T_L/(1-\eta)]\eta B_1 \quad (11.4.9)$$

where

$$B_1=(U_HA)/[1+(U_L/U_H)^{-1/2}]^2 \quad (11.4.10)$$

Eq.(9) is the optimal performance characteristics of the endoreversible Carnot heat engine. It indicates that P=0 when $\eta=0$ and $\eta=\eta_c=1-T_L/T_H$. Taking the derivative of P with respect to η and setting it equal to zero ($dP/d\eta=0$) gives

$$\eta=\eta_{C-A}=1-(T_L/T_H)^{1/2} \quad (11.4.11)$$

and the optimal power output delivered by the cycle is

$$P_{max}=B_1\,[(T_H)^{1/2}-(T_L)^{1/2}]^2 \quad (11.4.12)$$

The power versus efficiency characteristics of the endoreversible Carnot heat engine is a parabolic curve. The endoreversible heat engine is a simple model which considers the external heat transfer irreversibility between the heat engine and its surrounding heat reservoirs only.

The required optimum intermediate temperatures at maximum power condition are

$$T_W=\{(T_H)^{0.5}+[U_LA_L/(U_HA_H)](T_L)^{0.5}\}/\{1+[U_LA_L/(U_HA_H)]\}(T_H)^{0.5} \quad (11.4.13)$$

and

$$T_C=\{(T_H)^{0.5}+[U_LA_L/(U_HA_H)](T_L)^{0.5}\}/\{1+[U_LA_L/(U_HA_H)]\}(T_L)^{0.5} \quad (11.4.14)$$

By taking into account the rate of heat transfer associated with the endoreversible cycle, the upper bound of the power output of the cycle can be found. This bound provides a practical basis for a real power plant design. The industrial view is that the heat engine efficiency is secondary to the power output in power plants whose worth is constrained by economic consideration.

Another important industrial design objective function for power cycle design is net power per unit conductance of heat exchangers. The conductance of the high-temperature-side heat exchanger is U_HA_H in Eq. (11.4.1), and the conductance of the low-temperature-side heat exchanger is U_LA_L in Eq. (11.4.1). The net power per unit conductance of heat exchangers $[Wdot_{net}/(U_HA_H+U_LA_L)]$ represents the initial cost and operational cost of heat exchangers, which is a very important part of the power plant.

For many power plants such as waste-heat plants, geothermal plants and OTEC (Ocean thermal energy conversion) plants which have relatively small fuel cost, the industrial design objective function for these power cycle design is specific net power. The specific net power is defined as $Wdot_{net}/(A_H+A_L)$. The total cost of these plants is determined mainly by the construction cost. From the view points of cost and size, the most important components of these plants are the heat exchangers. The volume of supporting structure, the weight of buoyance-adjusting-ballast, the length of pipe, head loss, pump power and all other major components increases as the size of the heat exchangers increases. High performance of these plants is essential for producing power. For this reason, the performance evaluation objective function of these plants is taken to be the specific power.

Example 11.4.1. An endoreversible (Curzon and Ahlborn) cycle operates between a heat source at temperature T_H=1600 K and a heat sink at temperature T_L=400 K. Suppose U_H=100 kW/m^2 (overall heat transfer coefficient of the high-temperature side heat exchanger between the heat engine and the heat source), A_H=1 m^2 (heat transfer surface area of the high-temperature side heat exchanger between the heat engine and the heat source), and U_L=100 kW/m^2 (overall heat transfer coefficient of the low-temperature side heat exchanger between the heat engine and the heat sink). Determine the maximum power output of the cycle. Find the heat transfer added, heat transfer removed, heat transfer surface area of the low-temperature side heat exchanger between the heat engine and the heat sink, and efficiency of the cycle at the maximum power output condition.

Solution:

$Qdot_H=U_HA_H(T_H-T_W)=100(1)(1600-1200)=40000$ kW.
$Qdot_H/T_W=40000/1200=Qdot_L'/T_C=Qdot_L/600$ gives $Qdot_L=20000$ kW
$Qdot_L=20000=U_LA_L(T_C-T_L)=100A_L(600-400)$ gives $A_L=1$ m^2
$P=Qdot_H-Qdot_L=40000-20000=20000$ kW

and

$\eta=1-(T_C/T_W)=1-600/1200=50\%$.

Example 11.4.2. An endoreversible (Curzon and Ahlborn) steam cycle operates between a heat source at temperature T_H=640 K and a heat sink at temperature T_L=300 K. The following information is given:

> Heat source: fluid=water, T_5=640 K, x_5=1, T_6=640 K, and x_6=0
> Heat sink: fluid=water, T_7=300 K, x_7=0, T_8=300 K, and x_8=1
> Steam cycle: fluid=water, x_2=0, T_3=500 K, x_3=1, T_4=400 K, and mdot=1 kg/s.

Determine the rate of heat added from the heat source, rate of heat removed to the heat sink, power required by the isentropic pump, power produced by the isentropic turbine, net power produced and efficiency of the cycle.

To solve this problem by CyclePad, we take the following steps:

1. Build the cycle and its surroundings as shown in Figure 11.4.2.
2. Analysis: (A) Assuming the heat exchangers are isobaric (notice that isobaric is also isothermal in the saturated mixture region), and turbine and pump are isentropic. (B) Input Heat source fluid=water, T_5=640 K, x_5=1, T_6=640 K, and x_6=0; Heat sink fluid=water, T_7=300 K, x_7=0, T_8=300 K, and x_8=1; Steam cycle fluid=water, x_2=0, T_3=500 K, x_3=1, T_4=400 K, and mdot=1 kg/s.
3. Display results: The results are: rate of heat added from the heat source=1827 kW, rate of heat removed to the heat sink=-1462 kW, power required by the isentropic pump=-50.53 kW, power produced by the isentropic turbine=416.0 kW, net power produced=365.4 kW and efficiency of the cycle=20%.

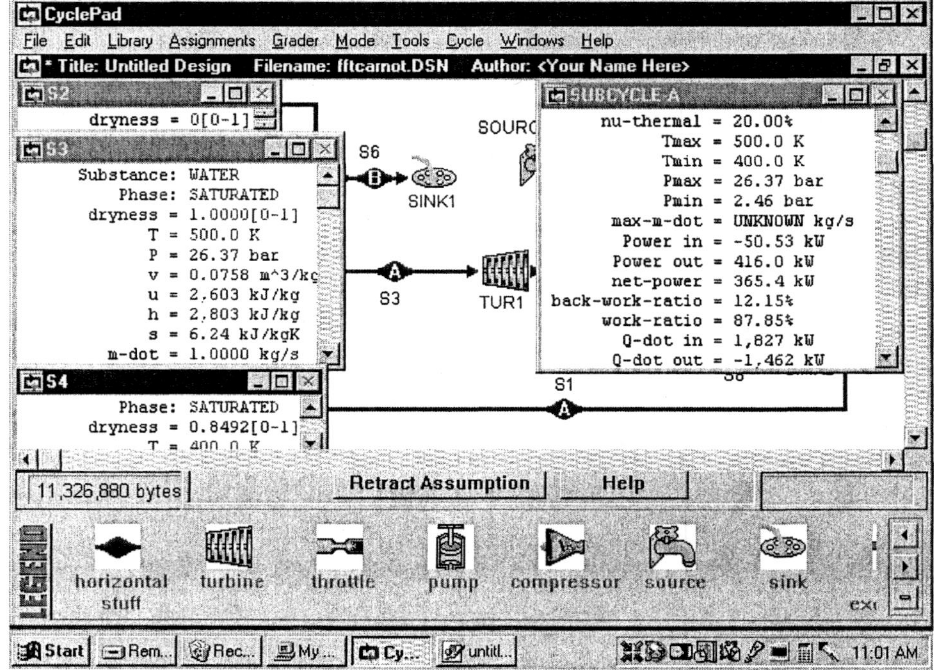

Figure Example 11.4.2. Endoreversible (Curzon and Ahlborn) steam cycle

Example 11.4.3. An endoreversible (Curzon and Ahlborn) steam cycle operates between a heat source at temperature T_H=640 K and a heat sink at temperature T_L=300 K. The following information is given:

> Heat source: fluid=water, T_5=640 K, x_5=1, T_6=640 K, and x_6=0
> Heat sink: fluid=water, T_7=300 K, x_7=0, T_8=300 K, and x_8=1
> Steam cycle: fluid=water, x_2=0, x_3=1, T_4=400 K, and mdot=1 kg/s.

Determine the maximum net power produced and working fluid temperature at the inlet of the turbine.

To solve this problem by CyclePad, we take the following steps:

1. Build the cycle and its surroundings as shown in Figure 11.4.2.
2. Analysis: (A) Assuming the heat exchangers are isobaric (notice that isobaric is also isothermal in the saturated mixture region), and turbine and pump are isentropic. (B) Input Heat source fluid=water, T_5=640 K, x_5=1, T_6=640 K, and x_6=0; Heat sink fluid=water, T_7=300 K, x_7=0, T_8=300 K, and x_8=1; Steam cycle fluid=water, x_2=0, T_3=500 K, x_3=1, T_4=400 K, and mdot=1 kg/s.
3. Sensitivity analysis: Plot net power vs T_3 diagram as shown in Figure Example 11.4.3. The answer are: maximum net power is about 427.0 kW and T_3 is about 558.6 K.

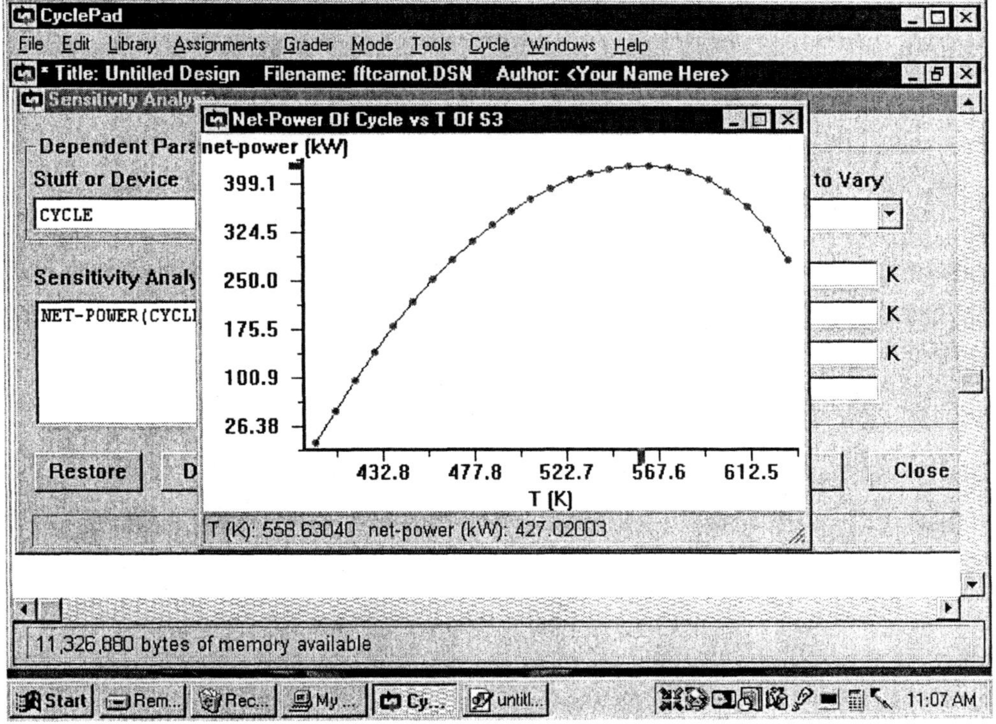

Figure Example 11.4.3. Endoreversible cycle Sensitivity analysis

Comment: The partial optimization is only for ∂(net power)/∂(T_3)=0. To have the full optimization, we must let ∂(net power)/∂(T_4)=0 also.

Example 11.4.4. An endoreversible (Curzon and Ahlborn) steam cycle as shown in Figure Example 11.4.4a operates between a heat source at temperature T_H=300°C and a heat sink at temperature T_L=20°C. The following information is given:

Heat source: fluid=water, T_1=300°C, x_1=1, T_2=300°C, and x_2=0
Heat sink: fluid=water, T_3=20°C, x_3=0, T_4=20°C, and x_4=1
Carnot cycle: fluid=water, T_5=100°C, s_5=3.25 kJ/kg(K), T_7=200°C, s_7=5.70 kJ/kg(K), and mdot=1 kg/s.

The heat exchangers are counter-flow type, U_H=0.4 kJ/(m^2)K and U_L=0.4 kJ/(m^2)K.

Determine the rate of heat added from the heat source, rate of heat removed to the heat sink, power required by the isentropic pump, power produced by the isentropic turbine, net power produced and efficiency of the cycle.

Taking the specific net power output (net output power per unit total heat exchanger surface area) as the design objective function, optimize the warm-side (heater or high-temperature-side heat exchanger) and cold-side (cooler or low-temperature-side heat exchanger) working fluid temperatures.

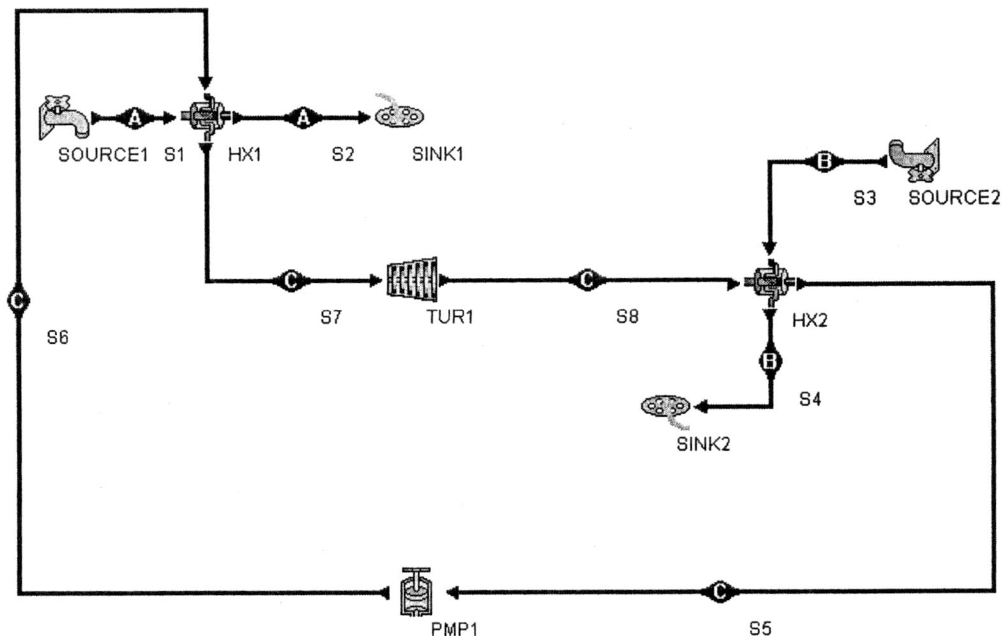

Figure Example 11.4.4a Finite-time Carnot cycle

To solve this problem by CyclePad, we take the following steps:

1. Build the cycle and its surroundings as shown in Figure 11.4.4a.

2. Analysis: (A) Assuming the heat exchangers are isobaric and counter-flow, and turbine and pump are isentropic. (B) Input Heat source: fluid=water, T_1=300°C, x_1=1, T_2=300°C, and x_2=0

   Heat sink: fluid=water, T3=20 C, x3=0, T4=20 C, and x4=1
   Carnot cycle: fluid=water, T5=100 C, s5=3.25 kJ/kg(K), T7=200 C, s7=5.70 kJ/kg(K), and mdot=1 kg/s.

3. Display results: The results are: $LMTD_H$=100 K, $LMTD_H$=80 K, rate of heat added from the heat source=1159 kW, rate of heat removed to the heat sink=-914.2 kW, power required by the isentropic pump=-143.2 kW, power produced by the isentropic turbine=388.2 kW, net power produced=245.0 kW and efficiency of the cycle=21.14% as shown in Figure Example 14.4.4b.
4. Calculate $U_HA_H=Q_H/LMTD_H$=1159/100=11.59 kW/K, $U_LA_L=Q_L/LMTD_L$=914.2/80= 11.43 kW/K, $U_HA_H+U_LA_L$=11.59+11.43=23.02 kW/K, and specific net power output= $Wdot_{net}/(A_H+A_L)$=245.0/(11.59/0.4+11.43/0.4)=kW/m^2.

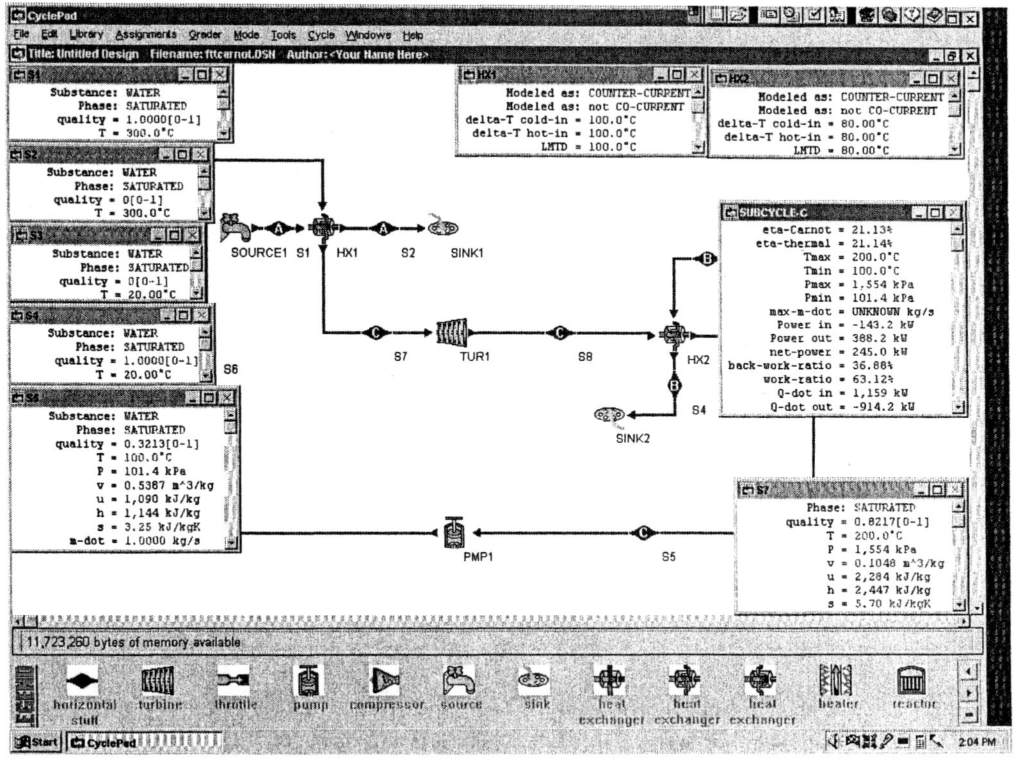

Figure Example 11.4.4a Finite-time Carnot cycle input and output

5. To optimize the specific power output of the cycle, we let ∂(specific power output)/∂T_7=0 first and then

∂(specific power output)/∂T_7=0 as shown in the following two tables.

It is seen that ∂(specific power output)/∂T_7=0 occurs at T_7=220°C as shown in Table Example 11.4.4a.

Table Example 11.4.4a Specific power optimization with respect to T_7

		T1=T2=300		T3=T4=20									
T5	T7	pnet/UA	pnet/A	efficiency	power in	power out	powernet	heat in	heat out	LMTDH	LMTDL	UHAH	ULAL
100	100.1	0.015276	0.019096	0.0267	-0.195	0.4395	0.2444	914.2	-914.2	200	80	4.571	11.4275
100	120	2.920981	3.651226	5.09	-36.72	85.74	49.01	963.2	-914.2	180	80	5.351111	11.4275
100	140	5.522602	6.903253	9.69	-69.22	167.3	98.04	1012	-914.2	160	80	6.325	11.4275
100	160	7.734371	9.667963	13.85	-97.68	244.7	147	1061	-914.2	140	80	7.578571	11.4275
100	180	9.478902	11.84863	17.66	-122.3	318.3	196	1110	-914.2	120	80	9.25	11.4275
100	200	10.64408	13.30509	21.14	-143.2	388.2	245	1159	-914.2	100	80	11.59	11.4275
100	**220**	**11.08284**	**13.85355**	**24.34**	**-160.6**	**454.6**	**294**	**1208**	**-914.2**	**80**	**80**	**15.1**	**11.4275**
100	240	10.59378	13.24222	27.28	-174.6	517.6	343	1257	-914.2	60	80	20.95	11.4275
100	260	8.893426	11.11678	30.01	-185.5	577.5	392	1306	-914.2	40	80	32.65	11.4275
100	280	5.569764	6.962205	32.54	-193.2	634.2	441	1355	-914.2	20	80	67.75	11.4275
100	299.9	0.034822	0.043528	34.86	-198.4	687.7	489.3	1404	-914.2	0.1	80	14040	11.4275
			assume UH=UL=0.4 kJ/m^2(K)										

6. Then let ∂(specific power output)/∂T_5=0

It is seen that ∂(specific power output)/∂T_5=0 occurs at T_5=80°C as shown in Table Example 11.4.4b.

Table Example 11.4.4b Specific power optimization with respect to T_5

T5	T7	pnet/UA	pnet/A	efficiency	power in	power out	powernet	heat in	heat out	LMTDH	LMTDL	UHAH	ULAL
20.1	220	0.068036	0.085045	40.54	-354.9	844.5	489.8	1208	-718.4	80	0.1	15.1	7184
40	220	8.249158	10.31145	36.5	-298.7	739.8	441	1208	-767.2	80	20	15.1	38.36
60	220	11.0407	13.80087	32.45	-247.8	639.8	392	1208	-816.2	80	40	15.1	20.405
80	**220**	**11.61924**	**14.52405**	**28.39**	**-201.8**	**544.9**	**343**	**1208**	**-865.2**	**80**	**60**	**15.1**	**14.42**
100	220	11.08284	13.85355	24.34	-160.6	454.6	294	1208	-914.2	80	80	15.1	11.4275
120	220	9.906194	12.38274	20.28	-123.9	368.9	245	1208	-963.2	80	100	15.1	9.632
140	220	8.328612	10.41076	16.22	-91.38	287.4	196	1208	-1012	80	120	15.1	8.433333
160	220	6.48189	8.102362	12.17	-62.92	209.9	147	1208	-1061	80	140	15.1	7.578571
180	220	4.446965	5.558707	8.11	-38.35	136.3	98	1208	-1110	80	160	15.1	6.9375
200	220	2.274955	2.843694	4.06	-17.4	66.41	49	1208	-1159	80	180	15.1	6.438889
219.9	220	0.011569	0.014461	0.0202	-0.0787	0.3233	0.2446	1208	-1208	80	199.9	15.1	6.043022

7. Let T_5=80°C and T_7=220°C, the optimized specific power output of the cycle is 14.53 kW/m^2. At the maximum optimized specific power output condition, $LMTD_H$=80 K, $LMTD_H$=60 K, rate of heat added from the heat source=1208 kW, rate of heat removed to the heat sink=-865.2 kW, power required by the isentropic pump=-201.8 kW, power produced by the isentropic turbine=544.9 kW, net power produced=343.0 kW and efficiency of the cycle=28.39% as shown in Figure Example 14.4.4c.

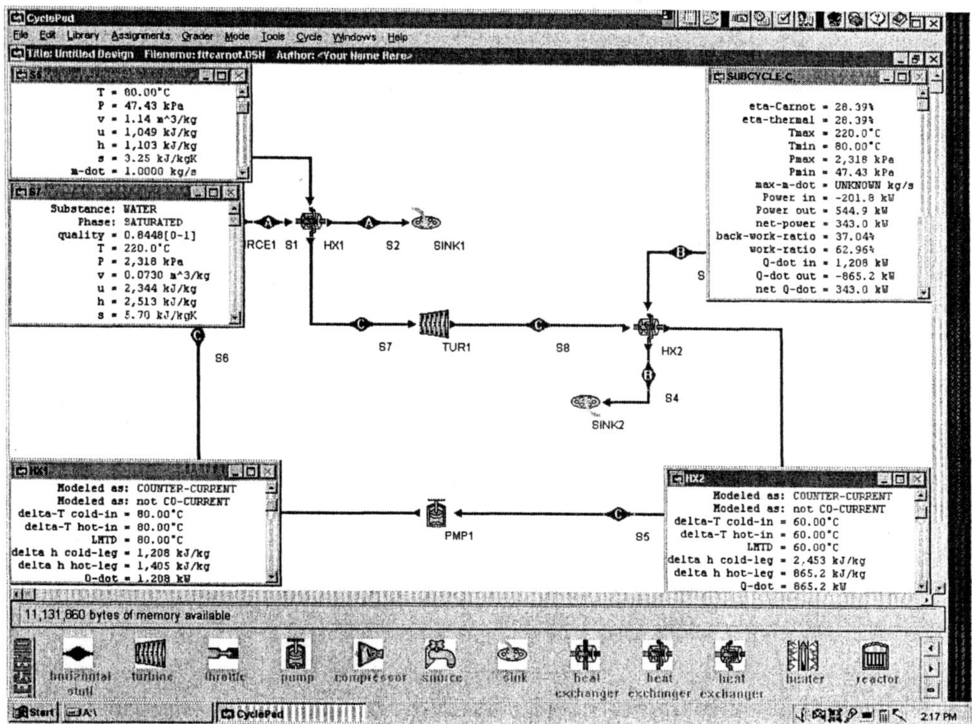

Figure Example 11.4.4a Finite-time Carnot cycle optimization result

Homework 11.4 Curzon and Ahlborn Cycle With Infinite Heat Capacity Heat Source and Sink

1. An endoreversible (Curzon and Ahlborn) cycle operates between a heat source at temperature T_H=3600 K and a heat sink at temperature T_L=300 K. The temperatures of the working fluid in the isothermal heat addition and heat rejection processes are T_W=2000 K and T_C=400 K. Suppose U_H=100 kW/m^2 (overall heat transfer coefficient of the high-temperature side heat exchanger between the heat engine and the heat source), A_H=1 m^2 (heat transfer surface area of the high-temperature side heat exchanger between the heat engine and the heat source), and U_L=100 kW/m^2 (overall heat transfer coefficient of the low-temperature side heat exchanger between the heat engine and the heat sink). Determine the heat transfer added, heat transfer removed, heat transfer surface area of the low-temperature side heat exchanger between the heat engine and the heat sink, power output and efficiency of the cycle.
2. Referring to problem 1and with fixed heat source and heat sink temperatures, determine the maximum power output of the cycle. Find the working fluid temperatures in the isothermal heat addition and heat rejection processes, heat transfer added, heat transfer removed, heat transfer surface area of the low-temperature side heat exchanger between the heat engine and the heat sink, and efficiency of the cycle at the maximum power output condition.

3. An endoreversible (Curzon and Ahlborn) steam cycle operates between a heat source at temperature T_H=640 K and a heat sink at temperature T_L=300 K. The following information is given:
   Heat source: fluid=water, T_5=600 K, x_5=1, T_6=600 K, and x_6=0
   Heat sink: fluid=water, T_7=290 K, x_7=0, T_8=290 K, and x_8=1
   Steam cycle: fluid=water, x_2=0, T_3=550 K, x_3=1, T_4=300 K, and mdot=1 kg/s.
   Determine the rate of heat added from the heat source, rate of heat removed to the heat sink, power required by the isentropic pump, power produced by the isentropic turbine, net power produced and efficiency of the cycle.
4. Referring to problem 1 and with fixed heat source and heat sink temperatures as well as working fluid temperatures in the isothermal heat rejection processes, determine the maximum power output of the cycle. Find the working fluid temperatures in the isothermal heat addition, heat transfer added, heat transfer removed, and efficiency of the cycle at the maximum power output condition. Draw the sensitivity diagram.
5. Referring to problem 3 and with fixed heat source and heat sink temperatures as well as working fluid temperatures in the isothermal heat addition processes, determine the maximum power output of the cycle. Find the working fluid temperatures in the isothermal heat removing, heat transfer added, heat transfer removed, and efficiency of the cycle at the maximum power output condition. Draw the sensitivity diagram.
6. An endoreversible (Curzon and Ahlborn) steam cycle as shown in Figure Example 11.4.4a operates between a heat source at temperature T_H=300°C and a heat sink at temperature T_L=40°C. The following information is given:
   Heat source: fluid=water, T_1=300°C, x_1=1, T_2=300°C, and x_2=0
   Heat sink: fluid=water, T_3=40°C, x_3=0, T_4=40°C, and x_4=1
   Carnot cycle: fluid=water, T_5=100°C, s_5=3.25 kJ/kg(K), T_7=200°C, s_7=5.70 kJ/kg(K), and mdot=1 kg/s.
   The heat exchangers are counter-flow type, U_H=0.4 kJ/(m^2)K and U_L=0.4 kJ/(m^2)K.
   Determine the rate of heat added from the heat source, rate of heat removed to the heat sink, power required by the isentropic pump, power produced by the isentropic turbine, net power produced and efficiency of the cycle.
   Taking the specific net power output (net output power per unit total heat exchanger surface area) as the design objective function, optimize the warm-side (heater or high-temperature-side heat exchanger) and cold-side (cooler or low-temperature-side heat exchanger) working fluid temperatures.

11.5 Curzon and Ahlborn Cycle with Finite Heat Capacity Heat Source and Sink

The T-s diagram and schematic diagram of the Curzon and Ahlborn (endoreversible Carnot) cycle are shown in Figure 11.5.1 and Figure 11.5.2, respectively. The cycle operates between a heat source and a heat sink with finite heat capacity. The fluid of the heat source enters the hot-side heat exchanger at T_5 and exits at T_6. The fluid of the heat sink enters the cold-side heat exchanger at T_7 and exits at T_8. The temperatures of the working fluid in the isothermal heat addition and heat rejection processes are T_W and T_C. The finite mean temperature difference $LMTD_H$ {$LMTD_H=(T_5-T_6)/[LN (T_5-T_W)/(T_6-T_W)]$} allows a finite heat transfer from the heat source to the working fluid in the heat addition process. Similarly, the finite mean temperature difference $LMTD_L$ {$LMTD_L=(T_8-T_7)/[LN (T_C-T_7)/(T_C-T_8)]$} allows a heat transfer from the working fluid to the heat sink in the heat rejection process. The cycle is a modified Carnot cycle. Other than the external irreversibilities due to the two heat transfer processes, the modified cycle is an internal reversible heat engine.

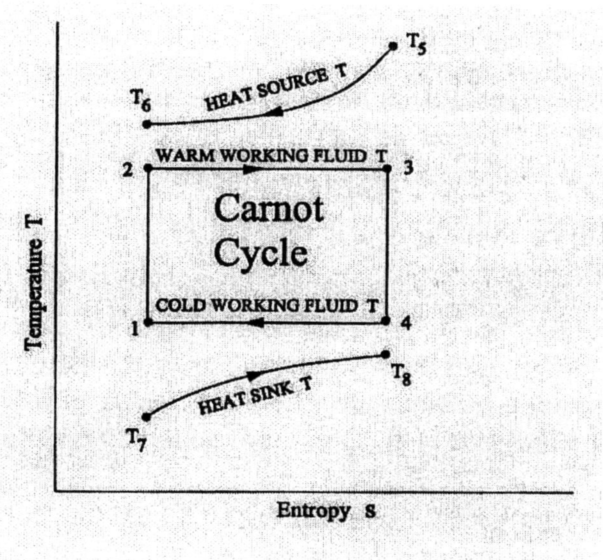

Figure11.5.1 Curzon and Ahlborn cycle with finite heat capacity source and sink T-s diagram

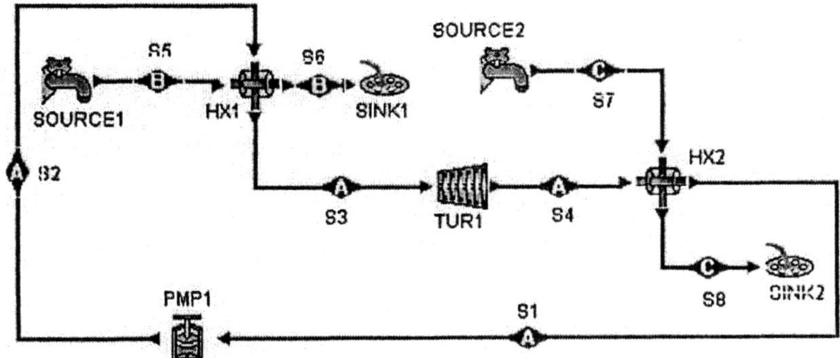

Figure11.5.1 Curzon and Ahlborn cycle with finite heat capacity source and sink schematic diagram

Assume that the working fluid flows through the heat engine in a steady-state fashion. Such a steady-state endoreversible Carnot heat engine is schematically illustrated in Figure 11.4.2. The rates of heat rejection and addition of the heat engine are:

$$Qdot_H=U_HA_HLMTD_H \quad (11.5.1)$$

$$Qdot_L=U_LA_LLMTD_L \tag{11.5.2}$$

where U_H is the heat transfer coefficient and A_H is the heat transfer surface area of the high-temperature side heat exchanger between the heat engine and the heat source; U_L is the heat transfer coefficient and A_L is the heat transfer surface area of the low-temperature side heat exchanger between the heat engine and the heat sink.

The total heat transfer surface area (A) of the two heat exchangers is assumed to be a constant.

$$A=A_H+A_L \tag{11.5.3}$$

The power output (P) of the heat engine according to the first law of thermodynamics is

$$P=Qdot_H-Qdot_L' \tag{11.5.4}$$

The second law of thermodynamics requires that

$$Qdot_H/T_W=Qdot_L'/T_C \tag{11.5.5}$$

The efficiency (η) of the heat engine is

$$\eta=P/Qdot_H \tag{11.5.6}$$

Example 11.5.1. An endoreversible (Curzon and Ahlborn) steam cycle operates between a finite heat capacity heat source and a finite heat capacity heat sink. The following information is given:

Heat source: fluid=air, T_5=1500 K, p_5=1 bar, T_6=650 K, and p_6=1 bar
Heat sink: fluid=air, T_7=290 K, p_7=1 bar, T_8=400 K, and p_8=1 bar
Steam cycle: fluid=water, x_2=0, T_3=500 K, x_3=1, T_4=420 K, and mdot=1 kg/s.

Determine the rate of heat added from the heat source, rate of heat removed to the heat sink, power required by the isentropic pump, power produced by the isentropic turbine, net power produced and efficiency of the cycle.

To solve this problem by CyclePad, we take the following steps:

1. Build the cycle and its surroundings as shown in Figure 11.5.2.
2. Analysis: (A) Assuming the heat exchangers are isobaric, and turbine and pump are isentropic. (B) Input Heat source fluid=air, T_5=1500 K, p_5=1 bar, T_6=650 K, and p_6=1 bar; Heat sink: fluid=air, T_7=290 K, p_7=1 bar, T_8=400 K, and p_8=1 bar; Steam cycle: fluid=water, x_2=0, T_3=500 K, x_3=1, T_4=420 K, and mdot=1 kg/s.
3. Display results: The results are: rate of heat added from the heat source=1827 kW, rate of heat removed to the heat sink=-1535 kW, power required by the isentropic pump=-32.83 kW, power produced by the isentropic turbine=325.1 kW, net power produced=292.3 kW and efficiency of the cycle=16%.

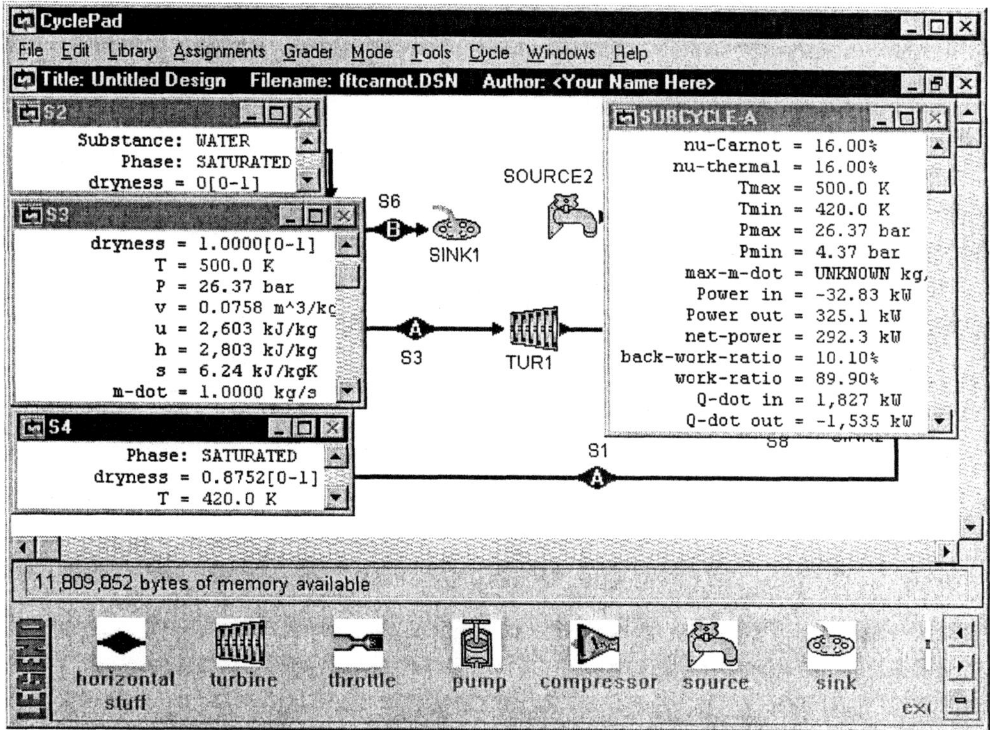

Figure Example 11.5.1. Curzon and Ahlborn cycle with finite heat capacity heat source and sink

Example 11.5.2. An endoreversible (Curzon and Ahlborn) steam cycle operates between a finite heat capacity heat source and a finite heat capacity heat sink. The following information is given:

Heat source: fluid=air, T_5=1500 K, p_5=1 bar, T_6=650 K, and p_6=1 bar
Heat sink: fluid=air, T_7=290 K, p_7=1 bar, T_8=400 K, and p_8=1 bar
Steam cycle: fluid=water, x_2=0, x_3=1, T_4=420 K, and mdot=1 kg/s.

Determine the maximum net power produced and working fluid temperature at the inlet of the turbine with fixed condenser temperature.

To solve this problem by CyclePad, we take the following steps:

1. Build the cycle and its surroundings as shown in Figure 11.5.2.
2. Analysis: (A) Assuming the heat exchangers are isobaric (notice that isobaric is also isothermal in the saturated mixture region), and turbine and pump are isentropic. (B) Input heat source: fluid=air, T_5=1500 K, p_5=1 bar, T_6=650 K, and p_6=1 bar; heat sink: fluid=air, T_7=290 K, p_7=1 bar, T_8=400K,and p_8=1 bar; steam cycle: fluid=water, x_2=0, x_3=1, T_4=420 K, and mdot=1 kg/s.
3. Sensitivity analysis: Plot net power vs T_3 diagram as shown in Figure Example 11.4.4. The answer are: maximum net power is about 375.2 kW and T_3 is about 563.7 K.

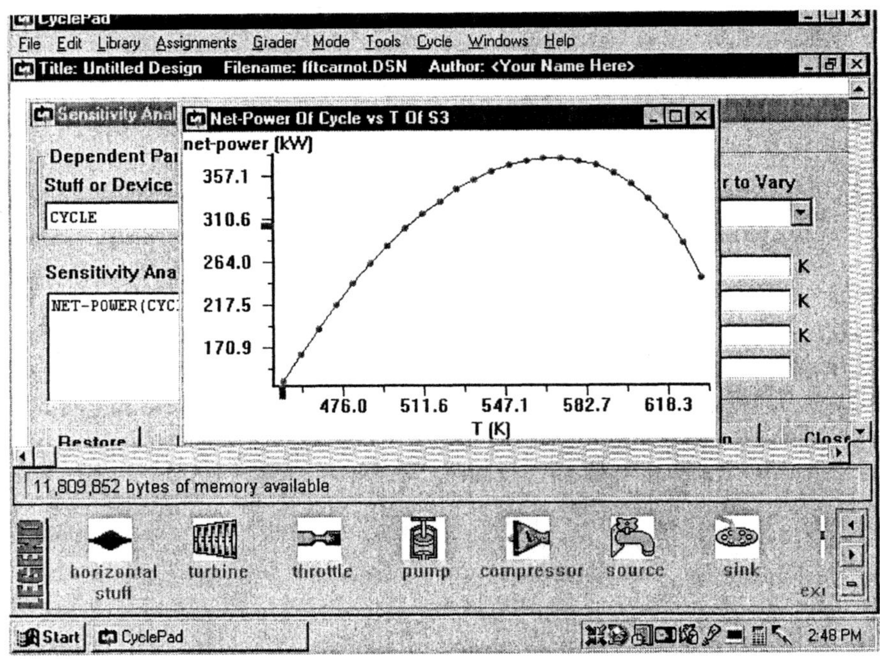

Figure Example 11.5.2. Endoreversible cycle with finite heat capacity heat source and sink sensitivity analysis

Comment: The partial optimization is only for ∂(net power)/∂(T_3)=0. To have the full optimization, we must let ∂(net power)/∂(T_4)=0 also.

Homework 11.5 Curzon and Ahlborn Cycle with Finite Heat Capacity Heat Source and Sink

1. An endoreversible (Curzon and Ahlborn) steam cycle operates between a finite heat capacity heat source and a finite heat capacity heat sink. The following information is given:
   Heat source: fluid=air, T_5=1500 K, p_5=1 bar, T_6=650 K, and p_6=1 bar
   Heat sink: fluid=air, T_7=290 K, p_7=1 bar, T_8=400 K, and p_8=1 bar
   Steam cycle: fluid=water, x_2=0, T_3=450 K, x_3=1, T_4=420 K, and mdot=1 kg/s.
   Determine the rate of heat added from the heat source, rate of heat removed to the heat sink, power required by the isentropic pump, power produced by the isentropic turbine, net power produced and efficiency of the cycle.
   ANSWER: rate of heat added from the heat source=2026 kW, rate of heat removed to the heat sink=-1891 kW, power required by the isentropic pump=-4.99 kW, power produced by the isentropic turbine=140.0 kW, net power produced=135.0 kW and efficiency of the cycle=6.67%.
2. An endoreversible (Curzon and Ahlborn) steam cycle operates between a finite heat capacity heat source and a finite heat capacity heat sink. The following information is given:
   Heat source: fluid=air, T_5=1500 K, p_5=1 bar, T_6=650 K, and p_6=1 bar

Heat sink: fluid=air, T_7=290 K, p_7=1 bar, T_8=400 K, and p_8=1 bar
Steam cycle: fluid=water, x_2=0, T_3=480 K, x_3=1, T_4=410 K, and mdot=1 kg/s.
Determine the rate of heat added from the heat source, rate of heat removed to the heat sink, power required by the isentropic pump, power produced by the isentropic turbine, net power produced and efficiency of the cycle.
ANSWER: rate of heat added from the heat source=1913 kW, rate of heat removed to the heat sink=-1634 kW, power required by the isentropic pump=-25.32 kW, power produced by the isentropic turbine=304.3 kW, net power produced=278.9 kW and efficiency of the cycle=14.58%.

3. An endoreversible (Curzon and Ahlborn) steam cycle operates between a finite heat capacity heat source and a finite heat capacity heat sink. The following information is given:
Heat source: fluid=air, T_5=1500 K, p_5=1 bar, T_6=650 K, and p_6=1 bar
Heat sink: fluid=air, T_7=290 K, p_7=1 bar, T_8=400 K, and p_8=1 bar
Steam cycle: fluid=water, x_2=0, T_4=480 K, x_3=1, T_4=420 K, and mdot=1 kg/s.
Determine the maximum net power produced and working fluid temperature at the inlet of the turbine with fixed condenser temperature.
ANSWER: maximum net power is about 401.2 kW and T_3 is about 564.3 K.

11.6 Finite Time Rankine Cycle

11.6.1 Ideal Rankine Cycle with Infinitely Large Heat Reservoirs

The ideal Finite time Rankine cycle and its T-s diagram are shown in Figure 11.6.1.1 and Figure 11.6.1.2. The cycle is an endoreversible cycle which is made of two isentropic processes and two isobaric heat transfer processes. The cycle exchanges heats with its surroundings in the two isobaric external irreversible heat transfer processes. The heat source and heat sink are infinitely large. Therefore, the temperature of the heat source and heat sink are unchanged during the heat transfer processes.

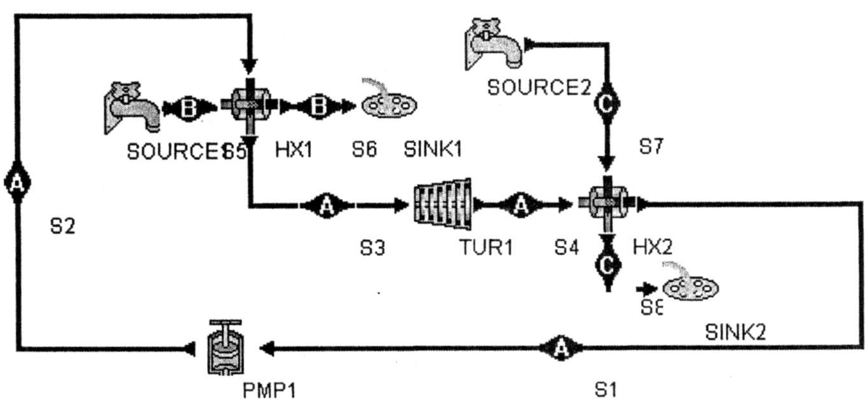

Figure 11.6.1.1 Finite time ideal Rankine cycle with infinitely large heat reservoirs

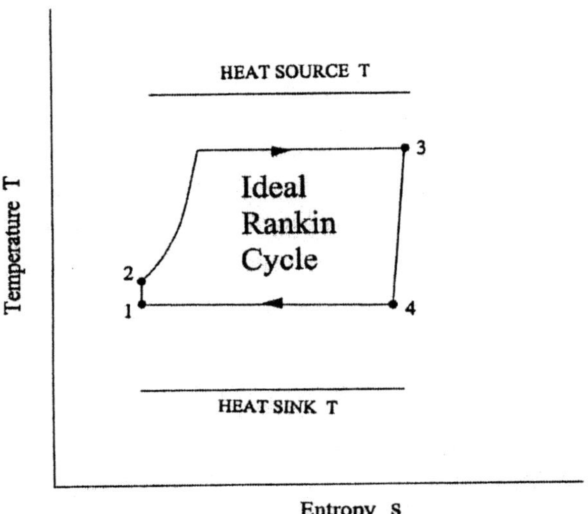

Figure 11.6.1.2. Finite-time ideal Rankine cycle T-s diagram

Assume that the working fluid flows through the heat engine in a steady-state fashion. The rates of heat rejection and addition of the heat engine are:

$$Qdot_H=U_HA_HLMTD_H \tag{11.6.1}$$

$$Qdot_L=U_LA_LLMTD_L \tag{11.6.2}$$

where U_H is the heat transfer coefficient and A_H is the heat transfer surface area of the high-temperature side heat exchanger between the heat engine and the heat source; U_L is the heat transfer coefficient and A_L is the heat transfer surface area of the low-temperature side heat exchanger between the heat engine and the heat sink.

The total heat transfer surface area (A) of the two heat exchangers is assumed to be a constant.

$$A=A_H+A_L \tag{11.6.3}$$

The power output (P) of the heat engine according to the first law of thermodynamics is

$$P=Qdot_H-Qdot_L' \tag{11.6.4}$$

The efficiency (η) of the heat engine is

$$\eta=P/Qdot_H \tag{11.6.5}$$

Example 11.6.1.1. An endoreversible Rankine steam heat engine with its infinitely large steam heat source and heat sink is shown in Figure 11.6.1.1. The following information is given:

> p_1=1 bar, x_1=0, $mdot_1$=1 kg/s, p_3=100 bar, x_3=1, p_5=200 bar, x_5=1, p_6=200 bar, x_6=0, p_5=639 K, x_5=1, T_6=639 K, x_6=0, p_7=0.02 bar, x_7=0, p_8=0.02 bar, and x_8=1.

Determine the power required by the pump, power produced by the turbine, net power produced by the cycle, rate of heat added by the heat source, rate of heat removed to the heat sink, and cycle efficiency.
Optimize the net power produced by the cycle with fix p_1. Draw the sensitivity diagram of net power vs p_3. Find the maximum net power and p_3 at the maximum net power condition.

To solve this problem by CyclePad, we take the following steps:

1. Build the cycle and its surroundings as shown in Figure 11.6.1.
2. Analysis: (A) Assuming the heat exchangers are isobaric, and turbine and pump are isentropic. (B) Input Heat source fluid=steam, T_5=639 K, x_5=1, T_6=639 K, and x_6=0; Heat sink: fluid=steam, p_7=0.02 bar, x_7=0, p_7=0.02 bar, p_8=0.02 bar, and x_8=1; Steam cycle: fluid=water, x_1=0, p_1=1 bar, x_3=1, p_3=200 bar, and mdot=1 kg/s.
3. Display results: The results shown in Figure Example 11.6.1.1a are: rate of heat added from the heat source=2297 kW, rate of heat removed to the heat sink=-1507 kW, power required by the isentropic pump=-10.30 kW, power produced by the isentropic turbine=699.9 kW, net power produced=689.6 kW and efficiency of the cycle=30.02%.

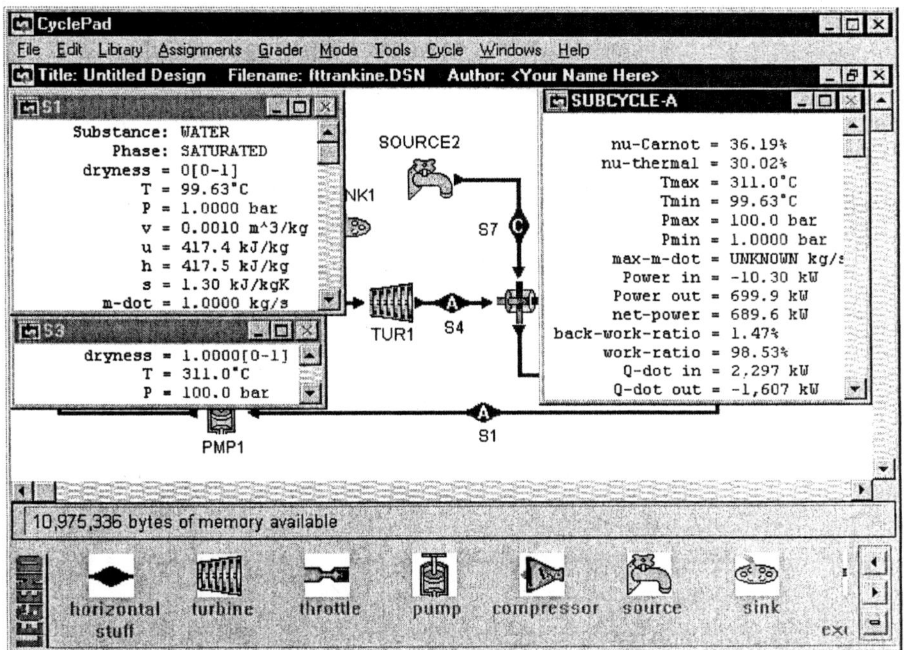

Figure Example 11.6.1.1a. Finite time ideal Rankine cycle with infinitely large heat reservoirs

4. Optimization
   Draw the sensitivity diagram of net power vs p_3 as shown in Figure Example 11.6.1.1b. The maximum net power is about 692.4 kW, and p_3 at the maximum net power condition is about 116.1 bar with fixed condenser pressure.

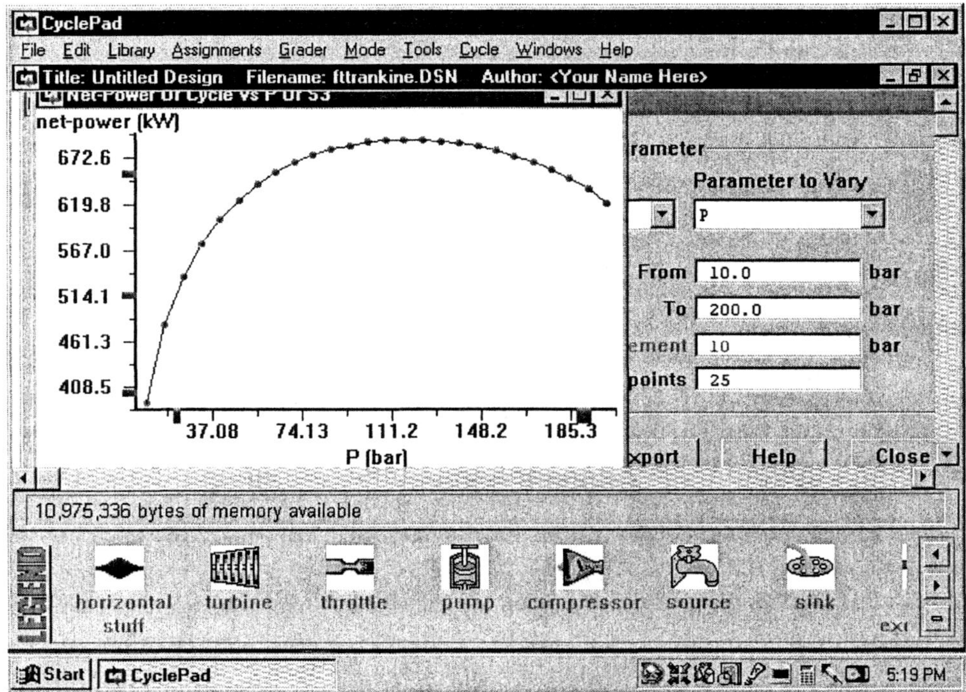

Figure Example 11.6.1.1b. Finite time ideal Rankine cycle with infinite largely heat reservoirs sensitivity diagram

Comment: The partial optimization is only for ∂(net power)/∂(p_3)=0. To have the full optimization, we must let ∂(net power)/∂(p_1)=0 also.

Homework 11.6.1.1. Finite Time Ideal Rankine Cycle with Infinitely Large Heat Reservoirs

1. An endoreversible Rankine steam heat engine with its infinitely large steam heat source and heat sink is shown in Figure 11.6.1.1. The following information is given:
   p_1=0.1 bar, x_1=0, $mdot_1$=1 kg/s, p_3=120 bar, x_3=1, p_5=200 bar, x_5=1, p_6=200 bar, x_6=0, p_5=639 K, x_5=1, T_6=639 K, x_6=0, p_7=0.02 bar, x_7=0, p_8=0.02 bar, and x_8=1.
   Determine the power required by the pump, power produced by the turbine, net power produced by the cycle, rate of heat added by the heat source, rate of heat removed to the heat sink, and cycle efficiency.
   Optimize the net power produced by the cycle with fix p_1. Draw the sensitivity diagram of net power vs p_3. Find the maximum net power and p_3 at the maximum net power condition.

ANSWER: power required by the pump=-12.13 kW, power produced by the turbine=948.0 kW, net power produced by the cycle=935.9 kW, rate of heat added by the heat source=2481 kW, rate of heat removed to the heat sink=-1545 kW, and cycle efficiency=37.73%.

2. An endoreversible Rankine steam heat engine with its infinitely large steam heat source and heat sink is shown in Figure 11.6.1.1. The following information is given:
p_1=0.1 bar, x_1=0, $mdot_1$=1 kg/s, p_3=150 bar, x_3=1, p_5=200 bar, x_5=1, p_6=200 bar, x_6=0, p_5=639 K, x_5=1, T_6=639 K, x_6=0, p_7=0.02 bar, x_7=0, p_8=0.02 bar, and x_8=1.
Determine the power required by the pump, power produced by the turbine, net power produced by the cycle, rate of heat added by the heat source, rate of heat removed to the heat sink, and cycle efficiency.
Optimize the net power produced by the cycle with fix p_1. Draw the sensitivity diagram of net power vs p_3. Find the maximum net power and p_3 at the maximum net power condition.
ANSWER: power required by the pump=-15.5 kW, power produced by the turbine=699.0 kW, net power produced by the cycle=683.5 kW, rate of heat added by the heat source=2177 kW, rate of heat removed to the heat sink=-1494 kW, and cycle efficiency=31.40%.

11.6.2 Actual Rankine Cycle with Infinitely Large Heat Reservoirs

The actual Finite time Rankine cycle is shown in Figure 11.6.2.1. The cycle is an external and internal irreversible cycle which is made of two irreversible internal adiabatic processes (pump and turbine) and two irreversible external isobaric heat transfer processes. The heat source and heat sink are infinitely large.

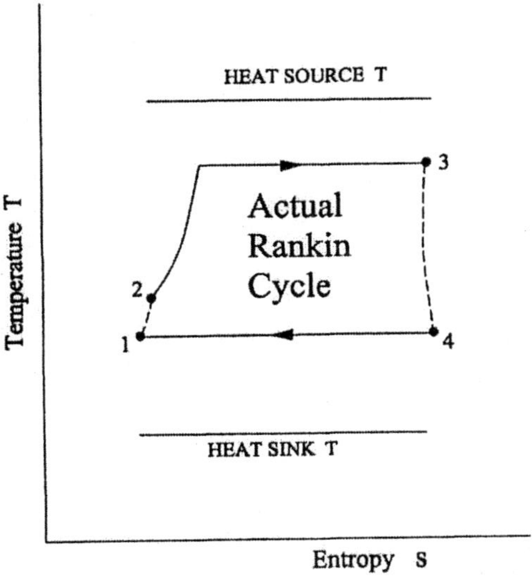

Figure 11.6.2.1. Finite time actual Rankine cycle with infinitely large heat reservoirs

Example 11.6.2.1. An endoreversible Rankine steam heat engine with its infinitely large steam heat source and heat sink is shown in Figure 11.5.1.1. The following information is given:

> $\eta_{turbine}$=85%, η_{pump}=100%, p_1=1 bar, x_1=0, $mdot_1$=1 kg/s, p_3=100 bar, x_3=1, p_5=200 bar, x_5=1, p_6=200 bar, x_6=0, p_5=639 K, x_5=1, T_6=639 K, x_6=0, p_7=0.02 bar, x_7=0, p_8=0.02 bar, and x_8=1.

Determine the power required by the pump, power produced by the turbine, net power produced by the cycle, rate of heat added by the heat source, rate of heat removed to the heat sink, and cycle efficiency.
Optimize the net power produced by the cycle with fix p_1. Draw the sensitivity diagram of net power vs p_3. Find the maximum net power and p_3 at the maximum net power condition.

To solve this problem by CyclePad, we take the following steps:

1. Build the cycle and its surroundings as shown in Figure 11.6.1.
2. Analysis: (A) Assuming the heat exchangers are isobaric, and turbine and pump are isentropic. (B) Input Heat source fluid=steam, T_5=639 K, x_5=1, T_6=639 K, and x_6=0; Heat sink: fluid=steam, p_7=0.02 bar, x_7=0, p_7=0.02 bar, p_8=0.02 bar, and x_8=1; Steam cycle: fluid=water, x_1=0, p_1=1 bar, x_3=1, p_3=200 bar, and mdot=1 kg/s.
3. Display results: The results shown in Figure Example 11.6.2.1a are: rate of heat added from the heat source=2297 kW, rate of heat removed to the heat sink=-1712 kW, power required by the isentropic pump=-10.30 kW, power produced by the isentropic turbine=594.9 kW, net power produced=584.6 kW and efficiency of the cycle=25.45%.

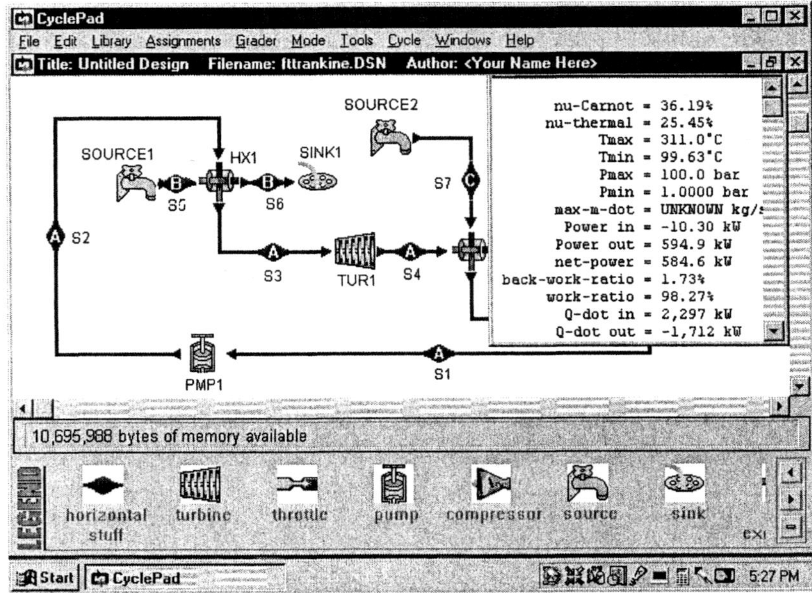

Figure Example 11.6.2.1a. Finite time actual Rankine cycle with infinitely large heat reservoirs

4. Optimization
   Draw the sensitivity diagram of net power vs p_3 as shown in Figure Example 11.5.1b. The maximum net power is about 424.8 kW, and p_3 at the maximum net power condition is about 107.4 kPa.

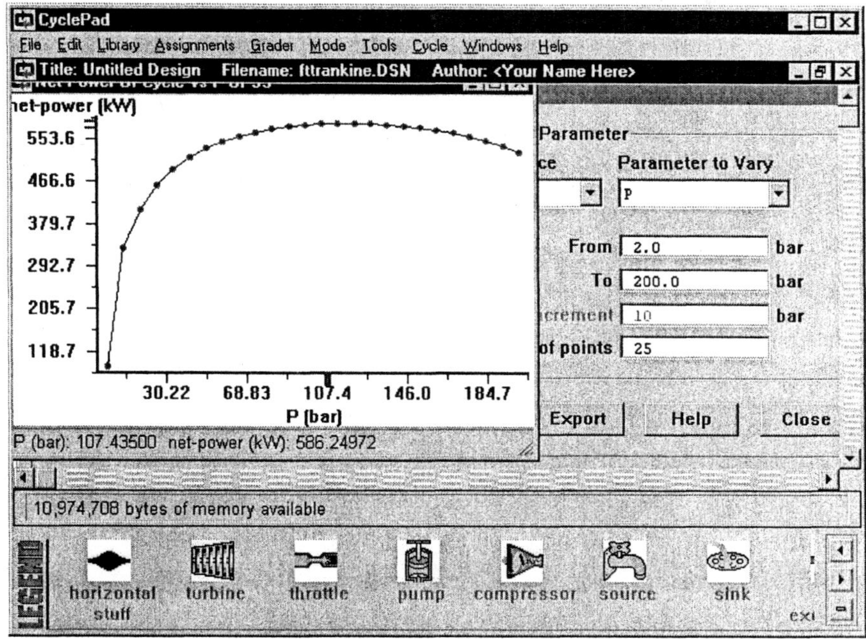

Figure Example 11.6.2.1b. Finite time actual Rankine cycle with infinite largely heat reservoirs sensitivity diagram

Comment: The partial optimization is only for ∂(net power)/∂(p_3)=0. To have the full optimization, we must let ∂(net power)/∂(p_1)=0 also.

Homework 11.6.2. Finite Time Actual Rankine Cycle with Infinitely Large Heat Reservoirs

1. An endoreversible Rankine steam heat engine with its infinitely large steam heat source and heat sink is shown in Figure 11.5.1.1. The following information is given:
   $\eta_{turbine}$=85%, η_{pump}=100%, p_1=1 bar, x_1=0, $mdot_1$=1 kg/s, p_3=150 bar, x_3=1, p_5=150 bar, x_5=1, p_6=200 bar, x_6=0, p_5=639 K, x_5=1, T_6=639 K, x_6=0, p_7=0.02 bar, x_7=0, p_8=0.02 bar, and x_8=1.
   Determine the power required by the pump, power produced by the turbine, net power produced by the cycle, rate of heat added by the heat source, rate of heat removed to the heat sink, and cycle efficiency.
   ANSWER: power required by the pump=-15.5 kW, power produced by the turbine=594.2 kW, net power produced by the cycle=578.7 kW, rate of heat added by the heat source=2177 kW, rate of heat removed to the heat sink=-1598 kW, and cycle efficiency=26.58%.

2. An endoreversible Rankine steam heat engine with its infinitely large steam heat source and heat sink is shown in Figure 11.5.1.1. The following information is given:
$\eta_{turbine}$=85%, η_{pump}=100%, p_1=0.5 bar, x_1=0, $mdot_1$=1 kg/s, p_3=100 bar, x_3=1, p_5=150 bar, x_5=1, p_6=200 bar, x_6=0, p_5=639 K, x_5=1, T_6=639 K, x_6=0, p_7=0.02 bar, x_7=0, p_8=0.02 bar, and x_8=1.
Determine the power required by the pump, power produced by the turbine, net power produced by the cycle, rate of heat added by the heat source, rate of heat removed to the heat sink, and cycle efficiency.
Optimize the net power produced by the cycle with fix p_1. Draw the sensitivity diagram of net power vs p_3. Find the maximum net power and p_3 at the maximum net power condition.
ANSWER: power required by the pump=-10.27 kW, power produced by the turbine=663.6 kW, net power produced by the cycle=653.3 kW, rate of heat added by the heat source=2374 kW, rate of heat removed to the heat sink=-1720 kW, and cycle efficiency=27.52%; maximum net power=approximately 653.7 kW and p_3=102.3 bar at the maximum net power condition

11.6.3 Ideal Rankine Cycle with Finite Capacity Heat Reservoirs

The ideal Finite time Rankine cycle is shown in Figure 11.6.3.1. The cycle is an endoreversible cycle which is made of two isentropic processes and two isobaric heat transfer processes. The cycle exchanges heats with its surroundings in the two isobaric external irreversible heat transfer processes. The heat source and heat sink are not infinitely large. Therefore, the temperature of the heat source and heat sink are unchanged during the heat transfer processes.

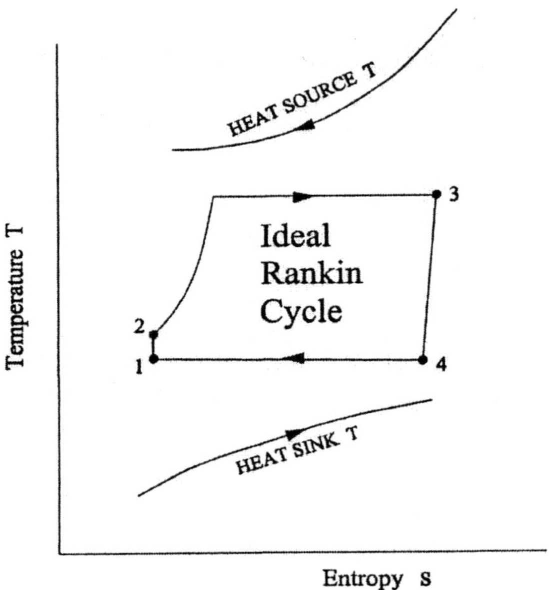

Figure 11.6.3.1. Finite time ideal Rankine cycle with finite heat reservoirs

Example 11.6.3.1. A finite time ideal Rankine cycle operates between a finite heat capacity heat source and a finite heat capacity heat sink. The following information is given:

> Heat source: fluid=air, T_5=2000°C, p_5=1 bar, T_6=800°C, and p_6=1 bar
> Heat sink: fluid=water, T_7=17°C, p_7=1 bar, T_8=30°C, and p_8=1 bar
> Steam cycle: fluid=water, x_2=0, p_3=150 bar, x_3=1, p_4=1 bar, and mdot=1 kg/s.

Determine the rate of heat added from the heat source, rate of heat removed to the heat sink, power required by the isentropic pump, power produced by the isentropic turbine, net power produced and efficiency of the cycle.
Optimize the net power produced by the cycle with fix p_1. Draw the sensitivity diagram of net power vs p_3. Find the maximum net power and p_3 at the maximum net power condition.

To solve this problem by CyclePad, we take the following steps:

1. Build the cycle and its surroundings as shown in Figure 11.6.1.
2. Analysis: (A) Assuming the heat exchangers are isobaric, and turbine and pump are isentropic. (B) Input Heat source fluid=air, T_5=2000°C, p_5=1 bar, T_6=800°C, and p_6=1 bar; Heat sink: fluid=water, T_7=17°C, p_7=1 bar, T_8=°C, and p_8=1 bar; and Steam cycle: fluid=water, x_2=0, p_3=150 bar, x_3=1, p_4=1 bar, and mdot=1 kg/s.
3. Display results: The results are: rate of heat added from the heat source=2177 kW, rate of heat removed to the heat sink=-1494 kW, power required by the isentropic pump=-15.5 kW, power produced by the isentropic turbine=699.0 kW, net power produced=683.5 kW and efficiency of the cycle=31.40%.

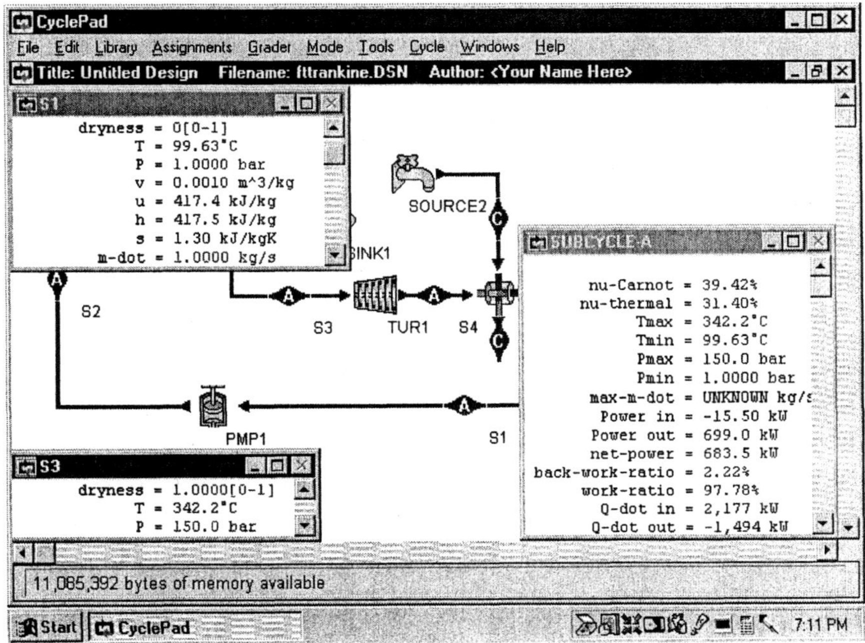

Figure Example 11.6.3.1a. Finite time ideal Rankine cycle with finite capacity heat reservoirs

4. Optimization
Draw the sensitivity diagram of net power vs p_3 as shown in Figure Example 11.6.3.1b. The maximum net power is about 868.7 kW, and p_3 at the maximum net power condition is about 90.0 bar.

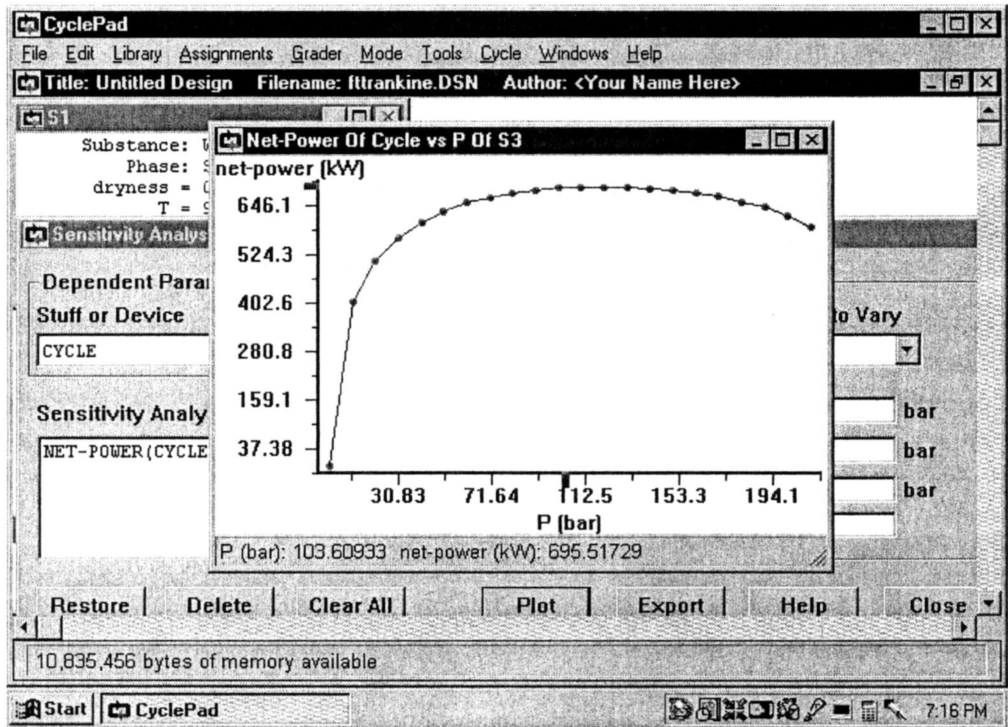

Figure Example 11.6.3.1b. Finite time ideal Rankine cycle with finite heat reservoirs sensitivity diagram

Comment: The partial optimization is only for ∂(net power)/∂(p_3)=0. To have the full optimization, we must let ∂(net power)/∂(p_1)=0 also.

Example 11.6.3.2. A finite time ideal Rankine cycle operates between a finite heat capacity heat source and a finite heat capacity heat sink. The following information is given:

Heat source: fluid=water, p_5=200 bar, x_5=1 bar, x_6=0, and p_6=200 bar,
Heat sink: fluid=water, x_7=0, p_7=0.02 bar, x_8=1, and p_8=0.02 bar,
Steam cycle: fluid=water, x_1=0, p_1=1 bar, x_3=1, p_3=117.6 bar, and mdot=1 kg/s,
Heat exchangers are counter-flow type.

Determine the rate of heat added from the heat source, rate of heat removed to the heat sink, power required by the isentropic pump, power produced by the isentropic turbine, net power produced and efficiency of the cycle.

Optimize the net power produced by the cycle with variable p_3 based on the criterion of (A) net power per unit conductance of heat exchangers, and (B) specific net power per unit surface area of heat exchangers with U_H=U_H=0.5 kW/[m^2(K)].

To solve this problem by CyclePad, we take the following steps:

1. Build the cycle and its surroundings as shown in Figure 11.6.1.
2. Analysis: (A) Assuming the heat exchangers are isobaric, and turbine and pump are isentropic. (B) Input Heat source fluid=water, p_5=200 bar, x_5=1 bar, x_6=0, and p_6=200 bar; Heat sink: fluid=water, x_7=0, p_7=0.02 bar, x_8=1, and p_8=0.02 bar; Heat exchangers are counter-flow type; and Steam cycle: fluid=water, x_1=0, p_1=1 bar, x_3=1, p_3=117.6 bar, and mdot=1 kg/s as shown in Figure Example 11.6.3.2a.
3. Display results: The results are: rate of heat added from the heat source ($Qdot_H$)=2260 kW, logarithm mean temperature difference of the high-temperature side heat exchanger ($LMTD_H$)=121.8°C, rate of heat removed to the heat sink ($Qdot_L$)=-1567 kW, logarithm mean temperature difference of the low-temperature side heat exchanger ($LMTD_L$)=82.13°C, net power produced=693.0 kW and efficiency of the cycle=30.66%.

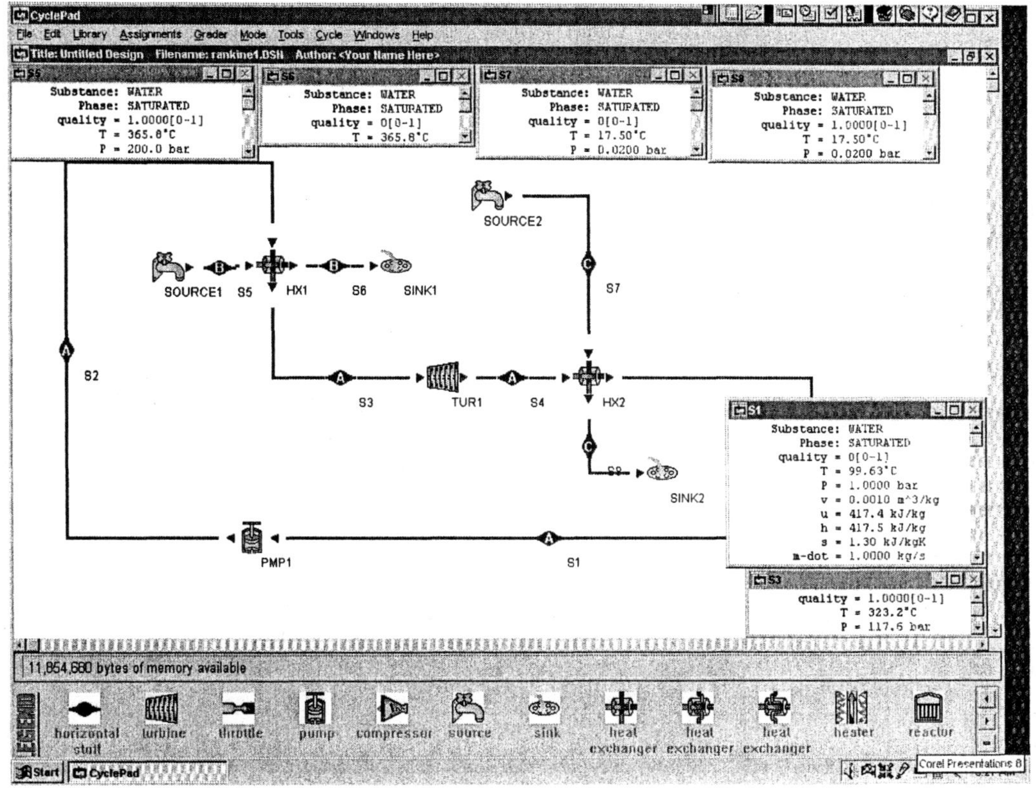

Figure Example 11.6.3.2a Finite time ideal Rankine cycle with finite heat capacity source and sink input

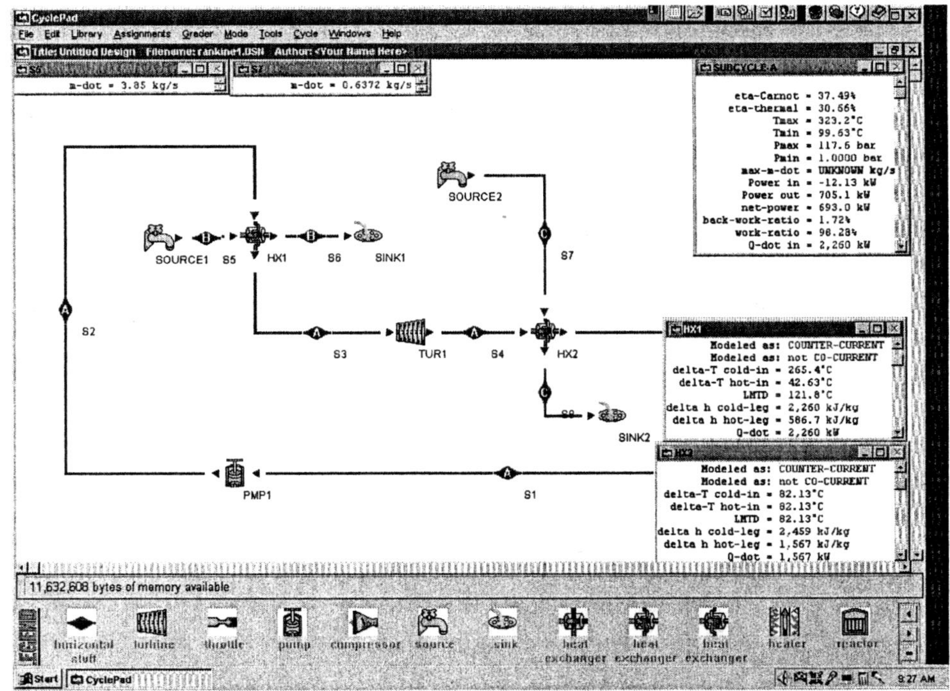

Figure Example 11.6.3.2b Finite time ideal Rankine cycle with finite heat capacity source and sink output

The conductances of the heat exchangers are U_HA_H=2260/121.8=18.56 kW/K and U_LA_L=1567/82.13= 19.08 kW/K. The total conductances of the heat exchangers is 18.56+19.08=37.64 kW/K. The net power output per unit conductance of heat exchangers is 693.0/37.64=18.41 kW/K.

The surface areas of the heat exchangers are A_H=18.56/0.5=37.12 m^2 and A_L=19.08/0.5=38.16 m^2. The total surface areas of the heat exchangers is 37.12+38.16=75.28 m^2. The specific power per unit total surface areas of the heat exchangers is 693.0/75.28=9.206 m^2.

p3 bar	LMTDH K	LMTDL K	QDOTH kW	QDOTL kW	UHAH kW/K	ULAL kW/K	SUM(UA) kW/K	PNET kW	sppnet kW/(kW/K)	
2	254.4	82.13	2277	2171	8.950472	26.4337	35.38417	105.6	2.984385	
10	223.6	82.13	2359	1969	10.55009	23.97419	34.52428	389.8	11.29061	
30	191.2	82.13	2383	1820	12.46339	22.15999	34.62338	562.6	16.24914	
50	170.9	82.13	2371	1741	13.87361	21.1981	35.07171	630.3	17.97175	
80	147.3	82.13	2332	1655	15.83164	20.15098	35.98262	676.8	**18.80908**	**MAXsppnet**
100	133.5	82.13	2297	1607	17.20599	19.56654	36.77253	689.6	18.75313	
110	126.8	82.13	2277	1584	17.95741	19.2865	37.24391	692.3	18.58827	
120	120.2	82.13	2255	1562	18.7604	19.01863	37.77903	**693**	18.34351	**MAXPNET**
130	113.6	82.13	2231	1539	19.63908	18.73859	38.37767	691.7	18.0235	
150	99.86	82.13	2177	1494	21.80052	18.19067	39.99119	683.5	17.09126	
180	75.15	82.13	2073	1418	27.58483	17.26531	44.85014	656	14.62649	

Figure Example 11.6.3.2c Net power per unit conductance of heat exchangers

Using p_3 as a design parameter, the following tables are made as shown in Figure Example 11.6.3.2c and d. Based on the criterion of (A) net power per unit conductance of heat exchangers, the optimization p_3=120 bar, and (B) specific net power per unit surface area of heat exchangers, the optimization p_3=80 bar.

p3 bar	LMTDH K	LMTDL K	QDOTH kW	QDOTL kW	AH m^2	AL m^2	SUM(A) m^2	PNET kW	sppnet kW/m^2	
2	254.4	82.13	2277	2171	17.90094	52.86741	70.76835	105.6	1.492193	
10	223.6	82.13	2359	1969	21.10018	47.94837	69.04855	389.8	5.645303	
30	191.2	82.13	2383	1820	24.92678	44.31998	69.24676	562.6	8.124568	
50	170.9	82.13	2371	1741	27.74722	42.3962	70.14342	630.3	8.985875	
80	147.3	82.13	2332	1655	31.66327	40.30196	71.96523	676.8	**9.404541**	**MAXsppnet**
100	133.5	82.13	2297	1607	34.41199	39.13308	73.54507	689.6	9.376564	
110	126.8	82.13	2277	1584	35.91483	38.57299	74.48782	692.3	9.294137	
120	120.2	82.13	2255	1562	37.5208	38.03726	75.55806	**693**	9.171755	**MAXPNET**
130	113.6	82.13	2231	1539	39.27817	37.47717	76.75534	691.7	9.011751	
150	99.86	82.13	2177	1494	43.60104	36.38135	79.98239	683.5	8.545631	
180	75.15	82.13	2073	1418	55.16966	34.53062	89.70028	656	7.313243	

Figure Example 11.6.3.2d Specific net power per unit surface area of heat exchangers

Example 11.6.3.3. A finite time ideal Rankine OTEC (Ocean thermal energy conversion) cycle as shown in Figure Example 11.6.3.3a operates between a finite heat capacity heat source and a finite heat capacity heat sink. The following information is given:

Heat source: fluid=warm ocean surface water, T_1=26°C, p_1=101 kPa, T_2=22°C, and p_2= 101 kPa

Heat sink: fluid=cold deep ocean water, T_3=5°C, p_3=101 kPa, T_4=9°C, and p_4=101 kPa,

Steam cycle: fluid=ammonia, x_5=0, T_5=12°C, x_7=1, T_7=20°C, and mdot=1 kg/s.

The heat exchangers are counter-flow type, U_H=0.4 kJ/(m^2)K and U_L=0.4 kJ/(m^2)K.

Determine the rate of heat added from the heat source, rate of heat removed to the heat sink, power required by the isentropic pump, power produced by the isentropic turbine, net power produced and efficiency of the cycle.

Since the fuel cost of the OTEC is free, the primary cost is the initial construction cost. The heat exchangers are the major concern of the initial construction cost. Let us take the specific net power output (net output power per unit total heat exchanger surface area) as the design objective function, optimize the warm-side (heater or high-temperature-side heat exchanger) and cold-side (cooler or low-temperature-side heat exchanger) working fluid temperatures.

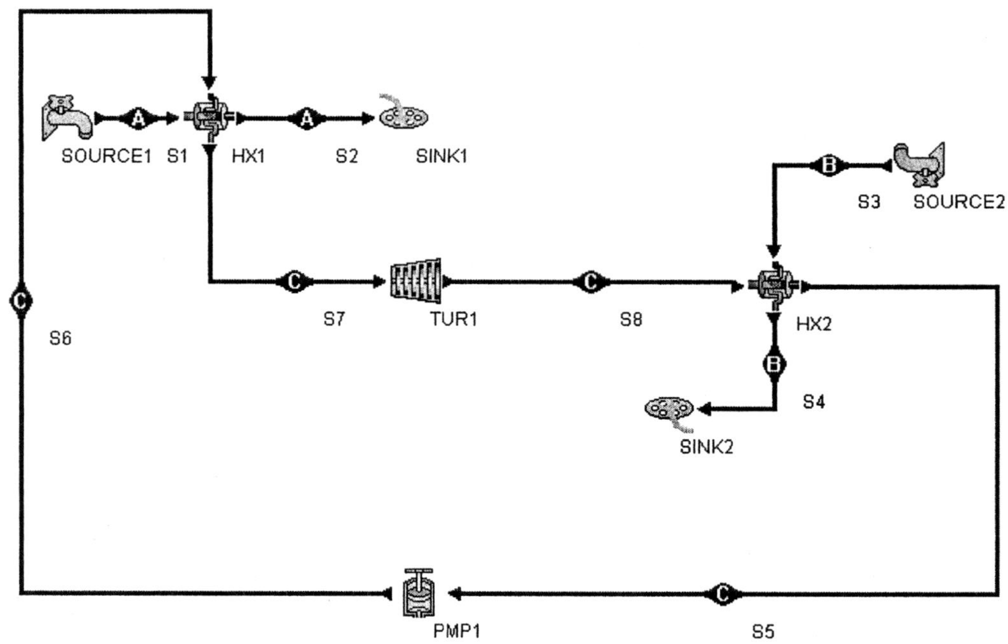

Figure Example 11.6.3.3aFinite-time OTEC cycle

To solve this problem by CyclePad, we take the following steps:

1. Build the cycle and its surroundings as shown in Figure 11.6.3.3a.
2. Analysis: (A) Assuming the heat exchangers are isobaric and counter-flow type, and turbine and pump are isentropic. (B) Input Heat source: fluid=warm ocean surface water, T_1=26°C, p_1=101 kPa, T_2=22°C, and p_2= 101 kPa
   Heat sink: fluid=cold deep ocean water, T_3=5°C, p_3=101 kPa, T_4=9°C, and p_4=101 kPa,
   Steam cycle: fluid=ammonia, x_5=0, T_5=12°C, x_7=1, T_7=20°C, and mdot=1 kg/s.
3. Display results: The results shown in Figure Example 11.6.3.3b are: $LMTD_H$=7.56 K, $LMTD_L$=4.72 K, rate of heat added from the heat source=1219 kW, rate of heat removed to the heat sink=-1191 kW, power required by the isentropic pump=-4.38 kW, power produced by the isentropic turbine=33.20 kW, net power produced=28.82 kW and efficiency of the cycle=2.36%.
4. Calculate $U_HA_H=Q_H/LMTD_H$=1219/7.56=161.2 kW/K, $U_LA_L=Q_L/LMTD_L$=1191/4.72= 252.3 kW/K, $U_HA_H+U_LA_L$=161.2+252.3=413.5 kW/K, and specific net power output=$Wdot_{net}/(A_H+A_L)$=28.82/(161.2/0.4+252.3/0.4)=0.1742 kW/m^2.
5. To optimize the specific power output of the cycle, we let ∂(specific power output)/∂T_7=0 first and then ∂(specific power output)/∂T_7=0 as shown in the following two tables.

It is seen that ∂(specific power output)/∂T_7=0 occurs at T_7=24°C as shown in Table Example 11.6.3.3a.

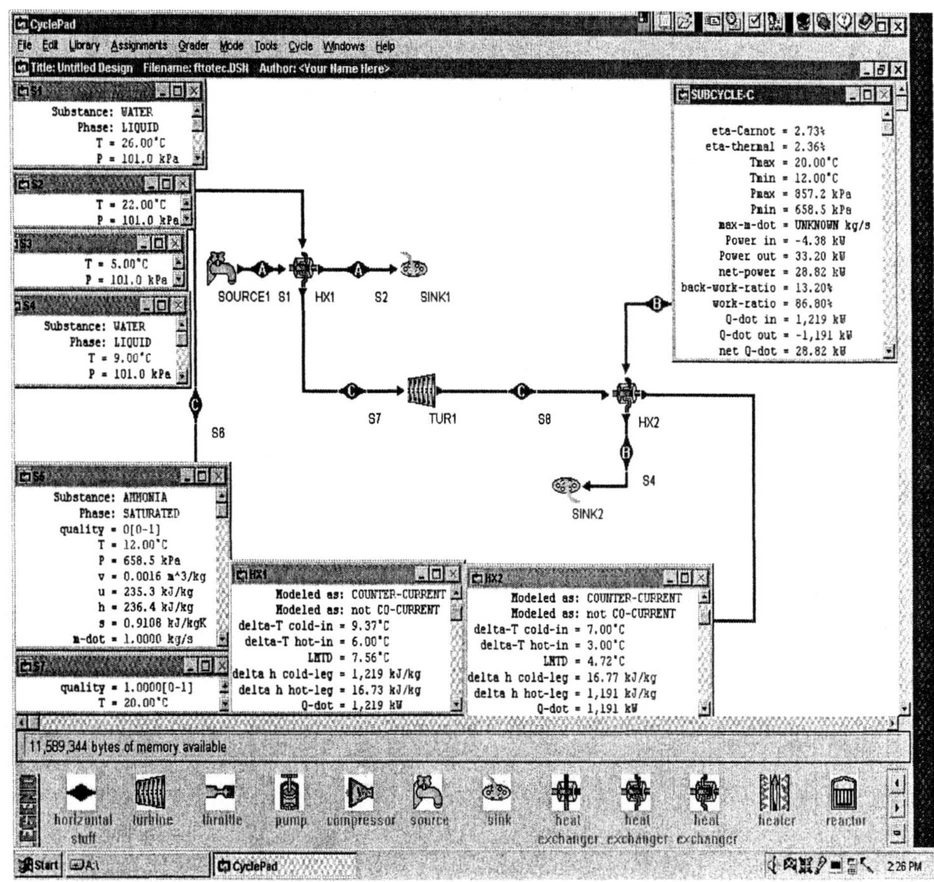

Figure Example 11.6.3.3b Finite-time OTEC cycle input and output

Table Example 11.6.3.3a Specific power optimization with respect to T_7

OTEC	T1=26	T2=22	T3=5	T4=9		COUNTER FLOW HX								
12	12.1	0.001028	0.001285	0.0304	-0.0561	0.4267	0.3706	1217	-1217	11.84	4.72	102.7872	257.839	
12	14	0.020189	0.025236	0.611	-1.09	8.53	7.44	1218	-1211	10.88	4.72	111.9485	256.5678	
12	16	0.038816	0.04852	1.21	-2.21	16.92	14.71	1219	-1204	9.84	4.72	123.8821	255.0847	
12	18	0.055613	0.069516	1.79	-3.28	25.14	21.86	1219	-1197	8.74	4.72	139.4737	253.6017	
12	20	0.069685	0.087107	2.36	-4.38	33.2	28.82	1219	-1191	7.56	4.72	161.2434	252.3305	
12	21.9	0.07957	0.099463	2.9	-5.42	40.73	35.32	1220	-1184	6.32	4.72	193.038	250.8475	
12	22	0.079947	0.099933	2.92	-5.47	41.13	35.66	1220	-1184	6.25	4.72	195.2	250.8475	
12	23	0.082671	0.103339	3.2	-6.03	45.02	38.99	1220	-1181	5.51	4.72	221.4156	250.2119	
12	**24**	**0.082867**	**0.103584**	**3.47**	**-6.58**	**48.91**	**42.33**	**1220**	**-1178**	**4.67**	**4.72**	**261.242**	**249.5763**	
12	24.5	0.081377	0.101722	3.6	-6.86	50.83	43.97	1220	-1176	4.19	4.72	291.1695	249.1525	
12	24.6	0.080922	0.101152	3.63	-6.91	51.21	44.3	1220	-1176	4.09	4.72	298.2885	249.1525	
12	24.7	0.080328	0.10041	3.66	-6.97	51.59	44.62	1220	-1175	3.98	4.72	306.5327	248.9407	
12	24.8	0.079672	0.09959	3.68	-7.02	51.97	44.95	1220	-1175	3.87	4.72	315.2455	248.9407	

6. Then let ∂(specific power output)/$\partial T_5=0$

It is seen that ∂(specific power output)/$\partial T_5=0$ occurs at $T_5=12°C$ as shown in Table Example 11.6.3.3b.

7. Let $T_5=12°C$ and $T_7=24°C$, the optimized specific power output of the cycle is 0.1036 kW/m^2. At the maximum optimized specific power output condition, $LMTD_H=4.67$ K,

$LMTD_H$=4.72 K, rate of heat added from the heat source=1220 kW, rate of heat removed to the heat sink=-1178 kW, power required by the isentropic pump=-6.58 kW, power produced by the isentropic turbine=48.91 kW, net power produced=42.33 kW and efficiency of the cycle=3.47% as shown in Figure Example 11.6.3.3c.

Table Example 11.6.3.3b Specific power optimization with respect to T_5

OTEC	T1=26	T2=22	T3=5	T4=9		COUNTER FLOW HX								
12	12.1	0.001028	0.001285	0.0304	-0.0561	0.4267	0.3706	1217	-1217	11.84	4.72	102.7872	257.839
12	14	0.020189	0.025236	0.611	-1.09	8.53	7.44	1218	-1211	10.88	4.72	111.9485	256.5678
12	16	0.038816	0.04852	1.21	-2.21	16.92	14.71	1219	-1204	9.84	4.72	123.8821	255.0847
12	18	0.055613	0.069516	1.79	-3.28	25.14	21.86	1219	-1197	8.74	4.72	139.4737	253.6017
12	20	0.069685	0.087107	2.36	-4.38	33.2	28.82	1219	-1191	7.56	4.72	161.2434	252.3305
12	21.9	0.07957	0.099463	2.9	-5.42	40.73	35.32	1220	-1184	6.32	4.72	193.038	250.8475
12	22	0.079947	0.099933	2.92	-5.47	41.13	35.66	1220	-1184	6.25	4.72	195.2	250.8475
12	23	0.082671	0.103339	3.2	-6.03	45.02	38.99	1220	-1181	5.51	4.72	221.4156	250.2119
12	**24**	**0.082867**	**0.103584**	**3.47**	**-6.58**	**48.91**	**42.33**	**1220**	**-1178**	**4.67**	**4.72**	**261.242**	**249.5763**
12	24.5	0.081377	0.101722	3.6	-6.86	50.83	43.97	1220	-1176	4.19	4.72	291.1695	249.1525
12	24.6	0.080922	0.101152	3.63	-6.91	51.21	44.3	1220	-1176	4.09	4.72	298.2885	249.1525
12	24.7	0.080328	0.10041	3.66	-6.97	51.59	44.62	1220	-1175	3.98	4.72	306.5327	248.9407
12	24.8	0.079672	0.09959	3.68	-7.02	51.97	44.95	1220	-1175	3.87	4.72	315.2455	248.9407
9.1	24	0.040529	0.050661	4.32	-7.8	61.09	53.29	1232	-1179	5.52	1.08	223.1884	1091.667
9.5	24	0.059103	0.073878	4.21	-7.64	59.4	51.76	1231	-1179	5.4	1.82	227.963	647.8022
10	24	0.0705	0.088125	4.06	-7.44	57.28	49.84	1228	-1179	5.26	2.49	233.4601	473.494
11	24	0.080891	0.101114	3.77	-7.01	53.11	46.1	1224	-1178	4.97	3.64	246.2777	323.6264
12	**24**	**0.082867**	**0.103584**	**3.47**	**-6.58**	**48.91**	**42.33**	**1220**	**-1178**	**4.67**	**4.72**	**261.242**	**249.5763**
15	24	0.066312	0.082889	2.6	-5.14	36.5	31.36	1207	-1176	3.74	7.83	322.7273	150.1916
18	24	0.037198	0.046497	1.72	-3.59	24.18	20.6	1195	-1174	2.68	10.88	445.8955	107.9044
20	24	0.018341	0.022927	1.15	-2.45	16.08	13.62	1186	-1173	1.82	12.9	651.6484	90.93023
21	24	0.010039	0.012549	0.8623	-1.86	12.05	10.19	1182	-1172	1.27	13.9	930.7087	84.31655
21.5	24	0.005931	0.007414	0.7177	-1.57	10.03	8.47	1180	-1171	0.8762	14.41	1346.724	81.26301

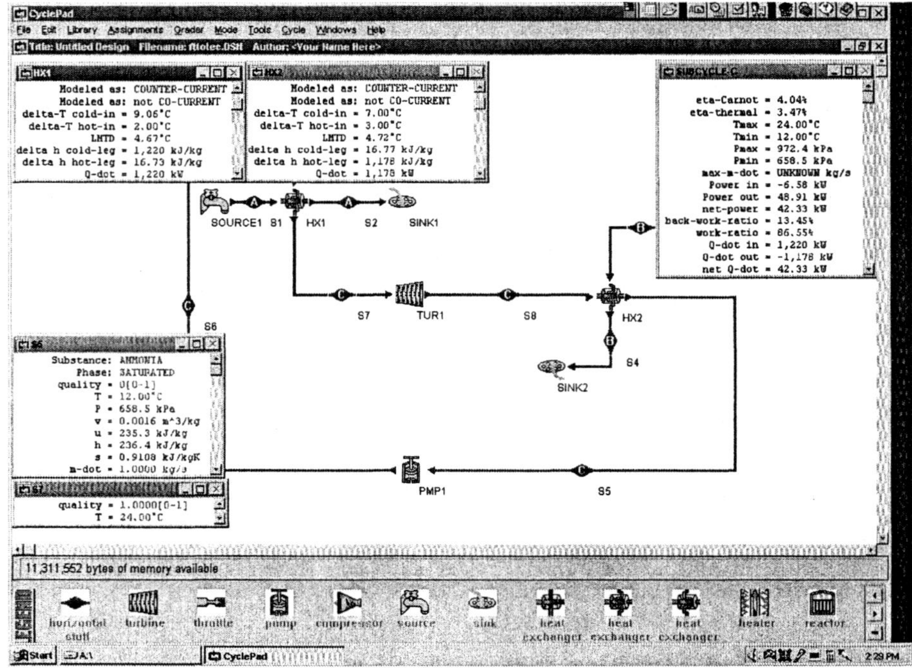

Figure Example 11.6.3.3c Finite-time OTEC cycle optimization

Homework 11.6.3. Finite Time Ideal Rankine Cycle with Finite Heat Capacity Reservoirs

1. An endoreversible (Curzon and Ahlborn) steam cycle operates between a finite heat capacity heat source and a finite heat capacity heat sink. The following information is given:
   Heat source: fluid=air, T_5=2000°C, p_5=1 bar, T_6=800°C, and p_6=1 bar
   Heat sink: fluid=water, T_7=17°C, p_7=1 bar, T_8=°C, and p_8=1 bar
   Steam cycle: fluid=water, x_2=0, p_3=200 bar, x_3=1, p_4=1 bar, and mdot=1 kg/s.
   Determine the rate of heat added from the heat source, rate of heat removed to the heat sink, power required by the isentropic pump, power produced by the isentropic turbine, net power produced and efficiency of the cycle.
   ANSWER: rate of heat added from the heat source=1975 kW, rate of heat removed to the heat sink=-1353 kW, power required by the isentropic pump=-20.68 kW, power produced by the isentropic turbine=642.7 kW, net power produced=622.0 kW and efficiency of the cycle=31.49%.
2. An endoreversible (Curzon and Ahlborn) steam cycle operates between a finite heat capacity heat source and a finite heat capacity heat sink. The following information is given:
   Heat source: fluid=air, T_5=2000°C, p_5=1 bar, T_6=800°C, and p_6=1 bar
   Heat sink: fluid=water, T_7=17°C, p_7=1 bar, T_8=°C, and p_8=1 bar
   Steam cycle: fluid=water, x_2=0, p_3=200 bar, x_3=1, p_4=0.2 bar, and mdot=1 kg/s.
   Determine the rate of heat added from the heat source, rate of heat removed to the heat sink, power required by the isentropic pump, power produced by the isentropic turbine, net power produced and efficiency of the cycle.
   ANSWER: rate of heat added from the heat source=2141 kW, rate of heat removed to the heat sink=-1366 kW, power required by the isentropic pump=-20.25 kW, power produced by the isentropic turbine=795.5 kW, net power produced=775.2 kW and efficiency of the cycle=36.20%.
3. An endoreversible (Curzon and Ahlborn) steam cycle operates between a finite heat capacity heat source and a finite heat capacity heat sink. The following information is given:
   Heat source: fluid=air, T_5=2000°C, p_5=1 bar, T_6=800°C, and p_6=1 bar
   Heat sink: fluid=water, T_7=17°C, p_7=1 bar, T_8=°C, and p_8=1 bar
   Steam cycle: fluid=water, x_2=0, p_3=200 bar, x_3=1, p_4=0.2 bar, and mdot=1 kg/s.
   Optimize the net power produced by the cycle with fixed p_4. Draw the sensitivity diagram of net power vs p_3. Find the maximum net power and p_3 at the maximum net power condition.
   ANSWER: The maximum net power is about 691.7 kW, and p_3 at the maximum net power condition is about 103.6 bar.
4. A finite time ideal Rankine OTEC (Ocean thermal energy conversion) cycle as shown in Figure Example 11.6.3.3a operates between a finite heat capacity heat source and a finite heat capacity heat sink. The following information is given:
   Heat source: fluid=warm ocean surface water, T_1=26°C, p_1=101 kPa, T_2=20°C, and p_2= 101 kPa

Heat sink: fluid=cold deep ocean water, T_3=5°C, p_3=101 kPa, T_4=10°C, and p_4=101 kPa, Steam cycle: fluid=ammonia, x_5=0, T_5=12°C, x_7=1, T_7=20°C, and mdot=1 kg/s. The heat exchangers are counter-flow type, U_H=0.4 kJ/(m^2)K and U_L=0.4 kJ/(m^2)K. Determine the rate of heat added from the heat source, rate of heat removed to the heat sink, power required by the isentropic pump, power produced by the isentropic turbine, net power produced and efficiency of the cycle.
Since the fuel cost of the OTEC is free, the primary cost is the initial construction cost. The heat exchangers are the major concern of the initial construction cost. Let us take the specific net power output (net output power per unit total heat exchanger surface area) as the design objective function, optimize the warm-side (heater or high-temperature-side heat exchanger) and cold-side (cooler or low-temperature-side heat exchanger) working fluid temperatures.

5. A finite time ideal Rankine OTEC (Ocean thermal energy conversion) cycle as shown in Figure Example 11.6.3.3a operates between a finite heat capacity heat source and a finite heat capacity heat sink. The following information is given:
   Heat source: fluid=warm ocean surface water, T_1=26°C, p_1=101 kPa, T_2=20°C, and p_2= 101 kPa
   Heat sink: fluid=cold deep ocean water, T_3=5°C, p_3=101 kPa, T_4=10°C, and p_4=101 kPa, Steam cycle: fluid=ammonia, x_5=0, T_5=12°C, x_7=1, T_7=20°C, and mdot=1 kg/s.
   The heat exchangers are cocurrent (parallel)-flow type, U_H=0.4 kJ/(m^2)K and U_L=0.4 kJ/(m^2)K.
   Determine the rate of heat added from the heat source, rate of heat removed to the heat sink, power required by the isentropic pump, power produced by the isentropic turbine, net power produced and efficiency of the cycle.
   Since the fuel cost of the OTEC is free, the primary cost is the initial construction cost. The heat exchangers are the major concern of the initial construction cost. Let us take the specific net power output (net output power per unit total heat exchanger surface area) as the design objective function, optimize the warm-side (heater or high-temperature-side heat exchanger) and cold-side (cooler or low-temperature-side heat exchanger) working fluid temperatures.

11.6.4 Actual Rankine Cycle with Finite Capacity Heat Reservoirs

The actul finite time Rankine cycle is shown in Figure 11.6.4.1. The cycle is an actual Rankine cycle which is made of two adibatic processes and two isobaric heat transfer processes. The cycle exchanges heats with its surroundings in the two isobaric external irreversible heat transfer processes. The heat source and heat sink are not infinitely large. Therefore, the temperature of the heat source and heat sink change during the heat transfer processes.

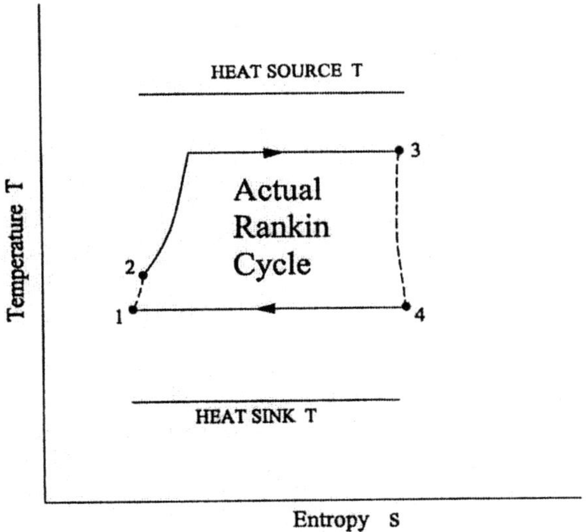

Figure 11.6.4.1. Finite time actual Rankine cycle with finite heat reservoirs

Example 11.6.4.1. A finite time actual Rankine cycle operates between a finite heat capacity heat source and a finite heat capacity heat sink. The following information is given:

Heat source: fluid=air, T_5=2000°C, p_5=1 bar, T_6=800°C, and p_6=1 bar
Heat sink: fluid=water, T_7=17°C, p_7=1 bar, T_8=30°C, and p_8=1 bar
Steam cycle: fluid=water, x_2=0, p_3=200 bar, x_3=1, p_4=0.2 bar, and mdot=1 kg/s, $\eta_{turbine}$=85%.

Determine the rate of heat added from the heat source, rate of heat removed to the heat sink, power required by the isentropic pump, power produced by the isentropic turbine, net power produced and efficiency of the cycle.
Optimize the net power produced by the cycle with fix p_1. Draw the sensitivity diagram of net power vs p_3. Find the maximum net power and p_3 at the maximum net power condition.

To solve this problem by CyclePad, we take the following steps:

1. Build the cycle and its surroundings as shown in Figure 11.6.1.
2. Analysis: (A) Assuming the heat exchangers are isobaric, and turbine and pump are isentropic. (B) Input Heat source fluid=air, T_5=2000°C, p_5=1 bar, T_6=800°C, and p_6=1 bar; Heat sink: fluid=water, T_7=17°C, p_7=1 bar, T_8=°C, and p_8=1 bar; and Steam cycle: fluid=water, x_2=0, p_3=200 bar, x_3=1, p_4=0.2 bar, and mdot=1 kg/s, $\eta_{turbine}$=85%.
3. Display results: The results are: rate of heat added from the heat source=2141 kW, rate of heat removed to the heat sink=-1486 kW, power required by the isentropic pump=-20.25 kW, power produced by the turbine=676.1 kW, net power produced=655.9 kW and efficiency of the cycle=30.63%.

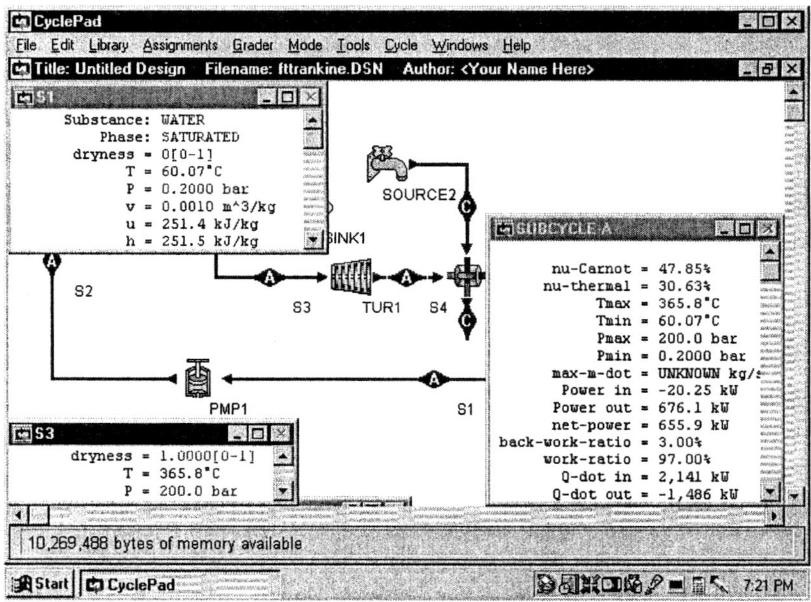

Figure Example 11.6.4.1a. Finite time actual Rankine cycle with finite capacity heat reservoirs

4. Optimization

   Draw the sensitivity diagram of net power vs p_3 as shown in Figure Example 11.6.4.1b. The maximum net power is about 733.8 kW, and p_3 at the maximum net power condition is about 90.69 bar.

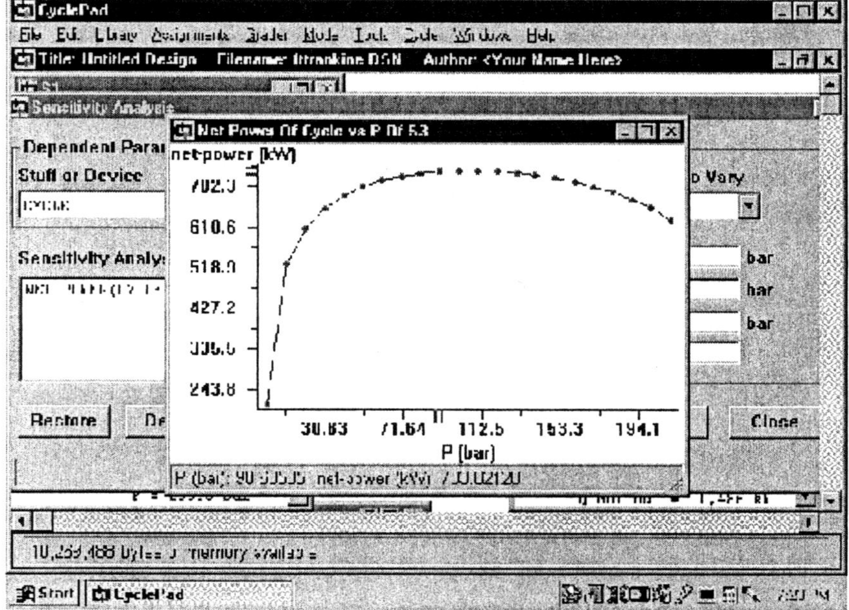

Figure Example 11.6.4.1b. Finite time actual Rankine cycle with finite heat reservoirs sensitivity diagram

Comment: The partial optimization is only for ∂(net power)/∂(p_3)=0. To have the full optimization, we must let ∂(net power)/∂(p_1)=0 also.

Example 11.6.4.2. A finite time actual Rankine cycle operates between a finite heat capacity heat source and a finite heat capacity heat sink. The following information is given:

Heat source: fluid=air, T_5=2000°C, p_5=1 bar, T_6=800°C, and p_6=1 bar
Heat sink: fluid=water, T_7=17°C, p_7=1 bar, T_8=30°C, and p_8=1 bar
Steam cycle: fluid=water, x_2=0, p_3=150 bar, T_8=400°C (superheated vapor), p_4=0.1 bar, and mdot=1 kg/s, $\eta_{turbine}$=85%.

Determine the rate of heat added from the heat source, rate of heat removed to the heat sink, power required by the isentropic pump, power produced by the isentropic turbine, net power produced and efficiency of the cycle.
Optimize the net power produced by the cycle with fix p_1. Draw the sensitivity diagram of net power vs p_3. Find the maximum net power and p_3 at the maximum net power condition.

To solve this problem by CyclePad, we take the following steps:

1. Build the cycle and its surroundings as shown in Figure 11.6.1.
2. Analysis: (A) Assuming the heat exchangers are isobaric, and turbine and pump are isentropic. (B) Input Heat source fluid=air, T_5=2000°C, p_5=1 bar, T_6=800°C, and p_6=1 bar; Heat sink: fluid=water, T_7=17°C, p_7=1 bar, T_8=30°C, and p_8=1 bar; and Steam cycle: fluid=water, x_2=0, p_3=150 bar, T_8=400°C (superheated vapor), p_4=0.1 bar, and mdot=1 kg/s, $\eta_{turbine}$=85%.
3. Display results: The results are: rate of heat added from the heat source=2768 kW, rate of heat removed to the heat sink=-1836 kW, power required by the pump=-15.14 kW, power produced by the turbine=947.3 kW, net power produced=932.1 kW and efficiency of the cycle=33.68%.
4. Optimization
   Draw the sensitivity diagram of net power vs p_3 as shown in Figure Example 11.6.4.1b. The maximum net power is about 947.0 kW, and p_3 at the maximum net power condition is about 89.32 bar.

Comment: The partial optimization is only for ∂(net power)/∂(p_3)=0. To have the full optimization, we must let ∂(net power)/∂(p_1)=0 also.

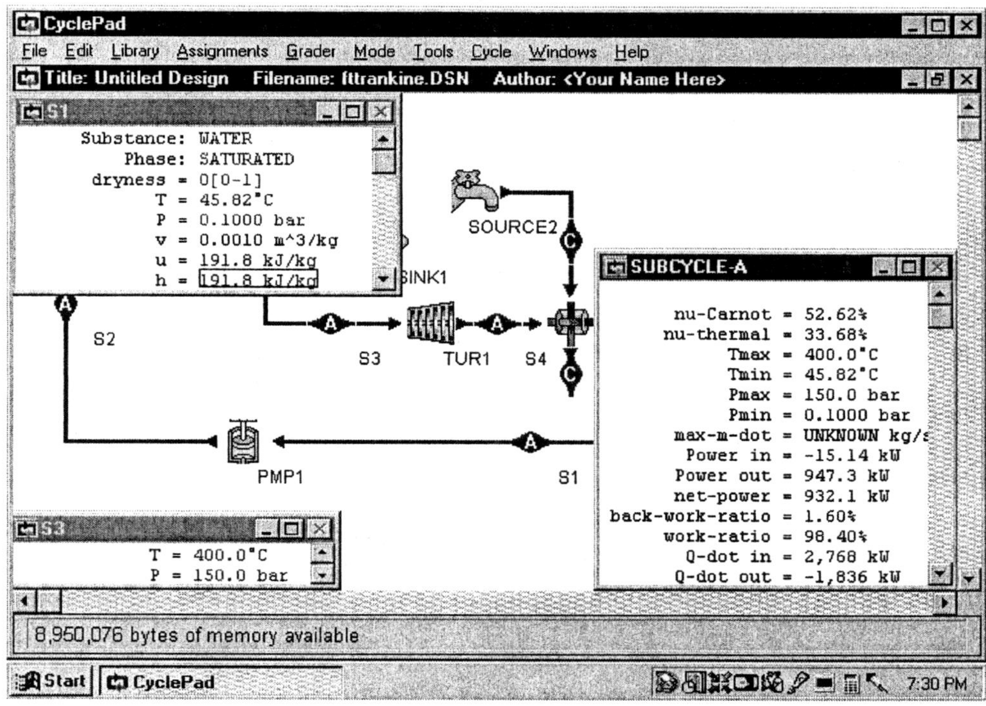

Figure Example 11.6.4.2a. Finite time actual Rankine cycle with finite capacity heat reservoirs

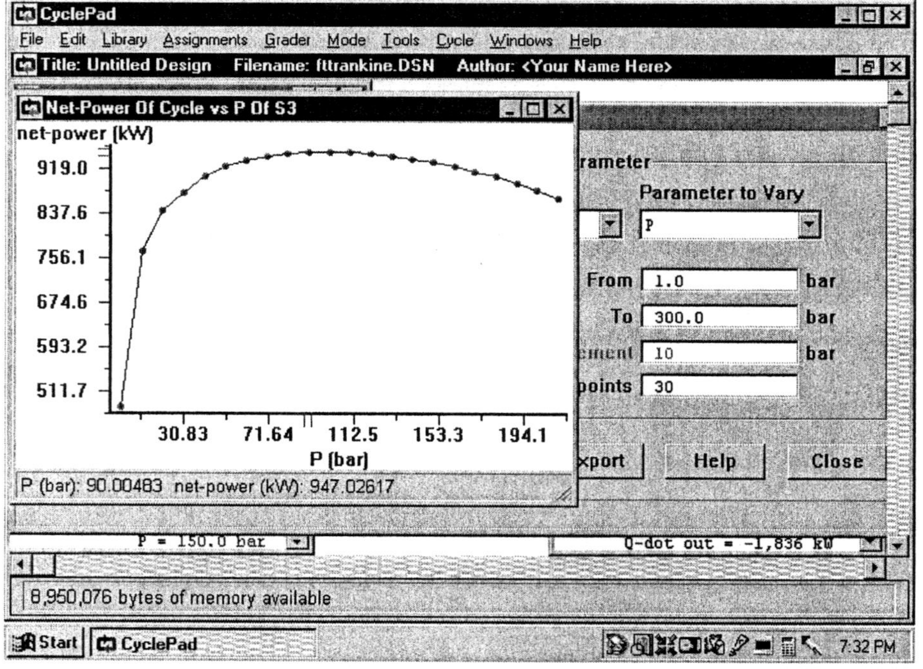

Figure Example 11.6.4.2b. Finite time actual Rankine cycle with finite heat reservoirssensitivity diagram

Homework 11.6.4. Finite Time Ideal Rankine Cycle with Finite Heat Capacity Reservoirs

1. A finite time actual Rankine cycle operates between a finite heat capacity heat source and a finite heat capacity heat sink. The following information is given:
   Heat source: fluid=air, T_5=2000°C, p_5=1 bar, T_6=800°C, and p_6=1 bar
   Heat sink: fluid=water, T_7=17°C, p_7=1 bar, T_8=30°C, and p_8=1 bar
   Steam cycle: fluid=water, x_2=0, p_3=120 bar, T_8=400°C (superheated vapor), p_4=0.1 bar, and mdot=1 kg/s, $\eta_{turbine}$=85%.
   Determine the rate of heat added from the heat source, rate of heat removed to the heat sink, power required by the isentropic pump, power produced by the turbine, net power produced and efficiency of the cycle.
   ANSWER: rate of heat added from the heat source=2847 kW, rate of heat removed to the heat sink=-1900 kW, power required by the pump=-12.13 kW, power produced by the turbine=959.2 kW, net power produced=947.1 kW and efficiency of the cycle=33.27%.
2. A finite time actual Rankine cycle operates between a finite heat capacity heat source and a finite heat capacity heat sink. The following information is given:
   Heat source: fluid=air, T_5=2000°C, p_5=1 bar, T_6=800°C, and p_6=1 bar
   Heat sink: fluid=water, T_7=17°C, p_7=1 bar, T_8=30°C, and p_8=1 bar
   Steam cycle: fluid=water, x_2=0, p_3=120 bar, T_8=400°C (superheated vapor), p_4=0.5 bar, and mdot=1 kg/s, $\eta_{turbine}$=85%.
   Determine the rate of heat added from the heat source, rate of heat removed to the heat sink, power required by the isentropic pump, power produced by the turbine, net power produced and efficiency of the cycle.
   ANSWER: rate of heat added from the heat source=2698 kW, rate of heat removed to the heat sink=-1908 kW, power required by the pump=-12.32 kW, power produced by the turbine=802.2 kW, net power produced=789.9 kW and efficiency of the cycle=29.28%.
3. A finite time actual Rankine cycle operates between a finite heat capacity heat source and a finite heat capacity heat sink. The following information is given:
   Heat source: fluid=air, T_5=2000°C, p_5=1 bar, T_6=800°C, and p_6=1 bar
   Heat sink: fluid=water, T_7=17°C, p_7=1 bar, T_8=30°C, and p_8=1 bar
   Steam cycle: fluid=water, x_2=0, p_3=120 bar, T_8=400°C (superheated vapor), p_4=0.5 bar, and mdot=1 kg/s, $\eta_{turbine}$=85%.
   Optimize the net power produced by the cycle with fixed p_1. Draw the sensitivity diagram of net power vs p_3. Find the maximum net power and p_3 at the maximum net power condition.
   ANSWER: The maximum net power is about 789.4 kW, and p_3 at the maximum net power condition is about 100.9 bar.
4. A finite time actual Rankine OTEC (Ocean thermal energy conversion) cycle as shown in Figure Example 11.6.3.3a operates between a finite heat capacity heat source and a finite heat capacity heat sink. The following information is given:
   Pump efficiency=85% and turbine efficiency=85%
   Heat source: fluid=warm ocean surface water, T_1=26°C, p_1=101 kPa, T_2=20°C, and p_2= 101 kPa
   Heat sink: fluid=cold deep ocean water, T_3=5°C, p_3=101 kPa, T_4=10°C, and p_4=101 kPa,

Steam cycle: fluid=ammonia, x_5=0, T_5=12°C, x_7=1, T_7=20°C, and mdot=1 kg/s.
The heat exchangers are counter-flow type, U_H=0.4 kJ/(m^2)K and U_L=0.4 kJ/(m^2)K.
Determine the rate of heat added from the heat source, rate of heat removed to the heat sink, power required by the isentropic pump, power produced by the isentropic turbine, net power produced and efficiency of the cycle.
Since the fuel cost of the OTEC is free, the primary cost is the initial construction cost. The heat exchangers are the major concern of the initial construction cost. Let us take the specific net power output (net output power per unit total heat exchanger surface area) as the design objective function, optimize the warm-side (heater or high-temperature-side heat exchanger) and cold-side (cooler or low-temperature-side heat exchanger) working fluid temperatures.

5. A finite time actual Rankine OTEC (Ocean thermal energy conversion) cycle as shown in Figure Example 11.6.3.3a operates between a finite heat capacity heat source and a finite heat capacity heat sink. The following information is given:
Pump efficiency=85% and turbine efficiency=85%
Heat source: fluid=warm ocean surface water, T_1=26°C, p_1=101 kPa, T_2=20°C, and p_2= 101 kPa
Heat sink: fluid=cold deep ocean water, T_3=5°C, p_3=101 kPa, T_4=10°C, and p_4=101 kPa,
Steam cycle: fluid=ammonia, x_5=0, T_5=12°C, x_7=1, T_7=20°C, and mdot=1 kg/s.
The heat exchangers are cocurrent (parallel)-flow type, U_H=0.4 kJ/(m^2)K and U_L=0.4 kJ/(m^2)K.
Determine the rate of heat added from the heat source, rate of heat removed to the heat sink, power required by the isentropic pump, power produced by the isentropic turbine, net power produced and efficiency of the cycle.
Since the fuel cost of the OTEC is free, the primary cost is the initial construction cost. The heat exchangers are the major concern of the initial construction cost. Let us take the specific net power output (net output power per unit total heat exchanger surface area) as the design objective function, optimize the warm-side (heater or high-temperature-side heat exchanger) and cold-side (cooler or low-temperature-side heat exchanger) working fluid temperatures.

11.7 FINITE TIME BRAYTON CYCLE

11.7.1 Ideal Brayton Cycle

The schematic and T-s diagrams of the ideal Finite time Brayton cycle are shown in Figure 11.7.1 and Figure 11.7.2. The cycle is an endoreversible cycle which is made of two isentropic processes and two isobaric heat transfer processes. The cycle exchanges heats with its surroundings in the two isobaric external irreversible heat transfer processes. By taking into account the rates of heat transfer associated with the cycle, the upper bound of the power output of the cycle can be found as illustrated in the following example.

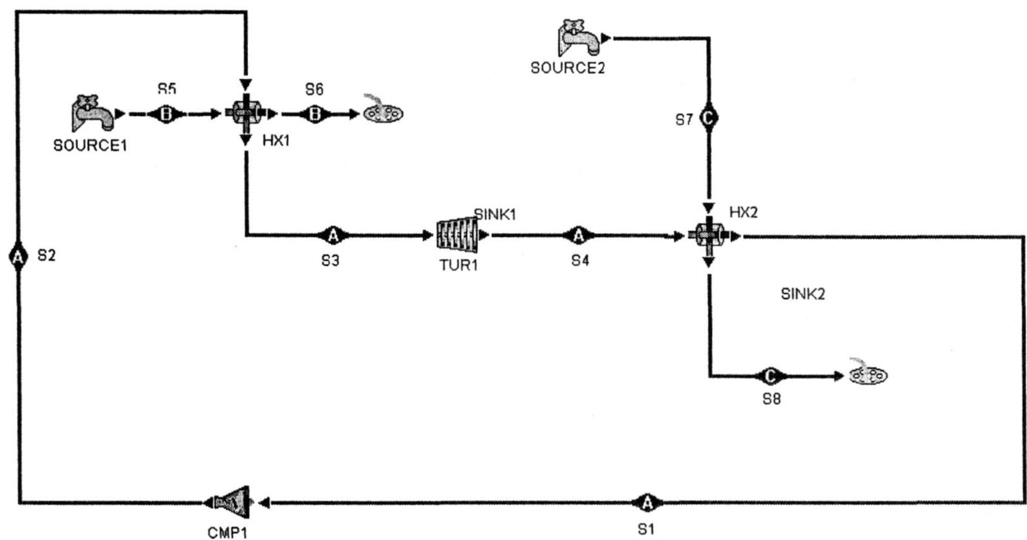

Figure 11.7.1 Schematic diagram of the ideal Finite time Brayton cycle

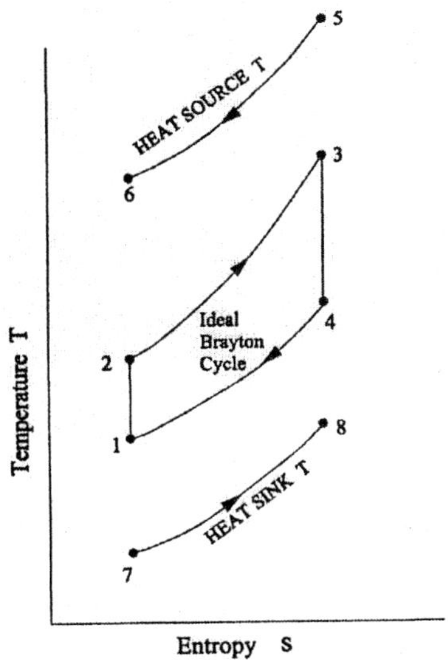

Figure 11.7.1 T-s diagrams of the ideal Finite time Brayton cycle

Example 11.7.1.1. A finite time ideal Brayton cycle operates between a heat source and a heat sink. The following information is given:

Heat source: fluid=air, T_5=2773 K, p_5=100 kPa, T_6=2473 K, and p_6=100 kPa
Heat sink: fluid=water, T_7=373.1 K, x_7=0, T_8=373.1 K, and x_8=1

Brayton cycle: fluid=helium, T_1=423 K, p_1=100 kPa, T_3=1500 K, p_3=800 kPa, and mdot=1 kg/s.

Determine the rate of heat added from the heat source, rate of heat removed to the heat sink, power required by the isentropic compressor, power produced by the isentropic turbine, net power produced and efficiency of the cycle.
Optimize the net power produced by the cycle with fixed p_1. Draw the sensitivity diagram of net power vs p_3. Find the maximum net power and p_3 at the maximum net power condition.
Optimize the net power produced by the cycle with fixed p_3. Draw the sensitivity diagram of net power vs p_1. Find the maximum net power and p_1 at the maximum net power condition.

To solve this problem by CyclePad, we take the following steps:

1. Build the cycle and its surroundings as shown in Figure 11.7.1.
2. Analysis: (A) Assuming the heat exchangers are isobaric, and turbine and compressor are isentropic. (B) Input Heat source: fluid=air, T_5=2773 K, p_5=100 kPa, T_6=2473 K, and p_6=100 kPa; Heat sink: fluid=water, T_7=373.1 K, x_7=0, T_8=373.1 K, and x_8=1; Brayton cycle: fluid=helium, T_1=423 K, p_1=100 kPa, T_3=1500 K, p_3=800 kPa, and mdot=1 kg/s.
3. Display results: The results are: rate of heat added from the heat source=2722 kW, rate of heat removed to the heat sink=-1182 kW, power required by the compressor=-2854 kW, power produced by the turbine=4394 kW, net power produced=1540 kW and efficiency of the cycle=56.58%.
4. Optimization
   Draw the sensitivity diagram of net power vs p_3 as shown in Figure Example 11.7.1.1b. The maximum net power is about 1697 kW, and p_3 at the maximum net power condition is about 491.6 kPa.

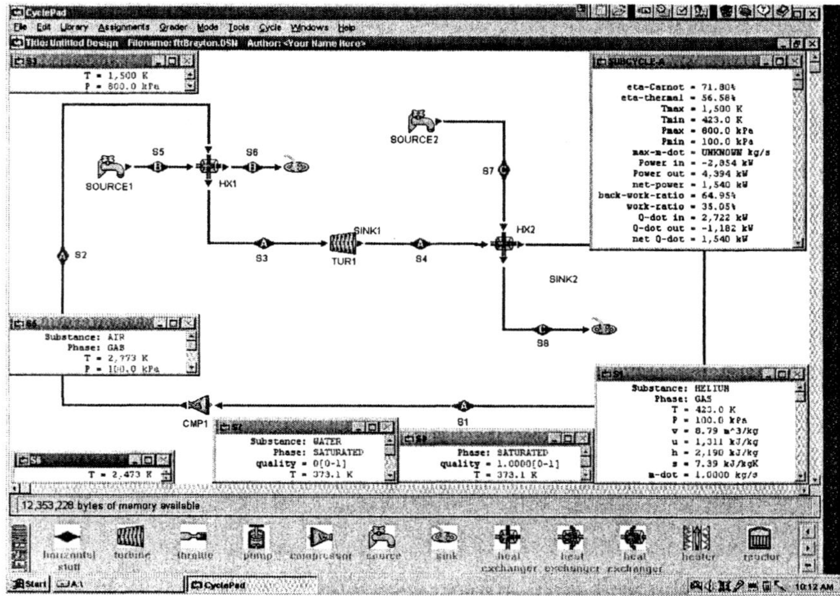

Figure Example 11.7.1.1a. Finite time ideal Brayton cycle

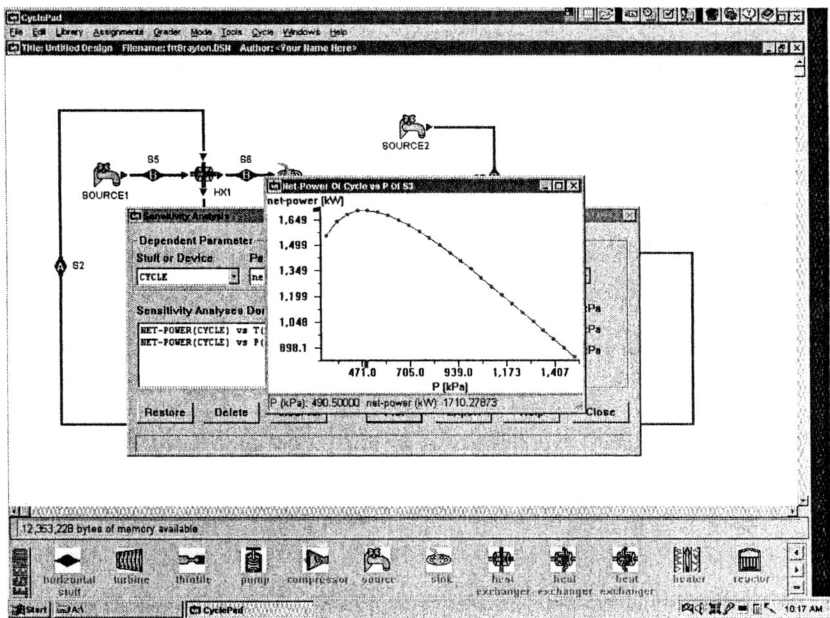

Figure Example 11.7.1.1b. Finite time ideal Brayton cycle Sensitivity diagram

Comment: The partial optimization is only for ∂(net power)/∂(p_3)=0. To have the full optimization, we must let ∂(net power)/∂(p_1)=0 also.

5. Optimization

   Draw the sensitivity diagram of net power vs p_1 as shown in Figure Example 11.7.1.1c. The maximum net power is about 1699 kW, and p_3 at the maximum net power condition is about 147.4 kPa.

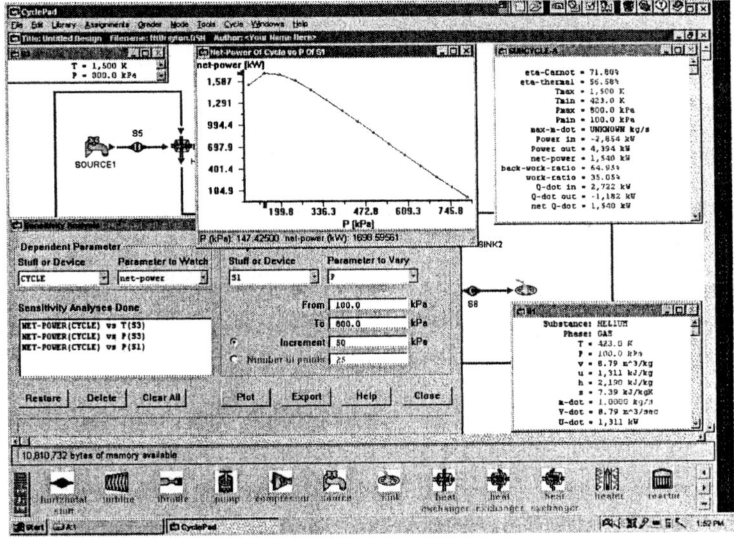

Figure Example 11.7.1.1c. Finite time ideal Brayton cycle Sensitivity diagram

Comment: The partial optimization is only for ∂(net power)/∂(p_3)=0. To have the full optimization, we must let ∂(net power)/∂(p_1)=0 also.

Example 11.7.1.2. A finite time ideal Brayton cycle operates between a heat source and a heat sink. The following information is given:

Heat source: fluid=air, T_4=2500°C, p_4=1 bar, T_5=1500°C, and p_5=1bar;
Heat sink: fluid=air, T_7=15°C, p_7=1 bar, T_8=90°C, and p_8=1 bar;
Brayton cycle: fluid=air, T_1=100°C, p_1=1 bar, T_3=1200°C, p_3=10 bar, and mdot=1 kg/s.
The heat exchangers are counter-flow type with U_H=1 kW/(m^{2}K) and U_L=1 kW/(m^{2}K).

Determine the rate of heat added from the heat source, rate of heat removed to the heat sink, power required by the isentropic compressor, power produced by the isentropic turbine, net power produced and efficiency of the cycle.
(A) Optimize the cycle based on cycle efficiency with respect to p_3, (B) Optimize the cycle based on net power with respect to p_3, (C) Optimize the cycle based on net power per unit conductance of heat exchangers with respect to p_3, and (D) Optimize the cycle based on net power per unit surface of heat exchangers with respect to p_3.

To solve this problem by CyclePad, we take the following steps:

1. Build the cycle and its surroundings as shown in Figure 11.7.1.
2. Analysis: (A) Assume the heat exchangers are isobaric and counter-flow type, and turbine and compressor are isentropic. (B) Input Heat source: fluid=air, T_4=2500°C, p_4=1 bar, T_5=1500°C, and p_5=1bar; Heat sink: fluid=air, T_7=15°C, p_7=1 bar, T_8=90°C, and p_8=1 bar; and Brayton cycle: fluid=air, T_1=100°C, p_1=1 bar, T_3=1200°C, p_3=10 bar, and mdot=1 kg/s as shown in Figure Example 11.7.1.2a.
3. Display results: The results are: rate of heat added from the heat source=755.3 kW, rate of heat removed to the heat sink=-391.2 kW, net power produced=364.1 kW, $LMTD_H$=1172 K, $LMTD_L$=203.3 K, and efficiency of the cycle=48.21% as shown in Figure Example 11.7.1.2b.

Change p_3=2, 3, 4, 5, 6, 7, 8, 9, 11, 12, 14 and 15 bar. The following table is made as shown in Figure Example 11.7.1.2c and we have: (A) Optimize the cycle based on cycle efficiency with respect to p_3, η_{max}=52.95% at p_3=14 bar; (B) Optimize the cycle based on net power with respect to p_3, net power$_{max}$=364.7 kW at p_3=11 bar; (C) Optimize the cycle based on net power per unit conductance of heat exchangers with respect to p_3, net power per unit conductance of heat exchangers$_{max}$=151.8 kW/(kW/K) at p_3=14 bar; and (D) Optimize the cycle based on net power per unit surface of heat exchangers with respect to p_3, net power per unit surface of heat exchangers$_{max}$= kW/m^2.

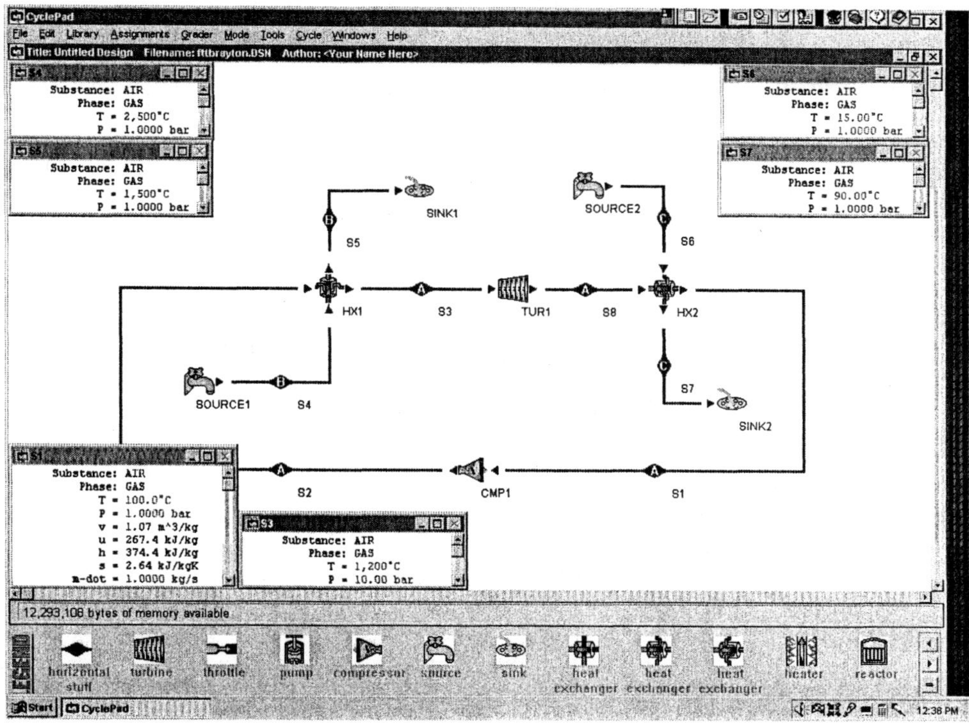

Figure Example 11.7.1.2a Finite time Brayton cycle input

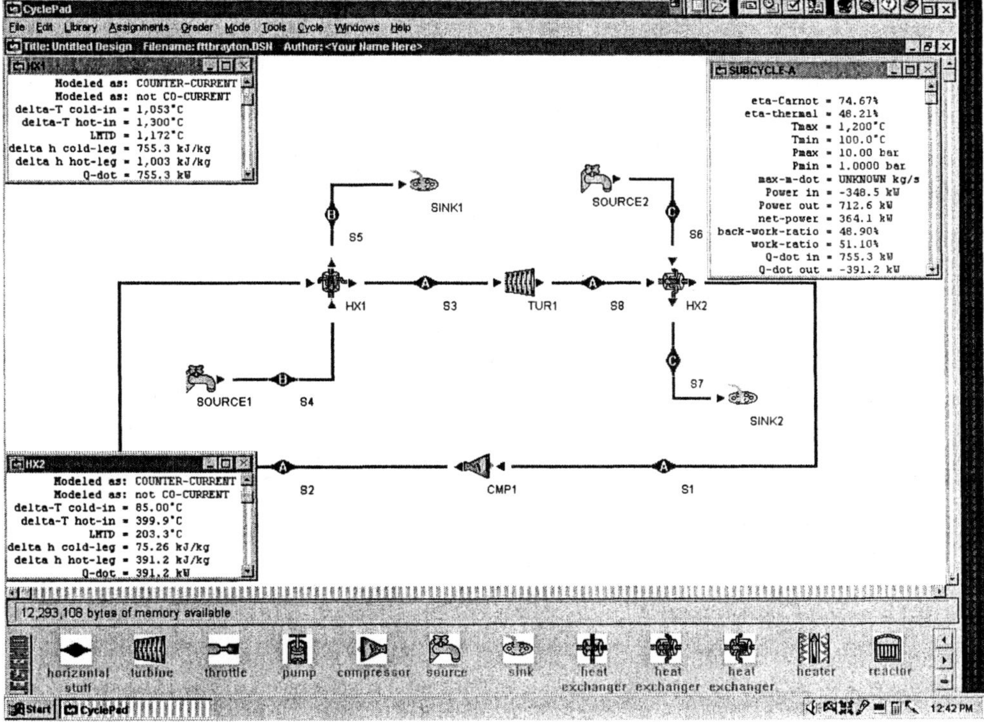

Figure Example 11.7.1.2b Finite time Brayton cycle output

p3 bar	CY EFF %	QDOTH kW	QDOTL kW	LMTDH K	LMTDL K	UHAH kW/K	ULAL kW/K	SUM(UA) kW/K	PNET kW	PNET/UA kW/kW/K	AH m^2	AL m^2	PNET/A kW/m^2
2	17.97	1022	838.2	1309	331	0.780749	2.532326	3.313075	183.6	55.4168	0.013793	5.064653	36.1528
3	26.94	965.7	705.5	1281	295.3	0.753864	2.389096	3.14296	260.2	82.7882	0.009182	4.778192	54.35131
4	32.7	921.8	620.3	1259	271.6	0.732168	2.283873	3.016042	301.5	99.96546	0.007575	4.567747	65.89701
5	36.86	885.2	558.9	1240	254	0.713871	2.200394	2.914265	326.3	111.9665	0.006743	4.400787	74.03238
6	40.07	853.4	511.5	1224	240.1	0.697222	2.130362	2.827585	341.9	120.9159	0.006231	4.260725	80.12739
7	42.65	825.3	473.3	1209	228.7	0.68263	2.069523	2.752154	352	127.8998	0.005879	4.139047	84.92311
8	44.8	799.9	441.6	1196	219	0.668813	2.016438	2.685251	358.3	133.4326	0.005628	4.032877	88.72096
9	46.62	776.7	414.6	1183	210.7	0.656551	1.967727	2.624278	362.1	137.9808	0.005434	3.935453	91.88286
10	48.21	755.3	391.2	1172	203.3	0.644454	1.92425	2.568704	364.1	141.7446	0.005285	3.8485	94.47855
11	49.6	735.5	370.6	1161	196.8	0.633506	1.88313	2.516636	**364.7**	144.9157	0.005164	3.76626	96.70088
12	50.83	716.6	352.3	1151	190.9	0.622589	1.845469	2.468058	364.3	147.6059	0.005066	3.690938	98.56593
14	**52.95**	682.3	321	1132	180.6	0.602739	1.777409	2.380147	361.3	**151.7973**	0.004919	3.554817	**101.4963**
15	51.95	699	335.9	1142	185.5	0.612084	1.810782	2.422866	363.1	149.8639	0.004987	3.621563	100.1227

Figure Example 11.7.1.2b Finite time Brayton cycle optimization

Homework 11.7.1 Finite Time Ideal Brayton Cycle

1. A finite time ideal Brayton cycle operates between a heat source and a heat sink. The following information is given:
   Heat source: fluid=air, T_5=2773 K, p_5=100 kPa, T_6=2473 K, and p_6=100 kPa
   Heat sink: fluid=water, T_7=373.1 K, x_7=0, T_8=373.1 K, and x_8=1
   Brayton cycle: fluid=helium, T_1=423 K, p_1=100 kPa, T_3=1800 K, p_3=800 kPa, and mdot=1 kg/s.
   Determine the rate of heat added from the heat source, rate of heat removed to the heat sink, power required by the isentropic compressor, power produced by the isentropic turbine, net power produced and efficiency of the cycle.
   ANSWER: rate of heat added from the heat source=4275 kW, rate of heat removed to the heat sink=-1856 kW, power required by the compressor=-2854 kW, power produced by the turbine=5272 kW, net power produced=2419 kW and efficiency of the cycle=56.58%.
2. A finite time ideal Brayton cycle operates between a heat source and a heat sink. The following information is given:
   Heat source: fluid=air, T_5=2773 K, p_5=100 kPa, T_6=2473 K, and p_6=100 kPa
   Heat sink: fluid=water, T_7=373.1 K, x_7=0, T_8=373.1 K, and x_8=1
   Brayton cycle: fluid=helium, T_1=423 K, p_1=100 kPa, T_3=1800 K, p_3=800 kPa, and mdot=1 kg/s.
   Optimize the net power produced by the cycle with fix p_3. Draw the sensitivity diagram of net power vs p_1. Find the maximum net power and p_1 at the maximum net power condition.
   ANSWER: The maximum net power is about 2457 kW, and p_1 at the maximum net power condition is about 145.2 kPa.
3. A finite time ideal Brayton cycle operates between a heat source and a heat sink. The following information is given:
   Heat source: fluid=air, T_5=2773 K, p_5=100 kPa, T_6=2473 K, and p_6=100 kPa
   Heat sink: fluid=water, T_7=373.1 K, x_7=0, T_8=373.1 K, and x_8=1
   Brayton cycle: fluid=helium, T_1=423 K, p_1=100 kPa, T_3=1800 K, p_3=800 kPa, and mdot=1 kg/s.

Determine the rate of heat added from the heat source, rate of heat removed to the heat sink, power required by the isentropic compressor, power produced by the isentropic turbine, net power produced and efficiency of the cycle.

ANSWER: rate of heat added from the heat source=3803 kW, rate of heat removed to the heat sink=-1510 kW, power required by the compressor=-3326 kW, power produced by the turbine=5619 kW, net power produced=2293 kW and efficiency of the cycle=60.30%.

4. A finite time ideal Brayton cycle operates between a heat source and a heat sink. The following information is given:
   Heat source: fluid=air, T_5=2773 K, p_5=100 kPa, T_6=2473 K, and p_6=100 kPa
   Heat sink: fluid=water, T_7=373.1 K, x_7=0, T_8=373.1 K, and x_8=1
   Brayton cycle: fluid=helium, T_1=423 K, p_1=100 kPa, T_3=1800 K, p_3=1000 kPa, and mdot=1 kg/s.
   Optimize the net power produced by the cycle with fix p_1. Draw the sensitivity diagram of net power vs p_3. Find the maximum net power and p_3 at the maximum net power condition.
   ANSWER: The maximum net power is about 2460 kW, and p_3 at the maximum net power condition is about 608.6 kPa.
5. A finite time ideal Brayton cycle operates between a heat source and a heat sink. The following information is given:
   Heat source: fluid=air, T_5=2773 K, p_5=100 kPa, T_6=2473 K, and p_6=100 kPa
   Heat sink: fluid=water, T_7=373.1 K, x_7=0, T_8=373.1 K, and x_8=1
   Brayton cycle: fluid=air, T_1=423 K, p_1=100 kPa, T_3=1500 K, p_3=800 kPa, and mdot=1 kg/s.
   Determine the rate of heat added from the heat source, rate of heat removed to the heat sink, power required by the isentropic compressor, power produced by the isentropic turbine, net power produced and efficiency of the cycle.
   Optimize the net power produced by the cycle with fix p_1. Draw the sensitivity diagram of net power vs p_3. Find the maximum net power and p_3 at the maximum net power condition.
   ANSWER: rate of heat added from the heat source=736.3 kW, rate of heat removed to the heat sink=-406.4 kW, power required by the compressor=-344.0 kW, power produced by the turbine=674.2 kW, net power produced=329.8 kW and efficiency of the cycle=44.80%.
   The maximum net power is about 330.7 kW, and p_3 at the maximum net power condition is about 836.8 kPa.

11.7.2 Actual Brayton Cycle

The actual finite time Brayton cycle as shown in Figure 11.7.2.1 is made of two adiabatic processes and two isobaric heat transfer processes. The cycle exchanges heats with its surroundings in the two isobaric external irreversible heat transfer processes. By taking into account the rates of heat transfer associated with the cycle, the upper bound of the power output of the cycle can be found as illustrated in the following example.

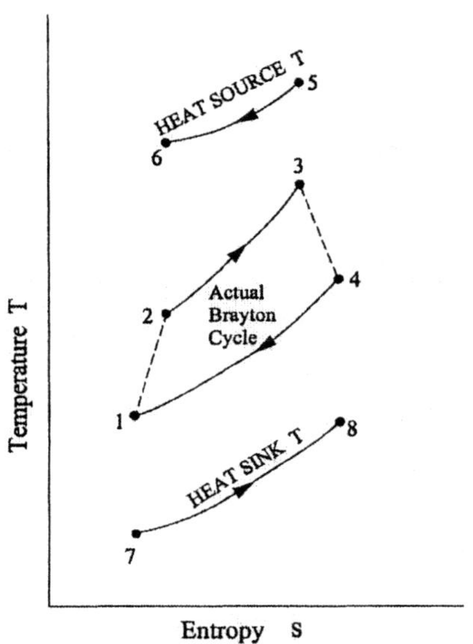

Figure 11.7.2.1 Actual finite time Brayton cycle

Example 11.7.2.1. A finite time actual Brayton cycle operates between a heat source and a heat sink. The following information is given:

Heat source: fluid=air, T_5=2773 K, p_5=100 kPa, T_6=2473 K, and p_6=100 kPa
Heat sink: fluid=water, T_7=373.1 K, x_7=0, T_8=373.1 K, and x_8=1
Brayton cycle: fluid=air, $\eta_{compressor}$=85%, $\eta_{turbine}$=85%, T_1=423 K, p_1=100 kPa, T_3=1500 K, p_3=800 kPa, and mdot=1 kg/s.

Determine the rate of heat added from the heat source, rate of heat removed to the heat sink, power required by the isentropic compressor, power produced by the isentropic turbine, net power produced and efficiency of the cycle.
Optimize the net power produced by the cycle with fixed p_1. Draw the sensitivity diagram of net power vs p_3. Find the maximum net power and p_3 at the maximum net power condition.
Optimize the net power produced by the cycle with fixed p_3. Draw the sensitivity diagram of net power vs p_1. Find the maximum net power and p_3 at the maximum net power condition.

To solve this problem by CyclePad, we take the following steps:

1. Build the cycle and its surroundings as shown in Figure 11.7.1.
2. Analysis: (A) Assuming the heat exchangers are isobaric, and turbine and compressor are isentropic. (B) Input Heat source: fluid=air, T_5=2773 K, p_5=100 kPa, T_6=2473 K, and p_6=100 kPa; Heat sink: fluid=water, T_7=373.1 K, x_7=0, T_8=373.1 K, and x_8=1; Brayton cycle: fluid=air, $\eta_{compressor}$=85%, $\eta_{turbine}$=85%, T_1=423 K, p_1=100 kPa, T_3=1500 K, p_3=800 kPa, and mdot=1 kg/s.

3. Display results: The results are: rate of heat added from the heat source=675.5 kW, rate of heat removed to the heat sink=-507.6 kW, power required by the compressor=-405.2 kW, power produced by the turbine=573.1 kW, net power produced=167.9 kW and efficiency of the cycle=24.86%.

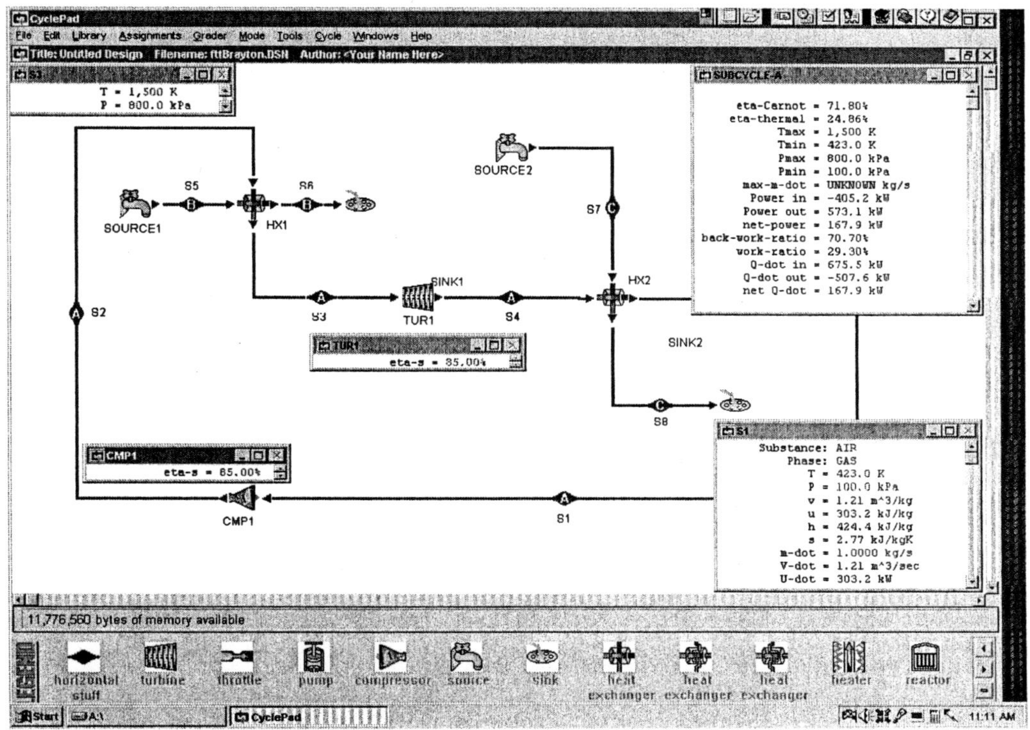

Figure Example 11.7.2.1a. Finite time actual Brayton cycle

4. Optimization
   Draw the sensitivity diagram of net power vs p_3 as shown in Figure Example 11.7.2.1b. The maximum net power is about 179.8 kW, and p_3 at the maximum net power condition is about 544.3 kPa.

Comment: The partial optimization is only for ∂(net power)/$\partial(p_3)$=0. To have the full optimization, we must let ∂(net power)/$\partial(p_1)$=0 also.

5. Optimization
   Draw the sensitivity diagram of net power vs p_3 as shown in Figure Example 11.7.2.1c. The maximum net power is about 179.8 kW, and p_1 at the maximum net power condition is about 147.4 kPa.

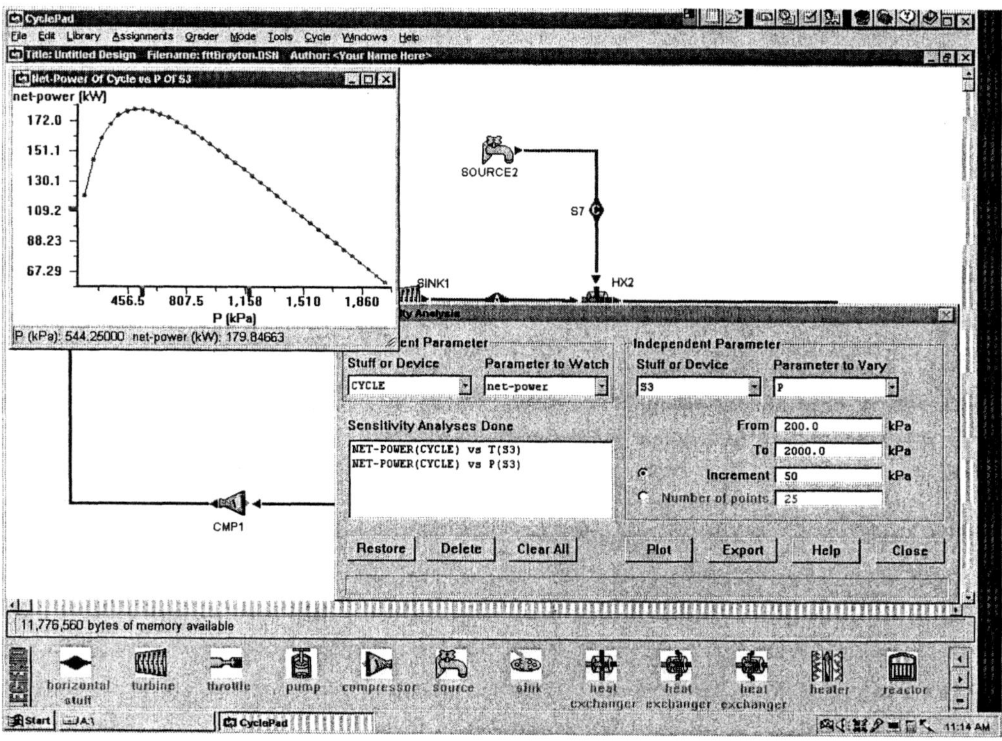

Figure Example 11.7.2.1b. Finite time actual Brayton cycle Sensitivity diagram

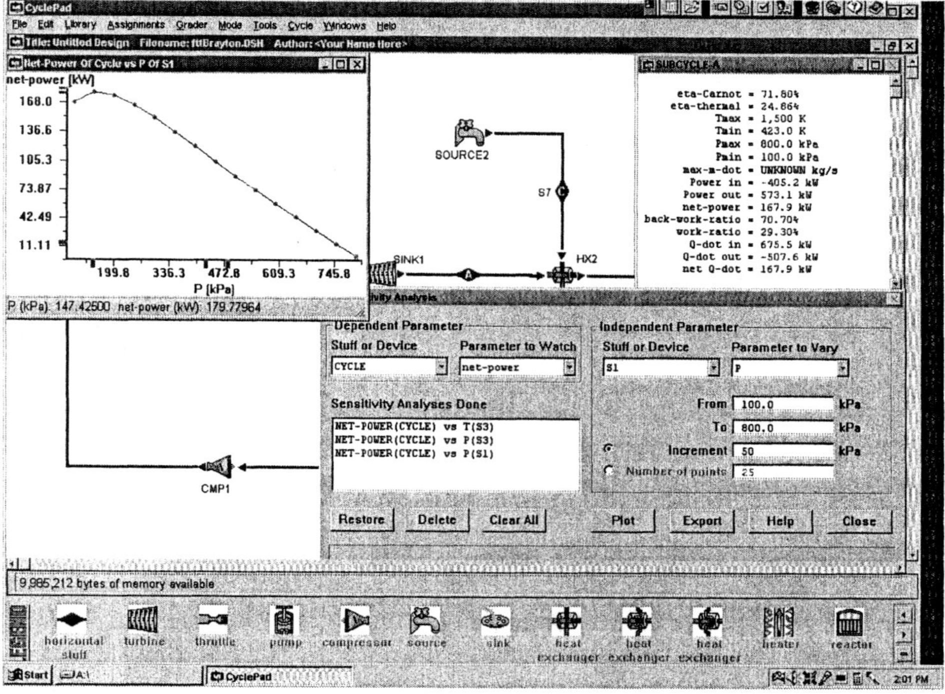

Figure Example 11.7.2.1c. Finite time actual Brayton cycle Sensitivity diagram

Comment: The partial optimization is only for ∂(net power)/∂(p_{31})=0. To have the full optimization, we must let ∂(net power)/∂(p_3)=0 also.

Homework 11.7.2 Finite Time Actual Brayton Cycle

1. A finite time actual Brayton cycle operates between a heat source and a heat sink. The following information is given:
   Heat source: fluid=air, T_5=2773 K, p_5=100 kPa, T_6=2473 K, and p_6=100 kPa
   Heat sink: fluid=water, T_7=373.1 K, x_7=0, T_8=373.1 K, and x_8=1
   Brayton cycle: fluid=air, $\eta_{compressor}$=80%, $\eta_{turbine}$=80%, T_1=423 K, p_1=100 kPa, T_3=1500 K, p_3=800 kPa, and mdot=1 kg/s.
   Determine the rate of heat added from the heat source, rate of heat removed to the heat sink, power required by the isentropic compressor, power produced by the isentropic turbine, net power produced and efficiency of the cycle.
   ANSWER: rate of heat added from the heat source=650.2 kW, rate of heat removed to the heat sink=-541.3 kW, power required by the compressor=-430.5 kW, power produced by the turbine=539.4 kW, net power produced=108.9 kW and efficiency of the cycle=16.74%.
2. A finite time actual Brayton cycle operates between a heat source and a heat sink. The following information is given:
   Heat source: fluid=air, T_5=2773 K, p_5=100 kPa, T_6=2473 K, and p_6=100 kPa
   Heat sink: fluid=water, T_7=373.1 K, x_7=0, T_8=373.1 K, and x_8=1
   Brayton cycle: fluid=air, $\eta_{compressor}$=80%, $\eta_{turbine}$=80%, T_1=423 K, p_1=100 kPa, T_3=1500 K, p_3=800 kPa, and mdot=1 kg/s.
   Optimize the net power produced by the cycle with fix p_1. Draw the sensitivity diagram of net power vs p_3. Find the maximum net power and p_3 at the maximum net power condition.
   ANSWER: The maximum net power is about 135.0 kW, and p_3 at the maximum net power condition is about 200.0 kPa.
3. A finite time actual Brayton cycle operates between a heat source and a heat sink. The following information is given:
   Heat source: fluid=air, T_5=2773 K, p_5=100 kPa, T_6=2473 K, and p_6=100 kPa
   Heat sink: fluid=water, T_7=373.1 K, x_7=0, T_8=373.1 K, and x_8=1
   Brayton cycle: fluid=air, $\eta_{compressor}$=90%, $\eta_{turbine}$=90%, T_1=423 K, p_1=100 kPa, T_3=1500 K, p_3=800 kPa, and mdot=1 kg/s.
   Determine the rate of heat added from the heat source, rate of heat removed to the heat sink, power required by the isentropic compressor, power produced by the isentropic turbine, net power produced and efficiency of the cycle.
   Optimize the net power produced by the cycle with fixed p_1. Draw the sensitivity diagram of net power vs p_3. Find the maximum net power and p_3 at the maximum net power condition.
   ANSWER: rate of heat added from the heat source=698.0 kW, rate of heat removed to the heat sink=-473.9 kW, power required by the compressor=-382.7 kW, power produced by the turbine=606.8 kW, net power produced=224.1 kW and efficiency of the cycle=32.11%.

The maximum net power is about 227.0 kW, and p_3 at the maximum net power condition is about 608.6 kPa.

4. A finite time actual Brayton cycle operates between a heat source and a heat sink. The following information is given:
   Heat source: fluid=air, T_5=2773 K, p_5=100 kPa, T_6=2473 K, and p_6=100 kPa
   Heat sink: fluid=water, T_7=373.1 K, x_7=0, T_8=373.1 K, and x_8=1
   Brayton cycle: fluid=helium, $\eta_{compressor}$=90%, $\eta_{turbine}$=90%, T_1=423 K, p_1=100 kPa, T_3=1500 K, p_3=800 kPa, and mdot=1 kg/s.
   Determine the rate of heat added from the heat source, rate of heat removed to the heat sink, power required by the isentropic compressor, power produced by the isentropic turbine, net power produced and efficiency of the cycle.
   Optimize the net power produced by the cycle with fixed p_1. Draw the sensitivity diagram of net power vs p_3. Find the maximum net power and p_3 at the maximum net power condition.
   ANSWER: rate of heat added from the heat source=2405 kW, rate of heat removed to the heat sink=-1621 kW, power required by the compressor=-3171 kW, power produced by the turbine=3954 kW, net power produced=783.6 kW and efficiency of the cycle=32.59%.
   The maximum net power is about 1158 kW, and p_3 at the maximum net power condition is about 357.1 kPa.
5. A finite time actual Brayton cycle operates between a heat source and a heat sink. The following information is given:
   Heat source: fluid=air, T_5=2773 K, p_5=100 kPa, T_6=2473 K, and p_6=100 kPa
   Heat sink: fluid=water, T_7=373.1 K, x_7=0, T_8=373.1 K, and x_8=1
   Brayton cycle: fluid=helium, $\eta_{compressor}$=90%, $\eta_{turbine}$=90%, T_1=423 K, p_1=100 kPa, T_3=1500 K, p_3=800 kPa, and mdot=1 kg/s.
   Optimize the net power produced by the cycle with fixed p_3. Draw the sensitivity diagram of net power vs p_1. Find the maximum net power and p_3 at the maximum net power condition.
   ANSWER: The maximum net power is about 1163 kW, and p_1 at the maximum net power condition is about 197.5 kPa.

11.8 Other Finite Time Cycles

Finite time thermodynamics is one of the newest and most challenging areas in thermodynamics. A book entitled *Recent Advances in Finite Time Thermodynamics* (Editors: Chih Wu, Lingen Chen and Jincan Chen, Nova Science Publishers, Inc., New York, USA, 1999, ISBN 1-56072-644-4)) provides results from research, which continues at an impressive rate. The book contains many academic and industrial papers that are relevant to current problems and practice. The numerous contributions from the international thermodynamic community are indicative of the continuing global interest in finite time thermodynamics.

The readers should find the following papers informative and useful for analysis and design of various finite time thermodynamic cycles. It is hoped that these papers will provide interest and encouragement for further study in the area of finite time thermodynamics.

Carnot, Brayton and Rankine finite cycles are the basic thermodynamic cycles.

Carnot Cycle

Wu, C., Power optimization of a finite-time Carnot heat engine, *Energy:The International Journal*, vol.13, no.9, pp681-687, 1988.

Wu, C. and R.L. Kiang, Finite time thermodynamic Analysis of a Carnot engine with internal irreversibility, *Energy:The International Journal*, vol.17, no.12, pp1173-1178, 1992.

Wu, C., Specific heating load of an endoreversible Carnot heat pump, *International Journal of Ambient Energy*, vol.14, no.1, pp25-28, 1993.

Wu, C., Maximum obtainable specific cooling load of a Carnot refrigerator, *Energy Conversion and Management*, vol.36, no.1, pp7-10, 1995.

Wu, C., General performance characteristics of a finite-speed Carnot refrigerator, *Applied Thermal Engineering*, vol.16, no.4, pp299-304, 1996.

Chen, Lingen and C.Wu, Performance of an endoreversible Carnot refrigerator, *Energy Conversion and Management,* vol.37, no.10, pp1509-1512, 1996.

Chen, Lingen and C.Wu, The influence of heat transfer law on the endoreversible Carnot refrigerator, *Journal of the Institute of Energy,* vol.69, no.480, pp96-100, 1996.

Chen, Lingen and C.Wu, Optimal performance of an endoreversible Carnot heat pump, *Energy Conversion and Management*, vol.38, no.14, pp1439-1444, 1997.

Chen, Lingen and C.Wu, Effect of heat transfer law on the performance of a generalized Carnot heat pump, *Journal of the Institute of Energy*, vol.72, no.2, pp64-68, 1999.

Chen, Lingen and C.Wu, Effect of heat transfer law on the performance of generalized irreversible Carnot refrigerator, *J. of Engineering Thermophysics,* vol.20, no.2, pp10-13, 1999.

Brayton Cycle

Wu, C., Work and power optimization of a finite-time Brayton cycle, *International Journal of Ambient Energy*, pp129-136, 1990.

Wu, C., Power optimization of an endoreversible Brayton gas heat engine, *Energy Conversion and Management*, vol.31, no.6, pp561-565, 1991.

Wu, C. and R.L. Kiang, Power performance of a nonisentropic Brayton cycle, *ASME Journal of Engineering for Gas Turbines and Power*, vol.113, pp501-504, 1991.

Wu, C., Performance of a regenerative Brayton heat engine, *Energy: The International Journal*, vol. 21, no.2, pp71-76, 1996.

Chen, Lingen, F. Sun and C. Wu, Performance analysis for a real closed regenerated Brayton cycle via methods of finite-time thermodynamics, *International Journal of Ambient Energy,* vol.20, no.2, pp95-104, 1999.

Chen, Lingen, F. Sun and C. Wu, Performance analysis of a closed regenerated Brayton heat pump with internal irreversibilities, *International Journal of Energy Research*, vol.23, pp1039-1050, 1999

Rankine cycle

Wu, C., Power optimization of a finite-time Rankine heat engine, *International Journal of Heat and Fluid Flow*, vol.10, no.2, pp134-138, 1989.

Wu, C., Intelligent computer aided design on optimization of specific power of finite-time Rankine cycle using CyclePad, *Journal of Computer Application in Engineering Education*, vol.6, no.1, pp9-13, 1998.

Other finite time thermodynamic cycle literature including Atkinson, Combined and Cascaded, Diesel, Dual, Ericsson, Otto, Rallis, and Stirling cycles are provided in the following:

Atkinson Cycle

Chen, Lingen, F. Sun and C. Wu, Efficiency of an Atkinson engine at maximum power density, *Energy Conversion and Management*, vol.39, no.3/4, pp337-342, 1998.

Combined and Cascaded Cycle

Wu, C., G. Karpouzian and R.L. Kiang, The optimal power performance of an endoreversible Combined cycle, *Journal of the Institute of Energy*, vol., pp41-45, 1992.

Wu, C., Power performance of a cascade Endoreversible cycle, *Energy Conversion and Management*, vol., no., pp, 1990.

Wu, C., Maximum obtainable power of a Carnot combined power plant, *Heat Recovery Systems and CHP*, vol.15, no.4, pp351-355, 1995.

Chen, Jincan and C. Wu, Maximum specific power output of a two-stage endoreversible Combined cycle, *Energy:The International Journal*, vol.20, no.4, pp305-309, 1995.

Wu, C., Performance of a cascade endoreversible heat pump system, *The Institute of Energy Journal*, vol.68, no.476, 137-141, 1995.

Wu, C., Finite-time thermodynamic analysis of a two-stage combined heat pump system, *International Journal of Ambient Energy*, vol.**16**, no.4, pp205-208, 1995.

Chen, Jincan and C. Wu, General performance characteristics of a N-stage endoreversible combined power cycle system at maximum specific power output, *Energy Conversion and Management*, vol.37,no.9, pp1401-1406, 1996.

Chen, Lingen, F. Sun and C.Wu, The equivalent cycles of an n-stage irreversible combined refrigeration system, *International Journal of Ambient Energy*, vol.18, no.4, pp.197-204, 1997.

Wu, C., Intelligent computer aided analysis of a Rankine/Rankine combined cycle, *International Journal of Energy, Environment and Economics*, vol.7, no.2, pp239-244, 1998.

Chen, Lingen, F. Sun and C.Wu, A generalized model of a combined refrigeration cycle and its performance, *International Journal of Thermal Sciences (Revue Generale de Thermique)*, vol.38, no.8, pp712-718, 1999.

Chen, Jincan and C.Wu, Thermoeconomic analysis on the performance characteristics of a multi-stage irreversible combined heat pump system, *ASME Journal of Energy Resources Technology*, vol.122, no.4, pp212-216, 2000.

Diesel Cycle

Wu, C. and D.A. Blank, The effect of combustion on a power optimised endoreversible Diesel cycle, *Energy Conversion and Management*, vol.34, no.6, pp493-498, 1993.

Chen, Lingen and C.Wu, Heat transfer effect on the net work and/or power versus efficiency characteristics for the air standard Diesel cycles, *Energy: The International Journal*, vol.21, no.12, pp1201-1205, 1996.

Dual Cycle

Wu, C. and D.A. Blank, The effect of combustion on a power optimised endoreversible Dual cycle, *International Journal of Power and Energy Systems*, vol.14, no.3, pp98-103, 1994.

Chen, Lingen, F. Sun and C.Wu, Finite thermodynamics performance of a Dual cycle, *International Journal of Energy Research*, vol.23, no.9, pp765-772, 1999.

Ericsson Cycle

Blank, D.A. and C. Wu, Performance potential of a terrestrial solar-radiant Ericsson power cycle from finite-time thermodynamics, *International Power and Energy Systems*, vol.15, no.2, pp78-84, 1995.

Blank, D.A. and C. Wu, Power limit of an endoreversible Ericsson cycle with regeneration, *Energy Conversion and Management*, vol.37, no.1, pp59-66, 1996.

Chen, Lingen, F. Sun and C.Wu, Cooling and heating rate limits of a reversed reciprocating Ericsson cycle at steady state, *Proceedings of the Institute of Mechanical Engineers, Part A, Journal of Power and Energy*, vol.214, pp75-85, 2000.

Otto Cycle

Wu, C. and D.A. Blank, The effect of combustion on a work optimised endoreversible Otto cycle, *Journal of the Institute of Energy*, pp86-89, 1992.

Chen, Lingen, F. Sun and C. Wu, Heat transfer effects on the net work output and efficiency characterisitic for an air-standard Otto cycles, *Energy Conversion and Management*, vol.39, no.7, pp643-648, 1998.

Rallis Cycle

Wu, C., Analysis of an endoreversible Rallis cooler, *Energy Conversion and Management*, vol.35, no.1, pp79-85, 1994.

Stirling Cycle

Wu, C., Analysis of an endoreversible Stirling cooler, *Energy Conversion and Management*, vol.34, no.12, pp1294-1253, 1993.

Blank, D.A. and C. Wu, Power optimization of an endoreversible Stirling cycle with regeneration, *Energy: The International Journal*, vol.19, no,1, pp125-133, 1994.

Blank, D.A. and C. Wu, Power optimization of an extra-terrestrial, solar radiant Stirling heat engine, *Energy:The International Journal*, vol.20, no.6, pp523-530, 1995.

Chen, Lingen, C.Wu and F. Sun, Optimum performance of irreversible Stirling engine with imperfect regeneration, *Energy Conversion and management*, vol.39, no.8, pp727-732, 1998.

Chen, Lingen, C.Wu and F. Sun, Optimal performance of an irreversible Stirling cryocooler, *International Journal of Ambient Energy*, vol.20, no.1, pp.39-44, 1999.

Chen, Lingen, C.Wu and F. Sun, Performance characteristic of an endoreversible Stirling refrigerator, *International Journal of Power and Energy Systems*, vol.19, no.1, pp79-82, 1999.

11.9 SUMMARY

Maximum efficiency and maximum coefficient of performance are not necessarily the primary concern in design of a real cycle. Net power output and specific net power output in a heat engine, cooling load and specific cooling load in a refrigerator, and heating load and specific heating load in a heat pump are probably more important in industrial design of thermodynamic cycles. A different criteria of real cycle performance is provided by finite time thermodynamics. The basic finite time thermodynamic cycles are Carnot, Brayton and Rankine cycles. Literature concerning other finite time thermodynamic cycles are also provided in this chapter.

REFERENCES

Balmer, Robert T., *Thermodynamics*, West Publishing Co., New York, 1990.

Balzhiser, Richard E. and Michael R. Samuels, *Engineering Thermodynamics*, Prentice-Hall Publishing Co., New Jersey, 1977.

Black, William Z, and James G. Hartley, *Thermodynamics*, Third Edition, Harper-Collins College Publishers, New York, 1998.

Cengel, Yunus A. and Michael A. Boles, *Thermodynamics: An Engineering Approach*, Third Edition, McGraw-Hill Publishing Co., New York, 1998.

Granet, Irving and Maurice Bluestein, *Thermodynamics and Heat Power*, Sixth Edition, Prentice-Hall Publishing Co., New Jersey, 2000.

Haberman, William, L. and James E.A. John, *Engineering Thermodynamics*, Allen and Bacon, Inc., Boston, 1980.

Huang, Francis F., *Engineering Thermodynamics, Fundamentals and Applications,* MacMillian Publishing Co., New York, 1976.

Moran, Michael, J. and Howard N. Shapiro, *Fundamentals of Thermodynamics*, Fourth Edition, John Wiley and Sons Publishing Co., New York, 2000.

Rogers, G.F.C. and Y.R. Mayhew, *Engineering Thermodynamics, Work and Heat Transfer*, Fourth Edition, Addison Wesley Longman Ltd, Essex, England, 1992.

Sonntag, Richard E., Claus Borgnakke, and Gordon J. Van Wylen, *Fundamentals of Thermodynamics*, Fifth Edition, John Wiley and Sons Publishing Co., New York, 1998.

Wark, Kenneth, Jr. and Donald E. Richards, *Thermodynamics*, Sixth Edition, McGraw-Hill Publishing Co., New York, 1999.

Wu, Chih, Lingen Chen and Jincan Chen, *Recent Advances in Finite Time Thermodynamics*, Nova Science Publishing Co., New York, 1999.

Index

C

D

E

F

G

H

I

K

L

M

N

O

P

Q

R

S

T

U

V

W

Z